T0406162

Geotechnical, Geological and Earthquake Engineering

Volume 52

Series Editor

Atilla Ansal, School of Engineering, Özyegin University, Istanbul, Turkey

Editorial Board Members

Julian Bommer, Imperial College, London, UK

Jonathan D. Bray, University of California, Berkeley, Walnut Creek, USA

Kyriazis Pitilakis, Aristotle University of Thessaloniki, Thessaloniki, Greece

Susumu Yasuda, Tokyo Denki University, Hatoyama, Japan

The book series entitled *Geotechnical, Geological and Earthquake Engineering* has been initiated to provide carefully selected and reviewed information from the most recent findings and observations in these engineering fields. Researchers as well as practitioners in these interdisciplinary fields will find valuable information in these book volumes, contributing to advancing the state-of-the-art and state-of-the-practice. This book series comprises monographs, edited volumes, handbooks as well as occasionally symposia and workshop proceedings volumes on the broad topics of geotechnical, geological and earthquake engineering. The topics covered are theoretical and applied soil mechanics, foundation engineering, geotechnical earthquake engineering, earthquake engineering, rock mechanics, engineering geology, engineering seismology, earthquake hazard, etc.

Prospective authors and/or editors should consult the **Series Editor Atilla Ansal** for more details. Any comments or suggestions for future volumes are welcomed.

EDITORS AND EDITORIAL BOARD MEMBERS IN THE GEOTECHNICAL, GEOLOGICAL AND EARTHQUAKE ENGINEERING BOOK SERIES:

Series Editor:

Atilla Ansal

Ozyegin University
School of Engineering
Alemdag Çekmeköy 34794, Istanbul, Turkey
Email: atilla.ansal@ozyegin.edu.tr

Advisory Board

Julian J. Bommer

Imperial College
Department of Civil & Environmental Engineering
Imperial College Road
London SW7 2AZ, United Kingdom
Email: j.bommer@imperial.ac.uk

Jonathan D. Bray

University of California, Berkeley
Department of Civil and Environmental Engineering
453 David Hall, Berkeley, CA 94720, USA
Email: jonbray@berkeley.edu

Kyriazis Pitilakis

Aristotle University Thessaloniki
Department of Civil Engineering
Laboratory of Soil Mechanics and Foundations
University Campus
54124 Thessaloniki, Greece
Email: pitilakis@civil.auth.gr

Susumu Yasuda

Tokyo Denki University
Department of Civil and Environmental Engineering
Hatoyama-Cho
Hiki-gun
Saitama 350-0394, Tokyo, Japan
Email: yasuda@g.dendai.ac.jp

More information about this series at https://link.springer.com/bookseries/6011

Lanmin Wang · Jian-Min Zhang ·
Rui Wang
Editors

Proceedings of the 4th International Conference on Performance Based Design in Earthquake Geotechnical Engineering (Beijing 2022)

Set 3

Editors
Lanmin Wang
Lanzhou Institute of Seismology
China Earthquake Administration
Lanzhou, China

Jian-Min Zhang
Department of Hydraulic Engineering
Tsinghua University
Beijing, China

Rui Wang
Department of Hydraulic Engineering
Tsinghua University
Beijing, China

ISSN 1573-6059 ISSN 1872-4671 (electronic)
Geotechnical, Geological and Earthquake Engineering
ISBN 978-3-031-11897-5 ISBN 978-3-031-11898-2 (eBook)
https://doi.org/10.1007/978-3-031-11898-2

© The Editor(s) (if applicable) and The Author(s), under exclusive license
to Springer Nature Switzerland AG 2022
This work is subject to copyright. All rights are solely and exclusively licensed by the Publisher, whether the whole or part of the material is concerned, specifically the rights of translation, reprinting, reuse of illustrations, recitation, broadcasting, reproduction on microfilms or in any other physical way, and transmission or information storage and retrieval, electronic adaptation, computer software, or by similar or dissimilar methodology now known or hereafter developed.
The use of general descriptive names, registered names, trademarks, service marks, etc. in this publication does not imply, even in the absence of a specific statement, that such names are exempt from the relevant protective laws and regulations and therefore free for general use.
The publisher, the authors, and the editors are safe to assume that the advice and information in this book are believed to be true and accurate at the date of publication. Neither the publisher nor the authors or the editors give a warranty, expressed or implied, with respect to the material contained herein or for any errors or omissions that may have been made. The publisher remains neutral with regard to jurisdictional claims in published maps and institutional affiliations.

This Springer imprint is published by the registered company Springer Nature Switzerland AG
The registered company address is: Gewerbestrasse 11, 6330 Cham, Switzerland

Preface

The 4th International Conference on Performance-based Design in Earthquake Geotechnical Engineering (PBD-IV) will be held on July 15–17, 2022, in Beijing, China. The PBD-IV conference is organized under the auspices of the International Society of Soil Mechanics and Geotechnical Engineering—Technical Committee on Earthquake Geotechnical Engineering and Associated Problems (ISSMGE-TC203). The PBD-I, PBD-II, and PBD-III events in Japan (2009), Italy (2012), and Canada (2017), respectively, were highly successful events for the international earthquake geotechnical engineering community. The PBD events have been excellent companions to the International Conference on Earthquake Geotechnical Engineering (ICEGE) series that TC203 has held in Japan (1995), Portugal (1999), USA (2004), Greece (2007), Chile (2011), New Zealand (2015), and Italy (2019). The goal of PBD-IV is to provide an open forum for delegates to interact with their international colleagues and advance performance-based design research and practices for earthquake geotechnical engineering.

The proceedings of PBD-IV is the outcome of more than two years of concerted efforts by the conference organizing committee, scientific committee, and steering committee. The proceedings include 14 keynote lecture papers, 25 invited theme lecture papers, one TC203 Young Researcher Award lecture paper, and 187 accepted technical papers from 27 countries and regions. Each accepted paper in the conference proceedings was subject to review by credited peers. The final accepted technical papers are organized into five themes and six special sessions: (1) Performance Design and Seismic Hazard Assessment; (2) Ground Motions and Site Effects; (3) Foundations and Soil–Structure Interaction; (4) Slope Stability and Reinforcement; (5) Liquefaction and Testing; (6) S1: Liquefaction experiment and analysis projects (LEAP); (7) S2: Liquefaction Database; (8) S3: Embankment Dams; (9) S4: Earthquake Disaster Risk of Special Soil Sites and Engineering Seismic Design; (10) S5: Special Session on Soil Dynamic Properties at Micro-scale: From Small Strain Wave Propagation to Large Strain Liquefaction; and (11) S6: Underground Structures.

The Chinese Institution of Soil Mechanics and Geotechnical Engineering - China Civil Engineering Society and China Earthquake Administration provided tremendous support to the organization of the conference. The publication of the proceedings was financially supported by the National Natural Science Foundation of China (No. 52022046 and No. 52038005).

We would like to acknowledge the contributions of all authors and reviewers, as well as members of both Steering Committee and Scientific Committee for contributing their wisdom and influence to confer a successful PBD-IV conference in Beijing and attract the participation of many international colleagues in the hard time of COVID pandemic.

March 2022

Misko Cubrinovski
ISSMGE-TC203 Chair

Jian-Min Zhang
Conference Honorary Chair

Lanmin Wang
Conference Chair

Rui Wang
Conference Secretary

Organization

Honorary Chair

Jian-Min Zhang

Conference Chair

Lanmin Wang

Conference Secretary

Rui Wang

Steering Committee

Atilla Ansal
Charles W. W. Ng
George Gazetas
Izzat Idriss
Jian Xu
Jian-Min Zhang
Jonathan Bray
Kenji Ishihara
Kyriazis Pitilakis
Liam Finn
Lili Xie
Misko Cubrinovski
Ramon Verdugo
Ross Boulanger
Takaji Kokusho
Xianjing Kong
Xiaonan Gong
Yunmin Chen

Scientific Committee

Anastasios Anastasiadis
A. Murali Krishna
Alain Pecker
Amir Kaynia
Andrew Lees
António Araújo Correia
Antonio Morales Esteban
Arnaldo Mario Barchiesi
Carl Wesäll
Carolina Sigarán-Loría
Christos Vrettos
Deepankar Choudhury
Dharma Wijewickreme
Diego Alberto Cordero Carballo

Duhee Park
Eleni Stathopoulou
Ellen Rathje
Farzin Shahrokhi
Francesco Silvestri
Gang Wang
George Athanasopoulos
George Bouckovalas
Ioannis Anastasopoulos
Ivan Gratchev
Jan Laue
Jay Lee
Jean-François Semblat
Jin Man Kim
Jorgen Johansson
Juan Manuel Mayoral Villa
Jun Yang
Kemal Onder Cetin
Louis Ge
M. A. Klyachko
Masyhur Irsyam
Mitsu Okamura
Nicolas Lambert
Paulo Coelho
Sjoerd Van Ballegooy
Rubén Galindo Aires
Ryosuke Uzuoka
Sebastiano Foti
Seyed Mohsen Haeri
Siau Chen Chian
Stavroula Kontoe
Valérie Whenham
Waldemar Świdziński
Wei F. Lee
Zbigniew Bednarczyk

Organizing Committee

Ailan Che
Baitao Sun
Degao Zou
Gang Wang
Gang Zheng
Guoxing Chen
Hanlong Liu
Hongbo Xin
Hongru Zhang
Jie Cui
Jilin Qi
Maosong Huang
Ping Wang
Shengjun Shao
Wei. F. Lee
Wensheng Gao
Xianzhang Ling
Xiaojun Li
Xiaoming Yuan
Xiuli Du
Yanfeng Wen
Yangping Yao
Yanguo Zhou
Yong Zhou
Yufeng Gao
Yunsheng Yao
Zhijian Wu

Acknowledgements

Manuscript Reviewers
The editors are grateful to the following people who helped to review the manuscripts and hence assisted in improving the overall technical standard and presentation of the papers in these proceedings:

Amir Kaynia
Atilla Ansal
Brady cox
Dongyoup Kwak
Ellen Rathje
Emilio Bilotta
Haizuo Zhou
Jiangtao Wei
Jon Stewart
Kyriazis Pitilakis
Lanmin Wang
Mahdi Taiebat
Majid Manzari
Mark Stringer
Ming Yang
Rui Wang
Scott Brandenberg
Tetsuo Tobita
Ueda Kyohei
Wenjun Lu
Wu Yongxin
Xiao Wei
Xiaoqiang Gu
Xin Huang
Yang Zhao
Yanguo Zhou
Yifei Cui
Yu Yao
Yumeng Tao
Zitao Zhang

Contents

Invited Theme Lectures

S5: Special Session on Soil Dynamic Properties at Micro-scale: From Small Strain Wave Propagation to Large Strain Liquefaction

S6: Special Session on Underground Structures

S1: Special Session on Liquefaction Experiment and Analysis Projects (LEAP)

Lessons Learned from LEAP-RPI-2020 Simulation Practice

Long Chen[1](✉), Alborz Ghofrani[2], and Pedro Arduino[3]

[1] Haley & Aldrich, Seattle, USA
lchen@haleyaldrich.com
[2] Google LLC, Mountain View, USA
[3] University of Washington, Seattle, USA

Abstract. In accordance with the Liquefaction Experiment and Analysis Projects (LEAP)-RPI-2020 guidelines, a model was built using finite element tool OpenSees for blind prediction of centrifuge experiments to investigate the seismic response of a sheet pile retaining structure supporting liquefiable soils. This paper discusses different aspects of a representative numerical model for the LEAP experiment, implications of the decisions made in the process, and presents the lessons learned from this effort. These aspects include modeling of the soil-fluid coupling, soil behavior, structure element, soil-structure interface, boundary conditions, and initial static conditions.

Keywords: Liquefaction · Centrifuge · Soil-structure interaction · Finite element method · Constitutive model

1 Introduction

A realistic numerical model of an advanced soil-structure interaction physical experiment involves a series of decisions to be made on different levels. The Liquefaction Experiment and Analysis Projects (LEAP)-RPI-2020 [1] provides an opportunity to validate and verify these decisions. The LEAP experiment was set up to investigate the seismic response of a sheet pile wall retaining liquefiable soils. In the process of building a finite element (FE) model[1] for a blind prediction of the results from the centrifuge experiments, we gathered a list of most influential decisions made to successfully capture the response of the physical experiment.

In this paper, we discuss different aspects of a representative numerical model for the LEAP experiment, implications of the decisions made in the process, and present the lessons we learned from this effort. This paper is not intended to be an exhaustive review on various ways a numerical model can be set up, but we hope that it covers most of the important aspects of such a model so that an interested reader would find valuable discussion on important decisions made to capture the parameters of interest realistically.

[1] We used OpenSees, a finite element framework developed by McKenna et al. [2].

© The Author(s), under exclusive license to Springer Nature Switzerland AG 2022
L. Wang et al. (Eds.): PBD-IV 2022, GGEE 52, pp. 1763–1771, 2022.
https://doi.org/10.1007/978-3-031-11898-2_159

2 LEAP and the Centrifuge Experiment Settings

LEAP [3] is a series of projects aiming to develop a high-quality centrifuge experiment database for assessment of the validity of numerical tools for predicting soil liquefaction and its consequences such as lateral spreading and failure of retaining structures.

During the RPI-2020 phase of LEAP, centrifuge experiments were performed at ten different facilities across the world to investigate the seismic response of a sheet pile wall supporting liquefiable soils. Numerical simulation exercises were also performed by several teams. Figure 1 illustrates the baseline schematic of the centrifuge model experiments. The model consisted of a (top) layer of medium-dense Ottawa F-65 sand supported by a sheet pile wall made of aluminum. This layer of soil was underlain by a dense sand layer. The total length of the model was about 20 m in prototype scale. The tip of the wall was embedded at mid-depth of the dense layer before the centrifuge acceleration was applied. The target motion was a ramped sine wave with a dominant frequency of 1 Hz and 5 consecutive uniform cycles at the PGA (Fig. 2). For more details regarding the model specifications and soil properties, the reader is encouraged to refer to [1].

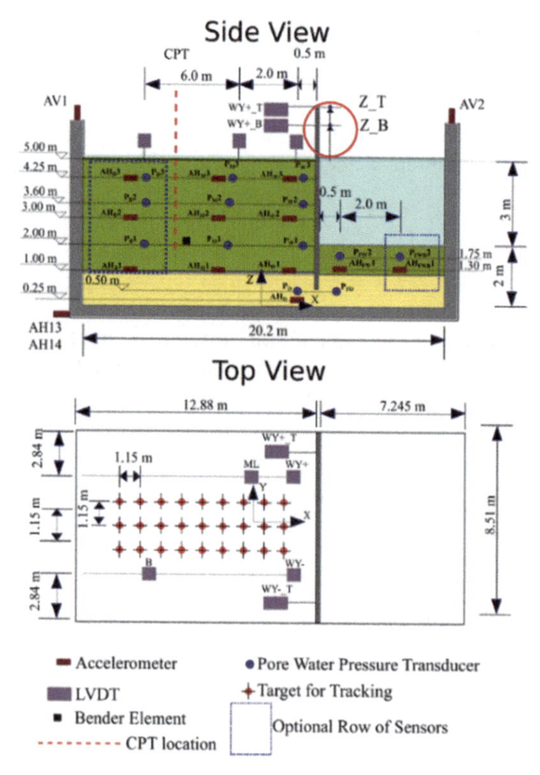

Fig. 1. Experimental setup and instrumentation for LEAP-2020 centrifuge tests (after [1]).

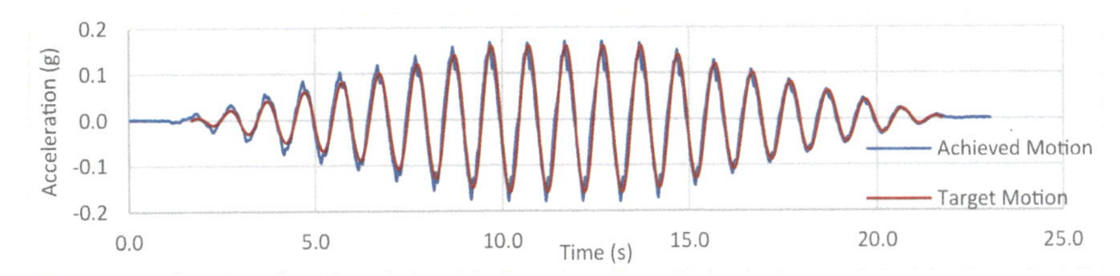

Fig. 2. Achieved and target base acceleration time histories from LEAP-2020 centrifuge experiments (RPI Test 9).

Several complex aspects of the numerical simulation of the experiment are discussed in the following sections. These aspects include but are not limited to:

- Nonlinear soil behaviors including contraction and dilation.
- Soil and structure interaction with water, as well as the free water.
- Soil-structure interaction during static and dynamic phases of the simulation. The embedment length is relatively short, which could pose a challenge to the constitutive model during the static phase which captures the experiment setup and spin up. Forming and closing of the gap between the soil and the structure is possible during shaking.

Other numerical modeling concerns such as the boundary conditions, element type, integration methods, dynamic damping, etc.

3 Modeling Decisions

3.1 General Model

We used OpenSees as the main FE tool in this study. Other numerical tools, such as FLAC [4] and PLAXIS [5], are capable of modeling this experiment. As the shaking was solely applied to the longitudinal direction, a 2D plane-strain model was a reasonable simplification. The main advantage of using a 2D model is the significant reduction in the computing time compared to a 3D model. Other benefits include simpler model set up and implementation.

A mesh size capable of resolving a minimum frequency of 25 Hz was used. The maximum element size was calculated based on the element order, mechanical wave velocity, and the target frequency. Where stress concentration may happen, e.g., zones with material contrast, local mesh refinement is usually required, e.g., around the wall tip. However, if mesh refinement is done closer to the surface, overly refined mesh could result in minimal effective confining pressure. When effective stress-based constitutive behavior is used, this may result in an ill-conditioned stiffness matrix which causes difficulty in numerical convergence.

Due to the radial distribution of the gravity field in a centrifuge experiment, the stress distribution in the scaled model could be different from the target model in prototypical scale. This difference is inversely proportional to the arm length of the centrifuge facility

[6]. In order to match the stress distribution between the scaled and prototype models, the geometry of the scaled model should be slightly changed with curved boundaries. However, considering the centrifuge arm length, the error encountered is negligible. In this study, the authors chose to model the prototype scale model with the FE mesh shown in Fig. 3.

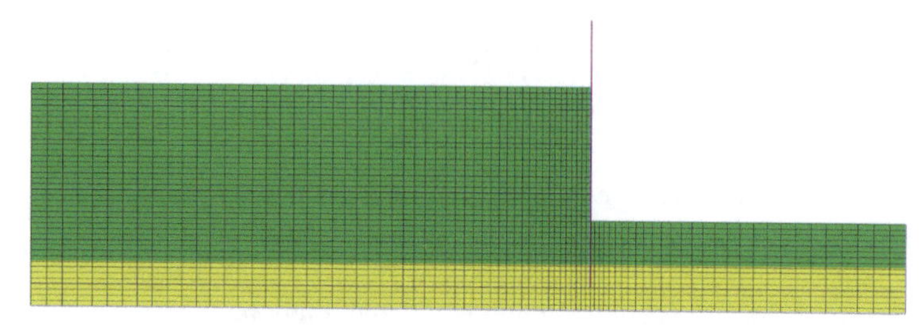

Fig. 3. Prototype scale 2D mesh for LEAP-2020 model.

3.2 Modeling Soil Behavior

Liquefaction of the medium dense sand was indicated by the recorded pore water pressure data. The constitutive model used in the numerical simulation needs to be able to represent the soil behaviors during shaking and have the capability to capture contraction and dilation behaviors. It is preferred to calibrate the model for conditions similar to the physical experiment conditions, e.g., cyclic stress ratio, initial confining pressure, initial static shear stress [6], and stress path (Fig. 4).

In this study, we used PM4Sand [7] and Manzari-Dafalias (MD) [8] to model the medium dense liquefiable layer. Both models are capable of capturing realistic liquefiable soil behaviors. The model calibration details can be found in Chen [6]. An example of calibrated model behavior is illustrated in Fig. 4.

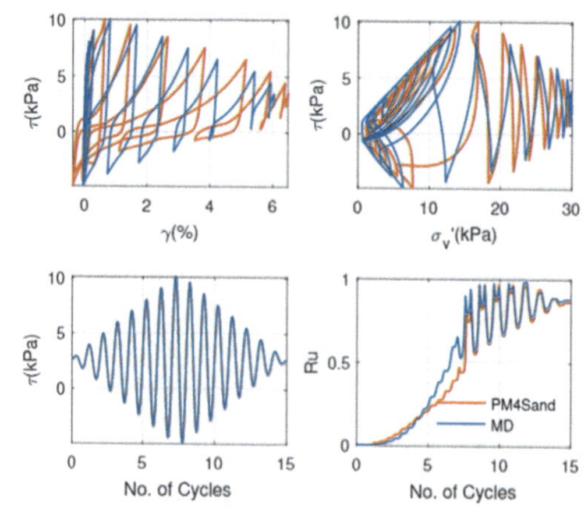

Fig. 4. Example calibrated model behaviors under nonuniform cyclic stress loading in undrained cyclic simple shear stress path.

3.3 Modeling the Soil-Fluid Coupling

In the experiment, water interacts with soil in two ways:

- excess pore pressure due to the interaction between the soil solid skeleton and the pore water,
- hydrodynamic pressure due to the dynamic interaction between the free water and the wall.

Generation of excess pore pressure can be modelled in different ways in different numerical tools. For example, in PLAXIS, this is achieved by defining stiffness and strength in terms of effective properties [5]. A large bulk stiffness for water is automatically applied to make the soil incompressible, then (excess) pore pressures are calculated, even above the phreatic surface. Dynamic and consolidation analyses are available based on u-P formulation. In FLAC, the soil and fluid coupling can be modelled with proper fluid bulk modulus and tension limit (to allow negative pore pressures due to soil dilation). In OpenSees, the fully coupled u-P elements can be used to add pore pressure degree(s) of freedom (DoF) to the nodes. Phreatic line needs to be modeled explicitly in OpenSees by applying explicit conditions on the pore pressure DoF. Stabilized single point quadrilateral element with mixed displacement-pressure formulation (SSPquadUP) developed by McGann et al. [9] was used in all simulations due to its efficiency. In this element, a stabilizing parameter is employed to permit the use of equal-order interpolation and its value was chosen to be on the same order as permeability.

It is worth mentioning that the so-called permeability input parameter in FLAC and OpenSees is defined as the conventional hydraulic conductivity (units: [L/T] - e.g., m/sec) divided by unit weight of water (units: [F/V] - e.g., Pa/m). Literature sometimes refers to it as dynamic permeability coefficient [10]. In PLAXIS, the conventional hydraulic conductivity is directly inputted. In this study, a constant permeability $k = 3 \times 10^{-5}$ m/s was applied to all soils based on reported values and author's previous experience with LEAP experiments [6].

The free water in front of the wall was not explicitly modelled in our analysis, which could have contributed to inaccurate prediction results. To properly account for this effect, one approach is to model water directly using an elastic solid material with negligible shear modulus, representative values of unit weight (e.g., 1 Mg/m^3) and bulk modulus (e.g., 2.2 GPa), and Poisson's ratio approaching 0.5. Another possible approach is to add additional mass to the structural nodes following Westergaard's method. Both PLAXIS and FLAC manuals [4, 5] offer explanations on how this method can be implemented. The distribution of added mass over depth of water is depicted in Fig. 5.

3.4 Modeling the Structure

The sheet pile wall can be modeled with either solid elements or beam elements. The flexibility of using section fibers with the beam elements, available in OpenSees, adds the major advantage of modeling nonlinear cross-sectional behavior with more accuracy. Beam elements provide higher accuracy compared to typical 2D bi-linear or 3D tri-linear solid elements, as well as easiness of extracting force distribution along the structure.

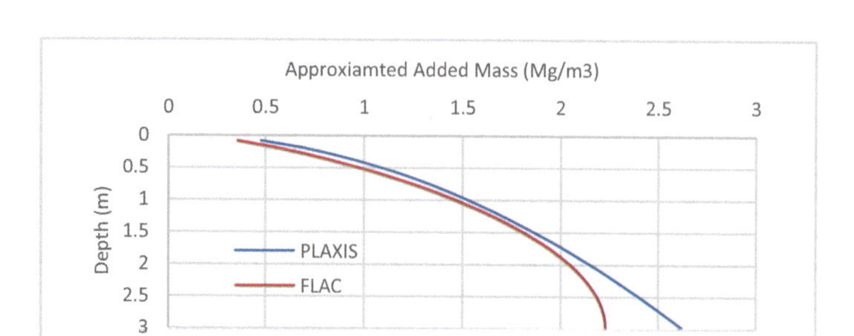

Fig. 5. Distribution of added mass to account for the hydrodynamic pressure.

In this study, beam elements were used to represent the wall. Since solid elements and beam elements typically have different DoFs, the choice of element will possibly affect the choice of technique to model soil structure interface. Modeling the structure of the container is discussed in the boundary condition section.

3.5 Modeling the Soil-Structure Interaction (SSI)

An important aspect of this experiment is the interaction between the soil and the sheet pile wall. The simplest way to model SSI is to assume perfect bonding between adjacent soil and structure nodes. It can be easily achieved by using tied (equal) DoF between the two adjacent nodes. This approach greatly simplifies the analysis and can make the solution easier to converge. However, this method lacks the ability to capture the separation of the soil body and the structure, modeling the frictional forces along the interface, and dynamic interaction forces.

Springs can also be used to represent the interface. Zero tension springs (elastic no tension (ENT) material in OpenSees) applied normally to the interface can be used to model the separation. These springs need to be reasonably stiff to prevent penetration while avoiding ill-conditioning of the stiff matrix. Springs parallel to the interface can be used to model the frictional behavior between the soil and the wall.

Contact (interface) elements can also be used to capture both parallel and perpendicular interface behaviors. In OpenSees, the *zeroLengthContact2D* element [11] is a node-to-node, frictional, 2D contact element and it allows to model the soil-structure interface in a more realistic way. Although the constrained nodes and retained nodes must have 2 DoF (soil node and beam node both have 3 DoF in this model), it can be used by introducing extra "shadow" 2 DoF nodes that are tied to the soil and beam nodes. However, it became computationally more expensive when contact had to be established on both sides of the beam. Other continuum soil-structure interface types (e.g., node-to-surface or surface- to-surface contact element) have been developed [12], but currently they can only be used in static analyses because the contact searching lags one step behind the analysis step, which could introduce unrealistic contact forces when the contact algorithm tries to push the soil back from a penetration state to a contact state.

In this study, we used ENT springs to model the interface between soil and the wall. The stiffness of the ENT springs was set to be ten times stronger than the wall to reduce the amount of penetration of the soil nodes into the wall. This simplified approach may contribute to inaccurate prediction results.

The interface between soil and the container was simplified by assuming perfect bonding.

3.6 Boundary Conditions

The centrifuge container defines the side boundary conditions. For an experiment held in a laminar container, the side boundaries can be tied at each level to represent the relative movement between container rings. While for a rigid container experiment, the soil nodes along the two vertical boundaries and the bottom boundary are typically assumed to move together with the rigid container. In OpenSees, this boundary condition can be applied by fixing horizontal DoF on the soil nodes and using *UniformExcitation* to apply uniform horizontal acceleration time history to all the nodes with fixed horizontal DoFs. However, consequently, the nodes along the base were fully fixed, which could result in inaccurate initial static shear stresses, compared to the case with roller fixity condition.

Pore pressure DoFs were fixed along the horizontal free surfaces to ensure zero excess pore water pressures at all times. Because the free water in front of the wall was not simulated, the static pore water pressures calculated using uP elements on nodes behind and in front of the wall are different. To ensure the correct initial effective stress field, the soil domain on both sides of the wall were separated. As shown in Fig. 6, there were three nodes located at the same location along the embedded portion of the wall: two soil nodes and one beam node that represents the wall.

No additional constraints were applied to other wall nodes during dynamic analysis, except that the bottom of the wall was fixed in rotation and vertical translation and free in horizontal translation, with the assumption that liquefaction would not be triggered in the dense sand layer. However, this condition could have led to overly constrained wall tip and limited the overall movement of the wall, which resulted in underprediction of simulation results.

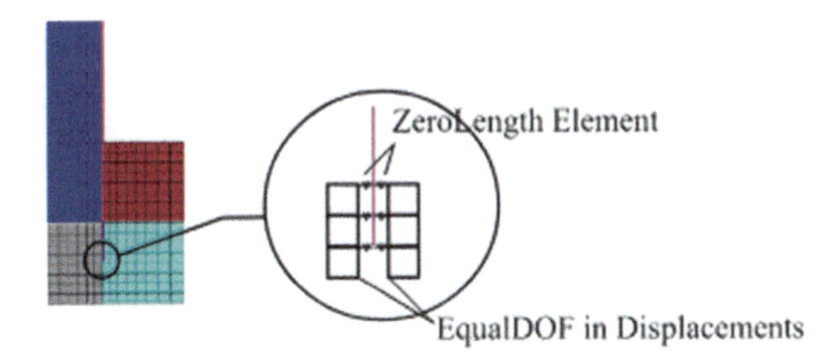

Fig. 6. Boundary condition around the toe of the embedded wall.

3.7 Initial Conditions

A gravity phase was required to generate a realistic initial stress field for effective stress based constitutive models before dynamic analysis and after the spin-up phase. It is important to note that when u-P elements are used in OpenSees, transient analysis is required for the static application of gravity. This is due to the fact that pore pressure DoFs are modeled in the velocity vector for computational efficiency. During this phase, both MD and PM4Sand were first set to behave elastically, and the soil permeability was set very high (k = 1 m/s) to facilitate the dissipation of generated excess pore water pressures. Newmark parameters $\gamma = 5/6$ and $\beta = 4/9$ was used to numerically damp out the waves generated due to the sudden application of gravity and change in stiffness when switching material from elastic to elastoplastic stage.

At the end of static analysis phase, stress concentration was observed at the corners and around the tip of the sheet pile wall (Fig. 7). The wall showed excessive deformation under static lateral pressures resulting from the retained soil (Fig. 8). Only minor lateral displacement was observed during centrifuge spin-up, while 5 to 20 cm of lateral displacements, depending on relative density and wall thickness, were predicted in OpenSees after releasing constraints on the wall at the beginning of shaking.

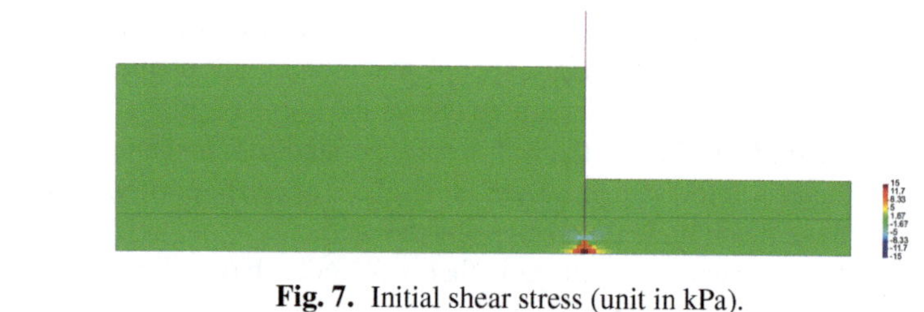

Fig. 7. Initial shear stress (unit in kPa).

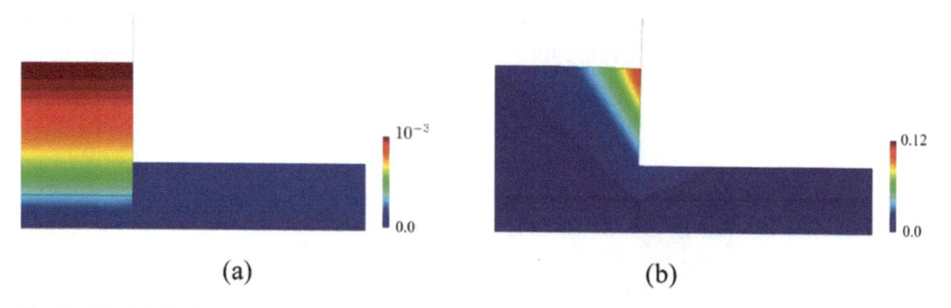

Fig. 8. Nodal displacement contours (m): (a) before and (b) after releasing the sheet pile wall.

As a result, the retained soil tended to 'collapse' on the wall and cause stress relaxation near the interface. Because the constitutive models behaved elastically during the gravity phase, some soil elements possibly had initial negative mean confining pressure, which is inadmissible for an effective stress-based model in the plastic stage. For bounding surface constitutive models, it was also possible that after gravity analysis, the initial stress state was outside the bounding surface, which needs to be corrected.

4 Conclusions

A numerical model was developed to simulate the LEAP-RPI-2020 centrifuge experiment in OpenSees. This paper summarizes modeling decisions and strategies adopted during the model building and analysis process, with particular emphasis on the shortcomings of the model, which lead to the inaccurate predictions.

It is recommended to calibrate the constitutive model under conditions similar to the physical experiment conditions, e.g., range of cyclic stress ratio, initial confining pressure, initial static shear stress, and stress path. Interaction of water and soil, including excess pore pressure and hydrodynamic pressure, should be accounted in the model. Special attention needs to be paid to the interface between soil and structure, especially for dynamic analysis, as different choices of interface could bring different degrees of complexity in the model.

References

1. LEAP-RPI-2020 Organizing Team: LEAP 2020 Simulation Exercise
2. McKenna, F., Scott, M.H., Fenves, G.L.: Nonlinear finite-element analysis software architecture using object composition. J. Comput. Civ. Eng. **24**(1), 95–107 (2010)
3. Manzari, M.T., et al.: Liquefaction experiment and analysis projects (LEAP): summary of observations from the planning phase. Soil. Dyn. Earthq. Eng. **113**, 714–743 (2017). https://doi.org/10.1016/j.soildyn.2017.05.015
4. Itasca Consulting Group, Inc.: FLAC—Fast Lagrangian Analysis of Continua, Ver. 8.1. Minneapolis (2019)
5. PLAXIS, PLAXIS 2D Reference Manual. Bentley Systems International Limited, Dublin (2021)
6. Chen, L., Ghofrani, A., Arduino, P.: Remarks on numerical simulation of the LEAP-Asia-2019 centrifuge tests. Soil Dyn. Earthq. Eng. **142**, 106541 (2021)
7. Boulanger, R.W., Ziotopoulou, K.: PM4Sand (Version 3.1): a sand plasticity model for earthquake engineering applications. UCD/CGM-17/01, Center for Geotechnical Modeling, University of California, Davis, CA (2017)
8. Dafalias, Y.F., Manzari, M.T.: Simple plasticity sand model accounting for fabric change effects. J. Eng. Mech. **130**(6), 622–634 (2004)
9. McGann, C.R., Aruino, P., MacKenzei-Helnwein, P.: Stabilized single-point 4-node quadrilateral element for dynamic analysis of fluid saturated porous media. Acta Geotech. **7**(4), 297–311 (2012). https://doi.org/10.1007/s11440-012-0168-5
10. Parra-Colmenares, E.J.: Numerical modeling of liquefaction and lateral ground deformation including cyclic mobility and dilation response in soil systems, Ph.D. thesis, Rensselaer Polytechnic Institute, Troy, NY (1996)
11. Wang, G., Sitar, N.: Nonlinear analysis of a soil-drilled pier system under static and dynamic axial loading, PEER Report No. 2006/06, Pacific Earthquake Engineering Research Center, University of California, Berkeley, CA (2006)
12. Petek, K.A.: Development and application of mixed beam-solid models for analysis of soil-pile interaction problems. Ph.D. Dissertation, University of Washington (2006)

Numerical Simulations of the LEAP 2020 Centrifuge Experiments Using PM4Sand

Renzo Cornejo(✉) and Jorge Macedo

Department of Civil and Environmental Engineering, Georgia Institute of Technology, Atlanta, USA
rcornejot@gatech.edu

Abstract. Soil liquefaction is a complex phenomenon that has been studied for several decades, and there is still much to learn about it and its consequences. In this context, as part of Liquefaction Experiments and Analysis Projects (LEAP), in 2020, eleven universities around the globe performed centrifuge tests of a model representing a sheet-pile retaining system supporting liquefiable soils. In this study, we document the numerical simulation performed to predict the 2020 LEAP centrifuge tests response of the Rensselaer Polytechnic Institute and the University of California Davis tests. The numerical simulations are performed using the PM4Sand model developed by Boulanger and Ziotopoulou (2017). Towards this end, we inspected the available laboratory information for the Ottawa sand, and, interestingly, we find discrepancies in the critical state line (CSL) definition, a key input for critical state-based numerical models. Independent of this finding, we selected a CSL to calibrate the PM4Sand model further using the information from cyclic tests. Once calibrated, the PM4Sand model is used in the context of a boundary value problem in finite-difference calculations using the software FLAC. The results show that displacements are well predicted, excess pore pressures are reasonably well predicted, and high-frequency dilation peaks cannot be predicted, which deserves further examination.

Keywords: Liquefaction · LEAP · Sheet-pile · PM4Sand · FLAC

1 Introduction

Liquefaction is a phenomenon that has been extensively studied for several years due to its complex mechanism of occurrence and the disastrous consequences it can cause (e.g., Ishihara and Yoshimine 1992; Bray and Dashti 2014; Popescu and Prevost 1993; Elgamal et al. 2005; Karimi and Dashti 2016a, 2016b; Travasarou et al. 2006; Luque and Bray 2015; Liu and Dobry 1997; Yoshimi and Tokimatsu 1977; Macedo and Bray 2018, Bray et al. 2017a, 2017b, 2017c). Even though there have important contributions to liquefaction engineering, there are still aspects that need further research, as pointed out by the National Academy of Sciences, Engineering and Medicine report (2016). In this context, the Liquefaction Experiments and Analysis Projects (LEAP) project started in 2011 to advance liquefaction engineering under a holistic approach that includes

© The Author(s), under exclusive license to Springer Nature Switzerland AG 2022
L. Wang et al. (Eds.): PBD-IV 2022, GGEE 52, pp. 1772–1784, 2022.
https://doi.org/10.1007/978-3-031-11898-2_160

laboratory tests (i.e., monotonic triaxial tests, cyclic direct shear tests), centrifuge tests, and numerical simulations.

As part of the 2020 LEAP project, eleven centrifuge experiments were tested in ten different laboratories using a pre-established configuration provided to all the participants. The system tested in the centrifuge experiments consists of a sheet-pile retaining system supported by liquefiable soils. The sheet pile was supported by a 4 m thick loose sand and 1 m-thick dense sand at prototype scale. The prototype/model scale was changed for each laboratory.

In this paper, the RPI and UCDavis' centrifuge experiments have been selected as the baseline centrifuge experiment to compare against the results from numerical simulations performed in this study.

2 Centrifuge Experiment

2.1 Experiment Description

Several centrifuge tests were performed as part of the LEAP 2020 at different universities. The dimensions at a prototype scale are 20 m in length and 5 m in height. These dimensions were reduced at each facility based on the equipment capacity. For instance, the University of California - Davis (UCDavis) used a model with a scale of 27, resulting on dimensions in the centrifuge test of 740 mm in length and 185 mm in height.

2.2 Sensor Distribution

The LEAP Project provided a detailed list of instruments with their coordinates to be incorporated within each centrifuge test. For the UCDavis' centrifuge experiment, the instrumentation was distributed as shown in Fig. 1: The sensor coordinates were also used to define control points in the numerical models.

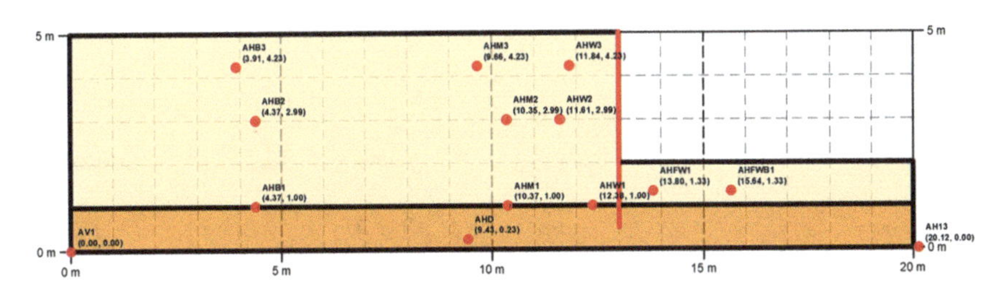

Fig. 1. Distribution of instrumentation within the centrifuge experimental model at UC Davis

3 Prediction Analysis

3.1 Analysis Platform

FLAC (Fast Lagrangian Analysis of Continua) is a numerical modeling software for advanced geotechnical analysis of soil, rock, groundwater, and ground support in two dimensions using either plane strain, plane stress, or axisymmetric conditions. FLAC is applicable to both static and dynamic conditions. In addition, FLAC utilizes an explicit finite difference formulation, and it solves the dynamic motion equation, even for static conditions using a so-called quasi-static scheme where a damping factor is added to the solution. More details on FLAC are available in (Itasca 2019).

3.2 Solution Algorithms and Assumptions

As opposed to the finite element method (FEM), in FLAC, there are no shape functions to capture the variation of field parameters (e.g., displacement, stresses, etc.). Instead, the derivatives of the governing equations are directly replaced by algebraic finite differences at discrete points in the problem domain. As shown in Fig. 2, within a calculation cycle of a boundary value problem, FLAC assumes that there are no interactions between different parts of the domain; hence, the updating of parameters in one cycle is isolated in the space domain. For example, the node velocities remain frozen while the stresses are updated. This characteristic enforces the time step to be sufficiently small so that the calculated information from one element cannot physically pass into neighboring elements. The convergence of the solution is checked after each iteration cycle (i.e., Fig. 2a) by calculating an unbalanced force for the problem domain. Figure 2b presents the numbering scheme for elements and grid points in FLAC. If the unbalanced force at a given grid point is below a predefined threshold, after many interactions, it is considered that the solution has been achieved. In the case of a dynamic excitation, the motion equation is solved within each time step, and the unbalanced force is checked accordingly before moving to the next time step. More detailed information on the FLAC operations can be found at (Itasca 2019).

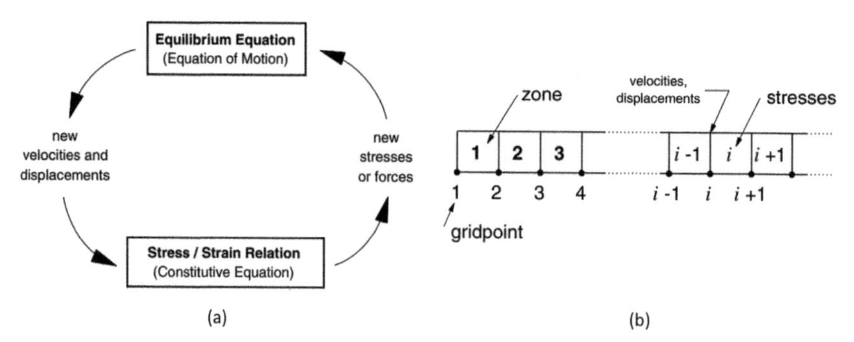

Fig. 2. a) Explicit calculation cycle in FLAC, b) Numbering scheme in FLAC illustrating that the ith zone is between the ith and $(i + 1)^{th}$ gridpoints. At each FLAC step, the cycle in a) is applied to the zones and gridpoints in b) under an explicit scheme

3.3 Model Description

The numerical model considers the prototype scale (i.e., the expected real dimensions) of the sheet-pile retaining structure and soil layers. The model has 300 squared zones of 0.5 m × 0.5 m. There is a 1 m-thickness layer of dense sand at the bottom, and the upper soil is a loose sand layer. Figure 3 also shows two important details considered in the modeling process namely, the addition of lateral boundaries to capture the rigid box's effect and the no relative movement between the soil and the sheetpile, the latter is a limitation of the current study and will be addressed in the future.

Fig. 3. Model geometry discretization

4 Constitutive Model

The model considered to perform the numerical analysis is PM4Sand, developed by Boulanger and Ziotopoulou (2017). The constitutive model PM4Sand is a stress-ratio controlled, critical state compatible, bounding surface plasticity model designed primarily for earthquake engineering applications.

The implementation of PM4Sand in FLAC has a set of 6 primary parameters and 21 secondary parameters. Three primary parameters were calibrated. These are the relative density (Dr), shear modulus coefficient (G_o) and contraction rate parameter (h_{po}). Some secondary parameters, such as the critical state parameter, were also considered for a better fit during calibration.

5 Type-B Simulations

5.1 Calibrations

The sheet-pile interacts with two layers of Ottawa F-65 sand. The deepest layer is a 90% relative density sand of 1 m thickness. The other soil material is a loose sand with a relative density of 60 to 70%, and its thickness is variable (4 m behind the sheet-pile and 1 m in front).

The Ottawa F-65 sand materials have been modeled using the PM4Sand constitutive model (Boulanger and Ziotopoulou 2015). The PM4sand model requires a critical state line (CSL) as an input; Boulanger and Ziotopoulou (2015) recommend using default parameters to represent the CSL in the PM4Sand model if there is no specific information.

However, in the case of the LEAP projects, there are several triaxial tests that can be used to inspect the CSL assessments. The information on triaxial tests from previous LEAP projects and from Ramirez et al. (2018) on Ottawa sand has been collected and processed. Figure 4 shows the information from different tests at instances expected to represent the critical state (i.e., at large strains). Using the information from Ramirez et al. (2018), we defined a CSL (labeled as CSL-this study in Fig. 4). We also plotted for reference the CSL defined by Ramirez et al. (2018) and the CSL that corresponds to the default parameters in PM4Sand. Interestingly, it can be observed that the data produced during the LEAP 2015 for loose sands (blue circles in Fig. 4) departs from the clustering of the data produced by Ramirez et al. (2018) for the same material. In addition, there are tests on quite dense samples that are far away from the CSL, perhaps due to localization in dense samples during the testing. The laboratory tests on loose samples during the LEAP 2015 program should be checked in future studies as they are not consistent with the rest of the data. In the context of this study, we will use the defined CSL in the numerical simulations. Of note, this CSL is significantly different from the PM4Sand default CSL.

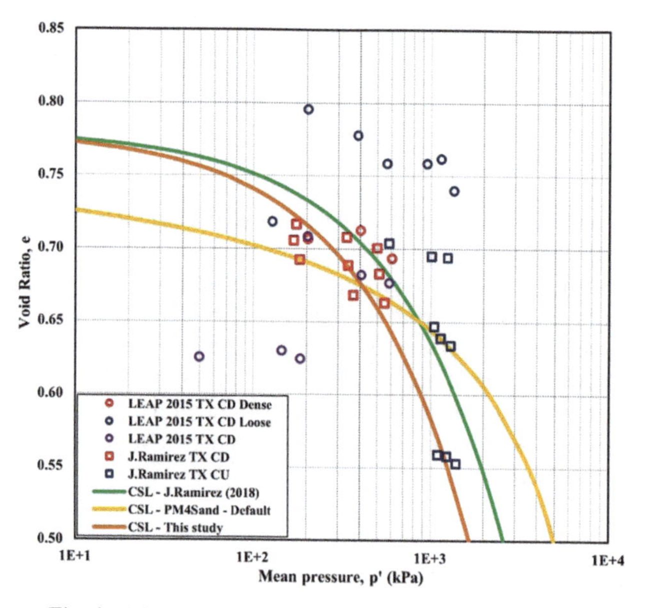

Fig. 4. Triaxial test data and critical state lines definitions

The calibrations of other parameters are performed considering two different stress levels, i.e., 40 and 100 kPa. Table 1 summarizes geotechnical parameters measured through laboratory tests for these stress levels.

Table 1. Ottawa F65 sand characteristics based on laboratory tests

Parameter	Symbol	Unit	Value	
			40 kPa	100 kPa
Specific gravity	G_s	–	2.65	2.65
Minimum void ratio	e_{min}	–	0.50	0.50
Maximum void ratio	e_{max}	–	0.78	0.78
Relative density	D_r	%	66.15	67.48
Shear wave velocity	V_s	m/s	145.98	159.23
Shear modulus	G_{max}	MPa	35.15	41.70

Two of the PM4Sand primary parameters (Dr, G_o) were estimated from the laboratory tests, the other primary parameter h_{po} and the secondary parameters were calibrated based on the observed response in cyclic simple shear tests. The initial calibrations showed a liquefaction curve (i.e., cyclic stress ratio- CSR, versus the number of cycles) with a steeper slope compared to the experimental data. We performed a parametric study with the different PM4Sand parameters to identify those that influence more the liquefaction curve's slope. The parameters identified are listed as follows (the definition of each parameter is provided in Table 2): plastic control parameter, n^b, the cyclic

Table 2. PM4Sand model parameter for Ottawa F-65 sand

Parameter	Symbol	Unit	Value	
			40 kPa	100 kPa
Primary parameters				
Relative density	D_r	%	66.2	67.5
Normalized shear modulus	G_o	–	950	950
Contraction rate	h_{po}	–	0.08	0.08
Atmospheric pressure	P_{atm}	kPa	101.3	101.3
Secondary parameters				
	h_o	–	**	**
Minimum void ratio	e_{min}	–	0.50	0.50
Maximum void ratio	e_{max}	–	0.78	0.78
	n^b	–	0.20	0.20
Phase transformation control	n^d	–	0.10*	0.10*
Fabric parameter	c_z	–	250*	250*
Cyclic mobility parameter	c_ε	–	2.0	2.0
Critical friction angle	ϕ_{cv}	°	30.0	30.0
Poisson's ratio	ν	–	0.3*	0.3*
Critical state line parameter 1	Q	–	10*	10*
Critical state line parameter 2	R	–	1.5*	1.5*
Thickness of yield surface	m	–	0.01*	0.01*

(*): default value.
(**): Calibrated based on the liquefaction resistance curves provided for the Ottawa F-65 sand and to reproduce the liquefaction triggering curves from Boulanger and Idriss (2016).

mobility parameter, c_ε, and the critical friction angle, ϕ_{cv}. Table 2 shows the final values of the calibrated parameters after iterations, and Fig. 5 shows a comparison of the numerical-based and experimental-based liquefaction curves.

Figure 5 shows a comparison of the liquefaction curves measured in the laboratory and simulated numerically for both applied pressures (40 and 100 kPa).

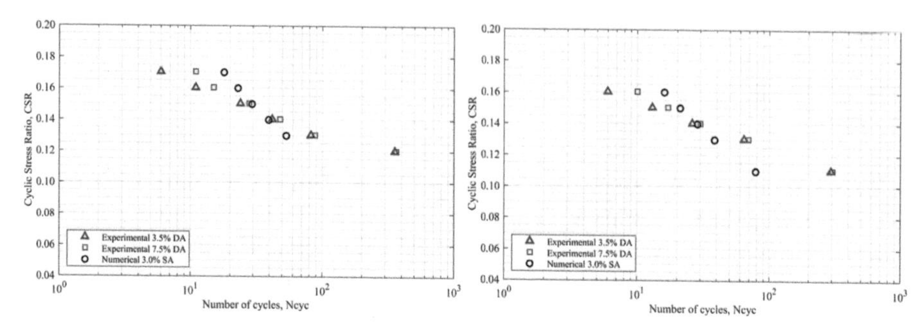

Fig. 5. Comparison between experimental and numerical Liquefaction Strength Curve for 40 kPa (left) and 100 kPa (right)

5.2 Modeling Process

For the boundary value problem simulation, the model was first assigned with elastic properties and subjected to gravity loads to reach static equilibrium. After reaching equilibrium, PM4Sand parameters, shown in Table 3, were assigned to the loose and dense layers.

Table 3. PM4Sand model parameter for each soil layer

Parameter	Symbol	Unit	Value	
			Loose sand	Dense sand
Primary parameters				
Relative density	D_r	%		
UCD 1			71.0	93.0
RPI 09			63.0	90.0
RPI 10			65.0	91.0
RPI 11			66.0	90.0
Normalized shear modulus	G_O	–	950	950
Contraction rate	h_{po}	–	0.08	0.08
Atmospheric pressure	P_{atm}	KPa	101.3	101.3

(continued)

Table 3. (*continued*)

Parameter	Symbol	Unit	Value	
			Loose sand	Dense sand
Secondary parameters				
	h_o	–	**	**
Minimum void ratio	e_{min}	–	0.50	0.50
Maximum void ratio	e_{max}	–	0.78	0.78
	n^b	–	0.20	0.20
Phase transformation control	n^d	–	0.10*	0.10*
Fabric parameter	c_z	–	250*	250*
Cyclic mobility parameter	c_ε	–	2.0	2.0
Critical friction angle	ϕ_{cv}	°	30.0	30.0
Poisson's ratio	ν	–	0.3*	0.3*
Critical state line parameter 1	Q	–	10*	10*
Critical state line parameter 2	R	–	1.5*	1.5*
Thickness of yield surface	m	–	0.01*	0.01*

(*): default value.
(**): Calibrated based on the liquefaction resistance curves provided for the Ottawa F-65 sand and to reproduce the liquefaction triggering curves from Boulanger and Idriss (2016).

The model is solved again to reach equilibrium after assigning PM4Sand properties. The dynamic loading is applied at the model's base, as an acceleration time history using a rigid base to simulate the conditions in a centrifuge test. In addition, a Rayleigh damping value of 0.2% has been incorporated into the model to avoid spurious frequencies. Finally, we defined history points at the sensor locations to perform comparisons against the experimental recorded observations.

5.3 Simulation Results

Figure 6 shows the comparisons between numerically-based and experimental-based response spectra at several locations indicated in Fig. 1. It can be observed that there is consistency for spectral periods higher than 0.5 s; however, the responses observed during the experiments for shorter periods are significantly higher. This significant contribution on large frequencies (i.e., short periods) was associated with the presence of dilation spikes, as shown in Fig. 7. The dilation spikes are enhanced when the sensors are closer to the surface, which is translated to larger differences on the response spectra (i.e., experimental versus numerical) at short periods. Even though we considered different sensitivities for the PM4Sand model parameters, we could not reproduce the dilation spikes, which should be further explored in future studies.

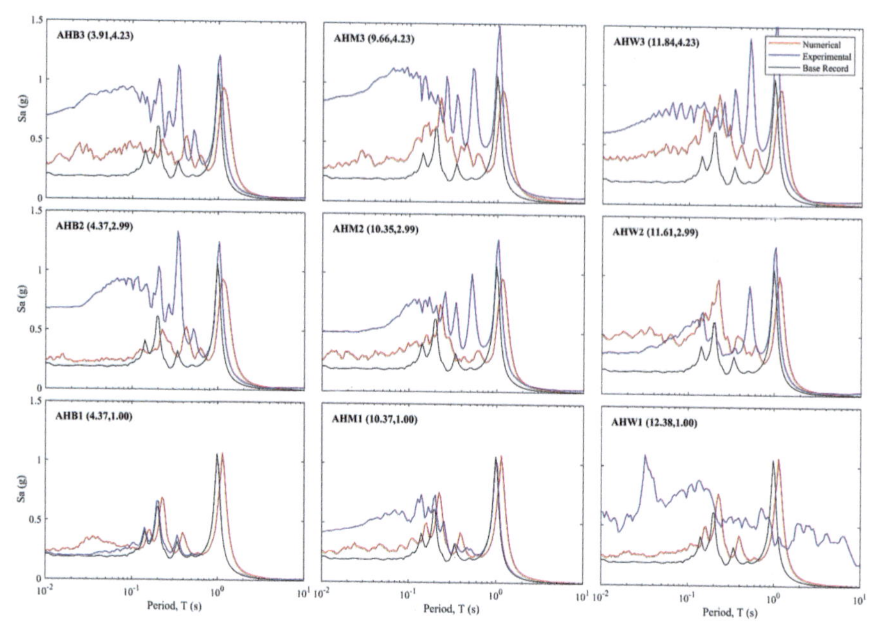

Fig. 6. Response spectrum comparison for the UCD-01 centrifuge test results and numerical model

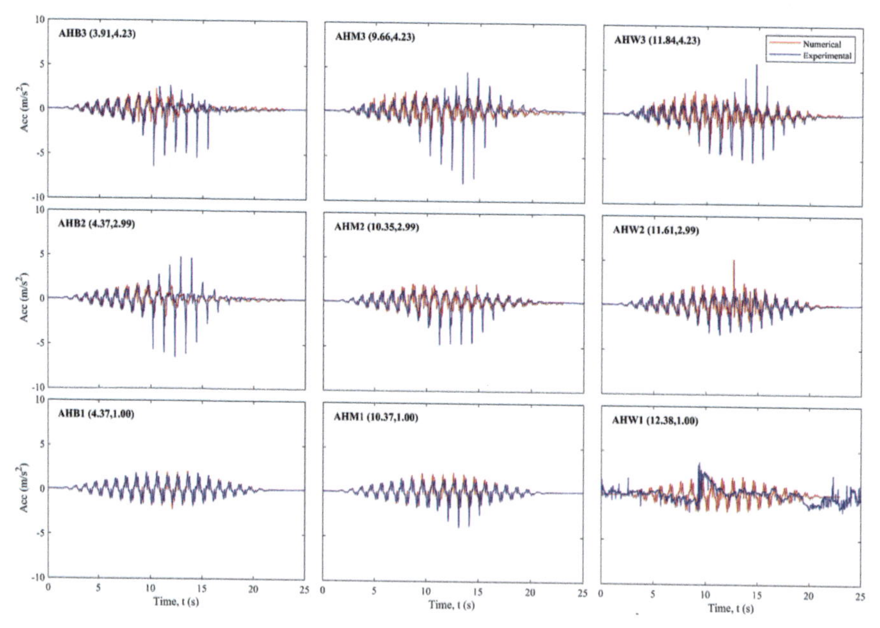

Fig. 7. Comparison of the predicted and measured acceleration response for the UCD-01 centrifuge test

Interestingly, the more significant differences in predicting the acceleration response are at the furthest control points (i.e., AHB2, AHB3). This might suggest an effect associated with the centrifuge boundary conditions.

In terms of settlements, the numerical prediction trends were similar (qualitatively) than the experimental responses (with a better performance in the middle points on the sensor array); however, there may be some significant quantitative differences with a maximum error in the order of 5 cm as observed in Fig. 8. We expect these differences be alleviated with complementary experimental testing and additional calibration, as suggested by Ghofrani and Arduino (2018).

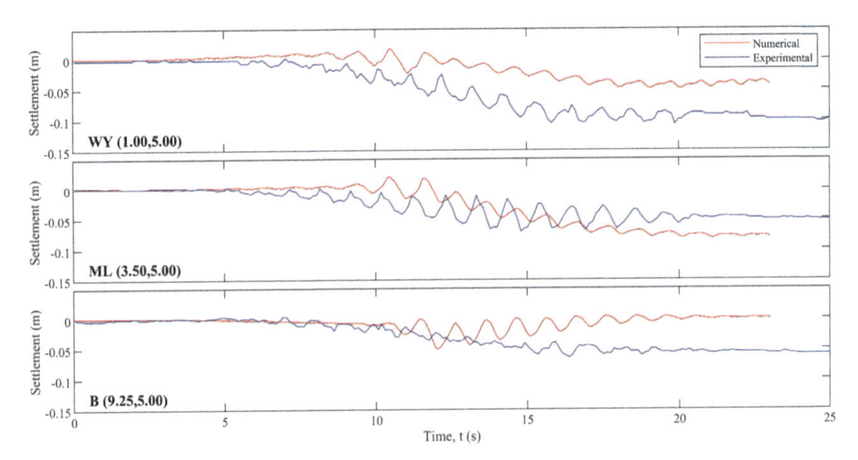

Fig. 8. Performance of predicted settlement compared to the experimental data of UCD-01 centrifuge test

Figure 9 shows the comparisons between numerical and experimental excess pore pressures. In general, the compassions show consistency for deeper sensors, with relatively lower values for the estimated excess pore pressures compared to those measured experimentally. In contrast, the excess pore pressures estimated numerically were quite different from those measured experimentally in the sensors installed in the profile's shallow parts. Finally, Fig. 10 shows the sheet-pile's deformed shape at the end of the seismic event (Fig. 10a for UCDavis experiments and Fig. 10b for RPI experiment). The deformed shape in the cases suggest similar patterns. The sheet-pile rotates clockwise because of the retained loose sand pressure. The sheet-pile rotation produces sand to go up at the toe. However, since the relative density of RPI models is lower, more deformation appears after the dynamic loading.

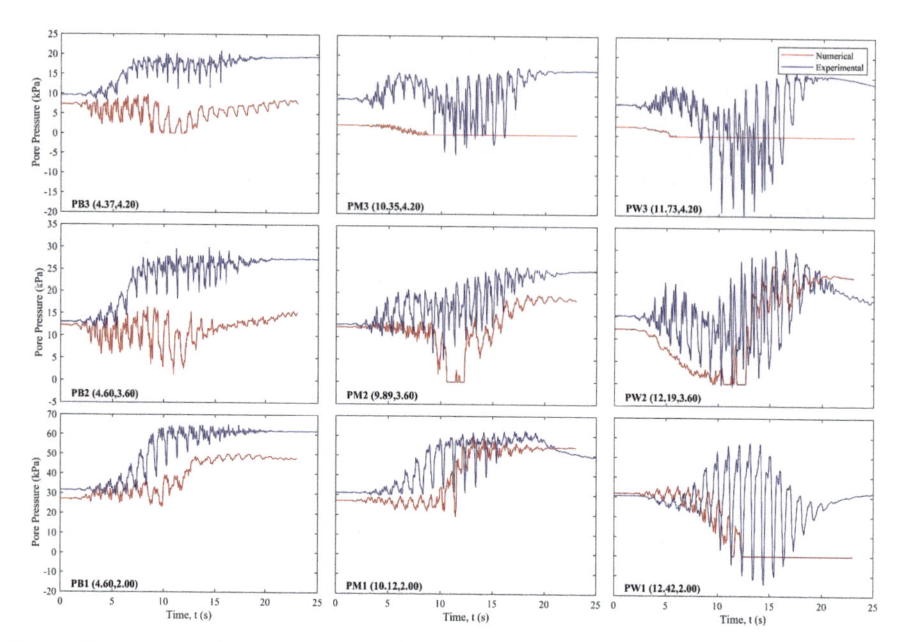

Fig. 9. Comparison of the predicted and measure pore pressure generation for the UCD-01 centrifuge test

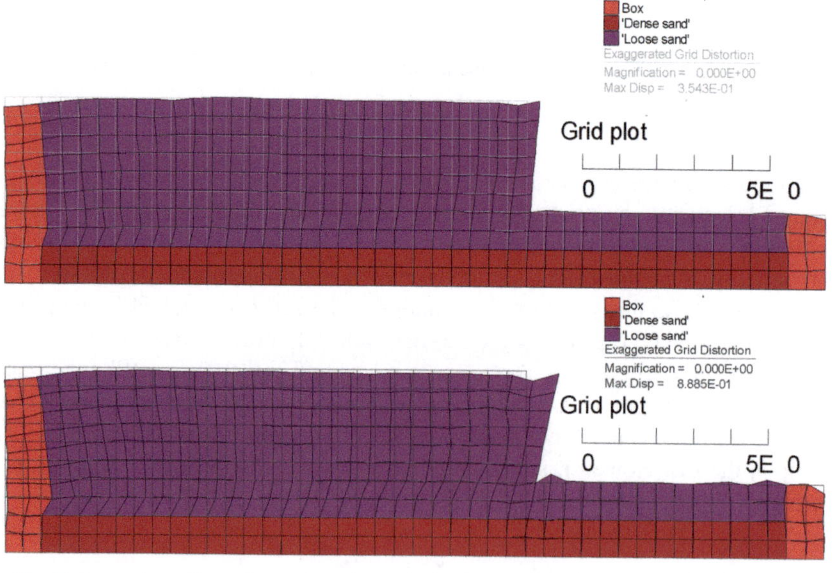

Fig. 10. Deformed configurations after dynamic loading corresponding for UCD-01 (top) and RPI-09 (bottom)

6 Conclusions

- We suggest inspecting further the experimental data for Ottawa sand to understand the differences in CSLs from different studies. This may be particularly crucial for constitutive models formulated in the critical state framework.
- In general, the PM4Sand model has been useful to inspect different aspects of the centrifuge tests, including the amplification of high, moderate, and large frequencies, the generation of excess pore pressures, and the displacements in areas close and far from the sheet pile.
- The estimated horizontal displacements numerically, unlike the settlements, were in general in good agreement with the experimentally measured displacements.
- Even though the trends in the estimated excess pore pressures are generally consistent with those measured experimentally, there are some important differences (depending on the sensor location), which may be associated with the calibration of the PM4Sand model. For example, we were not able to fully capture the slope of the experimental liquefaction curves with the PM4Sand model.
- We could not capture the dilation spikes observed during the experiments, which are translated in high-frequency content spectral amplitudes in some sensors. Future studies should examine this aspect further.

References

Ghofrani, A., Arduino, P.: Prediction of LEAP centrifuge test results using a pressure-dependent bounding surface constitutive model. Soil Dyn. Earthq. Eng. **113**, 758–770 (2018). https://doi.org/10.1016/j.soildyn.2016.12.001

Boulanger, R., Iddriss, I.M.: CPT-based liquefaction triggering procedure. J. Geotech. Geoenviron. Eng. **142**, 2 (2016)

Boulanger, R., Ziotopoulou, K.: PM4Sand (Version 3.1): A Sand Plasticity Model for Earthquake Engineering Application. Report No. UCD/CGM-17/01, Center for Geotechnical Modeling, University of California, Davis, CA, p. 112, March 2017

Bray, J.D., Dashti, S.: Liquefaction-induced building movements. Bull. Earthq. Eng. **12**(3), 1129–1156 (2014). https://doi.org/10.1007/s10518-014-9619-8

Bray, J., Macedo, J.: 6th Ishihara lecture: simplified procedure for estimating liquefaction-induced building settlement. Soil Dyn. Earthq. Eng. **102**, 215–231 (2017)

Bray, J., Macedo, J., Luque, R.: Key trends in assessing liquefaction-induced building settlement. In: Performance Based Design in Earthquake Geotechnical Engineering, pp. 16–19 (2017)

Bray, J., Markham, C., Cubrinovski, M.: Liquefaction assessments at shallow foundation building sites in the Central Business District of Christchurch, New Zealand. Soil Dyn. Earthq. Eng. J. **92**, 153–164 (2017)

Elgamal, A., Lu, J., Yang, Z.: Liquefaction-induced settlement of shallow foundations and remediation: 3D numerical simulation. J. Earthq. Eng. **9**, 17–45 (2005)

Ishihara, K., Yoshimine, M.: Evaluation of settlement in sand deposits following liquefaction during earthquakes. Soils Found. **32**(1), 173–188 (1992)

Itasca Consulting Group, Inc.: FLAC – Fast Lagragian Analysis of Continua, Version 8.1. Itasca Consulting Group, Minneapolis (2019)

Karimi, Z., Dashti, S.: Numerical and centrifuge modeling of seismic soil-foundation-structure interaction on liquefaction ground. J. Geotech. Geoenviron. Eng. **142**(1), 04015061 (2016)

Karimi, Z., Dashti, S.: Seismic performance of shallow-founded structures on liquefiable ground: validation of numerical simulation using centrifuge experiments. J. Geotech. Geoenviron. Eng. **142**(6), 04016011 (2016)

Liu, L., Dobry, R.: Seismic response of shallow foundation on liquefiable sand. J. Geotech. Geoenviron. Eng. **123**(6), 557–567 (1997). https://doi.org/10.1061/(ASCE)1090-0241(1997)123:6(557)

Luque, R., Bray, J.: Dynamic analysis of a shallow-founded building in Christchurch during the Canterbury earthquake sequence. In: Proceedings of the 6th International Conference on Earthquake Geotechnical Engineering, 1–4 November 2015, Christchurch, New Zealand (2015)

Luque, R.: Numerical analyses of liquefaction-induced building settlement. Ph.D. thesis, University of California, Berkeley (2017)

Macedo, J., Bray, J.D.: Key trends in liquefaction-induced building settlement. J. Geotech. Geoenviron. Eng. **144**(11), 04018076 (2018)

National Academies of Sciences, Engineering and Medicine: State of the Art and Practice in the Assessment of Earthquake-Induced Soil Liquefaction and Its Consequences. The National Academies Press, Washington, DC (2016). https://doi.org/10.17226/23474

Parra, A.: Ottawa F-65 sand characterization. Ph.D. thesis, University of California, Davis (2016)

Popescu, R., Prevost, J.: Centrifuge validation of a numerical model for dynamic soil liquefaction. Soil Dyn. Earthq. Eng. **12**, 73–90 (1993)

Ramirez, J., et al.: Site response in a layered liquefiable deposit: evaluation of different numerical tools and methodologies with centrifuge experimental results. J. Geotech. Geoenviron. Eng. **144**(10) (2018). https://doi.org/10.1061/(ASCE)GT.1943-5606.0001947

Travasarou, T., Bray, J., Sancio, R.: Soil-structure interaction analyses of building responses during the 1999 Kocaeli earthquake. In: Proceedings of the 8th US National Conference on Earthquake Engineering, 100th Anniversary Earthquake Conference Commemorating the 1906 San Francisco Earthquake. EERI, Paper 1887 (2006)

Yoshimi, Y., Tokimatsu, K.: Settlement of building on saturated sand during earthquakes. Soils Found. **17**(1), 23–38 (1977). https://doi.org/10.3208/sandf1972.17.23

Ziotopoulou, K., et al.: Cyclic strength of Ottawa F-65 sand: laboratory testing and constitutive model calibration. In: Proceedings of Geotechnical Earthquake Engineering and Soil Dynamics V (GEESDV 2018), Austin, Texas (2018)

LEAP-2021 Cambridge Experiments on Cantilever Retaining Walls in Saturated Soils

Xiaoyu Guan(✉), Alessandro Fusco, Stuart Kenneth Haigh, and Gopal Santana Phani Madabhushi

University of Cambridge, Cambridge, UK
xg257@cam.ac.uk

Abstract. Retaining walls are important geotechnical structures that are used worldwide. The lateral movement of retaining walls can cause damage to structures behind them or their collapse can endanger the infrastructure on the excavated side. This is particularly an issue when retaining structures are located in saturated soils in seismic areas. To address the need to understand the behaviour of retaining walls in saturated soils, an international collaborative project called LEAP (Liquefaction Experiments and Analysis Projects) was conducted. This paper describes the dynamic centrifuge tests on a retaining wall with two different embedment ratios conducted for LEAP-2020 and LEAP-2021. The behaviour of the retaining wall is studied using Particle Image Velocimetry (PIV). Additionally, the acceleration and the excess pore water pressure in the saturated Ottawa sand are presented. It is found that more dilation occurred in the test with the lower embedment ratio. However, the embedment ratio was still crucial in terms of the stability of the retaining wall.

Keywords: Soil-structure interaction · Retaining structures · Physical modelling · Liquefaction

1 Introduction

Retaining walls are commonly constructed in some coastal areas which are highly seismic. The excess pore pressure generation and ensuing soil liquefaction can exacerbate the lateral movements of the retaining walls. Due to the complex dynamic behaviour of retaining walls and different soil conditions in engineering practice, numerical modelling has been widely employed in the design of retaining walls. Despite numerous existing studies of retaining walls in saturated soils for numerical modelling, the accuracy of numerical predictions is often under question due to the required selection of input parameters, constitutive models, drainage, and boundary conditions. The determination of the inputs can be challenging if not supported by experimental data, especially when liquefaction is involved. Based on the findings from the VELACS project aiming to validate numerical implementations [1], further efforts continue in LEAP. LEAP-2020 and LEAP-2021 aim to provide experimental data for numerical validation of the behaviour of retaining walls in saturated soils. Therefore, dynamic centrifuge tests were conducted on a small-scale retaining wall model in saturated sand.

© The Author(s), under exclusive license to Springer Nature Switzerland AG 2022
L. Wang et al. (Eds.): PBD-IV 2022, GGEE 52, pp. 1785–1793, 2022.
https://doi.org/10.1007/978-3-031-11898-2_161

2 Experimental Set-Up

Two dynamic centrifuge tests at an enhanced gravity of 35-g were conducted at the Schofield Centre to investigate the dynamic behaviour of a cantilever retaining wall in saturated sand. The Turner beam centrifuge used for the tests has a capacity of 150-g ton and a radius of 4.125 m.

2.1 Model Preparation

The centrifuge tests were conducted in a rigid container with a transparent Perspex side that allowed digital image correlation. The inner dimension of the container is 730 mm × 250 mm × 398 mm (length × width × height). In other words, the container is large enough to ensure that the retaining wall is not close to the boundary. Additionally, the philosophy of centrifuge testing in LEAP-2020 and LEAP-2021 was to develop a database that can be used by numerical modellers to improve the FE codes. By using rigid boundaries that can be simulated easily in a FE model, the need to incorporate the non-reflecting boundaries in the FE analysis is avoided. The models were prepared with Ottawa F65 sand whose characteristics have been summarised by Kutter et al. [2]. For good uniformity in the soil deposit and repeatability of the tests, the automatic sand pourer available at the Schofield Centre was used [3].

Both models consisted of one thin layer of dense sand at the base. The rest of the sand at the backfill and in front of the wall was medium dense. The calculation of relative densities (D_r) in the tests referred to a maximum sand density of 1760 kg/m^3 and a minimum one of 1490 kg/m^3. Table 1 shows the relative densities of sand in both models. The sand was saturated with a highly viscous aqueous solution of hydroxypropyl methylcellulose (35 cSt), utilising the difference in the vacuum pressure between the container and the fluid tank. Such a pressure difference allowed the viscous fluid to flow into the model.

The cantilever wall made of aluminium represents a prototype wall with a thickness of 0.1 m and a height of 5.7 m. The flexural stiffness of the prototype is 5.8 MNm^2/m. The dimensions of the two models are shown in Fig. 1. The depth of the sand behind the retaining wall in either test is around 5 m. The model in LEAP-2020 has an embedment ratio (i.e., the ratio of the retained height h over the embedment depth d) of 2:1. This test aimed to promote large soil and structural deformation during seismic loading with the consideration of soil liquefaction. For comparisons with the results from LEAP-2020, the embedment ratio of the retaining wall was reduced to 1:1 in the model of LEAP-2021 in which less deformation and movement were expected subjected to seismic loading and soil liquefaction.

2.2 Instrumentation

Piezo-electric accelerometers and the pore pressure transducers (PPT) were placed at different elevations on the active and passive side of the retaining wall and the farther field to capture the dynamic response of saturated sand in different zones, as shown in

Fig. 1. Additionally, strains gauges were installed in the wall to measure the bending moments. In-flight cone penetration tests were conducted in the free field for the tip resistance to analyse the changes in soil stiffness. The Particle Image Velocimetry (PIV) technique was employed to capture the movement of the retaining wall by tracing the markers (see Fig. 4) on the wall during earthquakes.

2.3 Input Motions

The seismic input motions were generated by a servo-hydraulic earthquake actuator [4]. Figure 2 shows the acceleration time histories of the input motions. In LEAP-2020, two earthquakes with five peak cycles but different amplitudes were applied to the model. The amplitude of the second earthquake EQ2 was twice the amplitude of the previous earthquake EQ1. On the other hand, the model in LEAP-2021 experienced three earthquakes. The first specified earthquake is the same as the second earthquake in LEAP-2020, followed by two earthquakes with thirty peak cycles.

3 Behaviour of the Retaining Wall

The locations of the wall in the tests are presented in Fig. 3. The wall was approximately vertical at 1-g. However, during the swing-up phase of the centrifuge tests, the retaining wall accumulated some permanent displacements. The retaining wall with an embedment ratio of 2:1 experienced significant movement and an outward rotation of over 8 ° during swing-up, resulting in an increased actual embedment ratio before any seismic excitation. This result verified that for this investigated cantilever wall, such an embedment ratio could not even guarantee static equilibrium. For the retaining wall with a higher embedment ratio, its movement and rotation were much smaller when 35-g was reached.

Under earthquake loading, the wall moved towards the excavated side. During the stronger earthquake, the top of the wall horizontally moved ~0.7 m in the test with an h/d ratio of 2:1, much larger than that in the test with an h/d ratio of 1:1 (~0.15 m). Slight wall movements occurred during the weaker earthquake ($a_{max} =\sim 0.05$ g), but significant movements were observed when the amplitudes were ~0.1 g regardless of the number of peak cycles.

Figure 4 shows the displacement time histories of the marker at about the mid-height of the wall during earthquake loading. The displacement increments were substantial during the first several cycles and then dropped afterwards in both tests. The wall had less horizontal residual displacement and rotation when it was embedded deeper into the sand, showing that a higher embedment ratio contributes to a higher level of structural safety. When comparisons are made with the displacements accumulated during earthquake EQ1 in the test with an h/d ratio of 1:1, it should be noted that the wall with an h/d ratio of 2:1 had already experienced large permanent displacements before EQ2.

4 Behaviour of Saturated Sand

4.1 Soil Stiffness

In-flight cone penetration tests were performed at 35-g before and after earthquakes. Figure 5 shows the profiles of cone tip resistance measured in the two tests. A reasonably good match in the penetration resistance can be observed before earthquake loading, indicating similar relative densities of the sand deposits. The resistance also increased dramatically at the depth of around 3.75 m in both tests, when the penetrometer reached the dense layer near the base. Additionally, the measured resistance experienced a moderate increase after seismic loading due to the densification of the sand. However, the soil resistance was slightly smaller at the shallower layers (<2 m deep) in the'h/d = 2:1' test, compared to the soil resistance in the 'h/d = 1:1' test. The same amount of resistance, which was ~1.7 MPa, was seen at the depth of 2 m in the two tests.

Table 1. Relative densities of sand layers in the centrifuge models

Test number	D_r: dense sand layer	D_r: medium dense sand layer
LEAP-2020	86%	50%
LEAP-2021	98%	55%

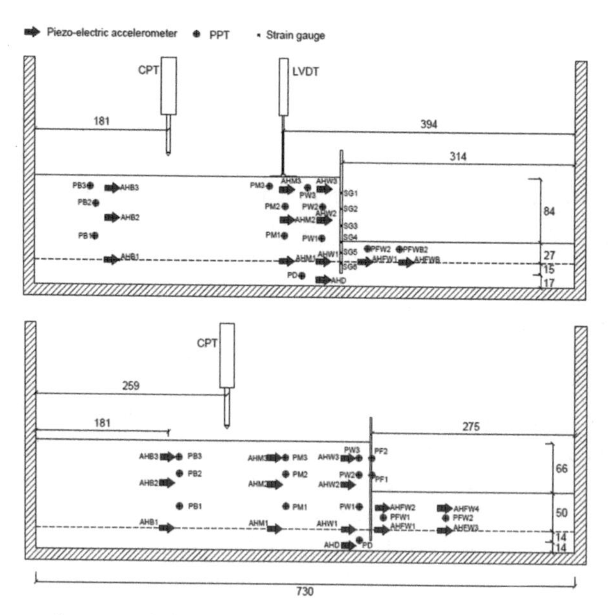

Fig. 1. Instrument layout and dimensions at model scale (mm) of the two centrifuge models

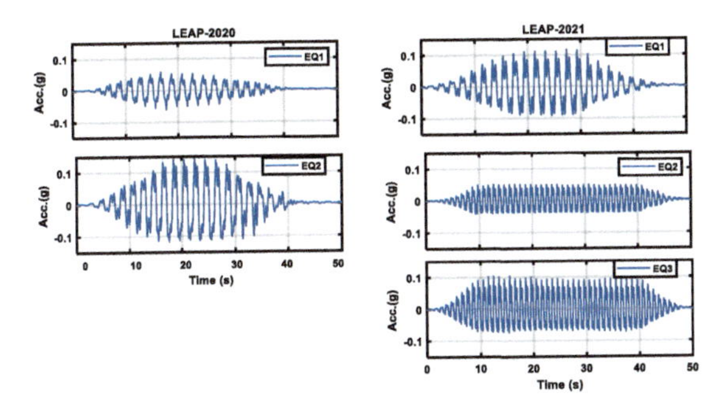

Fig. 2. Acceleration time histories applied in LEAP-2020 and LEAP-2021

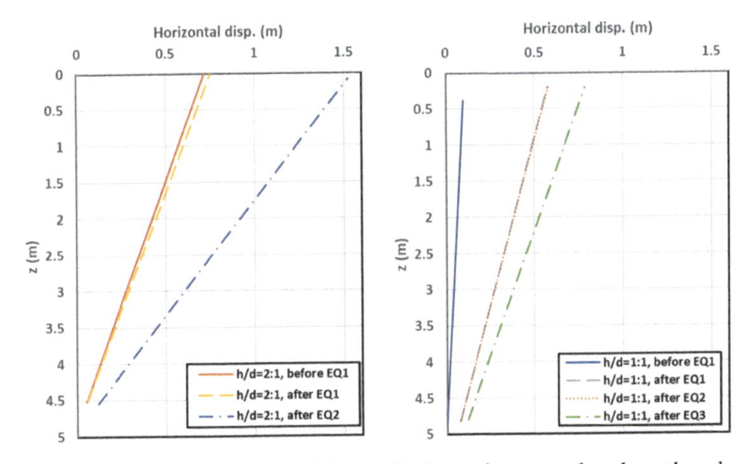

Fig. 3. The position of the retaining wall after swing-up and each earthquake

4.2 Accelerations

Since the acceleration time histories of EQ2 in LEAP-2020 (h/d = 2:1) and EQ1 in LEAP-2021 (h/d = 1:1) approximately coincided, the accelerations in the saturated sand during the mentioned earthquakes were compared and analysed. The accelerations in different areas, i.e. the free field and both sides of the retaining wall, are shown in Fig. 6 and Fig. 7. In the backfill, the accelerations time histories at the deeper sand layers were highly similar. Similar amplitudes and frequency contents were also seen on the excavated side, as shown in Fig. 7. However, after several shaking cycles, the signals showed a significant de-amplification in the shallower layers of soil, and high-frequency acceleration spikes were observed at several locations. Such phenomena resulted from liquefaction of the medium dense soil, preventing the seismic waves from travelling towards shallower soil layers. Interestingly, larger de-amplifications were observed in the test with a ratio h/d = 1:1. Possibly, this resulted from the soil densification occurring in the test with a ratio h/d = 2:1 during EQ1, and therefore partially inhibited the degree of liquefaction in the medium dense sand during the following earthquake. The other interesting observation is that larger de-amplifications occurred in the soil of the free field. This is because in the soil adjacent to the retaining wall the accelerations were transmitted to shallower soil layers by the wall itself, consequently allowing the seismic waves to reach the soil at shallow depths if liquefaction occurred.

4.3 Pore Water Pressure

During earthquake loading, excess pore water pressure generated. The excess pore water pressure on the active side of the wall during the stronger excitation with 5 peak cycles in both tests is presented in Fig. 8. The horizontal solid line represents the effective vertical stress in the 'h/d = 2:1' test, while the dash-dot line represents the effective vertical stress in the 'h/d = 1:1' test. In general, more excess pore water pressure generated in the test with an h/d ratio of 1:1. In the 'h/d = 2:1' test, the movement of the wall top before EQ2 was around three times more than that before EQ1 in the 'h/d = 1:1' test (see Fig. 3). This is possibly due to an increase in the relative density before the considered earthquake loading. Therefore, more dilation occurred, inhibiting the generation of excess pore water pressure in the 'h/d = 2:1' test.

At the shallow layers, the excess pore water pressure remained under the effective stress line during the first several cycles and then exceeded the effective stress line in the test with more embedment. However, the excess pore water pressure in the test with less embedment was below the effective stress line. This is consistent with the decrease in the amplitudes of the accelerations displayed in Fig. 7. For the deep layers, the excess pore water pressure never exceeded the effective stress in both tests, indicating that full liquefaction was not achieved.

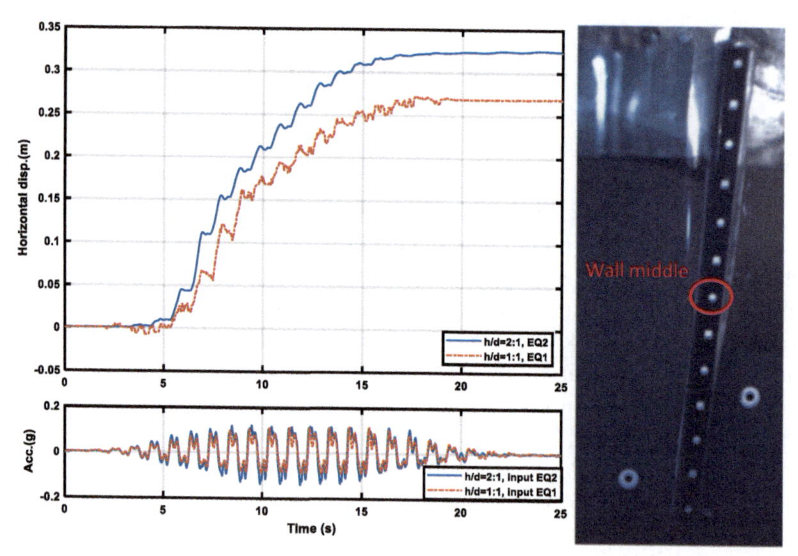

Fig. 4. Horizontal displacements time histories of the retaining wall at the mid-height of the wall

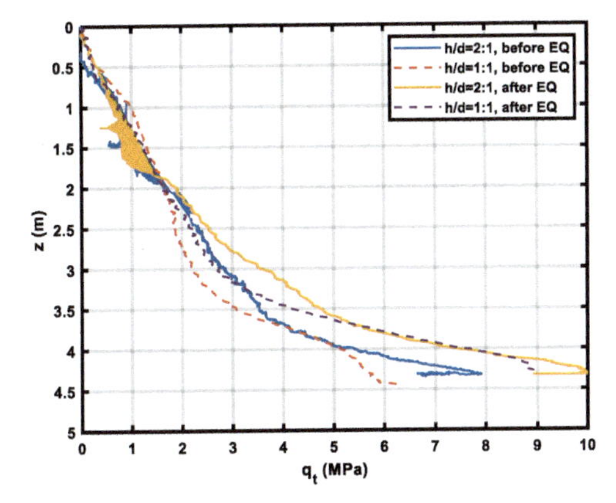

Fig. 5. Cone penetration resistance measured before and after earthquake loading

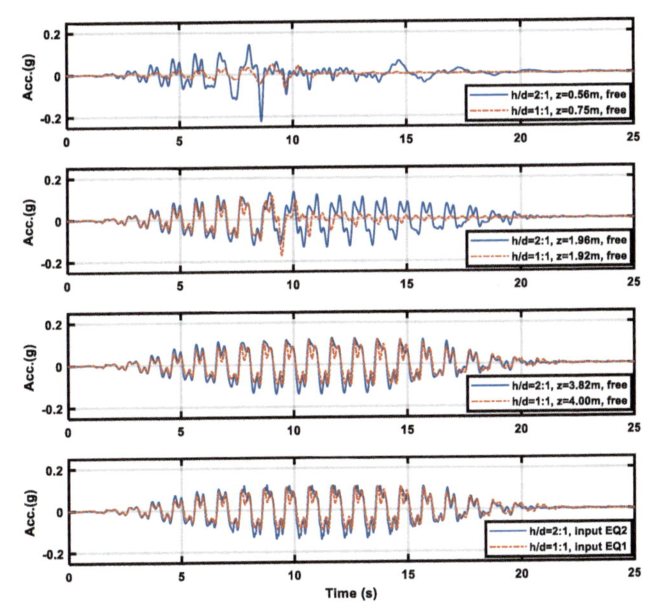

Fig. 6. Acceleration time histories in the free field of both tests recorded by accelerometers (top to bottom): AHB3, AHB2, AHB1

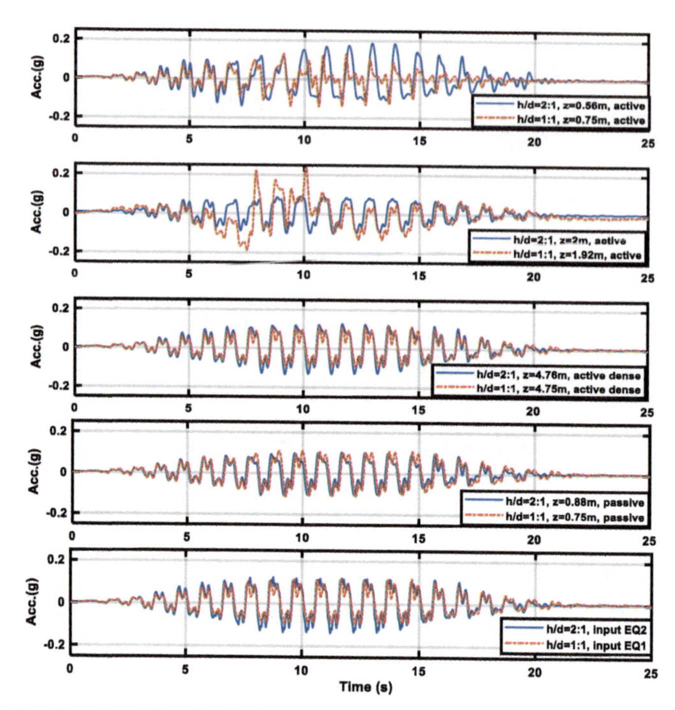

Fig. 7. Acceleration time histories in the sand on both sides of the retaining wall recorded by accelerometers (top to bottom): AHW3, AHW2, AHD and AHFW1/AHFW2

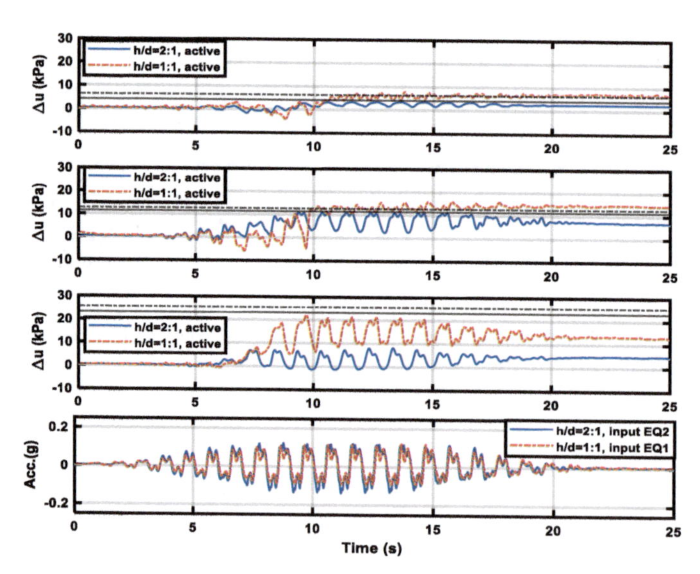

Fig. 8. Excess pore water pressure recorded by PPTs (top to bottom): PW3, PW2 and PW1

5 Conclusions

Two dynamic centrifuge tests were performed on a small-scale retaining wall model with an embedment ratio of 2:1 and 1:1 respectively. As expected, the impact of the embedment ratio on the safety of the retaining wall was significant. Compared to LEAP-2021 (h/d = 1:1), larger wall movement and rotation occurred during swing-up in LEAP-2020 (h/d = 2:1), likely resulting in more densification in the sand immediately behind the wall before seismic loading and consequently contributing to more dilation. Despite that, the displacements of the wall in the test of LEAP-2020 were still larger than in the test of LEAP-2021.

References

1. Arulanandan, K., Scott, R.F.: Verification of numerical procedures for the analysis of soil liquefaction problems. A.A. Balkema (1993)
2. Kutter, B.L., et al.: LEAP-GWU-2015 experiment specifications, results, and comparisons. Soil Dyn. Earthq. Eng. **113**, 616–628 (2018)
3. Madabhushi, S.P.G., Houghton, N.E., Haigh, S.K.: A new automatic sand pourer for model preparation at University of Cambridge. In: Proceedings of the 6th International Conference on Physical Modelling in Geotechnics, pp. 217–222. Taylor & Francis, London (2006)
4. Madabhushi, S.P.G., Haigh, S.K., Houghton, N.E., Gould, E.: Development of a servo-hydraulic earthquake actuator for the Cambridge Turner beam centrifuge. Int. J. Phys. Model. Geotech. **12**(2), 77–88 (2012)

Repeatability Potential and Challenges in Centrifuge Physical Modeling in the Presence of Soil-Structure Interaction for LEAP-2020

Evangelia Korre[1(✉)], Tarek Abdoun[2], and Mourad Zeghal[2]

[1] ETH Zurich, Stefano-Franscini-Platz 5, 8093 Zurich, Switzerland
evkorre@ethz.ch

[2] Rensselaer Polytechnic Institute, 110 8th Street, Troy, NY 12180, USA

Abstract. The LEAP (Liquefaction Experiment and Analysis Project) is a continuing international collaboration to create a reliable databank of high-quality experimental results for the validation of numerical tools. This paper investigates the response of a floating rigid sheet-pile quay wall under conditions of seismically induced liquefaction, embedded in dense sand and supporting a saturated liquefiable soil deposit. The experimental challenges related to repeatability in physical modeling in such a soil-structure-interaction regime are also discussed. To this end, three experiments performed at Rensselaer Polytechnic Institute (RPI) as part of the experimental campaign for the LEAP-2020 are discussed herein. Models RPI_REP-2020 and RPI10-2020 investigate the repeatability potential in centrifuge modeling in the presence of soil-structure-interaction. Model RPI_P-2020 is the pilot test of the LEAP-2020 experimental campaign at RPI and investigates the effect of the wall's initial orientation on the system's dynamic response and soil liquefaction, as a possible "defect" in the model construction procedure. The three models were built in a consistent way, employed comparable instrumentation layout while simulating the same prototype and comparable soil conditions. The three models were subjected to the same acceleration target input motion, which was repeated across all three models with high consistency.

Keywords: Centrifuge modeling · Liquefaction · Sheet-pile wall

1 Introduction

Several case histories in past earthquakes have underscored the severity of infrastructure damage in coastal areas due to seismically triggered liquefaction [1, 2]. Sheet-pile quay walls have shown to be particularly vulnerable to this phenomenon developing excessive lateral displacements towards the waterfront, as a combined result of the seismic excitation and the interaction with the backfill [3, 4].

Several experimental studies have been conducted by means of shake table tests as well as geotechnical centrifuge testing, mainly to investigate the effect of the sheet-pile wall response on the structures and their foundations resting on the backfill, under seismically induced liquefaction [5, 6]. Several experimental investigations utilizing a

© The Author(s), under exclusive license to Springer Nature Switzerland AG 2022
L. Wang et al. (Eds.): PBD-IV 2022, GGEE 52, pp. 1794–1801, 2022.
https://doi.org/10.1007/978-3-031-11898-2_162

geotechnical centrifuge have also been conducted to shed light on the sheet-pile wall response as a mitigation measure against earthquake-induced liquefaction [7–9].

This paper presents a summary of the results from three centrifuge tests performed at Rensselaer Polytechnic Institute (RPI) as part of the experimental campaign for the Liquefaction Experiments and Analysis Project (LEAP). The LEAP is an ongoing international collaboration aiming at generating a databank of reliable experimental data for seismically induced soil liquefaction, which can be utilized for the validation of numerical tools.

One of the experiments presented herein was the pilot test of the experimental campaign (RPI_P-2020). It aimed at shedding light on the mechanisms of the system's response and contributed to the design improvement for the subsequent experiments. The pilot test was followed by the benchmark test of the LEAP-2020 experimental campaign at RPI, RPI10-2020, which was later repeated in model test RPI_REP-2020. The overview of the experiments discussed herein is presented in Table 1. The main aspects of the response as well as the challenges and lessons learned from these experiments are discussed in the following sections.

Table 1. Overview of the centrifuge experiments.

Experiment	$Dr(\%)$	$a_{max}(g)$	Details
RPI_P-2020	65/90	0.17	Wall initial rotation $\approx 2°$
RPI_REP-2020	65/90	0.16	–
RPI10-2020	65/90	0.17	–

2 Experimental Methodology

The experimental layout for the benchmark test RPI10-2020 presented in Fig. 1, illustrates the soil model consisting of two layers, a dense one with soil relative density $D_r \approx 90\%$ and a medium dense one with $D_r \approx 65\%$. The dense layer provided the embedment of the floating rigid sheet-pile, which supported the 3m-deep excavated backfill. Practically the same layout was adopted also for the pilot test, with small variations in the elevations of some sensors.

A consistent experimental methodology was adopted for the construction of the model tests presented herein. The experimental methodology entailed preparation of the rigid aluminum container, construction and saturation of the model and application of the testing sequence. All models observed the scaling laws for centrifuge testing [10] and were conducted at 23g gravitational field. All dimensions henceforth are presented in prototype scale.

2.1 Container Preparation

The utilized aluminum container for the centrifuge experiments was practically rigid, and had a PMMA window along its longitudinal side (Fig. 2). Preparation of the container prior to model construction, established highly consistent boundary conditions for all model tests performed.

The boundary conditions entailed a rough surface at the base of the model and consistently smooth interface between the sheet-pile wall and the container sides. The former was facilitated by means of a water and tear resistant sanding sheet with cloth backing which was attached with glue on the bottom of the container, and minimized the sliding of the soil model relative to the container during shaking (Fig. 2a). The latter was achieved (a) by means of teflon tape, attached to the container sides at the installation location of the sheet-pile wall (Fig. 2a–b); and (b) by means of high viscosity silicone grease, which was applied along the sides of the wall. The tape layers contributed to effortless sliding of the sheet-pile wall inside the container, creating consistent boundary conditions on either side of the wall while at the same time minimizing the gap between the wall and the container boundaries. The role of the silicone grease was also instrumental in the prevention of grain migration from the backfill to the excavated side. Upon installation of the wall, clamps and tape were installed to prevent any vertical sliding or lateral rotation of the structure during model construction (Fig. 2b–c).

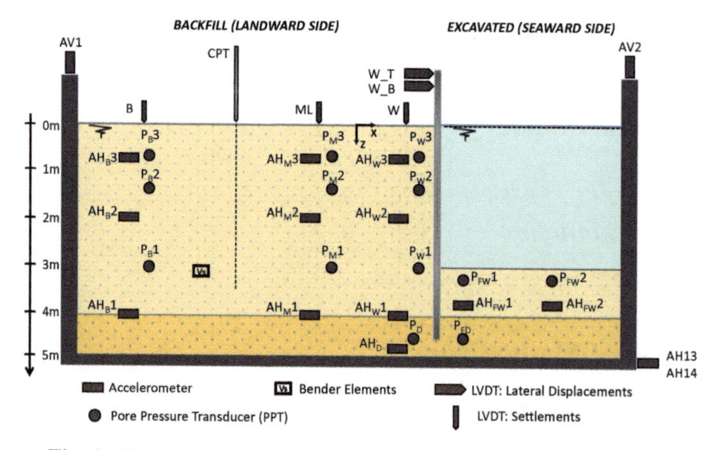

Fig. 1. Experimental layout of benchmark model test RPI10-2020.

2.2 Model Construction

The soil models were pluviated with dry Ottawa F-65 sand by means of air pluviation in horizontal layers, maintaining consistent drop-height and velocity to ensure consistent relative density along the depth. Compaction through tapping took place after pluviation of the bottom layers, to achieve the mass density of approximately 1.73 kg/m^3 ($D_r \approx 90\%$). The corresponding achieved mass density for the medium dense layers was approximately 1.66 kg/m^3, corresponding to $D_r \approx 65\%$.

Between the soil layers, the accelerometers and pore pressure transducers (PPTs) were embedded, observing the prescribed locations. After construction, the model was transferred to the centrifuge basket and was prepared for saturation as described in [11].

The clamps supporting the wall were removed, thus allowing the model container to be sealed. Saturation took place at a slow supply rate of viscous fluid at 23*cP* viscosity, to prevent hydraulic piping failure and avoid leaving dry pockets of soil inside the model. Due to technical limitations the free standing water in the excavated side could not be instrumented. Nevertheless, the influence of the water on the overall system's response is not expected to have influenced the results significantly.

Before spinning the centrifuge, the verticality of the sheet-pile was visually inspected with a spirit level in all model tests. For RPI_P-2020, it was found that the wall had tilted about 1° towards the excavated side, corresponding approximately to 0.03 m in prototype scale displacement at the soil surface. Since this was observed only during the pilot test, this "defect" was attributed to the installation technique of the wall in the pilot model. This was one of the main challenges during model construction. Further improvement of the installation technique in the subsequent tests included improving the attachment of the securing clamps and tape, and taking additional measurements during the wall installation.

During spin-up of the centrifuge, the rotated sheet-pile in the pilot model test, sustained further rotation of about 0.7°, as a result of the de-stabilizing component from the centrifugal acceleration field. Therefore, the total initial rotation of the sheet-pile wall in the pilot test prior to shaking was assumed to be approximately 1.7°, corresponding to displacement of about 0.06 m in prototype scale at the soil surface.

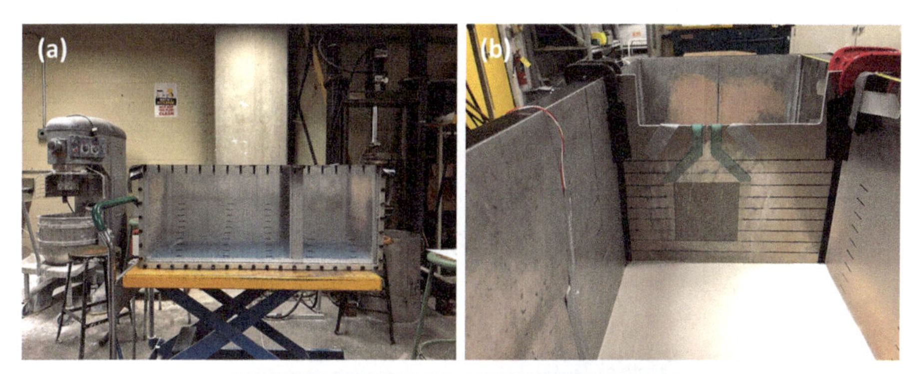

Fig. 2. Photos during model construction of the RPI10-2020 model.

2.3 Testing Sequence

All models discussed herein were subjected to a bender element trial before and after the destructive shaking, revealing consistent shear wave velocity measurements for all models. This further corroborated the consistency in the achieved relative density for all model tests presented herein.

The models were subjected to a tapered sinusoidal acceleration input motion along their longitudinal direction, with 5 cycles at the maximum target acceleration of $0.15g$. The achieved horizontal input motion was recorded by accelerometers AH13 and AH14, attached to the base of the rigid container (Fig. 1). Their average response is illustrated in Fig. 3a, while the vertical response (average of AV1 and AV2) is shown in Fig. 3b. Observe that the amplitude of the vertical acceleration response of the rigid container was minimal and is not expected to have had any major impact on the model response. Figure 3 also confirms the high degree of repeatability of the input motion for RPI_P-2020, RPI10-2020 and RPI_REP-2020.

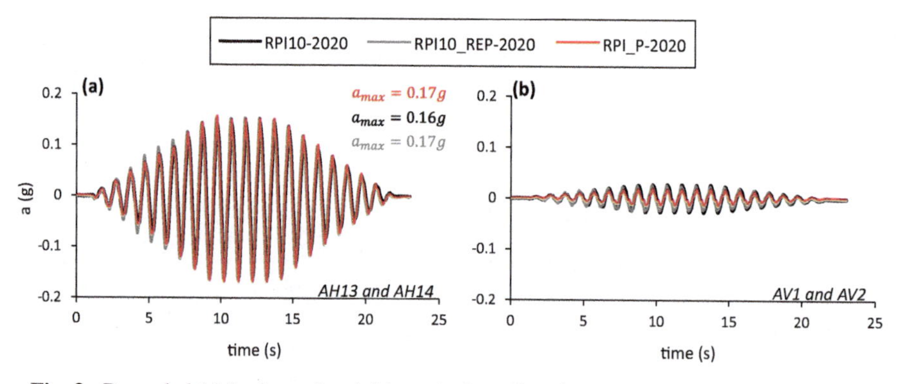

Fig. 3. Recorded (a) horizontal and (b) vertical acceleration response of the rigid container.

3 Experimental Results

The observed acceleration response in the dense stratum was practically identical to the input motion and the corresponding excess pore pressure ratio $R_u \leq 0.60$ in all performed experiments. Due to space limitations, Fig. 4 illustrates the comparison of the response in terms of accelerations only for location AHW3 and excess pore pressure ratio $R_u = \frac{\Delta u}{\sigma_v'}$, for location PW2 (PW3 malfunctioned for PRI_P-2020). A more detailed description of the experimental results will become available in an upcoming journal publication.

High consistency is observed among all model tests in the recorded acceleration of the AHW3 location (Fig. 4). Strong dilation spikes (negative acceleration peaks) reveal the re-stiffening of the soil as the sheet-pile rotated towards the excavated side (seawards). Slightly higher dilation peaks are observed in RPI_REP-2020 compared to RPI10-2020.

In terms of excess pore water pressure (EPWP), the comparison is satisfactory revealing good agreement in terms of the amplitude and accumulation rate of EPWP. As illustrated in Fig. 4, the soil in models in RPI10-2020 and RPI_REP-2020 liquefied ($R_u \approx 1$) approximately at $t \approx 13.5$ s, whereas in the pilot test soil liquefaction occurred approximately 3 s later. Moreover, discrepancies in the negative pore pressure peaks are observed, with RPI10-2020 exhibiting significantly stronger dilative response compared to RPI_P-2020. This may be partly attributed to the deeper embedment depth (approximately 0.5 m) of the PW2 sensor in the pilot test. An additional factor accounting for this difference may also be the initial seaward sheet-pile rotation in the pilot model, as will be discussed in the following sections.

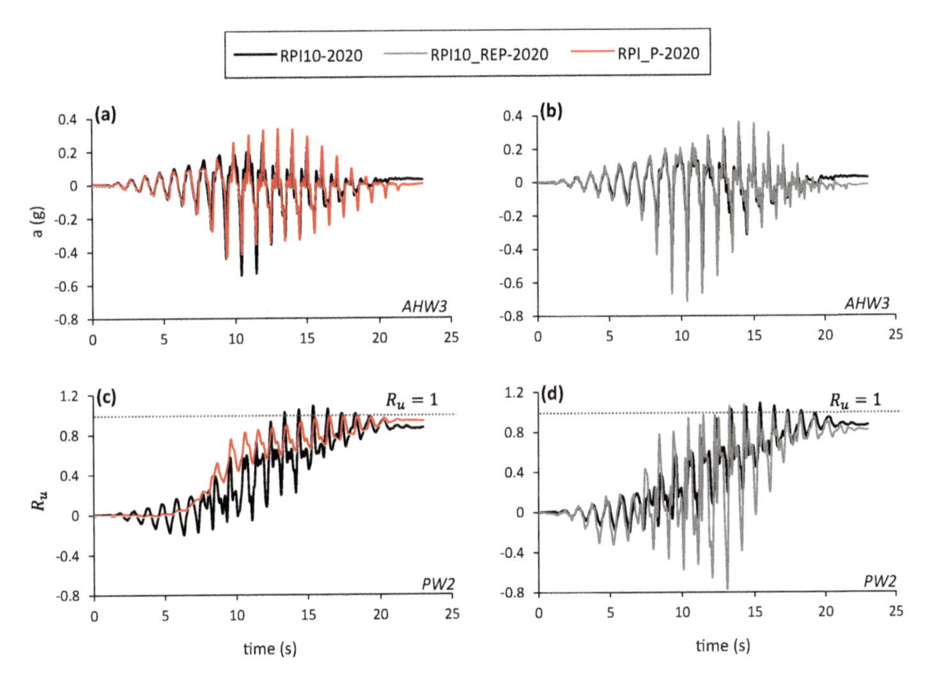

Fig. 4. Comparison of the response in terms of acceleration and excess pore pressure ratio (R_u).

Consistently with the acceleration results, RPI_REP-2020 developed significantly stronger dilation adjacent to the wall compared to RPI10-2020, which may be attributed to localized difference in relative density or soil disturbance. This discrepancy between the two seemingly identical models underscores the inherent challenge in physical modeling to accurately replicate localized soil conditions during model building, particularly in areas under strong soil-structure interaction (SSI) influence.

The pilot model test exhibits earlier and more rapid accumulation of settlements (Fig. 5a) and seaward wall lateral displacements (Fig. 5b) compared to RPI10-2020. It is postulated that the initial seaward rotation of the sheet-pile wall towards the excavated side in the pilot test created a bias in its response, thus favoring its outward rotation. As shown in Fig. 5b, the sheet-pile wall in all model tests did not re-center during its oscillation, but rather it accumulated permanent seaward displacements. Nevertheless in the pilot model test, it is evident that the range of oscillation for the sheet-pile is slightly smaller, which indicates its increased inability to rotate towards the backfill.

Both in terms of settlements (Fig. 5a) as well as in terms of lateral displacements for the sheet-pile wall (Fig. 5b), RPI10-2020 and RPI_REP-2020 exhibited high consistency both in terms of accumulation rate as well as ultimate displacement. This further corroborates the high degree of achieved repeatability between the two model tests.

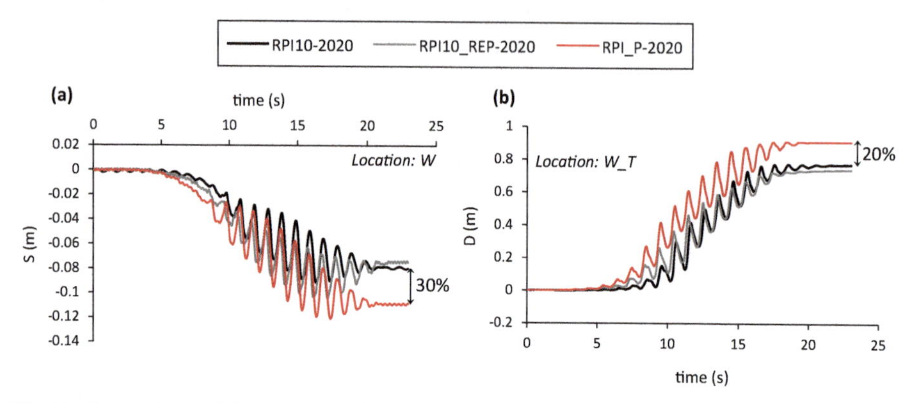

Fig. 5. Comparison of the response (a) in terms of settlements and (b) in terms of sheet-pile lateral displacements.

4 Conclusions

Three experiments were performed at RPI as part of the LEAP-2020 experimental campaign, investigating the system response of a floating rigid sheet-pile quay wall supporting a liquefiable sand deposit. Installation of the floating sheet-pile wall was especially challenging, and its installation in the pilot model test introduced an imperfection to the model, by allowing the wall to rotate towards the excavated side. This rotation further increased during spin-up, accentuating in this way the initial rotation of the wall prior to shaking.

Improvement of the model building technique minimized such imperfections, leading to the benchmark (RPI10-2020) and the repeatability model (RPI_REP-2020) to have highly consistent response both for the soil settlements as well as for the sheet-pile wall displacements. RPI_REP-2020 revealed a stronger dilative response compared to RPI10-2020, which further underlined the inherent limitations of accurately reproducing soil conditions in a model, especially close to the structure.

On the other hand, the initial rotation of the sheet-pile wall led to significantly milder dilative response compared to RPI10-2020, which was associated with larger settlements and larger lateral displacements.

References

1. DesRoches, R., Comerio, M., Eberhard, M., Mooney, W., Rix, G.J.: Overview of the 2010 Haiti Earthquake. Earthq. Spectra **27**(1_suppl1), 1–21 (2011). https://doi.org/10.1193/1.3630129
2. Sugano, T., Nozu, A., Kohama, E., Shimosako, K., Kikuchi, Y.: Damage to coastal structures. Soils Found. **54**(4), 883–901 (2014). https://doi.org/10.1016/j.sandf.2014.06.018
3. Hamamoto, Y.T., Nagano, T.: Analytical study of the sheet-pile quay damaged by 2005 Fukuoka Earthquake [Online]. (2005). https://iiirr.ucalgary.ca/files/iiirr/A5-4_.pdf
4. Ku, C.-Y., Jang, J.-J., Lai, J.-Y., Hsieh, M.-J.: Modeling of dynamic behavior for port structures using the performance-based seismic design. J. Mar. Sci. Technol. **25**(6), 732–741 (2017). https://doi.org/10.6119/JMST-017-1226-14
5. Motamed, R., Towhata, I., Honda, T., Yasuda, S., Tabata, K., Nakazawa, H.: Behaviour of pile group behind a sheet pile quay wall subjected to liquefaction-induced large ground deformation observed in shaking test in E-defense project. Soils Found. **49**(3), 459–475 (2009). https://doi.org/10.3208/sandf.49.459
6. Tazoh, T., Sato, M., Jang, J., Gazetas, G.: Centrifuge tests on pile foundation-structure systems affected by liquefaction-induced soil flow after quay wall failure. In: Kikuchi, Y., Otani, J., Kimura, M., Morikawa, Y. (eds.) Advances in Deep Foundations, p. 409. Taylor and Francis Group, London (2007)
7. Adalier, K., Elgamal, A.-W., Martin, G.R.: Foundation liquefaction countermeasures for earth embankments. J. Geotech. Geoenviron. Eng. **124**(6), 500–517 (1998). https://doi.org/10.1061/(ASCE)1090-0241(1998)124:6(500)
8. Hausler, E.A.: Influence of Ground Improvement on Settlement and Liquefaction: A Study on Field Case History Evidence and Dynamic Geotechnical Centrifuge Tests. University of California Berkeley (2002)
9. Saha, P., Horikoshi, K., Takahashi, A.: Performance of sheet pile to mitigate liquefaction-induced lateral spreading of loose soil layer under the embankment. Soil Dyn. Earthq. Eng. **139**, 106410 (2020). https://doi.org/10.1016/j.soildyn.2020.106410
10. Garnier, J., et al.: Catalogue of scaling laws and similitude questions in geotechnical centrifuge modelling. Int. J. Phys. Model. Geotech. **7**(3), 01–23 (2007). https://doi.org/10.1680/ijpmg.2007.070301
11. Korre, E., Abdoun, T., Zeghal, M.: Verification of the repeatability of soil liquefaction centrifuge testing at rensselaer. In: Kutter, B.L., Manzari, M.T., Zeghal, M. (eds.) Model Tests and Numerical Simulations of Liquefaction and Lateral Spreading, pp. 385–400. Springer, Cham (2020). https://doi.org/10.1007/978-3-030-22818-7_19

Numerical Modeling of the LEAP-RPI-2020 Centrifuge Tests Using the SANISAND-MSf Model in FLAC3D

Keith Perez, Andrés Reyes, and Mahdi Taiebat(✉)

Department of Civil Engineering, University of British Columbia, Vancouver, BC, Canada
{kcperez,reyespa}@mail.ubc.ca, mtaiebat@civil.ubc.ca

Abstract. The SANISAND-MSf model is an extension of a reference and established critical state bounding surface plasticity model for sands with two new constitutive ingredients: a memory surface and a semifluidized state. Each ingredient is designed to significantly enhance the reference model in capturing the progressive reduction of mean effective stress during undrained cyclic shearing for various cyclic stress ratios in the pre-liquefaction, and the subsequent development of large shear strains in the post-liquefaction, respectively. This paper presents the validation of this model using the results of ten centrifuge tests from LEAP-RPI-2020. These experiments consisted of submerged liquefiable soil deposits, supported by a 4.5 m sheet-pile wall, and subjected to ramped sine wave motions. Fully coupled dynamic analyses were conducted in the finite difference computational platform FLAC3D. Comparisons between the numerical simulations and experimental results reveal that the employed modeling approach is effective in reproducing several essential aspects of the system response.

Keywords: Numerical modeling · Centrifuge test · Liquefaction · Sheet-pile

1 Introduction

The Liquefaction Experiments and Analysis Projects (LEAP) [1] are recognized for providing a wealth of element level and centrifuge testing data for the validation of numerical models designed to simulate liquefaction response of sands. Predictability of such numerical models depends largely on the constitutive model used for cyclic response of sand, and therefore detailed validation of such model allows determining its capabilities and limitations, and the domain of reliable application. This approach is being used extensively to assess the performance of different constitutive models and numerical modeling frameworks. Within the context of LEAP, validation starts with calibrating model constants using as target the results of a large variety of element level cyclic tests specifically conducted for this project. Following the calibration, numerical models of centrifuge tests subjected to seismic motions are analyzed. This allows for an assessment of the calibrated model to reasonably capture various key features of response in the boundary value problem. In these liquefaction-related problems, special

© The Author(s), under exclusive license to Springer Nature Switzerland AG 2022
L. Wang et al. (Eds.): PBD-IV 2022, GGEE 52, pp. 1802–1811, 2022.
https://doi.org/10.1007/978-3-031-11898-2_163

consideration is placed on engineering demand parameters such as determining the onset of liquefaction and seismically-induced volumetric and shear deformations.

This paper presents the validation of the newly formulated SANISAND-MSf constitutive model [2], using the laboratory and centrifuge tests completed for LEAP-RPI-2020. The LEAP-RPI-2020 centrifuge model tests consisted of a 5 m-depth two-layered submerged deposit of Ottawa F65 sand, where the top 3 m backfill is supported by a 4.5 m sheet-pile wall, and the system is subjected to a ramped sine wave motion at the base. The SANISAND-MSf model constants were determined based on a set of hollow cylinder cyclic torsional shear tests conducted on specimens with an isotropic initial stress state, i.e., with a lateral stress coefficient $K_0 = 1.0$. Then, the centrifuge models were simulated in the finite difference computer platform FLAC3D. The dynamic analyses showed that SANISAND-MSf can reproduce the most important features of recorded response, provided the impact of the mean shear stress, namely K_0 and τ_{mean}, is taken into consideration in the calibration process.

2 SANISAND-MSf Model

The newly proposed SANISAND-MSf model by Yang et al. [2] extends the seminal work of the Dafalias and Manzari [3] by incorporating two new and key constitutive ingredients which improve its performance in modeling the undrained cyclic of sands. The first one is a memory surface (MS) designed to better control of the rate of plastic deviatoric and volumetric strains in the pre-liquefaction stage. The second new ingredient is the concept of the semifluidized state (Sf), which is used to degrade the plastic stiffness and dilatancy to model the progressive development of large cyclic shear strains in the post-liquefaction stage. These modifications come at the cost of at least two new model constants for each feature. In this work, the initial liquefaction is defined based on the first instance of the mean stress p reaching very small threshold value. Additional minor modifications were also introduced in the reference model. The combined use of the MS and Sf state components provides a significant improvement of the reference model for liquefaction related problems.

Model constants associated to the reference model for Ottawa F65 sand were obtained from Ramirez et al. [4] and Barrero et al. [5], who mainly followed the procedure outlined in Taiebat et al. [6]. Then, model constants associated to the MS, μ_0 and u, and the Sf state, x and c_l, were determined based on results of the hollow cylinder cyclic torsional shear (HCCTS) undrained tests performed by Vargas et al. [7]. The tests chosen for this initial round of calibration were conducted on reconstituted samples with a relative density $D_r = 50\%$, initial mean effective stress $p_0 = 100$ kPa, and coefficient of initial lateral pressure $K_0 = 1.0$, and sheared with cyclic stress ratios τ_{cyc}/p_0 (CSR) ranging 0.099–0.191. Figure 1 show the experiments and performance of the model for a HCCTS test with a CSR = 0.149, in which the MS component is responsible for the excellent fit of the stress-path response, while the Sf component independently provides the flexibility to capture the cyclic shear strains in post-liquefaction. Figure 2 presents the liquefaction strength curve from the experiments and simulations, showing the model capabilities in capturing the observed response for different levels of cyclic shear stresses. The model constants for the calibration are shown in Table 1.

Table 1. SANISAND-MSf model constants for Ottawa F65 sand.

Model constant	Symbol	Value	Model constant	Symbol	Value
Elasticity	G_0	125	Yield surface	m	0.01
	ν	0.05	Kinematic hardening	n^{b}	2.3
Critical state	M	1.26		$h_0^{'}$	4.6
	c	0.735		c_{h}	0.968
	$e_{\mathrm{c}}^{\mathrm{ref}}$	0.78	Fabric dilatancy	$z_{\max}$	15
	λ_{c}	0.0287		c_{z}	2000
	ξ	0.7	Memory surface	μ_0	1.99
				u	1.32
Dilatancy	n^{d}	2.5	Semifluidized state	x	3.5
	$A_0^{'}$	0.626		c_{l}	35
	n_{g}	0.9			

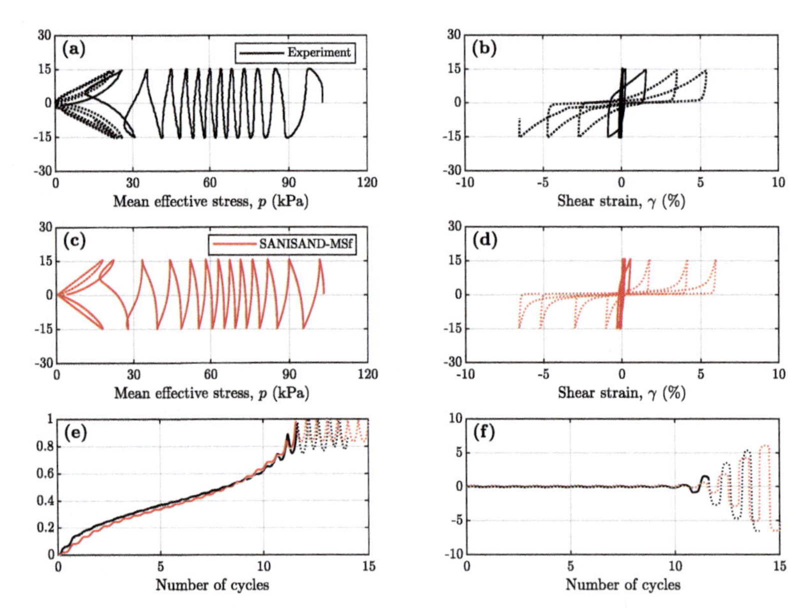

Fig. 1. Simulations compared with experiments in undrained HCCTS tests on isotropically consolidated ($K_0 = 1.0$) samples of Ottawa F65 sand with $D_r = 50\%$, $p_0 = 100$ kPa, and sheared with CSR = 0.149: (a,b) experimental data from Vargas et al. [7]; and (c,d) simulations of SANISAND-MSf, (e,f) comparisons between experiments and simulations in terms of pore pressure generation and shear strain development.

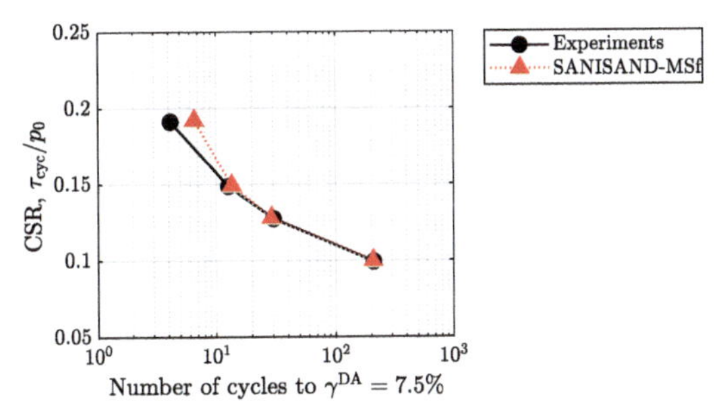

Fig. 2. Liquefaction strength curves of Ottawa F65 sand from undrained HCCTS tests [7] and corresponding simulations on samples consolidated to $D_r = 50\%$ and $p_0 = 100$ kPa at $K_0 = 1.0$.

3 Numerical Modeling of LEAP-RPI-2020 Centrifuge Tests

In LEAP-RPI-2020, the centrifuge models consisted of submerged two-layered deposits of Ottawa F65 sand. The top and bottom layers were deposited with relative densities D_r ranging from 55–76 and 65–93%, respectively. The deposits were supported by a 4.5 m-high sheet-pile wall embedded 0.5 m in the bottom layer. The centrifuge models were shaken at their base using a ramped sine wave with different effective peak ground accelerations (PGA_{eff}) [1]. Table 2 shows a summary of the ten experiments used in this study and their relevant characteristics.

Table 2. LEAP-RPI-2020 centrifuge tests.

Centrifuge model	Top layer D_r (%)	Bottom layer D_r (%)	PGA (g)	PGA_{eff} (g)
EU2	62.8	87.9	0.173	0.158
KAIST2	63.3	94.2	0.158	0.138
KyU3	64.7	64.7	0.158	0.135
RPI9	63.7	87.9	0.180	0.167
RPI10	65.9	91.3	0.169	0.156
RPI11	67.2	91.3	0.156	0.158
RPI12	56.0	91.3	0.158	0.153
RPI13	76.2	87.9	0.174	0.162
UCD1	70.8	93.2	0.187	0.170
ZJU1	74.0	74.0	0.190	0.167

The prototype scale numerical model of the centrifuge experiments was set up in the non-linear finite difference computer platform FLAC3D v7 [8], which allows for solving the dynamic equation of motion using an explicit time-integration scheme, as well as the use of coupled solid-pore fluid interaction. Three-dimensional brick elements were

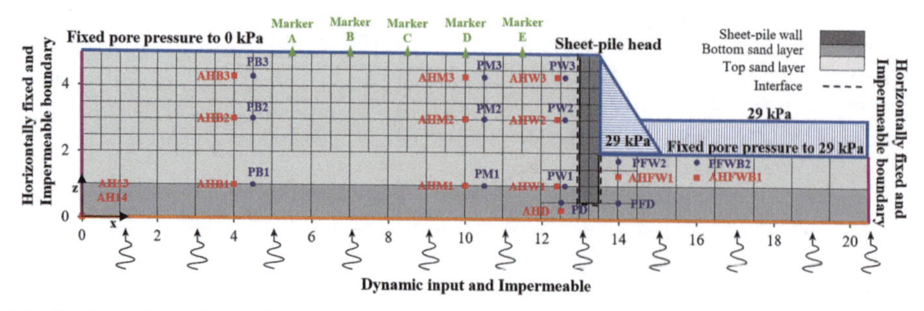

Fig. 3. Two-dimensional view of the centrifuge numerical model showing its spatial discretization, boundary conditions, and control points for horizontal accelerations (AH), pore water pressures (P), and horizontal displacements (markers).

employed to build the centrifuge numerical model, which is shown in a two-dimensional view in Fig. 3. For computational efficiency, the size of the zones in x and z directions was not uniform throughout the model, with sizes of 0.5 and 1 m depending on their proximity to the surface and the sheet-pile wall. In the y direction, only a single layer of zones with a thickness of 1.0 m was used. Furthermore, deformations along this direction were not allowed, effectively rendering the model to behave as in plane strain conditions. The boundary conditions on the sides of the model were constrained laterally, while the base was fully fixed.

The sheet-pile wall was embedded into the ground and modeled as a continuum with a width of 0.5 m. The soil layers were initially modeled with the linear elastic material model with a bulk modulus of 21.9×10^5 kPa and a shear modulus of 9.5×10^5 kPa, before switching the model to SANISAND-MSf. The sheet-pile wall was modeled with a linear elastic material model with Young's modulus of 25.0×10^5 kPa and a Poisson's ratio of 0.31. These values were chosen with the consideration of preserving the bending stiffness of the aluminum sheet-pile wall, whose prototype scale width was approximately 0.1 m and with a Young's modulus of 69×10^6 kPa. Additionally, interface elements were placed along the sides and base of the wall to model the contact between soil and structure, with shear and normal stiffnesses of 138.6×10^6 kPa and a friction angle of 20°.

The model was constructed sequentially, starting by considering dry continuum layers of elastic material. After bringing the model to equilibrium, the solid-pore fluid interaction module was activated in order to assign the built-in isotropic fluid model to the soil elements, with a water bulk modulus of 2.0×10^5 kPa and a soil hydraulic conductivity $k = 1.15 \times 10^{-4}$ kPa [1]. The sheet-pile wall was considered impermeable by using a null fluid model. Pore water pressures at the upper boundary of the model were fixed to 0 and the system was brought to equilibrium again. Then, the 3 m-high excess top sand layer downstream of the wall was removed in order to arrive to the final configuration of the system. Normal stress gradients representing the distribution of submerged water pressures were applied on the right face of the wall and the soil surface downstream of the sheet-pile, as shown in Fig. 3. Pore water pressures corresponding to the submerged conditions were fixed for the grid points of the newly uncovered soil surface. No provision was made for simulating the potential hydrodynamical effects of water. After bringing the updated system to equilibrium, the soil model was switched to SANISAND-MSf using the model constants specified in Table 1, and the system was brought to mechanical and fluid equilibrium.

The above approach was used for the ten centrifuge models in Table 2. Upon reaching the state of equilibrium, the dynamic module was activated. A Rayleigh damping of 1% with a central frequency of 2.5 Hz was adopted to reduce high-frequency numerical noise. Finally, acceleration records of the achieved base motions were applied at the base grid points of the models. Acceleration, excess pore water pressure and displacement time histories were recorded at the control points specified in Fig. 3.

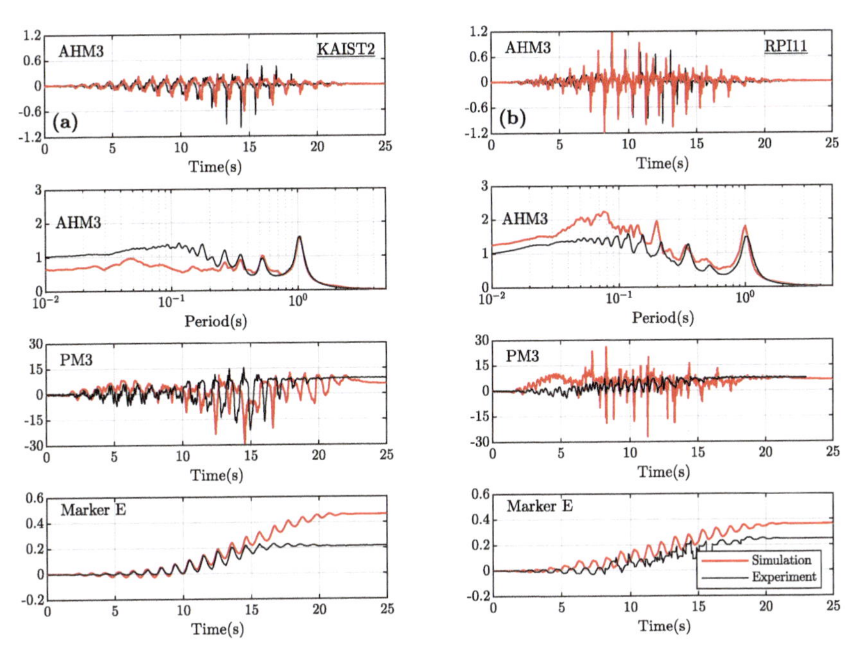

Fig. 4. Comparing the recorded and simulated acceleration histories and acceleration response spectra (AHM3), excess pore water pressure histories (PM3), and horizontal displacement histories (marker E) for the (a) KAIST2 and (b) RPI11 centrifuge models.

Experimental and simulated time-histories of horizontal accelerations at AHM3, excess pore water pressure ratios r_u at PM3 and horizontal displacements at marker E are shown in Fig. 4 for centrifuge models KAIST2 and RPI11. The results indicate the model can capture the acceleration response, resembling the recorded time-histories, presence of dilation spikes, and the spectral values for periods mainly above 0.2 s. The mismatch observed for spectral acceleration at high frequencies corresponds to the relative amplitude of the dilation spikes present in the time histories. The simulations also capture the maximum r_u measured in the experiments, although they also tend to overestimate spikes of negative r_u. Finally, the histories of horizontal displacements were simulated relatively well until 10 s into the shaking; what follows is an overestimation of the accumulated displacements, which results in larger values by around 0.1–0.2 m at the end of shaking. Evaluation of the simulations of the rest of the centrifuge models yields similar observations (not shown here), with an overall reasonable estimation of acceleration response and maximum excess porewater pressures. However, on average, the end of shaking surface horizontal displacements tended to be overestimated.

4 Influence of Mean Shear Stress

The overprediction of the surface displacements can attributed to the fact that the combined impact of having $K_0 \neq 1.0$ and non-zero τ_{mean} was not considered during the calibration of model constants. Recall that the target cyclic tests used for calibration did not reflect the initial stress conditions the soil experienced in the centrifuge tests. While the HCTSS from Vargas et al. [7] were reconstituted with a $K_0 = 1.0$, the soil in the centrifuge experiments had a $K_0 \neq 1.0$ and non-zero τ_{mean} as consequence of the depositional procedure followed to prepare the centrifuge model and the presence of the sheet-pile wall. To investigate the combined effect of non-zero mean shear stresses on the undrained cyclic response, El Ghoraiby and Manzari [9] conducted cyclic direct simple shear (CDSS) constant-volume tests on samples of Ottawa F65 sand. These experiments were conducted on samples with $D_r \approx 66\%$, initial vertical stress $\sigma_{v0} = 30$ kPa, $\tau_{mean} =$ 2.6 kPa, and subjected to 15 cycles of non-uniform shear waves with different maximum CSR. It should be noted that the CDSS tests not only had initial stress states similar to those of the soil deposit but were also subjected to non-uniform shear stresses resembling the ramped sine waves used as base input in the centrifuge tests.

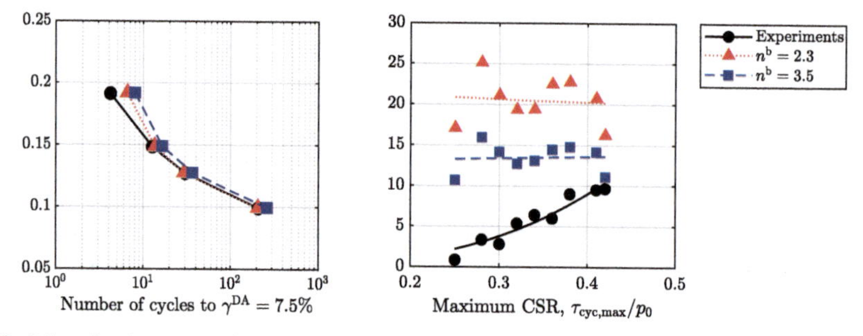

Fig. 5. Liquefaction strength curves from HCCTS [7] tests and final shear strain versus maximum CSR from CDSS [9] tests on samples of Ottawa F65 sand. Simulations of SANISAND-MSf are shown with n^b values of 2.3 and 3.5.

Using the CDSS tests, the calibration of the constitutive model was adjusted to achieve a closer match between final shear strains of the experiments and simulations by changing the n^b from 2.3 to 3.5. This model constant affects the kinematic hardening in the reference model and was found to be effective in controlling the accumulated shear strains in post-liquefaction. This simple approach was used by Reyes et al. [10], who identified a similar issue in modeling submerged slopes in an earlier stage of LEAP. In that work, the excessive accumulation of shear strains was also attributed to the impact of non-zero mean shear stresses. Using the adjusted value of $n^b = 3.5$ resulted in slightly compromising the response in the isotropic HCCTS tests with respect to the one obtained using $n^b = 2.3$ calibration, but the prediction of final shear strains in the CDSS tests was improved. Figure 5 shows a summary of the liquefaction strength curves from the HCCTS tests and final shear strain against the maximum CSR for the CDSS tests using the original and adjusted values of n^b.

Using the adjusted set of model constants, the results of the ten centrifuge models are presented in Fig. 6 in terms of end of shaking horizontal displacements measured at marker E and at the head of the sheet-pile wall. The complete simulated response with further details on other aspects, e.g., acceleration, pore water pressures, settlements, can be found in Pérez [11]. The results indicate the numerical modeling approach provides reasonable estimates of horizontal displacement of the soil surface at marker E for most centrifuge models, with the evident exception of RPI12 whose recorded displacements appear to be much higher than other cases. The simulated displacements at the head of the sheet-pile wall show a larger scatter with respect to the experiments, yet they fall within the lines of 0.5:1 and 2:1. The larger differences observed between simulation and experiments can be attributed to the larger impact of the non-zero mean shear stresses near the sheet-pile wall and the modeling of the interaction of the soil and the structural element. The hydrodynamic effects might also explain the larger-than-predicted horizontal displacements.

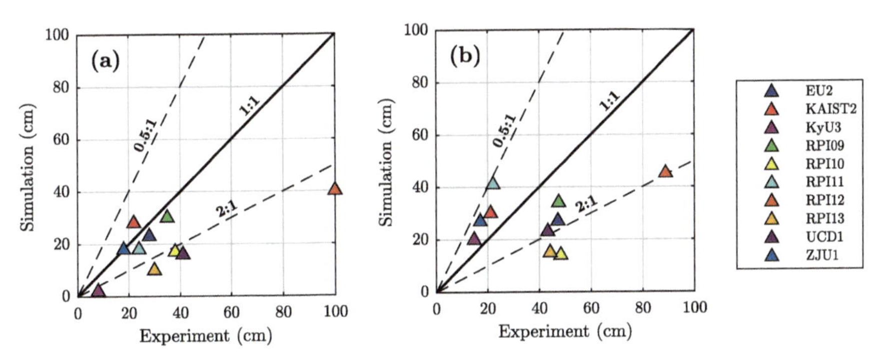

Fig. 6. (a) Comparison of experimentally measured and numerically simulated horizontal displacements at the end of shaking at: (a) marker E and (b) at sheet-pile head for all centrifuge models using the adjusted model constant n^b of 3.5.

The overall reasonable estimate of the system response was improved due to the adjustment of the model constant n^b to account for the effect of the mean shear stress on the undrained cyclic response, particularly on the modeling of the accumulated shear strains in post-liquefaction. Nevertheless, the simulated results of the CDSS tests from [9] and of the centrifuge models of LEAP-RPI-2020 indicate SANISAND-MSf requires a fundamental improvement to its formulation so that the effect of mean shear stresses on the resistance and deformation in undrained cyclic shearing is properly accounted for. This matter is the subject of ongoing research.

5 Conclusions

This study presented the validation of the newly formulated SANISAND-MSf model with the laboratory and centrifuge tests completed for LEAP-RPI-2020. This model extends that of Dafalias and Manzari [8] by introducing two constitutive ingredients which independently control the stiffness and rate of plastic strains in pre- and post-liquefaction. The centrifuge models consisted of a submerged liquefiable soil deposit supported by a sheet-pile wall and subjected at its base to ramped sine waves of varying intensity. The model constants of SANISAND-MSf were calibrated based on undrained HCTSS and constant-volume CDSS with isotropic and anisotropic initial stress conditions, respectively. The numerical simulations of the centrifuge models indicate that the constitutive model provides reasonable estimates of the overall response of the experiments provided the impact of the $K_0 \neq 1.0$ and non-zero τ_{mean} are taken into consideration in the determination of the model constants. The formulation of a new constitutive ingredient to allow for controlling of the effect of non-zero mean shear stresses in undrained cyclic response is the subject of ongoing research.

Acknowledgment. Support for this study was provided by the Natural Sciences and Engineering Research Council of Canada (NSERC). The authors are grateful to Professors B.L. Kutter, M. Manzari, and M. Zeghal, principal investigators of the Liquefaction Experiments and Analyses Project (LEAP), for the invitation to participate in the prediction exercise, and all the participating experimental facilities for providing the data used to conduct this investigation.

References

1. Kutter, B.L., Manzari, M.T., Zeghal, M. (eds.): Model Tests and Numerical Simulations of Liquefaction and Lateral Spreading: LEAP-UCD-2017. Springer, Cham (2020). https://doi.org/10.1007/978-3-030-22818-7
2. Yang, M., Taiebat, M., Dafalias, Y.F.: SANISAND-MSf: a sand plasticity model with memory surface and semifluidised state. Géotechnique **72**(3), 227–246 (2022)
3. Dafalias, Y.F., Manzari, M.T.: Simple plasticity sand model accounting for fabric change effects. J. Eng. Mech. **130**(6), 622–634 (2004)
4. Ramirez, J., et al.: Site response in a layered liquefiable deposit: evaluation of different numerical tools and methodologies with centrifuge experimental results. J. Geotech. Geoenviron. Eng. **144**(10), 04018073 (2018)
5. Barrero, A.R., Taiebat, M., Dafalias, Y.F.: Modeling cyclic shearing of sand in the semifluidized state. Intl. J. Num. Anal. Meth. Geomech. **44**(3), 371–388 (2020)
6. Taiebat, M., Jeremić, B., Dafalias, Y.F., Kanya, A.M., Cheng, Z.: Propagation of seismic waves through liquified soils. Soil Dynam. Earthq. Eng. **30**(4), 236–257 (2020)
7. Vargas, R., Ueda, K., Uemura, K.: Influence of the relative density and K0 effects in the cyclic response of Ottawa F65 sand - cyclic Torsional Hollow-Cylinder shear tests for LEAP-ASIA-2019. Soil Dynam. Earthq. Eng. **133**, 106111 (2020)
8. Itasca. FLAC3D 7.0 Documentation. Itasca Consulting Group, Minneapolis (2019)
9. El Ghoraiby, M.A., Manzari, M.T.: Cyclic behavior of sand under non-uniform shear stress waves. Soil Dynam. Earthq. Eng. **143**, 1106590 (2021)
10. Reyes, A., Yang, M., Barrero, A.R., Taiebat, M.: Numerical modeling of soil liquefaction and lateral spreading using the SANISAND-Sf model in the LEAP experiments. Soil Dynam. Earthq. Eng. **143**, 106613 (2021)

11. Pérez, K.: Numerical modeling of liquefaction sand deposits supported by a sheet-pile wall. Master's thesis. Department Civil Engineering, University of British Columbia, Vancouver, BC, Canada (in prep.)

Numerical Simulations of LEAP Centrifuge Experiments Using a Multi-surface Cyclic Plasticity Sand Model

Zhijian Qiu[1(✉)] and Ahmed Elgamal[2]

[1] Xiamen University, Xiamen 361005, Fujian, China
ZhijianQiu@xmu.edu.cn
[2] University of California San Diego, La Jolla, CA 92093-0085, USA
aelgamal@ucsd.edu

Abstract. Numerical simulations of LEAP-UCD-2017 centrifuge tests are conducted using a multi-surface cyclic plasticity sand model implemented with the characteristics of dilatancy, cyclic mobility and associated shear deformation. This model extends the OpenSees PDMY03 material to include the Lade-Duncan failure criterion as the yield function, thus allowing for considerable accuracy in three-dimensional shear response conditions. For this study, the model parameters are calibrated based on a series of available cyclic stress-controlled triaxial tests of Ottawa F-65 sand with relative density $D_{r.} = 65\%$. Using the calibrated model parameters, Finite Element (FE) simulations are performed for dynamic centrifuge tests of a liquefiable sloping ground. The computed results are systematically presented and directly compared to the centrifuge test data. An overall good match between the simulations and measurements demonstrated that the multi-surface cyclic plasticity model has capabilities for simulating the stress path of cyclic stress-controlled triaxial tests as well as the response of the liquefiable sloping ground subjected to seismically-induced liquefaction.

Keywords: LEAP · Centrifuge tests · Dynamic · Liquefiable · Multi-surface

1 Introduction

LEAP (Liquefaction Experiments and Analysis Projects) facilitates validation and verification of numerical techniques for liquefaction-induced lateral spreading analysis of a liquefiable sloping ground [1–3]. As one project within LEAP, LEAP-UCD-2017 [4–6] provides a number of reliable centrifuge test data conducted at various facilities about a liquefiable sloping ground.

Based on these high-quality centrifuge test data [4, 5], numerical simulations were undertaken during the LEAP-UCD-2017. All simulations are performed using a multi-surface cyclic plasticity sand model [7–9], which extends the OpenSees PDMY03 material [10] by employing the Lade-Duncan failure criterion as the yield function [11]. The model parameters are determined based on cyclic stress-controlled triaxial tests [6] and further employed in numerical simulations of the centrifuge tests.

© The Author(s), under exclusive license to Springer Nature Switzerland AG 2022
L. Wang et al. (Eds.): PBD-IV 2022, GGEE 52, pp. 1812–1820, 2022.
https://doi.org/10.1007/978-3-031-11898-2_164

The following sections of this paper outline: 1) specifics and calibration processes, 2) details of the employed numerical modeling techniques, and 3) computed results. Finally, conclusions are presented and discussed.

2 LEAP-UCD-2017 Centrifuge Tests

Figure 1 displays a schematic representation of the LEAP-UCD-2017 centrifuge tests [6]. The soil specimen is a 5° sloping layer of Ottawa F-65 sand with relative density $D_{r.} = 65\%$. The length and the height at the center of soil domain are 20 m and 4 m in prototype scale, respectively. The specimen is built in a container with rigid walls. All centrifuge models were subjected to a target motion of ramped, 1 Hz sine wave base motion with amplitude 0.15g [4, 5, 12, 13].

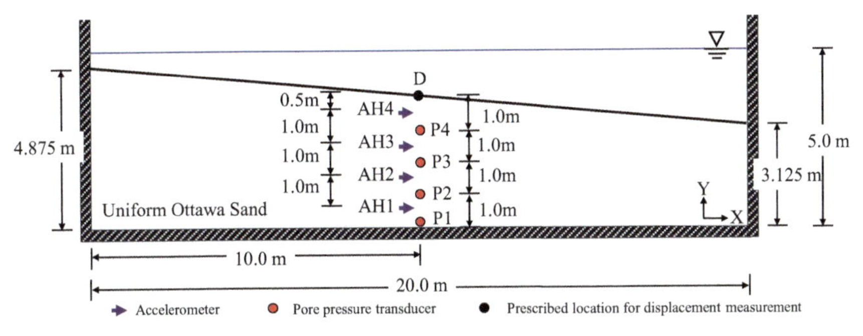

Fig. 1. A Schematic representation of the LEAP-UCD-2017 centrifuge test layout [6].

3 Finite Element Model

A two-dimensional (2D) Finite Element (FE) mesh (Fig. 2) is created to represent the centrifuge experiment model, comprising 4,961 nodes and 4,800 quadrilateral elements (maximum size = 0.2 m). More details of the boundary and loading conditions are discussed in [12, 13]. In this paper, all FE simulations are performed using the computational platform OpenSees [14]. For simulating the saturated soil response, quadrilateral Four-node plane-strain elements (quadUP) with two-phase material following the *u-p* [15] formulation were employed, where u is the displacement of the soil skeleton and p is the pore water pressure. Brief descriptions of the multi-surface cyclic plasticity sand model are included below.

4 Soil Constitutive Model

A multi-surface cyclic plasticity sand model [7–9] implemented with the characteristics of dilatancy, cyclic mobility and associated shear deformation were employed. To allow for further accuracy in capturing 3D shear response [7], the OpenSees PDMY03 material

[10] was extended with the Lode angle effect by employing the Lade-Duncan failure criterion as the yield function [11]. This failure criterion is represented by [16]:

$$J_3 - \frac{1}{3}I_1 J_2 + \left(\frac{1}{27} - \frac{1}{k_1}\right)I_1^3 = 0 \tag{1}$$

where, I_1 is volumetric stress, J_2 and J_3 are second and third deviatoric stress invariants, respectively, parameter k_1 (>27) is related to soil shear strength (or friction angle ϕ). A typical yield surface f_m (Fig. 3) is defined by [7]:

$$f_m = \overline{J}_3 - \frac{1}{3}(\eta_m I_1)\overline{J}_2 + a_1(\eta_m I_1)^3 = 0 \tag{2}$$

in which, η_m is normalized yield surface size ($0 < \eta_m < 1$) and $a_1 = \frac{1}{27} - \frac{1}{k_1}$. In Eq. 2, $\overline{J}_2 = \frac{1}{2}\overline{s} : \overline{s}$ and $\overline{J}_3 = \frac{1}{3}(\overline{s} \cdot \overline{s}) : \overline{s}$, where, $\overline{s} = \boldsymbol{s} - p'\boldsymbol{a}$ and $\boldsymbol{s} = \boldsymbol{\sigma} - p'\boldsymbol{\delta}$, $\boldsymbol{\delta}$ is the second-order identity tensor, $p' = \frac{1}{3}I_1$ is mean effective stress, deviatoric tensor $\boldsymbol{a}$ is back stress (yield surface center), and the operators "·" and ":" denote single and double contraction of two tensors, respectively.

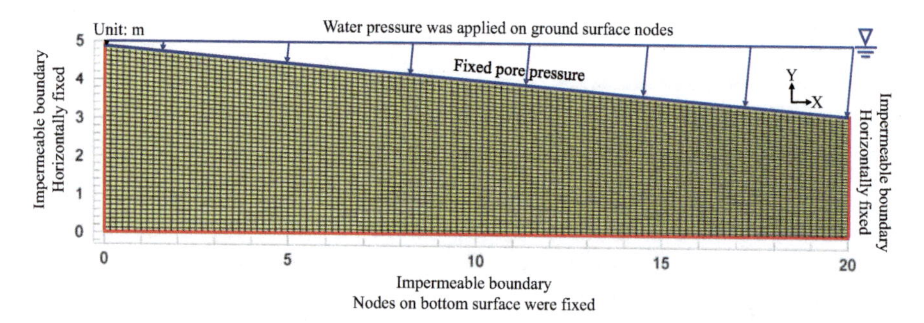

Fig. 2. Finite Element mesh (maximum size = 0.2m).

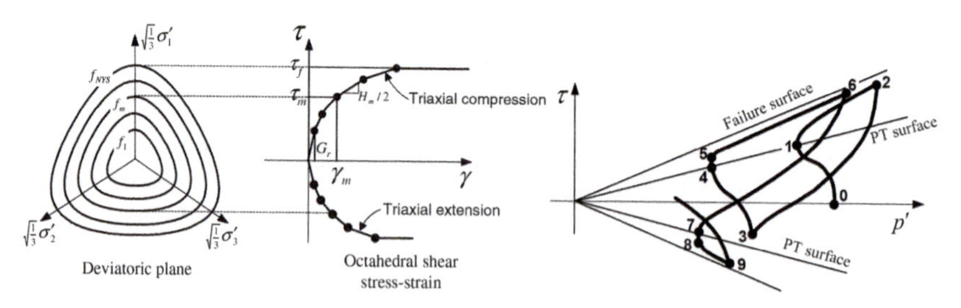

Fig. 3. A multi-surface cyclic plasticity sand model.

4.1 Contractive Phase

Shear-induced contraction occurs inside the phase transformation (PT and Fig. 3) surface ($\eta < \eta_{PT}$), as well as outside ($\eta > \eta_{PT}$) when $\dot{\eta} < 0$, where, η is the deviatoric stress ratio. The contraction rule is defined as:

$$P'' = (c_1 + c_2\gamma_c)\left(p'/p_a\right)^{c_3}(c_4\eta_v)^{c_5}$$
$$\eta_v = \tau_{oct}/p' \tag{3}$$

where c_1, c_2, c_3, c_4 and c_5 are non-negative calibration constants, γ_c is octahedral shear strain accumulated during previous dilation phases, p_a is atmospheric pressure and τ_{oct} is octahedral shear stress. The parameter c_3 is used to represent the dependence of pore pressure buildup on initial confinement (i.e., K_σ effect).

4.2 Dilative Phase

Dilation appears only due to shear loading outside the PT surface ($\eta > \eta_{PT}$ with $\dot{\eta} > 0$), and is defined as:

$$P'' = \left(d_1 + \gamma_d^{d_2}\right)\left(p'/p_a\right)^{-d_3}(\eta_v)^{d_4} \tag{4}$$

where d_1, d_2 and d_3 are non-negative calibration constants, and γ_d is the octahedral shear strain accumulated from the beginning of a particular dilation cycle as long as there is no significant load reversal. Parameter d_3 reflects the dependence of pore pressure buildup on initial confinement (i.e., K_σ effect).

4.3 Neutral Phase

When the stress state approaches the PT surface ($\eta = \eta_{PT}$) from below, a significant amount of permanent shear strain may accumulate prior to dilation, with minimal changes in shear stress and confinement. For simplicity, $P'' = 0.$ is maintained during this highly yielded phase until a boundary defined in deviatoric strain space is reached, and then dilation begins. This yield domain will enlarge or translate depending on load history. In deviatoric strain space, the yield domain is a circle with the radius $\gamma = (\gamma_s + \gamma_{rv})/2$ defined as [17]:

$$\gamma_s = y_1\left\langle\frac{p'_{max} - p'_n}{p'_{max}}\right\rangle^{0.25}(\eta_v)^{y_2}\int_0^t d\gamma_c$$
$$\gamma_{rv} = y_3\left\langle\frac{p'_{max} - p'_n}{p'_{max}}\right\rangle^{0.25} oct\left(\boldsymbol{e} - \boldsymbol{e}_p\right) \tag{5}$$

where, y_1, y_2 and y_3 are non-negative calibration constants, p'_{max} is maximum mean effective confinement experienced during cyclic loading, p'_n is mean effective confinement at the beginning of current neutral phase, and $< >$ denotes MacCauley's brackets (i.e., $\langle a\rangle = max(a, 0)$). Parameter y_1 is used to define the accumulated permanent shear strain γ_s as a function of dilation history $\int_0^t d\gamma_c$ and allow for continuing enlargement of the domain. The y_2 parameter is mainly used to define the biased accumulation of permanent shear strain γ_{rv} as a function of load reversal history and allows for translation of the yield domain during cyclic loading. In Eq. 5, $oct(\boldsymbol{e} - \boldsymbol{e}_p)$ denotes the octahedral shear strain of tensor $\boldsymbol{e} - \boldsymbol{e}_p$, where $\boldsymbol{e}$ is current deviatoric shear strain, and $\boldsymbol{e}_p$ is pivot strain obtained from previous dilation on load reversal point.

5 Determination of Soil Model Parameters

With the modeling parameters in Table 1, Fig. 4 shows the liquefaction strength curve of Ottawa F-65 sand with relative density $D_{r.} = 65\%$ (i.e., number of cycles to reach 2.5% single amplitude of axial strain) obtained from the FE simulation and measurement [6]. As seen in this figure, a good match is reached between the computed results and laboratory data. For illustration, an example of undrained cyclic triaxial stress-controlled test with $CSR = 0.2$ is displayed in Figs. 4b–c. It can be seen that the computed and experimental response show similar cycle-by-cycle permanent axial strain accumulation pattern.

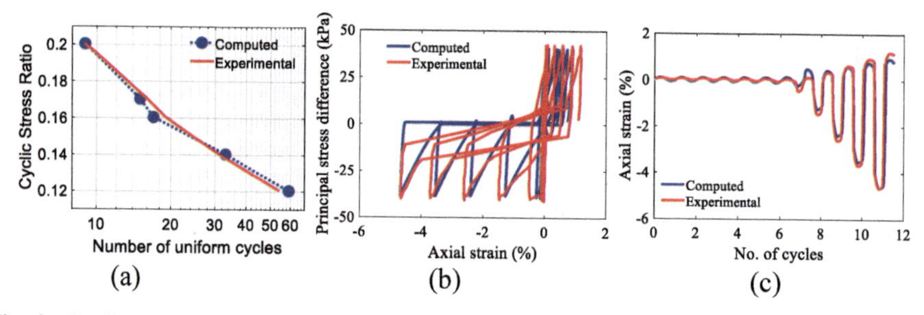

Fig. 4. Cyclic stress-controlled triaxial tests [6]: (a) Liquefaction strength curve; (b)–(c) Representative soil response with $CSR = 0.2$.

Table 1. Calibrated sand model parameters.

Model parameters	Value
Reference mean effective pressure, p'_r (kPa)	101
Mass density, ρ (t/m^3)	2.04
Maximum shear strain at reference pressure, $\gamma_{max,r}$	0.1
Shear modulus at reference pressure, Go (MPa)	25
Stiffness dependence coefficient d, $G = Go\ (p'/p'_r)^d$	0.5
Poisson's ratio ν (for dynamics)	0.4
Shear strength at zero confinement, c (kPa)	0.3
Friction angle and Phase transformation angle, ϕ/ϕ_{PT}	36°/26°
Contraction coefficients, $c_1/c_2/c_3/c_4/c_5$	0.012/3.0/0.1/2.0/1.1
Dilation coefficient, $d_1/d_2/d_3/d_4$	0.15/1.0/0.3/2.0
Damage parameter, $y_1/y_2/y_3$	1.0/0.3/0.0
Permeability (m/s)	1.1×10^{-4}

6 Computed Results of LEAP-UCD-2017 Centrifuge Tests

6.1 Acceleration

Figure 5 depicts the computed and experimental acceleration time histories at the locations AH1-AH4 (Fig. 1). It can be seen that the computed accelerations at deeper depths (AH1–AH2) are in good agreement with those from the measurements (Fig. 5). For shallower depths (AH3–AH4), both the computed results and measurements showed a consistent trend for acceleration spikes due to dilation. Nevertheless, the accelerations near ground surface (AH4) were not captured successfully in KAIST-1, KAIST-2, NCU-3, and KyU-3 tests. All these experimental tests showed very high acceleration spikes due to strong dilation. In tests of CU-2 and Ehime-2, the computed results displayed higher dilation spikes in contrast to measurements. These unintended inconsistencies are mainly due to the different relative densities in centrifuge experiments conducted at various facilities.

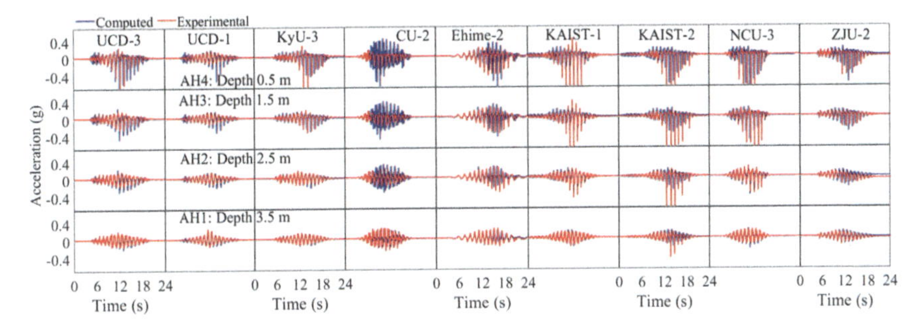

Fig. 5. Measured and computed acceleration time histories.

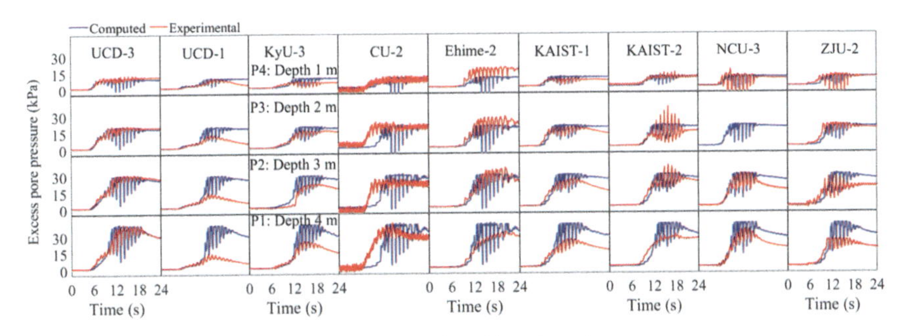

Fig. 6. Measured and computed time histories of excess pore pressure.

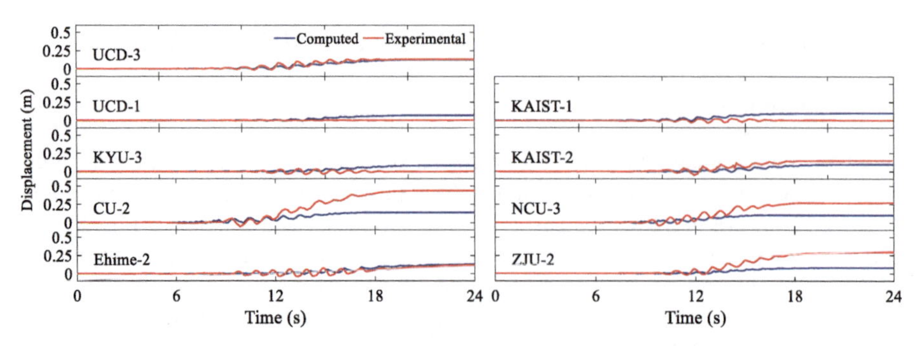

Fig. 7. Measured and computed displacement time histories.

6.2 Excess Pore Pressure

Figure 6 illustrates the time histories of excess pore pressure at transducers P1–P4. In general, the computed results reasonably match those from measurements. In accordance with the accelerations (Fig. 5), both the computed results and measurements showed a consistent trend of negative spikes due to dilation. However, the peak values of excess pore pressure at transducer P1 are overpredicted by numerical simulations in UCD-1, KyU-3, KAIST-1, KAIST02, ZJU-2 tests, might be due to the soils at deeper depths are not fully liquefied.

6.3 Displacement

Figure 7 displays the computed and experimental horizontal displacement time histories at the midpoint D near ground surface. It can be seen that the computed results of UCD-3 and KAIST-1 are in good agreement with those from the measurements. However, the horizontal displacements of CU-2, NCU-3, KAIST-2, ZJU-2 tests are underpredicted, and UCD-1, KAIST-1, KyU-3 are overpredicted, mainly due to the different relative densities in centrifuge experiments conducted at various facilities.

7 Conclusions

The FE simulation results of centrifuge tests about a liquefiable sloping ground in LEAP-UCD-2017 are presented. All simulations are performed using a multi-surface cyclic plasticity sand model with the Lade-Duncan failure criterion as the yield function. The sand model parameters are determined based on cyclic stress-controlled triaxial tests of Ottawa F-65 sand with relative density $D_{r.} = 65\%$. Using the calibrated model parameters, FE simulations are performed for dynamic centrifuge tests of a liquefiable sloping ground. The computed results are systematically presented and directly compared to the centrifuge data. The primary conclusions can be drawn as follows:

(1) It is shown that the employed sand model can reasonably capture the liquefaction strength curve of Ottawa F-65 sand with relative density $D_{r.} = 65\%$ and the corresponding soil response including stress path and cyclic-by-cycle permanent axial strain accumulation of cyclic triaxial stress-controlled tests.

(2) An overall good match between the simulations and measurements demonstrated that the multi-surface cyclic plasticity sand model as well as the overall employed computational framework have the potential to predict the dynamic response of the liquefiable sloping ground, and subsequently realistically evaluate the performance of an equivalent soil system subjected to seismically-induced liquefaction.
(3) The unintended inconsistencies of centrifuge test results make it challenging to match all experiments using the same model parameters. To better capture overall dynamic response of a particular test, contraction parameters are suggested to be adjusted to simulate the inconsistent contractive behavior exhibited by the Ottawa F-65 sand with different relative densities at various facilities.

Acknowledgements. The authors are grateful for the kind invitation by Professors Majid T. Manzari, Mourad Zeghal and Bruce L. Kutter to participate in LEAP-UCD-2017. Partial funding of this effort was provided by Caltrans through a project in which Ottawa sand is being used for experimentation, and Fundamental Research Funds for the Central Universities of China (22070103963, 20720220070).

References

1. Manzari, M.T., et al.: LEAP projects: concept and challenges. In: Proceedings, 4th International Conference on Geotechnical Engineering for Disaster Mitigation and Rehabilitation, 16–18 September 2014, Kyoto, Japan, pp. 109–116 (2014)
2. Kutter, B.L., et al.: Proposed outline for LEAP verification and validation processes. In: Proceedings, 4th International Conference on Geotechnical Engineering for Disaster Mitigation and Rehabilitation, 16–18 September 2014, Kyoto, Japan, pp. 99–108 (2014)
3. Kutter, B.L., et al.: LEAP Databases for verification, validation, and calibration of codes for simulation of liquefaction. In: 6th International Conference on Earthquake Geotechnical Engineering, Christchurch, New Zealand (2015)
4. Kutter, B.L., Zeghal, M., Manzari, M.T.: LEAP-UCD-2017 Experiments (Liquefaction Experiments and Analysis Projects). DesignSafe-CI [publisher], Dataset (2018). https://doi.org/10.17603/DS2N10S
5. Kutter, B.L., Manzari, M.T., Zeghal, M. (eds.): Model Tests and Numerical Simulations of Liquefaction and Lateral Spreading. Springer, Cham (2020). https://doi.org/10.1007/978-3-030-22818-7
6. Ghoraiby, M., Park, H., Manzari, M.T.: Physical and mechanical properties of Ottawa F65 sand. In: Kutter, B.L., Manzari, M.T., Zeghal, M. (eds.) Model Tests and Numerical Simulations of Liquefaction and Lateral Spreading, pp. 45–67. Springer, Cham (2020). https://doi.org/10.1007/978-3-030-22818-7_3
7. Yang, Z., Elgamal, A.: Multi-surface cyclic plasticity sand model with Lode angle effect. Geotech. Geol. Eng. **26**(3), 335–348 (2008)
8. Qiu, Z., Lu, J., Elgamal, A., Su, L., Wang, N., Almutairi, A.: OpenSees three-dimensional computational modeling of ground-structure systems and liquefaction scenarios. Comput. Model. Eng. Sci. **120**(3), 629–656 (2019)
9. Qiu, Z.: Computational modeling of ground-bridge seismic response and liquefaction scenarios. Ph.D. Thesis, UC San Diego (2020)
10. Khosravifar, A., Elgamal, A., Lu, J., Li, J.: A 3D model for earthquake-induced liquefaction triggering and post-liquefaction response. Soil Dyn. Earthq. Eng. **110**, 43–52 (2018)

11. Lade, P.V., Duncan, J.M.: Elastoplastic stress-strain theory for cohesionless soil. J. Geotech. Eng. Div. **101**(10), 1037–1053 (1975)
12. Qiu, Z., Elgamal, A.: Numerical simulations of LEAP centrifuge tests for seismic response of liquefiable sloping ground. Soil Dyn. Earthq. Eng. **139**, 106378 (2020)
13. Qiu, Z., Elgamal, A.: Numerical simulations of LEAP dynamic centrifuge model tests for response of liquefiable sloping ground. In: Kutter, B.L., Manzari, M.T., Zeghal, M. (eds.) Model Tests and Numerical Simulations of Liquefaction and Lateral Spreading, pp. 521–544. Springer, Cham (2020). https://doi.org/10.1007/978-3-030-22818-7_26
14. McKenna, F., Scott, M.H., Fenves, G.L.: Nonlinear finite-element analysis software architecture using object composition. J. Comput. Civ. Eng. **24**(1), 95–107 (2010)
15. Chan A.H.C.: A unified finite element solution to static and dynamic problems in geomechanics. Ph.D. Thesis, University College of Swansea (1988)
16. Chen, W.F., Mizuno, E.: Nonlinear Analysis in Soil Mechanics, Theory and Implementation. Elsevier, New York, NY (1990)
17. Yang, Z., Elgamal, A., Parra, E.: Computational model for cyclic mobility and associated shear deformation. J. Geotech. Geoenviron. Eng. **129**(12), 1119–1127 (2003)

Validation of Numerical Predictions of Lateral Spreading Based on Hollow-Cylinder Torsional Shear Tests and a Large Centrifuge-Models Database

R. Vargas[1,3](✉), Z. Tang[2,3], K. Ueda[2,3], and R. Uzuoka[2,3]

[1] Penta-Ocean Construction Co., Ltd., Tokyo, Japan
ruben.vargas@mail.penta-ocean.co.jp
[2] Department of Civil and Earth Resources Engineering, Kyoto University, Kyoto, Japan
[3] Disaster Prevention Research Institute, Kyoto University, Kyoto, Japan

Abstract. This paper aims to present a complete validation exercise that explores the capabilities of numerical predictions to simulate the lateral spreading phenomena in clean sands under a diverse range of densities and input motions. The validation exercise used the "Strain Space Multiple Mechanism Model" to simulate the lateral spreading phenomena (although the methodology presented might be used for the validation of other numerical tools as well), and was based on multiple, cross-checked, and high-quality physical models (Centrifuge Models) and element tests (Hollow Cylinder Cyclic Shear Tests), developed for LEAP Project.

The validation exercise showed that the numerical model is able to predict the displacements for the median trend and the 95% probability confidence bounds for PGA < 0.25 g.

Keywords: Validation · LEAP project · Liquefaction

1 Introduction

During the last decades, important efforts and developments of numerical tools for liquefaction modeling have contributed to increasing the accuracy of prediction of the liquefaction phenomena.

In order to enhance its reliability, continue evolving, and be included in the design practice, these numerical tools need to be assessed and validated [1] by taking into account the median response and its associate variability [2, 3].

In that context, the main objective of this paper is to present a summary of a complete validation exercise [4] that was developed to explore the capabilities of the numerical predictions to simulate the lateral spreading phenomena under a diverse range of densities and input motions, placing special attention on the quantification of the median response and the associated variability of the physical and numerical models. The validation exercise used the "Strain Space Multiple Mechanism Model" [5] to simulate the lateral spreading phenomena (although the methodology presented might be used for the validation of other numerical tools as well), and was based on multiple, cross-checked, and

© The Author(s), under exclusive license to Springer Nature Switzerland AG 2022
L. Wang et al. (Eds.): PBD-IV 2022, GGEE 52, pp. 1821–1828, 2022.
https://doi.org/10.1007/978-3-031-11898-2_165

high-quality physical models (Centrifuge Models) and element tests (Hollow Cylinder Cyclic Shear Tests), developed for LEAP (refer to Sect. 1.2).

1.1 Methodology

According to the ASME's "Guide for Verification and Validation for Computational Solid Mechanics" [2, 3], the goal of a validation exercise is to determine the predictive capability of a computational model for its intended use; and, for a meaningful validation, the next three prerequisites are required.

1. Having a clear definition of the model's intended use

   The intended use of the model is the simulation of the lateral spreading phenomena of clean sands under a diverse range of densities and input motions.
2. Having already conducted code verification

   The model intended to be validated has been implemented in the commercial finite element program "FLIP ROSE," and the verification was developed as part of the internal activities of FLIP Consortium INC. Hence, the verification step is considered out of the scope of this study.
3. Quantifying uncertainties in both the simulation outcomes and the experimental outcomes

   The uncertainty quantification in the experimental outcome is analyzed by means of the observed variability in the results of a large database of centrifuge models (i.e., centrifuge experiments developed for LEAP); on the other hand, the uncertainty quantification in the simulation outcomes is analyzed based on the observed variability of the liquefaction resistance curve (LRC).

   Figure 1 shows a schematic diagram of the main steps as part of the validation exercise.

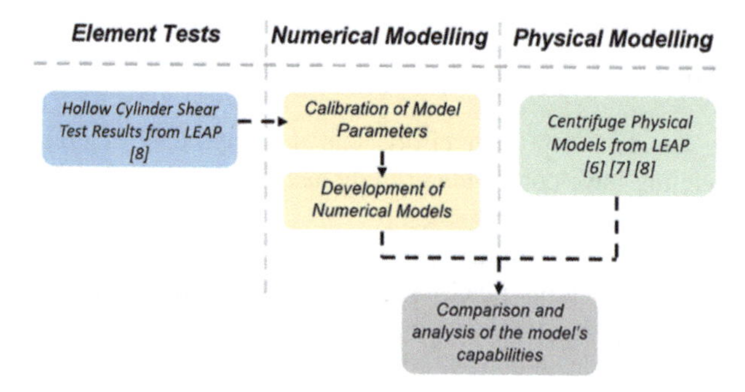

Fig. 1. Schematic diagram of the main steps of the validation exercise.

1.2 LEAP Project

In order to promote the assessment of the reliability of modern numerical techniques in the analysis of liquefaction-related problems, an international collaborative project named "Liquefaction Exercises and Analysis Project" (LEAP) was developed [1]. As part of this project, fifty-four centrifuge models [6–8] were developed in several centrifuge facilities around the world, aiming to perform a sufficient number of experiments to characterize the dynamic behavior of a saturated sloping deposit of Ottawa F-65 Sand. Figure 2 shows the geometry, dimensions, and instrumentation of the target models (in prototype scale). The target input motion consisted of a ramped sinusoidal 1 Hz wave.

Additionally, in order to study the mechanical characteristics of Ottawa F-65 Sand under cyclic loading, Vargas et al. [9] developed twenty-three Torsional Hollow Cylinder Cyclic Shear Tests that covered a wide range of Relative Densities (Dr) and Cyclic Stress Ratios (CSR).

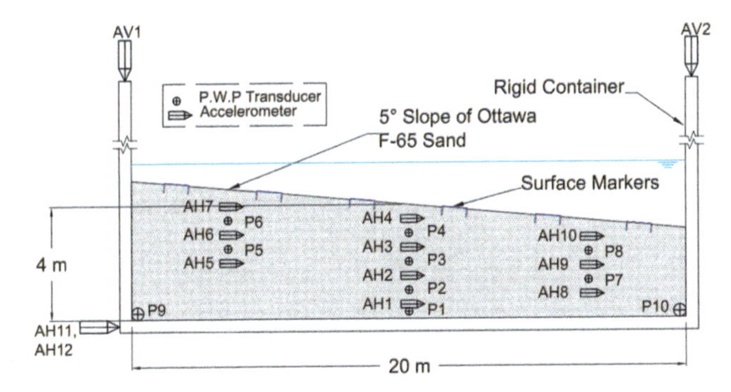

Fig. 2. Schematic diagram of the main steps of the validation exercise.

2 Mean Response and Variability of the Physical Models

Kutter et al. [10] found that, for the lateral spreading phenomena, the residual surface displacements (U_X) are primarily a function of the intensity of shaking and the relative density of the sand. In order to estimate the mean response and the associated variability of U_X, a correlation between the U_X, PGA, and Dr (based on the LEAP centrifuge physical models [6–8]) was developed through the implementation of a hybrid MCMC based Bayesian estimation [4]. Figure 3 illustrates the aforementioned correlation.

3 Mean Response and Variability of the Element Tests

In order to study the mechanical properties of Ottawa F-65 Sand under cyclic loading, Vargas et al. [9] conducted a series of Hollow Cylinder Dynamic Torsional Shear Tests for four different relative densities (Dr = 50%, 60%, 70%, and 85%), under a wide range of CSR values. In order to estimate the median trend and the associated variability of the torsional test results (specific, the liquefaction resistance curves), a hybrid MCMC based Bayesian correlation was developed [4]. Figure 4 illustrates the correlation showing the mean trend and the associated variability of the liquefaction resistance curves.

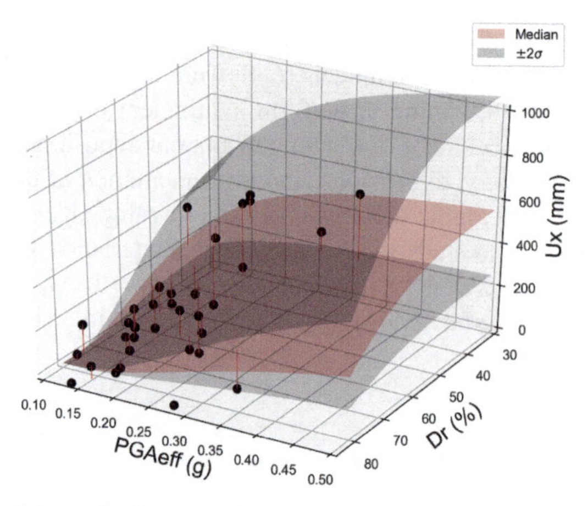

Fig. 3. Schematic diagram of the main steps of the validation exercise.

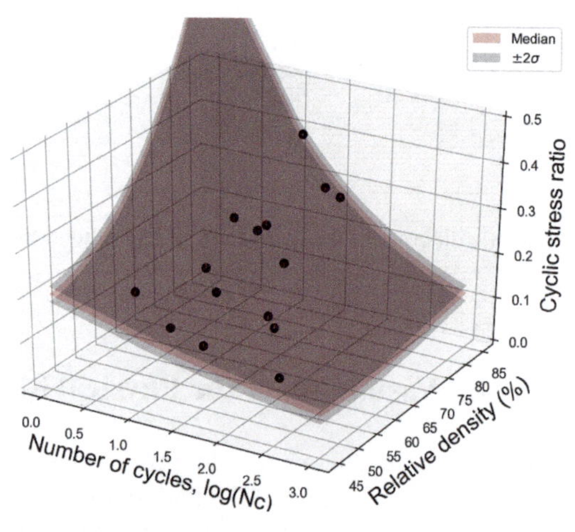

Fig. 4. Correlation between the liquefaction strength, the number of cycles (Nc), and the relative density (Dr) based on MCMC based Bayesian modeling

4 Numerical Model

As previously stated, the main objective of this paper is the validation of the capabilities of the "Strain Space Multiple Mechanism Model" [5] to simulate the lateral spreading phenomena.

For a plane-strain application of the model, and by assuming an isotropic texture of the material, the strain space multiple mechanism model has 17 primary parameters for the analysis of liquefaction; among them, five specify the volumetric mechanism, three specify the shear mechanism, and nine control the dilatancy.

4.1 Element Test Simulations

By aiming to replicate the mean trend of the LRC (obtained through the hybrid MCMC Bayesian simulation, Fig. 3), and focusing on the N = 10–100 cycles interval, the model parameters of the four densities under study (50%, 60%, 70%, and 85%) were estimated through an iterative procedure based on a parametric study.

A well-known approach (and suggested by the ASME's guide [2, 3]) for the model uncertainty quantification corresponds to estimating the uncertainties by defining the model parameters as random variables (i.e., defining the uncertainty of each parameter) and propagating it through the model. This approach seems reasonable and effective for cases in which the model parameters can be independently estimated, and the correlation between them can be established. Unfortunately, regardless that each model parameter has a physical interpretation, the current state of the art of the liquefaction modeling and the soil element testing does not allow an independent estimation of all the parameters.

During the numerical simulation process, it has been found that even if different sets of parameters are used but similar LRC (and stress-strain behavior) are achieved, the simulated lateral spreading displacements of clean sands become almost identical. This suggests that the numerical modeling results (for lateral spreading) depend mostly on the simulated LRC rather than on a specific set of parameters. Therefore, in this paper, the model input variability will be expressed as changes in the simulated LRCs, rather than individual changes in the parameters; in this sense, the variability estimated through the hybrid MCMC based Bayesian correlation (Fig. 3) was used as the base to define the input variability of the Numerical Models.

Figure 5 shows the median trend and the associated variability (95% and 50% probability boundaries) of the simulated LRC, compared with the mean trend of the hybrid MCMC analysis (Fig. 4).

It is important to mention that since the models were developed under laboratory conditions using precise/calibrated instruments, other sources of uncertainty (such as spatial variability and the uncertainty in the input wave) were not taken into account.

4.2 Physical Models Simulations

The simulations of the physical models were developed by using the "Strain Space Multiple Mechanism Model" [5]. The analysis was carried out under 2-D plane-strain conditions, aiming to simulate the models in prototype scale.

Figure 6 shows the mesh and the boundary conditions used in the numerical simulation; as seen in the figure, 384 4-node quadrilateral elements (including the pore water elements) were used.

First, a self-weight analysis was carried out to obtain the initial stress distribution before shaking; after finishing this step, a dynamic response analysis was performed for 130 s, considering pore water flow migration (during and after the shaking). Additional details of the numerical model implementation can be found in Vargas et al. [4].

5 Comparison Between Physical and Numerical Models

Figure 7 shows a comparison between the simulated displacements (obtained by the numerical simulations), and the physical models of the LEAP exercises (expressed as a correlation through the hybrid MCMC based Bayesian modeling, shown in Fig. 3), for Dr = 50%, 60%, 70%, and 85%, respectively. It is important to mention that the figures include a comparison between the mean values and the upper/lower boundaries for a 95% probability.

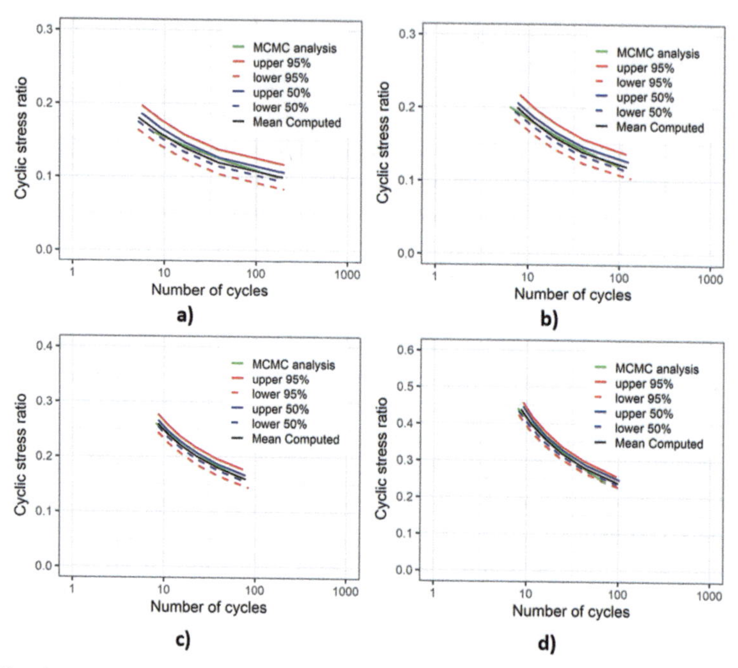

Fig. 5. Simulated LRC's - Comparison between the median trend of the hybrid MCMC analysis (i.e. target value) and the upper/lower (2.5%) bounds for 95% and 50% probability (a) Dr = 50%, (b) Dr = 60%, (c) Dr = 70%, (d) Dr = 80%

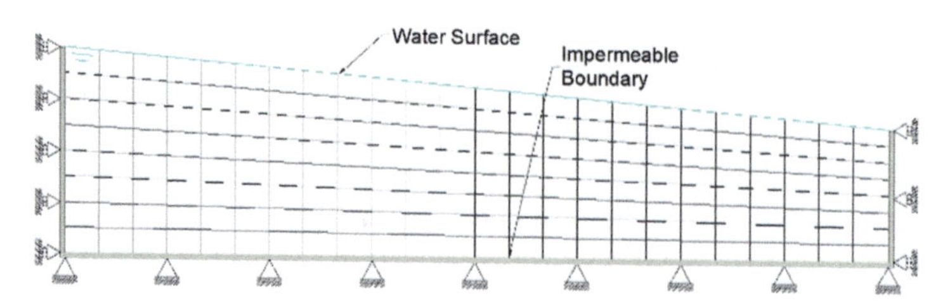

Fig. 6. FEM mesh and boundary conditions

Based on the comparison between the main trend and the associated variability of the physical modeling, and the numerical simulations, the main outcomes of the validation exercise can be summarized as follows:

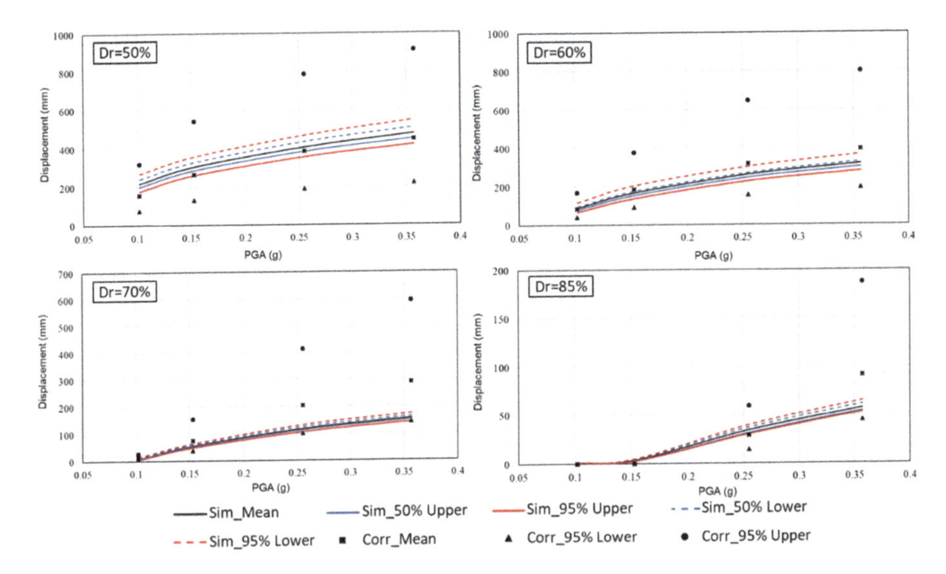

Fig. 7. Comparison between the displacements obtained in physical and numerical models for Dr = 50%, Dr = 60%, Dr = 70%, and Dr = 85%.

- As for the median response, a good agreement between the computed displacements and the estimated correlation can be seen for PGA values lesser than 0.25 g.
- For a 95% of probability, the confidence bounds of the computed displacements are located between the confidence bounds of the experimental outcomes (with few exceptions between 0.14–0.16 g at Dr85%); validating the model for the three variability conditions considered in this exercise, for PGA values lesser than 0.25 g.
- The variability of the experimental results is significantly higher than the variability of the numerical simulations. This might be explained by the fact that the element tests were developed in a unique facility, while the physical models were developed in several institutions around the world.
- As for PGA values higher than 0.25 g, important variations were found between the experimental and numerical results; this may be explained due to the fact that the induced CSR values cannot be replicated in the element tests (due to the instability of the sample), and also, few physical models were developed in this range of accelerations (which also partially explains the increment in the variability in comparison to the PGA < 0.25 range); so, additional research efforts would be required to explore the validity of the "Strain Space Multiple Mechanism Model" in these range of accelerations.

6 Conclusions

- A complete validation exercise has been developed to assess the capabilities of the numerical methods to simulate the lateral spreading phenomena. Although the validation was applied using the "Strain Space Multiple Mechanism Model," the authors believe that the methodology can be used to validate other numerical tools as well.

- In the comparison between the computed displacements and the results obtained in the estimated correlation, the "Strain Space Multiple Mechanism Model" has shown a good performance among the tested densities (Dr = 50, 60, 70, and 85%), for PGA values lesser than 0.25 g.
- Important variations were found between the experimental and numerical results for PGA values higher than 0.25 g; additional research efforts would be required to explore the validity of the "Strain Space Multiple Mechanism Model" in this range of accelerations.

References

1. Manzari, M.T., et al.: LEAP Projects: Concept and Challenges. Geotechnics for Catastrophic Flooding Events, pp. 109–116. CRC Press, New York (2015)
2. American Society of Mechanical Engineers: Guide for Verification and Validation in Computational Solid Mechanics. ASME Standard V&V 10-2016 (2006)
3. American Society of Mechanical Engineers: An Illustration of the Concepts of Verification and Validation in Computational Solid Mechanics. ASME V&V 10.1-2012 (2012)
4. Vargas, R., Tang, Z., Ueda, K., Uzuoka, R.: Validation of numerical predictions for liquefaction phenomenon – lateral spreading in clean sands. Soils Found. **62**(1), 101101 (2022)
5. Iai, S., Tobita, T., Ozutsumi, O., Ueda, K.: Dilatancy of granular materials in a strain space multiple mechanism model. Int. J. Numer. Anal. Methods Geomech. **35**(3), 360–392 (2011)
6. Kutter, B., et al.: LEAP-GWU-2015 experiment specifications, results, and comparisons. Soil Dyn. Earthq. Eng. **113**, 616–628 (2018)
7. Kutter, B., et al.: LEAP-UCD-2017 comparison of centrifuge test results. In: Kutter, B.L., Manzari, M.T., Zeghal, M. (eds.) Model Tests and Numerical Simulations of Liquefaction and Lateral Spreading, pp. 69–103. Springer, Cham (2020). https://doi.org/10.1007/978-3-030-22818-7_4
8. Tobita, T., et al.: LEAP-ASIA-2019: validation of centrifuge experiments and generalized scaling law on liquefaction-induced lateral spreading. In: Proceedings of the 16th Asian Regional Conference on Soil Mechanics and Geotechnical Engineering, 14–18 October 2019, Taipei, Taiwan, Paper No. TC104-001 (2019)
9. Vargas, R., Ueda, K., Uemura, K.: Influence of the relative density and K0 effects in the cyclic response of Ottawa F-65 sand - cyclic Torsional Hollow-Cylinder shear tests for LEAP-ASIA-2019. Soil Dyn. Earthq. Eng. **133**, 106111 (2019)
10. Kutter, B., et al.: Twenty-four centrifuge tests to quantify sensitivity of lateral spreading to Dr and PGA. Geotechnical Earthquake Engineering and Soil Dynamics V: Slope Stability and Landslides, Laboratory Testing, and In Situ Testing (2018)

Centrifuge Modeling on the Behavior of Sheet Pile Wall Subjected Different Frequency Content Shaking

Yi-Hsiu Wang(✉), Jun-Xue Huang, Yen-Hung Lin, and Wen-Yi Hung

National Central University, Taoyuan, Chinese Taipei
wang841215@gmail.com

Abstract. Sheet pile wall is often used as a retaining system at the riverbank owing to its economy, convenience, and constructability. The soil deposit nearby river is composed of alluvium soil with a high ground water level. Therefore, the soil deposit usually has high potential of liquefaction. The shaking may induce soil liquefaction when the earthquake occurs, causing the sheet pile wall damage or failure. Each earthquake has different frequency content and acceleration amplitude in real conditions. Thus, it would lead to different behavior of the wall-soil system. In this study, three dynamic centrifuge tests were carried out by NCU geotechnical centrifuge and shaking table under 24 g centrifugal acceleration field. Ottawa sand was used to prepare the liquefiable ground with a prototypical excavation depth of 3 m. The models were subjected to the input motions with different frequency content of 1 Hz and 3 Hz. The horizontal displacement of the sheet pile wall and ground surface were measured by the linear variable differential transformers and surface markers. Test results indicate that the model subjected to input motion with higher 3 Hz content and higher peak base acceleration has higher excess pore water pressure excitation and excitation rate. The shallow layer soil in the backfilled area of it achieved initial liquefaction. Moreover, it also has larger lateral displacement of sheet pile wall and ground surface as compared to the others.

Keywords: Sheet pile wall · Different frequency content · Liquefaction · Dynamic centrifuge test

1 Introduction

Liquefaction Experiments and Analysis Projects (LEAP) is a series of collaborative research projects, and LEAP aims to produce reliable experimental data for assessment, calibration, and validation of constitutive models and numerical modeling techniques [1]. In LEAP-UCD-2017, nine different centrifuge facilities conducted twenty-four separate model tests to obtain the meaningful assessment of the sensitivity and variability of the tests [2]. In LEAP-ASIA-2019, the centrifuge modeling tests to validate the generalized scaling law. For LEAP-RPI-2020, NCU conducted three dynamic centrifuge tests by the geotechnical centrifuge at the NCU centrifuge laboratory. to observe the effect of input motion frequency content on the behavior of sheet pile wall at liquefiable ground.

© The Author(s), under exclusive license to Springer Nature Switzerland AG 2022
L. Wang et al. (Eds.): PBD-IV 2022, GGEE 52, pp. 1829–1836, 2022.
https://doi.org/10.1007/978-3-031-11898-2_166

2 Equipment and Materials

The experiments were conducted by NCU geotechnical centrifuge with a capacity of 100g-ton in National Central University. The NCU geotechnical centrifuge has a nominal radius of 3 m and carried a 1-D servo-hydraulic actuator shaking table. NCU shaking table can operate under 80 g centrifugal acceleration field with maximum payload of 400 kg. The maximum displacement is 6.4 mm, and the nominal operating frequency range is 0–250 Hz. A rigid container is composed of aluminum alloy plates with inner dimensions of 767 mm (L) × 355 mm (W) × 400 mm (H).

Ottawa sand F65 was used to prepare the liquefiable sandy ground. A series of laboratory testing was conducted to find the physical properties of the soil. The minimum and maximum dry weights are 1445.96 kg/m^3 and 1723.32 kg/m^3, respectively. Ottawa sand F65 is classified as poorly graded sand in the Unified Soil Classification System (USCS). The specific gravity is 2.65 and the mean grain size is 0.20 mm. The sheet pile wall model (Fig. 1) was made of aluminum alloy to simulate the same dynamic bending moment behavior in prototype. The dimension of the sheet pile wall without flange is 245 mm (H) × 340 mm (W), and the thickness is 4.75 mm.

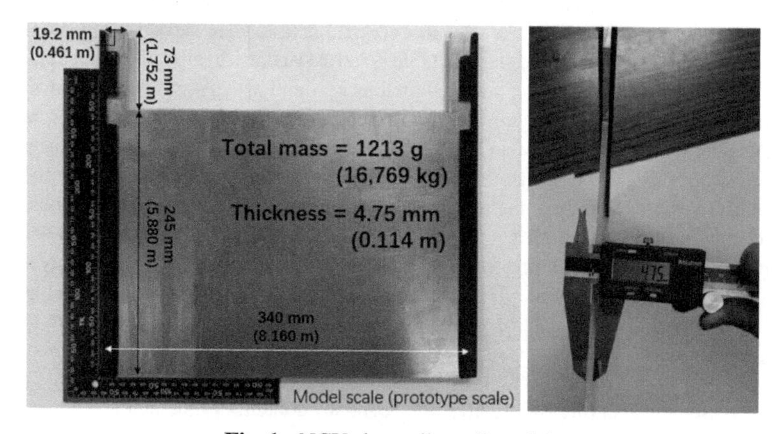

Fig. 1. NCU sheet pile wall model.

3 Model Preparation and Test Procedure

The air pluviation method was adopted to make the sandy deposits model. A sieve box with No.16 mesh is equipped on the pluviator. The bottom of the sieve box is a grille with three rectangular slots (Fig. 2). The slot dimensions are 10.3 mm × 100 mm for dense sand strata (target relative density of 90%) and 1.2 mm × 100 mm for medium dense strata (target relative density of 65%), respectively.

Before installing the sheet pile wall, the vacuum grease was glued on the Teflon tape. In addition, the rubbers covered two sides of the sheet pile wall to reduce the friction between the wall and the container. After installing the sheet pile wall on the sand bed, the adjustable bars were temporarily fixed beside the wall model. The accelerometers and pore water pressures transducers were embedded at prescribed locations as shown in Fig. 3.

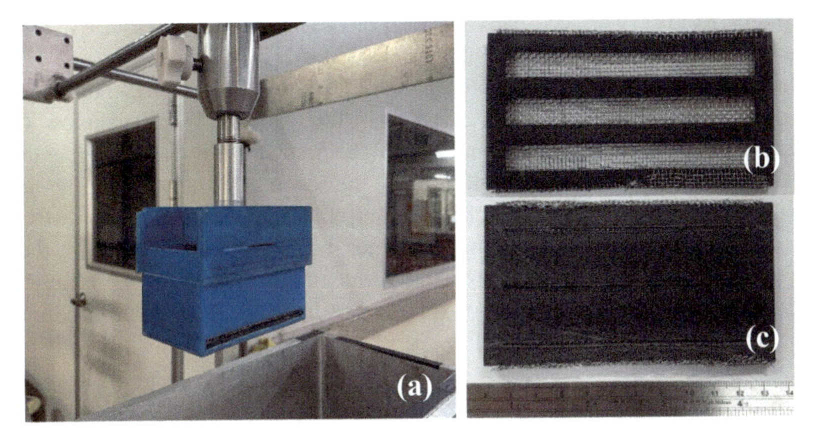

Fig. 2. Pluviation facilities (a) Sieve box (b) Grille with three rectangular slots for medium dense strata (c) Grille with three rectangular slots for dense strata.

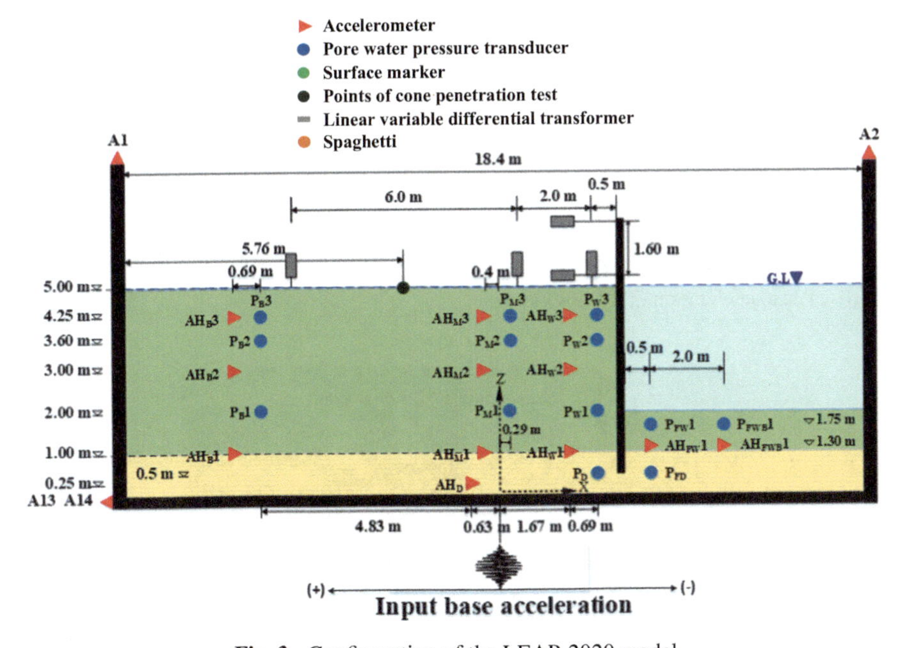

Fig. 3. Configuration of the LEAP-2020 model

From the previous research [3, 4], the dissipation time of excess pore water pressure should be taken into consideration under soil liquefaction conditions. To simulate the prototype fluid flow rate in a dynamic test, the methylcellulose viscous fluid was used to saturate the model by dropping fluid on the model carefully. During model saturation, the air inside the container was simultaneously and continuously vacuumed out. The target viscosity for NCU tests is 24 cSt. After saturation finished, the location and elevation of markers were measured by using digital vernier caliper. Four linear variable differential transformers (LVDTs) were vertically installed on the soil surface and four LVDTs were horizontally installed on the hinge of the sheet pile wall.

After the instruments were connected to the acquisition system, NCU centrifuge spun from 1 g to 24 g-level. The tests were carried out by the sequence described below; (1) the first shaking event, a nondestructive motion; (2) the first CPT test was implemented; (3) second shaking event, a destructive motion; (4) the second CPT test was implemented; (5) the third shaking event, a nondestructive motion. The cone penetration test system (Fig. 4) is controlled by the stepping motor. After the centrifuge was stopped, the final elevation of makers was measured by digital Vernier caliper and the soil profile was cut to observe the deformation behavior of soil deposit from spaghetti's deformation. The testing conditions are listed in Table 1.

Table 1. Conditions of models

Test No	Density (kg/m^3)	PBA (g)	I_A (m/s)	CAV (m/s)	PBA1Hz (g)	PBA3Hz (g)
NCU 1	1610.72	0.15	1.05	8.68	0.08	0.08
NCU 2	1623.52	0.08	0.39	5.30	0.06	0.02
NCU 3	1627.35	0.16	1.09	8.80	0.10	0.03

PBA: Peak base acceleration
I_A: Arias intensity
CAV: Cumulative absolute velocity
PBA_{1Hz}: Peak acceleration of 1Hz wave
PBA_{3Hz}: Peak acceleration of 3 Hz wave

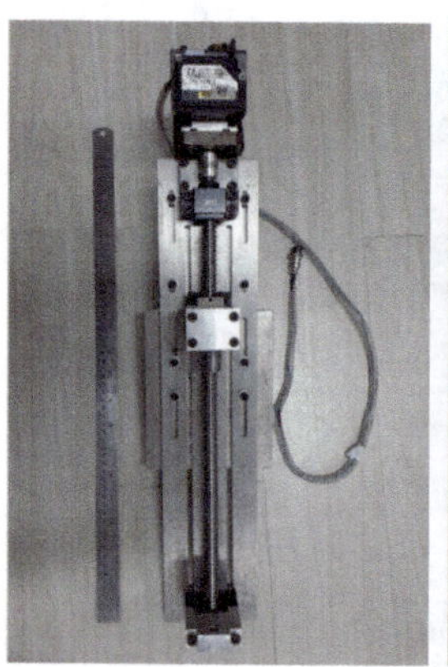
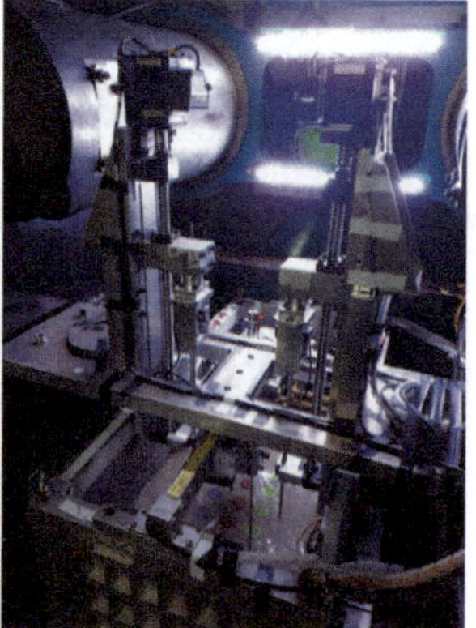

Fig. 4. Cone penetration test system

4 Test Results

4.1 Acceleration Response

Figure 5 shows the acceleration time histories of destructive motion. In three NCU tests, the acceleration responses of the B, M, and W arrays are the same at backfill zone (depth 2 m–4.75 m), and spike waves are observed at backfill zone (depth 0.75 m). Comparing NCU 1 and NCU 3, the shallow soil after wall (W array) has less excess pore water pressure as compared to B and M arrays, and the positive acceleration is higher than negative values because the inertial force of wall and would easier to move to excavation area. The soil close to free filed (B and M arrays) liquefies with symmetric spikes because of soil dilatancy. On the other hand, the FW and FWB array's acceleration responses at the excavation zone are similar to the input motion.

Based on the same frequency content, the test results show that the larger peak base input motion leads to the larger acceleration response. Two times of peak input motion would lead 4 times of response. The soil close to free filed (B and M arrays with depth 0.75 m) liquefies with symmetric spikes.

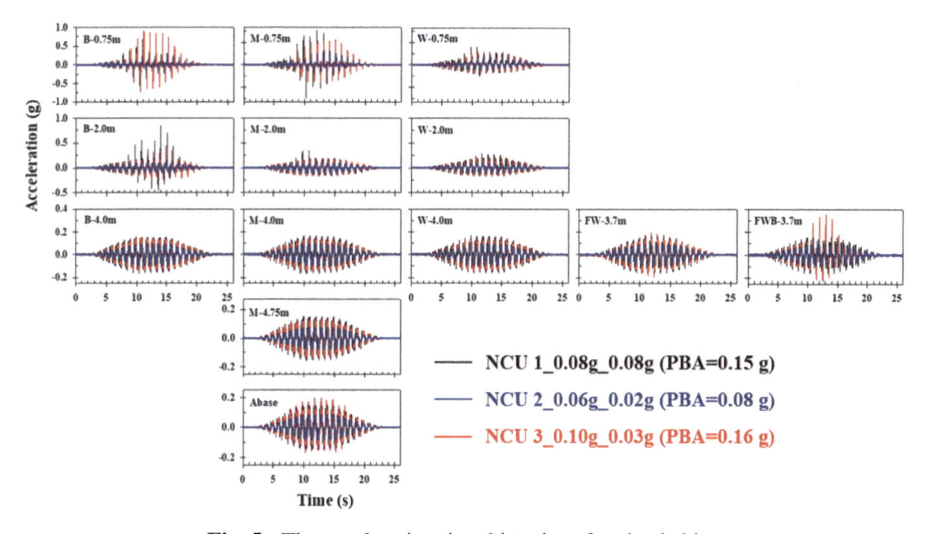

Fig. 5. The acceleration time histories of main shaking

4.2 Excess Pore Water Pressure Response

Figure 6 shows the excess pore water pressure generated behavior during destructive motion, where the black dashed line represents the initial effective overburden pressure. When the excess pore water pressure reaches the initial effective overburden stress, the initial liquefaction occurred. In NCU 1, the excess pore water pressure at depth 3 m, 1.4 m, and 0.75 m in B, M, and W arrays reach the initial effective stress; the initial liquefaction occurred at the backfill zone. At the depth 4.5 m in the W array, after the excess pore water pressure dissipates completely, the water pressure is lower than the initial condition. The model subjected to motion with more predominant frequency of 3 Hz has higher excess pore water pressure excitation.

Based on the same frequency content, the test results indicated that the larger peak base acceleration, the larger excess pore water pressure excitation. In NCU 3, the initial liquefaction could be observed at the backfill zone (depth 3 m). At the excavation zone (FW and FWB), the excess pore water pressure of shallow soil reaches the initial effective stress and it liquefied.

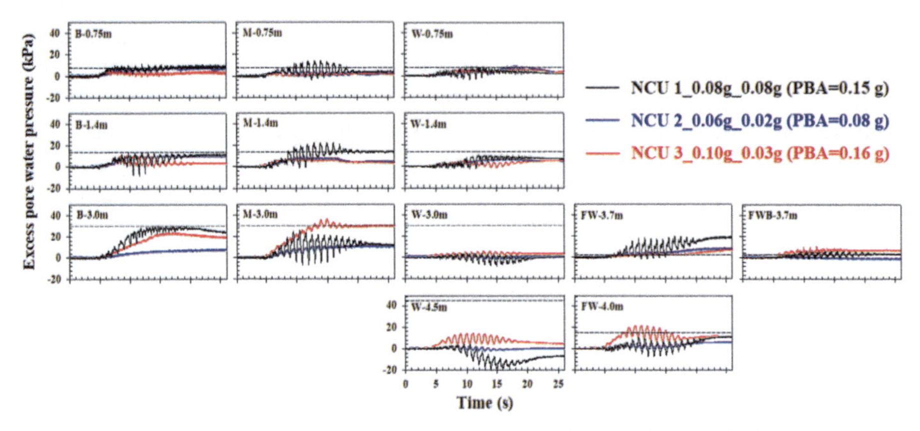

Fig. 6. The excess pore water pressure time histories of main shaking

4.3 Lateral Displacement and Rotation Angle of Wall

The lateral wall displacement and rotation angle of the models are listed in Table 2. The sheet pile wall moves toward to excavation zone after shaking for all the tests. From NCU 1 and NCU 3, the wall subjected to the input motion with more frequency of 3 Hz content (from 6% to 45%) has a larger lateral displacement and rotation angle about 1.15 times and 1.5 times. The wall subjected to double peak base acceleration with the same percent of frequency content has a larger lateral displacement and rotation angle about 2.8 times and 1.8 times. The largest rotation angle changes and lateral displacement is under the peak base input motion of 0.15 g with frequency content of 1 Hz-52% and 3 Hz-45%.

Table 2. The lateral wall displacement and rotation angle of the models

Test No.	Lateral wall displacement		Rotation angle
	Top of the flange	Bottom of the flange	
NCU 1	0.43 m	0.31 m	4.4°
NCU 2	0.14 m	0.09 m	1.6°
NCU 3	0.36 m	0.28 m	2.9°

4.4 Surface Marker Movement and Settlement

Figure 7 shows the surface marker movement and settlement. The red arrows indicate the direction and movement of the measuring point, and the different colors of the circle represented the settlement. Comparing NCU 1 and NCU 3 with B2 and B9 markers, the model subjected to the motion with more predominant frequency of 3 Hz has a smaller settlement about 0.5 times and 0.75 times. For B2 and B9 markers, the model subjected to the motion with more frequency of 3 Hz has a larger surface movement. As compared with NCU 2 and NCU 3 at B array, the model subjected to the larger acceleration amplitude results to a larger surface movement and settlement about 1.5 times.

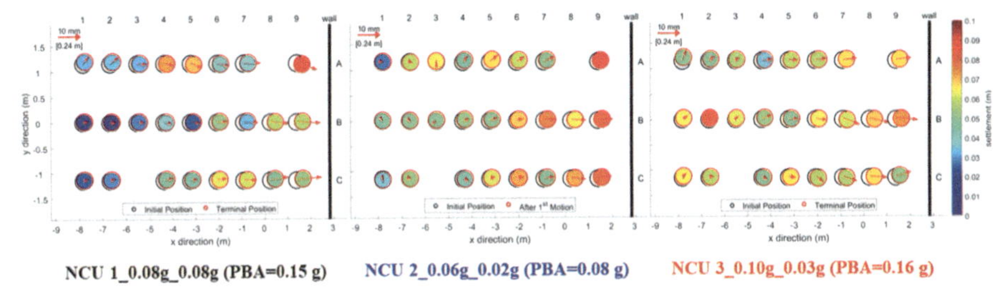

Fig. 7. Surface marker movement and settlement before and after main shaking

4.5 Soil Strength

The distribution of q_c along the depth is plotted in Fig. 8. The wall subjected to the input motion with more frequency of 3 Hz content (from 6% to 45%) would lead to the more change of soil resistance along the depth. The wall subjected to double peak base acceleration with the same percent of frequency content would lead to the more change of soil resistance along the depth.

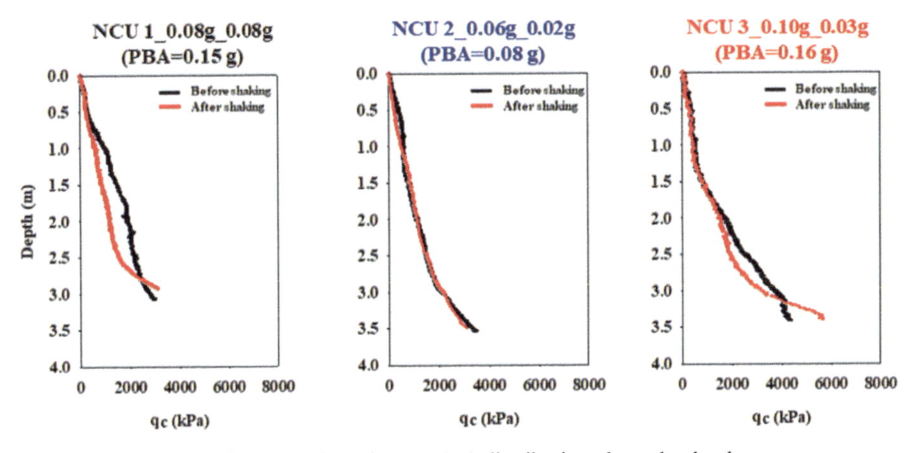

Fig. 8. Cone tip resistance (q_c) distribution along the depth

5 Conclusions

Three centrifuge modeling tests were conducted to observe the effect of input motion frequency content on the behavior of sheet pile wall during liquefaction. Initial liquefaction was observed at backfill zone in all the NCU tests. Several conclusions could be drawn as below.

Due to the soil dilatancy, a spike signal was obtained from acceleration histories at backfill zone. The larger peak base input motion leads to the larger acceleration response. Two times of peak input motion would lead 4 times of response. The soil close to free filed (B and M arrays with depth 0.75 m) liquefies with symmetric spikes.

The larger peak of input motion would lead to the more change of soil resistance along the depth. The wall subjected to the input motion with more frequency of 3 Hz content (from 6% to 45%) would lead to the more change of soil resistance along the depth.

Under the same input energy conditions, the wall subjected to the input motion with more frequency of 3 Hz content (from 6% to 45%) would lead to a larger lateral displacement and rotation angle about 1.15 times and 1.5 times. Under the same frequency content of input motion with two times of the different of PBA, the model subjected to higher PBA would lead to a larger lateral displacement and rotation angle about 2.8 times and 1.8 times.

Acknowledgements. The authors would like to express our gratitude for the financial and technical support from the Ministry of Science and Technology, Taiwan (R.O.C.) (MOST 106-2628-E-008-004-MY3), National Center for Research on Earthquake Engineering, and Geotechnical Centrifuge and Shaking Table Laboratory of Precious Instrument Utilization Center at National Central University. These supports made this study and further research possible and efficient.

References

1. Kutter, B.L., Manzari, M.T., Zeghal, M. (eds.): Model Tests and Numerical Simulations of Liquefaction and Lateral Spreading. Springer, Cham (2020). https://doi.org/10.1007/978-3-030-22818-7
2. Kutter, B.L., et al.: LEAP-UCD-2017 comparison of centrifuge test results. In: Kutter, B.L., Manzari, M.T., Zeghal, M. (eds.) Model Tests and Numerical Simulations of Liquefaction and Lateral Spreading, pp. 69–103. Springer, Cham (2020). https://doi.org/10.1007/978-3-030-22818-7_4
3. Ueda, K., Iai, S.: Numerical predictions for centrifuge model tests of a liquefiable sloping ground using a strain space multiple mechanism model based on the finite strain theory. Soil Dyn. Earthq. Eng. **113**, 771–792 (2018)
4. Stewart, D.P., Chen, Y.-R., Kutter, B.L.: Experience with the use of methylcellulose as a viscous pore fluid in centrifuge models. Geotech. Test. J. **21**(4), 365–369 (1998)

S2: Special Session on Liquefaction Database

Liquefaction Cases and SPT-Based Liquefaction Triggering Assessment in China

Longwei Chen[1,2](✉), Gan Liu[1,2], Weiming Wang[3], Xiaoming Yuan[1,2], Jinyuan Yuan[1,2], Zhaoyan Li[1,2], and Zhenzhong Cao[4]

[1] Key Laboratory of Earthquake Engineering and Engineering Vibration, Institute of Engineering Mechanics, China Earthquake Administration, Harbin 150080, China
chenlw@iem.ac.cn

[2] Key Laboratory of Earthquake Disaster Mitigation, Ministry of Emergency Management, Harbin 150080, China

[3] Heilongjiang Institute of Technology, Harbin 150050, China

[4] Guilin University of Technology, Guilin 541006, China

Abstract. In the 60s to 70s of last century, major damaging earthquakes hit China and trigged tremendous liquefaction. The post-earthquake in-situ investigation and site tests on liquefied sites and comparative non-liquefied sites were conducted that relevant data have been collected and studied. Using the data, liquefaction evaluation methods have been formulized and accepted by Chinese codes, in which the standard penetration test (SPT) based formula is widely used in engineering practice. This paper introduces liquefaction data collected in China mainland and the revision process of SPT-based liquefaction evaluation formulation. Adopting the established procedures, the SPT blows were corrected on trial by influence factors. Comparison with liquefaction evaluation methods, which base on cyclic-stress-ratio (CSR) framework, indicates the CSR-based liquefaction formulae basically overestimated the liquefaction data in low CSR range less than 0.1. Further investigation has to be performed on the data analysis to study the consistency of data collected from China with those from worldwide.

Keywords: Liquefaction · Standard penetration test · Evaluation method · Data correction

1 Introduction

Site liquefaction is a typical earthquake-induced geotechnical hazard that gives rise to a scientific concern globally [1–3]. In recent earthquakes, liquefaction still remains a problem and even made a dominant cause for earthquake loss [4–6]. Since engineers and scientists focused on the liquefaction issues after Niigata earthquake and Alaska earthquake in 1964, many a liquefaction mitigation technique has been proposed, in which, liquefaction evaluation becomes the priority for aseismic design and fortification of an engineering site. To establish a reliable liquefaction evaluation technique is the

© The Author(s), under exclusive license to Springer Nature Switzerland AG 2022
L. Wang et al. (Eds.): PBD-IV 2022, GGEE 52, pp. 1839–1847, 2022.
https://doi.org/10.1007/978-3-031-11898-2_167

first step for liquefaction mitigation and engineering safety protection. The technique, on one hand, guarantees the seismic safety of engineering structures yet closely relate to engineering cost; on the other hand, the reliability of liquefaction evaluation technique is so important that the technique itself needs continuously updating.

Fig. 1. Sand ejecta caused by liquefaction in 1976 Tangshan earthquake buried a farmland [9]

Fig. 2. A water pit caused by liquefaction ejection in 2003 Bachu earthquake in Xinjiang province (photo by Lu)

Liquefaction data from post-earthquake liquefaction survey and in-situ site investigation set the base of liquefaction mitigation, giving insight into liquefaction knowledge and liquefaction understanding. On the other hand, liquefaction data are the essential for constructing liquefaction evaluation methods like those in seismic design codes [7, 8]. Site liquefaction occurred in almost every damaging earthquake. In China mainland, the noticeable liquefaction cases were those from Tangshan, Haicheng earthquakes etc., in which the liquefaction data after Tangshan earthquake were abundant, stays dominant in Chinese liquefaction dataset. Figure 1 pictured the sand ejecta in a farmland caused by Tangshan earthquake [9]. The tremendous liquefaction field investigation data from earthquakes in 60s and 70s of last century have been used to construct the liquefaction evaluation methods, in which the standard penetration test (SPT) is the most popular method. The SPT is widely used for measurement of soil penetration resistance and subsequent correlation with soil properties such as relative density, shear strength, bearing capacity, and liquefaction resistance (Fig. 2).

Over decades, the SPT-based liquefaction evaluation formula has been revised several times by changing the functional forms. However, the original data used for regression were not updated, even though the liquefaction data have been collected from recent earthquakes both in China mainland and overseas. Figure 3 shows liquefaction generated a water pit surrounded by sand ejecta in 2003 Bachu earthquake in Xinjiang. Li et al. [10] conducted in-situ SPT tests and compared the measured blow counts with predictions by the code formula, then concluded that the code formulae was not feasible for Bachu cases. Therefore, liquefaction evaluation methods have to be updated by cumulated data from recent earthquakes.

Another issue concerns the quality of liquefaction data. The SPT setups have been improved through standardization and measurement of energy transferred from the hammer to the drill rod. Cetin et al. [11] and Idriss and Boulanger [12] have collected and compiled the SPT data of liquefaction from worldwide, following the work by Seed et al. [13], and used correction factors on the SPT data to make the data standardized. However, the database hardly encompasses the liquefaction data from China mainland. The main reason could possibly result from the correction factors. Hereinafter, the liquefaction data, i.e., SPT data, from China mainland are briefly presented, and then SPT-based liquefaction evaluation methods are introduced, and finally that trials of SPT data correction are performed to see the consistency of the SPT data with those from other countries.

2 Liquefaction Data

Xie [14] has compiled liquefaction data of SPT from earthquakes occurred in the 60s and 70s of last century and presents the methodology of building the liquefaction evaluation formula which was adopted in Chinese seismic design code (TJ 10-74) which was the former version of the current Seismic Design Code for Buildings (GB50011-2010). The liquefaction data herein were collected from Xie [14] and included Bachu earthquake [10]. Table 1 lists earthquake information. Totally, 160 effective data in seismic intensity VII, VIII and IX (*Chinese Intensity Scale*) are obtained, including 99 liquefaction data and 61 non-liquefaction data. Statistically, 85% of the liquefaction data are from Tangshan earthquake, Tonghai earthquake and Haicheng earthquake. Nearly 60% data are from the 1976 Tangshan earthquake.

Figure 3 displays the histograms of the data of sand-layer buried depth, with respect to liquefaction and non-liquefaction. Most sand-layer buried depth of liquefaction cased stays less than 6 m while those of non-liquefaction keeps roughly uniform distribution. Figure 4 portrays the distribution of SPT blows and ground motion intensity measured by PGA values. The measured SPT blows of liquefied sites were mostly less than 20 blows while those of non-liquefaction remains much larger. The ground shaking intensity, measured by peak ground motion acceleration (PGA), of the data ranges widely from less than 0.05 g to larger than 0.4 g, where g is the unit of gravity acceleration.

Table 1. Earthquakes of liquefaction data collected in China mainland [10, 14]

No.	Earthquake	Date	Magnitude
1	Heyuan	3/19/1962	6.4
2	Xingtai 1	3/08/1966	6.7
3	Xingtai 2	3/22/1966	7.2
4	Hejian	3/27/1967	6.3
5	Bohai	7/18/1969	7.4
6	Yangjiang	7/26/1969	6.4
7	Tonghai	1/05/1970	7.8
8	Haicheng	2/04/1975	7.3
9	Tangshan	7/28/1976	7.8 (M_w7.6)
10	Bachu	2/24/2003	6.8 (M_w6.3)

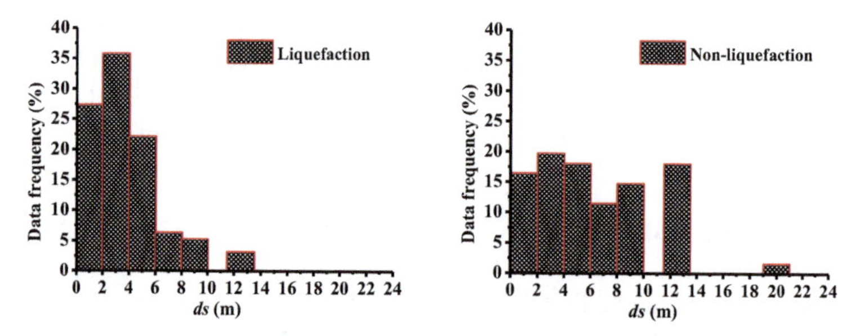

Fig. 3. Histogram of sandy-layer buried depth versus data frequency

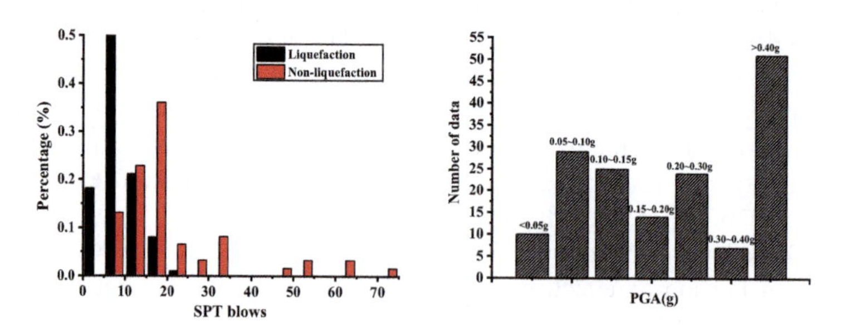

Fig. 4. Distribution of liquefaction data of SPT blows and PGA

3 Liquefaction Evaluation Formula

The liquefaction evaluation in Chinese seismic design codes consists of two steps. In the first step, site geological age and sandy-layers buried conditions are evaluated. In the second step, a liquefaction evaluation formula, which uses SPT blows as an index, is adopted to judge liquefaction possibility.

In the latest code (GB 50011-2010) [7], the provision is written as, "*Geological age of a soil unit is and before late quaternary-age (Q3) [greater than 2.6 million years old], the unit is judged as non-liquefaction for intensity VII (PGA = 0.1* g*) and VIII (PGA = 0.2* g*)*".

If the site is potentially capable of liquefying, then it moves to the second step which uses an empirical liquefaction evaluation formula. The formula has been revised several times that the functional forms have been changed from linearity to piecewise-linearity and to current logarithmic forms. In 1989 Chinese code, the liquefaction evaluation formula is linear and written as,

$$N_{\mathrm{cr}} = N_0[0.9 + 0.1(d_{\mathrm{s}} - d_{\mathrm{w}})]\sqrt{3/\rho_{\mathrm{c}}} \tag{1}$$

where, N_{cr} is the critical SPT blows, while N_0 is the reference SPT values seen in Table 2; d_w is the values of underground water tables in meter, and d_s is the sandy-layer buried depth in meter, herein $d_s \leq 15$ m, and ρ_c is the fine content which set to be 3 if ρ_c is less than 3 or the soil is classified as sand.

In 2001 Chinese code method (GB50011-2001) [15] d_s was extended to 20 m. The formula becomes,

$$N_{cr} = N_0[0.9 + 0.1(d_{\mathrm{s}} - d_{\mathrm{w}})]\sqrt{3/\rho_{\mathrm{c}}}\,(d_{\mathrm{s}} \leq 15) \tag{2}$$

$$N_{\mathrm{cr}} = N_0(2.4 - 0.1d_{\mathrm{w}})\sqrt{3/\rho_{\mathrm{c}}}\,(15 \leq d_{\mathrm{s}} \leq 20) \tag{3}$$

Considering that the critical line should be continuous instead of segments, the evaluating formula was modified in the current formula [7] which is written as,

$$N_{\mathrm{cr}} = N_0\beta[\ln(0.6d_{\mathrm{s}} + 1.5) - 0.1d_{\mathrm{w}}]\sqrt{3/\rho_{\mathrm{c}}} \tag{4}$$

In which, β is an adjusting coefficient and set values of 0.8, 0.95 and 1.05 with respect to the first design earthquake group, the second design earthquake group and the third design earthquake group provided in seismic design codes.

Table 2. Reference N_0 values under different design basic PGA values

PGA/g	0.10	0.15	0.20	0.30	0.40
N_0	7	10	12	16	19

Besides, Yang et al. [16] proposed a SPT-based hyperbolic formula for liquefaction evaluation to overcome the disadvantages of the code method, which makes conservative evaluation for sandy-layer buried depth of 10 m to 20 m in intensities VIII and IX. The liquefaction evaluation formula is written as,

$$N_{\mathrm{cr}} = 0.79N' \cdot (1 - 0.02d_{\mathrm{w}})\left(0.27 + \frac{d_{\mathrm{s}}}{d_{\mathrm{s}} + 6.2}\right) \tag{5}$$

In which, N' is taking the values suggested in Table 3.

Table 3. Reference N' values under different design basic PGA values

PGA/g	0.10	0.15	0.20	0.30	0.40
N'	16	20	23	31	37

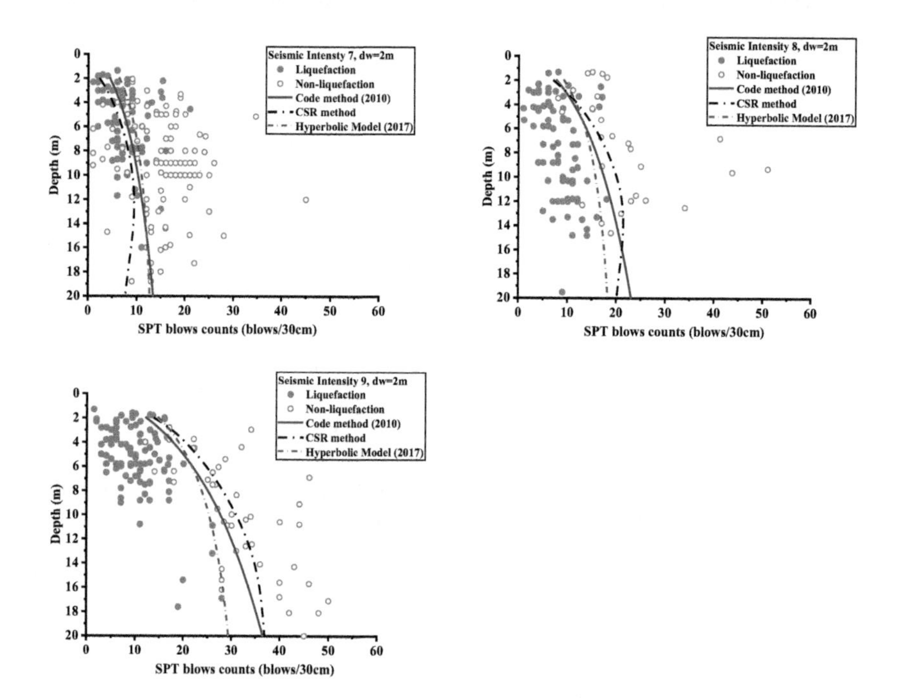

Fig. 5. Comparison of critical lines predicted by different methods

To be consistent with PGA values, an empirical equation to predict N' has been proposed, which is,

$$N' = 69\frac{PGA}{PGA + 0.4} \tag{6}$$

in which the unit of PGA value is in g. Figure 5 displays comparatively the critical lines predicted by different formulae on the data collected for seismic intensity (SI) 7, 8 and 9, respectively. To be emphasized, the data are not only collected from China.

4 Liquefaction Evaluation Methods Comparison

The liquefaction evaluation methods based on SPT blows mostly are formulated within the simplified cyclic stress ratio framework (CSR) formulated by Seed et al. (1971) [17]. The most representatives have been systematically introduced in NCEER (2001) report and Idriss and Boulanger (2010) report. The essential difference between Chinese SPT-based liquefaction evaluation formula with CSR methods comes from the SPT blow index (*N* values). In the CSR framework, the blow counts are corrected by influence factors such as overburden pressure, energy transfer ratio, rod length, borehole diameter, and types of samplers. The SPT index used in Chinese codes is the uncorrected measured blow values.

Apart from the difference of *N* values, the Chinese SPT-based liquefaction evaluation formula, takes sandy layer depths as a variable, is very convenient for engineering practice. By comparing the critical *N* values predicted with the measured SPT blow counts at depth according to seismic intensity, the liquefaction judgement can be conducted instead that the cyclic stress ratio and cyclic resistance ratio are not needed to be calculated.

To fit the CSR framework, the Chinese data could be corrected as well according to the established procedures. However, the data information was not well documented and the setups of the SPT have been improved ever since. Decades have passed that the availability of data records was unfortunately limited, which makes the correction factors uneasy to be determined. To clear the obstacles, the correction coefficients were presumed and the ground shaking intensity was assessed by ground-motion prediction equations or simply conversion relations presented in Chinese code between PGA values with respect to seismic intensities [7]. Figure 6 displays CSR critical curves of different methods comparing with the corrected data $(N_1)_{60}$. The correction coefficients are referring to Youd et al. (2001) [3]. Meanwhile, the liquefaction critical lines predicted by NCEER (2001) and Idriss and Boulanger (2010) are almost identical but stay offset

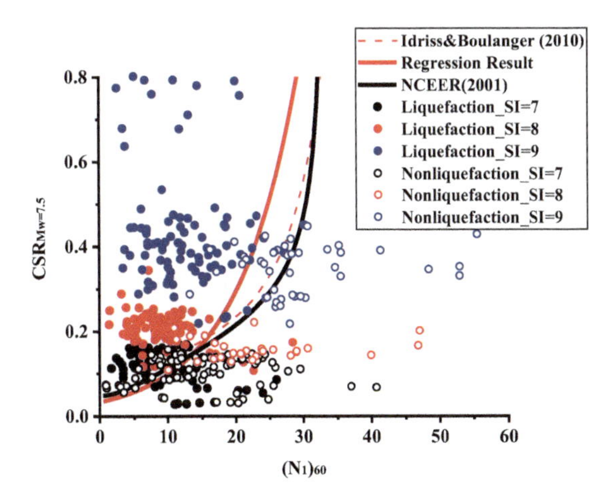

Fig. 6. The critical lines predicted by liquefaction evaluation methods comparing with Chinese SPT data of different seismic intensity values

from the direct regression line when CSR values are larger than 0.2. To be noticed, the critical lines predicted by NCEER (2001) and Idriss and Boulanger (2010) can fit well the liquefaction data when CSR stays larger than 0.1. But when low CSR values than 0.1, the liquefaction data are mostly misjudged. As for non-liquefaction data, there are some data intervening with the liquefaction data, that the critical lines cannot distinguish them. There are reasons for the liquefaction data of low CSR. For instance, underestimation of ground-motion intensity and the SPT values were, in some cases, converted from blows of un-standard samplers. Therefore, the liquefaction data have to be further investigation and evaluation formula within CSR framework has to be proposed.

5 Conclusive Remarks

Liquefaction hazard has been widely recognized by China geotechnical community, and that the liquefaction evaluation techniques have been proposed accordingly. The Chinese SPT-based liquefaction evaluation methods in the codes are different from the CSR-based methods which are widely used worldwide. The evaluation formulae adopted in Chinese code are unique but user-friendly. The SPT blow data, however, that are used to establish the formulae are uncorrected, which makes the methods hardly can compare with others and the theoretical background weak. To compensate these limitations, correction factors which are recognized and empirically recommended have been borrowed to correct the SPT blow values and then the data are compared with the predictions. The results indicate the corrected SPT blow values are not quite consistent with predictions especially in low CSR range of less than 0.1. Further investigation is needed to correct the data and liquefaction data from recent earthquakes have to be incorporated. With these data, the revision of the current liquefaction evaluation formulae will be benefited.

Acknowledgement. Financial support of Scientific Research Fund of Institute of Engineering Mechanics, China Earthquake Administration (Grant No. 2019C07) is sincerely acknowledged.

References

1. Ishihara, K.: Liquefaction and flow failure during earthquake. Geotechnique **43**(3), 351–415 (1993)
2. Berrill, J., Yasuda, S.: Liquefaction and piled foundations: some issues. J. Earthquake Eng. **6**(S1), 1–41 (2002)
3. Youd, T.L., et al.: Liquefaction resistance of soils: summary report from the 1996 NCEER and 1998 NCEER/NSF workshops on evaluation of liquefaction resistance of soils. J. Geotech. Geoenviron. Eng. **127**(10), 817–833 (2001)
4. Chen, L., et al.: Liquefaction macrophenomena in the great Wenchuan earthquake. Earthq. Eng. Eng. Vib. **8**(2), 219–229 (2009)
5. Cubrinovski, M., et al.: Soil liquefaction effects in the central business district during the February 2011 Christchurch earthquake. Seismol. Res. Lett. **82**(6), 893–904 (2011)
6. Papathanassiou, G., Mantovani, A., Tarabusi, G., Rapti, D., Caputo, R.: Assessment of liquefaction potential for two liquefaction prone areas considering the May 20, 2012 Emilia (Italy) earthquake. Eng. Geol. **189**, 1–16 (2015)

7. Design Code: Code for Seismic Design of Buildings GB 50011-2010. China Architecture and Building Press, Beijing (2010). (in Chinese)
8. Building Seismic Safety Council: NEHRP recommended Seismic Provisions for new buildings and other structures (FEMA P-750), Washington D C (2009)
9. Liu, H.X. (ed.): The Great Tangshan Earthquake. Seismic Press, Beijing (1985)
10. Li, Z.Y., Yuan, X.M., Cao, Z.Z., Sun, R., Dong, L., Shi, J.H.: New evaluation formula for sand liquefaction based on survey of Bachu earthquake in Xinjiang. Chin. J. Geotech. Eng. **34**(3), 483–489 (2012). (In Chinese with English abstract)
11. Cetin, K.O., et al.: Standard penetration test-based probabilistic and deterministic assessment of seismic soil liquefaction potential. J. Geotech. Geoenviron. Eng. **130**(12), 1314–1340 (2004)
12. Idriss, I.M., Boulanger, R.W.: SPT-based liquefaction triggering procedures, Geotechnical Engineering Report No. UCD/CGM-10-02, University of California at Davis (2010)
13. Seed, H.B., Tokimatsu, K., Harder, L.F., Chung, R.M.: The influence of SPT procedures in soil liquefaction resistance evaluations, Earthquake Engineering Research Center Rep. No. UCB/EERC-84/15, University of California, California (1984)
14. Xie, J.F.: Some comments on the formula for estimating the liquefaction of sand in revised aseismic design code. Earthq. Eng. Eng. Vib. **4**(2), 95–125 (1984). (In Chinese with English abstract)
15. Design Code: Code for Seismic Design of Buildings, GB 50011-2001. China Architecture and Building Press, Beijing (2001). (in Chinese)
16. Yang, Y., Chen, L., Sun, R., Chen, Y., Wang, W.: A depth-consistent SPT-based empirical equation for evaluating sand liquefaction. Eng. Geol. **221**, 41–49 (2017)
17. Seed, H.B., Idriss, I.M.: Simplified procedure for evaluating soil liquefaction potential. J. Soil Mech. Found. Div. **97**(SM8), 1249–1274 (1971)

Hammer Energy Measurement of Standard Penetration Test in China

Longwei Chen[1,2](✉), Tingting Guo[1,2], Tong Chen[1,2], Gan Liu[1,2], and Yunlong Wang[1,2]

[1] Key Laboratory of Earthquake Engineering and Engineering Vibration, Institute of Engineering Mechanics, China Earthquake Administration, Harbin 150080, China
chenlw@iem.ac.cn

[2] Key Laboratory of Earthquake Disaster Mitigation, Ministry of Emergency Management, Harbin 150080, China

Abstract. Hammer efficiency, which is defined as the energy transfer ratio, is an important index in standard penetration test which is widely used for measurement of soil penetration resistance and subsequent correlation with soil properties. In the routine procedure, the energy measurement should be made to correct penetration blow counts. However, the energy correction for standard penetration test are not considered in Chinese codes. To circumvent the shortcoming, in-situ tests at three selected sites in Xichang city region are performed to measure the energy transmitted into rod during hammer fall in the current commonly used standard penetration test setups. The energy transfer ratio data recorded are well consistent, that most energy transfer ratio values range in 60% to 90% with mean ratios exceeding 70%. The dependence of energy transfer ratio on testing depth is weak. The energy transfer ratio increases by about 10% when the penetration depth increases down to 20 m beneath surface. The analysis results can prove the reliability of current standard penetration test setups and provide a good reference for energy correction of blow counts.

Keywords: Standard penetration test · Energy measurement · Energy transfer ratio · In-situ test

1 Introduction

Automatic trip hammers have advantages for standard penetration test (SPT) of consistent drop height and low friction loss during hammer fall [1]. The SPT is widely used for measurement of soil penetration resistance and subsequent correlation with soil properties such as relative density, shear strength, bearing capacity, and liquefaction resistance [e.g., 2–4]. Over the years, the SPT has been improved through standardization and measurement of energy transferred from the hammer to the drill rod. The development and wide deployment of automatic trip hammers has improved test consistency and eliminated operational variables that previously plagued the test such as maintaining a constant drop height and creating near frictionless hammer fall.

© The Author(s), under exclusive license to Springer Nature Switzerland AG 2022
L. Wang et al. (Eds.): PBD-IV 2022, GGEE 52, pp. 1848–1856, 2022.
https://doi.org/10.1007/978-3-031-11898-2_168

When SPT is performed in engineering practice, energy measurements should be made to measure the energy transfer ratios (ETR) which are determined for correction of measured SPT blow counts (N_m) to a standard energy ratio of 60% [4]. The energy correction factor, denoted as C_E, is defined as,

$$C_E = \frac{ETR}{60\%} \tag{1}$$

$$N_{60} = C_E \cdot N_m \tag{2}$$

where N_m is the measured SPT resistance in the field in blows per 30 cm, that N_m is measured in whole integers. And N_{60} values which are corrected by energy measurement can be sensitive to the tested ETR.

Apart from energy correction, other factors like overburden pressure, rod length, borehole diameter etc., also have to be considered [4–6]. However, in Chinese codes [7, 8], even the energy correction is not considered instead that the raw measured blows N_m is used for representation of soil resistance. Consequently, empirical relations which are based N_m have been established and used in engineering practice. For example, liquefaction evaluation formulae take N_m as the index as the soil resistance against liquefaction [7]. Under such circumstance, the Chinese liquefaction evaluation methods can hardly be compared with other methods even though it is engineering friendly and convenient. One explanation is that no energy measure device was ever used in field tests during the site investigation. That is, when field data were compiled to build the empirical formulae and that the energy correction for SPT data had to be ignored.With the demanding for energy correction in blow counts of penetration tests, attention has paid on energy transmitted in to rod of the Chinses hammer using the current penetration test setups. The energy transfer ratio starts to be considered in correcting penetration test blow counts [9, 10]. Also, Ge et al. [11] has compared Chinese code and ASTM standard penetration tests by penetration energy analysis and correlation of blow counts. In another words, the importance of hammer efficiency is more and more recognized, that the energy correction gradually will be an inevitably necessary step in determination of soil penetration resistance.

To respond to the reliability and hammer efficiency of the current SPT setups in China, SPT tests were performed at three sites in Xichang city region of China Seismic Experimental Site (CSES). Energy measurement was employed to test how the Chinese SPT setups performed and to affirm the confidence of the SPT blows.

2 In-situ Testing

2.1 Energy Measurement

A Pile Dynamic Analyzer (PDA) was used to measure the dynamic energy transmitted into the drilling rod. The energy measurement equipment basically is equipped with two highly sensitive acceleration sensors and two strain gauges, that are used to measure the acceleration and strain, respectively, during hammer impact. The recorded rod strain and the acceleration time histories can be used to converted to axial force and velocity time histories which are processed straightforwardly by the PDA program as shown in Fig. 1. In theory, the maximum energy (EMX) transfer into the rod measured by the gauges can be calculated using force and velocity records as,

$$EMX = \int F(t)du = \int F(t)V(t)dt \tag{3}$$

where,

- F(t): the time history of impacting force;
- V(t): the time history of velocity

Using PDA, the EMX can be directly read from the screen, that the energy transfer ratio (ETR) can be obtained by,

$$ETR = \frac{EMX}{E_R} \tag{4}$$

where E_R is the theoretical total energy which is equal to the potential energy of the hammer lifted by 76 cm. In this case, the potential energy of the donut hammer, can be calculated as, $E_R = 63.5\ \text{kg} * 0.76\ \text{m} * 9.8\ \text{m/s}^2 = 63.5\ \text{kg} = 0.473\ \text{kN}\cdot\text{m}$. In the SPT test, ETR values are directed processed by the PDA that no need is required to calculate ETR values by Eq. (4). However, Eqs. (3) and (4) explain how the ETR values are obtained.

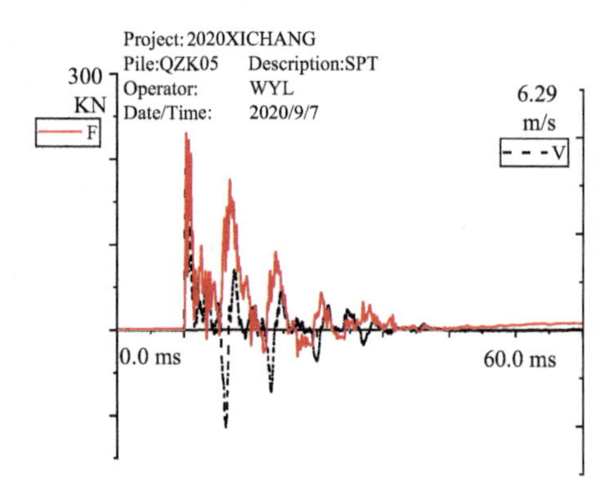

Fig. 1. The axial force time history and velocity time history measured during hammering

2.2 In-situ SPT Test

Three testing sites, which were denoted as QZK01, QZK05 and QZK08, were selected to carry out SPT test in Xi-Chang city region. The test layout is displayed in Fig. 2. The SPT apparatus is a routinely used in China using a standard sampler (Fig. 3). The donut hammer was driven by a diesel engine via a steel rope. During the SPT test, the 63.5 kg donut hammer (Fig. 3) was lifted by the diesel engine up to 76 cm high and released automatically. The hammer freely drops on the anvil and drives the rod penetrating the soil. The number of blows that required to penetrate 30 cm is defined as N_m, which is widely used as an index of soil resistance against penetration and consequently to determine the soil properties such as relative density, bearing capacity and liquefaction resistance etc. [12].

Fig. 2. In-situ layout of SPT testing with a 63.5 kg donut hammer falling to anvil above PDA energy measurement.

Fig. 3. The 63.5 kg donut hammer and the standard penetration test sampler

3 Data Analysis

3.1 ETR

Figure 4 displays the ETR values collected during performing SPT test with respect to depth by PDA. Figure 4(a) to Fig. 4(c) present ETR values with respect to individual QZK01, QZK05 and QZK08 sites. Most of the ETR values range in 60% to 90%, with exception that the ETR values at QZK08, at shallow depth, stayed less than 60% but greater than 50%. No clear difference can be discerned among the tested sites. Figure 4(d) displays the lumped ETR data inclusively of the three sites, showing a well matching of the data even though with variability in some extend. Neglecting the variation of ETR data varying with depth, ETR values of an individual site were lumped for statistical analysis. Table 1 presents the mean and variation of the ETR values. The means of ETR values of the three sites were very close, ranging between 70% to 80%. However, the standard deviation of QZK08 stayed a little larger than those of QZK01 and QZK05 by about 3–4%. Nevertheless, the hammering efficiency of the SPT apparatus are stable that approximately 60% to 90% of energy can transmitted into the rod.

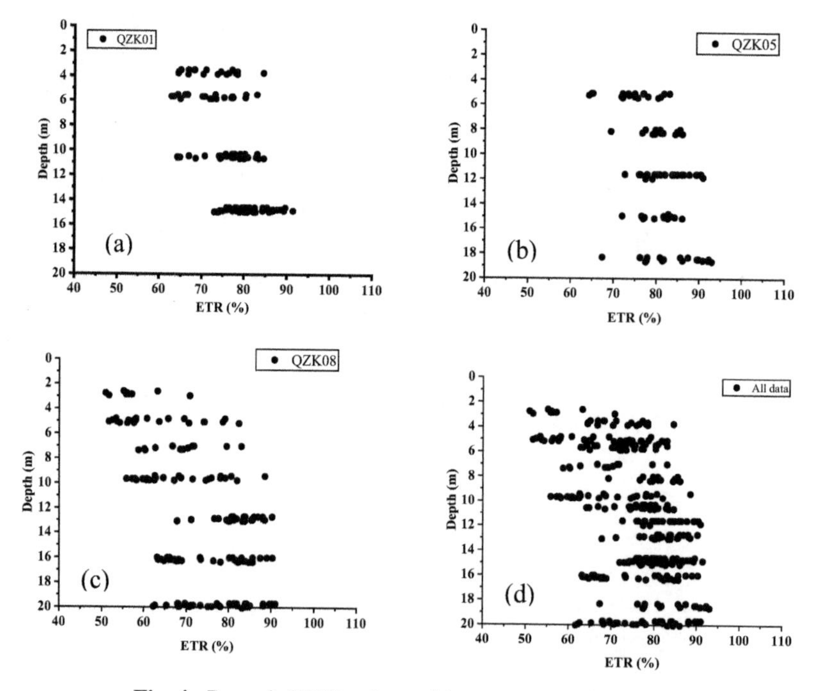

Fig. 4. Recorded ETR values with respect to testing depth

To see the statistical distribution of the data, Fig. 5 demonstrates the histograms of the ERT values. The black lines are normal distribution probability function prediction curves, which mean and standard deviation values are equal to those in Table 1. The ETR data from QZK01 and QZK05, which were shown in Fig. 5(a) and Fig. 5(b), respectively,

Table 1. The statistics of ETR values collected during in-situ test

Tested sites	ETR			
	Mean	Mean + Std.	Mean – Std.	Std.
QZK01	77.03%	83.50%	70.56%	6.47%
QZK05	80.61%	86.90%	74.32%	6.29%
QZK08	72.99%	83.83%	62.15%	10.84%

somehow stay consistent with the normal distribution function while that of QZK08 stays quite dispersive. Lumping all the data as displaying in Fig. 5(d), the normal distributed probability function models slightly-biased fit the data. The bias mainly stems from the data of QZK08.

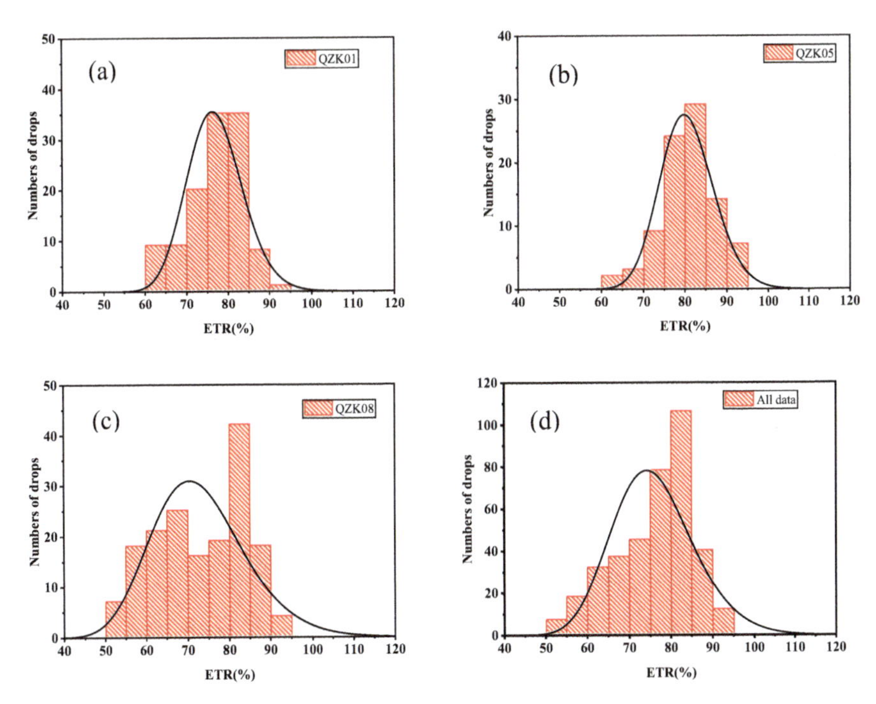

Fig. 5. Histograms of ETR values representing the statistics

The reasons for scattering of the ETR data can be, from the testing point of view, (1) the stability of the rod above level ground; (2) the rebound of the hammer on the anvil; (3) the rate of lifting and releasing of the hammer; (4) the artificial skill of the operator/driver. The first reason could be obvious. When the rod swayed back and forth during hammering, the ETR values monitored usually were small. Because, the swaying of the rod can dissipate hammering energy, resulting into less energy transmitted into the rod. The operating rate of hammer, i.e., drops per minute, also plays an important role in ETR measurement. The high testing rate usually causes the ETR values unexpectedly high. It was observed that, if the hammer was lifted fast and released automatically, the hammer would not drop immediately instead that it would fly further and then dropped. In this case, the "real" potential energy of the hammer is larger than the theoretical, that the energy transmitted into the rod stays larger than the expected.

3.2 Dependence of ETR on Depth

To investigate the dependence of ETR values on testing depth, the lumped data of three sites, to circumvent paucity of data, were subdivided into depth bins, i.e., <5.0 m, 5–10 m, 10–15 m, >15 m, according to depths where SPT were performed.

Figure 6 presents the histograms of ETR values in each depth bin and the black solid curves represent normal distributed probability density function prediction, which took the mean and standard deviation values from Table 2.

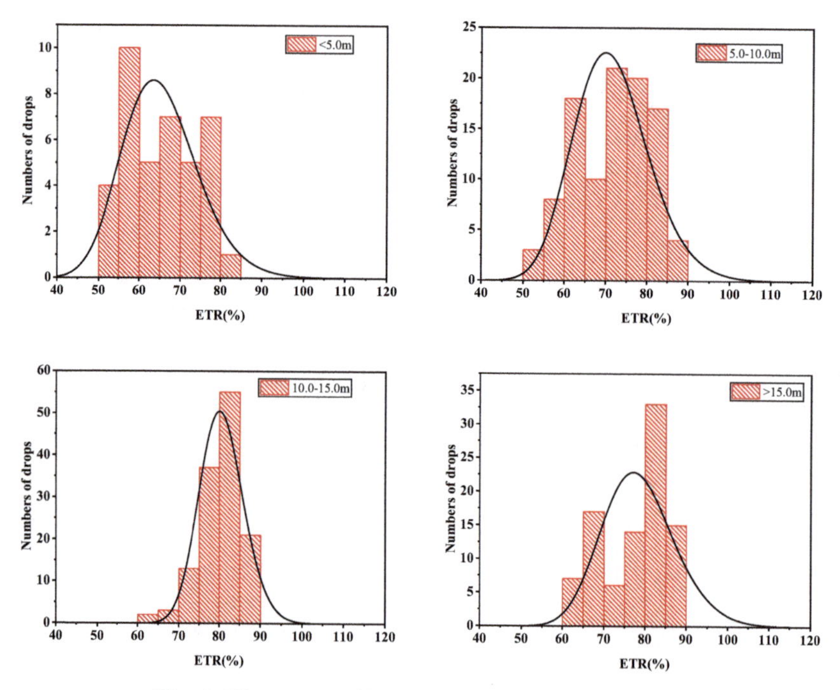

Fig. 6. Histograms of ETR values in different depth bins

As indicated in Table 2, the mean values of ETR keep a slight tendency of increasing with increasing depth by about 10%. Moreover, the standard deviations remain less than 10%. The dependence of ETR is not strong that, for the sake of simplicity and convenience, it can be neglected. In Fig. 6, in shallow depth, e.g., less than 10m, the ETR values appear to be dispersive in range of 50% to 90%. The uncertainty is greater than those in other depth bins. The variability of data in depth ranging in 10 m to 15 m stays the smallest. Majority of the data range in 80–85%.

Table 2. Mean and variation of ETR values in different depth ranges

Depth (m)	ETR			
	Mean	Mean + Std.	Mean – Std.	Std.
<5.0	65.36%	74.57%	56.15%	9.21%
5.0–10.0	71.60%	80.41%	62.79%	8.81%
10.0–15.0	80.57%	85.86%	75.28%	5.29%
>15.0	78.55%	87.04%	70.06%	8.49%

4 Conclusions and Remarks

Through in-situ test and energy measurement, the energy transfer of the hammering in the commonly used SPT procedures in China was evaluated. The results indicate that the current SPT test is quite stable and robust. The efficiency of the hammering stays high with ETR values mainly ranging in 60% to 90%. The mean ERT values are more than 70% and standard deviations stays less than 10%. The dependence of ERT on depth appears to be weak.

In engineering, when SPT measurement is performed, hammer energy measurement should have been made that energy correction factors could be used to correct SPT blow counts. Otherwise, the raw blow counts would bias the soil properties and cause large uncertainty.

In future work, more tests will be conducted on various categories of soil layers, since clay soil layers were performed SPT tests so far. The dependence of ETR values on soil types has to be explored.

Acknowledgement. Financial support of Scientific Research Fund of Institute of Engineering Mechanics, China Earthquake Administration (Grant No. 2020B01) is sincerely acknowledged.

References

1. Youd, T.L., Bartholomew, H.W., Steidl, J.H.: SPT hammer energy ratio versus drop height. J. Geotech. Geoenviron. Eng. **134**(3), 397–400 (2008)
2. Seed, H.B., Tokimatsu, K., Harder, L.F., et al.: Influence of SPT procedures in soil liquefaction resistance evaluations. J. Geotech. Eng. **111**(12), 1425–1445 (1985)
3. Liao, S.C., Whitman, R.V.: Overburden correction factors for SPT in sand. J. Geotech. Eng. **112**(3), 373–377 (1986)
4. Youd, T.L., Idriss, I.M., Ronald, D., et al.: Liquefaction resistance of soils: summary report from the 1996 NCEER and 1998 NCEER/NSF workshops on evaluation of liquefaction resistance of soils. J. Geotech. Geoenviron. Eng. **127**(10), 817–833 (2001)

5. Valiquette, M., Robinson, B., Borden, R.H.: Energy efficiency and rod length effect in standard penetration test hammers. Transp. Res. Rec. **2**(186), 47–56 (2010)
6. Lee, C., Lee, J.S., An, S., et al.: Effect of secondary impacts on SPT rod energy and sampler penetration. J. Geotech. Geoenviron. Eng. **136**(3), 522–526 (2010)
7. Code, C.: Seismic Design of Buildings (GB 50011-2010). China Architecture and Building Press, Beijng (2010)
8. Code, C.: Code for Investigation of Geotechnical Engineering (GB50021-2010). China Architecture and Building Press, Beijng (2002)
9. Cao, Z.Z., Youd, T.L., Yuan, X.M.: Chinese dynamic penetration test for liquefaction evaluation in gravelly soils. J. Geotech. Geoenviron. Eng. **139**(8), 1320–1333 (2013)
10. Chen, L.W., Wang, Y.L., Chen, Y.X.: Stability of DPT hammer efficiency and relationships of blow-counts obtained by different DTP apparatuses. Chin. J. Geotech. Eng. **42**(6), 1041–1049 (2020) (in Chinese with English Abstract)
11. Ge, Y.X., Zhang, J., Zhu, L.W., et al.: Chinese and ASTM standard penetration tests at sand site: penetration energy analysis and correlation of blow counts. J. Eng. Geol. (2021) (in Chinese with English Abstract)
12. National Academies of Sciences: Engineering, and Medicine: State of the Art and Practice in the Assessment of Earthquake-Induced Soil Liquefaction and Its Consequences. The National Academies Press, Washington, DC (2016)

Empirical Magnitude-Upper Bound Distance Curves of Earthquake Triggered Liquefaction Occurrence in Europe

Mauro De Marco[1], Francesca Bozzoni[1](✉), and Carlo G. Lai[1,2]

[1] European Centre for Training and Research in Earthquake Engineering, Eucentre, Pavia, Italy
francesca.bozzoni@eucentre.it

[2] Department of Civil and Architectural Engineering, University of Pavia, Pavia, Italy

Abstract. This article presents empirical magnitude – maximum distance threshold curves developed starting from the European interactive Catalogue of earthquake-induced soil Liquefaction phenomena (ECLiq). Based on the latter, European regressions were computed to predict the maximum distance of liquefaction occurrence starting from the main seismological information of an earthquake. The interactive catalogue ECLiq is a unique digital archive, which includes documented historical information regarding earthquake-induced manifestations of soil liquefaction occurred in Europe in the latest 1000 years or so. ECLiq is publicly available as web-based Geographical Information System (GIS) platform at the link http://ecliq.eucentre.it/. The catalogue includes both the main seismological data of the earthquakes that triggered liquefaction in Europe and the features of liquefaction manifestations. Based on ECLiq, new empirical European relationships between earthquake magnitude and maximum distance for liquefaction were developed and presented hereinafter. ECLiq was used to identify magnitude – maximum threshold distance pairs above which liquefaction is unlikely to occur. It is important to emphasize that since these models were developed based on historical data of liquefaction occurrence, their use to predict the occurrence/non occurrence of liquefaction at a given site and for a given magnitude and location of an earthquake in Europe, must be linked to the actual ground conditions of the site of interest and in particular of its susceptibility/non susceptibility to liquefaction.

Estimating the location where soil liquefaction can possibly occur in the immediate aftermath of a strong earthquake is valuable for rapid loss estimation, post reconnaissance surveys and site investigations. The proposed equations can also be adopted to estimate the magnitude of an earthquake in paleoseismic studies.

Keywords: Liquefaction · Database · Empirical models · Magnitude-distance threshold curves · Europe

1 Introduction

Early studies on functional relationships linking earthquake magnitude and maximum distance at which liquefaction occurs were published by Kuribayashi and Tatsuoka

© The Author(s), under exclusive license to Springer Nature Switzerland AG 2022
L. Wang et al. (Eds.): PBD-IV 2022, GGEE 52, pp. 1857–1864, 2022.
https://doi.org/10.1007/978-3-031-11898-2_169

(1975) [1], Youd (1977) [2], and Youd and Perkins (1978) [3]. However, Kuribayashi and Tatsuoka (1975) were the first authors to develop magnitude – maximum distance threshold relationships for liquefaction using data from 44 historic Japanese earthquakes occurred between 1872 and 1968. Youd (1977) and Youd and Perkins (1978) proposed correlations in terms of maximum distance from the fault rather than from the epicenter using seismological data from several earthquakes in the United States. Keefer (1984) [4] collected data from 40 historical earthquakes and presented threshold curves of magnitude versus maximum epicentral distance. Since 1988, Ambraseys [5] started to develop two new types of correlations, which linked the moment magnitude with both the maximum epicentral distance and the fault distance, based on a dataset that included a description of liquefaction manifestations occurred during 137 earthquakes worldwide.

In the nineties and early 2000s, efforts were spent in Europe to build databases of liquefaction manifestations at national level and then compute threshold curves linking earthquake magnitude and maximum liquefaction distance. In Italy, Galli and Meloni (1993) [6] used data on liquefaction manifestations occurred during several historical earthquakes and computed an epicentral intensity versus maximum epicentral distance threshold for liquefaction occurrence. Few years later Galli (2000) [7] published for Italy a catalogue of historical manifestations of liquefaction occurred in the period 1117–1990 and, based on the latter, proposed empirical relationships between magnitude and maximum distance for liquefaction. In Greece, Papadopoulos and Lefkopoulos (1993) [8] extended the Ambraseys's database with 30 new cases of liquefaction occurred in Greece and updated the maximum distance thresholds as a function of the magnitude. In Turkey, Aydan et al. (2000) [9] re-evaluated seismological data of Turkish earthquakes and developed magnitude-distance relationships for liquefaction occurrence.

Initiatives to update and improve the pioneering studies previously mentioned were carried by several authors, e.g. Papathanassiou et al. (2005) [10] with reference to the Aegean territory, Pirrotta et al. (2007) [11] focusing on the Central-Eastern Sicily in Southern Italy, and Martino et al. (2014) [12] for the overall Italian territory.

From a literature review of available correlations linking earthquake magnitude and maximum distance of liquefaction occurrence, it turns out that specific threshold curves for the whole European territory is currently missing. This paper is a first attempt to bridge this gap by presenting novel empirical magnitude – maximum distance thresholds based on using a dataset from a catalogue of earthquake-induced soil liquefaction phenomena specifically built for Europe. Those activities were carried out within the framework of LIQUEFACT, a 3.5-year research project funded under the Horizon 2020 Research and Innovation Programme. LIQUEFACT is the largest research project on the assessment and mitigation of risks associated with earthquake-induced soil liquefaction ever funded by the European Commission. The article includes a brief description of the *European interactive Catalogue of earthquake-induced soil Liquefaction phenomena*, ECLiq, fully illustrated in Bozzoni et al. (2021) [13]. Based on ECLiq, novel European correlations of moment magnitude versus both maximum epicentral and hypocentral distance are then proposed. Finally, comparison between the proposed empirical threshold curves and correlations available in the literature for specific countries (i.e. Italy and Turkey) are illustrated.

2 ECLiq, the European Interactive Catalogue of Earthquake-Induced Soil Liquefaction Phenomena: A Brief Description

A composite, homogeneous, and well-documented catalogue of earthquake-induced manifestations of soil liquefaction occurred in Europe was built by Bozzoni et al. (2021) as digital archive, named "*European interactive Catalogue of earthquake-induced soil Liquefaction phenomena*" ECLiq. It contains documented historical information regarding liquefaction-related phenomena (e.g. sand ejecta and boils, soil settlements and lateral spreading, ground and structural failures) triggered by seismic activity in continental Europe.

Data and metadata were gathered within the time window 1117–2019 AD and include: the main seismological characteristics of the earthquake (e.g. UTC date, epicentre coordinates, magnitude, etc.), location of the site where liquefaction phenomena were documented, a description of the features of ground failure. Data and information were retrieved, collected, critically reviewed and harmonized to compile the catalogue of earthquake-induced liquefaction occurrences in Europe.

ECLiq is fully accessible as web-based GIS platform and it is publicly available at the link http://ecliq.eucentre.it/. ECLiq platform is structured as an interactive chart the user can interrogate (Fig. 1a). He/she can also get access to the complete datasets and metadata archive. The map can be panned and zoomed. The locations of liquefaction occurrences are shown by using three different types of symbols (from light to dark blue) based on the level of accuracy associated to the definition of the coordinates of each liquefied site. Indeed, the identification of the location of liquefaction sites carries various levels of uncertainty and three typical conditions were distinguished namely "*A. Georeferenced coordinates*", "*B. Coordinates obtained from maps*" and "*C. Generic description of the site of liquefaction manifestation*". These categories are shown in the legend of the interactive map in Fig. 1b. The earthquake epicenters that triggered soil liquefaction are represented as red circles. The size and color of the circles are proportional to the earthquake magnitude according to the scale reported in the map.

3 Novel Magnitude-Upper Bound Distance Curves for Europe

Starting from the dataset of ECLiq which has been briefly described in Sect. 2, empirical regressions to predict liquefaction occurrence from the main seismological data of an earthquake are proposed hereinafter. Indeed, empirical correlations of moment magnitude versus both maximum epicentral distance and maximum hypocentral distance (Sect. 3.1) were computed. A comparison of the proposed threshold curves and the correlations available in the literature is illustrated in Sect. 3.2.

3.1 Correlations of Moment Magnitude Versus Maximum Epicentral and Hypocentral Distance

The functional relationship of the correlation between magnitude and maximum distance of liquefaction occurrence adopted hereinafter is the one originally proposed by Youd

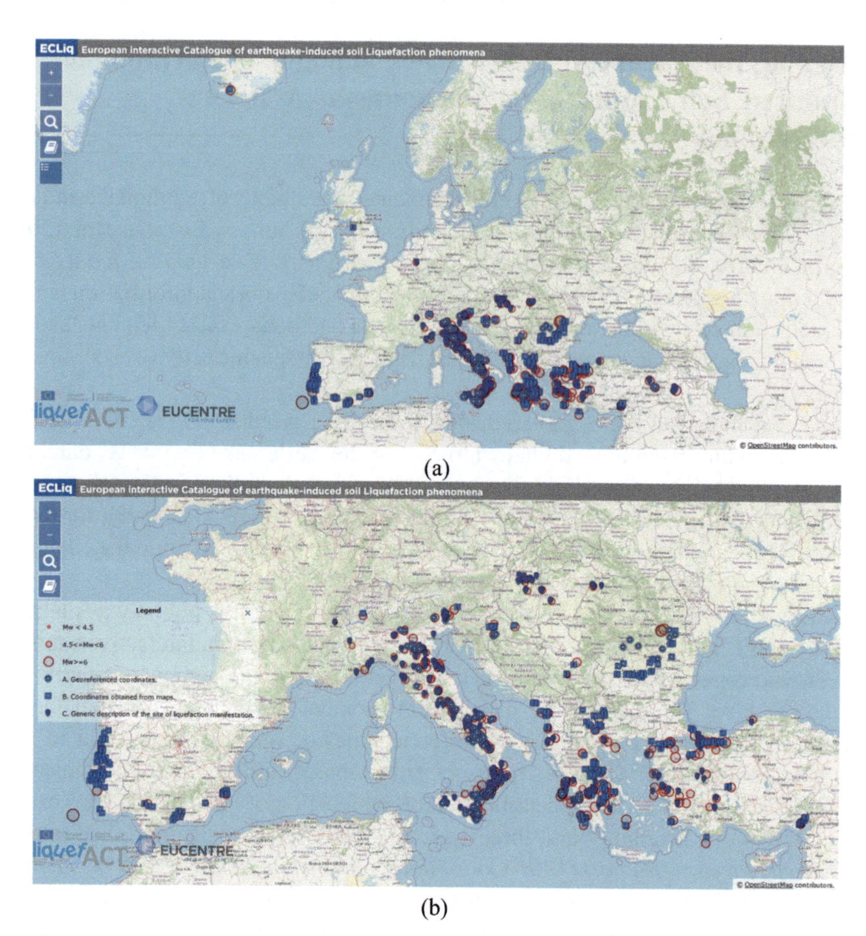

Fig. 1. European interactive catalogue of earthquake-induced liquefaction occurrences, ECLiq [13] throughout the time period 1117–2019 AD, publicly available at the link: http://ecliq.eucentre.it/. Locations of liquefaction occurrences are shown by using three different types of symbols (from light to dark blue) based on the level of accuracy associated to the location of each liquefaction site. Earthquake epicentres that triggered liquefaction are plotted as red circles whose size and color are proportional to earthquake magnitude.

and Perkins (1978). The input data in terms of magnitude-distance pairs are extracted from the ECLiq dataset and split into bins of increasing magnitude. Within each bin, the magnitude-distance pair referred to the maximum distance is selected. The size of the magnitude bin can influence data selection and, thus, the regression parameters, thus a sensitivity analysis was been carried out by considering magnitude bin of 0.25, 0.5, 1, and 2, as illustrated in Lai et al. (2018) [14]. Once the magnitude – maximum distance pairs were identified, the threshold curve is calculated using a standard nonlinear least squares algorithm. By this approach, an empirical correlation of moment magnitude M_W versus maximum epicentral distance R_{epi} was developed using 929 magnitude – maximum distance pairs available for the European territory as follows:

$$M_W = 1.377 + 2.394 * log\left(R_{epi}\right) \tag{1}$$

Figure 2a shows the proposed curve of moment magnitude versus maximum epicentral distance computed for Europe. Starting from the 467 magnitude – maximum distance pairs for which the earthquake depth was available, empirical correlation of moment magnitude versus maximum hypocentral distance was computed for Europe and shown in Fig. 2b:

$$M_W = 1.492 + 2.335 * log\left(R_{hypo}\right) \tag{2}$$

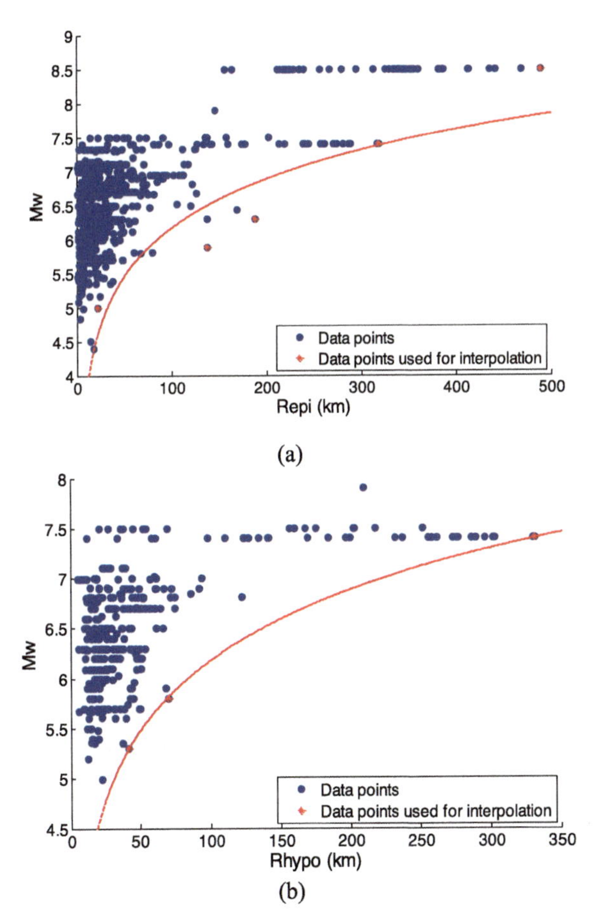

Fig. 2. Empirical correlations computed for Europe of moment magnitude versus maximum epicentral distance (a) and maximum hypocentral distance (b).

3.2 Comparison with Available Correlations

Comparisons between novel threshold curves presented in Sect. 3.1 and analogous correlations available in the literature, i.e. for Italy (Galli 2000) and for Turkey (Aydan et al. 2000), are shown hereinafter. First, specific threshold curves for the Italian and Turkey territories have been developed and defined in Eq. 3 and Eq. 4, respectively:

$$M_W = 0.921 + 2.596 * log(R_{epi})\text{for Italy} \tag{3}$$

$$M_W = 0.4554 + 3.232 * log(R_{hypo})\text{for Turkey} \tag{4}$$

It is worth remarking that the correlations originally proposed by Galli (2000) were not expressed in terms of M_W, therefore the modified version of Galli's curve provided in [14] (i.e. moment magnitude versus maximum epicentral distance), was adopted in this study for comparison. Indeed, Fig. 3 shows the comparison of the Italian threshold curve of Eq. 3 and the available correlation by Galli (2000). Galli (2000)'s curve seems to be less conservative than the one calculated herein and this is probably due to the post-1990 cases included in the updated version of the catalogue of liquefaction occurrences used for Italy. Furthermore, the uncertainty associated to the conversion of earthquake magnitude might have had an impact [14].

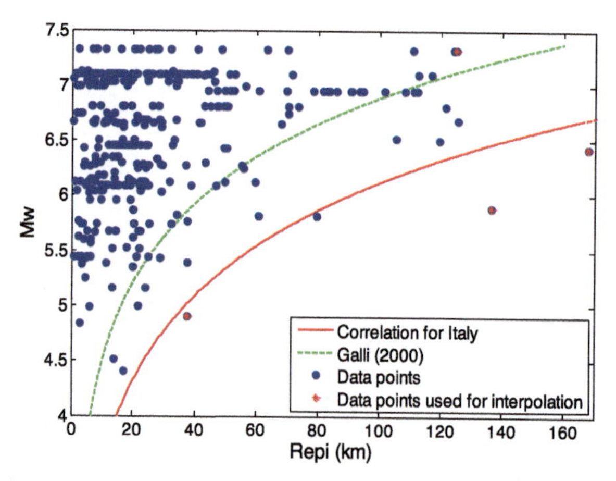

Fig. 3. Empirical correlation of moment magnitude versus maximum epicentral distance computed for Italy and comparison with a correlation from literature, i.e. Galli (2000).

With regard to Turkey, Aydan et al. (2000) developed a relationship in terms of surface-wave magnitude and maximum hypocentral distance. A modified version of Aydan et al. (2000) from [14], is adopted in Fig. 4 for comparison. It turns out that that the curve developed in this study seems to be more conservative especially for magnitude greater than 7.

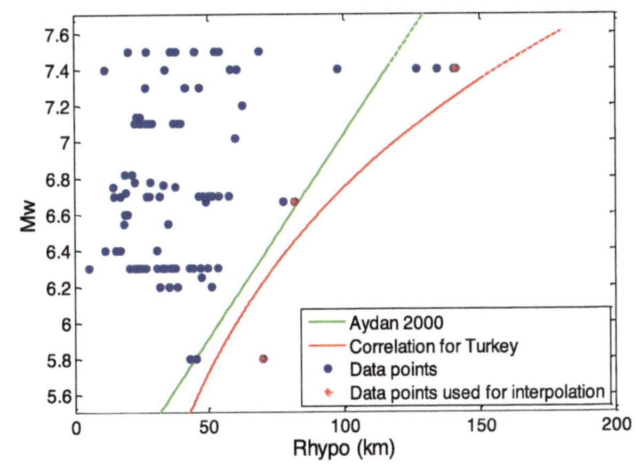

Fig. 4. Empirical correlation of moment magnitude versus maximum epicentral distance computed for Turkey and comparison with a correlation from the literature, i.e. Aydan et al. (2000).

4 Concluding Remarks

This article presents, for the first time, empirical models for the European territory to predict the maximum distance from the epicenter (or from the hypocenter) of an earthquake where liquefaction was historically observed. These correlations were constructed by using information and metadata contained in a composite yet homogenized earthquake catalogue named ECLiq dating back over 900 years, i.e. from 1117 AD to 2019. It must be remarked that since these models were developed based on historical data of liquefaction occurrence, their use to predict the occurrence/non occurrence of liquefaction at a given site and for a given magnitude and location of an earthquake in Europe, must be linked to the actual ground conditions of the site of interest and in particular of its susceptibility/non susceptibility to liquefaction.

A comparison between the empirical curves developed in this study and other types of correlations available in the literature was possible only for specific countries namely Italy and Turkey. It is worth highlighting that the proposed models appear to be more conservative in both cases. In the current version of these empirical models, the uncertainty associated to the location of the epicentre of the earthquakes (which is large in ancient events) and the corresponding liquefaction manifestations is not addressed. Efforts are underway to account for this uncertainty. Yet, despite this limitation, the empirical models presented in this article could be used for a quick zoning of the territory where soil liquefaction could possibly take place in the immediate aftermath of an earthquake. Therefore, they represent a useful tool in the framework of rapid loss estimation and planning of post-event reconnaissance and site investigations. Moreover, the proposed empirical models can be adopted to estimate the magnitude of an earthquake in paleoseismic studies.

Acknowledgements. This research has been carried out within the framework of the European LIQUEFACT Project. The LIQUEFACT Project has received funding from the European Union's

Horizon 2020 Research and Innovation Programme under Grant Agreement No. 700748. This support is gratefully acknowledged by the authors.

References

1. Kuribayashi, E., Tatsuoka, F.: Brief review of liquefaction during earthquakes in Japan. Soils Found. **15**, 81–92 (1975)
2. Youd, T.L.: Discussion of "Brief review of liquefaction during earthquakes in Japan" by E. Kuribayashi and F. Tatsuoka. Soils Found. **17**(1), 82–85 (1977)
3. Youd, T.L., Perkins, D.M.: Mapping liquefaction-induced ground failure potential. J. Geotech. Eng. Div. **104** (1978)
4. Keefer, D.K.: Landslides caused by earthquakes. Geol. Soc. Am. Bull. **95**, 406–421 (1984)
5. Ambraseys, N.N.: Engineering seismology. Earthq. Eng. Struct. Dynam. **17**, 1–105 (1988)
6. Galli, P., Meloni, F.: Liquefazione Storica. Un catalogo nazionale. Quat. Ital. J. Quat. Sci. **6**, 271–292 (1993)
7. Galli, P.: New empirical relationships between magnitude and distance for liquefaction. Tectonophysics **324**, 169–187 (2000)
8. Papadopoulos, G.A., Lefkopoulos, G.: Magnitude-distance relations for liquefaction in soil from earthquakes. Bull. Seismol. Soc. Am. **82**(3), 925–938 (1993)
9. Aydan, Ö., Ulusay, R., Kumsar, H.: Liquefaction phenomenon in the earthquakes of Turkey, including recent Erzincan, Dinar and Adana-Ceyhan earthquakes. In: 12WCEE, 12th World Conference on Earthquake Engineering. Paper 0207 (2000)
10. Papathanassiou, G., Pavlides, S., Christaras, B., Pitilakis, K.: Liquefaction case histories and empirical relations of earthquake magnitude versus distance from the broader Aegean region. J. Geodyn. **40**, 257–278 (2005)
11. Pirrotta, C., Barbano, M.S., Guarnieri, P., Gerardi, F.: A new dataset and empirical relationships between magnitude/intensity and epicentral distance for liquefaction in central-eastern Sicily. Ann. Geophys. **50**(6) (2007)
12. Martino, S., Prestininzi, A., Romeo, R.W.: Earthquake-induced ground failures in Italy from a reviewed database. Nat. Hazards Earth Syst. Sci. **14**, 799–814 (2014)
13. Bozzoni, F., Cantoni, A., De Marco, M.C., Lai, C.G.: ECLiq: European interactive catalogue of earthquake-induced soil liquefaction phenomena. Bull. Earthq. Eng. **19**(12), 4719–4744 (2021). https://doi.org/10.1007/s10518-021-01162-5
14. Lai, F., Bozzoni, M., De Marco, C., Zuccolo, E., Bandera, S., Mazzocchi G.: GIS database of the historical liquefaction occurrences in Europe and European empirical correlations to predict the liquefaction occurrence starting from the main seismological information v. 1.0. LIQUEFACT Project, Deliverable 2.4 (2018). http://www.liquefact.eu/

Laboratory Component of Next-Generation Liquefaction Project Database

Kenneth S. Hudson[1](✉), Paolo Zimmaro[1,2], Kristin Ulmer[3], Brian Carlton[4], Armin Stuedlein[5], Amalesh Jana[5], Ali Dadashiserej[5], Scott J. Brandenberg[1], John Stamatakos[3], Steven L. Kramer[6], and Jonathan P. Stewart[1]

[1] University of California Los Angeles, Los Angeles, CA 90095, USA
kenneth.s.hudson@gmail.com
[2] University of Calabria, Via Pietro Bucci, 87036 Arcavacata di Rende, Italy
[3] Southwest Research Institute, San Antonio, TX 78238, USA
[4] Norwegian Geotechnical Institute, Oslo, Norway
[5] Oregon State University, Corvallis, OR 97331, USA
[6] University of Washington, Seattle, WA 98195, USA

Abstract. Soil liquefaction and resulting ground failure due to earthquakes presents a significant hazard to distributed infrastructure systems and structures around the world. Currently there is no consensus in liquefaction susceptibility or triggering models. The disagreements between models is a result of incomplete datasets and parameter spaces for model development. The Next Generation Liquefaction (NGL) Project was created to provide a database for advancing liquefaction research and to develop models for the prediction of liquefaction and its effects, derived in part from that database in a transparent and peer-reviewed manner, that provide end users with a consensus approach to assess liquefaction potential within a probabilistic framework. An online relational database was created for organizing and storing case histories which is available at http://nextgenerationliquefaction.org/ (https://www.doi.org/10.21222/C2J040, [1]). The NGL field case history database was recently expanded to include the results of laboratory testing programs because such results can inform aspects of liquefaction models that are poorly constrained by case histories alone. Data are organized by a schema describing tables, fields, and relationships among the tables. The types of information available in the database are test-specific and include processed-data quantities such as stress and strain rather than raw data such as load and displacement. The database is replicated in DesignSafe-CI [2] where users can write queries in Python scripts within Jupyter notebooks to interact with the data.

Keywords: Liquefaction · Database · Laboratory

© The Author(s), under exclusive license to Springer Nature Switzerland AG 2022
L. Wang et al. (Eds.): PBD-IV 2022, GGEE 52, pp. 1865–1874, 2022.
https://doi.org/10.1007/978-3-031-11898-2_170

1 Introduction

Quantifying liquefaction susceptibility, triggering, and effects requires datasets that span a wide parameter space, and a modeling framework that is founded in first principles known to control soil response to undrained shear. The combination of a physically meaningful modeling framework and a significantly large data set is required to develop robust semi-empirical models regressed from the data. To date, the emphasis of the Next Generation Liquefaction (NGL) project has been on field case histories of liquefaction and its effects, as well as no-ground failure cases [1, 3]. A major goal of NGL is to support this model building process by providing objective data to modeling teams, along with results of additional supporting studies to constrain effects that cannot be established solely from data.

While the NGL database will support model development over a certain parameter space, it is not currently adequate to constrain models over the parameter space required for application. As one example, liquefaction models need to be applicable over a wide range of vertical effective stresses (also known as K_σ affects), ranging from effectively zero up to perhaps 6 atm. The available case histories involve relatively shallow soils, and hence do not include high-overburden pressure cases. Extending models across broad parameters spaces requires additional information, which can often be provided by laboratory studies of soil behavior. As a result, the NGL database schema of Brandenberg et al. [3] was expanded to allow for this information; this manuscript describes this work.

2 Laboratory Database Schema

2.1 Database Structure

A thorough description of the field case history portion of the NGL database structure can be found in Brandenberg et al. [1]. The laboratory component is built into the NGL relational database framework and is a structured database that can be queried using structured query language (SQL). A relational database comprises tables linked to one another by means of identifiers called keys. Each table has a primary key that uniquely identifies table entries. If two tables are linked, the primary key of a table is used as a foreign key in another table. Primary-foreign key relationships produce the organized hierarchical structure of a database. Such organizational structure is called schema. The NGL laboratory component was developed in consultation with the NGL database working group (S.J. Brandenberg, K.O. Cetin, R.E.S. Moss, K.W. Franke, K. Ulmer, and P. Zimmaro). The schema presented here is mostly complete, but population of the database is ongoing and should continue indefinitely as more testing is performed and researchers share data. The schema may have fields and/or tables added in the future if there is sufficient interest in additional types of datasets.

Twenty-four tables were added to the NGL database for the laboratory component with a laboratory table (LAB) at the top of the hierarchy. The field case history component is joined to the sample table via the sample-test table allowing samples to be associated with a test (under the field case history schema) or not (under the laboratory component schema). The hierarchy of the laboratory component schema is shown in Fig. 1. Table 1 contains descriptions for each database table. There are 140 fields contained within the tables defined in Table 1 and shown in Fig. 1.

The table names in Table 1 below also correspond to the primary keys of those tables (for example, table SPEC has a primary key SPEC_ID). SPEC_ID is used as a foreign key in the following tables: TXG, DSSG, PLAS, RDEN, OTHR, INDX, GRAG, and CONG. The TXG table primary key, TXG_ID is used as a foreign key in the TXS table, which has a primary key TXS_ID that is used as a foreign key in the TXD table. Similarly, the DSSG table primary key, DSSG_ID, is used as the foreign key in DSSS which has a primary key DSSS_ID that is used as a foreign key in DSSD1D and DSSD2D. The FILE table with a primary key FILE_ID is used as a foreign key in the OTHR table, which therefore has two foreign keys, SPEC_ID and FILE_ID. GRAG_ID is used as a foreign key in the GRAT table. CONG_ID is used as a foreign key in the CON_STGE table which has its primary key (CON_STGE_ID) used in the COND table.

To illustrate database functionality, consider the following example data entry for a triaxial cyclic shear test:

- First, the laboratory where the testing was performed needs to be created as an entry in the LAB table where information such as the lab name, location in latitude and longitude coordinates, and any description of the laboratory are entered. The specific testing program that the triaxial test was performed within also needs to have an entry created in the LAB_PROGRAM table and associated with the LAB entry using the LAB_ID foreign key in the LAB_PROGRAM table. Any personnel who worked on the testing program and are to be associated with the testing program can be linked to it through the LAB_PROGRAM_USER junction table.
- The sample used in the testing is assigned an identifier (SAMP_ID) and its name (SAMP_NAME), sample type (SAMP_TYPE), depth to the top and base of the sample within a boring if associated with a boring (SAMP_TOP, SAMP_BASE) these would be left blank if the material was a synthetic mixture created in the lab), the diameter of the sample (SAMP_SDIA), the date the sample was obtained (SAMP_DATE), the recovery rate for the sample (SAMP_REC), description of the sample (SAMP_DESC), and any remarks (SAMP_REM) are entered. This sample entry can be associated with the testing program via the LAB_PROGRAM_SAMP table and associated with a field test if it was not synthesized in the lab via the SAMP_TEST table.
- A specimen obtained from the sample (associated via the SAMP_ID foreign key) is assigned an ID (SPEC_ID), name (SPEC_NAME), and other metadata such as (1) the depth to the top and bottom of the specimen (SPEC_TOP, SPEC_BASE) if the specimen is from a boring (these would be left blank if the material was a synthetic mixture created in the lab), (2) name of the person or organization who did the testing (SPEC_CREW), and (3) other remarks about the specimen (SPEC_REM).
- Results of index testing, relative density measurements, grain size distribution analysis, or other testing are provided in tables INDX, RDEN, GRAG/GRAT, PLAS, and OTHR, respectively. That data is connected via the SPEC_ID foreign key to the SPEC table. If consolidation tests were performed separate from triaxial or direct shear tests, then metadata from each stage of the consolidation tests such as final effective vertical stress and final height of the specimen (CONG_STGE_SIGV and CONG_HI, respectively) is entered into the CONG and CON_STGE tables and the consolidation

data – time and displacement – are entered into the COND table. The COND table is linked via CON_STGE_ID as the foreign key which is linked to the CONG table using the CONG_ID foreign key.

- Triaxial data is entered by first entering the general metadata for the triaxial test (TXG table) such as initial void ratio, water content, specimen diameter, initial height, and any descriptive information (TXG_E0, TXG_W0, TXG_DIAM, TXG_H0, and TXG_DESC, respectively). The triaxial test stage table (TXS) contains a foreign key to the TXG table and also contains fields for the stage number, type of stage (i.e. consolidation, monotonic loading, or cyclic loading), drainage (i.e. drained, undrained, or neither), and a description of the stage (TXS_ST, TXS_TY, TXS_DR, and TXS_DESC, respectively). The triaxial test data (TXD) table has a foreign key connecting it to the TXS table (TXS_ID) and has fields for time, deviator stress, cell pressure, pore pressure, axial strain, radial strain, and volumetric strain vectors (TXD_TIME, TXD_SD, TXD_CP, TXD_PP, TXD_EA, TXD_ER, TXD_EV, respectively).

Direct simple shear tests are entered similarly to triaxial tests, however there is an option for entering data for 1- or 2-directional loading.

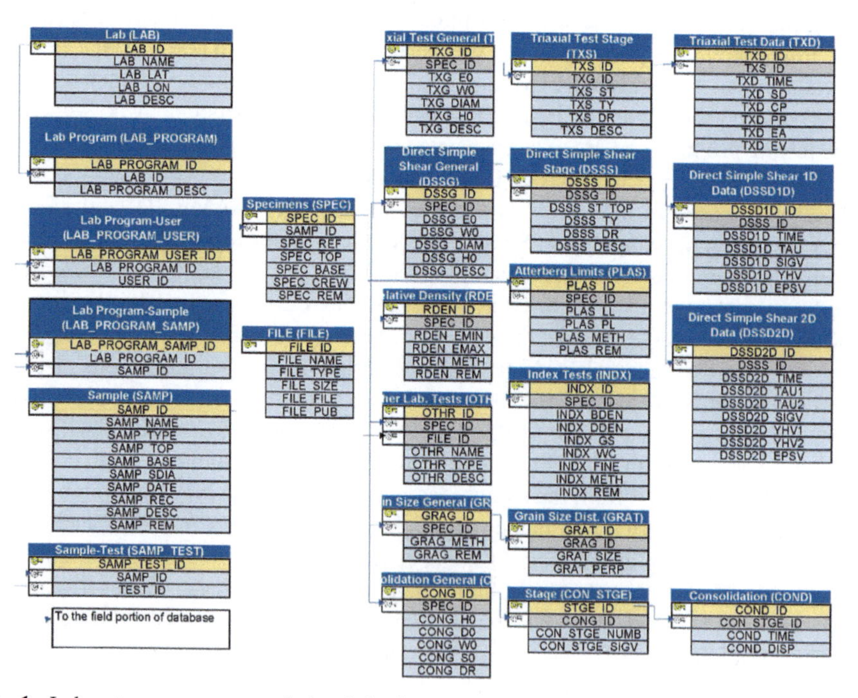

Fig. 1. Laboratory component relational database schema showing relationships between tables using keys.

Table 1. List of tables in the laboratory component of the NGL database.

Table name	Table description	Number of fields
LAB	Laboratory information	5
LAB_PROGRAM	Testing Program information	3
LAB_PROGRAM_USER	Junction table between testing program and users	3
LAB_PROGRAM_SAMP	Junction table between testing program and sample	3
SAMP	General information for laboratory or field samples	10
SAMP_TEST	Junction table between sample and specimen	3
SPEC	General information for laboratory specimens taken from samples	7
INDX	Index tests include:	9
	dry and bulk density (ASTM D7263-09),	
	water (moisture) content (ASTM D2216-10), and	
	fines content (ASTM D21140-17)	
	Standards recommended for each test are in parentheses	
RDEN	Relative density measurement	6
PLAS	Plasticity test (i.e., Liquid limit and plasticity limit) information (ASTM D4318-10e1)	6
GRAG	General information for particle size distribution analysis	4
GRAT	Test results (percent passing for a specific sieve) from particle size distribution analysis	4
OTHR	Other tests not specified above. Any format of test results can be uploaded	6
FILE	Table storing supplemental files	5
DSSG	Direct simple shear test general information	7
DSSS	Information about each direct simple shear test stage	6
DSSD1D	One-dimensional direct simple shear test data	7
DSSD2D	Two-dimensional direct simple shear test data	9
TXG	Triaxial test general information	7
TXS	General information for triaxial test stages	6

(*continued*)

Table 1. (*continued*)

Table name	Table description	Number of fields
TXD	Triaxial test data	8
CONG	Consolidation test general information	7
CON_STGE	Consolidation test stage information	5
COND	Consolidation test data	4

2.2 Data Querying and Visualization

Currently the laboratory component of the NGL database cannot be accessed in the same manner as the field case history component via the interactive website (https://nextgenerationliquefaction.org; https://www.doi.org/10.21222/C2J040) because there has not been adequate time or budget to add that capability. However, the database is replicated daily to DesignSafe [2], where it can be queried by any user using Python scripts in Jupyter notebooks. A Jupyter notebook is a server-client application that allows editing and running notebook documents in a web browser and combines rich text elements and computer code executed by a Python kernel [4]. Jupyter notebooks are published and available on DesignSafe in the NGL project partner data apps (Brandenberg et al. [2] and references therein).

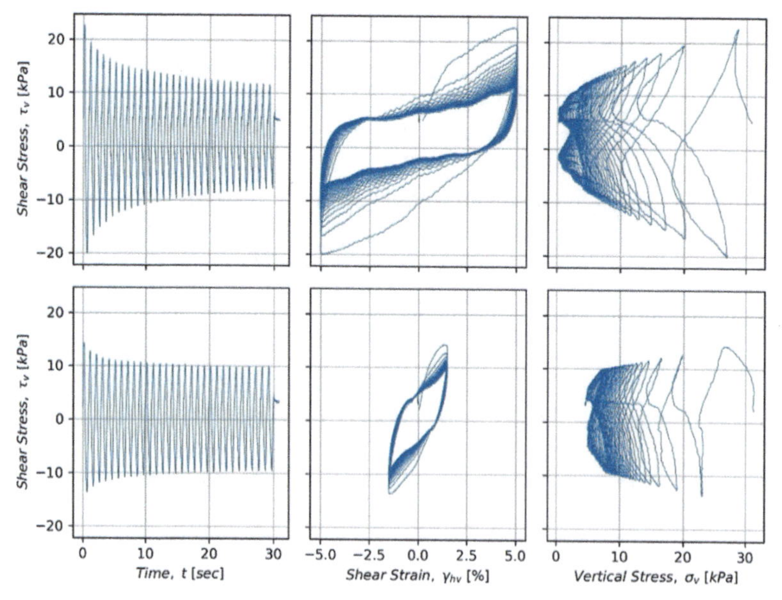

Fig. 2. Example strain-controlled cyclic direct simple shear test results developed at OSU and plotted for viewing with the Jupyter notebook tool (a) specimen that undergoes liquefaction and (b) specimen that did not liquefy.

Users can create their own custom Jupyter notebooks to query and visualize data from the NGL database for use in model development. The published notebooks are a good starting place to base new custom notebooks on and to learn how to write SQL queries in Python.

Figure 2 illustrates the use of a Jupyter notebook for visualizing direct simple shear test data. The user can select a laboratory from the first dropdown menu and the tool populates the Program dropdown menu with all the testing programs at that particular lab. Based on the selection from the Program dropdown, the Sample dropdown is populated with all samples within that testing program. The user can then select a specimen from the Specimen dropdown menu and the tool populates the DSSG_ID dropdown menu with the direct simple shear general table IDs that are performed on that specimen (from the DSSG table). Based on the selection from that dropdown, the tool populates the DSSS_ID dropdown with the stages for that particular test. The tool plots the data from the selected stage in nine separate plots to help visualize it. Figure 2 shows some subplots from the tool for a cyclic simple shear test performed at Oregon State University (OSU).

3 Status of Database

As of this writing, the NGL database has 347 sites, 147 of which have been fully reviewed (meaning independently reviewed by two database working group members) and 135 have been partially review (reviewed once). The database includes 843 cone penetration tests (CPTs), 696 borehole tests with 7559 standard penetration test (SPT) measurements, 125 invasive shear wave velocity measurements (such as downhole), and 30 surface wave method tests (such as spectral analysis of surface waves (SASW)).

Table 2 shows the labs testing programs currently in the database. The counts for laboratory tests are shown in Table 3. There are 6 laboratories and 8 laboratory testing programs. Tests other than direct simple shear, triaxial, and consolidation may be entered into the field database without specifying a laboratory, which is why the number of laboratories and test programs is low relative to the number of tests.

4 Application of NGL Laboratory Database

There are many potential problem-solving capacities within NGL such as the issues with adjustment factors (drainage effects (K_d), partial saturation, path correction (K_P), 2-dimensional loading (K_{2D}), initial effective stress (K_σ), initial static shear stress (K_α)), and the difficulty with liquefaction susceptibility criteria. The NGL database is also uniquely suited to addressing the issues with fine-grained soil susceptibility.

4.1 Contributions Addressing Transitional Silt Soils

Ongoing research on the seismic response and liquefaction susceptibility of transitional silts conducted at OSU has led to the development of a significant laboratory dataset currently being added to the NGL database. Targeted sites have been investigated using mud-rotary drilling with thin-walled tube sampling, downhole shear wave velocity tests,

Table 2. Lab testing programs currently in the NGL database.

ID	Lab program description
1	Testing of samples from sites associated with the Canterbury Earthquake Sequence [5]
2	Cyclic and monotonic direct simple shear on Orange Co. Silica Sand [6]
3	Testing on samples from Mihama Ward associated with 2011 Tohoku earthquake [7]
4	Lab testing associated with [8] and [9] "https://doi.org/10.1016/j.soildyn.2018.09.012" & "http://dx.doi.org/10.1016/j.dib.2018.08.043"
5	Lab testing associated with Graded area east of New River at SW edge of Brawley
6	Lab testing associated with 1979–1981 with CPT retesting in 2003 [10]
7	Cyclic testing on clay-silt blends [6]
8	Testing of remolded samples from Mihama Ward

Table 3. Counts of laboratory tests within the NGL database.

Tests	Total
Index (specific gravity, water content, and/or percent passing number 200 sieve)	3847
Relative density	77
Plasticity (Atterberg limits)	1385
Gradation	4495
Direct simple shear	53
Triaxial	63
Consolidation	4

and cone penetration tests. Laboratory investigation consists of soil characterization (e.g., Atterberg limits, gradation), quantification of stress history, and evaluation of monotonic and cyclic strength of natural, intact specimens. At present, six distinct study sites have been developed. The dataset consists of tens of constant-rate-of-strain consolidation and constant-volume, mono-tonic direct simple shear (DSS) tests on soils from each site, with the goal of establishing SHANSEP parameters suitable for the low and moderate plasticity silts. The dataset also includes over 150 stress-controlled and tens of strain-controlled, constant-volume cyclic DSS tests most of which are accompanied with post-cyclic reconsolidation of monotonic shearing phases. Representative oedometric compression specimens were used to judge sample quality using compressions ratios (e.g.,[11]) and provided the basis for selecting recompression consolidation techniques for DSS test specimens in view of the generally high quality of the samples. With fines contents and plasticity indices ranging from 25 to 100% and 0 to 38, this dataset will serve to refine thresholds in the transition between sand-like and intermediate, and intermediate and clay-like behavior. Portions of the OSU dataset have been and are continuing to be uploaded as sponsored projects progress towards closure.

4.2 Stress Effects (K_σ and K_α) Work

In cyclic stress-based liquefaction triggering evaluations (e.g. [12]), the overburden stress correction factor (K_σ) is used to modify the cyclic resistance ratio of the soil (*CRR*) for confining stresses (σ'_c) other than one atmosphere (atm), and the initial shear stress correction factor (K_α) is used to modify the *CRR* for when the initial static shear stress (τ_s) is not equal to zero. One approach to account for these effects de-pends on laboratory test data from tests such as cyclic triaxial, cyclic direct simple shear, or cyclic torsional shear. Tests are performed to develop relationships between cyclic stress ratio (*CSR*) and the number of cycles to liquefaction (*N*) for a given soil and similar relative density (D_r), σ'_c, and τ_s. Typically, this is done first using a base-line condition, such as $\sigma'_c =$ 1 atm and $\tau_s = 0$. *CRR* can then be computed from the *CSR*-*N* curve assuming a value of *N* associated with a given magnitude event. To compute K_σ or K_α, the same soil is tested using the same set of conditions but changing either σ'_c or τ_s. The ratio of the *CRR* of the second test to the *CRR* of the baseline test is the K_σ or K_α correction factor. Results from laboratory tests have shown that an increase in σ'_c leads to a reduced *CRR*, and the presence of a non-zero τ_s can either increase or decrease the *CRR* depending on the state of the soil.

As part of the NGL project, an ongoing study is investigating the effects of σ'_c and τ_s on liquefaction triggering [13]. The study requires *CSR* vs *N* relationships from many laboratory tests to develop K_σ and K_α models that apply to a variety of soils under a wide range of stress conditions. The laboratory component of the NGL database provides a centralized, open-source location to store the data from these published laboratory tests to facilitate the development of these K_σ and K_α models. There is a significant advantage to storing stress and strain relationships throughout the duration of the cyclic tests rather than providing only the summary statistics that the original authors reported (e.g., *CRR*, K_σ, K_α). For example, the computation of *CRR* requires the selection of a liquefaction triggering criterion. Some studies choose strain-based criteria, while others choose pore pressure-based criteria. The computation of *CRR* also requires the selection of an *N* that corresponds to the magnitude or duration of interest (e.g., **M**7.5). The value of *N* has typically been between 10 and 20 in published studies. Directly providing the stress and strain relationships through-out the duration of the cyclic tests, as the laboratory component of the NGL database does, circumvents these issues and allows model developers to consider a single, consistent interpretation and/or alternative frameworks and triggering criteria.

Acknowledgements. The Next-Generation Liquefaction project was originated in the Pacific Earthquake Engineering Research (PEER) center, and remains an active PEER project. This paper describes work performed by the Center for Nuclear Waste Regulatory Analyses (CNWRA) and its contractors for the U.S. Nuclear Regulatory Commission (USNRC) under Contract No. 31310018D0002 and through an inter-agency agreement with the U.S. Bureau of Reclamation (USBR) under Agreement Number R20PG00126. The activities reported here were performed on behalf of the USNRC Office of Nuclear Regulatory Research. This paper is an independent product of the CNWRA and does not necessarily reflect the view or regulatory position of the USNRC or the USBR.

References

1. Brandenberg, S., Kwak, D., Zimmaro, P., Bozorgnia, Y., Kramer, S., Stewart, J.: Next-generation liquefaction (NGL) case history database structure. In: 5th Decennial Geotechnical Earthquake Engineering and Soil Dynamics Conference, Austin, TX (USA), 10–13 June, vol. 290, pp. 426–433. Geotechnical Special Publication (2018)
2. Pinelli, J.-P., et al.: Disaster risk management through the designsafe cyberinfrastructure. Int. J. Disaster Risk Sci. **11**(6), 719–734 (2020). https://doi.org/10.1007/s13753-020-00320-8
3. Brandenberg, S., et al.: Next-generation liquefaction database. Earthq. Spectra **36**(2), 939–959 (2020)
4. Perez, F., Granger, B.: IPython: a system for interactive scientific computing. Inst. Electr. Electron. Eng. **9**(3), 21–29 (2007)
5. Beyzaei, C.: Fine-Grained Soil Liquefaction Effects in Christchurch, New Zealand, dissertation, UC Berkeley (2017)
6. Eslami, M.: Experimental Mapping of Elastoplastic Surfaces for Sand and Cyclic Failure of Low-Plasticity Fine-Grained Soils, dissertation, UCLA (2017)
7. Kwak, D.Y., Zimmaro, P., Nakai, S., Sekiguchi, T., Stewart, J.P.: Case study of liquefaction susceptibility from field performance of hydraulic fills. In: 11th U.S. National Conference on Earthquake Engineering Proceedings, 25–29 June 2018 (2018)
8. Shengcong, F., Tatsuoka, F.: Rep. of Japan-China cooperative research on engineering lessons from recent Chinese earthquakes, including the 1976 Tangshan earthquake (part 1). In: Tamura, C., Katayama, T., Tatsuoka, F. (eds.) Institute of Industrial Science, University of Tokyo, Tokyo (1983)
9. Cetin, K.O., et al.: SPT-based probabilistic and deterministic assessment of seismic soil liquefaction triggering hazard. Soil Dyn. Earthq. Eng. **115**, 698–709 (2018). https://doi.org/10.1016/j.soildyn.2018.09.012
10. Moss, R.E., Seed, R.B., Kayen, R.E., Stewart, J.P., Tokimatsu, K.: Probabilistic Liquefaction Triggering Based on the Cone Penetration Test, Geotechnical Special Publication (2005). https://doi.org/10.1061/40779(158)23
11. DeJong, J.T., Krage, C.P., Albin, B.M., DeGroot, D.J.: Work-based framework for sample quality evaluation of low plasticity soils. J. Geotech. Geoenviron. Eng. **144**(10), 04018074 (2018)
12. Seed, H.B., Idriss, I.M.: Simplified procedure for evaluating soil liquefaction potential. J. Geotech. Engrg. Div. ASCE **97**(9), 1249–1273 (1971)
13. Ulmer, K., Carlton, B.: Review of available data on the effects of confining pressure and initial static shear stress for use in liquefaction triggering analyses. In: Proceedings of the 12th National Conference in Earthquake Engineering, Earthquake Engineering Research Institute, Salt Lake City, UT (2022, under review)

Liquefaction Characteristics of 2011 New Zealand Earthquake by Cone Penetration Test

Zhao-yan Li(✉), Si-yu Zhang, and Xiao-Ming Yuan

Institute of Engineering Mechanics, China Earthquake Administration, Harbin 150080, China
hkjlizhaoyan@163.com

Abstract. The liquefaction-induced damage is one of the most serious damage in the earthquake. At present, the discrimination of sand liquefaction is one of the most effective and direct methods to prevent liquefaction-induced damage. Investigation of liquefied sites and data collection are the basis of the formation of the liquefaction discrimination method. The 2011 Christchurch Mw 6.3 earthquake caused large-scale liquefaction. The earthquake is the first earthquake in which sand liquefaction is the main cause of seismic damage and provides a mass of liquefaction data. This paper collects liquefaction data of 132 survey points after the earthquake. Through the analysis of the collected static cone penetration test data, groundwater level, peak acceleration and other data, the liquefaction characteristics of the earthquake are obtained. At the same time, it was discovered that multiple deep liquefaction phenomena occurred in the earthquake. By testing the existing CPT liquefaction evaluation method in Chinese code using the collected data, it is found that the method has obvious errors in the evaluation of deep sand liquefaction.

Keywords: Liquefaction · New Zealand earthquake · Liquefaction characteristics · Cone penetration test · Evaluation method

1 Introduction

The 2011 Christchurch Mw 6.3 earthquake is the first earthquake in which sand liquefaction is the main cause of seismic damage.[1]. The liquefaction zone of the earthquake occupied about one-third of the urban area, which was the most serious within 50 km^2 east of the city center [2, 3]. The area covered by sand ejecta reached 70%–80%. The amount of sand and silt reached 400,000 t, which brought huge difficulties to the post-earthquake reconstruction work. Liquefaction caused various degree of damage to local buildings, including the settlement and differential settlement to buildings. It also caused lateral spreading near the rivers leading to the damage of roads, bridges and lifelines facilities [4, 5].

The establishment of liquefaction discrimination methods is mainly based on the seismic field reconnaissance data. The liquefaction discrimination method in Chinese code is based on field reconnaissance data of the four earthquakes in China. The seismic field reconnaissance data play an important role in the formation and revision of the

© The Author(s), under exclusive license to Springer Nature Switzerland AG 2022
L. Wang et al. (Eds.): PBD-IV 2022, GGEE 52, pp. 1875–1883, 2022.
https://doi.org/10.1007/978-3-031-11898-2_171

liquefaction discrimination method [6, 7]. The 2011 New Zealand earthquake provided a lot of liquefaction data for testing and developing liquefaction discrimination methods. In this paper, the liquefaction data of the earthquake are used to test the liquefaction discrimination formula in Chinese code. Moreover, most of the basic data forming the liquefaction discrimination formula in Chinese code are in shallow layers, and there is a lack of high-intensity data. The data of the 2011 New Zealand earthquake are mostly in high-intensity areas, and there are liquefaction data in deep layers, which makes up for the vacancy in Chinese code (Fig. 1).

Fig. 1. Liquefaction-induced damage of the 2011 New Zealand earthquake [11].

2 Liquefaction Data

After the 2011 New Zealand earthquake, many scientific institutions and universities conducted investigations on the area [8]. This paper collects the field reconnaissance data of 132 investigation sites from the local scientific institutions and the distribution of the sites is shown in Fig. 2 [9]. The data include location, groundwater level, cone penetration test (CPT) (see Fig. 3), etc. Most of the CPT data are in picture format and this paper obtained data from all reports by image information extraction technology.

This paper obtained the distribution of the liquefaction area from the website of Canterbury Maps [10], as shown in Fig. 4 (the red area is the liquefaction area). There are 115 sites located in the liquefied area and 17 sites in the non-liquefied area.

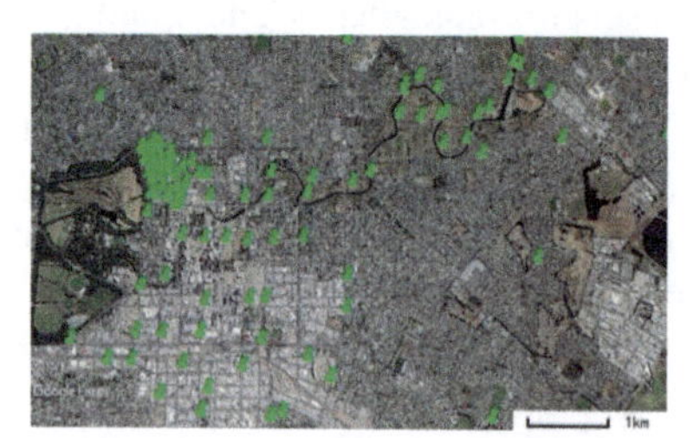

Fig. 2. Distribution of investigation sites.

This paper collected the peak accelerations recorded by 27 seismic stations within 50km of the epicenter [11, 12]. The distribution is shown in Fig. 5. Based on these, the peak accelerations of the 132 sites were calculated by the ArcGIS (see Fig. 6). The peak accelerations of all sites were converted into the intensities commonly used in Chinese code. There are 12 sites in the 8-degree region, 117 sites in the 9-degree region, and 3 sites in the 10-degree region.

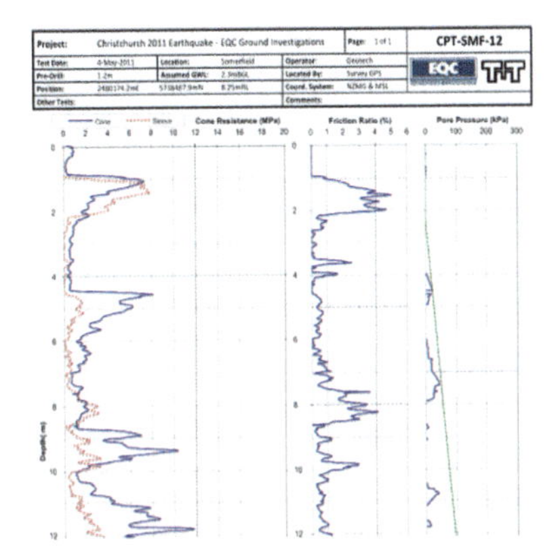

Fig. 3. Static cone penetration test report [9].

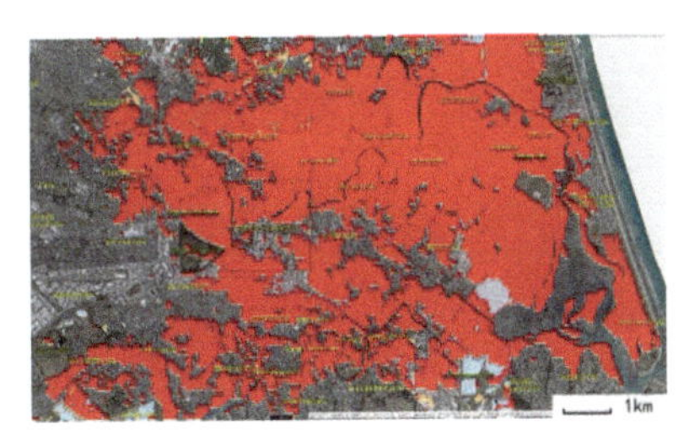

Fig. 4. Distribution of liquefaction area in and around Christchurch [10].

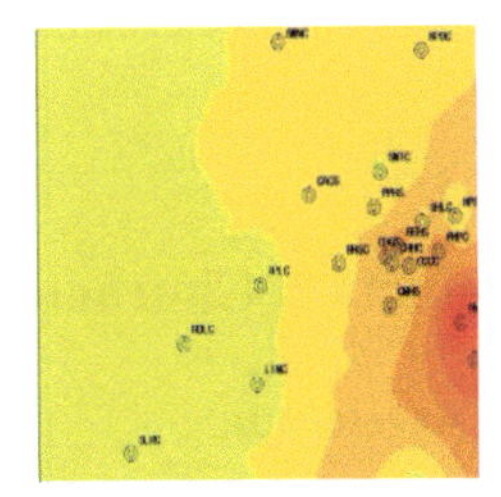

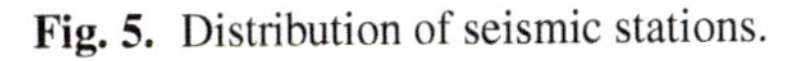

Fig. 5. Distribution of seismic stations.

Fig. 6. PGA distribution map after calculation by ArcGIS.

The geological interpretative report provided by the Christchurch City Council analyzed the distribution of the liquefied layers in 40 investigation sites. Part of it is shown in Fig. 7. It can be seen from the geological sectional view of the report that the liquefied layers are distributed in strips and there are multiple liquefied layers in the liquefaction sites.

According to the experience, the layers of small and stable cone tip resistance values were chosen as the liquefied layers and the layers of large and stable cone tip resistance values were chosen as the non-liquefied layers in other sites, as shown in Fig. 8 [13].

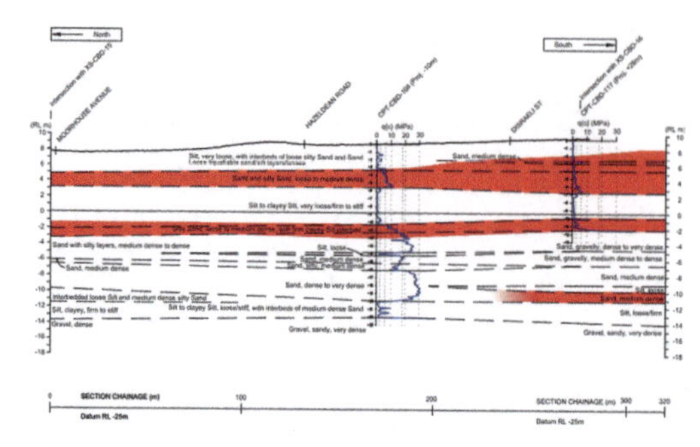

Fig. 7. Geological sectional view of Christchurch [9].

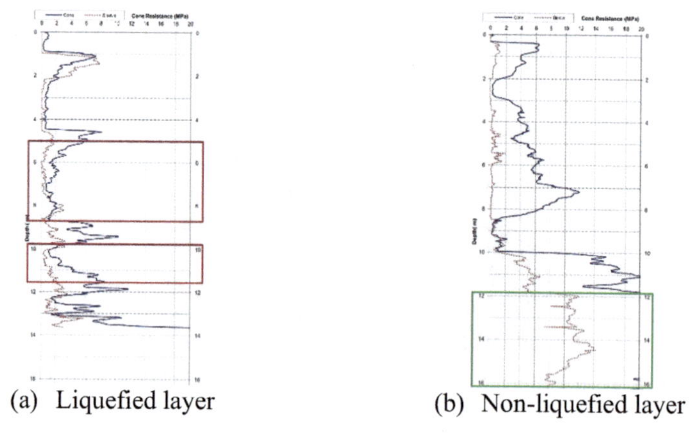

(a) Liquefied layer (b) Non-liquefied layer

Fig. 8. Example of evaluation between liquefied and non-liquefied layers in a typical site.

3 Liquefaction Characteristics

3.1 Peak Acceleration

The peak acceleration distributions of liquefied sites and non-liquefied sites are shown in Fig. 9. The peak accelerations of the liquefied sites are in the range of 0.2g–0.8g, most of which are in the range of 0.5g–0.7g, accounting for 63% of the total; There are few data in non-liquefied sites. The peak accelerations are in the range of 0.4g–0.8g, most of which are in the range of 0.5g–0.6g, accounting for 59% of the total.

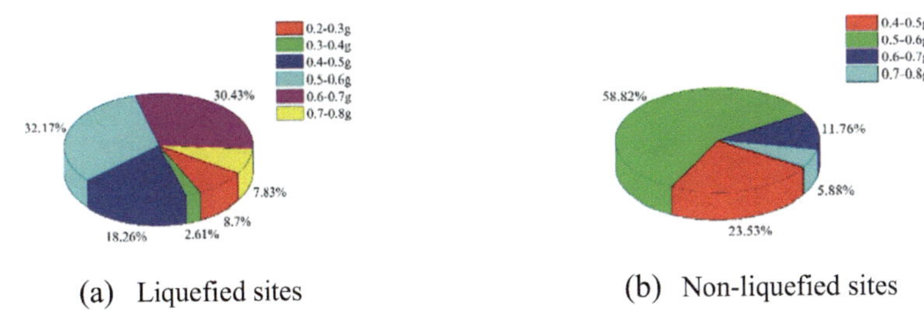

(a) Liquefied sites (b) Non-liquefied sites

Fig. 9. Distribution of PGA in liquefied and non-liquefied sites.

The intensity distribution of liquefied and non-liquefied sites is shown in Table 1 and Fig. 10. It can be seen that the liquefied sites are mainly located in the 9-degree region. There are 100 sites, accounting for 87% of the total. There are only 12 sites in the 8-degree region and 3 sites in the 10-degree region. The non-liquefied sites are all located in the 9-degree region.

Table 1. Seismic intensity distribution of investigation sites.

Intensity	8	9	10
Liquefied	12	100	3
Non-liquefied	0	17	0
Total	12	117	3

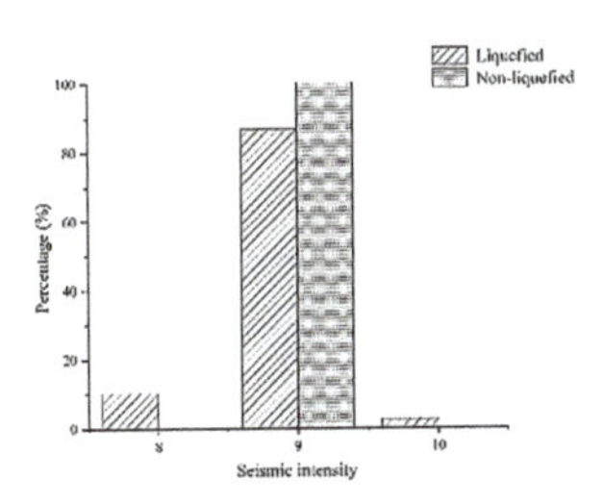

Fig. 10. Seismic intensity distribution of liquefied sites and non-liquefied sites.

3.2 Groundwater Level

The groundwater level distributions of liquefied and non-liquefied sites are shown in Fig. 11. The groundwater levels of the liquefied sites are in the range of 0.2 m–4.0 m, most of which are about 1.9 m, accounting for 35% of the total. The groundwater levels of the non-liquefied sites are in the range of 0.9 m–3.5 m, most of which are about 2.0 m, accounting for 23% of the total.

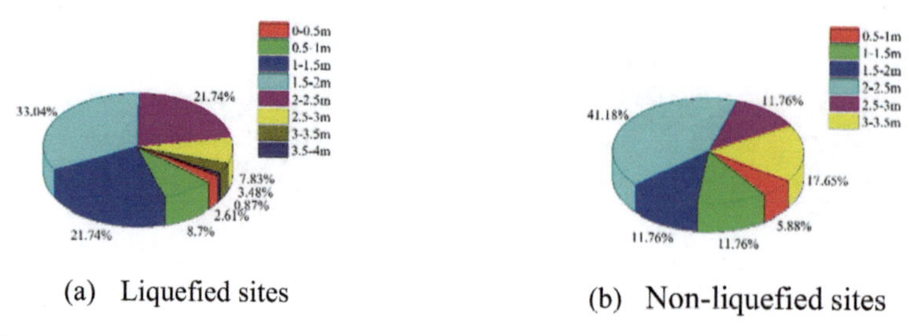

(a) Liquefied sites (b) Non-liquefied sites

Fig. 11. Distribution of groundwater level in liquefied sites and non-liquefied sites.

3.3 Sand Layer Burial Depth

The burial depth distributions of liquefied and non-liquefied layers are shown in Fig. 12. The burial depth of the liquefied layers is in the range of 1.6 m–22 m, most of which is within 5 m, accounting for 57% of the total. In this earthquake, there were liquefied layers buried below 20 m in 6 liquefied sites. The burial depth of the non-liquefied layers is in the range of 3.9 m–17.3 m, most of which is about 15m, accounting for 20% of the total, and the distribution is relatively even. Table 2 shows the statistical results of the burial depth of the liquefied and non-liquefied layers with the boundaries of 10 m and 15 m.

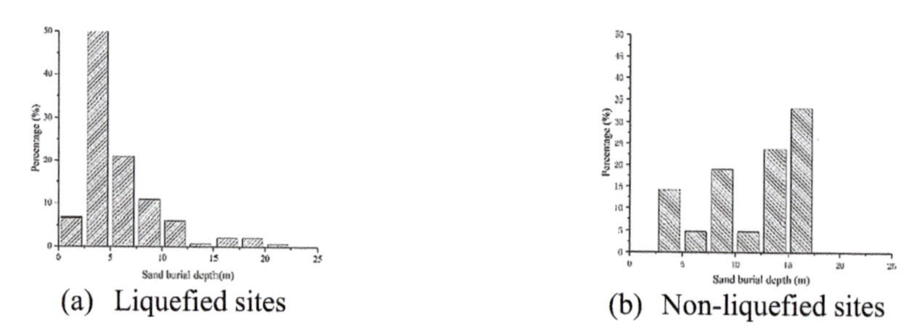

(a) Liquefied sites (b) Non-liquefied sites

Fig. 12. Distribution of sand burial depth in liquefied sites and non-liquefied sites.

Table 2. The statistical results of sand burial depth.

	<10 m	10–15 m	>15 m
Liquefied	131	10	7
Non-liquefied	8	6	7
Total	139	16	14

There are 159 cases of seismic field reconnaissance data in mainland China, which are the basic data that form the liquefaction evaluation method in Chinese code. It can be seen that the data of New Zealand have higher intensities than those of the mainland, and the depth of the liquefied layers is deeper.

4 Test

The CPT liquefaction discrimination formula in "Code for Investigation of Geotechnical Engineering" (GB50021-2001) can be expressed as:

$$q_{ccr} = q_{c0} a_w a_u a_p \tag{1}$$

$$a_w = 1 - 0.065(d_w - 2) \tag{2}$$

$$\alpha_u = 1 - 0.05(d_u - 2) \tag{3}$$

where q_{ccr} is the critical value of cone tip resistance of saturated soil (MPa), q_{c0} is the standard value of cone tip resistance, α_w is the modification coefficient of groundwater level buried depth, α_u is the modification coefficient of non-liquefied soil layers, α_p is the modification coefficient of soil property, d_w is the depth of the groundwater level, d_u is buried depth for saturated soil.

This paper tests the CPT liquefaction discrimination method in Chinese code by using the obtained data from 132 investigation sites of the New Zealand earthquake, as shown in Fig. 13 and Fig. 14. For the 8-degree region, there are a little data in liquefied sites. The discrimination success rate is 100%. For the 9-degree region, the discrimination success rate is about 93.5% for liquefied sites, and 100% for non-liquefied sites. However, it can be seen that in the sand layers below 10m, the success rate for liquefied sites is only about 53%, which greatly reduced. The success rate for the 9-degree intensity region is shown in Table 3.

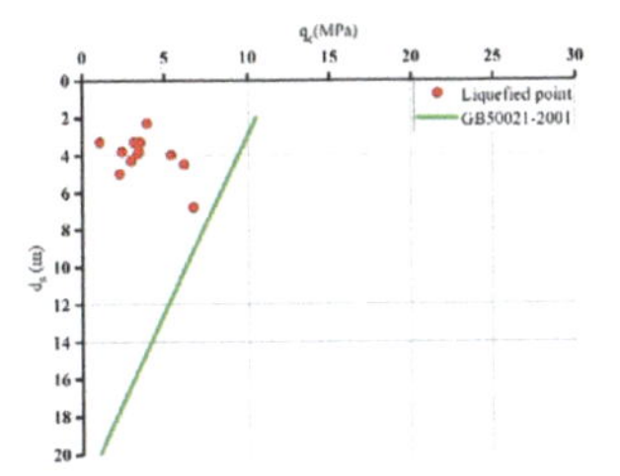

Fig. 13. Test of CPT liquefaction discrimination method in Chinese code (8-degree).

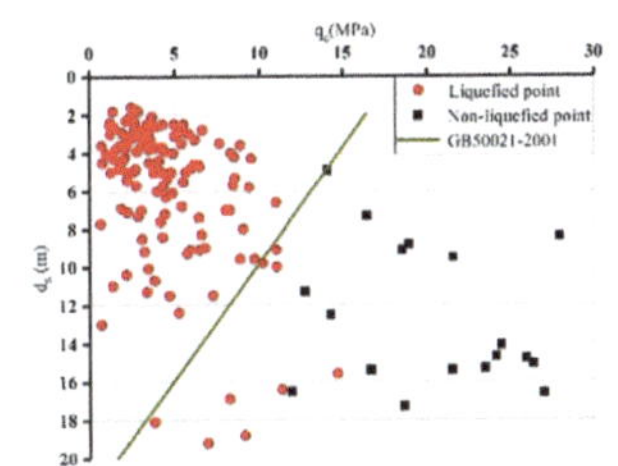

Fig. 14. Test of CPT liquefaction discrimination method in Chinese code (9-degree).

Table 3. Success rate of the discrimination method from Chinese code in the 9-degree intensity region.

	Liquefied	Non-liquefied
<10 m	98.0%	83.3%
10–20 m	53.0%	100.0%

5 Conclusions

(1) This paper collected the field reconnaissance data of 132 investigation sites in the 2011 New Zealand earthquake, including the geological sectional view and CPT data, and also obtained the distribution of liquefaction area and PGA of the earthquake.
(2) The liquefaction characteristics of the 2011 New Zealand earthquake were analyzed. The PGA of the liquefied sites was in the range of 0.2 g–0.8 g, most of which was in the range of 0.5 g–0.7 g; the PGA of the non-liquefied sites was in the range of 0.4 g–0.7 g, most of which was in the range of 0.5 g–0.6 g.
(3) The groundwater levels of the liquefied sites are in the range of 0.2 m–4.0 m, most of which are about 1.9 m. The groundwater levels of the non-liquefied sites are in the range of 0.9 m–3.5 m, most of which are about 2.0 m.
(4) The burial depth of the liquefied layers is in the range of 1.6 m–22 m, most of which is within 5 m. There are some liquefied layers buried below 20 m. The burial depth of the non-liquefied layers is in the range of 3.9 m–17.3 m, most of which is about 15 m, and the distribution is relatively even.
(5) The liquefaction discrimination method in Chinese code has a higher success rate within 10 m, but it greatly reduced below 10 m. In addition, there were many liquefied sand layers below 20 m in the 2011 New Zealand earthquake, which exceeded the discrimination scope of Chinese code, which will definitely promote the expansion of the understanding of the depth of sand liquefaction. In further study, we will propose a new method for evaluating liquefaction, which is suitable for sand layers of different buried depths.

Acknowledgments. This work was supported by Scientific Research of Institute of Engineering Mechanics, China Earthquake Administration (2019EEEVL0201), the National Natural Science Foundation of China (51508532) and Heilongjiang Provincial Natural Science Foundation of China (LH2019E099).

References

1. Chen, L.W., Yuan, X.M., Sun, R.: Review of liquefaction phenomena and geotechnical damage in the 2011 New Zealand Mw6.3 earthquake. World Earthq. Eng. **29**(03), 1–9 (2013)
2. Orense, R.P., Kiyota, T., Yamada, S.: Comparison of liquefaction features observed during the 2010 and 2011 Canterbury earthquakes. Seismol. Res. Lett. **82**(6), 905–918 (2011)
3. Cubrinovski, M., Bray, J.D., Taylor, M.: Soil liquefaction effects in the central business district during the February 2011 Christchurch earthquake. Seismol. Res. Lett. **82**(6), 893–904 (2011)
4. Omer, A., Resat, U., Masanori, H.: Geotechnical aspects of the 2010 Darfield and 2011 Christchurch earthquakes, New Zealand, and geotechnical damage to structures and lifelines. Bull. Eng. Geol. Env. **71**(4), 637–662 (2012)
5. Cubrinovski, M., Henderson, D., Bradley, B.A.: Liquefaction impacts in residential areas in the 2010–2011 Christchurch earthquakes. In: One Year after 2011 Great East Japan Earthquake: International Symposium on Engineering Lessons Learned from the Giant Earthquake, vol. 183, pp. 3–4. University of Canterbury Civil and Natural Resources Engineering, Tokyo (2012)

6. Yuan, X.M., Sun, R.: Proposals of liquefaction analytical methods in Chinese seismic design provisions. Rock Soil Mech. **32**(S2), 351–358 (2011)
7. Li, Z.Y., Wang, Y.L., Yuan, X.M.: Research on applicability of characteristic depth in liquefaction evaluation. World Earthq. Eng. **30**(02), 1–6 (2014)
8. Earthquake Engineering Research Institute, Pacific Earthquake Engineering Research Center. Learning from Earthquakes. The M 6.3 Christchurch, New Zealand, Earthquake of February 22, 2011. EERI Special Earthquake Report, May 2011
9. Tonkin & Taylor Ltd., Christchurch City Council. Christchurch central city geological interpretative report (2011)
10. CanterburyMaps. https://mapviewer.canterburymaps.govt.nz
11. Cubrinovski, M., Bradley, B., Wotherspoon, L.: Geotechnical aspects of the 22 February 2011 Christchurch earthquake. Bull. N. Z. Soc. Earthq. Eng. **44**(4), 205–226 (2011)
12. Sun, R., Yuan, X.M., Tang, F.H.: Blind detection of liquefaction sites by existing methods for the 22 Feb. 2011 Ms 6.3 New Zealand earthquake. J. Earthq. Eng. Eng. Vibr. **31**(03), 1–10 (2011)
13. Li, Z.Y., Wang, Y.Z., Yuan, X.M.: New CPT-based prediction method for soil liquefaction applicable to Bachu region of Xinjiang. Chin. J. Geotech. Eng. **35**(S1), 140–145 (2013)

Insights from Liquefaction Ejecta Case Histories from Christchurch, New Zealand

Z. Mijic[1(✉)], J. D. Bray[1], and S. van Ballegooy[2]

[1] University of California, Berkeley, CA 94720-1710, USA
zorana.mijic@berkeley.edu

[2] Tonkin and Taylor, Ltd., 105 Carlton Gore Road, Newmarket, Auckland 1023, New Zealand

Abstract. Liquefaction ejecta were a primary contributor to the damage of land and light-weight structures in Christchurch from the 2010–2011 Canterbury earthquakes. It occurred predominantly in east Christchurch, which is typically characterized by thick, clean sand deposits. By contrast, ejecta tended to be absent in the stratified silty soil swamp deposits of west Christchurch. To advance understanding of ejecta production, 235 well-documented case histories of ejecta occurrence, its quantity, and its effects on infrastructure were developed. The ejecta database includes 61 sites that underwent four major earthquakes that produced no-to-extreme ejecta. The case histories take advantage of numerous CPTs, boreholes, pre- and post-earthquake airborne LiDAR surveys, aerial photographs, liquefaction land damage documentation from insurance claims, and earthquake-specific PGAs and groundwater estimates. Ejecta coverage and amounts for each of the four main Canterbury earthquakes were extracted using photographic- and LiDAR-based approaches because direct measurements of ejecta quantities were not made. The database provides the basis for the development of procedures to evaluate the occurrence and quantity of ejecta.

Keywords: Liquefaction ejecta · Case histories · Canterbury earthquakes

1 Introduction

The 2010–2011 Canterbury earthquake sequence (CES) resulted in virtually unprecedented levels of liquefaction ejecta in a modern urban setting. Ejecta were a key mechanism of liquefaction-induced land damage and light-weight residential house damage in Christchurch, New Zealand [1]. Over 15,000 of 140,000 residential properties were damaged beyond economic repair [1]. A database with detailed case histories that could be used to develop a procedure or gain insights into the complex mechanism of ejecta, ground conditions, and seismic demand leading to the occurrence or non-occurrence of ejecta and the differing degrees of ejecta-induced settlement does not exist either. Therefore, this study takes advantage of the well-documented CES to develop detailed ejecta case histories for the four main Canterbury earthquakes: the 4 Sep 2010 (M_w 7.1), 22 Feb 2011 (M_w 6.2), 13 Jun 2011 (M_w 6.2), and 23 Dec 2011 (M_w 6.1) events.

© The Author(s), under exclusive license to Springer Nature Switzerland AG 2022
L. Wang et al. (Eds.): PBD-IV 2022, GGEE 52, pp. 1884–1892, 2022.
https://doi.org/10.1007/978-3-031-11898-2_172

Sites throughout Christchurch experienced various quantities of ejecta (Fig. 1). The degree of ejecta-induced damage varied from site to site and from earthquake to earthquake. Although direct measurements of ejecta were not made, ejecta coverage and amounts for each of the four major Canterbury earthquakes can be characterized with access to the comprehensive T+T (2015) [2] and LDAT (2021) [3] databases.

In this paper, the methodology for developing detailed ejecta case histories is described for two sites in Christchurch and key insights are shared. The Bideford Pl site manifested ejecta for the Sep 2010, Feb 2011, and Jun 2011 earthquakes, whereas the Tonks St site had no ejecta for any earthquake event.

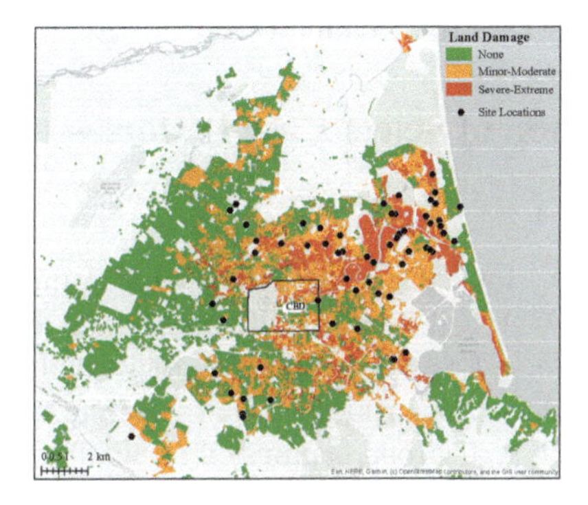

Fig. 1. The maximum liquefaction ejecta-induced damage map [2] with site locations.

2 Available Data and Methodology

Liquefaction-induced land damage for the CES was documented for more than 60,000 residential lots [4]. Over 25,000 CPTs and 5,000 boreholes, many with piezometers installed, were conducted. High-resolution aerial photographs were acquired after each of the four main earthquakes to identify areas with ejecta [5]. Bradley and Hughes (2012) [6] developed robust estimates of PGA with uncertainties for each earthquake using the recordings at 19 strong motion stations and an empirical ground motion model. The ground surface elevation was surveyed by airborne LiDAR prior to the CES and after each of the main earthquakes in the CES [7]. The surveys were typically acquired a month after each event (except for the Sep 2010 earthquake) when liquefaction ejecta were removed from most of the land to allow for the actual ground surface level recordings [7]. The LiDAR-derived surface elevation model and the water level measurements from wells were used to develop the event-specific groundwater depths [8].

In this study, 27 of 55 sites developed by the NZ-US researchers were investigated in detail [9]. These sites are not well captured by liquefaction triggering procedures. An additional 34 high-quality sites that are generally captured well by liquefaction triggering procedures and are with good observations (i.e., aerial and ground photographs and property inspection reports), reliable LiDAR-based settlement estimates, closely spaced CPTs with investigation depths of 15–20 m, and a nearby borehole were selected to form an unbiased database [9]. Figure 1 illustrates the locations of the 61 sites.

Each site was centered on a CPT or shear wave velocity measurement and encompassed an area within a 50-m radius of its center (termed 50-m buffer). However, due to the spatial variation in soil deposits and ejecta distribution, the 10-m and 20-m buffers were primarily used in the analysis. An area selected for detailed settlement assessment excluded vegetation, buildings, and significant anthropogenic changes as they could affect the LiDAR survey measurements. The ejecta-induced settlement was estimated using the LiDAR- and photographic evidence-based approaches. The best estimate of ejecta-induced settlement was provided as the weighted average of these two estimates based on the completeness of visual evidence, LiDAR measurement uncertainties, and liquefaction severity misestimates using the liquefaction triggering procedures. The details of the methodology are explained in Mijic et al. (2022) [9].

3 Detailed Evaluation of Select Case Histories

3.1 Bideford Pl

The Bideford Pl site is situated in the NE geologic quadrant of Christchurch (172.675071°, −43.512497°). It is a level site in residential area that did not undergo lateral spreading during the CES. Vegetation, buildings, anthropogenic changes, and spatial distribution of ejecta were considered when selecting the settlement assessment areas (Fig. 2). Unlike Patch A, the road manifested ejecta for the Sep 2010 earthquake. For the Feb and Jun 2011 earthquakes, ejecta occurred within Patch A and on the road, although the quantity of ejecta across the site was greater for the Feb 2011 earthquake. No ejecta were observed for the Dec 2011 earthquake.

The shape of ejected soil within Patch A was approximated by a prism with curvilinear bases due to the interconnectedness of soil deposits originating from different fissures and forming irregularly shaped ejecta "blankets." Different shades of gray of the ejecta were interpreted as different ejecta thicknesses. The total areas of the outlined thick ejecta layers and the outlined thin ejecta layers ($A_{E,thick}$ and $A_{E,thin}$, respectively) were measured in Google Earth™. The height range for the thick and thin ejecta layers ($H_{E,thick}$ and $H_{E,thin}$, respectively) was estimated based on the available ground photographs (Fig. 3) and detailed property inspection reports. The volume of ejecta was estimated as

$$V_{E,thick+thin} = \sum_{i=1}^{m} A_{E,thick,i} * H_{E,thick,i} + \sum_{j=1}^{n} A_{E,thin,j} * H_{E,thin,j} \quad (1)$$

and was divided by the total assessment area, A_T, to obtain the areal ejecta-induced free-field settlement, S_{E,P_areal} (Table 1). The localized ejecta-induced settlement, $S_{E,P\_localized}$, was evaluated, too, as the ratio of $V_{E,thick+thin}$ and the area of Patch A covered in ejecta, A_E (Table 1). Further, the volume of ejecta on the road was approximated as a prism with curvilinear bases (see Eq. 1) and a prism with right triangular bases

$$V_{E,prism} = \frac{1}{2}\sum_{k=1}^{p} W_{E,prism,k} * H_{E,prism,k} * L_{E,prism,k} \quad (2)$$

where $W_{E,prism,k}$ is the width of a rectangle and is perpendicular to the curb, while $L_{E,prism,k}$ is the length of a rectangle and aligns with the curb. The lower and upper estimates of the height of ejecta at the curb, $H_{E,prism,k}$, were based on the typical cross-slopes of normal crown of 2% and 4%. The volumes of the ejecta shapes on the road were summed and normalized by A_T of the road to obtain S_{E,P_areal} and by A_E within A_T to obtain $S_{E,P\_localized}$ (Table 1).

To estimate the LiDAR-based liquefaction-induced settlement, S_T, the average earthquake-induced change in ground surface elevation within the assessment area was evaluated for individual LiDAR points and adjusted for the LiDAR flight error, global offset, and vertical tectonic movements. The average volumetric settlement, S_{V1D}, was subtracted from S_T to obtain the ejecta-induced settlement, $S_{E,L}$ (Table 2). Figure 4 illustrates the q_t and I_c profiles. Table 2 summarizes the PGA and groundwater table depths.

The $S_{E,L}$ values were used in combination with the S_{E,P_areal} values to provide the best estimate of ejecta-induced free-field settlement, $S_{E,final}$ (Table 3). The best estimate of the ejecta-induced settlement at the Bideford Pl site for the Sep 2010, Feb 2011, Jun 2011, and Dec 2011 earthquakes is < 5 mm, 90 ± 30 mm, 25 ± 20 mm, and 0 mm, respectively, considering that the 20-m buffer is the most representative buffer.

The Bideford Pl site is a thick, clean sand site with a 4.5-m thick layer of fine to medium sand, SP, of marine/estuarine origin in the upper 10 m (from the 5.5- to 10-m depth) and below the average groundwater table depth of 2 m BGS ($q_{t,avg} = 14$ MPa). The top 5.5 m of the soil profile consist of the 0.4-m thick silty, ML, topsoil with organics ($q_{t,avg} = 4$ MPa), alluvial sandy silt, ML, to a depth of 3 m ($q_{t,avg} = 3$ MPa), and alluvial gravelly fine to coarse sand, SW, to a depth of 5.5 m ($q_{t,avg} = 14$ MPa). The average LPI (Iwasaki et al. (1982) [12]) = 0, 5, 0, and 1 and LSN (van Ballegooy et al. (2014) [13]) = 1, 14, 2, and 6 for the Sep 2010, Feb 2011, Jun 2011, and Dec 2011 earthquakes, respectively. Considering the observations of liquefaction ejecta (Fig. 2), the severity of

Fig. 2. Aerial photographs acquired for Bideford Pl in Sep 2010, Feb 2011, Jun 2011, and Dec 2011 [5] with ejecta outlines for the 10-, 20-, and 50-m buffers delineated by white circles. Patch A and the road are outlined in red.

liquefaction manifestation at the ground surface was higher than estimated by LPI or LSN for the Feb 2011 and Jun 2011 earthquakes, only slightly underestimated for the Sep 2010 earthquake, and well estimated for the Dec 2011 earthquake.

Fig. 3. Ground photographs of ejecta remnants (date: 16 Jun 2011) [3].

Table 1. Comparison of photographic-based areal and localized ejecta-induced settlement.

EQ event	Patch A (20- and 50-m b.)		Road (20-m b.)	
	S_{E,P_areal} (mm)	$S_{E,P\_localized}$ (mm)	S_{E,P_areal} (mm)	$S_{E,P\_localized}$ (mm)
Sep 2010	0	0	<5	15 ± 5
Feb 2011	70 ± 15	80 ± 20	35 ± 5	35 ± 5
Jun 2011	20 ± 5	50 ± 10	10 ± 5	25 ± 10
Dec 2011	0	0	0	0

Table 2. Ejecta-induced settlement for the top 20 m of the soil profile using the Boulanger and Idriss (2016) [10] and Zhang et al. (2002) [11] procedures ($P_L = 50\%$, $C_{FC} = 0.13$, and $I_{c,cutoff} = 2.6$).

EQ event	PGA (g)	GWT (m)	Patch A (20- and 50-m b.)			Road (20-m b.)		
			S_T (mm)	S_{V1D} (mm)	$S_{E,L}$ (mm)	S_T (mm)	S_{V1D} (mm)	$S_{E,L}$ (mm)
Sep 2010	0.19	2.5	–	4 ± 20	–	5 ± 50	2 ± 20	3 ± 54
Feb 2011	0.43	1.7	161 ± 25	52 ± 50	109 ± 56	101 ± 25	38 ± 50	63 ± 56
Jun 2011	0.26	2.5	38 ± 25	10 ± 25	28 ± 35	43 ± 25	5 ± 25	38 ± 35
Dec 2011	0.28	2.0	–	26 ± 50	–	–	9 ± 50	–

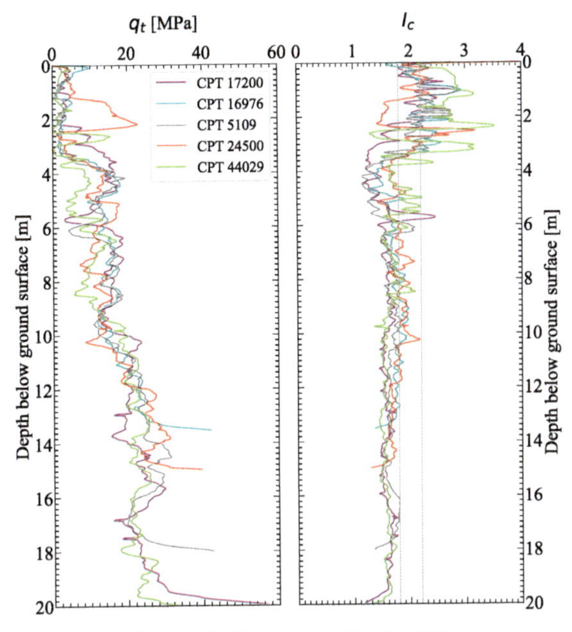

Fig. 4. CPT traces for Bideford Pl.

Table 3. Best final estimates of ejecta-induced settlement for Bideford Pl.

EQ event	$S_{E,final}$ (mm)	
	Patch A (20- and 50-m b.)	Road (20-m b.)
Sep 2010	0	<5
Feb 2011	90 ± 30	50 ± 30
Jun 2011	25 ± 20	25 ± 20
Dec 2011	0	0

3.2 Tonks St

The Tonks St site (172.724500°, −43.493746°) is in the residential area of the NE quadrant of Christchurch. Lateral spreading was not observed for any earthquake. The site is mostly level apart from the properties in the western portion of the 50-m buffer that typically have sloped driveways as they are approximately 0.5 m above the elevation of the road and their front lawns are retained by 0.5-m tall walls. No specific area was outlined for settlement assessment at this site because ejecta were absent for all earthquakes (Fig. 5). Thus, it was not necessary to consider the LiDAR surveys to conclude that the ejecta-induced settlement at the Tonks St site for the Sep 2010, Feb 2011, Jun 2011, and Dec 2011 earthquakes is 0 mm, 0 mm, 0 mm, and 0 mm, respectively.

The average groundwater depth at the Tonks St site is 2.8 m BGS. The soil profile at the site consists primarily of fine to medium sand, SP, of marine/estuarine origin to a depth of 20 m ($q_{t,avg}$ = 17 MPa) (sec Fig. 6). Thus, the soil profile belongs to the thick, clean sand category. The PGA for the Sep 2010, Feb 2011, Jun 2011, and Dec 2011 earthquakes was 0.19, 0.56, 0.22, and 0.41 g, respectively, while the respective groundwater table depths were 3.3, 3.3, 2.5, and 2.3 m BGS. For the Sep 2010, Feb 2011, Jun 2011, and Dec 2011 earthquakes, LPI = 0, 2, 0, and 1, respectively, and LSN = 0, 2, 0, and 2, respectively. Thus, the severity of liquefaction manifestation at the ground surface was captured well by liquefaction triggering procedures for all earthquake events.

Fig. 5. Aerial photographs for the Tonks St site acquired in Sep 2010, Feb 2011, June 2011, and Dec 2011 [5] with no ejecta observed within the 10-, 20-, and 50-m buffers delineated by white circles.

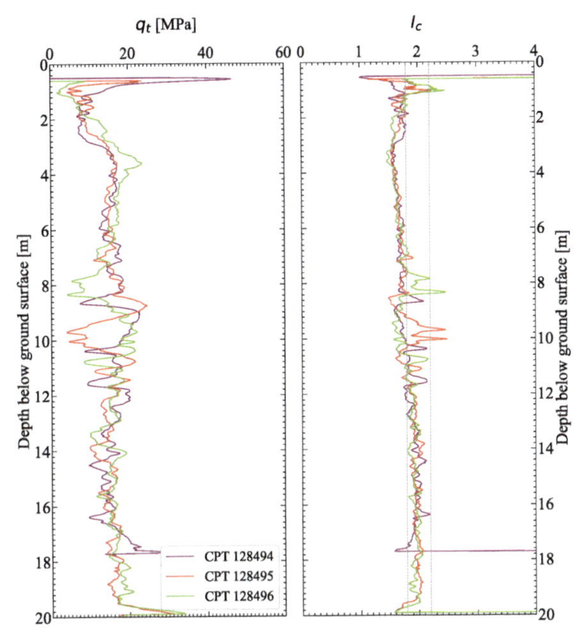

Fig. 6. CPT traces for Tonks St.

4 Conclusions

The liquefaction ejecta-induced free-field settlement at 61 sites in Christchurch was estimated for each of the four major Canterbury earthquakes because direct measurements of the ejecta-induced settlement were not available. The Bideford Pl and Tonks St sites were presented to describe the methodology for estimating the ejecta-induced free-field settlement using the photographic- and LiDAR-based approaches. Both sites had a saturated clean sand layer thicker than 3 m in the upper 10 m of the soil profile, but they experienced strikingly different levels of ejecta-induced land damage. The Feb 2011 earthquake caused the most intense shaking at both sites; the Tonks St site experienced more severe PGA yet had no ejecta, whereas the Bideford Pl site settled 90 ± 30 mm due to ejecta. Liquefaction ejecta were also manifested at the Bideford Pl site for the Sep 2010 and Jun 2011 earthquakes, whereas the Tonks St site had no ejecta for either of these events. Several factors could have contributed to the differing observed responses of these two sites, and additional studies are ongoing to identify the most likely factors. As an example, the lack of ejecta at the Tonks St site may be due to the absence of an impermeable cap or the greater depth of groundwater table at the site.

This unique database of 235 insightful case histories is summarized in a flatfile and documented in detail in Mijic et al. [9]. It provides a reasonable basis for the development of procedures to evaluate when liquefaction ejecta will or will not occur and to estimate the quantity of ejecta in earthquakes. In the future, post-earthquake reconnaissance teams should directly measure ejecta volume at sites by terrestrial LiDAR, structure-from-motion, or conventional land surveys, photographs, and hand measurements to provide more confident estimates of ejecta quantities.

Acknowledgments. This research was supported primarily by the U. S. Geological Survey (USGS), Department of Interior, under USGS award number G20AP00079. The views and conclusions contained in this document are those of the authors and should not be interpreted as necessarily representing the official policies, either expressed or implied, of the U.S. Government. The authors would also like to extend gratitude to James Russell and Oliver Hay of Tonkin and Taylor, Ltd., New Zealand, for their help on this project.

References

1. Rogers, N., van Ballegooy, S., Williams, K., Johnson, L.: Considering post-disaster damage to residential building construction - is our modern building construction resilient? In: 6th International Conference on Earthquake Geotechnical Engineering, Christchurch, New Zealand (2015)
2. Tonkin and Taylor, Ltd.: Tonkin and Taylor Geotechnical Database: Canterbury Maps (Database) (2015). https://canterburygeotechnicaldatabase.projectorbit.com/
3. Land Damage Assessment Team (LDAT): LDAT Reports Data Entry (Database) (2021). https://tracker.projectorbit.com/Sites/LDAT/EQCFieldReportFormExtra.aspx
4. Tonkin and Taylor Ltd. (T+T): Liquefaction vulnerability study - Report to Earthquake Commission (Report 52020.0200) (2013)
5. Canterbury Geotechnical Database: Aerial Photography, Map Layer CGD0100, 1 June 2012. https://canterburygeotechnicaldatabase.projectorbit.com/. Accessed Jul 2021
6. Bradley, B., Hughes, M.: Conditional peak ground accelerations in the Canterbury earthquakes for conventional liquefaction assessment. Technical Report prepared for the Department of Building and Housing, New Zealand (2012)
7. Russell, J., van Ballegooy, S.: Canterbury earthquake sequence: increased liquefaction vulnerability assessment methodology. T+T Report 0028-1-R-JICR-2015 prepared for the Earthquake Commission (2015)
8. Canterbury Geotechnical Database: Event Specific Groundwater Surface Elevations, Map Layer CGD0800, 10 June 2014. https://canterburygeotechnicaldatabase.projectorbit.com/. Accessed Jul 2021
9. Mijic, Z., Bray, J.D., van Ballegooy, S.: Liquefaction ejecta case histories for 2010-11 Canterbury Earthquakes. Int. J. Geoengineering Case Histories **6**(3), 73–93 (2022). https://doi.org/10.4417/IJGCH-06-03-04
10. Boulanger, R.W., Idriss, I.M.: CPT-based liquefaction triggering procedure. J. Geotech. Geoenviron. Eng. **142**(2), 04015065-1–04015065-11 (2016). https://doi.org/10.1061/(asce)gt.1943-5606.0001388
11. Zhang, G., Robertson, P.K., Brachman, R.W.I.: Estimating liquefaction-induced ground settlements from CPT for level ground. Can. Geotech. J. **39**, 1168–1180 (2002). https://doi.org/10.1139/T02-047
12. Iwasaki, T., Arakawa, T., Tokida, K.: Simplified procedures for assessing soil liquefaction during earthquakes. In: Conference on Soil Dynamics and Earthquake Engineering, Southampton, UK, pp. 925–939 (1982)
13. van Ballegooy, S., et al.: Assessment of liquefaction-induced land damage for residential Christchurch. Earthq. Spectra **30**(1), 31–55 (2014)

Chilean Liquefaction Surface Manifestation and Site Characterization Database

Gonzalo A. Montalva[1](✉), Francisco Ruz[2], Daniella Escribano[1], Felipe Paredes[1], Nicolás Bastías[3], and Daniela Espinoza[1]

[1] Universidad de Concepcion, Concepcion, Chile
gmontalva@udec.cl
[2] RyV Engineers, Ñuñoa, Chile
[3] Gensis Geotechnical Earthquake Engineering, Chiguayante, Chile

Abstract. Seismic sources differ significantly from great megathrust to shallow crustal earthquakes in several aspects, yet we analyze them in many engineering applications as if they were similar. This work is framed within the study of the liquefaction behaviour of sites subjected to subduction events. To pursue the analysis of subduction triggered liquefaction, we must first have a database with observations of surface manifestation of liquefaction on such events, seismic demand, and site characterization. We present a database focused on the last three large events that affected Chile since 2010, but include some sites from other subduction events as well. This database has more than 200 sites, including surface manifestation of liquefaction; geotechnical data such as CPT and SPT logs, grain size distribution, and Atterberg limits; geophysical data including shear-wave velocity profiles, dispersion curves, and horizontal to vertical spectral ratios; surface intensity estimations at each site including PGA, PGV, and spectral ordinates.

Keywords: Subduction · Liquefaction · Chile · NGL · Next Generation Liquefaction

1 Introduction

Chile has a very prominent seismic activity, because it's located near the subduction boundary of the Nazca and South American plates. During its history, it has been subjected to earthquakes of great magnitude, like the 1960 Mw 9.5 Valdivia earthquake and the 2010 Mw 8.8 Maule earthquake, in which substantial liquefaction was observed.

The semi-empirical simplified methodologies, introduced by Seed and Idriss (1971) and then updated by several authors (e.g., Boulanger and Idriss 2014; Cetin et al. 2018), are the most widely used methodologies to evaluate liquefaction hazards. Nevertheless, studies made by Montalva and Leyton (2014), Montalva and Ruz (2017) and Espinoza (2018) showed that this methodologies perform poorly in the chilean subduction zone. This is caused, in part, by the significant under-representation of subduction events in the currently employed liquefaction databases.

To address this issue, we compiled a database with 209 Chilean sites with or without liquefaction manifestation. These sites were subjected to one of the following subduction

© The Author(s), under exclusive license to Springer Nature Switzerland AG 2022
L. Wang et al. (Eds.): PBD-IV 2022, GGEE 52, pp. 1893–1900, 2022.
https://doi.org/10.1007/978-3-031-11898-2_173

events: 2010 Mw 8.8 Maule, 2014 Mw 8.2 Iquique, 2015 Mw 8.3 Illapel and 2016 Mw 7.6 Melinka (Fig. 1). The seismic performance and geotechnical tests were compiled from research papers, engineering and site reconnaissance reports, undergraduate theses from Chilean universities, geotechnical reports and fieldwork from the geotechnical group of the University of Concepción. Finally, ground motion intensity measures were estimated with event-corrected GMPE's, developed specifically for the Chilean subduction zone.

This database is currently open to the public, and can be downloaded from DesignSafe (Montalva et al. 2021a).

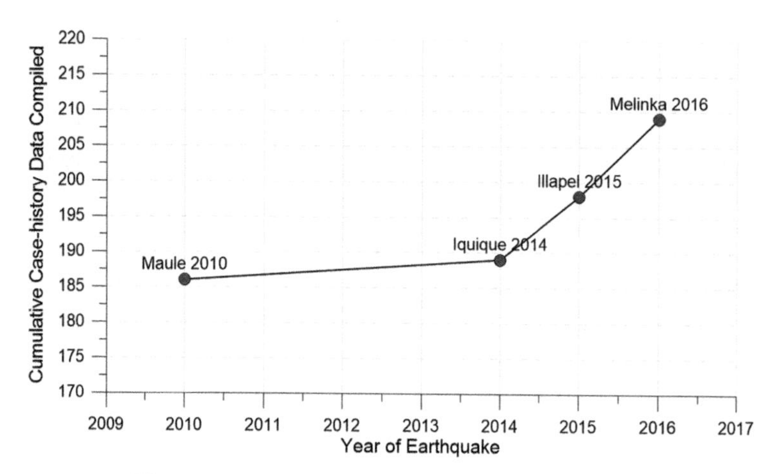

Fig. 1. Evolution of case history for megathrust events

2 Geotechnical Data

Following the databases by Geyin et al. (2021) and Next Generation Liquefaction project (Brandenberg et al. 2020), we employed a relational model to organize the database.

There is a main table that contains the metadata, the associated seismic event and the estimated intensity measures for the 209 sites. Also, there is a separate file for each site, and each file is organized into nine main tables (Fig. 2) that contain, in detail, the site information [SITE], geotechnical tests [TEST], and seismic parameters [SPAR].

2.1 Site Background and Seismic Performance

There are two tables associated with general site information: site background [SITE_SBGND] and evidence of liquefaction [SITE_LE]. The table [SITE_SBGND] contains the following information: (1) Site identification (site ID); (2) Latitude and longitude (WGS84, decimal degrees); (3) City and region; (4) Structure type (i.e., building, bridge, free-field level ground); (5) Geological age.

On the other hand, the table [SITE_LE] contains the following: (1) Site performance; (2) Seismic event associated with the site; (3) Observed features of liquefaction; (4) Photographies and field comments.

The performance is divided in the following categories: (1) Surface manifestations of liquefaction; (2) No visible surface manifestations of liquefaction; (3) Marginal.

Sites with liquefaction evidence described as Marginal correspond to areas without liquefaction manifestation but that are close to sites with liquefaction evidence (~500 m or less).

The evidence of liquefaction is divided in the following features: (1) Sand Boils; (2) Lateral Spreading; (3) Flow Failure; (4) Excessive Settlement; (5) Structures Flotation; (6) Surface Cracks; (7) Structural Damage.

Every site in the database has, at least, the aforementioned information available.

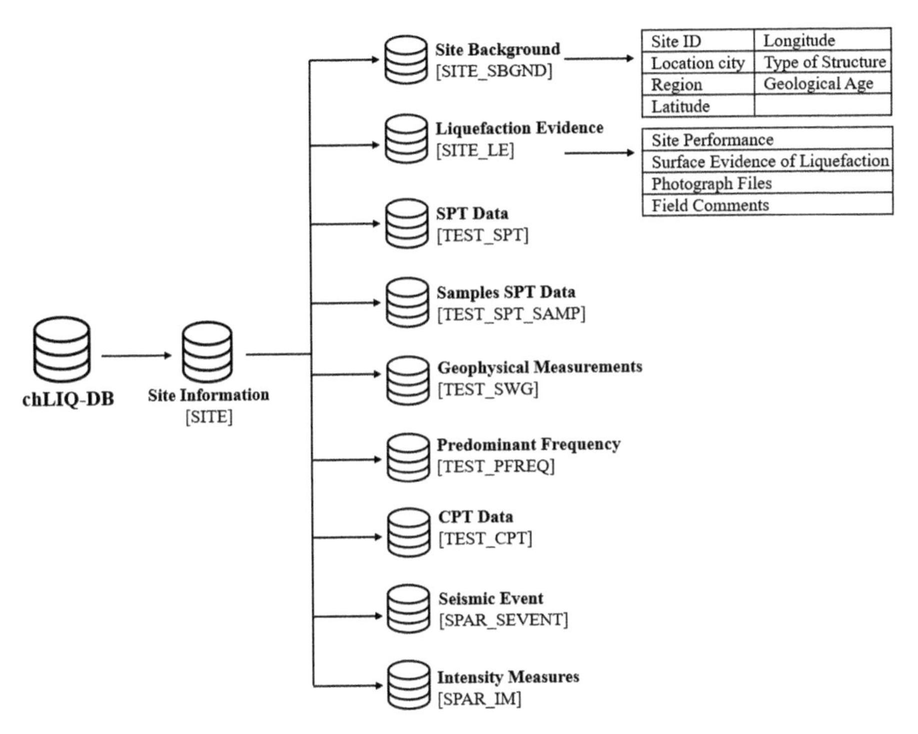

Fig. 2. Chilean liquefaction database: main fields of structure (Montalva et al. 2021a).

2.2 Geotechnical Tests

The geotechnical test tables [TEST] contains information from SPT's [TEST_SPT], laboratory tests [TEST_SPT_SAMP], and CPTu measurements [TEST_CPT]. The available information from geotechnical tests is summarized in Table 1.

SPT Data. The [TEST_SPT] table contains the following: (1) ID for each SPT on the site; (2) Location of each test (WGS84, decimal degrees); (3) Reported groundwater depth; (4) Type and average energy of the hammer; (5) Bar diameter; (6) Presence or absence of liners; (7) Depth range of each sample; (8) Blow count of each sample.

Table 1. #Tests from case histories (Compiled from Montalva et al. 2021a).

Surface liquefaction evidence	#SPT boreholes	#Vs measurements	#HVSR measurements	#CPT tests
Yes	204	109	115	11
No	100	98	108	11
Marginal	8	6	20	0

The database contains a total of 3130 SPT samples. The table [TEST_SPT_SAMP] contains detailed information for each sample, corresponding to: (1) SPT ID; (2) Sample ID; (3) Depth range of each sample; (4) Percentage passing the N°4, N°10, N°40 and N°200 sieves; (5) USCS classification; (6) Moisture content; (7) Liquid and plastic limits; (8) Plasticity index; (9) Specific gravity of solid particles; (10) An estimation of the saturated unit weight. Histograms of the blow count and sand content in the Chilean liquefaction database are presented in Fig. 3.

In total, the database contains 781 samples with a measured plasticity index greater than zero, 1955 non-plastic samples and 394 samples without measured Atterberg limits.

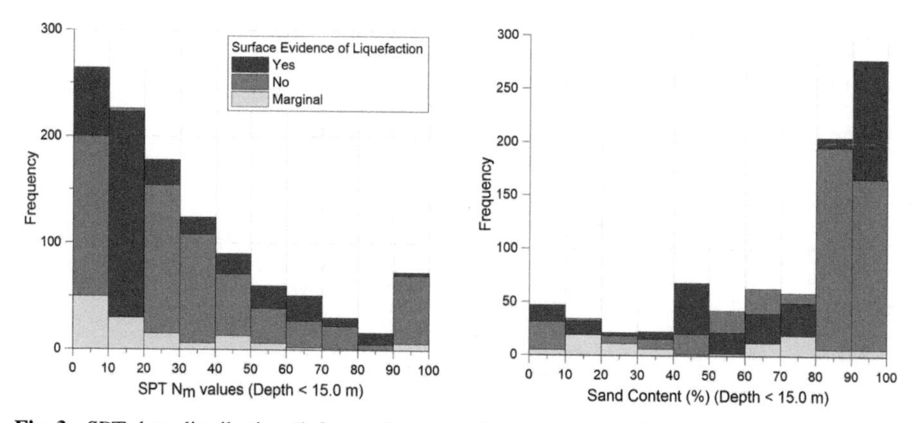

Fig. 3. SPT data distribution (left panel) and sand content on database (right panel) (Montalva et al. 2021a).

CPT Data. The [TEST_CPT] tables contains the following information: (1) Test ID; (2) Test location (WGS84, decimal degrees); (3) Groundwater table depth; (4) Cone resistance (qc), sleeve friction resistance (fs) and pore pressure (u) over depth (z).

2.3 Geophysical Data

The geophysical test tables correspond to shear velocity [TEST_SWG] and predominant frequency [TEST_PFREQ] measurements.

Surface Shear-Wave Measurements. In the majority of cases, the shear velocity measurements were derived from non-invasive methods, and in only a few cases, the shear velocity profiles were obtained from downhole tests.

The [TEST_SWG] tables contains the following information: (1) Site ID; (2) Vs18 (m/s); (3) Vs30 (m/s); (4) Shear velocity profile ID; (5) Depth range; (6) Shear velocity; (7) Measurement method; (8) Dispersion curve.

HVSR Measurements. The predominant frequency (f_0) and amplitude (A_0) were obtained from the HVSR curves from ambient noise measurements. These measurements were recorded with triaxial seismometers.

The [TEST_PFREQ] tables contains the following information: (1) Site ID; (2) f_0 (Hz); (3) T_0 (sec); (4) HVSR measurement ID; (5) Test location (WGS84, decimal degrees); (6) Record duration; (7) Clear peak; (8) HVSR curve (Frequency and amplitude). The variation of site, performance with respect to the predominant frequency and the HVSR amplitude, is shown in Fig. 4.

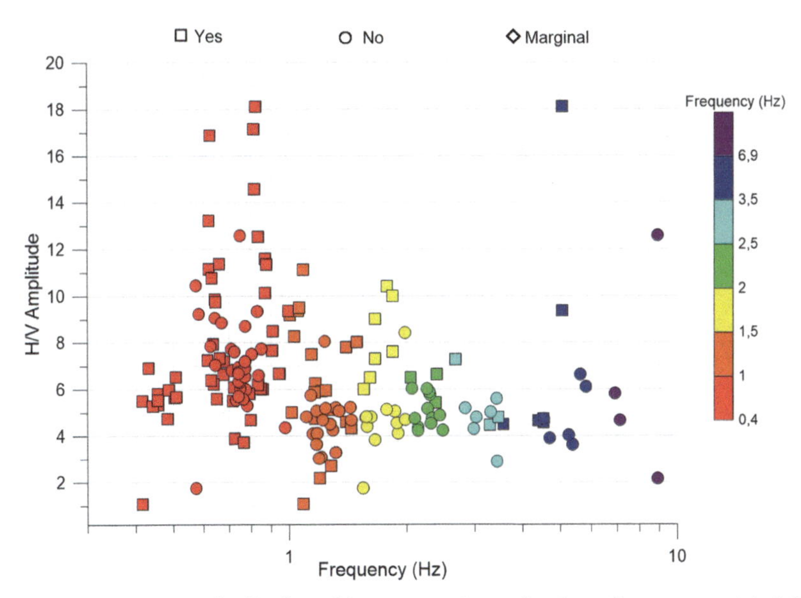

Fig. 4. Field performance distribution with respect to the predominant frequency and their HVSR amplitude (Montalva et al. 2021a).

3 Surface Intensity Estimations

In principle, the liquefaction potential models are developed based on cyclic demand versus cyclic resistance of the soil. In this context, the description of the earthquake's parameters and the estimation of the seismic demand, with their uncertainties, are key topics on liquefaction databases.

The metadata of the seismic event corresponds to the hypocenter location (i.e., latitude, longitude, and depth) and the moment magnitude, which is obtained from the Chilean National Seismological Center (CSN, 2018), and is supplemented with the information of the Global CMT project (Ekstrom et al. 2012). The finite-fault rupture planes for each earthquake are available from the SRCMOD repository (Mai and Thingbaijam 2014; Hayes 2017) and are used to estimate the source-to-site distances: the closest distance to the earthquake rupture plane (Rrup), epicentral (Repi), and hypocentral (Rhyp) distances. In the Chilean liquefaction database, the nearest site with evidence of liquefaction is located at 25 km and the farthest site at 103 km from the fault plane (i.e., Rrup).

Unfortunately, in the Chilean liquefaction database, no case histories of liquefaction are recorded by a strong ground motion station. Therefore, the estimation of the seismic intensities of each site are obtained from the latest strong Ground Motion Prediction Models (GMPMs) available from the Chilean subduction zone (Montalva et al. 2017; 2021b). Due to the event-term of the partitioned residuals of the GMPMs being known, the intensities are event-corrected. The use of the native residuals obtained directly from the regression process allows an unbiased estimate of intensities (Stafford 2015). The ground motion intensity measures (IM) included in the database are the peak ground acceleration (PGA), peak ground velocity (PGV), and the spectral ordinates PSA and SV at 1.0 s, 0.2 s and at the site's predominant frequency. The intensity distribution with respect to the distance is presented in Fig. 5.

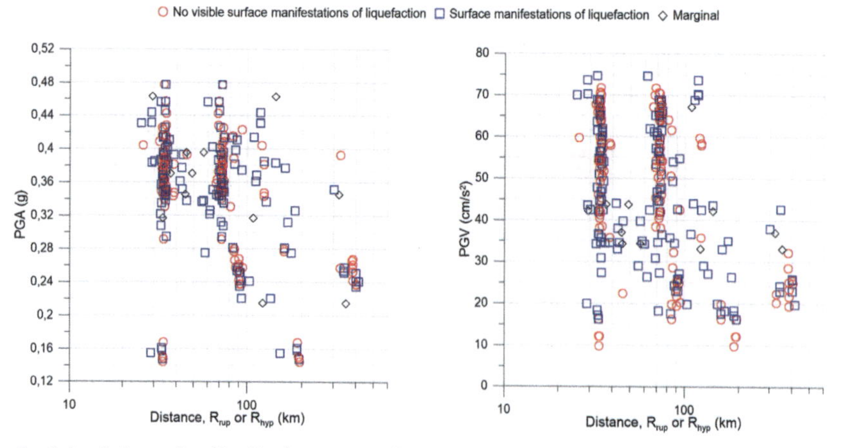

Fig. 5. Seismic intensity distribution versus distance on Chilean liquefaction database (Montalva et al. 2021a).

These IM are typically used to define the seismic demand on the liquefaction potential schemes, and also allow the estimation of the other parameters like S_1 (Kishida and Tsai 2014) which reflects the amplitude ratio between the velocity- and acceleration-controlled zone of the spectrum. To account for the variability on the seismic intensity we also provided to the database a minimum and maximum estimation obtained from the median minus and plus one standard deviation (respectively). The standard deviation is also event-corrected, i.e., is reduced the event term variability (τ) from total sigma (see Montalva et al. 2021a).

4 Conclusions

Based on previous research (Montalva and Ruz 2017) there is a significant mismatch between the observed performance on subduction environments and the estimation of liquefaction potential based on simplified methodologies, which is the main reason for the development of this database.

The database shown herein is the most extensive on subduction tectonic environments and includes 209 case histories, with known field performance, from the four latest megathrust events on the Chilean subduction zone: 2010 Mw 8.8 Maule, 2014 Mw 8.2 Iquique, 2015 Mw 8.3 Illapel, and 2016 Mw 7.6 Melinka.

The main aim is the compilation and integration of liquefaction data on the subduction zone, for the evaluation, training, and development of new models to quantify the liquefaction potential and the risk associated with civil structures.

The dataset contains both unprocessed (e.g., SPT blow counts) and processed data (e.g., shear wave velocity profiles). Each user can select the appropriate processing schema for their own investigations.

An aspect worth nothing is that the information on the database is organized around sites, not case histories. In particular, some sites correspond to bridges that can span hundreds of meters. So, the selection of case histories from this dataset should take into account the objectives of each site investigation.

Currently, the investigation team is focused on the expansion of the database, particularly by making CPT soundings in sites with no or scarce information, in order to improve the completeness and reliability of the database. Also, as part of a collaboration with the NGL project, an homologation of the dataset to the NGL schema is currently taking place.

Acknowledgments. We thank the Roads Department of the Ministry of Public Works (MOP) for providing essential data for the development of this work. We also acknowledge Ruz and Vukasovic Engineers for providing valuable data on many sites. Finally, special thanks go to the geotechnical research group at University of Concepci6n for the fieldwork performed to collect and process data.

This work was partially funded by the National Commission for Scientific and Technological Research (ANID) through the projects FONDEF ID16i10157, FONDEF ID16i20157, the Millennium Nucleus CYCLO, grant NC160025, funded by the Millennium Scientific Initiative (ICM) this support is greatly acknowledged. Site Response Observatory on Alluvial Basins funded by FONDEQUIP under grant EQM160015, provided valuable data for some of the sites.

References

Bolton Seed, H., Idriss, I.M.: Simplified procedure for evaluating soil liquefaction potential. J. Soil Mech, Found. Div. **97**(9), 1249–1273 (1971)

Boulanger, R., Idriss, I.: CPT and SPT Based Liquefaction Triggering Procedures. University of California, Department of Civil and Environmental Engineering, Davis, California (2014)

Cetin, K., et al.: Dataset on SPT-based seismic soil liquefaction. Data Brief **20**, 544–548 (2018)

Montalva, G., Leyton, F.: Discussion to: "shear-wave velocity-based probabilistic and deterministic assessment of seismic soil liquefaction potential." J. Geotech. Geoenviron. Eng. **140**(4), 407–419 (2014)

Montalva, G., Ruz, F.: Liquefaction Evidence in the Chilean Subduction Zone. PDB-III Earthquake Geotechnical Engineering, Vancouver, Cánada (2017)

Espinoza, D.: Evaluación del Potencial de Licuación en Zonas de Subducción. Universidad de Concepción, Departamento de Ingeniería Civil, Concepción (2018). (in Spanish)

Montalva, G., Ruz, F., Escribano, D., Bastias, N., Espinoza, D., Paredes, F.: Chilean liquefaction case history database. Earthq. Spectra **38**, 2260–2280 (2021)

Geyin, M., Maurer, B.W., Bradley, B.A., Green, R.A., van Ballegooy, S.: CPT-based liquefaction case histories compiled from three earthquakes in Canterbury, New Zealand. Earthq. Spectra **37**, 2920–2945 (2021)

Brandenberg, S., Zimmaro, P., Stewart, J.: Next-generation liquefaction database. Earthq. Spectra **36**(2), 939–959 (2020)

Ekström, G., Nettles, G., Dziewonski, A.: The global CMT project 2004–2010: centroid-moment tensors for 13,017 earthquakes. Phys. Earth Planet. Inter. **200–201**, 1–9 (2012)

Mai, P., Thingbaijam, K.: SRCMOD: an online database of finite-fault rupture models. Seismol. Res. Lett. **85**(6), 1348–1357 (2014)

Hayes, G.: The finite, kinematic rupture properties of great-sized earthquakes since 1990. Earth Planet. Sci. Lett. **486**, 94–100 (2017)

Montalva, G., Bastías, N., Rodríguez-Marek, A.: Ground motion prediction equation for the Chilean subduction zone. Bull. Seismol. Soc. Am. **107**, 901–911 (2017)

Montalva, G.A., Bastías, N., Leyton, F.: Strong ground motion prediction model for PGV and spectral velocity for the Chilean subduction zone. Bull. Seismol. Soc. Am. **112**(1), 348–360 (2021)

Stafford, P.J.: Extension of the random-effects regression algorithm to account for the effects of nonlinear site response. Bull. Seismol. Soc. Am. **105**(6), 3196–3202 (2015)

Kishida, T., Tsai, C.-C.: Seismic demand of the liquefaction potential with equivalent number of cycles for probabilistic seismic hazard analysis. J. Geotech. Geoenviron. Eng. **140**(3), 04013023.1-04013023.14 (2014)

S3: Special Session on Embankment Dams

Deterministic Seismic Hazard Analysis of Grand Ethiopia Renaissance Dam (GERD)

Mohammed Al-Ajamee[1](✉), Abhishek Baboo[2], and Sreevalsa Kolathayar[2]

[1] Building and Road Research Institute, University of Khartoum, Khartoum, Sudan
mohammedalajamee@gmail.com

[2] Department of Civil Engineering, National Institute of Technology Karnataka, Surathkal, India

Abstract. This paper presents the Deterministic Seismic Hazard Analysis for Grand Ethiopia Renaissance Dam (GERD) (11.2183°N, 35.0941°E). This area is of special importance because GERD is the largest hydroelectric power plant in Africa and the seventh-largest in the world. The Hazard analysis was done using linear sources present near the dam site (Inferred faults and Indicated faults); 713 faults were used for the analysis. Earthquake data was collected for the dam region, and homogenization of data was done to moment magnitude followed by declustering foreshock and aftershock events. 1375 earthquakes of moment magnitude 4–7 were considered and represented on a seismotectonic map. Ground Motion Prediction Equations (GMPEs) were identified for the dam site, and suitability was checked by comparing the Data Support Index (DSI) value of various prediction equations. Considering the selected GMPEs and input data, DSHA was performed by using a MATLAB program developed for this purpose. The dam site was further divided into a grid cell of 100 m × 100 m, and hazard parameters were obtained at the midpoint of the grid cells, bearing in mind all seismic sources within a 500 km radius. Obtained PGA values were compared with USGS developed Instrumental Intensity scale. The hazard map showing the spatial variation of the risk associated with the dam site is presented.

Keywords: DSHA · Grand Ethiopia Renaissance Dam (GERD) · Seismicity · East African Rift System (EARS)

1 Introduction

The Eastern part of Africa near the East African Rift System (EARS) and the Red Sea region close to the Afro-Arabian faults are mostly associated with high levels of seismic activity. However, a thorough examination of both recent and historical instrumental data has shown that other regions of Africa are likely to have major intraplate earthquakes, necessitating the need for conduction Seismic Hazard (SH) studies. Moreover, the sparse number of earthquakes recording stations and absence of sufficient earthquake catalogs is a significant impediment to carrying out SH assessments and calculating related hazard parameters. Conducting SH and risk studies is crucial and will help in minimizing the harmful effects that earthquakes may have on human beings, and infrastructure. Thus,

© The Author(s), under exclusive license to Springer Nature Switzerland AG 2022
L. Wang et al. (Eds.): PBD-IV 2022, GGEE 52, pp. 1903–1913, 2022.
https://doi.org/10.1007/978-3-031-11898-2_174

improving regional resilience in the presence of such disastrous events (Atalay 2017). Plethora of attempts have been made to assess SH in this reign by various researchers, (Abdalla et al. 2001) (Sudan and South Sudan and its vicinity); (Goitom et al. 2017) (Eritrea, and the surrounding regions); (Atalay 2017) (Ethiopia and the neighboring region). When designing infrastructure in such seismic region, it is essential to understand the ground motion distribution affecting the facility throughout its economic life, especially critical structures such as dams, bridges, nuclear power plants, roads, and railway lines (Baker 2008). As such, GERD is surrounded by many faults; earthquake events, and it is not far away from earthquake threats. There are diverse geological structures within the African continent which comprises regions of active deformation. Volcanic fields, transform faults, rift zones, thrust and fold mountain belts are all seismically active zones (A. Atalay 2021).

SH Analysis is the likelihood of an earthquake occurring in a specified geographic zone and with ground motion intensity surpassing a certain level. This paper aims to get an estimate of SH and risk around GERD location. Earthquake consequences lead to a risk of ground shaking, ground displacement, landslide, soil liquefaction, etc. (Al-Ajamee 2022a; 2022b). Evaluation of SH can be done by 1) Deterministic Seismic Hazard Assessment (DSHA) and, 2) Probabilistic Seismic Hazard Assessment (PSHA). In this paper, DSHA procedure will be followed for SH estimation in GERD location because hazard parameters obtained using DSHA will be maximum for a region, whereas PSHA considers the technical aspects and can be used for comparison (Baboo 2022). DSHA values are called the cap of PSHA (Kolathayar 2012a; 2012b), which explains why DSHA is still used for very important structures like large dams, nuclear power plants, hazardous waste containment facilities, and large bridges, which gives quantitative estimation of ground shaking threats at a specific location.

2 Tectonic Setting and Seismic Sources

The tectonics of the study area are governed by deviation of the Somali, Arabian and Nubian plates which connect at the Afar triple junction, near the international borders of Djibouti, Eritrea, and Ethiopia. The EARS mark the boundary between the Somali plate to the east and the Nubian plate to the west; both are parts of what was previously identified as the African plate. EARS is classified as a continental rift, and it's part of the Afro Arabian Rift System (see Fig. 1), the Gulf of Aden which the separate the Arabian plate from the Somali plate (Chu and Gordon 1988).

While this study focuses on estimating SH near the GERD region, the study area is situated between (11.2183ºN, 35.0941ºE), which covers parts of Ethiopia, Sudan, and South Sudan (see Fig. 1). There are two kinds of physiographic units that can be labeled within this region that are linked to several tectonic structures: the Red Sea rift, and the Afar rift Fig. 1; (Ogubazghi et al. 2004). The Afar depression is characterized by salt plains and graben in the south and north, respectively. The Afar depression is situated in northeastern Ethiopia and southeastern Eritrea. It is bounded by the Danakil horst, the Somali plate, the Red Sea, and the plateau, further details can be found elsewhere (Chu and Gordon 1988).

From the various sources generated, magnitude estimation is done. The magnitude of an earthquake that can be generated from a particular fault is based on the source geometry (USGS 2021). For moment magnitude estimation, various empirical relations are available, which uses the geometry of the faults. In this paper, for estimation, various vulnerable seismic sources considered are linear sources, and for their estimation, length relations were being used, which is shown in Fig. 2. For the length of each fault, start and end, latitudes and longitudes were determined using field calculator in QGIS, and lengths were determined using the same.

For All types of fault

$$M_w = 5.08 + 1.16\log(L) \tag{1}$$

where L = Length of faults

3 DSHA Methodology

Determination of SH for a given location involves a certain procedure that will lead to hazard estimation. First, regional geology and seismologic settings are inspected for patterns and sources of earthquakes, which can be obtained from the records of the seismometer. The effects of these sources are evaluated proportionally to the local geologic rock and groundwater conditions, soil types, and slope angles. Finally, a hazard indicator is required for an earthquake site and distance. Hazard indicators can be 1)

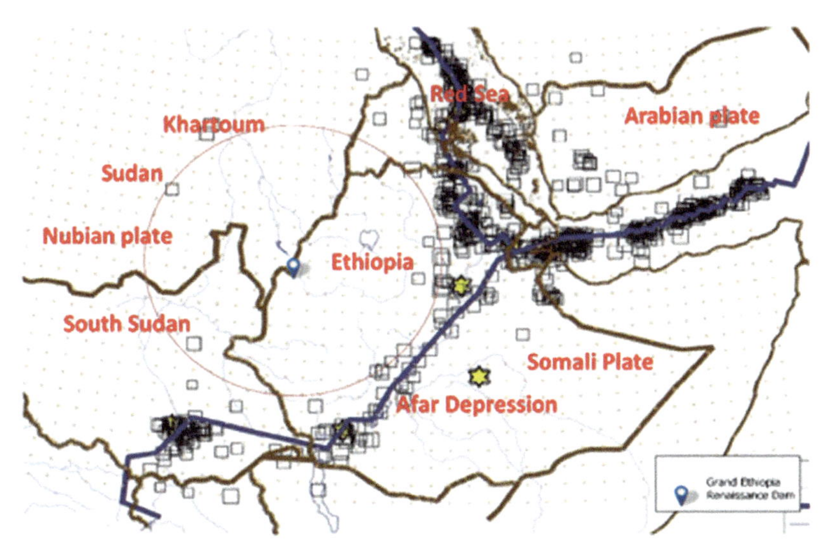

Fig. 1. Seismicity of Ethiopia and neighboring countries. The blue marker shows the location of Grand Ethiopia Renaissance Dam, and the circle around it covers the neighboring location in a radius of 300 km. Square marks on the map show the location of the earthquake event, where each square is proportional to the corresponding magnitude. The yellow star mark shows a location of magnitude above 6.5. Brown Line marks the boundary of the borders, and the Dark blue line shows the Plate boundaries, known as the Afar triple junction (Nubian, Somali and Arabian Plate).

Peak Ground Acceleration PGA, 2) Peak Velocity PV, 3) Response Spectrum RS ordinates. In this paper, PGA values are obtained for the dam site, where PGA refers to the maximum acceleration of the ground occurring during an earthquake at a specific site. PGA is equivalent to the amplitude of the largest absolute acceleration recorded on an accelerogram at a site during a particular earthquake.

In DSHA ground motion parameters are estimated for maximum credible earthquake, and it is expected to happen at the nearest possible location from the location of interest. DSHA utilizes established seismic sources sufficiently close to the location and accessible historical geological and seismic data. DSHA analysis can be categories into four steps, namely 1) identification and characterization of all sources, 2) selection of source site distance boundaries, 3) selection of controlling earthquake, 4) definition of hazard by means of controlling earthquake (Kolathayar 2012a; 2012b). Similarly, for the GERD location, all the above steps were followed, ending with the performance of the seismic hazard analysis. In respect to the DSHA analysis, certain assumptions are considered, which are: - 1) All sources have the ability to create substantial ground motion at the location, 2) large sources at long distances, 3) small sources at a short distance. DSHA produces a scenario or design earthquake, which can be a worst-case scenario. Hazard parameters obtained through the deterministic approach are always greater than the probabilistic approach, thus preferred for very important structures. The only drawback of the deterministic approach is that it offers no sign of how probable the design earthquake is to happen throughout the structure's life.

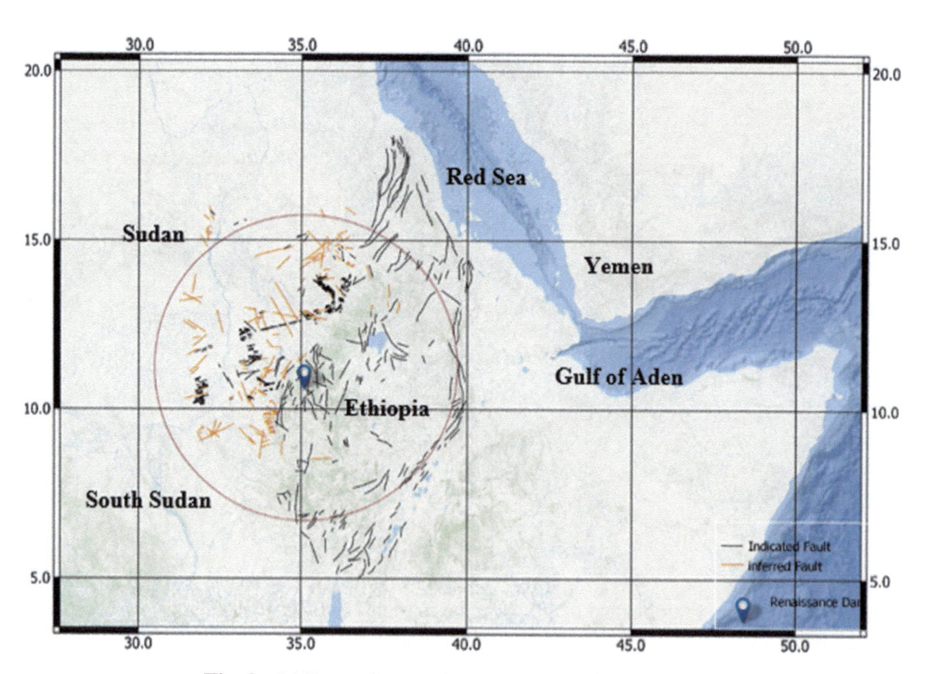

Fig. 2. Different faults identified for the GERD region

3.1 Earthquake Database

A collection of the spatial distribution of epicenters of earthquake occurrences was done, and the data was obtained from USGS (USGS 2021) along with fault details. The compilation of the database was a key objective associated with the seismotectonic map. An instrumentally recorded earthquake catalog was prepared to extend from the year 1906 to 2021, with a magnitude range of 4 to 7. 1327 earthquake data were recorded, and the details of past earthquakes were collected in a radius of about 500 km, covering the dam sites.

3.2 Unification of Magnitude

The magnitude of the earthquake is one of the most used parameters for the classification of seismicity of any region or area. Earlier, the most commonly used earthquake magnitude representation was in Richter magnitude, which was well-known but not well understood. Richter magnitude scale was developed for measuring the size of an earthquake in southern California only; thus, a better magnitude scale representation was required. For the GERD location, the data of the earthquake was reported in separate magnitude scales, and for the analysis, these magnitude scales need to be homogenized. Data was stated in body-wave magnitude scale (M_b), surface wave magnitude scale (M_s), local magnitude (M_L), and moment magnitude (M_w). Out of the reported magnitude scales, Moment Magnitude is considered to give a more quantitative measure of earthquake magnitude and gives the most reliable estimate of earthquake size (Hanks and Kanamori 1979). Magnitude is a function of energy release, which is the best representation by moment magnitude and for all sizes of magnitude. For magnitude conversion, Scordilis relations were identified (Scordilis 2006), which provides different relations for different magnitude scales. Scordilis given the moment conversion relations for a particular magnitude range and error value, which matches the magnitude range of our earthquake catalog. Magnitude conversion relations in this paper are represented below.

Conversion M_b to M_w (Body wave magnitude into moment magnitude)

$$M_w = 0.85 \pm .04M_b + 1.03 \pm 0.23 \quad 3.5 \leq M_b \leq 6.2 \tag{2}$$

Conversion Ms to M_w (Surface wave magnitude into moment magnitude)

$$M_w = 0.67(\pm 0.005)M_S + 2.07(\pm 0.03) \quad 3.0 \leq M_S \leq 6.1 \tag{3}$$

$$M_w = 0.99(\pm 0.02)M_S + 0.08(\pm 0.13) \quad 6.2 \leq M_S \leq 8.2 \tag{4}$$

Data reported in the local magnitude scale were removed from the earthquake catalog as there is no globally valid relation available between M_L and M_S. Total earthquake events reported after the unification of magnitude are 1275 events.

3.3 Declustering

In estimating the earthquake hazard, declustering of earthquake events was required. Decluttering is the process in which foreshocks and aftershocks are removed with respect to mainshocks (Zhuang et al. 2013). Declustering gives a better understanding of the earthquake events. All the dependent events that fall inside the time and space interval of the mainshock are disregarded to attain a data set of mainshocks presumed to illustrate a Poisson distribution. Declustering an earthquake catalog results in a catalog composed of independent events. All the unified earthquake catalog data (moment magnitude) was decluttered based on a software zmap, which runs on MATLAB. Declustering is done based on the assumption that time and distance depend on magnitude. The algorithm used for declustering is shown below:

$$\text{Distance} = \exp\left(-1.024 + 0.804\text{M}_\text{w}\right) \tag{5}$$

$$\text{Time} = \exp(-2.87 + 1.235\text{M}_\text{w}) \tag{6}$$

4 Estimation of Hazard by DSHA

The DSHA technique considers a specific earthquake scenario, either a presumed or realistic one. The DSHA method utilizes established seismic sources which are close to the site and accessible historical geological and seismic data to create models of ground motion at the site. It is assumed that the earthquakes occurred more often at the nearest plausible potential source. The DSHA technique is regarded as an upper bound estimate for seismic risk, and it has therefore been used to estimate SH for key structures such as dams, nuclear power plants (NPP), and wastewater contamination systems (Kolathayar 2012a; 2012b).

4.1 Attenuation Relations

For the estimation of various hazard parameters of earthquake-like peak ground velocity (PGV), peak ground acceleration (PGA), response spectrum ordinates, one requires Ground Motion Prediction Equations (GMPEs) or Attenuation Relations. There are several ground motion equations available all over the world, but most are region-specific, which makes the prediction for a particular earthquake difficult. Strong motion recording in Europe and the Middle East was first installed much later than U.S. and Japan, due to which there are no region-specific ground motion equations available for the GERD location.

From the seismotectonic map of the study region, we can conclude that GERD lies in a stable continental crust and can be considered a moderate seismic region. Since no region-specific equations are known, the best GMPEs identified for the study area which shares similar seismic attenuation characteristics are 1) Atkinson and Boore for Eastern North America (Atkinson and Boore 2006); 2) Grazier for Central and Eastern North America (Graizer 2016); and 3) Akkar and Bommer for Mediterranean region (Akkar and Boomer 2010), 4) Douglas (Douglas 2018). Attenuation relations were used considering rock sites, defined as $Vs > 750$ m/s, and prediction is made for 5% damping to obtain pseudo-spectral acceleration.

4.2 Selection of the Attenuation Model

The selection of the attenuation equation for the GERD dam site is done by calculating the Log-Likelihood value (LLH), for all attenuation equations for the maximum occurred magnitude in that region.

$$\mathrm{LLH} = -\frac{1}{\mathrm{n}}\Sigma_{\mathrm{i}}^{\mathrm{n}} = \log_2(\mathrm{t}(\mathrm{x_i})) \tag{7}$$

where

n = number of earthquakes.

t(x) = {xi}, i = 1,.... n = probability density function of ground motion model.

(Delavaud et al. 2009) suggested a theoretic approach for the selection of GMPEs and have proposed the efficacy test to quantitatively assess the suitability of the GMPE for the region of interest. The average value of log-likelihood has been used for the ranking purpose in the efficacy test. LLH value for a site is obtained by comparing the predicted ground motion from attenuation relations (identified earlier) with the actual ground motion data recorded in that region. Although finding the ground motion data is difficult for the African continent and we could not find any promising data near Ethiopia. Thus, for checking the suitability of the GMPEs, the recorded ground motion data used was of Greece country in Southeastern Europe. An empirical equation is used to compare the recorded value and predicted value is shown as follows

$$t(x) = \frac{1}{\sqrt{2\pi}\sigma}\exp\left(-\frac{1}{2}\left(\frac{x-u}{\sigma}\right)^2\right) \tag{8}$$

where

x = observations from past earthquakes.

u = predicted observation.

The lower value of LLH denotes that the GMPE selected is close to the unknown original model (accurate equation). Based on the LLH value, certain weights must be given to each equation. For this purpose, we required Data Support Index (DSI) value. If the DSI value is negative, the weightage given to that particular equation is zero. And if the DSI value is positive, an estimated weightage is given to the equations. Data Support Index value calculations are represented in Table 1 below.

Table 1. Calculations of Data Support Index values (DSI)

S. no	GMPE	LLH	2^(−LLH)	wt (i)	DSI	Wt (f)
A	(Atkinson and Boore 2006)	1.975065	0.254358425	0.414844747	65.9379	**0.6**
B	(Akkar and Boommer 2010)	2.296569	0.203546577	0.331973388	32.78936	**0.4**
C	(Graizer 2016)	3.80561	0.071514993	0.116637061	−53.3452	**0**
D	(Douglas 2018)	3.578262	0.083721251	0.136544804	−45.3821	**0**

The above table represents the DSI value of different GMPEs, where we can observe that the DSI value obtained for (Graizer 2016) and (Douglas 2018) is negative. Hence, the weightage given to those equations is zero.

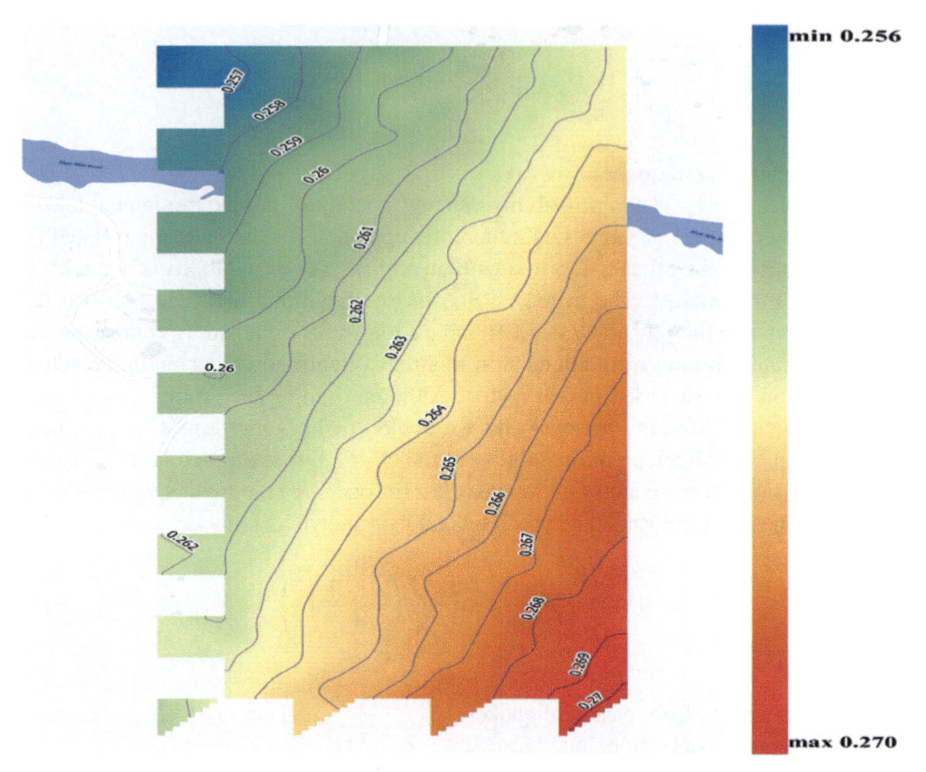

Fig. 3. Contour map of the spatial variation of the Hazard in the GERD dam site

5 Results and Discussions

DSHA analysis for the Dam site based on a MATLAB code, resulted in a maximum PGA value. The average focal depth considered for the GERD location based on different linear sources is 11.87 m, and a sum of 713 faults was considered for the analysis. Analysis was performed on the two selected equations. Performing DSHA using MATLAB code on the equation given by (Akkar and Boomer 2010) resulted in a controlling magnitude of 7.3 and a maximum PGA of 0.140g. While using the equation given by (Atkinson and Boore 2006), the maximum PGA value obtained was 0.266g having the same controlling magnitude of 7.3. Final PGA value for the site evaluated by giving weightage to the obtained values, i.e., of 0.6 and 0.4. The final PGA reported based on the analysis is 0.096g which represents the maximum PGA value for the dam site.

Further analysis was carried for the whole dam site, where the whole GERD dam, approximately having a surface area of 100 m × 1780 m, was divided in a grid of size 0.001 × 0.001 (about 100 m × 100 m). A sum of 90 grid points was considered for the analysis. For the grid analysis, the ground motion equation used was (Atkinson and Boore 2006), and for each grid point, the PGA value is obtained at the center of each grid, on the bedrock level, using MATLAB code. All sources were considered inside a radius of 500 km for every grid point.

Based on the hazard parameters obtained for the dam site in a grid format, a contour map is prepared for the GERD showing the variation of ground motion around the dam location (see Fig. 3). United States geological survey developed relation of PGA with other hazard parameters in Table 2. PGA value from the grid analysis (0.256–0.270) lies in the highlighted column, where a description of the Hazard associated with GERD location can be seen in Table 2.

Table 2. PGA and hazard parameters based on the USGS

Instrumental intensity	Acceleration (g)	Velocity (cm/s)	Perceived shaking	Potential damage
I	<0.0017	<0.1	Not felt	None
II–III	0.0017–0.014	0.1–1.1	Weak	None
IV	0.014–0.039	1.1–3.4	Light	None
V	0.039–0.092	3.4–8.1	Moderate	Very light
VI	0.092–0.18	8.1–16	Strong	Light
VII	**0.18–0.34**	**16–31**	**Very strong**	**Moderate**
VIII	0.34–0.65	31–60	Severe	Moderate to heavy
IX	0.65–1.24	60–116	Violent	Heavy
X+	>1.24	>116	Extreme	Very heavy

6 Conclusion

While the generation of seismotectonic framework and mapping at a level such as the continent of Africa is a tough task, numerous past and present local projects provided an ample amount of information for this paper. Seismic Hazard Analysis has been performed for the Grand Ethiopia Renaissance Dam, whose construction has been recently completed. There has been a lot of conflicts associated with the dam, and one of the concern matters was the seismic stability of the dam. Through this paper, an attempt has been made to get a closer view of what are the hazards associated with GERD. A seismotectonic map prepared considering the earthquake data from 1906 to 2021 concluded that GERD lies in a stable region which was one of the concern points related to GERD location. Also, the deterministic approach provided us with the worst-case scenario, derived from the latest earthquake data. From the maximum PGA obtained based on the logic tree weighing of 0.096g, it can be concluded that GERD lies in a moderate zone where the potential damage to the structure is moderate, whereas the perceived ground shaking can be very strong, as stated in the table from USGS. The maximum PGA value in is study is lower than the obtained by (Atalay 2017), in which the whole region of Ethiopia was considered, including the Afar region. The ground motion parameters studied in this paper provide an idea of the risk associated with the dam location. Further, conducting site responses studies are recommended for better understanding and gaining more knowledge about the role of surficial soils on ground motions (Al-Ajamee 2022a; 2022b).

References

Abdalla, J.A., Mohamedzein, Y.E., Wahab, A.A.: Probabilistic seismic hazard assessment of Sudan and its vicinity. Earthq. Spectra **3**(17), 399–415 (2001)

Akkar, S., Bommer, J.J.: Empirical equations for the prediction of PGA, PGV, and spectral accelerations in Europe, the Mediterranean region, and the Middle East. Seismol. Res. Lett. **81**(2), 195–206 (2010)

Al-Ajamee, M., Mahmoud, M.M.M., El Sharief, A.M.: Site-specific seismic ground response analysis for typical soil sites in Central Khartoum, Sudan. In: Sitharam, T.G., Kolathayar, S., Jakka, R. (eds.) Earthquake Geotechnics. LNCE, vol. 187, pp. 529–546. Springer, Singapore (2022). https://doi.org/10.1007/978-981-16-5669-9_43

Al-Ajamee, M., Mahmoud, M.M.M., Ali, A.M.: Khartoum Geohazard: an assessment and a future warning. In: Adhikari, B.R., Kolathayar, S. (eds.) Geohazard Mitigation. LNCE, vol. 192, pp. 87–98. Springer, Singapore (2022). https://doi.org/10.1007/978-981-16-6140-2_8

Atalay, A.: Probabilistic seismic hazard analysis (PSHA) for Ethiopia and the neighboring region. J. Afr. Earth Sc. **134**, 257–267 (2017)

Atkinson, G.M., Boore, D.M.: Earthquake ground-motion prediction equations for eastern North America. Bull. Seismol. Soc. Am. **96**(6), 2181–2205 (2006)

Atalay, A.: Dams in Stable Continental Interiors: The Case for the Great Ethiopian Renaissance Dam (2021). www.hornafricainsight.org. Accessed 6 June. https://www.hornafricainsight.org/post/gerd-on-seismically-stable-ground

Baboo, A., Al-Ajamee, M., Kolathayar, S.: Regional probabilistic seismic hazard assessment of the grand Ethiopian renaissance dam and its vicinity. J. Afr. Earth Sci. (2022)

Baker, W.J.: An introduction to probabilistic seismic hazard assessment (PSHA). US Nuclear Regulatory Commission (2008)

Chu, D., Gordon, R.G.: Current plate motions across the Red Sea. Geophys. J. Int. **135**(2), 313–328 (1988)

Delavaud, E., Scherbaum, F., Kuehn, N., Riggelsen, C.: Information-theoretic selection of ground-motion prediction equations for seismic hazard analysis: an applicability study using Californian data. Bull. Seismol. Soc. Am. **99**(6), 3248–3263 (2009)

Douglas, J.: Ground motion prediction equations 1964–2020. Department of Civil Environment Engineering, University of Strathclyde, Glaskow, UK (2018)

Goitom, B., et al.: Probabilistic seismic-hazard assessment for Eritrea. Bull. Seismol. Soc. Am. **107**(3), 1478–1494 (2017)

Graizer, V.: Ground-motion prediction equations for Central and Eastern North America. Bull. Seismol. Soc. Am. **106**(4), 1600–1612 (2016)

Hanks, T.C., Kanamori, H.: A moment magnitude scale. J. Geophys. Res. **84**, 3 (1979)

Kolathayar, S., Sitharam, T.G.: Comprehensive probabilistic seismic hazard analysis of the Andaman-Nicobar regions. Bull. Seismol. Soc. Am. **102**(5), 2063–2076 (2012)

Kolathayar, S., Sitharam, T.G., Vipin, K.S.: Deterministic seismic hazard macrozonation of India. J. Earth Syst. Sci. **1351–1364**, 13–14 (2012)

Ogubazghi, G., Ghebreab, W., Havskov, J.: Some features of the 1993 Bada earthquake swarm of southeastern Eritrea. J. Afr. Earth Sc. **2**(38), 135–143 (2004)

Scordilis, E.M.: Empirical global relations converting $\mathbf{M_S}$ and $\mathbf{m_b}$ to moment magnitude. J. Seismol. **10**, 225–236 (2006)

USGS. 2021. U.S. Geological Survey. https://www.usgs.gov/. Accessed 28 Sept 2021

Zhuang, J., Ogata, Y., Vere-Jones, D.: Stochastic declustering of space-time earthquake occurrences. J. Am. Stat. Assoc. **97**, 369–380 (2013)

Parametric Study of Seismic Slope Stability of Tailings Dam

T. S. Aswathi(✉) and Ravi S. Jakka

Indian Institute of Technology, Roorkee, Uttarakhand 247667, India
as@eq.iitr.ac.in

Abstract. Tailings are the byproduct of the mining industry. The minerals mined by these mining industries are only about 3 to 5% pure, so the rest, 97 to 95%, become the tailings. It is being said that the increasing population has increased the demand for minerals for various uses. This demand produces a massive volume of tailings that, when disposed of inadequately, causes several failures and has cost lives in some cases. Therefore, a proper study of the design and safety of these structures is needed. This study aims to understand the stability variation of different tailings construction methods with different seismic loads. In this paper, the seismic slope stability of tailings dam is done by varying the slope angle of each dyke and material properties for three methods of construction – upstream, centerline, and downstream method of construction under different seismic loading. The numerical study is done in GeoStudio software.

Keywords: Seismic slope stability · Numerical analysis · Tailings dam

1 Introduction

In the mining industry, tailings are residue after extracting minerals from the mineral ore. In most cases, the mineral extracted is only 1 to 3% of the mineral ore; the rest is tailings deposited or pumped to a site nearby in a slurry form. Tailings contain water, rock fragments of varying sizes, metal (in trace quantities), and other chemicals used in ore processing. The most common disposing or distributing methods of tailings are subaqueous discharge, subaerial discharge, and thickened discharge. The cycloning principle separates the whole tailings slurry to coarse sand fraction called tailing sand and fine fraction with water called slimes. These slimes are impounded by the raised embankment of dyke using the tailing sand. These raised embankments are primarily built by the following three methods: upstream, downstream, or centerline. (Vick [11]). Among these three, the downstream method of construction is most stable. (Vick [11]; Psarropoulos and Tsompanakis [6] and Jakka [5]).

The literature has reported that there are 18000 tailings dams across the globe, out of which approximately 3500 are active. The primary difficulty in handling tailings dam is its instability during and after the mining operations. Before the year 2000, a total of 198 tailings dam failures were reported (Rio et al. [10]). Between the year 2000 - 2010 (Azam and Li [1]) and 2010 – 2021 (WISE [12]), a total of 20 and 36 tailings dam

© The Author(s), under exclusive license to Springer Nature Switzerland AG 2022
L. Wang et al. (Eds.): PBD-IV 2022, GGEE 52, pp. 1914–1928, 2022.
https://doi.org/10.1007/978-3-031-11898-2_175

failure has been reported. The failure of the water retention dam (0.01%) is lower than the failure of the tailings dam (1.2%) (ICOLD [7]; Azam and Li [1]). The high failure rate of tailing dams has led to increasing awareness of the need for enhanced safety in the design and operation of tailing dams. Higher the tailings dam height higher the risk of failure (Klohn [8]; Azam and Li [1]; Davis [2]; Psarropoulos and Tsompanakis [6]; Ferdosi et al. [3, 4]).

In this paper, an attempt is made to study the effect of strength of foundation, tailings sand, and slimes on the factor of safety on three raised embankment types under static and dynamic conditions. The dynamic study is done by pseudo-static analysis considering that the tailings are constructed in India considering all four zones: zone 2, 3, 4 and 5. The parametric study is carried out with four different foundations, slimes, and tailings sand materials. Thus, Combination and permutation give a total of 64 cases. Each case is designated by nomenclature slime material - tailings material - foundation material. For example, S1_T1_F1 means that slime material is S1, tailings sand T1, and Foundation F1 is used in the case.

2 Methodology

This paper discusses the static and pseudo-static factors of safety of tailings dam constructed by upstream, downstream, and centerline methods. The geometry of these three tailings construction is shown in Fig. 1, 2, and 3. For this study, the construction of 3 types of the tailings dam is carried out with individual dyke of height 5 m, slope 4H:1V, and crest width 3 m to have a total height of 50 m. The analysis is carried out using GeoStudio software. The in-situ stresses are determined using an in-situ model of the QUAKE/W module. These results are then imported to the SLOPE/W module to find the static and pseudo-static factors of safety of the downstream side of the tailings dam in all three cases. The FEM static factor of safety (FOS_{static}) is carried out, and pseudo-static factor of safety (FOS_{ps}) is carried out by LEM Morgenstern – price method using effective stress parameters.

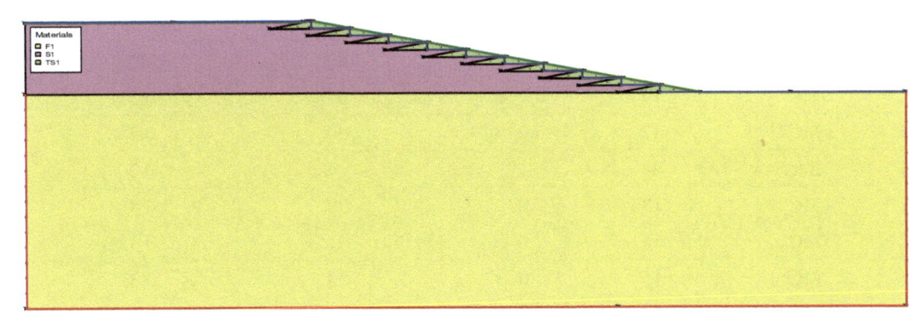

Fig. 1. The geometry of upstream method of construction with drainage condition

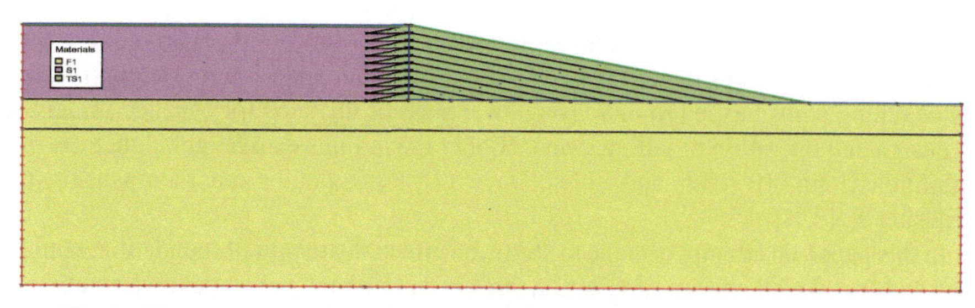

Fig. 2. The geometry of centerline method of construction with drainage condition

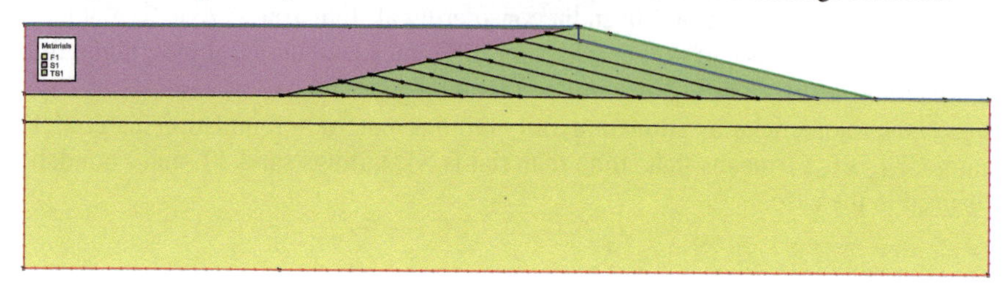

Fig. 3. The geometry of downstream method of construction with drainage condition

For the parametric study, the material properties are taken as Table 1 and pseudo-static coefficient as Table 2. Four sets of values are considered for the material properties of tailings sand (T1, T2, T3, T4), slimes (S1, S2, S3, S4), and foundation (F1, F2, F3, F4). Therefore, there are 64 cases for each tailings dam type. The horizontal seismic coefficient is calculated per IITKGSDMA [9] by considering the importance factor, I, as 1.5; site amplification factor S1 for foundations F1, F2, and F4 whereas S2 for foundation F3. The vertical seismic coefficient is neglected as its effect on the factor of safety is negligible.

Table 1. Material properties considered in the analysis

Material	Material model	Unit weight $\gamma\left(kN/m^3\right)$	Cohesion c' (kPa)	Frictional angle (°)	Poisson's ratio
F1	MC	21	0	40	0.28
F2	MC	19	0	35	0.3
F3	MC	17	0	30	0.3
F4	Bedrock – LE	24			0.2
S1	MC	13	0	29	0.33
S2	MC	14	0	31	0.33
S3	MC	15	0	33	0.33
S4	MC	16	0	35	0.33
T1	MC	16.5	0	33	0.3
T2	MC	17.5	0	35	0.3
T3	MC	18.5	0	37	0.3
T4	MC	19.5	0	39	0.3

* MC – Mohr-Coulomb and LE – Linearly Elastic

Table 2. Horizontal seismic coefficient

Zone	Zone factor, Z	Site amplification factor		Importance factor, I	Horizontal seismic coefficient α_H considering site factor	
		S1	S2		S1	S2
2	0.1g	1	2	1.5	0.05g	0.1g
3	0.16g	1	1.5	1.5	0.08g	0.12g
4	0.24g	1	1.2	1.5	0.12g	0.144g
5	0.36g	1	1.0	1.5	0.18g	0.18g

3 Result and Discussion

3.1 Effect of Foundation

To study the effect of foundation on the factor of safety of tailings dam, the cases with constant tailings sand for dyke and slimes are considered. In this study, the order of strength of foundation considered is F4 > F1 > F2 > F3. The FOS_{static} and FOS_{ps} follows the order of strength of the foundation in all the three cases names upstream, centerline, and downstream methods of construction of the tailings dam. The variation of FOS_{ps} is shown in Fig. 4, 5, and 6 for the upstream, centerline, and downstream tailings method in zone 2. Similar charts are also drawn for static and other pseudo-static cases.

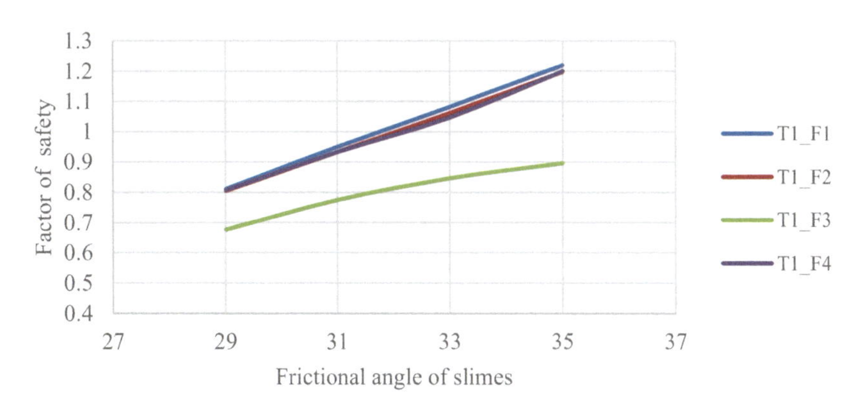

Fig. 4. The factor of safety of upstream tailings dam with constant tailings in zone 2

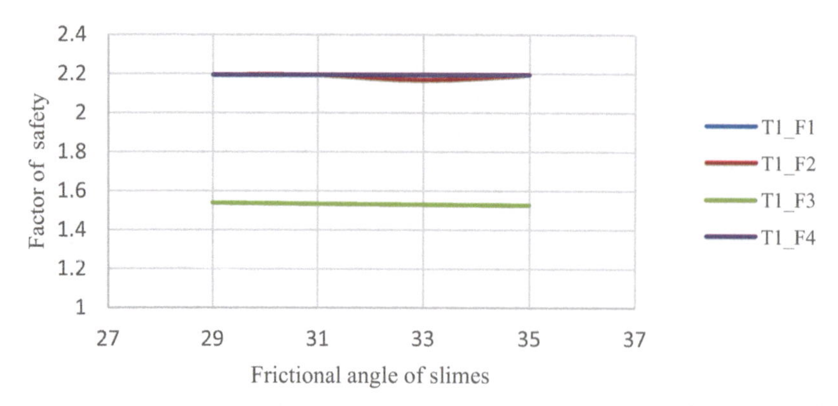

Fig. 5. The factor of safety of centerline tailings dam with constant tailings in zone 2

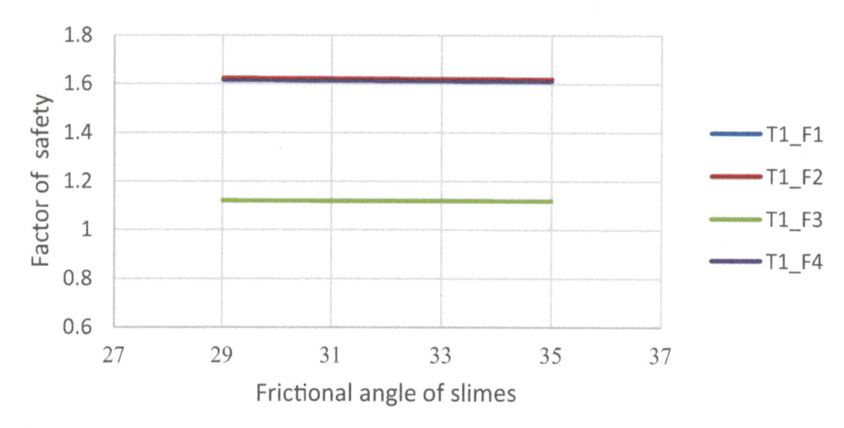

Fig. 6. The factor of safety of downstream tailings dam with constant tailings in zone 2

The slip surface changes from the shallow slope failure for F4, F1, and F2 to a deep base failure for F3, as shown in Figs. 8 and 9 in the case of centerline and downstream construction methods. However, there is not much change in slip surface in the case upstream method of construction constructed on the different foundation (Fig. 7).

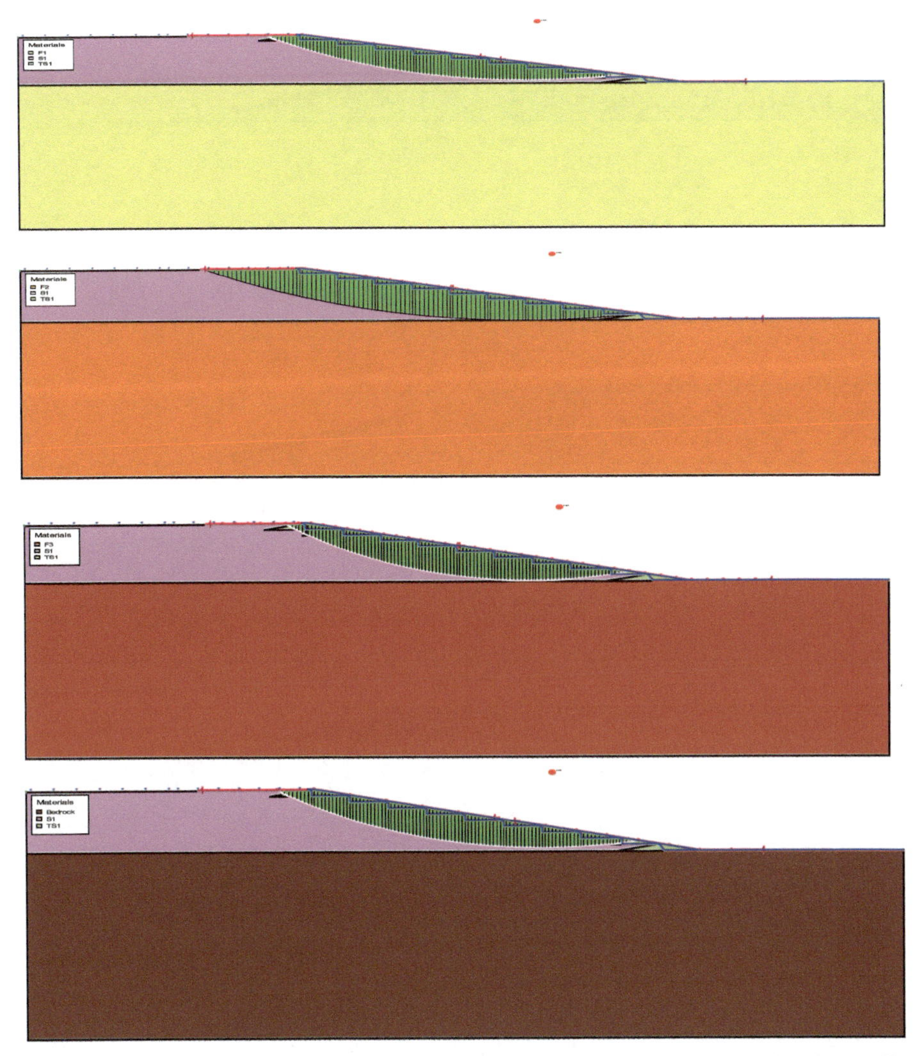

Fig. 7. Slip surface of upstream tailings under static analysis having different foundations F1, F2, F3, and F4 in order. (The results shown are for the cases S1_T1_F1, S1_T1_F2, S1_T1_F3, S1_T1_F4)

3.2 Effect of Tailings

To study the effect of tailings on the factor of safety of the tailings dam, cases with constant foundation and slimes are analyzed, and the results are plotted (Fig. 10, 11, and 12) for pseudo-static analysis in zone 2 of upstream, centerline, and downstream tailings on Foundation F1. Similar graphs are drawn from the rest of the sets. Therefore, the factor of safety is in the order of strength of the tailings for all the three types: upstream,

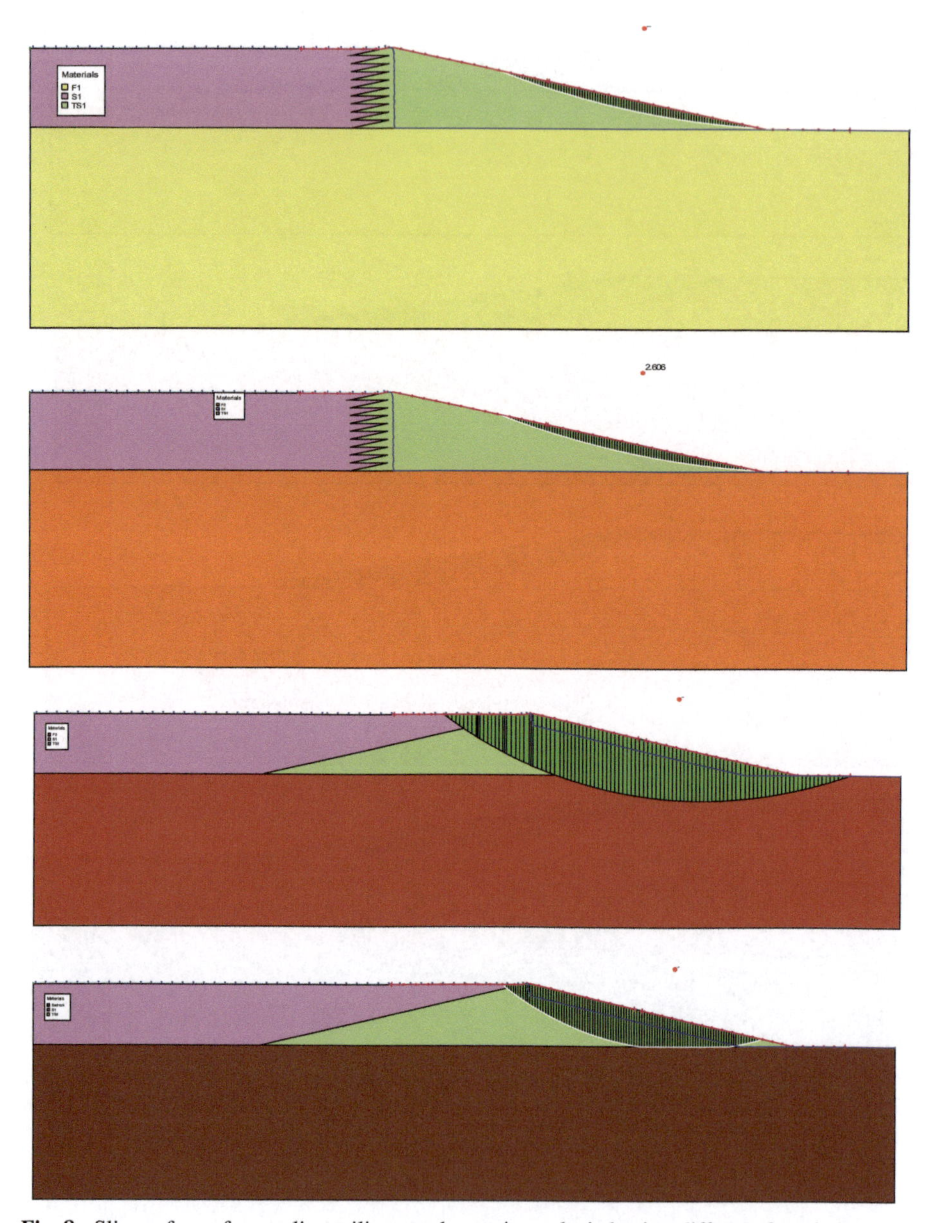

Fig. 8. Slip surface of centerline tailings under static analysis having different foundations F1, F2, F3, and F4 in order. (The results shown are for the cases S1_T1_F1, S1_T1_F2, S1_T1_F3, S1_T1_F4)

downstream, and centerline for both static and pseudo-static cases. The critical slip surface shape remains the same for the same foundation combination cases (Fig. 13).

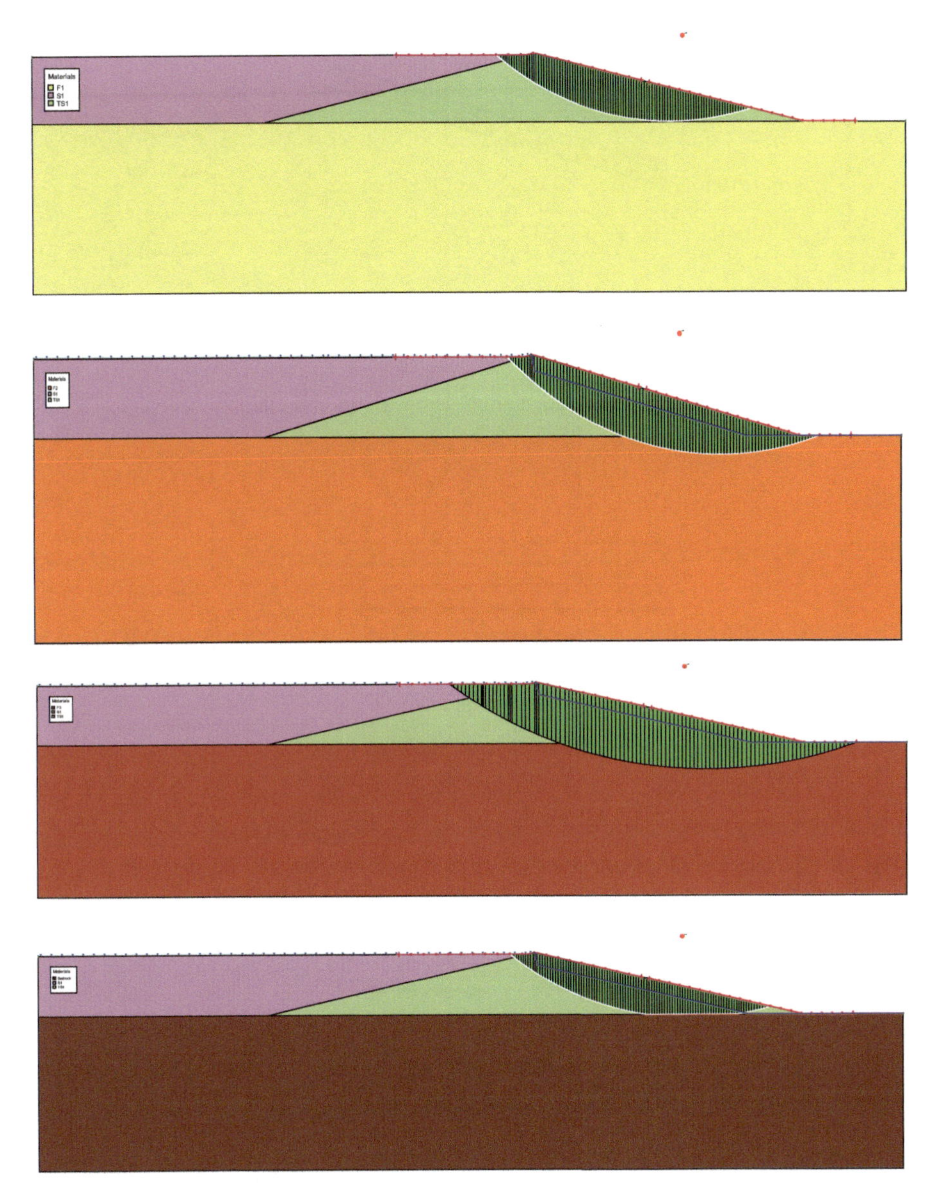

Fig. 9. Slip surface of downstream tailings under static analysis having different foundations F1, F2, F3, and F4 in order. (The results shown are for the cases S1_T1_F1, S1_T1_F2, S1_T1_F3, S1_T1_F4)

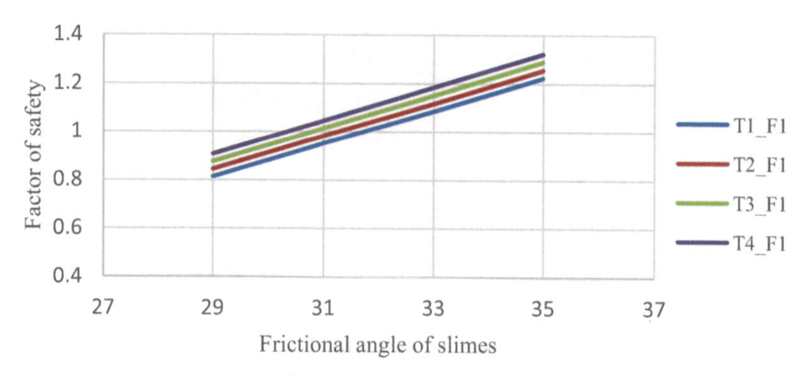

Fig. 10. The factor of safety upstream tailings dam with the constant foundation in zone 2

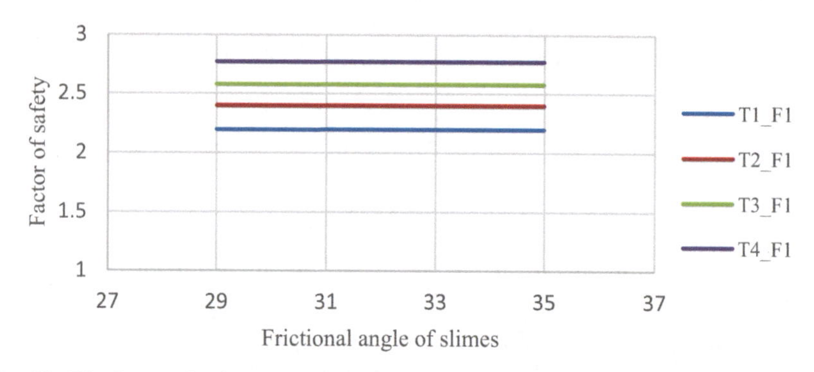

Fig. 11. The factor of safety centerline tailings dam with the constant foundation in zone 2

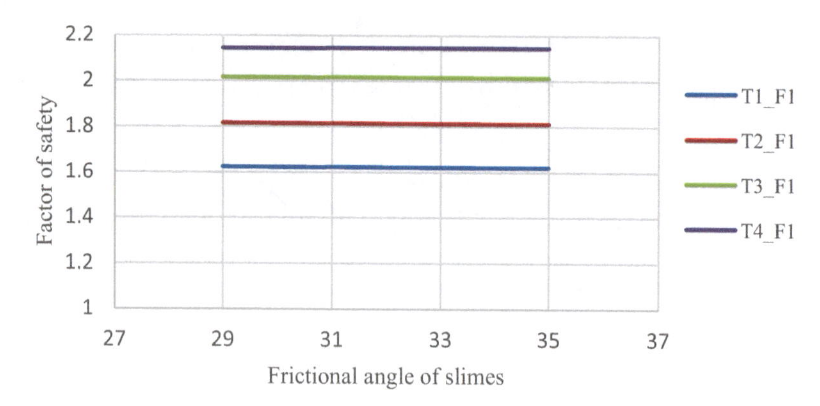

Fig. 12. The factor of safety downstream tailings dam with the constant foundation in zone 2

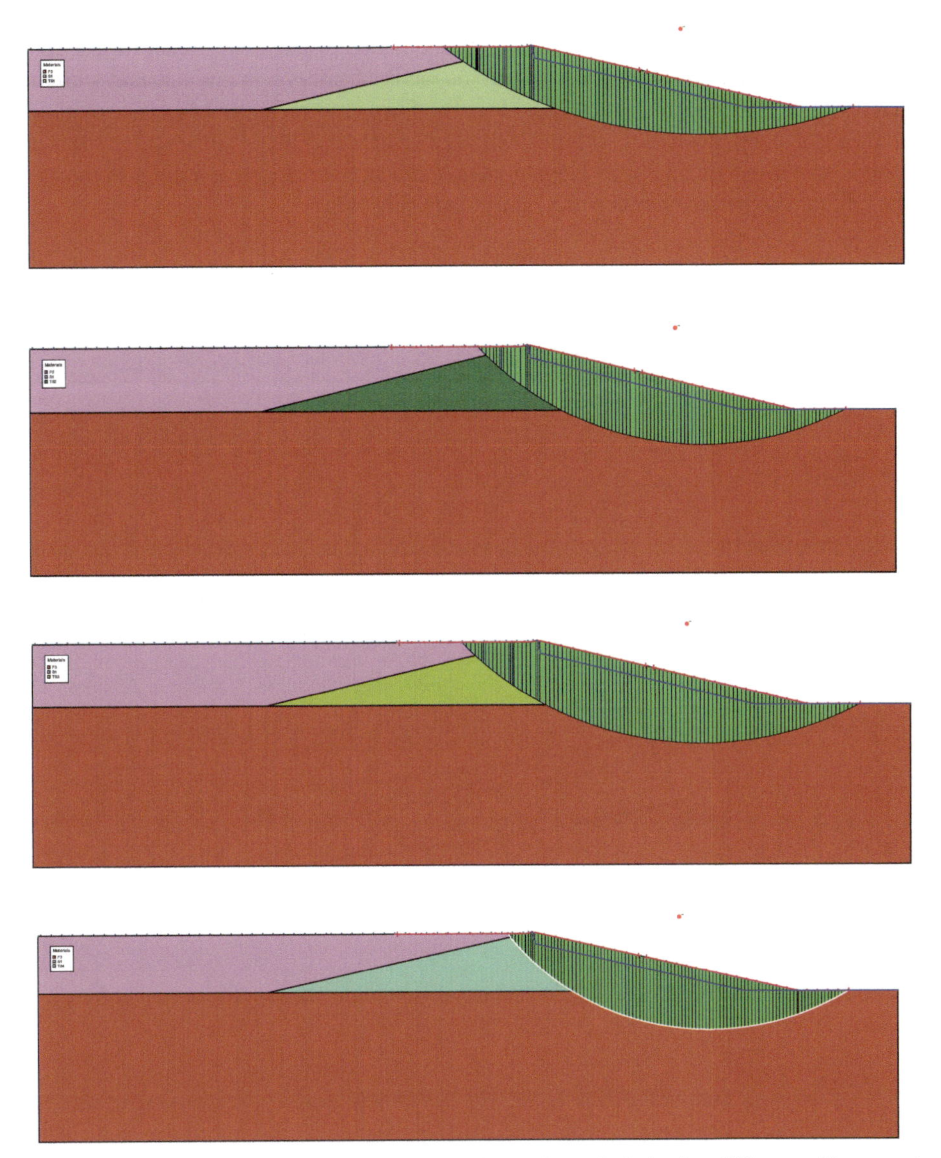

Fig. 13. Slip surface of downstream tailings under static analysis having different tailings sand TS1, TS2, TS3, and TS4 in order. (The results shown are for the cases S1_T1_F3, S1_T2_F3, S1_T3_F3, S1_T4_F4)

3.3 Effect of Slimes

To study the effect of slimes on the factor of safety of tailings dam, cases are grouped having the same foundation and tailings sand for dyke, and graphs are plotted for the factor of safety variation with frictional angle of slimes. Figure 14 shows that the factor of safety increases as the angle of slimes increases in the case of upstream for pseudo-static in zone 2. In the case of downstream, the factor of safety remains constant with the increase in the frictional angle of slimes (Fig. 16). Furthermore, for the centerline method, the factor of safety decreases with the increased frictional angle of slimes (Fig. 15). The same is true for static and other pseudo-static cases in the respective tailings dam type. The critical slip surface remains the same shape (Fig. 17).

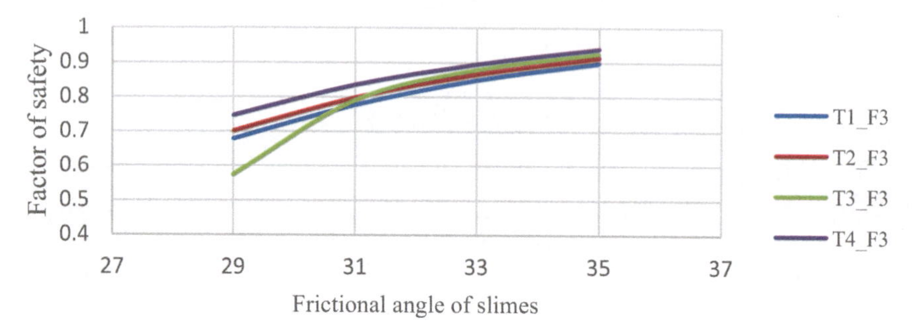

Fig. 14. The factor of safety upstream tailings dam with the constant foundation in zone 2

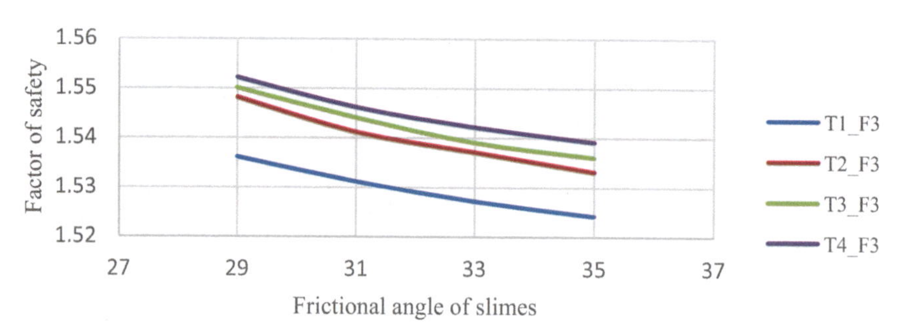

Fig. 15. The factor of safety centerline tailings dam with the constant foundation in zone 2

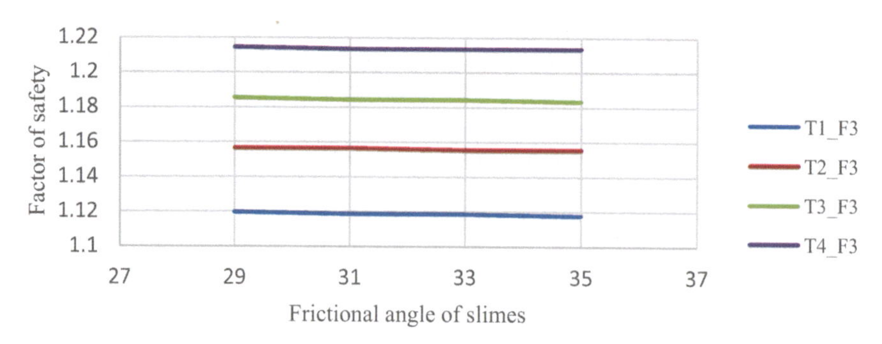

Fig. 16. The factor of safety downstream tailings dam with the constant foundation in zone 2

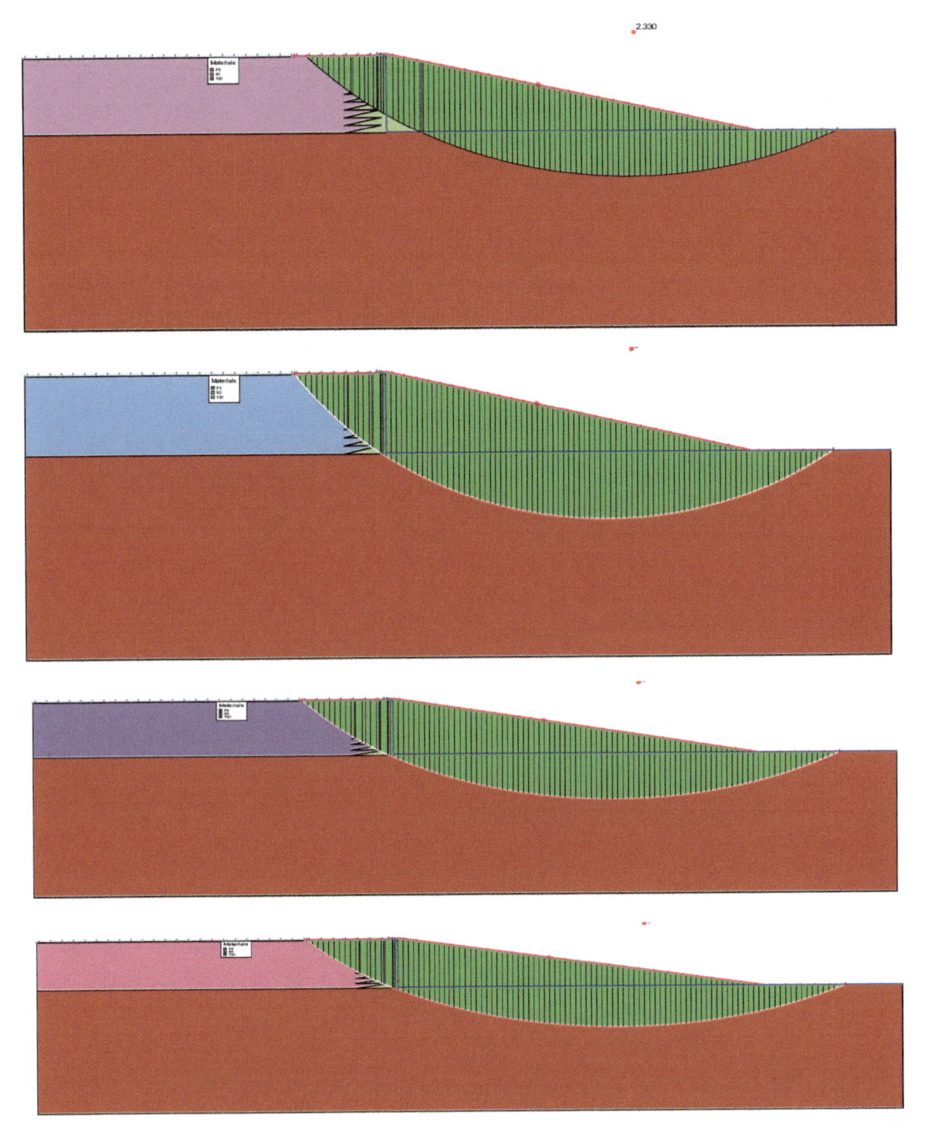

Fig. 17. Slip surface of centerline tailings under static analysis having different slimes S1, S2, S3, and S4 in order. (The results shown are for the cases S1_T1_F3, S2_T1_F3, S3_T1_F3, S4_T1_F3)

3.4 Construction Method

Comparing factors of safety between the three methods of construction, Fig. 21, it can be said that the centerline method is the safest and upstream method least safe for both static and pseudo-static cases. This difference in the order that the centerline method has more safety than downstream is because of the drainage condition considered in this study. Usually, the downstream method gets inclined drainage towards the upstream side of the dyke. The upstream tailings dam is unsafe, especially in the active seismic area. From Fig. 18, 19 and 20, it can be said that the factor of safety decreases with an increase in seismic loading, and the trend of the safety remains the same with respect to the slimes, tailings sand, and foundation combination.

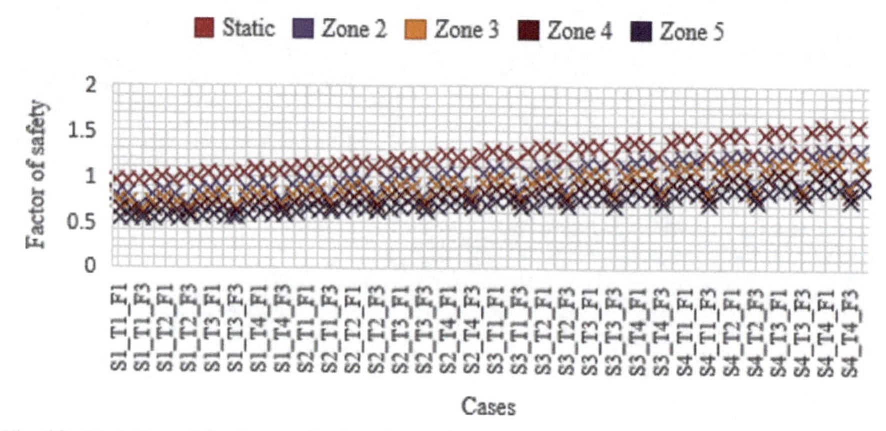

Fig. 18. Variation of the factor of safety for static and pseudo-static cases for zone 2, 3, 4, and 5 for upstream tailings dam for 64 cases.

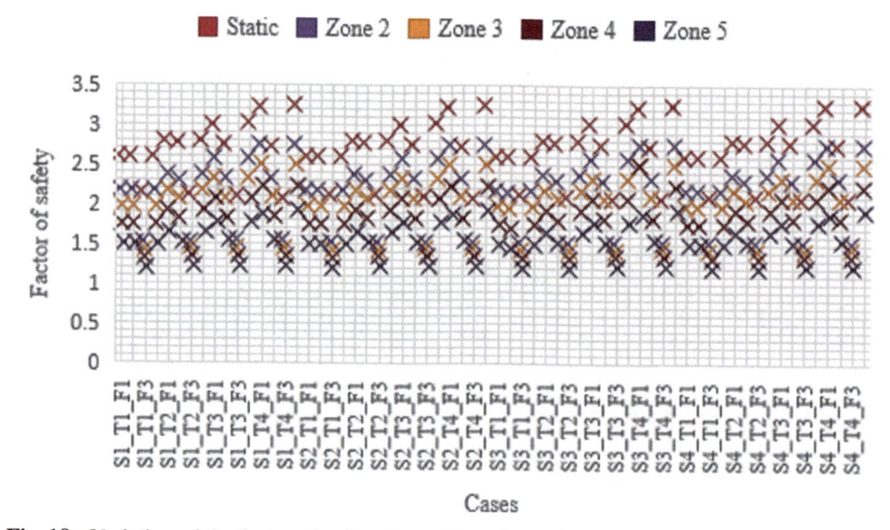

Fig. 19. Variation of the factor of safety for static and pseudo-static cases for zone 2, 3, 4, and 5 for centerline tailings dam for 64 cases.

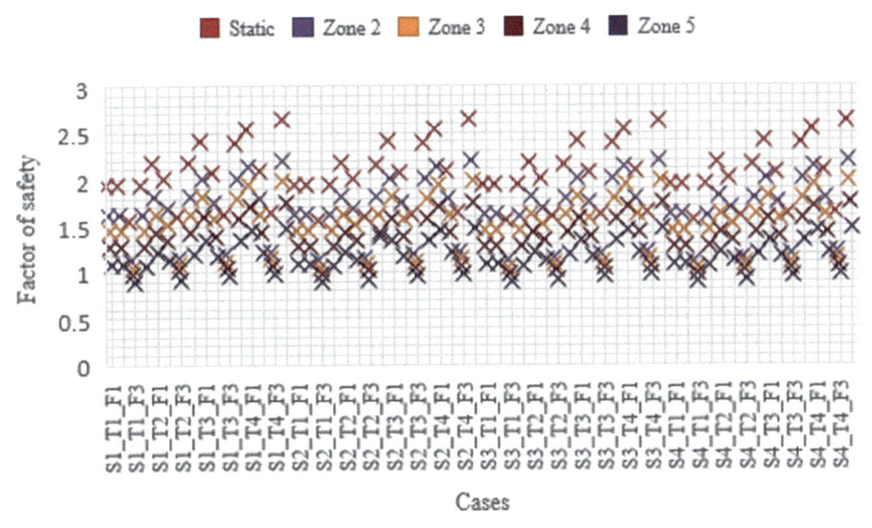

Fig. 20. Variation of the factor of safety for static and pseudo-static cases for zone 2, 3, 4, and 5 for downstream tailings dam for 64 cases.

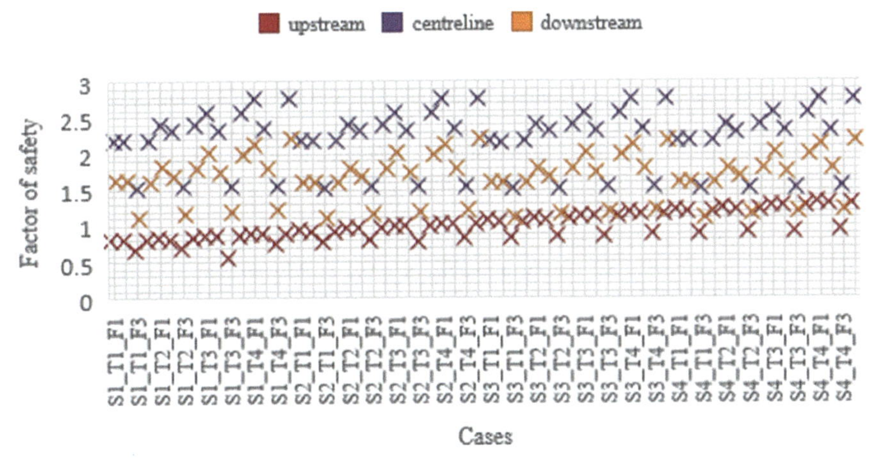

Fig. 21. Variation of the factor of safety for pseudo-static zone 2 for upstream, centerline, and downstream tailings dam for 64 cases.

4 Conclusions

In this paper, a parametric study is carried out with four sets of material properties for foundation, tailings sand, and slimes (64 cases) for three raised embankment construction methods, namely upstream, downstream, and centerline methods of construction with the same dike geometry for constructing a total height of 50 m height tailings dam. The following conclusions are drawn. The upstream method is the most unsafe construction in seismic conditions. According to this study, the centerline construction method is most stable, but this is due to the drainage condition considered in downstream construction.

The foundation effect is linear and more profound, followed by tailings sand and finally slimes. The effect of slimes has a nonlinear relationship to safety in the upstream and centerline construction method. However, there is no effect of slimes in downstream tailings. The factor of safety decreases with the increase in seismic loading irrespective of the tailings type.

References

1. Azam, S., Li, Q.: Tailings dam failures: a review of the last one hundred years. Geotech. News **28**(4), 50–54 (2010)
2. Davies, M.P.: Tailing's impoundment failures Are geotechnical engineers listening? Geotech. News-Vancouver **20**(3), 31–36 (2002)
3. Ferdosi, B., James, M., Aubertin, M.: Investigation of the effect of waste rock inclusions configuration on the seismic performance of a tailings impoundment. Geotech. Geol. Eng. **33**(6), 1519–1537 (2015)
4. Ferdosi, B., James, M., Aubertin, M.: Numerical simulations of seismic and post-seismic behavior of tailings. Can. Geotech. J. **53**(1), 85–92 (2015)
5. Jakka, R.S., Ramana, G.V., Datta, M.: Seismic slope stability of embankments constructed with pond ash. Geotech. Geol. Eng. **29**(5), 821–835 (2011)
6. Psarropoulos, P.N., Tsompanakis, Y.: Stability of tailings dams under static and seismic loading. Can. Geotech. J. **45**(5), 663–675 (2008)
7. ICOLD U. Tailings dams–risk of dangerous occurrences, lessons learnt from practical experiences (bulletin 121). Commission Internationale des Grands Barrages, Paris (2001)
8. Klohn, E.J.: Tailings dams in Canada. Geotechnical News 117-23 (1997)
9. Roy, D., Dayal, U., Jain, S.K.: IITKGSDMA guidelines for seismic design of earth dams and embankments, provisions with commentary and explanatory examples. Indian Institute of Technology Kanpur (2007)
10. Rico, M., Benito, G., Salgueiro, A.R., Díez-Herrero, A., Pereira, H.G.: Reported tailings dam failures: a review of the European incidents in the worldwide context. J. Hazard. Mater. **152**(2), 846–852 (2008)
11. Vick, S.G.: Planning, Design, and Analysis of Tailings Dams. Wiley, New York (1983)
12. WISE (World Information Service on Energy Uranium Project). Chronology of major tailings dam failures. https://www.wise-uranium.org/mdaf.html. Accessed 16 Dec 2021

Seismic Performance of a Zoned Earth Dam

Orazio Casablanca, Andrea Nardo, Giovanni Biondi, Giuseppe Di Filippo, and Ernesto Cascone(✉)

Department of Engineering, University of Messina, 98166 Messina, Contrada di Dio, Italy
ecascone@unime.it

Abstract. 2D dynamic numerical analyses of a zoned earth dam located in a high seismic hazard area of Southern Italy have been carried out. In the analyses soil behaviour has been described through an isotropic hardening elasto-plastic hysteretic model. A set of 47 real accelerograms recorded on rock outcrop has been used as input motion. The activation of plastic mechanisms induced by the seismic loading and the amplification of the horizontal accelerations in the dam are investigated. Also, assuming that the vertical settlement w_c of the crest of the dam can be regarded as an index of the seismic performance of the dam, the effects of various parameters of the input motion on w_c have been examined.

Keywords: Dynamic analysis · Earth dam · Seismic performance

1 Introduction

The seismic stability of existing earth dams located in zones characterized by moderate to high seismicity is a critical issue for the safety of the downstream areas that could be jeopardized by potential sudden flooding. In fact, earthquake shaking causes additional loads and possibly excess pore water pressures to the dam that can lead to large permanent displacements (e.g. [1]) and even failure (e.g. [2]). In this regard, the selection and modeling of the seismic excitation for engineering applications play a relevant role [3, 4].

It is usual to assess the seismic performance of earth dams through the value of the maximum crest settlement w_c [5, 6], since it is considered a good proxy of the dam response and of the seismic-induced damage; this approach has been recently applied in the seismic evaluation of some Italian earth and rockfill dams [7, 8] also pinpointing the difference in seismic response of homogeneous and zoned earth dams [9].

Performance based design and evaluation of the stability of slopes and earth dams can be carried out according to simplified methods based on Newmark-type displacement analysis [10–13]. Numerical analyses have also been performed focusing on the ground motion characteristics, the effect of the vertical component of ground motion, the bedrock compliance and the excess pore pressure build-up [14–16].

In this article, the seismic performance of an earth dam has been studied focusing on the possible activation of plastic mechanisms and amplification of the horizontal accelerations induced by a large set of earthquake records characterised by seismic parameters (maximum acceleration, predominant frequency, duration, Arias intensity,

© The Author(s), under exclusive license to Springer Nature Switzerland AG 2022
L. Wang et al. (Eds.): PBD-IV 2022, GGEE 52, pp. 1929–1936, 2022.
https://doi.org/10.1007/978-3-031-11898-2_176

number of equivalent loading cycles [17]) varying in wide ranges, with the aim of developing empirical formulas correlating the crest settlement w_c to some relevant seismic parameters.

The San Pietro dam has been considered in the analyses for which the results of an extensive geotechnical investigation were available. The dam is a 49 m high zoned earth dam located in a high seismicity area of Southern Italy. The core is made of low plasticity clayey silts, the shells are made of granular soils and the foundation soil consists of a layer of alluvial gravels overlying a deep stiff overconsolidated flysch deposit (Fig. 1). The seismic performance of the dam was checked through finite element elasto-plastic dynamic analyses with reference to the main cross section of the dam, using the code Plaxis 2D. Dynamic analyses were conducted considering 47 real accelerograms recorded on rock outcrop and arranged into four groups according to the values of the horizontal peak ground acceleration *PGA*:

- A (18 accelerograms) with *PGA* about 0.1g (±5%)
- B (13 accelerograms) with *PGA* about 0.2g (±10%)
- C (9 accelerograms) with *PGA* about 0.3g (±10%)
- D (7 accelerograms) with *PGA* in the range 0.33g–0.67g.

The selected accelerograms were imposed at the base of the numerical model in both directions, this resulting in 94 dynamic analyses.

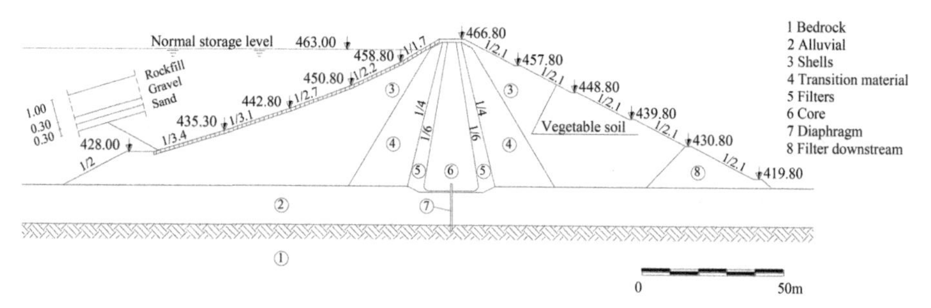

Fig. 1. Main cross section of the San Pietro dam.

2 Numerical Model

The numerical analyses presented in the paper were performed in plane strain condition, discretizing the dam and the foundation soils through a finite element mesh of 5000 triangular 15-noded elements. The model used in the analyses is depicted in Fig. 2. To avoid numerical distortion of the propagating seismic waves, the maximum height of elements was assumed smaller than 1/6 of the wavelength λ_{min} associated with the highest frequency f_{max} of the input motion. The mesh extended laterally to about three times the width of the base of the embankment (where the influence of the earth dam is negligible) and at the bottom to a depth of 25 m. The nonlinear soil behaviour is taken into account through an isotropic hardening elasto-plastic hysteretic model known as Hardening Soil model with small strain stiffness (*HSsmall*), capable to describe the nonlinear behaviour of soil in the small strain range. The model parameters have been calibrated using oedometer tests and drained and undrained compression triaxial tests available for the specific case, while to model the behaviour of the soil in the small strain range the results of two resonant column (RC) tests carried out at confining pressures of 150 and 300 kPa on samples retrieved from the core of the dam (Fig. 3a) have been used.

No cyclic or dynamic tests were performed on the materials of the shells and of the foundation soils during the geotechnical characterization, so in the analyses it was necessary to use data selected from the literature (Fig. 3).

Figure 3 shows only the pre-yield response of the *HSsmall* model used for the calibration of the parameters that describe the small-strain behaviour. However, when irreversible strains become predominant, a decrease of the shear modulus and an increase of the damping ratio more similar to experimental data is obtained.

To reproduce the state of effective stress at the end of the dam construction, a preliminary static analysis was carried out simulating the staged construction of the dam as a drained process, via the progressive activation of 24 layers of mesh elements (each about 2 m thick), the impoundment of the reservoir and, finally, the ensuing steady state seepage flow. The values of the small strain shear modulus G_0 were evaluated starting from the results of the cross-hole tests *CH3*, *CH4* and *CH5* shown in Fig. 4. For the dam body and for the shells, the *CH* data were fitted by a power relationship describing G_0 as a function of the minimum principal effective stress σ'_3 obtained in the static analyses. As shown in Fig. 4, G0 exhibits a small variability with depth in the alluvial deposit; thus, a constant value equal to $G_0 = 1100$ MPa, in the soils directly underlying the dam, and $G_0 = 580$ MPa, in the soils aside the foundation of the dam, were assumed in the dynamic analyses. The larger value assumed for the soils under the dam can be ascribed to the state of stress induced by the presence of the embankment. Finally, for the flysch deposit and the bedrock, a constant value of G_0 equal to 1100 MPa and 1900 MPa respectively, were considered (Fig. 4). To simulate radiation damping, a compliant base was introduced at the bottom of the mesh, while free-field boundaries were applied at lateral sides. Finally, for all the materials a viscous Rayleigh damping was introduced assuming $f_1 = 1$ Hz and $f_2 = 12$ Hz as control frequencies, and $\xi_1 = \xi_2 = 1.5\%$, to account for damping at small shear strain levels.

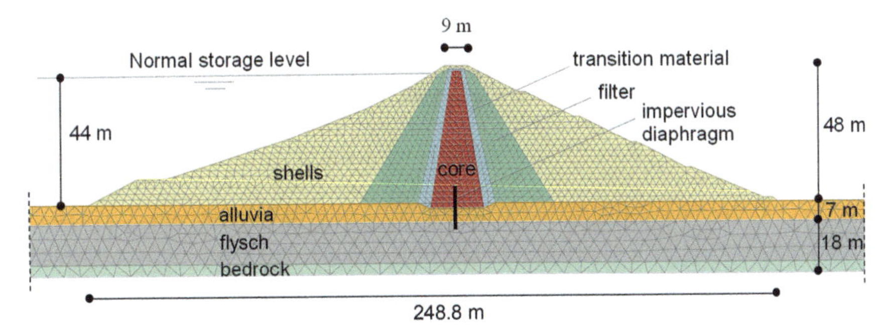

Fig. 2. Finite element numerical model adopted in the analyses.

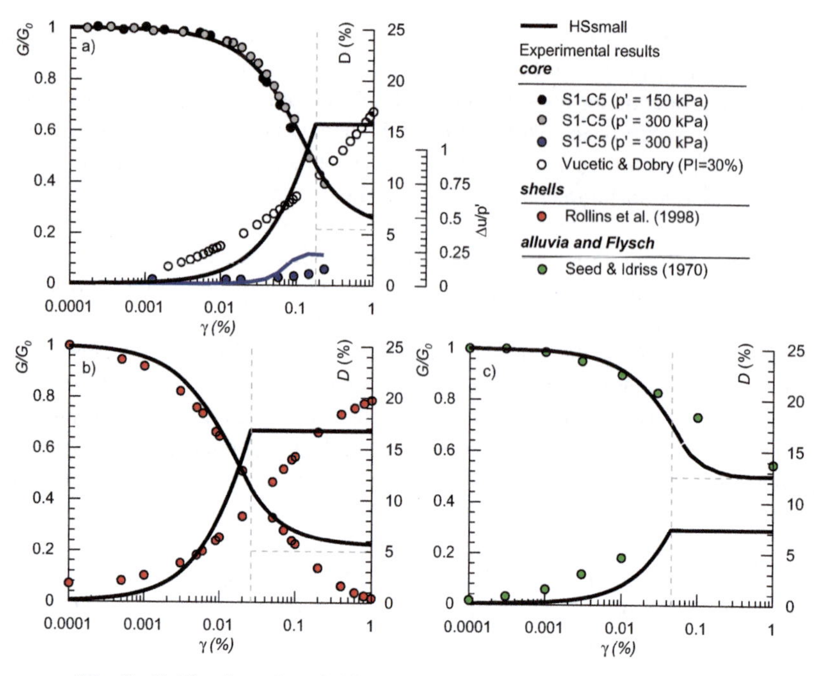

Fig. 3. Calibration of model for the core, the shells, alluvia and flysch.

3 Analysis Results

Some of the results of the dynamic analyses are shown in Fig. 5 in terms of contours of deviatoric strain as obtained using the records with $PGA = 0.39$g and 0.43g, which resulted the most severe for the group *D* of input motions.

For the cases of Fig. 5, as for all the other analyses carried out with the whole set of input motions, deviatoric strains are concentrated in shallow portions of the dam, the largest strains generally affecting the upstream shell, without intersecting the core. The maximum value of the crest settlement w_c is about 37 cm and is much smaller than the service freeboard of the dam. It was obtained for the record with lower *PGA* (Fig. 5a, $PGA = 0.39$g), characterised by a large energy content.

Figure 6 shows the average profile of the maximum horizontal acceleration ($a_{h,max}$) along the axis of the dam, normalized to the peak horizontal acceleration computed at its base ($a_{h,base}$), for all the considered groups of input motions; the shaded area represents the envelope of the normalized acceleration profiles. It is noted that small values of the ratio $a_{h,max}/a_{h,base}$, lower than 1.7, are generally observed in the average profiles obtained for the accelerograms of the *C* and *D* groups (Fig. 6c, d), whereas larger amplification ratios, up to 3, were evaluated for the *A* and *B* groups (Fig. 6a, b), revealing, as expected, smaller amplifications associated to stronger motions, capable to activate the non-linear response of the dam.

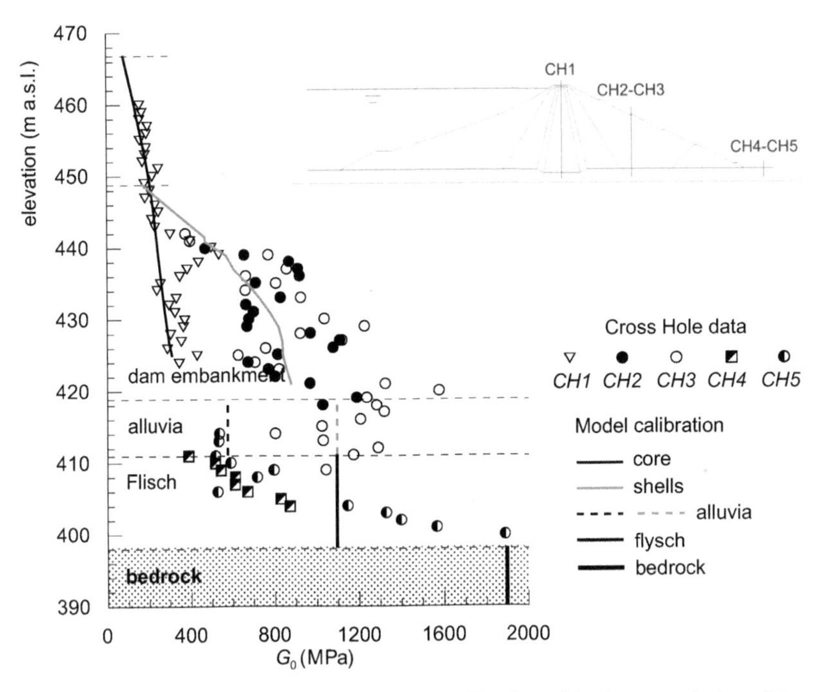

Fig. 4. Profiles of small strain shear modulus G_0 adopted in the numerical model.

The dam crest settlement w_c, is plotted in Fig. 7 versus the period ratios T_0/T_m (Fig. 7a) and T_0/T_p (Fig. 7b), T_0, T_m and T_p being the natural period of elastic horizontal vibration of the dam and the mean and predominant period of the input motion, respectively; w_c, is plotted also against the Arias intensity of the input motion I_a (Fig. 7c) and the product $I_a \cdot T_m$ (Fig. 7d). Figure 7a shows that the highest crest settlements occur for values of the period ratio T_0/T_m close to unity for which remarkable amplification phenomena occur; conversely larger values of the ratio T_0/T_p correspond to the highest crest settlements. This result is consistent with literature where T_m is generally recognized as a more robust indicator of the frequency content than T_p.

From Figs. 7c and 7d it is apparent that the dam crest settlement w_c of the San Pietro dam is well correlated to I_a and to the values of the product $I_a \cdot T_m$. Preliminary analyses revealed also that the product $I_a \cdot T_p$ does not represent a reliable predictive variable. By regression of the data relative to all groups of accelerograms, empirical best-fit formulas were developed for an approximate prediction of the expected maximum crest settlement w_c. A linear equation describes the relationship w_c-I_a though a slight scattering can be observed for values of I_a larger than about 0.75 m/s (Fig. 7c); a better approximation is attained when the product $I_a \cdot T_m$, accounting for both energy and frequency content of the input motion, is assumed as an independent variable and a grade 2 polynomial equation is used to fit numerical results (Fig. 7d). The two correlations are shown as dashed lines in Figs. 7c and 7d and are given in the following equation together with the corresponding regression coefficient R^2:

$$w_c(\text{cm}) = 19{,}56 \cdot I_a, \quad R^2 = 0.93 \tag{1}$$

$$w_c(\text{cm}) = -23{,}642 \times (I_a \cdot T_m)^2 + 63{,}242 \cdot (I_a \cdot T_m), \quad R^2 = 0.98 \tag{2}$$

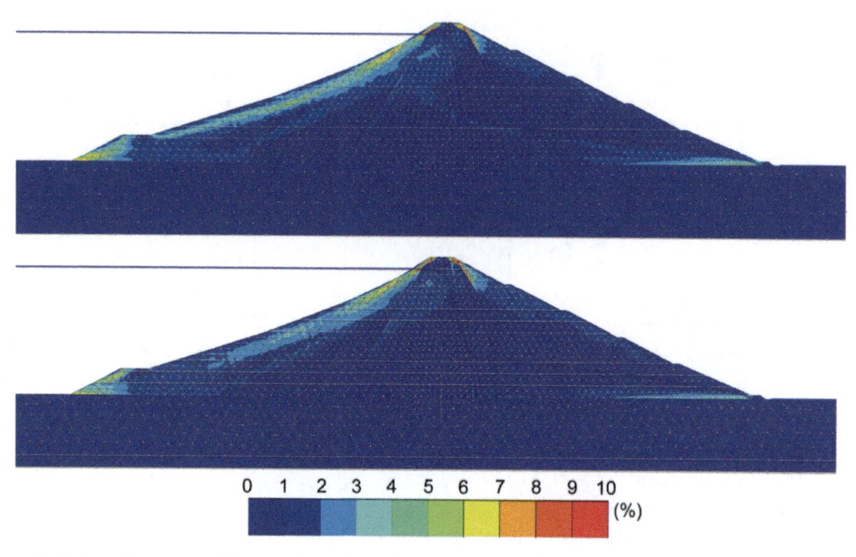

Fig. 5. Contours of the deviatoric strains relative to the most intense input motions.

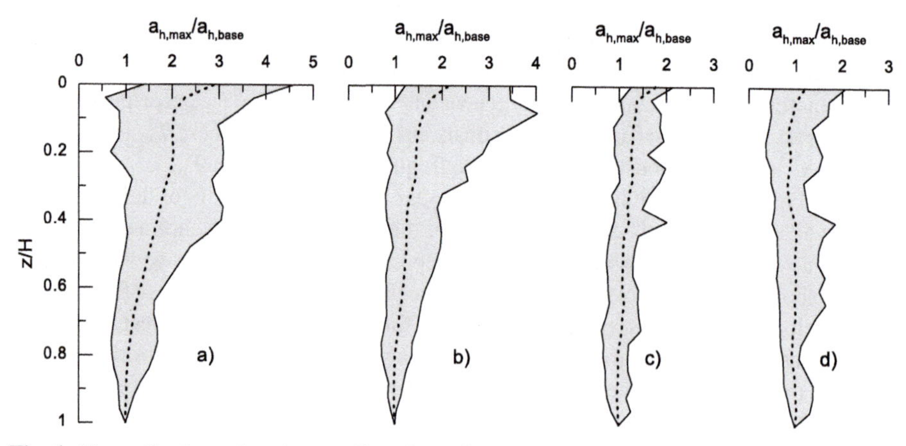

Fig. 6. Normalized acceleration profiles along the dam axis obtained for each group of input motions: a) group A, b) group B, c) group C, d) group D.

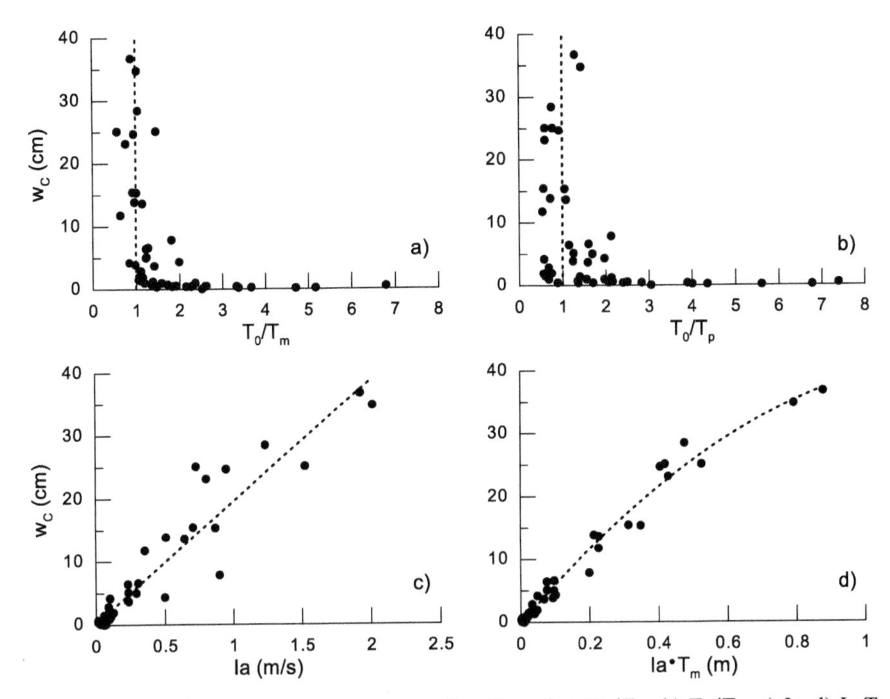

Fig. 7. Maximum dam crest settlement w_c as a function of: a) T_0/T_m, b) T_0/T_p, c) Ia, d) $I_a{\cdot}T_m$.

4 Conclusions

The seismic performance of a zoned earth dam has been studied through 2D finite element analyses and using the Hardening Soil model with small strain stiffness, capable to describe the nonlinear behaviour of soil in the small strain range. 94 dynamic analyses have been carried out using 47 real accelerograms recorded on rock outcrop. The activation of plastic mechanisms induced by the seismic loading has been investigated examining the distribution of deviatoric strains at the end of shaking. The results point out that strains mostly affect shallow portions of the upstream shell of the dam. Amplification of the horizontal accelerations in the dam showed that stronger seismic excitations lead to smaller amplifications in the dam body due to the activation of the nonlinear response of the soils constituting the dam.

Finally, assuming that the crest settlement w_c can be regarded as an index of the seismic performance of the dam, empirical equations relating w_c to some parameters of the input motion have been developed that can be used to predict a possible expected seismic-induced damage in the dam.

Acknowledgements. This work is part of the research activities carried out in the framework of the research project of major national interest, PRIN n. 2017YPMBWJ, on "Risk assessment of Earth Dams and River Embankments to Earthquakes and Floods (REDREEF)" funded by the Italian Ministry of Education University and Research (MIUR).

References

1. Catalano, A., Caruana, R., Del Gizzi, F., De Sortis, A.: Observed behaviour of Italian dams under historical earthquakes. In: 9th ICOLD European Club Symposium, Venice (2013)
2. Wieland, M., Chen, H.: Lessons learnt from the Wenchuan earthquake. Int. J. Water Power Dam Construct., 36–40 (2009)
3. Muscolino, G., Genovese, F., Biondi, G., Cascone, E.: Generation of fully non-stationary random processes consistent with target seismic accelerograms. Soil Dyn. Earthq. Eng. **141**, 106467 (2021)
4. Genovese, F.: Influence of soil non-linear behaviour on the selection of input motion for dynamic geotechnical analysis. In: Barla, M., Di Donna, A., Sterpi, D. (eds.) IACMAG 2021. LNCE, vol. 126, pp. 588–596. Springer, Cham (2021). https://doi.org/10.1007/978-3-030-64518-2_69
5. Swaisgood, J.: Embankment dam deformations caused by earthquakes. In: Proceedings of the 2003 Pacific Conference on Earthquake Engineering, Christchurch, New Zealand. New Zealand Society for Earthquake Engineering Inc. (2003)
6. Ishihara, K.: Performance of rockfill dams during recent large earthquakes. In: Prakash, S. (ed.) 5th International Conference on Recent Advances in Geotechnical Earthquake Engineering and Soil Dynamics, San Diego, California (2010)
7. Aliberti, D., Vecchiotti, M., Cascone, E., Biondi, G.: Numerical analysis of the seismic behavior of the menta BFR dam. In: Bolzon, G., Sterpi, D., Mazzà, G., Frigerio, A. (eds.) ICOLD-BW 2019. LNCE, vol. 91, pp. 419–438. Springer, Cham (2019). https://doi.org/10.1007/978-3-030-51085-5_23
8. Aliberti, D., Biondi, G., Cascone, E., Rampello, S.: Performance indexes for seismic analyses of earth dams In: Silvestri, S., Moraci, N. (eds.) 7th ICEGE, Rome, Italy, pp. 1066–1073 (2019)
9. Masini, L., Rampello, S., Donatelli, R.: Seismic performance of two classes of earth dams. Earthq. Eng. Struct. Dynam. **50**(2), 692–711 (2021)
10. Ingegneri, S., Biondi, G., Cascone, E., Di Filippo, G.: Influence of cyclic strength degradation on a Newmark-type analysis. In: Silvestri, S., Moraci, N. (eds.) 7th ICEGE, Rome, Italy, pp. 2996–3004 (2019)
11. Biondi, G., Cascone, E., Rampello, S.: Evaluation of seismic stability of natural slopes. Rivista Italiana di Geotecnica **45**(1), 11–34 (2011)
12. Meehan, C.L., Vahedifard, F.: Evaluation of simplified methods for predicting earthquake-induced slope displacements in earth dams and embankments. Eng. Geol. **152**, 180–193 (2013)
13. Biondi, G., Cascone, E., Aliberti, D., Rampello, S.: Screening-level analyses for the evaluation of the seismic performance of a zoned earth dam. Eng. Geol. **280**, 105954 (2021)
14. Aliberti, D., Cascone, E., Biondi, G.: Seismic performance of the San Pietro dam. Procedia Eng. **158**, 362–367 (2016)
15. Masini, L., Rampello, S.: Influence of input assumptions on the evaluation of the seismic performance of earth dams. J. Earthq. Eng. **26**(9), 4471–4495 (2020)
16. Cascone, E., Biondi, G., Aliberti, D., Rampello, S.: Effect of vertical input motion and excess pore pressures on the seismic performance of a zoned dam. Soil Dyn. Earthq. Eng. **142**, 106566 (2021)
17. Biondi, G., Cascone, E., Di Filippo, G.: Affidabilità di alcune correlazioni empiriche per la stima del numero di cicli di carico equivalente. Rivista Italiana di Geotecnica **46**(2), 9–39 (2012)

Evaluation of the Seismic Performance of Small Earth Dams

Andrea Ciancimino[1(✉)], Renato Maria Cosentini[1], Francesco Figura[2], and Sebastiano Foti[1]

[1] Department of Structural, Building and Geotechnical Engineering, Politecnico di Torino, Turin, Italy
andrea.ciancimino@polito.it

[2] Laboratory of Experimental Rock Mechanics, École Polytechnique Fédérale de Lausanne, Lausanne, Switzerland

Abstract. The evaluation of the seismic response of earth dams is a nontrivial task, which generally requires the use of advanced soil models able to accurately reproduce the material behaviour under dynamic loadings. The reliability of the numerical simulations is however constrained by the level of knowledge of the geotechnical model of the dams. The issue is particularly relevant when small earth dams, characterized by a reduced height and a limited reservoir volume, are considered. Such structures indeed frequently lack proper characterization of the materials constituting the dam body and its foundation. The implementation of dynamic numerical analyses is therefore limited in the common practice and the seismic performance of the dams are frequently assessed through simplified empirical methods. This study investigates the seismic behaviour of two small earth dams for which a reliable geotechnical model, based on both laboratory and *in situ* tests, is available. The seismic responses of the dams are analyzed through fully coupled effective stress dynamic analyses. The analyses are developed within the context of the ReSba European project for the French-Italian Alps area. The results have allowed comparing the seismic performance of the structure as predicted by simplified and advanced approaches in terms of stability conditions and seismic-induced settlement of the crest.

Keywords: Earth dam · Seismic response · Seismic performance · Effective stress analysis · Earthquake

1 Introduction

The evaluation of the risk associated with existing dams is a major issue in high seismicity regions. The seismic-induced inertial forces may compromise the stability of the embankment, leading to the accumulation of permanent crest settlement and reducing, in turn, the freeboard. As a consequence, several studies have been devoted in the past to the assessment of the seismic performance of earth dams (e.g. [1]). Such an assessment can be carried out by employing different approaches, namely pseudo-static numerical analyses, simplified empirical relationships, displacement-based approaches derived from

© The Author(s), under exclusive license to Springer Nature Switzerland AG 2022
L. Wang et al. (Eds.): PBD-IV 2022, GGEE 52, pp. 1937–1945, 2022.
https://doi.org/10.1007/978-3-031-11898-2_177

Newmark's rigid block method [2], and more accurate dynamic numerical simulations. These methods vary from simplified to highly sophisticated and, thus, require different levels of knowledge of the geotechnical model of the site.

According to the Italian National Code [3], the seismic response of existing dams should be assessed by following the principle of *gradualness*, thus selecting the method of analysis according to the amount of available information. Sophisticated methods may be suitable for studying the performance of *large dams* (i.e. dams with a height > 15 m or/and a reservoir volume > 10^6 m^3 [3]) for which a comprehensive characterization is usually available. Conversely, *small dams*, characterized by modest height and reservoir volume, frequently lack proper geotechnical information, therefore limiting the implementation of advanced numerical simulations. As a consequence, a first-level screening is usually carried out by using empirical relationships to identify the small dams for which more refined analyses are required. Despite their limited size, the risk associated with the potential failure of the small dams is indeed considerable due to their proximity to populated areas.

The ReSba (Resilience of Dams) project, funded by the European fund for regional development (Interreg-ALCOTRA), fits in this context intending to improve our ability to assess the natural risks associated with dams located in the French-Italian Alps area. Within the project, a simplified methodology has been firstly developed to classify the embankments according to the associated seismic risk [4]. The methodology has been then applied to select the small earth dams classified from medium to high seismic risk [5]. This study presents the results of the dynamic numerical simulations performed to analyze the seismic performance of two of them, namely the Arignano and Briaglia dams. The fully coupled analyses are carried out with due consideration of the possible pore water pressure build-up due to the coupled shear-volumetric response of granular soils. The results of the simulations are used as a benchmark for testing the ability (and the drawbacks) of the simplified empirical relationship proposed by Swaisgood (2003) to estimate the seismic-induced settlements of dams.

2 Geophysical and Geotechnical Survey

The Arignano and the Briaglia dams are two small earth dams located in the Piedmont region in Italy. Both the dams have been the subject of an extensive geophysical and geotechnical survey carried out within the ReSba project. In the following, the main results of this survey are briefly summarized. Further details can be found in [5].

2.1 Arignano Dam

The dam was built in the second half of the 19th century in Arignano (Turin, Italy) and dikes the course of the Rio del Lago stream forming a reservoir with a volume of about 640,000 m^3. The embankment has a trapezoidal cross-section with a crest 380 m long and 5 m wide, and a height of about 7 m.

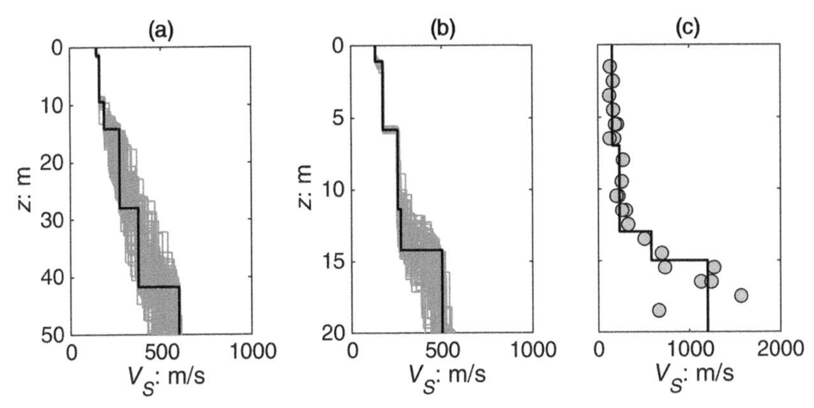

Fig. 1. Results of the MASW tests carried out along the crest of the Arignano (a) and Briaglia (b) dams (the thick lines are the minimum misfit profiles); (c) results of the DH test performed on the Briaglia dam.

A first geotechnical survey was conducted in 2004, comprising 12 Standard Penetration tests, 5 boreholes, and 4 Lefranc variable head permeability tests. Additionally, direct simple shear, unconfined undrained triaxial and oedometric tests were performed on undisturbed samples retrieved from the boreholes. The geophysical characterization was subsequently carried out within the framework of the ReSba project, involving the execution of active Multi-Channel Analysis of Surface (MASW) tests and single-station Horizontal-to-Vertical Spectral Ratio (HVSR) tests. Passive 2D array measurements were also carried out downstream from the dam. The experimental data were processed in the *f-k* domain using the Surface Wave. The main results of the geophysical tests performed along the dam crest are presented in Fig. 1a in terms of a statistical sample of the shear wave velocity V_S profile.

A geotechnical model of the site was built based on the results of the laboratory and *in situ* tests. According to the model, the embankment is constituted by a 2 m thick layer of sand and silt overlying 5 m of plastic clays and silts. The foundation consists instead of 2 m of clayey silts, followed by 10 m of fine silty sand overlying two layers of soft-to-hard clays with stiffness increasing with depth. The seismic bedrock is a deep layer of marlstones with sandstone inclusions. A summary of the mechanical parameters of the materials is given in Table 1.

Table 1. Geotechnical parameters adopted for the numerical model of the Arignano dam.

Layer	z:m	V_S:m/s	k:cm/s	φ':°	c':kPa	MRD curves	Finn-Byrne parameters
Dam 1	$0.0 \div 2.0$	150	$4 \cdot 10^{-7}$	21	12	Seed et al. (1986) [6]	$C_1 = 0.3 \div 0.7$ $C_2 = 0.6 \div 0.3$
Dam 2	$2.0 \div 7.0$	160	$4 \cdot 10^{-7}$	21	12	Sun et al. (1988) [7]	–
Found 1	$7.0 \div 9.0$	160	$4 \cdot 10^{-7}$	21	12	Sun et al. (1988) [7]	–
Found 2	$9.0 \div 19$	165	$8 \cdot 10^{-7}$	21	12	Seed et al. (1986) [6]	$C_1 = 0.09 \div 0.5$ $C_2 = 0.4 \div 2.2$
Found 3	$19 \div 40$	350	10^{-9}	30	1	Sun et al. (1988) [7]	–
Found 4	$40 \div 51$	600	10^{-9}	30	1	Sun et al. (1988) [7]	–
Bedrock	>51	1050	10^{-9}	–	–	–	–

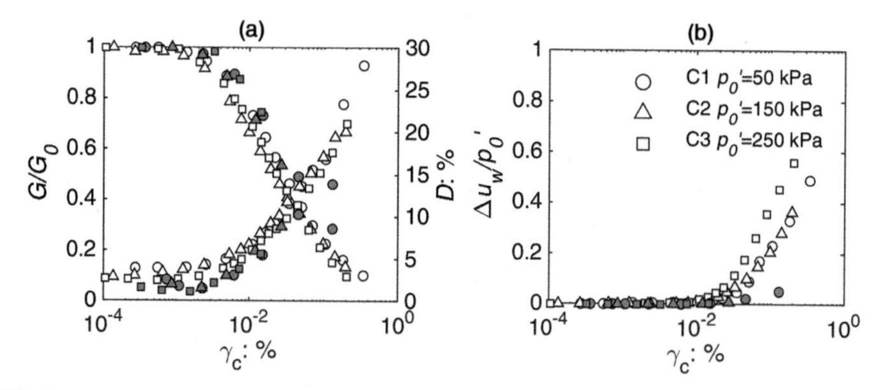

Fig. 2. MRD curves from RC/TS tests (a, the filled markers are the TS tests) along with r_u (b).

2.2 Briaglia Dam

The dam is located within the municipality of Briaglia (Cuneo, Italy) and it was built at the beginning of the 1990s to retain a reservoir volume of about 59,500 m^3 from the Rio del Frocco river. The embankment is characterized by a height of about 11 m and a crest 88.5 m long and 4 m wide.

No prior information was available regards the foundation and structure material, the site was therefore object of an extensive geotechnical and geophysical survey, which included: a borehole on the crest of dam instrumented to conduct Down Hole tests; active MASW and passive HVSR tests; a seismic cone penetration (SPCTU) test and a flat-dilatometer (DMT) test. The main results of the geophysical investigations are presented in Fig. 1b in terms of a statistical sample of the V_S profiles from MASW test, whereas Fig. 1c present the results of the DH test.

Laboratory tests were also conducted on undisturbed samples. Specifically, isotropically consolidated undrained triaxial tests and oedometric tests were carried out to define the mechanical properties of the materials. In addition, combined Resonant Column and Torsional Shear (RC/TS) tests were performed to investigate the soil dynamic behaviour. The results of the tests are presented in Fig. 2a in terms of Modulus Reduction and Damping ratio (MRD) curves along with the pore water pressure build-up normalized to the initial confining pressure $r_u = \Delta u_w/p_O'$ (Fig. 2b).

Table 2. Geotechnical parameters adopted for the numerical model of the Briaglia dam.

Layer	z: m	V_S:m/s	k: cm/s	φ':°	c':kPa	MRD curves	Finn-Byrne parameters
Dam 1	0.0 ÷ 4.5	160	10^{-9}	32	1	RC test - C1	$C_1 = 0.09$ $C_2 = 2.2$
Dam 2	4.5 ÷ 6.8	190	10^{-10}	32	1	Darendeli (2001) [8]	$C_1 = 0.08$ $C_2 = 2.4$
Dam 3	6.8 ÷ 7.6	230	10^{-8}	38	1	Rollins et al. (1998) [9]	–
Dam 4	7.6 ÷ 14	270	10^{-8}	35	1	Seed et al. (1986) [6]	$C_1 = 0.10 \div 0.12$ $C_2 = 1.7 \div 2.0$
Found 1	14 ÷ 20	500	$5 \cdot 10^{-10}$	38	1	RC test - C2	–
Found 2	20 ÷ 30	800	10^{-11}	34	1	RC test - C3	–
Bedrock	>30	1500	10^{-11}	–	–	–	–

The embankment is constituted by silts and clayey silts over a thin layer of weathered sandstones overlying a 6.4 m thick layer of clayey silty sand. The dam is built on a 6 m thick layer of medium-hard clays overlying a stiffer deposit of marlstone. The latter constitutes the seismic bedrock, reaching high V_S values (≈1500 m/s) at a depth of about 30 m. The geotechnical parameters of the dam are reported in Table 2.

3 Numerical Modelling

The fully coupled effective stress analyses are performed using the Finite-Difference code Flac 2D (Itasca C.G.). The soil stress-strain behaviour is modelled through an elastic-perfectly plastic constitutive model with a Mohr-Coulomb failure criterion, coupled with a hysteretic formulation to capture the nonlinear soil response. The parameters of the hysteretic model are calibrated based on the results of RC/TS tests (Fig. 2) and widely-used empirical models. Additionally, a small amount ($<1\%$) of Rayleigh viscous damping is added to consider the energy dissipation at small strains.

The simulations are carried out by taking into account the eventual pore water pressure build-up due to the coupling between shear and volumetric strains above the volumetric shear strain threshold. Such feature is considered for the sandy and non-plastic silty soils by employing the empirical model proposed by Byrne [10], with constitutive parameters defined according to the following relationships:

$$C_1 = 7600D_r^{-2.5}\ C_2 = 0.4/C_1 \tag{1}$$

being D_r the relative density of the soil, defined through the CPT and SCPT tests.

The numerical field is discretized through four-sided mesh elements with sizes defined according to the Kuhlemeyer and Lysmer [11] criteria to ensure the accuracy of the wave propagation phenomena. The geotechnical models developed to study the Arignano and Briaglia dams responses are presented in Fig. 3. The simulations comprise three numerical steps: (i) definition of the geostatic stress-state; (ii) hydro-mechanical seepage analysis; (iii) fully coupled dynamic analysis. During the static steps, elementary boundary conditions are employed. Conversely, free-field lateral conditions and base absorbing dashpots are introduced under dynamic conditions to avoid undesired waves reflections [12].

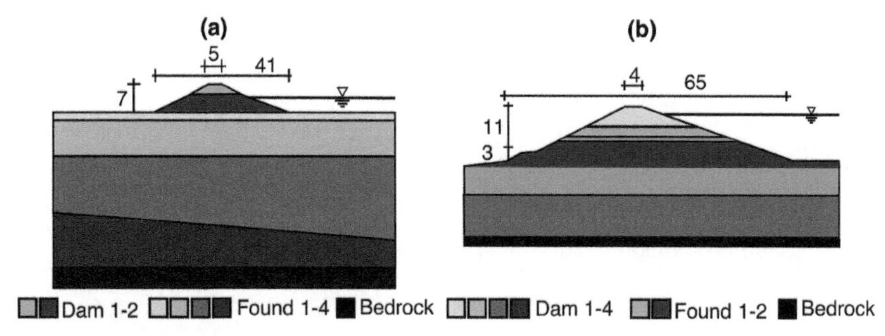

Fig. 3. Geotechnical models for the Arignano (a) and Briaglia dams (b) (dimensions in m).

The dynamic analyses are performed by applying input strong motions recorded on reference rock-like outcropping conditions. The latter were selected from the ITACA archive [13] in compliance with the seismicity of the Piedmont region and, specifically, of the sites of the dams. Figure 4 reports, as an example, the maximum acceleration a_{max} profiles recorded along the axes of the models for some selected input motions. For the Arignano model, the acceleration profiles slightly increase up to a depth of about 7 m, after which there is an abrupt increase of a_{max} due to the 2D topographic effects induced by the dam body (Fig. 4a). Conversely, a reduction of a_{max} can be observed within the Briaglia dam (Fig. 4b) as a consequence of the large amount of energy dissipated by the first 6 m of soil characterized by significant pore water pressure build-up.

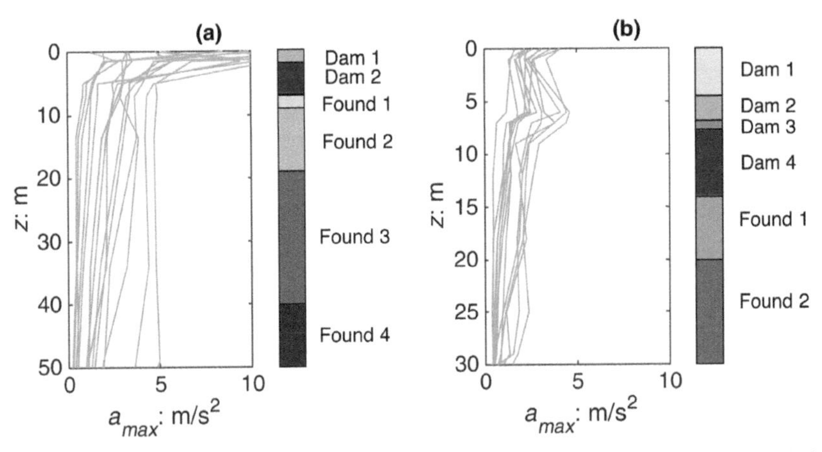

Fig. 4. Maximum acceleration profiles recorded along the axes of the (a) Arignano and (b) Briaglia dams.

4 Empirical Relationships for Predicting the Crest Settlement

The results of the numerical simulations are here employed to assess the performance of the empirical relationship proposed by Swaisgood [14] to estimate the permanent settlement of the crest w normalized to the height of the dam H_d plus the thickness of the alluvium H_f in percentage:

$$w(\%) = \frac{\Delta H}{H_d + H_f} = e^{6.07 \cdot PGA + 0.57 \cdot M - 8} \tag{2}$$

The numerical w are compared to Swaisgood's equation [14] in Fig. 5a as a function of the PGA. The empirical relationship seems to be in good accordance with w obtained for the Arignano dam, although it slightly overpredicts w at low PGA (Fig. 5a). Conversely, the equation significantly underestimates w for the Briaglia dam. Similar trends can be observed by looking at the comparison between numerical w and predicted w_{pred} settlement in Fig. 5b.

The data does not follow the diagonal of the plot, revealing a systematic bias coming from the application of Eq. (2). Figures 5 also report the predictions from relationships calibrated through a nonlinear least-squares procedure on the results of the simulations. In particular, the equations here adopted are expressed in the following, general, form:

$$w(\%) = A \cdot e^{\alpha \cdot PGA} \tag{3}$$

where A and α are calibration parameters. It is worth noting that although the calibrations were initially performed separately for the two dams, the obtained values of the parameter α, defining the slope of the lines, resulted to be in very good accordance. It was therefore decided to adopt the same value, equal to 27.1, for the two dams. Conversely, the value of A (equal to $5 \cdot 10^{-4}$ and 0.11, respectively for the Arignano and Briaglia dams), seems to be characteristic of each dam. The differences in this parameter are attributed mainly to the specific geometric features of the dams, as well as to the different thicknesses of

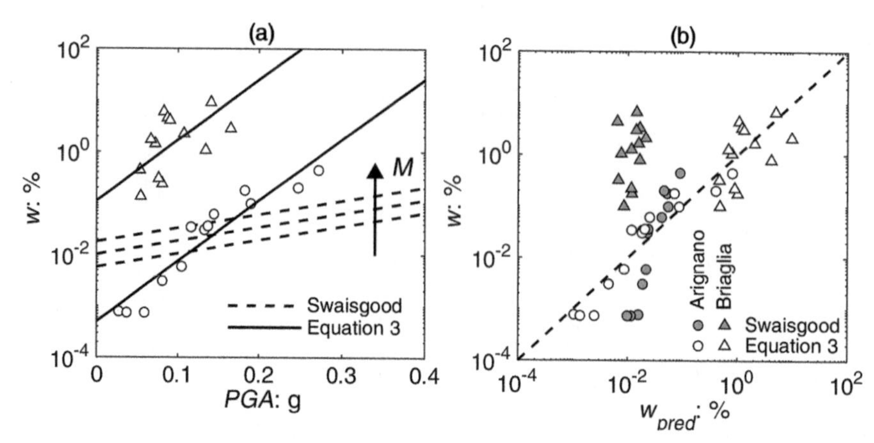

Fig. 5. (a) Simulated permanent settlement versus *PGA* and (b) simulated versus predicted settlement as obtained from Swaisgood's equation [14] and Eq. 3.

the soil layers affected by pore water pressure build-up. Further parametrical analyses on dams characterized by different geometric features are thus required to derive an empirical relationship for linking A to the specific characteristic of the structure.

5 Conclusions

This study has investigated the ability of the most used empirical equations for predicting the seismicity-induced permanent settlement of the crest of small earth dams. For this purpose, fully coupled dynamic numerical simulations were performed to analyze the seismic response of two embankments for which reliable geotechnical models were available based on laboratory and *in situ* tests.

The results of the analyses have shown that care should be exercised when such equations are employed in the common practice, as the seismic damage of the dams may be significantly underestimated. Nevertheless, the results of the analyses also suggest that the estimate may be substantially improved by explicitly including in the formulations of the empirical relationships the influence of the geometry of the dam and the thickness of the soil layers prone to pore water pressure build-up.

Acknowledgement. This study was supported by the European Union for regional development (Interreg-ALCOTRA) for the French-Italy Alps, within the ReSba project, for which funding is gratefully acknowledged. The Authors thank the Piedmont Region Administration for their availability and support. The Authors wish also to thank F. Passeri, Ph.D., and L. Fazio, M.Sc., who worked actively on the project.

References

1. Makdisi, F.I., Seed, H.B.: Simplified procedure for estimating dam and embankment earthquake-induced deformations. J. Geotech. Eng. Div. **104**(7), 849–867 (1978)

2. Newmark, N.M.: Effects of earthquakes on dams and embankments. Geotechnique **15**(2), 139–160 (1965)
3. Ministero delle Infrastrutture e dei Trasporti: Norme Tecniche per la Progettazione e la Costruzione degli Sbarramenti di Ritenuta (dighe e traverse) (2014)
4. Cosentini, R.M., Passeri, F., Foti, S.: A simplified methodology for the assessment of the seismic risk associated with small earth dams. In: Calvetti, F., Cotecchia, F., Galli, A., Jommi, C. (eds.) Geotechnical Research for Land Protection and Development: Proceedings of CNRIG 2019, pp. 92–100. Springer, Cham (2020). https://doi.org/10.1007/978-3-030-21359-6_10
5. Cosentini, R.M., Foti, S., Passeri, F.: Geophysical and geotechnical characterization of small earth dams in the Piedmont region for seismic risk assessment. In: 6th International Conference on Geotechnical and Geophysical Site Characterization (2021)
6. Seed, H.B., Wong, R.T., Idriss, I., Tokimatsu, K.: Moduli and damping factors for dynamic analyses of cohesionless soils. J. Geotech. Eng. **112**(11), 1016–1032 (1986)
7. Sun, J.I., Golesorkhi, R., Seed, H.B.: Dynamic moduli and damping ratios for cohesive soils. Earthquake Engineering Research Center, University of California, Berkeley (1988)
8. Darendeli MB: Development of a new family of normalized modulus reduction and material damping curves. The University of Texas at Austin (2001)
9. Rollins, K.M., Evans, M.D., Diehl, N.B., Daily III, W.D.: Shear modulus and damping relationships for gravels. J. Geotech. Geoenviron. Eng. **124**(5), 396–405 (1998)
10. Byrne, P.M.: A cyclic shear-volume coupling and pore pressure model for sand (1991)
11. Kuhlemeyer, R.L., Lysmer, J.: Finite element method accuracy for wave propagation problems. J. Soil Mech. Found. Div. **99**(5), 421–427 (1973)
12. Lysmer, J., Kuhlemeyer, R.L.: Finite dynamic model for infinite media. J .Eng. Mech. Div. **95**(4), 859–877 (1969)
13. Luzi, L., Pacor, F., Puglia, R.: ITalian ACcelerometric Archive (ITACA), version 3.0. INGV (2019). https://doi.org/10.13127/itaca.3.0
14. Swaisgood, J.: Embankment dam deformations caused by earthquakes. In: Pacific Conference on Earthquake Engineering (2003)

Liquefaction Potential for In-Situ Material Dams Subjected to Strong Earthquakes

Hong Nam Nguyen(✉)

Division of Geotechnical Engineering, Faculty of Civil Engineering, Thuyloi University, 175 Tay Son Street, Dong Da, Hanoi, Vietnam
hongnam@tlu.edu.vn

Abstract. Dam safety assessment is a crucial task to ensure the safety of the existing or the new dam which can be subjected to high risks such as heavy floods and/or strong earthquakes. In some old existing dams in Vietnam, the filling and foundation materials were consisted of sand which could possibly trigger a liquefaction when subjected to strong earthquakes. Thus, liquefaction potentials of some in-situ material large dams in the west-north of Vietnam were analyzed based on the seismic data, dynamic soil properties and finite element modeling with various scenarios regarding earthquake return periods and reservoir water levels. The simulation results suggested that the sand zones in the upstream dam body foundation and under the downstream drainage rockfill could be liquefied when subjected strong earthquakes. The extent of liquefied zone increased with the increase of the peak ground acceleration and reservoir water levels.

Keywords: In-situ dam · Earthquake · Liquefaction · Cyclic loading

1 Introduction

In-situ material dam is a type of dam using local materials such as soils, rocks, which are widely used in the world and in Vietnam for the construction of reservoirs for irrigation and hydropower. Currently, Vietnam has a total of about 6648 hydraulic reservoirs with a total storage capacity of about 13.5 billion m^3, including 702 large reservoirs and 5,946 small reservoirs [10].

Earthquakes are one of the major disasters for dam safety. The territory of Vietnam is located in a land with a complex geological - tectonic structure. Areas at risk of earthquakes from 6.0 to 7.0 on the Richter scale in Vietnam include: fault zones on the Red and Chay river systems; Lai Chau - Dien Bien fault zone; regions of Ma, Son La and Da rivers; Cao Bang and Tien Yen regions; Rao Nay - Ca river zone; Dakrong - Hue zone; Truong Son zone; Ba river zone; Central coastal zone… Some strong earthquakes have occurred in the northwest region of the country such as the earthquake ($M = 7$) in Lai Chau in 1914, Dien Bien in 1920 occurring in the Lai Chau-Dien Bien fault, The 1926 Son La earthquake on Son La fault, the 6–7 level Ta Khoa earthquake ($M = 5$) and a number of level 7 earthquakes were aftershocks of the 1983 Tuan Giao earthquake. A map of earthquake-generating areas in the territory of Vietnam has been

© The Author(s), under exclusive license to Springer Nature Switzerland AG 2022
L. Wang et al. (Eds.): PBD-IV 2022, GGEE 52, pp. 1946–1958, 2022.
https://doi.org/10.1007/978-3-031-11898-2_178

made [29]. Earthworks such as reservoir dams, river dikes and road embankments are often damaged in strong earthquakes around the world [1, 17, 20]. This damage is mainly due to liquefaction of the embankment and/or foundation soil.

Seed and Idriss [22] published a "simple procedure for assessing the liquefaction potential of soils", which was improved over the next decade and half the time was updated information and improved methods [16].

The most common method for assessing the liquefaction activation potential is the cyclic stress method [30]. There are also cyclic deformation methods [7] and others.

Dam seismic design has been included in design standards in several countries [8, 11, 15, 26, 28]. According to TCVN 9386-2012 "Design of earthquake-resistant buildings" [27], for strong earthquakes with design ground acceleration greater than 0.08g, it is necessary to calculate the seismic resistance structure. However, the seismic design standard for local material dams has not yet been released.

Note that some of the local material dams in the country are large dams that are filled with water-saturated gravel and sand and/or use coarse-grained backfill material that has a risk of liquefaction during strong earthquakes.

2 Project Study

The Loong Luong 1 reservoir is located in Loong Luong village, Muong Phang commune, Dien Bien district, Dien Bien province. The project aims to provide stable water for 150 ha of rice-growing land, create an ecological environment for Muong Phang historical site, and at the same time create a source of water for aquaculture; supply water for daily life; irrigate farm produce and fruit trees. Project has grade III [19]. The earth-fill dam has the maximum height of 29 m. The reservoir volume at the normal water level is 1.081 million m^3. The construction was completed in 2013.

Within the scope of the work under the state funded study, the topographical survey, seismic and geological investigation of Loong Luong 1 dam were conducted [14]. The survey was to draw maps of the existing plan and cross-sections of the existing dam. The geotechnical work involved in making 16 boreholes at the dam site, taking in-situ soil samples in the dam body and foundation for testing the soil index and dynamic soil properties of the dam body and its foundation. The seismic work dealt with calculation of peak ground acceleration PGA and acceleration times histories at site. The seismic data and dynamic soil properties were then employed in a geotechnical software to assess the liquefaction potential and slope stability of the dam when subjected to strong earthquakes in the area with different earthquake scenarios. Figure 1 shows the cross section of Loong Luong 1 dam. Figure 2 shows the layout of boreholes, in which 16 boreholes (HK1 to HK16) implemented in four cross Sects. 1, 2, 3 and 4 and some excavations (HĐ1 and HĐ2) arranged in the downstream. Since the investigation time, the reservoir water level was at high elevation, the boreholes in the upstream slope could only be arranged near the top of the dam (Fig. 1).

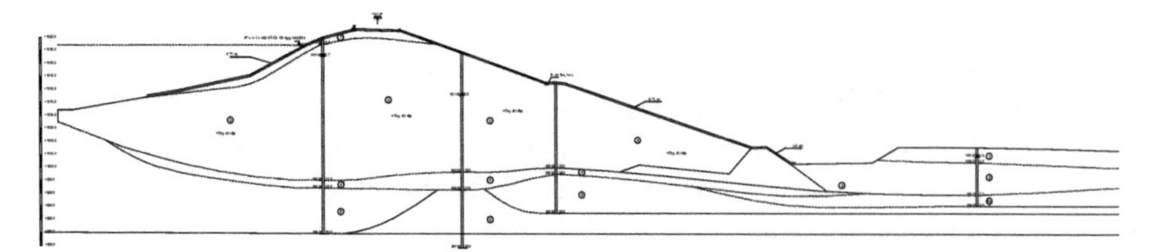

Fig. 1. Cross Sect. 1 of Loong Luong 1 dam.

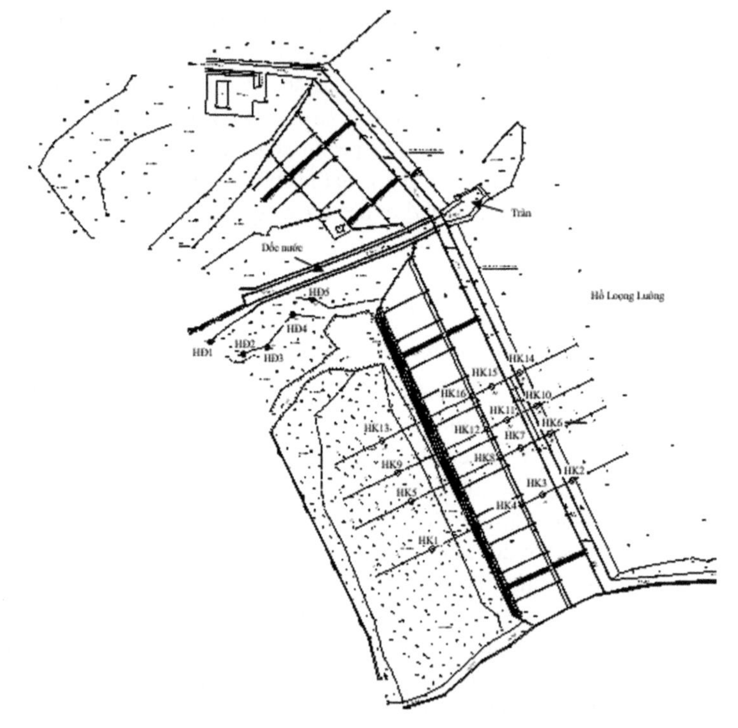

Fig. 2. Layout of boreholes at Loong Luong 1 dam.

3 Liquefaction Modeling of Loong Luong Dam Subjected to Strong Earthquakes

3.1 Seismic Data

The seismic data were collected and constructed from the earthquake observation data, the field geological data and the model calculation [25] as described follows. The data includes peak ground acceleration (PGA) values, spectral acceleration, and acceleration time history records at the dam site.

The geological, tectonic, and engineering geological data from different sources were collected. The seismic data was collected up to the year 2012. To calculate the PGA values with earthquake return periods, T = 475 years and T = 2475 years, the method of probabilistic seissmic hazard assessment (PSHA) [6] was employed with using the software CRISIS99 [18] and empirical attenuation relations [3–5] since the number of earthquake observation stations at site were limited.

The spectral accelerations SA of the dam were also calculated by the same probabilistic method as done for the PGA values.

To calculate the acceleration times history at the bedrock, the transform method in the frequency domain based on the random oscillation theory was employed with using the RASCAL program [24]. Based on the similarities in magnitudes, focal depths of the site, three observed earthquake motions, namely Campano, Italy [2] (called 1a) (Bagnoli-Irpino station, E-W direction, Repi = 23 km, M = 6.87, PGA = 177 cm/s^2); Lancang, China [13] (called 2a) (YNBA0004 station, N-S direction, Repi = 4 km, M = 6.7, PGA = 508 cm/s^2); and Dien Bien, Vietnam (called 3a) (Dien Bien station, N-S direction, Repi = 18 km, M = 5.3, PGA = 109 cm/s^2) were selected to calculate the time history acceleration records for the Loong Luong 1 dam.

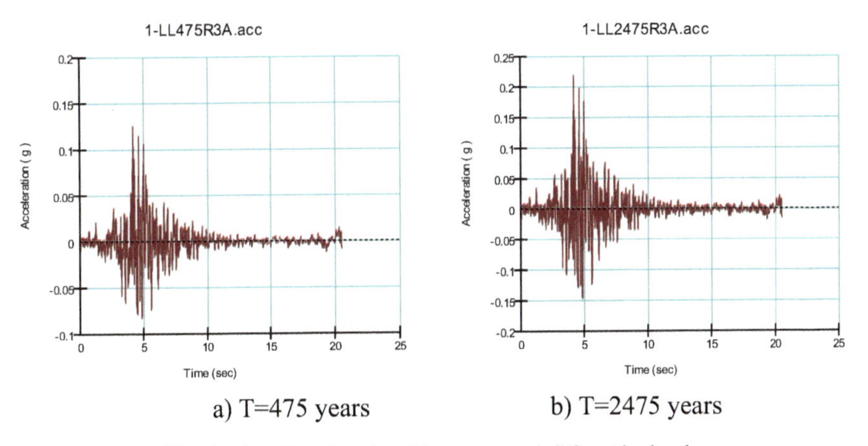

Fig. 3. Acceleration time history records R3a at bedrock.

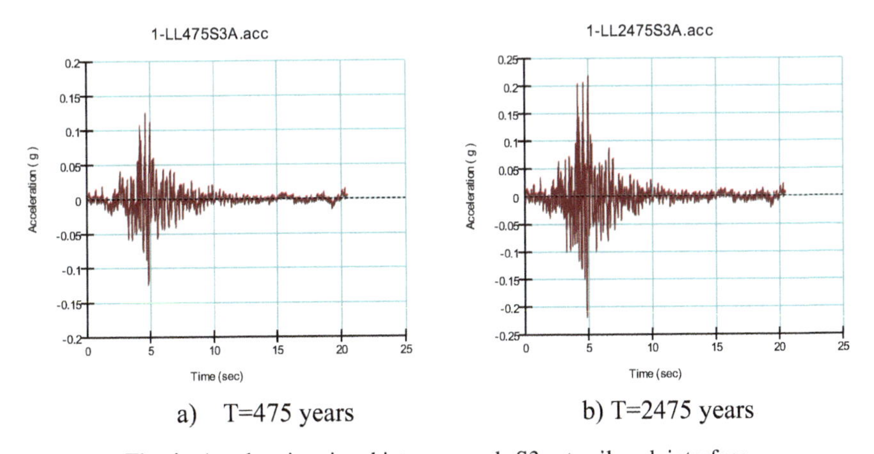

Fig. 4. Acceleration time history records S3a at soil-rock interface.

To calculate the acceleration time histories at the boundary between soil and rock as requested by the dynamic software QUAKE/W [9], the problem of wave propagation in an elastic medium with horizontal layering ground was analyzed, using the program SHAKE91 [23]. Earthquake activity, faults were considered within a radius of approximately 300 km from Dien Bien Phu city. Refer to ICOLD [12] for the guidelines of selecting seismic parameters for large dams.

The seismic calculation results yielded the average $Vs_{,30} = 261.7$ m/s for the dam site. This value was obtained based on the N-SPT values, which were measured at 6 boreholes with the depth of 30 m. $Vs_{,30}$ is the average value of shear wave propagation velocity in the upper 30 m of the ground section where the shear strain, $\gamma \leq 10^{-5}$ [27]. The dam foundation belongs to type C foundation [27]. The seismic calculation revealed the average PGA values at the dam site of being 0.125g (T = 475 years) and 0.218g (T = 2475 years).

Three acceleration time history records at the bedrock R1a, R2a, R3a (Fig. 3) and three at the soil–rock interface S1a, S2a, S3a (Fig. 4) corresponding to T = 475 years and 2475 years at the Loong Luong 1 dam were used for dam liquefaction analysis.

3.2 Geotechnical Data

The stratigraphy of the dam site from top to bottom to a depth of 31.0 m can be classified into 8 layers of soil, specifically described as follows:

- Layer 1: Filling soil.
- Layer 2: Dark gray clay loam, organic humus, soft plastic state.
- Layer 3: Red-brown, gray, yellow-gray, yellow-brown clay loam, mixed with gravel, hard plastic state; some place with semi-hard state clay loam.
- Layer 4A: Sand mixed with small pebbles, dark gray and gray clay loam.
- Layer 4: Sandy gravel, gray-brown clay, gray-white clay.
- Layer 5: Clay loam of gray, white spots, yellow brown, greenish gray patchy, semi-hard state, with some hard plastic clay loam.
- Layer 6: Sand mixed with grit, gray-green, yellow-gray, white spots, plastic state, with some hard plastic clay loam.
- Layer 7: Strong to very strong weathered granodiorite, gray-brown, gray, medium soft rock.

Table 1. Soil properties of the dam filling and foundation layers

Soil layer	w (%)	γ (g/cm^3)	e	S (%)	LL (%)	PL (%)	PI (%)	LI (%)	C (kPa)	ϕ (deg)
2	44.9	1.77	1.219	99.9	49.2	36.2	13.0	0.67	18.3	11°23′
3	25.37	1.82	0.861	79.6	33.3	20.9	12.4	0.35	22.6	16°25′
5	23.1	1.94	0.728	86.4	33.1	20.2	12.9	0.22	24.5	18°22′
6	18.9	1.88	0.710	71.6	22.1	15.6	6.5	0.50	17.2	22°20′

Table 1 shows the tested index properties of soil layers from boreholes. Note that it was very difficult to take the sand samples of layers 4 and 4A in an undisturbed state since the layer thickness is very thin in the foundation. Thus, sand samples of the same layer were collected in HĐ1 and HĐ2 excavation holes at the downstream (Figs. 1 and 2).

3.3 Simulation of Liquefaction

For liquefaction analysis, we consider the special calculation case of earthquake loading with the intensity grade VIII (MSK-64 scale). The corresponding water load combination is as follows: the upstream reservoir water level is at the normal water level (+1022.1), there is no water at downstream. The input data for seismic calculation include topographic, geological and seismic data.

Table 2. Model parameters for the soil layers

Soil layer	γ (kN/m^3)	ϕ (deg.)	c (kPa)	K (m/s)	E (kPa)	ν	G (kPa)	h
1	20.00	20.00	10.0	1.0E−8	15000	0.25	6000.0	0.1
2	17.70	11.38	18.3	1.5E−7	10000	0.30	3846.2	0.1
3	19.23	16.42	22.6	1.0E−8	20000	0.30	7692.3	0.1
4	20.00	30.00	0.0	1.0E−5	110000	0.20	45833	0.1
5	20.01	18.37	24.5	3.5E−7	50000	0.30	19231	0.1
6	19.95	22.33	17.2	3.0E−7	40000	0.30	15385	0.1
7	20.00	30.00	0.0	3.0E−7	200000	0.20	83333	0.1
Sand	20.00	30.00	0.0	1.0E−4	100000	0.20	41667	0.1
Sand filter	20.00	30.00	0.0	1.0E−4	100000	0.20	41667	0.1
Rockfill	22.00	30.00	0.0	1.0E−4	150000	0.20	62500	0.1

Two cases with T = 475 years and T = 2475 years were implemented with three given acceleration time histories on bedrock and three on the subsoil-rock interface. The analyses were carried out by the finite element method for the plane strain problem. Four cross Sects. 1, 2, 3 and 4 were employed for the simulation. The total computation cases were 48 for one load combination regarding the reservoir water level as above mentioned. Note that the sand layer 4 exists in the dam foundation as seen in all sections; however, in Sect. 4, it only exists below the downstream rockfill. Note that the existence of sand layers 4 and 4a in the dam foundation can trigger a liquefaction if the strong earthquake happens. The thin sand layer 4 exists in the dam body foundation which can cause seepage loss as well as liquefaction potential when subjected to strong earthquakes.

The finite element software GeoStudio [9] was employed. The linear elastic model was applied to all soil layers for simplicity and saving calculation time. Table 2 shows the model parameters of the soil layers.

The initial porewater pressures in the dam were simulated under the steady state seepage condition with the reservoir's normal water level by SEEP/W module. After that, module QUAKE/W was employed to calculate the initial stress condition. Next, the dynamic analysis was activated to simulate the earthquake loading by applying the input acceleration time history record on the problem boundary. In the seepage modeling, water flow was prevented at the bottom boundary, while the constant upstream water head was assigned to the nodes on the upstream face. Either water head or flow values were assigned to the nodes on the downstream face/drainage rockfill. In calculating the in-situ stresses of soils, the horizontal displacements were fixed along the vertical boundaries and both horizontal and vertical displacements were fixed in the bottom boundary. During the seismic loading, all displacements were fixed on the bottom boundary, while the horizontal displacements were allowable on the vertical boundaries. The FE mesh

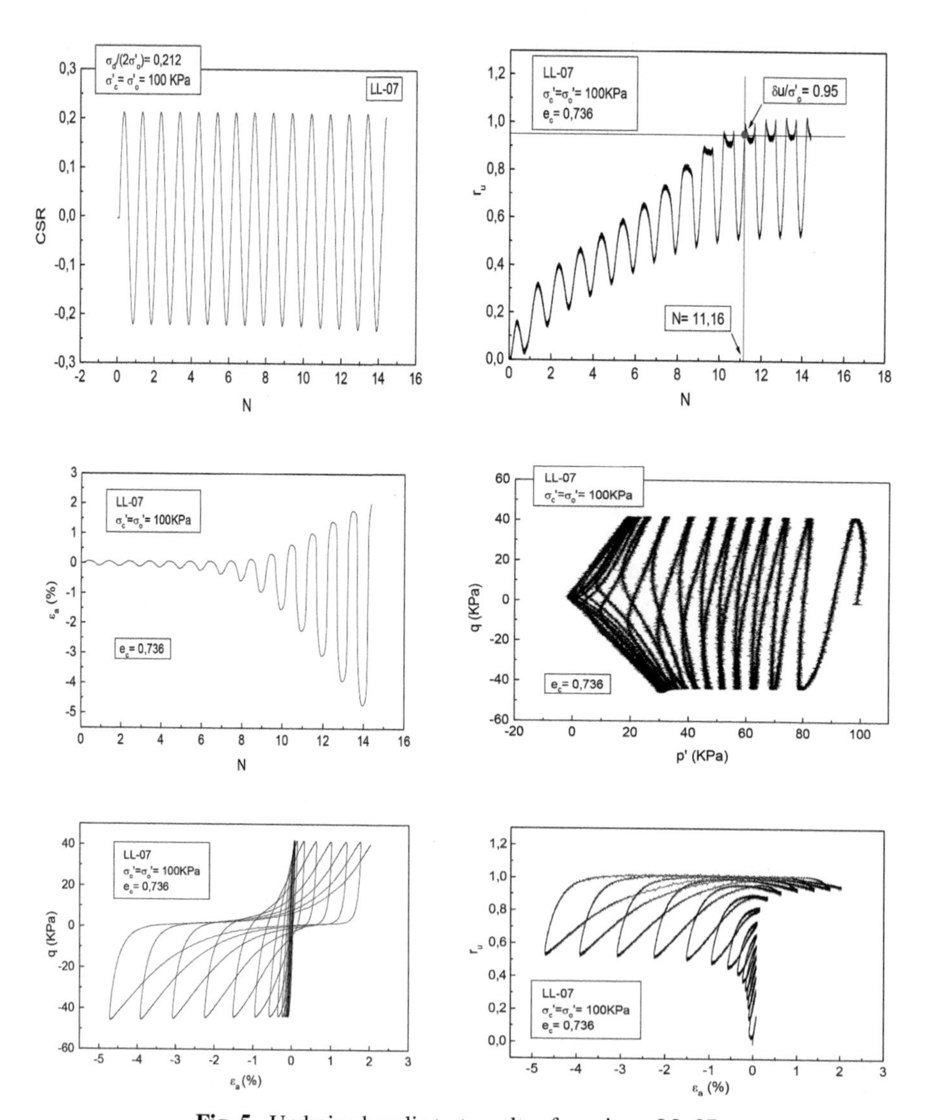

Fig. 5. Undrained cyclic test results of specimen LL-07.

consisted of 2627 elements (Sect. 1, average element size of 2 m) or even finer (with average size of 1 m). The results from QUAKE/W analysis were then employed for dam slope stability analysis by finite element method.

Note that the time durations of all initial acceleration records were trimmed to 10 s, generally, for saving the time calculation. For coarse grain materials (4 and 4a), their dynamic soil properties such as small strain vertical Young's modulus, damping ratio, liquefaction curve were obtained by running laboratory cyclic triaxial tests. Figure 5 shows typically undrained cyclic test results of specimen LL-07. To construct the liquefaction curve (Fig. 6), four sand specimens were reproduced with initial similar void ratios and tested with different stress ratios until they became liquefied, in terms of liquefaction triggering criteria either on the excess pore water pressure ($r_u = 0.95$) or the double strain amplitude (DA = 5%).

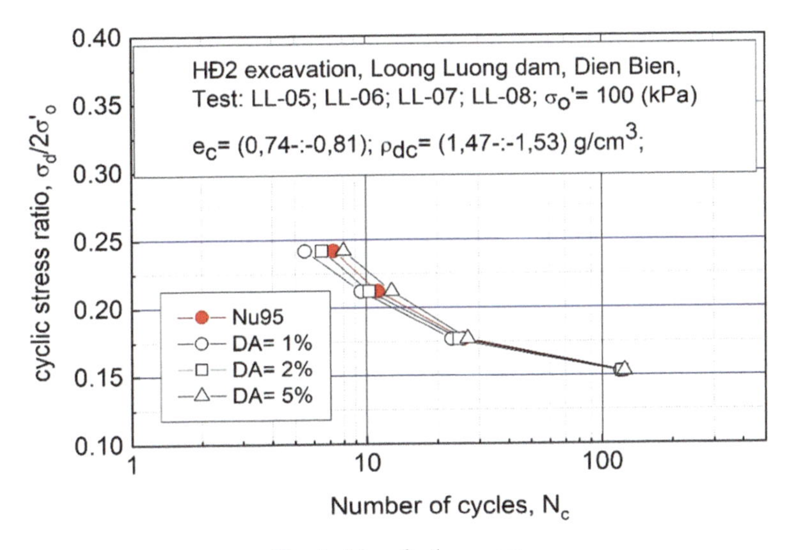

Fig. 6. Liquefaction curve.

Figure 7 shows the relation between normalized vertical Young's modulus versus axial strain. Note that the vertical Young's modulus was evaluated from 11 unload-reload cycles with the single strain amplitude changed from less than 10^{-5} to 10^{-4}.

The dynamic simulation results show that, there existed liquefied zones in the sand layer 4 in the upstream foundation and under the drainage rockfill at downstream. In cross Sect. 1, for T = 475 years, there existed the liquefied zones in the filter under the rockfill (with R1a, S1a records). For T = 2475 years, liquefied zones occurred in the filter under rockfill and in the layer 4 at the right end (with R1a, S1a, S2a records), see Fig. 8. For S3a record, only liquefied zone occurred in the filter under rockfill. No liquefaction zoned were observed with R2a, R3a records for both return periods. In cross Sect. 2, there existed the liquefied zones in layer 4 at the left end, in the upstream dam foundation, for all acceleration records, and in the filter under the rockfill for T = 2475 years, S3a record (Fig. 9).

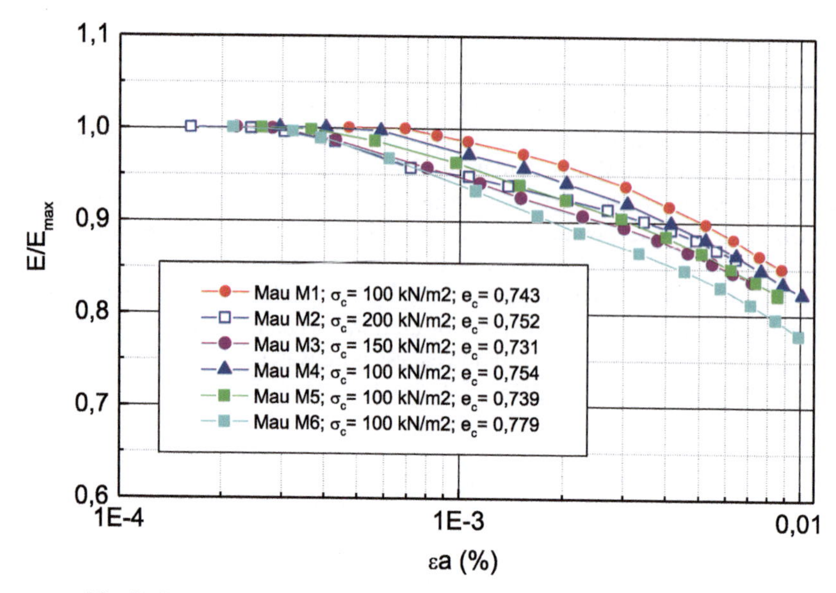

Fig. 7. Relation between normalized Yong' modulus versus axial strain.

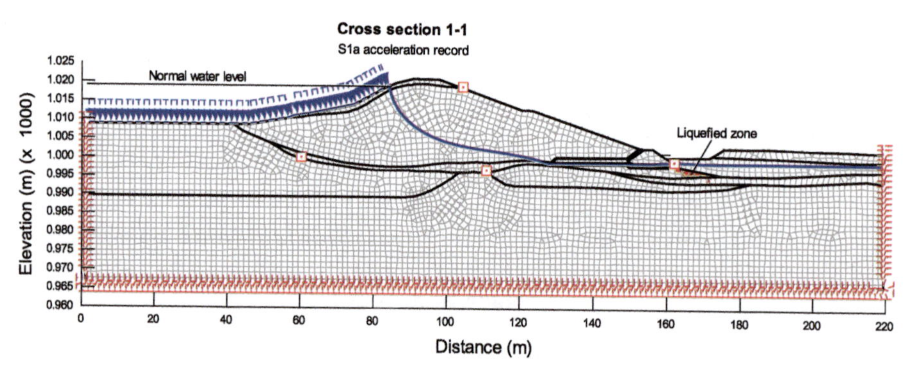

Fig. 8. Liquefied zones in cross Sect. 1, T = 2475 years, S1a acceleration record.

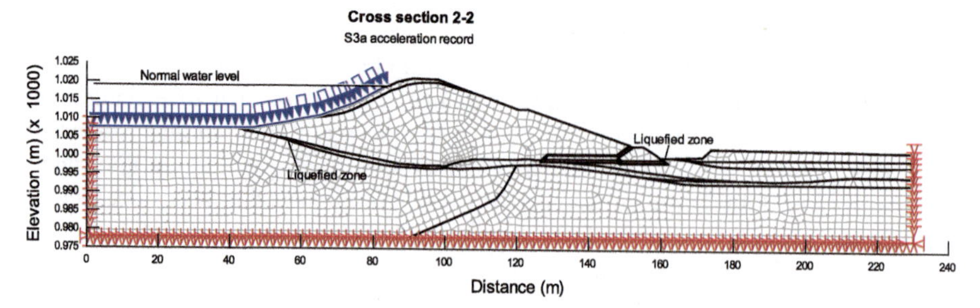

Fig. 9. Results of liquefaction at cross Sect. 2, T = 2475 years, S3a acceleration record.

From above consideration, effect of acceleration time history record on the liquefaction potential was noticeable.

No liquefied zones were observed in the cross Sects. 3 and 4 from the dynamic simulations implemented. Note that in Sect. 4, sand layer 4 does not exist in the foundation position under the dam body; it only exists in the rockfill foundation.

Note also from the above simulation results that the size of liquefied zone simulated with T = 2475 years was generally bigger than that simulated with T = 475 years. This is because the PGA values at T = 2475 years is bigger than the values at T = 475 years, possibly causing bigger earthquake load.

The slope stability analysis results show that the factors of safety of all upstream and downstream dam slopes were satisfied by the existing standard in the event of an earthquake with the return period of 475 years. For the return period of 2475 years, the minimum factor of safety decreased approximately to 1.1. Thus, it is necessary to pay much attention to safety measures for the dam when the strong earthquakes occur.

Note that the existing Vietnamese standard [27] only provides seismic accelerations for 1 in 475 year earthquake. The safety evaluation earthquake (SEE) with a return period 1 in 10,000 year [12] was not implemented in the current study.

The results of deformation analysis show that the largest horizontal strain (0.636 m) appeared at the cross Sect. 3 with a period of 2475 years, using the R1a acceleration record. This can be considered a large strain with high potential for dam instability. However, one of main limitations in the simulation is that the elastic model employed does not consider the degradation of soil characteristics with the increase of strain.

Given the event of a 1 in 475 year earthquake, effect of raising reservoir water level on liquefaction potential was implemented by taking into account three reservoir water levels, namely, the normal water level, the design flood level (probability 1%) and the check flood level (probability 0.2%). The simulation was carried out with the cross Sect. 2, considering the earthquake intensity grade VIII (MSK-64 scale), acceleration record R3a, the time duration of 10 s. The simulation results show that the liquefied zone size increases gradually with the water level. The largest liquefied zone occurs when the reservoir reached the checked flood level. Note that the probability that the checked flood water level happens at the same time with the strong earthquake is not mentioned in the existing standard.

Regarding the effect of material model, the analysis results from the same cross Sect. 2 show that in general, the size of liquefied zone simulated by the linear elastic model was larger than that by the linear equivalent model. The simulation difference could be due to the fact that in the linear elastic model the deformation characteristics are assumed to be constant, whereas in the linear equivalent model the dynamic soil properties such as shear modulus (deduced from Young's modulus with an assumption of Poisson's ratio) and the damping ratio vary with the strain level (Fig. 7), so the analysis results using equivalent linear model can be more realistic.

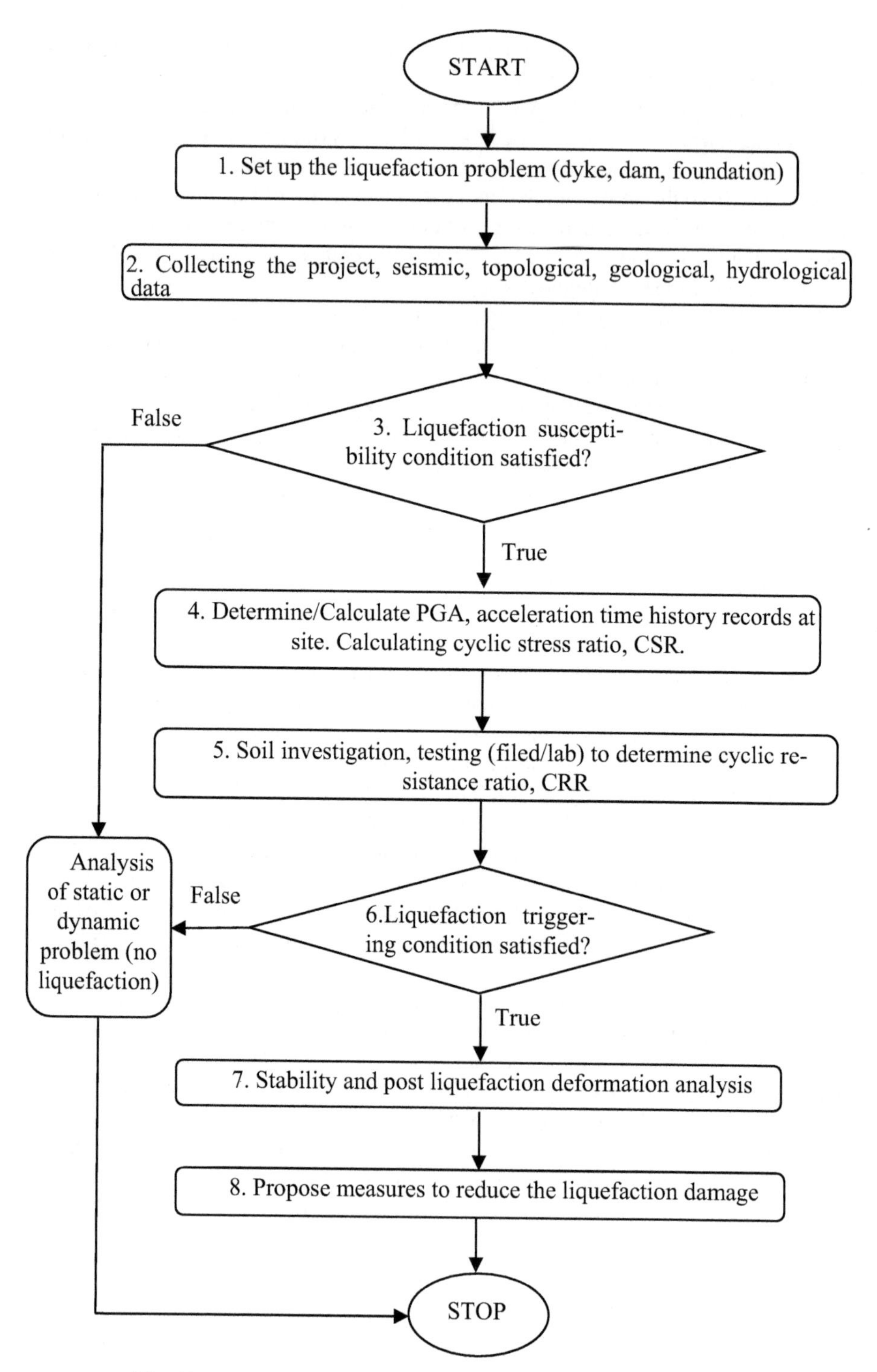

Fig. 10. Procedure for screening, analysis, prediction of liquefaction.

The summary procedure for defining, screening and predicting the liquefaction problem of in situ material dams subjected to strong earthquakes is plotted in Fig. 10, referring to the development of simple procedure [21, 22, 30]. The application of the above procedure needs to ensure synchronous steps such as building a seismic database, soil dynamic data, calculating seismic resistance, promulgating seismic design standards for in-situ material dams. The procedure needs to be continuously studied and put into practical application in the design of seismic resistance for local material dams and other construction works.

4 Conclusions and Recommendations

The analysis results of Loong Luong 1 dam show that the foundation sand layer in the upstream and under the downstream drainage rockfill could be highly liquefied when subjected to strong earthquakes with the return periods of 475 years and 2475 years.

The liquefaction potential of the dam depends on some main factors such as seismic, geotechnical data, and reservoir water level. The liquefied zone becomes bigger subjected to stronger earthquake, and higher the reservoir water level.

It is necessary to develop a liquefaction safety factor mapping system for in-situ material dams to proactively respond to strong earthquake disasters. The design standards for seismic resistance and liquefaction prevention for local material dams must be issued in the coming time. It is necessary to take measures to prevent dam instability in the event of strong earthquakes, especially during the flood season.

The author would like to acknowledge the Ministry of Science and Technology, Vietnam for providing the financial support within the framework of the state-funded research project No. KC08.23/11-15.

References

1. Adalier, K., Sharp, M.K.: Embankment dam on liquefiable foundation—dynamic behavior and densification remediation. J. Geotech. Geoenviron. Eng. **130**(11), 1214–1224 (2004)
2. Ambrasey, N., Smith, P., Berardi, R., Rinaldis, D., Cotton, F., Berge-Thierry , C.: Dissemination of European Strong - Motion Data, CD-ROM collection. European council, Environment and Climate Research Programme (2000)
3. Boore, D., Atkinson, G.: Next generation attenuation relations to be published in Earthquake Spectra (2007)
4. Campbell, K., Bozorgnia, Y.: Next generation attenuation relations to be published in Earthquake Spectra (2007)
5. Chiou, B., Youngs, R.: Next generation attenuation relations to be published in Earthquake Spectra (2007)
6. Cornell, C.A.: Engineering seismic risk analysis. Bull. Seismol. Soc. Am. **58**(5), 1583–1606 (1968)
7. Dobry, R., Ladd, R. S., Yokel, F. Y., Chung, R. M., Powell, D.: Prediction of pore water pressure buildup and liquefaction of sands during earthquake by the cyclic strain method. National Bureau of Standards, Gaithersburg, MD, NBS Building Science Series 138 (1982)
8. FEMA: Federal Guidelines for dam safety. Earthquake analyses and design of dams (2005)
9. Geo-slope International Ltd.: Dynamic modeling with Quake/w 2007, An engineering methodology, 4th edn. (2010)

10. General Department of Irrigation: Report on the management of reservoir dam safety, Proceedings of conference on safety management of water resource reservoirs, MARD, Hanoi 28/3/2018 (9 p.) (2018). (in Vietnamese)
11. Golder Associates: Design Guidelines for Dikes, second edition, prepared for Ministry of Forests, Lands and Natural Resource, Operations Flood Safety Section, British Columbia, Canada (2014)
12. ICOLD: Selecting seismic parameters for large dams. Guidelines, Revision of Bulletin 72, Committee on Seismic Aspects of Dam Design, International Commission on Large Dams, Paris (2010)
13. Institute of Engineering Mechanics (China) Chinese Strong - Motion Data, CD-ROM collection
14. Hong Nam, N.: Study on liquefaction potential of insitu material dams and dykes subjected earthquake loading and structural stabilization measures. State funded project No. KC.08.23/11-15, Final report (2016). (in Vietnamese)
15. Koester, J.P.: Seismic activity and vulnerabilities to levees. In: 2010 Missouri River/Texoma Regional and Midwest Levee Conference (2010)
16. Marcuson, W.F., Hynes, M.E., Franklin, A.G.: Seismic Design and Analysis of Embankment Dams - The State of Practice. The Donald M. Burmister Lecture, Columbia University (2007)
17. Matsuo, O.: Damage to river dikes. Soils Found. **36**(Special issue), 235–240 (1996)
18. Ordaz, M. Aguilar, A., Arboleda, J.: CRISIS99: a computer code to evaluate seismic hazard. Engineering Institute, National Autonomous University of Mexico, Mexico (1999)
19. QCVN 04-05:2012/BNNPTNT: National technical regulation on hydraulic structures - The basic stipulation for design (2012). (in Vietnamese)
20. Seed, H.B., Lee, K.L., Idriss, I.M., Makadisi, F.I.: The slides in the San Fernando dams during the earthquake of February 9, 1971. ASCE. J. Geotech. Eng. Div. **101**, 651–688 (1975)
21. Seed, R.B., et al.: Recent advances in soil liquefaction engineering: a unified and consistent framework. Keynote Presentation, 26th Annual ASCE Los Angeles Geotechnical Spring Seminar, Long Beach. Report No. EERC 2003-06 (2003)
22. Seed, H.B., Idriss, I.M.: Simplified procedure for evaluating soil liquefaction potential. J. GED, ASCE **97**(9), 1249–1273 (1971)
23. Schnable, P.B., et. al.: SHAKE: a computer program for earthquake response analysis of horizontally layered site. Earthquake Engineering Research Center, College of Engineering, University of California, Berkeley, California, USA (1972)
24. Silva, W.J.. Lee, K.: State-of-the-Art for Assessing Earthquake Hazards in the United States. Report 24. WES RASCAL Code for Synthesizing Earthquake Ground Motions (1987)
25. Son, L.T.: Calculation of PGA, acceleration time histories for in-situ material dams in Dien Bien province, with return periods of 475 years and 2475 years, Report No. 2-2., State funded Project No. KC.08.23/11-15, 140 p. (2015). (in Vietnamese)
26. Sugita, H., Tamura, K.: Development of seismic design criteria for river facilities against large earthquakes. Earthquake Disaster Prevention Research Group, Japan (2007)
27. TCVN 9386-2012: Design of structures for earthquake resistances (2012). (in Vietnamese)
28. USBR: Design Standards: Embankment dams, No. 13, Chap. 13. Seismic design and analysis. US Bureau of Reclamation, Denver, Co. (2015)
29. Xuyen, N.D.: Earthquake prediction and ground motion in Vietnam, state funded project, final report (2004). (in Vietnamese)
30. Youd, T.L., et al.: Liquefaction resistance of soils: summary report from the 1996 NCEER and 1998 NCEER/NSF workshops on evaluation of liquefaction resistance of soils. J. Geotech. Geoenviron. Eng. **127**(10), 817–833(2001)

Seismic Assessment of a Dam on a Clayey Foundation

Franklin R. Olaya[1](✉) and Luis M. Cañabi[2]

[1] University of California, Berkeley, CA 94720, USA
folaya@berkeley.edu
[2] Itasca Peru SAC Lima, Lima, Peru

Abstract. The seismic performance of a rockfill dam founded on a 20 m clayey foundation layer is presented. This stratum is the foundation of a 45 m high rockfill dam. A specific site seismic hazard study indicates that a Mw 8.0 event with peak ground acceleration of about 0.45 g is appropriate for design. The final aim of this ongoing study is to study the fragility of the dam-foundation system. The geotechnical characterization of the foundation involved the in-situ measurement of index properties (e.g., water content and PI) and the retrieval of undisturbed samples. Next, a battery of compressibility and strength testes were performed. The inferred compressibility characteristics of the clay layer and the measured monotonic stress-strain curves allowed the calibration of the modified Cam-Clay model. For the dynamic evaluation, a series of monotonic and cyclic direct simple shear tests results representative of this clay layer were collected to calibrate the PM4Silt model. Shear wave velocity profiles allowed the dynamic characterization of the core material and rockfill shells for which the UBCHyst model was employed. The static and dynamic analysis results are presented in terms of permanent deformations for design level motions.

Keywords: Clay foundation · Seismic response · Dynamic analyses

1 Introduction

Typically, embankment and rockfill dams are founded on rock or hard soil. However, the case of a 45 m high clayey gravel core rockfill dam founded on a relatively soft clayey dam is illustrated. Preliminary Limit Equilibrium (LE) analyses indicate that the structure may be stable under gravity forces but that under seismic demands, the factors of safety (FoS) are close to unity. The integrity of the dam is evaluated in terms of the induced permanent deformations near the crest and downstream face and their relative size compared to the width of the seepage control compacted clayey gravel core and shells.

A series of index tests (in-situ density, grain size, water content and plasticity) were performed to classify relevant soil units (see Fig. 1). The rock basement is composed of reddish marls of low to medium strength (10 to 20 MPa) overlaid by a 4m thick clayey gravel layer with boulder content of 10% to 20%. On top of this gravel stratum, there is

© The Author(s), under exclusive license to Springer Nature Switzerland AG 2022
L. Wang et al. (Eds.): PBD-IV 2022, GGEE 52, pp. 1959–1967, 2022.
https://doi.org/10.1007/978-3-031-11898-2_179

a thick clay (fines content from 60% to 80%) layer with traces of sand and gravel. The thickness of this clayey soil ranges from 12 m to 20 m. The superstructure is made of a compacted clayey core (GC), two rockfill shells (GW) and a filter (GW).

The performance of the clay layer in the dam foundation, under static and dynamic loadings plays a key role in the performance of the dam-foundation system, specially if softening is induced and large deformations are triggered. To reliably evaluate deformations, appropriate nonlinear constitutive models are calibrated and used for both the static and dynamic analyses. The software FLAC 2D V. 8.1 [1] was used for all calculations.

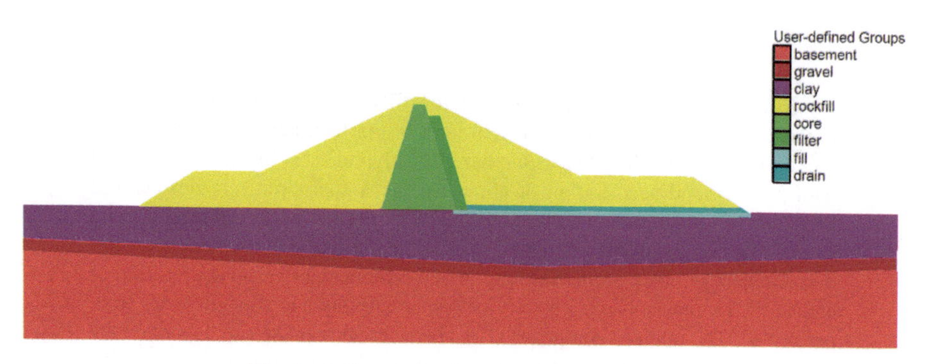

Fig. 1. Soil units conforming the dam model.

2 Site Conditions and Materials

The dam under evaluation is located in a high-seismicity area of the Andes. The controlling seismic sources are interface and intraplate with historic events as high as $M_w = 8$. A specific seismic hazard deaggregation evaluation indicates that a characteristic magnitude M_w 8 and PGA = 0.45g event contributes most to the hazard for a return period of 2500 yr. Therefore, a suite of 18 ground motions recorded in rock conditions (Site classes B and A) was compiled from which an initial subset of four $M_w \geq 8.0$ seed motions was selected for analysis. The motions were scaled and adjusted to fit the 2500 uniform hazard spectrum (UHS). Variability was not completely eliminated by loosening the matching criteria so that the scaled spectral values are bound by the ± 10% band around the UHS. The selected seed motions and the 2500 yr UHS band are shown in Fig. 2.

The clayey layer is mainly composed of lacustrine clays (60% - 80%) with some gravel and sand traces (20% - 40%). The fine matrix has water content (w_c) ranging from 12% to 36% and a plastic limit (PL) of 20%. Undisturbed samples were obtained using 3-in and 3 1/2-in diameter Shelby tube samplers from which 2.8-in diameter specimens were trimmed for testing. One-dimensional consolidation tests indicate that the in-situ pre-consolidation pressure of this clay is about 100 kPa with an OCR close to 1.1. Isotropically consolidated undrained triaxial tests (CU TX) indicate an average undrained shear strength ratio, S_u/σ'_v, of 0.23. The dam shells are composed of GW-GC materials with average grain size composition of 50% Gravels, 40% Sand and 10%

Fines and characterized in-situ by normalized SPT values, $(N_1)_{60}$, ranging from 20 to 28, whereas the core is composed of compacted clayey gravel moraine. Measured average shear wave velocities for the rockfill and clay layer are 360 m/s and 250 m/s respectively. Table 1 gives a summary of material properties.

Table 1. Classification and material properties.

Material	Classification	Unit weight (kN/m^3)	Friction angle (°)	S_u/σ'_v	G_{max}
Basement	Marl	22.0	--	--	5.6 E9
Gravel	GC	18.0	35	--	4.5 E8
Clay	CL - SC	17.0	--	0.23	1.1 E8
Rockfill	GW - GC	21.0	38[1]	--	3.0 E8
Filter	GW-SW	20.0	38	--	4.0 E8
Fill	SW	20.0	38	--	4.0 E8
Drain	GW, GP	20.0	38	--	4.0 E8

1. This is an average value. A loglinear variation of the friction angle was considered in the analysis

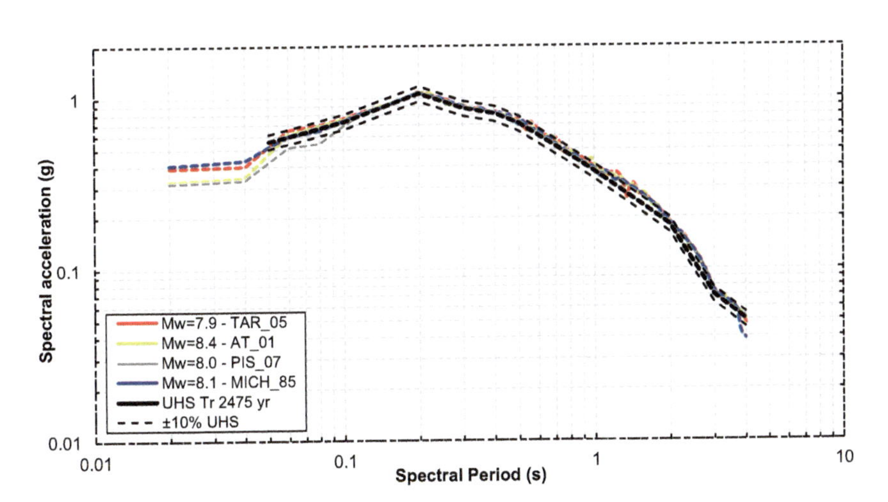

Fig. 2. Target response spectrum and suite of candidate seed motions.

3 Analysis Procedure

3.1 Static Response

The static response is concerned with the foundation response to the stresses induced by the construction of the dam and with the deformations developed in the dam body. Particular attention is paid to deformation of the clay layer and around the rockfill and core interface so that continuity is assured. The rock foundation is considered as an elastic half space; granular materials are model as Mohr-Coulomb materials and the clay layer which is expected to respond normally consolidated was modeled using the Modified Cam-Clay (MCC) model. A description of the model and the calibration process is shown as follow.

The MCC as proposed by [2] emanates from a balance of virtual work and internal energy dissipated by the soil. The model parameters capture the compressibility and strength of the soil. Therefore, information from the 1D consolidation and CU TX tests was employed to calibrate the model. The set of model parameters are shared in Table 2 and the calibration results are presented in Fig. 3 where the test data is shown in black solid lines (CU1 and CU2) and the MCC stress-strain curves in marked blue lines (MCC-CU1 and MCC-CU2).

Table 2. Modified Cam-clay parameters.

MCC parameter	Value
M	1.10
λ	0.10
κ	0.03
N	2.09
ν	0.20

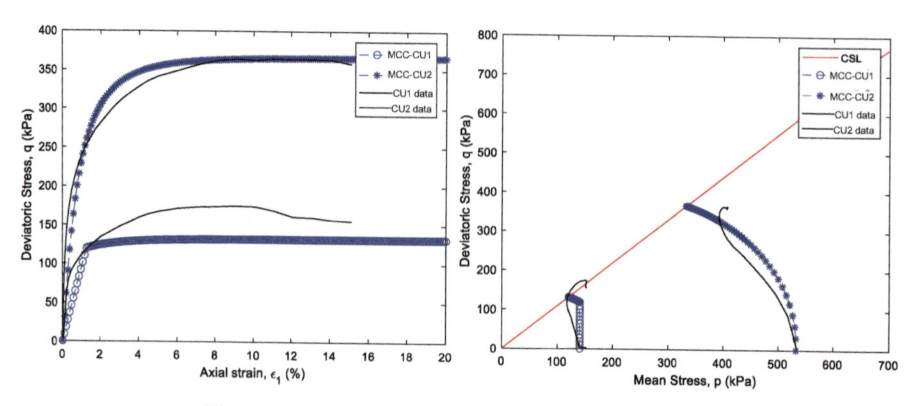

Fig. 3. Calibration results of MCC using CU TX tests.

The piezometric water level in the dam was estimated from piezometric readings from 2009 to 2018 which indicate that the vertical and horizontal drain in the downstream face of the dam function correctly. The resulting steady-state distribution of pore water pressure is shown in Fig. 4. The horizontal displacements after construction were estimated with FLAC and shown in Fig. 5a. The deformations induced by construction concentrate along the interface between the clay and the dam body. The downstream portion of the dam slides around 35 cm near the toe and the rockfill deforms between 15 cm to 35 cm. Vertical settlements of about 1.8 m concentrate in the top of the core/clay contact as shown in Fig. 5b. No records of construction-induced movements are available.

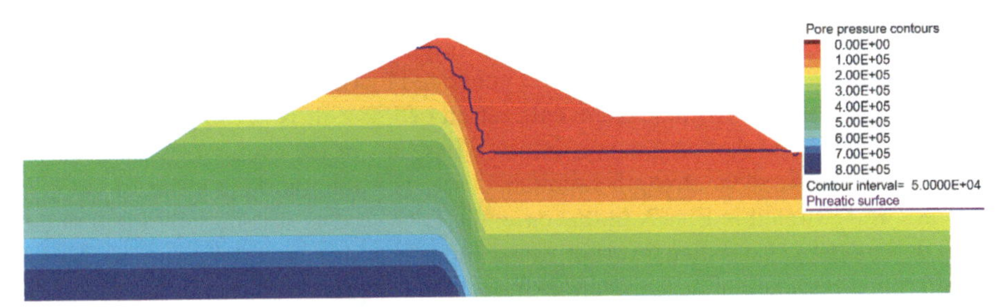

Fig. 4. Distribution of pore water pressures in the dam.

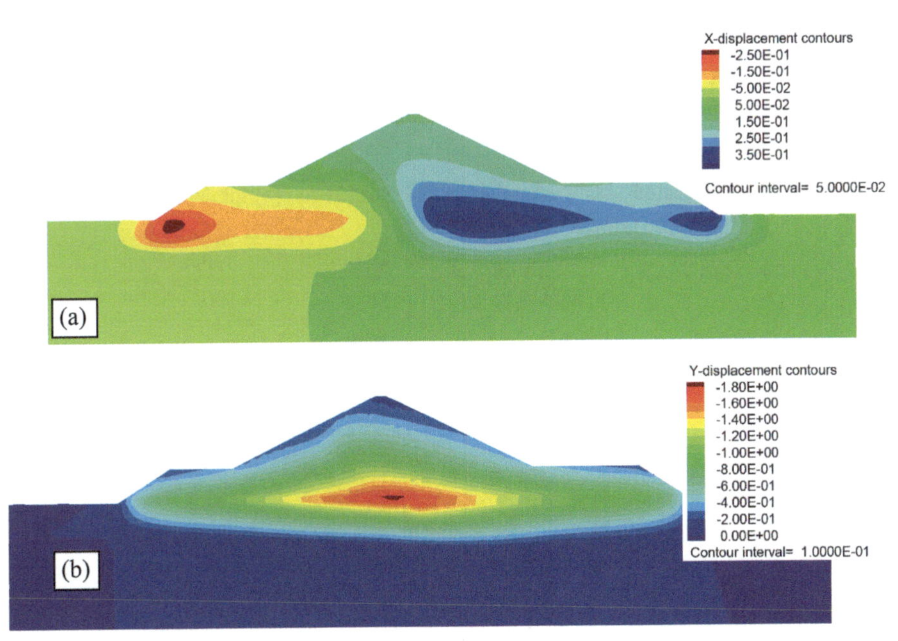

Fig. 5. (a) X-direction construction-induced displacement. (b) Y-direction construction-induced displacements

3.2 Dynamic Response

The dynamic response of the dam and clay materials were captured by means of the UBCHyst [3] and PM4Silt [4] models. UBCHyst can be seen as an extension of the Mohr-Coulomb model that incorporates a nonlinear hyperbolic stress-strain relationship and damping generation following the Masing rule, thus it provides an approximation of hysteretic response to cyclic loading while PM4Silt is an advanced critical-state-based bounding surface model developed for silts and clays. According to the grain size distribution of the rockfill material discussed in Sect. 2 and the N-SPT values, the G/G_{max} and damping curves from Rollins et al. [5] were assumed to be representative for the rockfill as a function of the effective confinement. UBCHyst have been used to model coarse grained materials that are expected to respond fully drained (e.g., Armstrong et al. [6]). The G/G_{max} and damping curves generated by UBCHyst on top pf the Rollins data are shown in Fig. 6 whereas the calibration of PM4Silt is performed not only in terms of G/G_{max} and damping curves but also in terms of the cyclic resistance (CSR) curve and the static shear strength simultaneously. The calibration in terms of G/G_{max} and CSR is illustrated in Fig. 7. In this case, the G/G_{max} data was initially fitted to the Darendeli model [7] and then PM4Silt was compared to this smooth curve. The degree of calibration obtained in both hysteretic models is adequate considering the degree of variability expected for this soil's responses. In addition, the overestimation of damping at $\gamma_{shear} > 1\%$ is expected in these model as reported in [3] and [4]. The model parameters used in this analysis are listed in Table 3. UBCHyst does not consider the increase of confinement with depth automatically, hence a pressure dependance was imposed so that key properties of the rockfill shell were captured. For example, the change in G_{max} with pressure is shown in Fig. 8

Table 3. UBCHyst and PM4silt main parameters. See [3] and [4] for detailed description

UBCHyst	Value	PM4Silt	Value
h_n	Variable(Fig. 8)	G_o	500
ϕdil	0	Su_{Rat}	0.3
R_f	0.98	h_{po}	1.2
h_{rm}	0.5	G_{exp}	0.5
hd_{fac}	0	$n^{b,wet}$	0.75
H_{modf1}	1	ϕcv	28

UBCHyst: hn: tangent stiffness exponent, ϕdil: dilation angle, Rf: hysteretic parameter; hrm: stress ratio pa-rameter; hdfac: Gmax reduction factor; Hmodf1: mod-ulus reaction. PM4Silt: Go: shear modulus coefficient, Surat: undrained shear strength ratio; hpo:contraction rate parameter; Gexp: exponent for shear modulus; nb,wet: strength reduction factor; ϕcv: critical state friction angle.

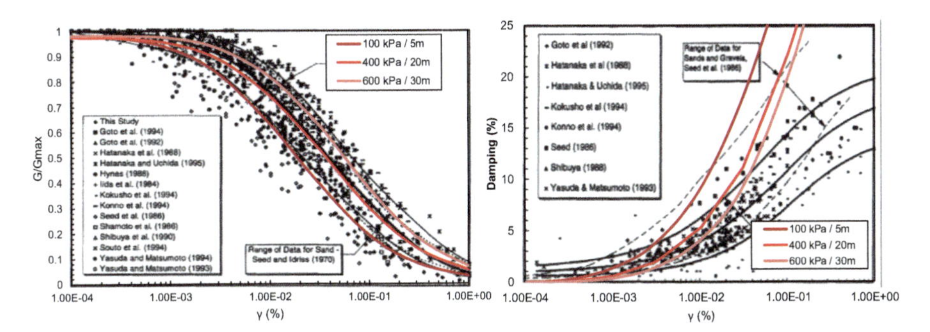

Fig. 6. G/G$_{max}$ and damping curves by UBCHyst. Rollins' data shown in the background

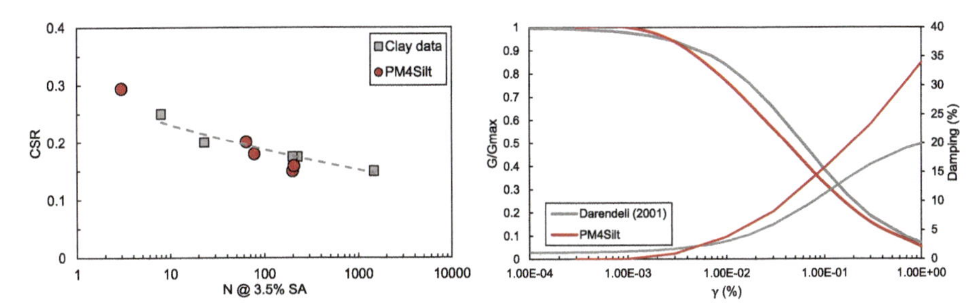

Fig. 7. CSR and G/G$_{max}$ calibration by PM4Silt

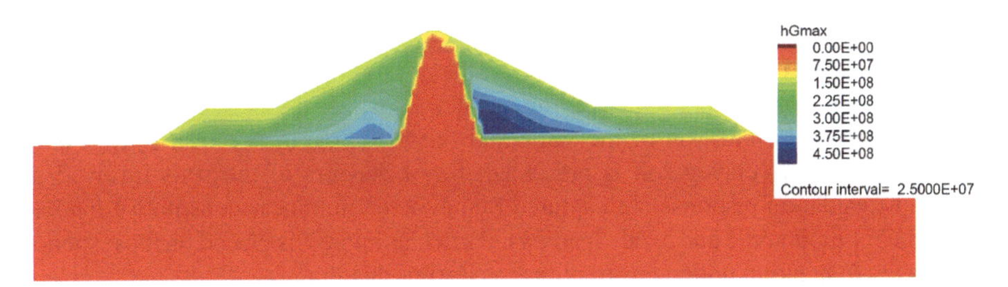

Fig. 8. UBCHyst hG$_{max}$ variation for dam shells

The dynamic response of the dam is ilustrated in Fig. 9 for time history Mw8.4 – AT_01. Similar to the static deformations, the relatively soft clay layer underlying the dam influences the pattern of deformation. In the horizontal direction, displacements of about 60 cm to 80 cm are expected in the downstream shell. It is expected that the deformation is caused by uniform sliding of the dam shell along the clay which in turn induces diferential deformations on the rockfill. The upstream shell of the dam that is in contact with the body of water develops a more rotational pattern triggerd by sliding along the clay. In terms of vertical deformation, the area near the upstream face of the crest settles around 90 cm. This amount of settlement is smaller than the impoundment free borad (3 m) and is located above the considered water level. However, the magnitude is not small and can be indicative of an engineering demand parameter (EDP) to consider in further analyses.

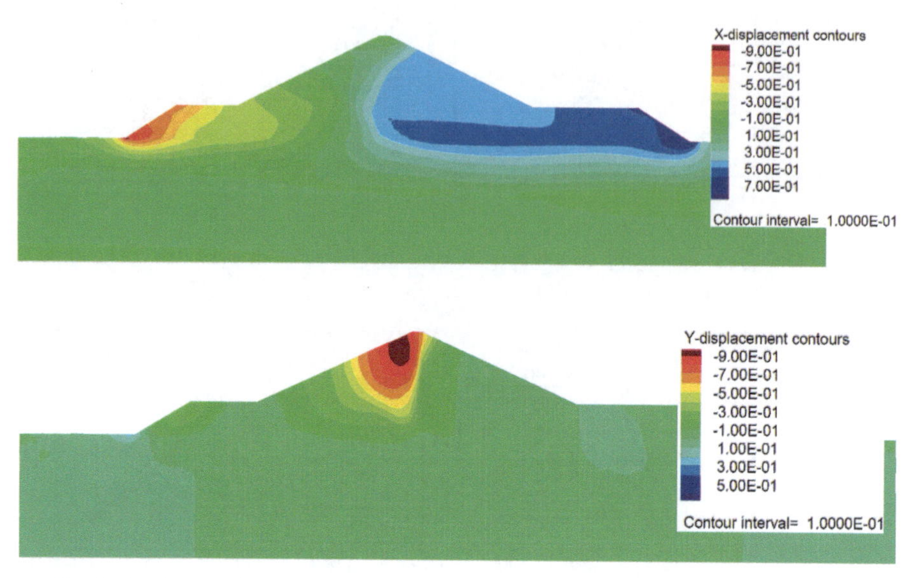

Fig. 9. Earthquake-induced Horizontal and Vertical displacements induced in the dam

4 Conclusion

A comprehensive evaluation of the static and seismic performance of a rockfill dam founded on mainly clayey soil is presented. Key behaviors of the soils involved in the dam-foundation response are captured by advanced nonlinear constitutive models adequate for monotonic and cyclic loadings. Particular interest is placed on the response of the clay foundation under construction and during shaking.

Different aspects of the constitutive relations are shown to be reasonably well captured by the Modified Cam-Clay, PM4Silt and UBCHyst models. After construction, settlements of about 1.8 m are expected in the foundation, whereas settlements of about 90 cm are expected to be induced in the crest of the dam by the ground shaking. Therefore, crest settlement is identified as a potential EDP for further analyses.

Next steps involve the use of the larger set of seed ground motions illustrated in Fig. 2 to evaluate the behavior of the selected EDP. The main objective is to capture the fragility of this earth structure and its dependance on the relatively soft foundation.

References

1. Itasca. Fast Lagrangian Analysis of Continua. Version 8.1 (2021)

2. Roscoe, K.H., Burland. J.B.: On the generalized stress-strain behaviour of 'wet' clay. In: J. Heyman, F., Leckie. (eds.) Engineering Plasticity, pp. 535–609. Cambridge University Press (1968)
3. Byrne, P.M.: UBCHsyt Model, University of British Columbia (2010)
4. Boulanger, R.W., and Ziotopoulou, K.: PM4Silt (Version 1): a silt plasticity model for earthquake engineering applications. Report No. UCD/CGM-18/01, Center for Geotechnical Modeling, Department of Civil and Environmental Engineering, University of California, Davis, CA (2018)
5. Rollins. K.M., Evans. M.D., Diehl, N.B., Daily III, W.D.: Shear modulus and damping relationships for gravels. J, Geotech. Geoenviron. Eng. **124** (5) (1998)
6. Armstrong, R., Tadahiro, K., Park, D.: Efficiency of ground motion intensity measures with earthquake-induced earth dam deformations. Earthq. Spectra **37**, 1–11 (2020)
7. Darendeli, M.B.: Development of a new family of normalized modulus reaction and material damping curves. The University of Texas at Austin (2001)

Seismic Fragility Analysis of Two Earth Dams in Southern Italy Using Simplified and Advanced Constitutive Models

Gianluca Regina[1(✉)], Paolo Zimmaro[1,2], Katerina Ziotopoulou[3], and Roberto Cairo[1]

[1] University of Calabria, Rende, Italy
gianluca.regina@unical.it
[2] University of California, Los Angeles, USA
[3] University of California, Davis, USA

Abstract. The seismic performance of earth dams can be assessed using various procedures spanning from simple deterministic methods to fully-probabilistic approaches, such as fragility analyses. In this study, analytical seismic fragility functions (i.e., based on results of numerical analyses) for two earth dams (the Farneto del Principe and the Angitola dams) in the Calabria region (Southern Italy) are developed. The Farneto del Principe dam does not have liquefaction-related issues, while the Angitola dam is founded on soils susceptible to liquefaction. The analyses are performed using numerical simulations based on the so-called multiple-stripe method. This framework takes inputs from site-specific probabilistic seismic hazard analysis results and it is used here to develop fragility functions for intensity measure (IM) values at the same return period of the seismic action. Separate calculations are presented, using the relatively simple hysteretic soil model implemented in the 2D finite difference method software FLAC and two advanced constitutive models (PM4Sand and PM4Silt) specifically developed to simulate the stress-strain response of sands and silts in geotechnical earthquake engineering applications. Fragility functions for several damage mechanisms and IMs are obtained. The analyses show that the IMs with the highest predictive power (i.e., those with the lowest fragility function standard deviations) are the peak ground acceleration and velocity (PGA and PGV) for the Farneto del Principe dam and the cumulative absolute velocity (CAV), the cumulative absolute velocity after application of a 5 cm/s^2 threshold acceleration (CAV5), and PGV for the Angitola dam.

Keywords: Fragility functions · Earth dams · PM4Sand · PM4Silt · Multiple-stripe analysis

1 Introduction

Earth dams are strategic and essential infrastructure systems for the economy and the society. In Italy, there are 165 large earth dams, with most of them being in operation for over 30 years. The seismic safety assessment of these structures has become critical in the last decade because of three main reasons: (1) the geotechnical properties of the

© The Author(s), under exclusive license to Springer Nature Switzerland AG 2022
L. Wang et al. (Eds.): PBD-IV 2022, GGEE 52, pp. 1968–1975, 2022.
https://doi.org/10.1007/978-3-031-11898-2_180

embankment may have changed over the years as a consequence of ageing; (2) most of the earth dams were designed using old standards in which seismic hazard and dynamic loading were either neglected or treated in a simplified manner; (3) the liquefaction potential of soils was typically ignored.

The evaluation of the seismic assessment of a strategic earth dam is typically performed using input motions derived from site-specific probabilistic seismic hazard analysis (PSHA). There are several procedures to define the target spectra and select hazard-consistent ground motions [1]. In this study, a PSHA model [2] developed for Southern Italy is adopted. Once the input motions to be used for response history analyses (RHA) are selected, two approaches can be followed to evaluate the seismic performance of a structure: (1) the uncoupled and (2) the coupled approaches. In the former, the seismic input is derived from probabilistic analyses, while the dynamic response of the system is based on a certain number of deterministic RHA [3]. However, this procedure comprises two issues: (1) the information gathered from the PSHA are essentially lost; (2) the results from deterministic analyses cannot be used to estimate the fragility and risk related to the structure. These issues can be solved using a fully probabilistic coupled approach by means of fragility functions, which, for any given level of seismic intensity, provide the probability that the structure reaches (or exceeds) a certain damage level or limit state. Presently, the number of seismic fragility functions for earth dams is very limited and based on input selections performed using uniform hazard spectra [4].

In this study, seismic fragility analyses consistent with a conditional target spectrum [5] are performed for two earth dams in Southern Italy: (1) the Farneto del Principe dam, (2) the Angitola dam. Both dams are located in the same area, characterized by high seismicity. However, they are built using different soil types. Thus, they are potentially vulnerable to different failure types and mechanisms. The former is non-susceptible to liquefaction, whereas the latter is founded on material susceptible to liquefaction (more details in subsequent sections). As a result, different constitutive models are used for these dams [6, 7]. Fragility functions (FFs) for these dams are obtained for multiple intensity measures (IMs) and damage measures (DMs) using numerical simulations based on the so-called multiple-stripe analysis (MSA) [8]. The fragility functions showed in this study can be used for structure-specific applications (e.g., to develop early-warning systems for these dams) and/or give information on the limit state that is most likely to be exceeded in earth dams similar to those analyzed herein.

2 The Case Studies

2.1 The Farneto Del Principe Dam

The Farneto del Principe dam is a zoned earth dam located in the Calabria region (Southern Italy, latitude 39.6515°N, longitude 16.1627°E) designed in the '60s and built during the late '70s and early '80s and it is used for flow balancing and irrigation purposes. The dam is located in a seismically active region characterized by the presence of shallow crustal faults and deep seismicity related to the subduction zone of the Calabrian Arc. The dam height is 30 m and its length is more than 1200 m; the crest elevation is at 144.20 m above sea level. The geotechnical characterization of the Farneto del Principe dam is based on a comprehensive field investigation program [9]. The results of the field investigation tests indicate that both the dam and the foundations are not susceptible to liquefaction. The water tightness of the dam system is ensured by a central core made of compacted, low permeability clay and silt. Its static stability is ensured by the shells, which comprise sand and gravel. Two filters (with thickness of one meter each) protect the dam core, with grain size distributions similar to that of the adjacent soil, both upstream and downstream. The foundation of the dam comprises a shallower layer of high-permeability alluvial material (i.e., sand and gravel) which is in turn founded on a thick clay layer. The exact location of a compliant bedrock is unknown. A cut-off wall is present throughout the length of the dam within the alluvial material and embedded for 3 m into the clay bed, to prevent groundwater flow and possible seepage. Figure 1a shows a representative cross section of the dam.

2.2 The Angitola Dam

The Angitola dam (latitude 38.7486°N and longitude 16.2319°E) is a zoned earth dam located in the Calabria region, Southern Italy, in the same seismic context of the Farneto del Principe dam (the two dams are 130 km apart). It was built in the period 1960–1966 and it has irrigation purpose. The structure comprises two dams located on the two sides of mount Marello. One of this dam is founded on soils that are potentially susceptible to liquefaction and thus it was selected to be the objective of this study. This decision was made to compare fragility results between a dam without liquefaction issues and another one with potential liquefaction prone-soils. The dam is 27.75 m high and about 195 m long. The core of the dam comprises medium plasticity silty sands, while the shells are composed of sandy gravel. There are two filters between the core and the shells; the internal filter, adjacent to the core, is 60 cm thick and comprises sand and gravel, while the external filter is 140 cm thick.

The foundation layers are characterized by a complex geology. On the downstream side, there is a 5m-thick layer of sandy silts, which comprises recent alluvial material and it is susceptible to liquefaction [10]. Beneath this layer, there is an old alluvial material comprising gravel with sandy silt, together with tiny fractions of clay material. The dam lays on a pliocenic formation of consistent silty clay, whose thickness rises longitudinally from downstream to upstream. Beneath those layers, a heterogeneous sand layer (with soil particles comprising gravel, sand, silt, and clay) is present, coming from the weathering of the underlying rock foundation (sandy layer), which is made up of fractured gneiss. The intact rock is about 40 m below the free surface. A representative cross section of the dam is shown in Fig. 1b.

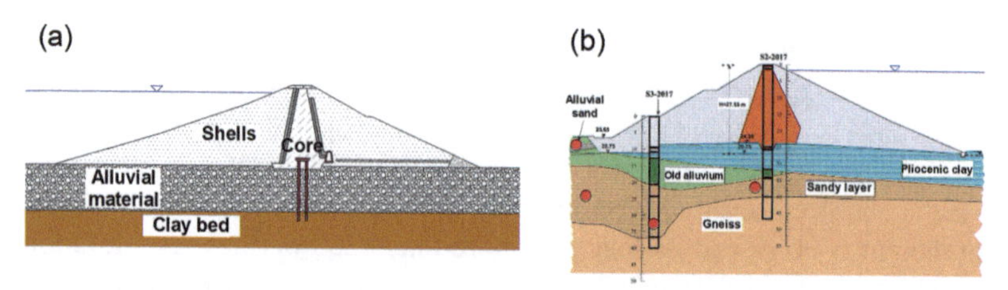

Fig. 1. Representative cross sections of: (a) Farneto del Principe dam and (b) Angitola dam.

3 Numerical Modelling of the Dams

The numerical modeling of the two case studies is performed with the 2D finite difference commercial software platform FLAC [11], which is based on the explicit, time-marching method to solve the equations of motion. The size of the mesh has been chosen based on the shear wave velocities and the geometrical details of the dams (i.e. filters zone). At the base of the model, quiet boundaries are used to simulate the elastic half-space. On the lateral boundaries of the model, the free-field boundary condition is adopted to ensure radiation damping towards infinity.

Four different constitutive models are used in this study. For the material susceptible to liquefaction of the Angitola dam, the PM4Sand V3.1 model is adopted [6]. The behavior of the foundation bedrock is assumed to be linear elastic, while all the other materials are characterized by the Mohr-Coulomb (MC) failure criterion coupled with a simplified hysteretic procedure [11]. For the core of the Farneto del Principe dam, the PM4Silt model [7] was also chosen to analyze the possible phenomenon of cyclic softening. PM4Sand and PM4Silt are plane strain stress ratio-controlled, critical state compatible, bounding surface plasticity models. Their calibrations are performed with single-element simulations for the range of loading paths important for the two dams. In particular, the PM4Sand and PM4Silt parameters are calibrated to obtain a cyclic resistance curve to liquefaction and cyclic softening, respectively, which matched those estimated from empirical procedures [12, 13]. For both dams, a 0.5% Rayleigh damping is used to remove high frequency noise.

3.1 Selection of Hazard Consistent Ground Motions

The choice of the input ground motions is conducted using a conditional spectrum as target [5] and the fundamental periods of the dams as the conditioning period. These types of response spectra are also called scenario target spectra, which are consistent with a scenario characterized by a certain magnitude, distance, and rate of occurrence. The controlling scenarios for the sites of interest are identified with the disaggregation of the seismic hazard, for six different return periods, T_R (75, 475, 710, 1460, 1950, and 2475). For each T_R, seven to ten inputs were selected to capture the uncertainties related to ground motions variability. More details about the ground motion selection and scaling process used in this study are provided by Regina [10].

4 Seismic Fragility Analysis

Fragility analyses are a popular tool to perform rapid assessment of the probability of damage induced by a stressing event on structure and infrastructural systems, including dams. FFs relate the probability of exceeding a certain DM given the occurrence of a seismic event with IM. Mathematically, they are typically defined as lognormal cumulative distribution functions:

$$P(\mathrm{d} > \mathrm{DM}|IM = x) = \Phi\left(\frac{\ln\frac{x}{\theta}}{\beta}\right) \tag{1}$$

In Eq. (1), θ is the IM that corresponds to 50% probability of exceedance (i.e., the median value), and β is the standard deviation of the natural logarithm of the IM.

FFs can be built using three approaches: (1) analytical (i.e., based on the results of numerical models); (2) empirical (i.e., based on empirical data); and (3) based on expert judgement. In this study, analytical fragility functions are developed. The procedure adopted to build FFs is the MSA, which is based on the method of maximum likelihood [8]. This approach is performed at discrete IM levels, typically corresponding to given T_R of the seismic action (i.e. those reported in Sect. 3.1). For the Farneto del Principe dam the following IMs were considered: peak ground acceleration and velocity (PGA and PGV, respectively) and Arias Intensity (AI). For the Angitola dam, in addition to those used for the Farneto del Principe dam, the following IMs were also used: cumulative absolute velocity (CAV) and the cumulative absolute velocity after application of a 5 cm/s^2 threshold acceleration (CAV5). A larger number of IMs were used for the Angitola dam because it is founded on materials susceptible to liquefaction, and cumulative velocity-based IMs were found to correlate well with liquefaction damage [14].

The DM for which FF are constructed are: filters displacement, global instability, free board reduction, excess pore water pressure ratio, r_u (for the Angitola dam only), Fell damage classes [15], and Swaisgood normalized crest settlement (NCS) [16]. The DM threshold adopted are both dam-specific (e.g. maximum displacement of the filters equal to the filter thickness) or taken from the literature.

4.1 Nonlinear Deformation Analysis Results and Fragility Functions

For the Farneto del Principe dam we ran nonlinear deformation analyses (NDAs) using two different models: (1) a model adopting the hysteretic model implemented in FLAC in combination with MC model as failure criterion (referred as MC dam model hereafter) and (2) a model where the PM4Silt constitutive model is used for the core, whereas all other materials are still modeled using the hysteretic MC model (this model is referred to as PM4Silt dam model hereafter). For the Angitola dam, two models were considered: (1) the simple MC dam model (where all materials are modelled using the hysteretic MC model) and (2) a model where sands susceptible to liquefaction were modelled using the PM4Sand model (this model is referred to as PM4Sand dam model hereafter). In this study, we show results from models #1-2 for the Farneto del Principe dam and model #2 for the Angitola dam. Regina [10] shows results for all models.

Figure 2 shows strain patterns produced by the two models for the Farneto del Principe dam and the r_u ratio in four zones of the core, near the upstream filter (upper, middle, lower, and central zone) using the PM4Silt dam model. As expected, excess pore water pressures are mostly negative. This indicates a dilative behavior of the compacted core material. Only in the lower portion of the core, where pore pressures and confining stresses are greater, there is a slight tendency to develop positive r_u values. Overall, differences between the MC and PM4Silt models in the amplitude and spatial distribution of strains in the Farneto del Principe dam are typically negligible, with the exception of few ground motions characterized by long return periods for which some differences are visible.

For this reason, all the FFs for the Farneto del Principe dam were derived using the MC model.

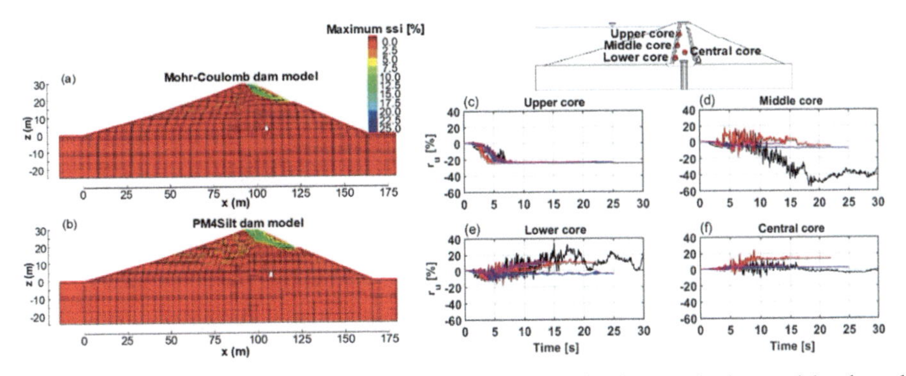

Fig. 2. Shear strains pattern for the Farneto del Principe dam for the constitutive models adopted (a) and (b), and excess pore water pressure ratio in some zones of the core (c-e).

Figure 3 shows typical results of MSA and the FFs obtained for the Farneto del Principe dam using the PGA as IM. It is interesting to note that, within the MSA framework, higher values of the IM does not necessarily produce an increasing fraction of exceeding a given DM. This is because each set of ground motions are scaled at the same value of the selected intensity measure. This scaling process can cause changes in the energy content of the input motion if compared against the original one. If the FFs are compared at the same IM, it is possible to identify the DM that is most likely to be exceeded. For the Farneto del Principe dam, this DM is the NCS, while the free board reduction is the least likely DM that is to be exceeded (Fig. 4a).

Figure 4c–f shows FFs obtained for the Angitola dam using the PM4sand dam model and the r_u as DM. The soil behavior is contractive only in some zones of the model (i.e. pit zone) and dilatant elsewhere (i.e. free-field zones). Thus, it is possible to build well-defined FFs for liquefaction only if the soil presents an overall contractive behavior. The IM with the least β are PGA and PGV for the Farneto del Principe dam, and CAV, CAV5, and PGV for the Angitola dam.

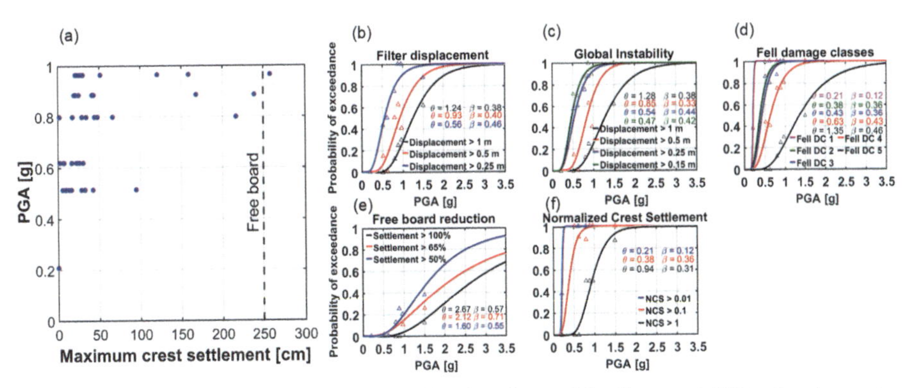

Fig. 3. Typical results of MSA (a) and fragility functions of the Farneto del Principe dam using the PGA as IM.

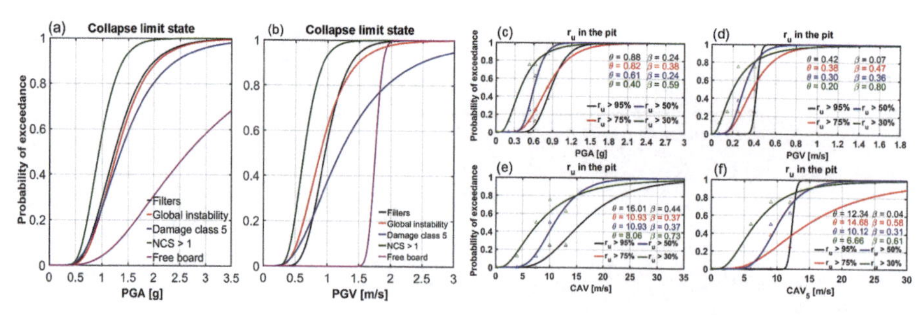

Fig. 4. Fragility functions comparison (a–b) and fragility functions for the excess pore water pressure ratio in the Angitola dam (c–f).

5 Conclusions and Discussion

In this paper, the results of seismic fragility analyses using a fully probabilistic approach for two earth dams in Southern Italy were illustrated. Ground motions were chosen to be hazard-consistent with a site-specific PSHA, while the seismic performance was addressed using fragility functions based on simplified and advanced constitutive models.

The Farneto del Principe dam did not present liquefaction- or cyclic softening-related issues. As a result, the comparison between simplified and advanced constitutive models were similar. Some soil layers in the foundations of the Angitola dam are susceptible to liquefaction. Thus, the more advanced PM4Sand constitutive model was used to derive FFs for this dam. The MSA approach was used to build FFs for several IMs and DMs. Results from this study show that the IMs with the highest predictive power are the PGA and PGV for the Farneto del Principe dam and the CAV, CAV5, and PGV for the Angitola dam.

Fragility functions can be used to determine the DMs that are most likely to be exceeded, thus creating a hierarchy of limit states. For both dams, the most important DM was found to be NCS followed by the global instability of the shells, and filters displacement. These outcomes are structure-specific and strictly speaking, only valid for the dams for which they were obtained. However, they can be used as screening tools for similar dams.

FFs may also be a useful tool for rapid post-earthquake damage evaluation. Such assessment can be performed using predicted IMs at the dam site, perhaps from readily available products such as ShakeMaps [17], or recorded IMs if the dam is instrumented with a monitoring system that includes ground motion recording stations. If the latter is available, fragility functions may be implemented as part of early warning systems [18]. If a strong ground motion occurs, the IM of the event is recorded and the probability that a certain damage will be exceeded can be promptly estimated and used in combination with other data available as part of the monitoring system (e.g., pore water pressures, crest settlements) to identify the performance of the structure.

Acknowledgements. We thank Professor Giuseppe Lanzo at University of Rome "La Sapienza", Mr. Pasqualino Cimbalo, and the Consorzio di Bonifica Tirreno-Catanzarese for kindly providing the geotechnical characterization data of the Angitola dam.

References

1. Ministero delle Infrastrutture e dei Trasporti. Aggiornamento delle Norme Tecniche per le Costruzioni, D.M.17-1-2018. Gazzetta Ufficiale della Repubblica Italiana, vol. 42 (2018)
2. Zimmaro, P., Stewart, J.P.: Site-specific seismic hazard analysis for Calabrian dam site using regionally customized seismic source and ground motion models. Soil Dyn. Earthq. Eng. **94**, 179–192 (2017)
3. Krawinkler, H., Miranda, E.: Chapter 9 - Performance-based earthquake engineering. Bozorgnia, Y., Bertero, V.V., (eds.) Earthquake Engineering. CRC Press (2004)
4. Jin, C., Chi, S.: Seismic fragility analysis of high earth-rockfill dams considering the number of ground motion records. Math. Probl. Eng. **2019**, 1–12 (2019)
5. Baker, J.W.: Conditional mean spectrum: tool for ground motion selection. ASCE J. Struct. Eng. **137**(3): 322–333 (2911)
6. Boulanger, R.W., Ziotopoulou K.: PM4Sand (Version 3.1): A sand plasticity model for earthquake engineering applications. Report No. UCD/CGM-17/01 University of California, Davis, CA (2017)
7. Boulanger, R.W., Ziotopoulou, K.: PM4Silt (version 1): a silt plasticity model for earthquake engineering applications. Rep. No. UCD/CGM-18/01. Center for Geotechnical Modeling, Department of Civil and Environ. Engineering, University of California, Davis, CA (2018)
8. Baker, J.W.: Efficient analytical fragility function fitting using dynamic structural analysis. Earthquake Spectra **31**(1), 579–599 (2015)
9. Regina, G., Ausilio, E., Dente, G., Zimmaro, P.: A critical overview of geophysical investigation and laboratory test results used in the seismic re-evaluation of the Farneto del Principe dam in Italy. In: 6th International Conference on Geotechnical and Geophysical Site Characterization Budapest (2021)
10. Regina, G.: Probabilistic assessment of the seismic performance of two earth dams in Southern Italy using simplified and advanced constitutive models. Ph.D. Dissertation (2021)
11. Itasca - FLAC – Fast Lagrangian Analysis of Continua – Version 8.1, User's Guide, Itasca Consulting Group, Minneapolis, USA (2019)
12. Boulanger, R.W., Idriss, I.M.: CPT and SPT based liquefaction triggering procedures. Rep. No. UCD/CGM-14/01, Univ. of California, Davis, CA (2014)
13. Boulanger, R.W., Idriss, I.M.: Evaluation of cyclic softening in silts and clays. J. Geotech. Geoenviron. Eng. ASCE **133**(6), 641–52 (2007)
14. Kramer, S.L., Mitchell, R.A.: An efficient and sufficient scalar intensity measure for soil liquefaction. Earthq. Spectra **22**(2), 1–26 (2006)
15. Fell, R., MacGregor, P., Stapledon, D., Bell, G., Foster, M.: Geotechnical Engineering of Dams, 2nd edn. Taylor & Francis Group (2014)
16. Swaisgood, J.R.: Behavior of embankment dams during earthquake. J. Dam Safety, ASDSO **12**(2), 35–44 (2014)
17. Worden, C.B., Wald, D.J.: ShakeMap manual online: technical manual, user's guide, and software guide. U.S.1022 Geological Survey (2016)
18. Pagano, L., Sica, S.: Earthquake early warning for earth dams: concepts and objectives. Nat. Hazards **66**, 303–318 (2012)

Comparison of the Monotonic and Cyclic Response of Tailings Sands with a Reference Natural Sand

David Solans(✉), Stavroula Kontoe, and Lidija Zdravković

Imperial College London, London, UK
david.solans16@imperial.ac.uk

Abstract. This article considers the monotonic and cyclic behaviour of a tailings sand material, tested over a wide range of confining pressures and relative densities. The collated experimental data are used to interpret the behaviour of this material in the framework of Critical State Soil Mechanics (CSSM) and to calibrate an advanced bounding surface plasticity model. The response of this man-made material is then contrasted, within the same framework, to that of Toyoura sand, which is a well-characterised reference natural sand.

This process aims to identify the principal behavioural differences between the two types of geo-materials and the implications that these may have on the calibration process of an advanced constitutive model. Ultimately, the process also supports the numerical modelling of Tailings Storage Facilities under both monotonic and cyclic/earthquake loading.

Keywords: Numerical modelling · Mine tailings · Bounding surface plasticity

1 Introduction

Several recent failures (e.g. [6, 11, 15]) in tailings storage facilities (TSFs) spotlight the need for further research in these complex man-made structures. Their extensive use in countries of high seismicity, also renders necessary a good understanding of their seismic response. A systematic study on the response of TSFs requires advanced numerical tools and material characterisation data for their calibration. In practice laboratory strength tests (e.g. triaxial, simple shear) in tailings materials are usually restricted due to the mining companies' policies and therefore there is a scarcity of published test data (e.g. [2, 5, 8]). Site-specific calibrations of numerical models are rarely attempted due to the lack of data. Analysts of TSFs often adopt a "default" set of parameters, usually only adjusted to be compatible with CPT/SPT site data, following, for example, the process proposed by Boulanger & Ziotopoulou (2013) [1] for the PM4Sand model. A major limitation of this type of generic calibration processes is that they are based on natural sands. Hence, to understand the static and seismic behaviour of tailings dams, it is important to develop consistent calibration procedures for these artificial materials based on experimental data. Through the comparison of calibrations for a tailings sand and for a natural sand, this article aims to explore their most relevant characteristics, intending to assess the likely impact for subsequent seismic simulations.

© The Author(s), under exclusive license to Springer Nature Switzerland AG 2022
L. Wang et al. (Eds.): PBD-IV 2022, GGEE 52, pp. 1976–1983, 2022.
https://doi.org/10.1007/978-3-031-11898-2_181

2 Sands Descriptions

Two sandy soils are analysed herein: El Torito tailings sand and Toyoura sand. The former is a cycloned tailings sand material employed in the construction of the El Torito tailings dam (Chile), while the latter is a well-known reference laboratory sand.

Due to various mining processes, mainly crushing of rock blocks, tailings can be treated as artificial soils where the particle shapes are typically sub-angular to angular, with presence of fines contents. The El Torito tailings sands cover a wide range of particle sizes with non-plastic fines content reported from 15% to 23%, which is mainly attributed to the mineralogy of copper tailings. On the other hand, Toyoura sand is a natural well-characterised sand consisting of rounded to subangular particles formed by natural processes. Toyoura sand shows a uniform particle size distribution without fines contents.

Figure 1 plots the particle size distribution (PSD) for both sands, also including the coarse fraction of the tailings sand which is very similar to Toyoura sand in terms of PSD.

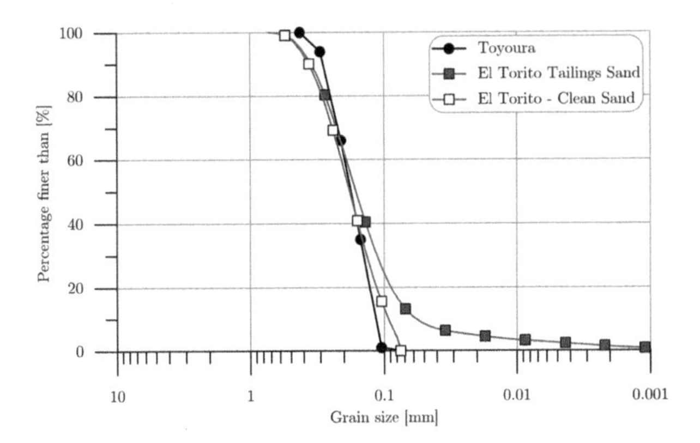

Fig. 1. Particle size distribution for Toyoura sand, the El Torito tailings sand and clean fraction

The characterisation of the El Torito tailings sands, which includes critical state strength, stiffness at small strains and cyclic strength, is presented and discussed in detail in Solans et al. (2019) [19]. Toyoura sand is well-known reference sand, with a large set of references available (e.g. [5, 7, 23, 24]).

3 Constitutive Model

A state parameter-based bounding surface plasticity model (BSPM) is employed herein, implemented in the Imperial College Finite Element Program ICFEP ([13]) by Taborda et al. (2014) [20] who introduced several modifications to the model originally proposed by Manzari & Dafalias (1997) [9]. The cyclic behaviour of the soil is simulated by means of a fabric variable influencing the hardening modulus ([12]), as well as by modifications introduced by Tsaparli (2017) [21] to the spherical part of the flow rule and the fabric tensor aiming to improve the prediction of the cyclic strength. Further details of the model formulation can be found in Taborda et al. (2014) [20] and Tsaparli (2017) [21]. A summary of all model equations and a recent application can be found in Tsaparli et al. (2019) [22]. Table 1 summarises the model parameters obtained through the calibration process for both materials.

Table 1. Model parameters BSPM for Toyoura sand and El Torito tailings sand.

Component	Toyoura sand	El Torito tailings sand
Critical State	$p'_{ref} = 100$ kPa; $e_{o,ref} = 0.9325$; $\lambda = 0.019$; $\xi = 0.70$	$p'_{ref} = 100$ kPa; $e_{o,ref} = 1.105$; $\lambda = 0.2395$; $\xi = 0.205$
Small-strain parameters	$B = 840$; $n_g = 0.5$; $\nu = 0.15$	$B = 458$; $n_g = 0.55$; $\nu = 0.15$
Surface parameters	$M^c_{cs} = 1.25$; $M^e_{cs} = 0.9$; $k^b_c = 1.1$; $k^d_c = 3.4$; $A_o = 0.9$	$M^c_{cs} = 1.50$; $M^e_{cs} = 1.05$; $k^b_c = 2.0$; $k^d_c = 4.0$; $A_o = 0.80$ For flow rule mod: $A_o = 0.80$; $A_{o,min} = 0.0$; $b_d = 0.10$; $d_1 = 0.07821$; $d_2 = 0.00388$; $d_3 = 0.0$
Hardening modulus	$h_o = 0.1025$; $\nu = 0.02$; $\alpha = 1.0$; $\beta = -0.5$; $\gamma = 0.92$	$h_o = 0.13$; $\nu = -0.10$; $\alpha = 1.0$; $\beta = -0.5$; $\gamma = 0.92$ $d_1 = 3.83861$; $d_2 = 0.00538$
Fabric Tensor	Fabric: Ho = 55000; $\zeta = 1.55$; $C_f = 50$	Fabric: Ho = 14890; $\zeta = 0.317$; $C_f = 75$; For flow rule mod: Ho = 15000; $\zeta = 0.0$; $C_f = 75$

4 Calibration Results

The position of the CSL in the $\log p' - e$ plane is presented in Fig. 2 for both Toyoura sand and the El Torito tailings sand, along with the initial states (red and blue symbols) for the selected monotonic and cyclic tests to be simulated. For Toyoura sand, the position of the CSL is clear, with an only minor scattering in the data interpretation. Conversely, the El Torito tailings sand data presents high scatter with values of $\Delta e \sim 0.05$ to 0.1, which is consistent with the observed range reported for tailings by Fourie & Papageorgiou (2003) [4] and Reid et al. (2020) [14]. It is noteworthy that for both sands the void ratio was determined at the end of each test, following the procedure recommended by Verdugo & Ishihara (1996) [23].

In terms of the state parameter, ψ, , the monotonic tests selected include contractive ($\psi > 0$) and dilative ($\psi < 0$) conditions; while for the cyclic tests, only the dilative cases ($\psi < 0$) have been taken into account, due to the model formulation of the fabric tensor[1].

[1] The fabric tensor is affected by the H parameter defined as: $H = H_o\left(\sigma'_{1,0}/p'_{ref}\right)\langle-\psi_o\rangle$, where $\langle\rangle$ corresponds to the Macauley brackets ($\langle x\rangle = x$ if $x > 0$ and $\langle x\rangle = 0$ if $x < 0$). (Taborda et al. 2014) [20]

Both sets of experimental data, Toyoura and El Torito tailings sands, were obtained for reconstituted samples formed using the moist tamping technique. The initial state conditions, $p'_o - e_o$, were defined aiming to cover a wide range of densities and mean effective stresses for Toyoura sand. For the El Torito tailings sand, the initial conditions aimed to reproduce the in-situ compaction state with an initial relative density of $D_{Ro} =$ 60%, a typical value for constructing cycloned tailings dams, and high p' values which are expected in TSFs for dams over 80 m height.

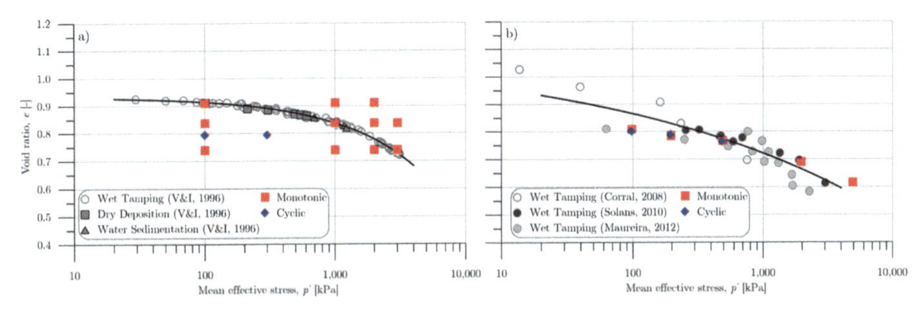

Fig. 2. Critical State Line and selected tests for: a) Toyoura sand ([23]), and b) El Torito tailings sand ([3, 10, 17])

Examining the two materials, the observed scatter evidences a bigger challenge to establish the CSL for the tailings than for the natural sand. It is noteworthy that no significant particle breakage was detected after the tests on the tailings sand, regardless of the high mean effective stresses applied ([19]).

4.1 Monotonic Response

The result of the model calibration is presented first for the monotonic response for both Toyoura and the El Torito tailings sands.

Figure 3 compares the experimental data ([23]) and the model simulations for undrained triaxial compression tests of Toyoura sand. The experimental data covers a wide range of mean effective stresses (*p'* from 100 kPa to 3 MPa) and void ratios (*e* from 0.735 to 0.907). The model simulations, indicated in solid lines, show a good agreement with the experiments, indicated with dotted lines, reproducing a variety of responses from highly dilatant to highly contractant. Also, the model reproduces well the phase transformation state and peak strength, typically observed in sandy soils.

A set of undrained triaxial tests for the El Torito tailings sand obtained from Solans (2010) [17] and Maureira (2012) [10] has been chosen for the numerical simulations, see Fig. 4. The selected tests cover a wide range of mean effective stresses (*p'* from 98 kPa to 4.9 MPa) and void ratios (*e* from 0.613 to 0.805). A major challenge in the calibration process is capturing the experiments for a wider range of mean effective stresses with a single set of parameters. The simulations provide a good agreement with the experimental data, reproducing the characteristic aspects (e.g. phase transformation) of the sand response. Better simulation results are obtained if the calibration is separated for a low and a high range of p' ([18]). Complementary simulations, for both materials, under drained triaxial compression conditions can be found in Solans (2022) [18].

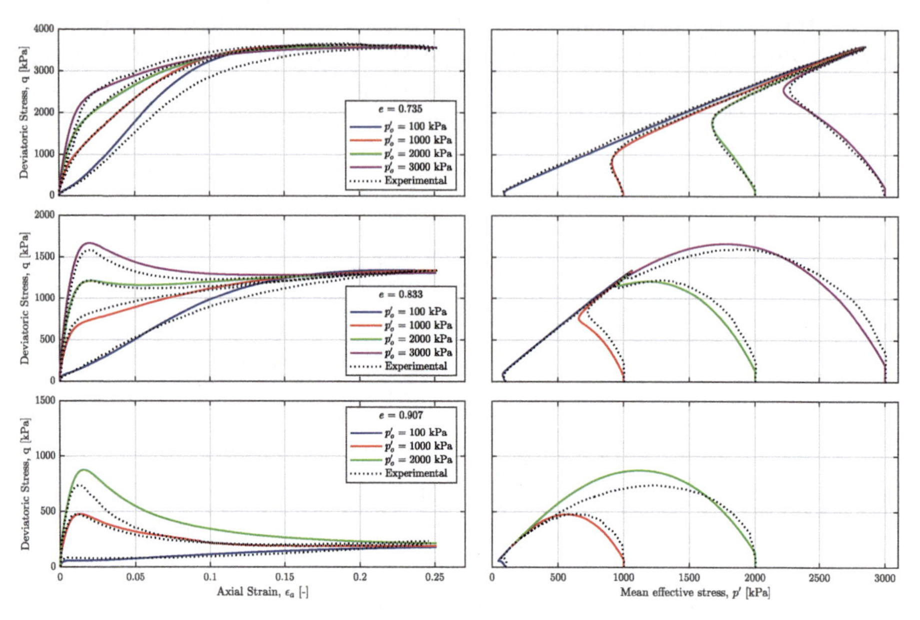

Fig. 3. Toyoura sand. Simulations compared with experiments in undrained triaxial tests ([23])

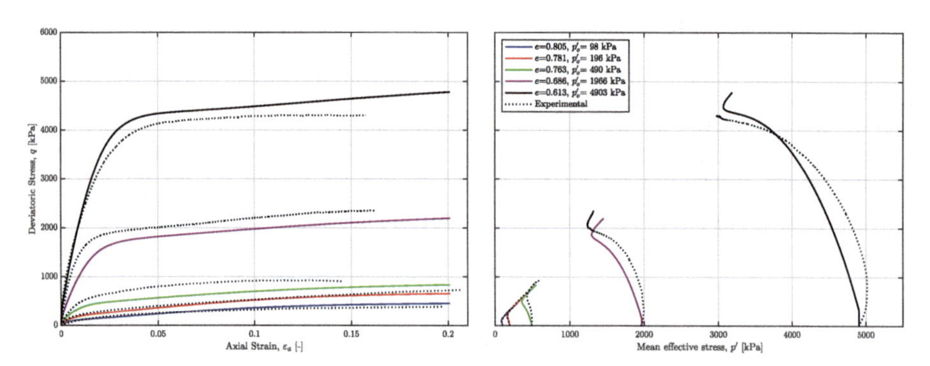

Fig. 4. El Torito tailings sand. Simulation compared with experiments in undrained triaxial tests based on ([17]) and ([10])

4.2 Cyclic Response

The cyclic response is assessed by undrained cyclic triaxial tests for both studied sands. The experimental data considered for Toyoura sand is published by Yang & Sze (2011) [24], while the tests of Solans (2010) [17] were used for the El Torito tailings sand.

The experimental data and the numerical simulations are summarised in Fig. 5 in terms of cyclic resistance curves. The calibration aims to reproduce the cyclic curves with reasonable agreement at around 10 to 15 cycles, considered as the relevant range for seismic applications. Another essential aspect to consider is the effect of mean effective stress in the cyclic response (often referred to as the overburden correction factor or K_σ ([16])).

The Toyoura sand simulations (Fig. 5.a) were performed employing the fabric tensor formulation described by Taborda et al. (2014) [20]. The experimental data uses the 5% double-amplitude (DA) axial strain as a liquefaction criterion. As some of the simulations did not reach the 5% DA criterion, it was necessary to include 90% of the excess of pore water pressure (PWP) build-up as a complementary criterion, typically employed in liquefaction assessment. The numerical simulations, plotted for 5%DA and 90% PWP criteria, provide satisfactory results compared to the experimental data, reproducing very well the K_σ effect for 300 kPa of confining stress.

Figure 5.b summarises the cyclic simulations for the El Torito tailings sand, where two calibration approaches were considered: (i) a fabric tensor formulation and (ii) a flow rule modification ([21]). Both the experimental data and the simulation employed the criterion of 90% PWP to define liquefaction. The original fabric tensor formulation provides good agreement for p' = 98 kPa, but as the mean effective stress increases, the simulated curves depart from the experimental data. For the flow rule modification, a good match is achieved at around 10 cycles, overestimating the cyclic strength for lower cycles and underestimating it for higher than 10 cycles, especially at $p' = 490$ kPa.

The differences in the cyclic strength curve simulations and the experiments may be attributed to the inherent scatter for the CSL and to very low initial state parameter values, ψ_0, observed for the El Torito tailings sands. If these conditions are controlled, as it is the case for Toyoura sand, it is possible to improve the model performance achieving low B_{CRR} values ($CRR = A_{CRR}N^{B_{CRR}}$).

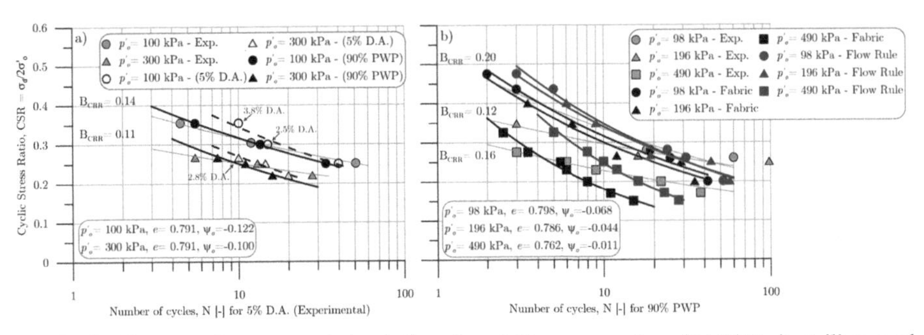

Fig. 5. Cyclic strength curves and simulations for: a) Toyoura sand, and b) El Torito tailing sand

Figure 6 shows the excess of pore water pressure (PWP) build-up of two cyclic tests. The experimental data in both tests indicate approximately 7 cycles to reach the 90% PWP liquefaction criterion, while the simulations predict higher number of cycles. In the case of Toyoura sand, the model simulation follows a PWP rate consistent with the experiment. The PWP rate obtained from the flow rule approach is closer to the experiment for the El Torito tailings sands simulation.

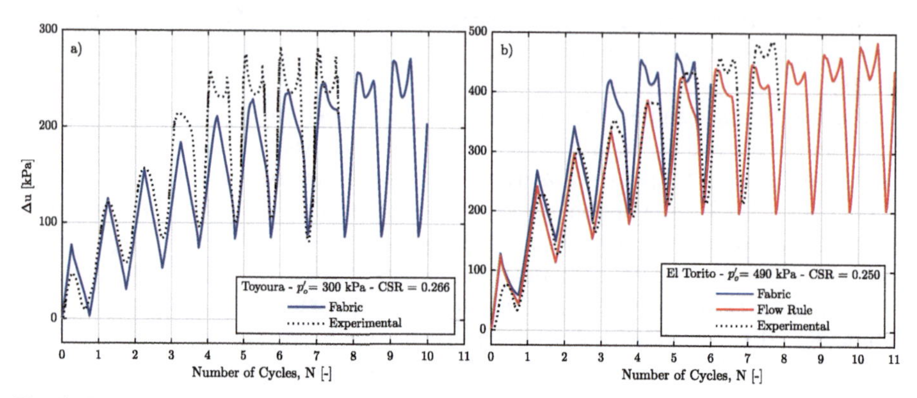

Fig. 6. Excess of pore water pressure build-up for: a) Toyoura sand, and b) El Torito tailings sand

5 Conclusions

This article presents and contrasts the calibration of an advanced state parameter-based bounding surface plasticity model for the El Torito tailings sand and for the well-known natural Toyoura sand.

Despite the fact that tailings sands are inherently variable, the model simulations for monotonic conditions compare well with the experimental data in the wide range of mean effective stress considered. For the cyclic response, reasonable simulation results are obtained for mean effective stress around 100 to 200 kPa. Important differences are observed for $p' = 490$ kPa, which may be explained by the low ψ_o values (proximity to the CSL), and the inherent scatter of the CSL, expected for this kind of material.

Finally, regardless of the variability of the tailings sand studied, the BSPM employed reproduces most of the behavioural features of this kind of material with reasonable accuracy. However, the use of this model formulation in boundary value problems should take into account the challenging calibration process at element level, to predict reasonable results in dynamic analysis of tailings sands dams.

Acknowledgements. The first author would like to express his gratitude to ANID – Chile (formerly CONICYT) (grant number: 72180052) and Dixon Scholarship from Imperial College London for the funding to his PhD studies.

References

1. Boulanger, R.W., Ziotopoulou, K.: Formulation of a sand plasticity plane-strain model for earthquake engineering applications. Soil Dyn. Earthq. Eng. **53**, 254–267 (2013)

2. Carrera, A., Coop, M., Lancellotta, R.: Influence of grading on the mechanical behaviour of stava tailings. Geotechnique **61**(11), 935–946 (2011)
3. Corral, G.: Efecto del esfuerzo de corte estático inicial en la resistencia cíclica de arenas (in Spanish) [MSc. dissertation]. University of Chile (2008)
4. Fourie, A.B., Papageorgiou, G.: Defining an appropriate steady state line for Merriespruit gold tailings. Can. Geotech. J. **40**(2), 484–486 (2003)
5. Jefferies, M., Been, K.: Soil Liquefaction: A Critical state Approach. 2nd edn. (2016)
6. Jefferies, M., Morgenstern, N.R., Van Zyl, D.J.A., Wates, J.: Report on the NTSF embankment failure, Cadia Valley Operations for Ashurst Australia (2019)
7. Kokusho, T.: Cyclic triaxial test of dynamic soil properties for wide strain range. Soils Found. **20**(2), 45–60 (1980)
8. Li, W., Coop, M.R.: Mechanical behaviour of panzhihua iron tailings. Can. Geotech. J. **56**(3), 420–435 (2019)
9. Manzari, M.T., Dafalias, Y.F.: A critical state two-surface plasticity model for sands. Géotechnique **47**(2), 255–272 (1997)
10. Maureira, S.: Respuesta Cíclica de Arena de Relaves en un Amplio Rango de Presiones (in Spanish) [MSc. dissertation]. University of Chile (2012)
11. Morgenstern, N.R., Vick, S.G., Viotti, C.B., Watts, B.D.: Fundão Tailings Dam Review Panel. Report on the Inmediate Causes of the Failure of the Fundão Dam (2016)
12. Papadimitriou, A.G., Bouckovalas, G.D.: Plasticity model for sand under small and large cyclic strains: A multiaxial formulation. Soil Dyn. Earthq. Eng. **22**(3), 191–204 (2002)
13. Potts, D.M., Zdravković, L.: Finite element analysis in geotechnical engineering. Theory. Thomas Telford (1999)
14. Reid, D., et al.: Results of a critical state line testing round robin programme. Géotechnique, 1–15 (2020)
15. Robertson, P.K., De Melo, L., Williams, D.J., Wilson, G.W.: Report of the Expert Panel on the Technical Causes of the Failure of Feijão Dam I Expert Panel (2019)
16. Seed, H.B.: Earthquake resistant design of earth dams. Symposium on Seismic Design of Embankments and Caverns (1983)
17. Solans, D.: Equipo Triaxial Monótono y Cíclico de Altas Presiones y su Aplicación en Arenas de Relaves (in Spanish) [MSc. dissertation]. University of Chile (2010)
18. Solans, D.: Seismic performance of tailings sand dams (under preparation) [PhD Thesis]. Imperial College London (2022)
19. Solans, D., Kontoe, S., Zdravković, L.: Monotonic and Cyclic Response of Tailings Sands. In: 2019 SECED Conference. Earthquake Risk and Engineering towards a Resilient World, pp. 1–10, Sep 2019
20. Taborda, D.M.G., Zdravković, L., Kontoe, S., Potts, D.M.: Computational study on the modification of a bounding surface plasticity model for sands. Comput. Geotech. **59**, 145–160 (2014)
21. Tsaparli, V.: Numerical modelling of earthquake-induced liquefaction under irregular and multi-directional loading. (Issue May) [PhD Thesis]. Imperial College London (2017)
22. Tsaparli, V., Kontoe, S., Taborda, D.M.G., Potts, D.M.: A case study of liquefaction: demonstrating the application of an advanced model and understanding the pitfalls of the simplified procedure. Géotechnique **70**(6), 538–558 (2020)
23. Verdugo, R., Ishihara, K.: The Steady State of Sandy Soils. Soils Found. **36**(2), 81–91 (1996)
24. Yang, J., Sze, H.Y.: Cyclic behaviour and resistance of saturated sand under non-symmetrical loading conditions. Geotechnique **61**(1), 59–73 (2011)

Seismic Performance of a Bituminous-Faced Rockfill Dam

M. Tretola, E. Zeolla, and S. Sica(✉)

University of Sannio, Benevento, Italy
{mariagrazia.tretola,stefsica}@unisannio.it

Abstract. Assessing the seismic safety of an existing earth dam is much more cumbersome than the design of a new structure since several actions and internal processes might have occurred over time. A detailed knowledge of both dam and foundation soils together with a proper characterization of the embankment response under the actual past static loading history, is a prerequisite before handling the dam seismic assessment through a performance-based approach. The paper describes the analysis procedures developed for analyzing the static and seismic response of a bituminous upstream-faced rockfill dam placed in Sicily (IT). The outcomes of the numerical study, performed through a complete dynamic formulation, are presented and discussed focusing on the embankment settlements at the end of the construction stage and on the seismic performance of the dam under the Collapse Limit State scenario. The paper tries to shed light on the role of a strong heterogeneity (limestone block) detected within the foundation soil to ascertain if its presence, which was completely disregarded during the original design of the dam, is beneficial or detrimental for the dam overall response.

Keywords: Earth dam · Bituminous liner · Earthquake

1 Introduction

For many decades, the seismic performance of earth dams has been overwhelmed due to the motiveless belief that these structures are intrinsically safe against earthquakes. Actually, plenty of case studies worldwide have demonstrated that, when they are properly constructed, earth and rockfill dams may withstand significantly strong shaking without collapse or severe damage. A few cases of complete failure reported in literature regard very old dams, small tailings or hydraulic fill dams built at the beginning of the last century with none or insufficient compaction of the construction materials. Consequently, liquefaction of the sandy soils of the embankment or foundation deposit occurred, e.g. the iconic case-history of the Lower San Fernando Dam (USA) in 1971 [1].

The assessment of earth dam seismic safety through performance-based criteria, as required by the more updated technical codes worldwide, is not a simple task in case of existing earth dams due to plenty of uncertainties about the actual state of the construction soils, the past loading history (static and seismic) experienced by the dam and the internal

© The Author(s), under exclusive license to Springer Nature Switzerland AG 2022
L. Wang et al. (Eds.): PBD-IV 2022, GGEE 52, pp. 1984–1994, 2022.
https://doi.org/10.1007/978-3-031-11898-2_182

processes occurred during the long operation of the reservoir. Most existing earth dams, moreover, were designed without accounting for seismic loading or with an inadequate characterization of the seismic hazard at the dam site.

The paper illustrates the numerical study carried out to investigate the static and seismic response of an Italian bituminous-faced rockfill dam (BFRD), Olivo Dam (Fig. 1), placed in Sicily (IT) and built in the late Seventies. The embankment has a maximum height of 49.5 m in correspondence of the central cross-section (Fig. 1a, b) and is made of coarse-grained soils (evaporitic limestone) with watertightness assured by an upstream-bituminous liner. The foundation deposit is made of two clay layers, an upper (AM1) and a lower (AM2) Miocenic clay, which englobe a big limestone block (CM), so that only the downstream part of the embankment is directly founded on it. Finally, a clayey breccia (MP) layer, called "*trubi*", has been found on the downstream side of the soil deposit (Fig. 1a).

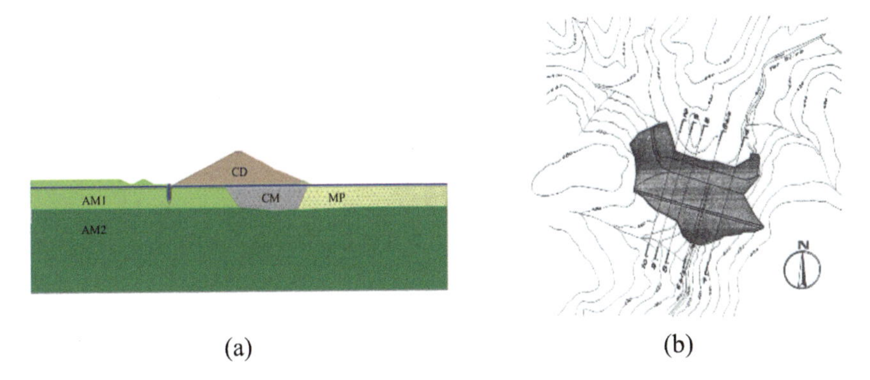

Fig. 1. Main cross Sect. 6 bis (a) and plan view (b) of the olivo dam.

The seismic response of the dam has been investigated through 2D numerical analyses of increasing complexity. For the sake of succinctness, only the outcomes of the advanced dynamic analyses, carried out through the finite element (f.e.) software Plaxis 2D [5], will be discussed hereinafter. According to technical codes (NTC2018, NTD2014) and indications (DGD, 2019) in force in Italy for dams, the seismic safety of these structures should be evaluated against two Serviceability Limit States (Operational and Damage) and two Ultimate Limit States (Life Safety and Collapse). In addition, for each limit state, at least five spectrum-compatible acceleration time histories consistent with the seismicity of the dam site, need to be considered (NTD2014).

The paper describes the analysis procedure developed for analyzing the static and seismic response of the Olivo Dam, trying to unravel the beneficial or detrimental role of the limestone block (CM) on the overall performance of the structure.

2 Input Signal Selection

With regard to the most severe scenario of the Collapse Limit State (CLS), corresponding to a return period T_R equal to 1462 years, the maximum horizontal acceleration excepted at the dam site is 0.141g. Six spectrum-compatible accelerogram groups (i.e., horizontal and vertical components) recorded on horizontal rock outcrop (soil type A), were selected from the European Strong Motion database [11]. Details on the adopted accelerograms are reported in Table 1. All input signals were applied at the bottom boundary of the f.e. model shown in Fig. 1a, after a linear deconvolution analysis within the clay formation AM2 (reference bedrock).

Table 1. Input signals adopted for the seismic analyses of the CLS scenario

OLIVO DAM					
Signal	**Earthquake**	**Mw**	**Epicentral Dist. [km]**	**PGA [g]**	**PGA NTC [g]**
55	Friuli (06/05/1976)	6.5	23	0.306	0.141
234	Montenegro (24/05/1979)	6.2	30	0.069	0.141
290	Campano Lucano (23/11/1980)	6.9	32	0.216	0.141
198	Montenegro (15/04/1979)	6.9	21	0.181	0.141
294	Campano Lucano (23/11/1980)	6.9	26	0.092	0.141
5819	Kalamata (13/10/1997)	6.4	48	0.121	0.141

3 Numerical Analyses

3.1 Constitutive Model

The constitutive law used for the advanced dynamic analyses of the dam is the *Hardening Soil with Small Strain Stiffness* (HS small) model implemented in Plaxis 2D. This model is able to describe with sufficient accuracy the non-linear, hysteretic and plastic response of the soil even at small strain levels. It is an evolution of the original hardening soil model developed by Schanz et al. [4] and Benz et al. [2], with two types of hardening (Fig. 2), one controlling the deviatoric response (shear hardening with conic yield surface) and another the volumetric behavior (compression hardening with cap yield surface). The model is able to predict soil stiffness decay with increasing strain amplitude by adding two further material parameters: G_0^{ref}, representing the initial shear modulus of the soil, and $\gamma_{0.7}$, corresponding to the strain level at which the soil shear modulus is about 70% of the initial value [5]. Table 2 summarizes the parameters of the *Hardening Soil with Small Strain Stiffness* model used in the computation.

3.2 Geometry and Boundary Conditions

As stated above, the static and seismic response of the dam was investigated in 2D by modelling its maximum cross section by finite elements. A suitable portion of the

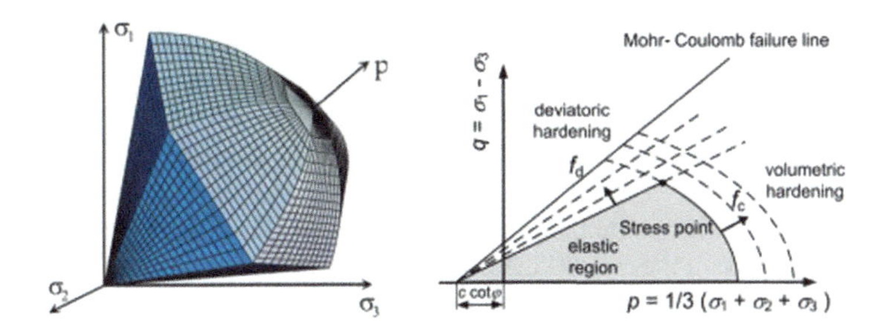

Fig. 2. Yield locus and hardening response of the HS-small constitutive law [6]

Table 2. Parameters of the *Hardening Soil with Small Strain Stiffness* model.

	c'	φ'	ψ	E_{50_ref}	E_{oed_ref}	E_{ur_ref}	G_{0_ref}	$\gamma_{0.7}$	m	k_0^{nc}
	[kPa]	[°]	[°]	[MPa]	[MPa]	[MPa]	[MPa]	[-]	[-]	[-]
CD	0.02	35	0	70	68	230	475	8E-5	0.5	0.43
AM1	50	25	0	20	20	100	200	4E-4	0.5	0.58
AM2	50	25	0	35	35	170	333	1	0.5	0.58
MP	50	25	0	20	20	100	333	1.5E-4	0.5	0.58
CM	200	45	0	30	30	170	360	1	0.5	0.29

foundation soil was included underneath the dam embankment to properly simulate the seismic wave propagation from the bedrock to the ground surface. The Miocenic clay AM2 was considered as the reference bedrock for the dam site since shear wave velocity higher than 1000 m/s were measured through Down-hole and MASW tests. To avoid wave reflection at the boundary of the analysis domain during the seismic stage, special absorbing elements (free-field boundaries) have been imposed along the lateral sides (Lysmer and Kuhlemeyer [7]) while a compliant base condition was adopted at the bottom of the soil domain. Conversely, elementary boundaries were used in simulating the static stages of the dam. The overall analysis procedure consisted of three main phases. First, the embankment construction was simulated through 23 loading steps with a total duration of 16 months and the activation of about 2 m-high layer for each step. A post-construction consolidation was also included in this stage. Second, the first filling and the operating conditions of the reservoir were simulated. Third, the seismic analyses were carried out to predict dam performance under the different seismic scenarios prescribed by the Italian code.

3.3 Static Response

Among plenty of results regarding the static stage simulation, contours of the vertical effective stress and settlements at the end of the embankment construction stage were selected (Fig. 3). With reference to Fig. 3 a, it could be observed that the olistolitis CM, which is much stiffer than formation AM, induced a stress discontinuity in the foundation deposit since the initial stages of the construction process. Consistently with what stated above, the settlement distribution is not symmetric with respect to the embankment centerline but shifted towards the left (upstream side) (Fig. 3 b). To better clarify this aspect, Fig. 4 shows the foundation settlement distribution along the x-axis (i.e. base of the embankment).

The maximum settlement at the foundation level was found to be 22 cm and it was located on the left side of the embankment vertical axis. In order to evaluate the influence of the limestone block CM on the deformed shape of the dam foundation at the end of the construction stage, a further analysis was carried out in which the olistolitis CM was ignored and soil layering was assumed the same of the upstream side (AM1 and AM2). As expected, the settlement distribution along the dam base turned to be symmetric with an increase of about 45% in the maximum value with respect to the former prediction. From this point of view, the role of CM could be considered beneficial as it reduced the amount of foundation settlements during the construction stage of the embankment.

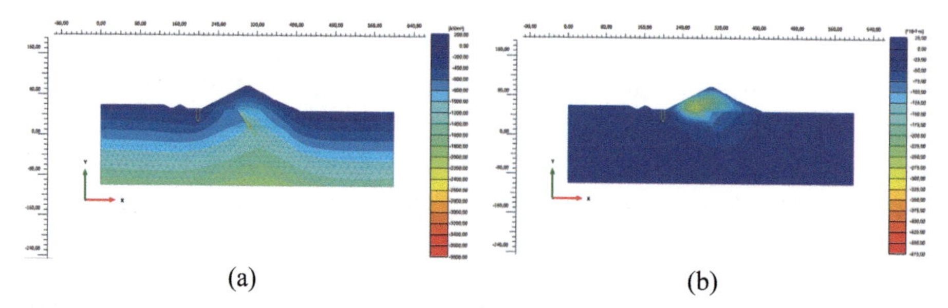

Fig. 3. Contour of (a) vertical effective stress (kPa) and (b) settlements at the end of the embankment construction

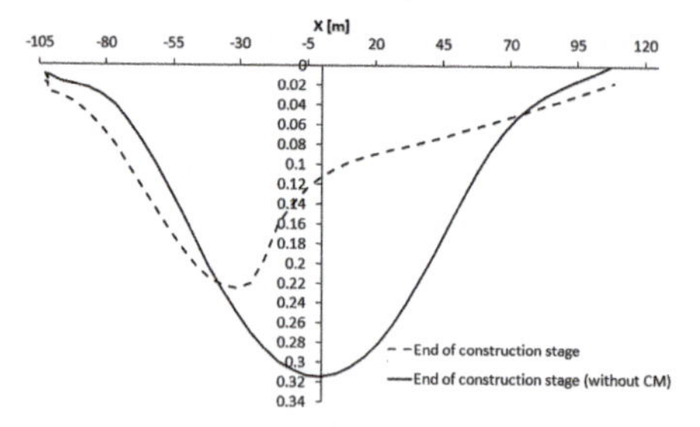

Fig. 4. Settlements (m) predicted along the embankment base with and without modeling the limestone block CM.

3.4 Dynamic Response Under the CLS Seismic Scenario

The seismic performance of the dam was predicted in both full and empty reservoir conditions, with and without modeling the limestone block CM. Figure 5 shows the permanent vertical displacements computed at the end of the seismic stage along the vertical axis of the embankment in all the analyzed conditions (full vs empty reservoir, with and without the block CM). Independently from the presence of the olistolitis CM in foundation, the empty reservoir condition (Fig. 5 b-d) revealed to be the most dangerous since the average (on the six CLS input signals) settlement at the dam crest was found to be about 18 cm (0.36% of H), i.e. higher than the value of 12 cm (0.24% of H) computed with full height of the reservoir (Fig. 5 a-b). This response should not be considered surprising for this type of dam as the reservoir water could have a stabilizing effect at least on the upstream shell. In any case, the crest permanent settlement resulted always lower than the critical threshold (1% of H) that, according to Hynes-Griffin and Franklin [9], might be used as a reference value for most earth dam typologies. Furthermore, the expected damage under the CLS seismic scenario corresponds to a moderate level according to the indications of Pells & Fell [8]. To unravel the contribution of the block CM on the seismic response of the dam, comparison among results shown in Fig. 5 a-c or in Fig. 5 b-d enhances that no significant difference emerged in terms of permanent settlements along the vertical axis of the embankment. The same statement could be extended also the whole embankment domain as shown in Fig. 6 and Fig. 7 for the input signal groups #290 e #294 that provided for higher permanent displacements at the end of the seismic analysis.

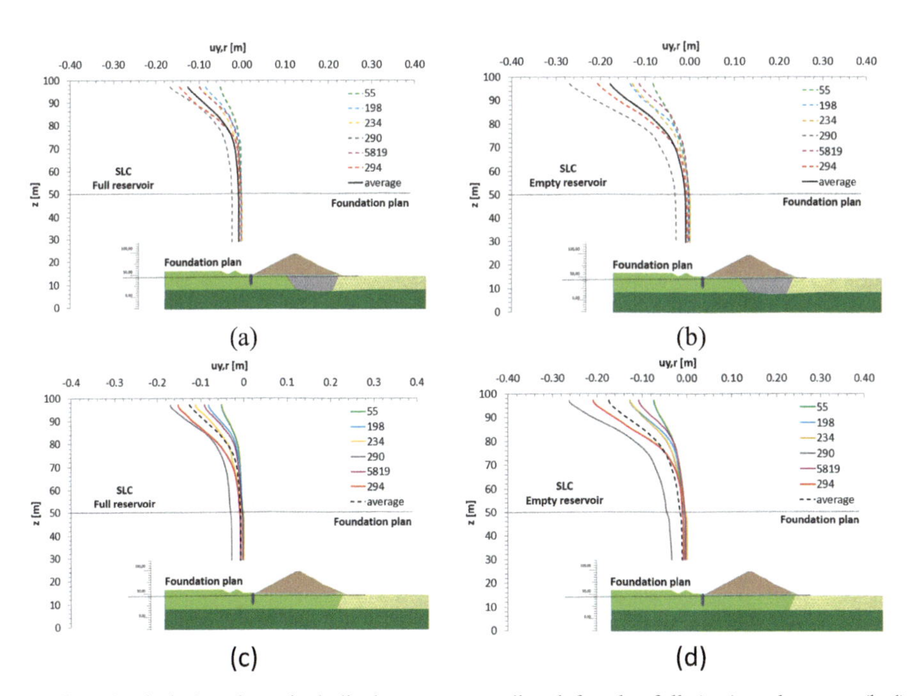

Fig. 5. Seismic-induced vertical displacements predicted for the full (a-c) and empty (b-d) reservoir conditions with (top) and without (bottom) the limestone block CM.

Some slight differences between the dynamic analyses with and without the block CM were found in terms of maximum accelerations along the dam base (Fig. 8 and Fig. 9). The presence of the block CM generally was beneficial as lower PGAs were obtained along the dam base, especially in the full reservoir condition. The main differences in PGA along the embankment base were found for signal 55× in the full reservoir condition (Fig. 8 a-c). Moreover, the presence of the olistolitis CM (Fig. 8 a–b) induces higher amplification of the ground motion on the left side of CM due to wave reflection caused by the stiffer limestone material.

Also at the dam crest, the maximum acceleration in the horizontal direction resulted similar in case of considering or not the block CM (Fig. 9). This finding was also corroborated by the PGA contours shown in Fig. 10 and Fig. 11 for the signal groups #290 and #294, i.e. the same signals of Fig. 6 and Fig. 7.

The above slight differences in acceleration, however, did not influence the final performance of the embankment in terms of seismic-induced permanent settlements and freeboard loss.

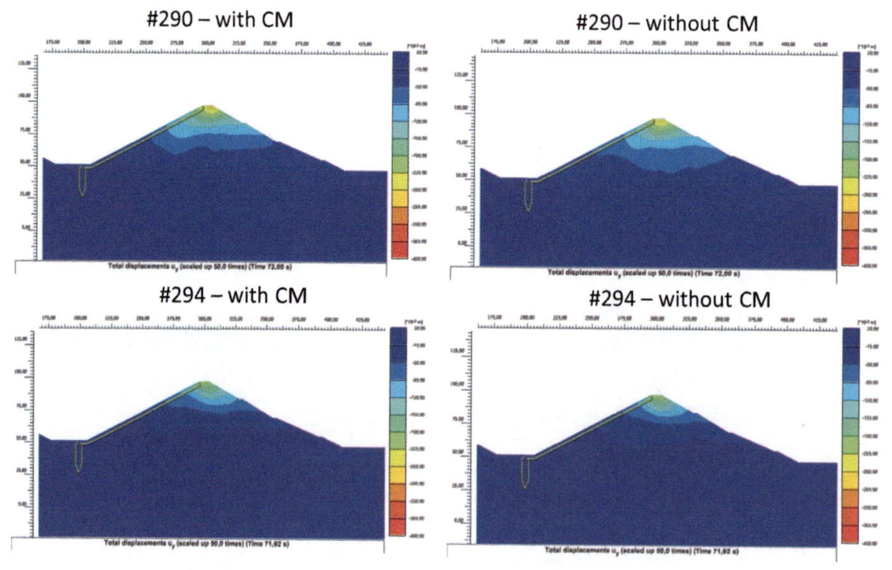

Fig. 6. Contour of maximum permanent settlements for the empty reservoir condition with and without the limestone block CM (input signal groups #290 and #294).

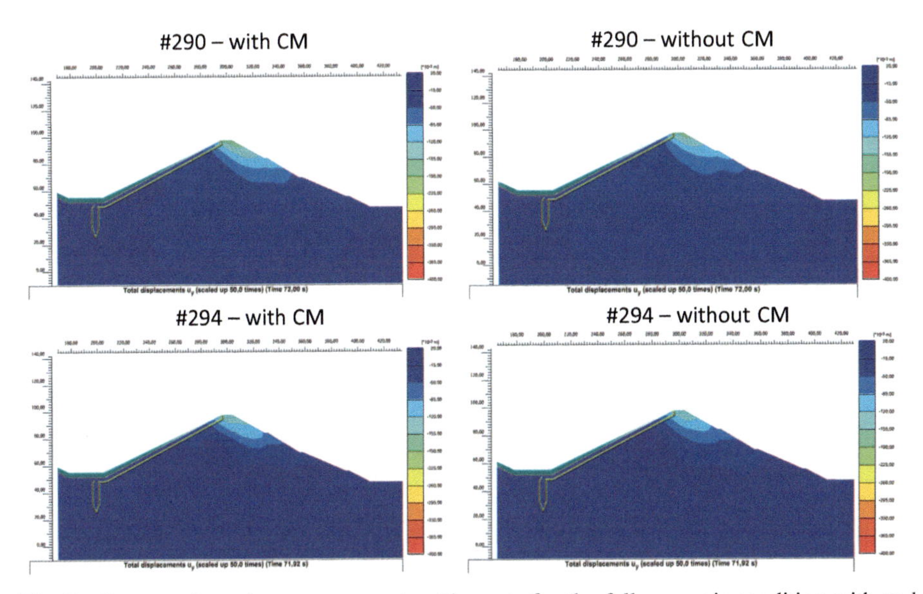

Fig. 7. Contour of maximum permanent settlements for the full reservoir condition with and without the limestone block CM (input signal groups #290 and #294).

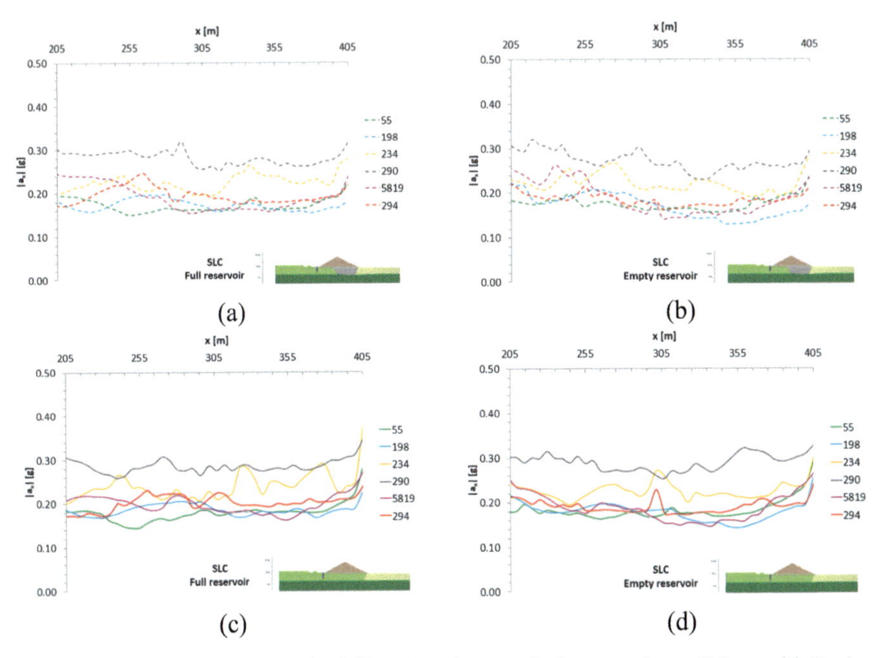

Fig. 8. PGA along the dam base for full (a-c) and empty (b-d) reservoir conditions with (top) and without (bottom) the limestone block CM.

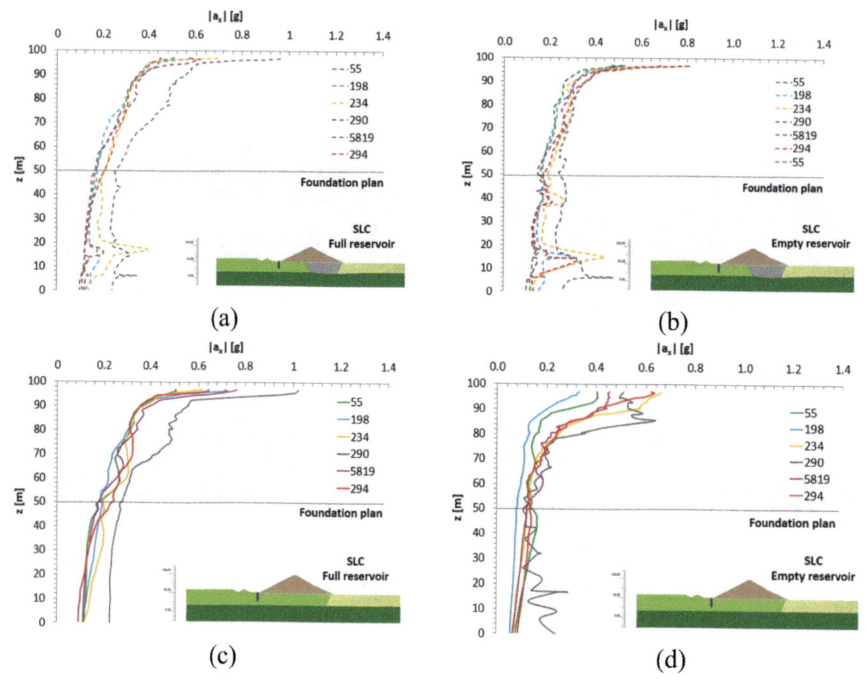

Fig. 9. Maximum horizontal accelerations along the vertical axis of the embankment for full (a-c) and empty (b-d) reservoir conditions with (top) and without (bottom) the limestone block CM.

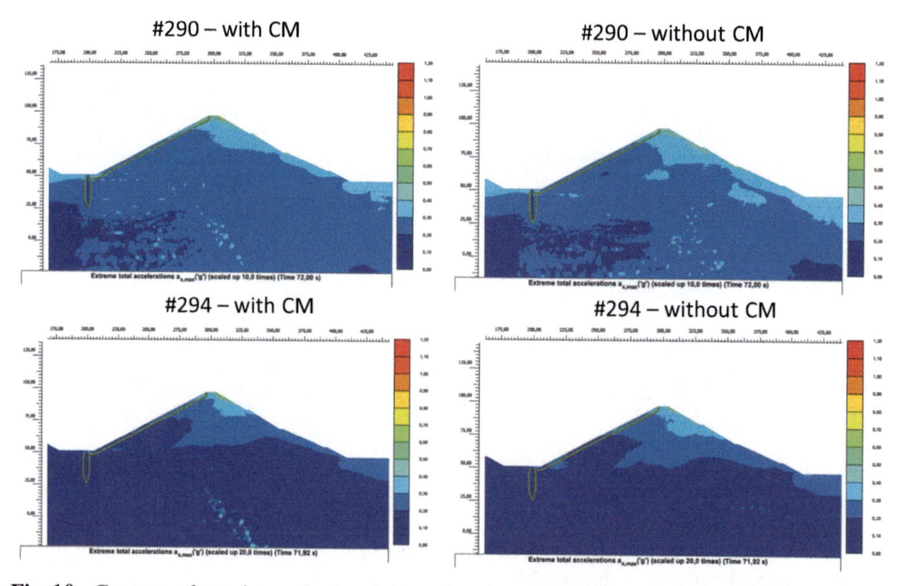

Fig. 10. Contour of maximum horizontal accelerations for empty reservoir condition with and without the limestone block CM (input signals #290 and #294).

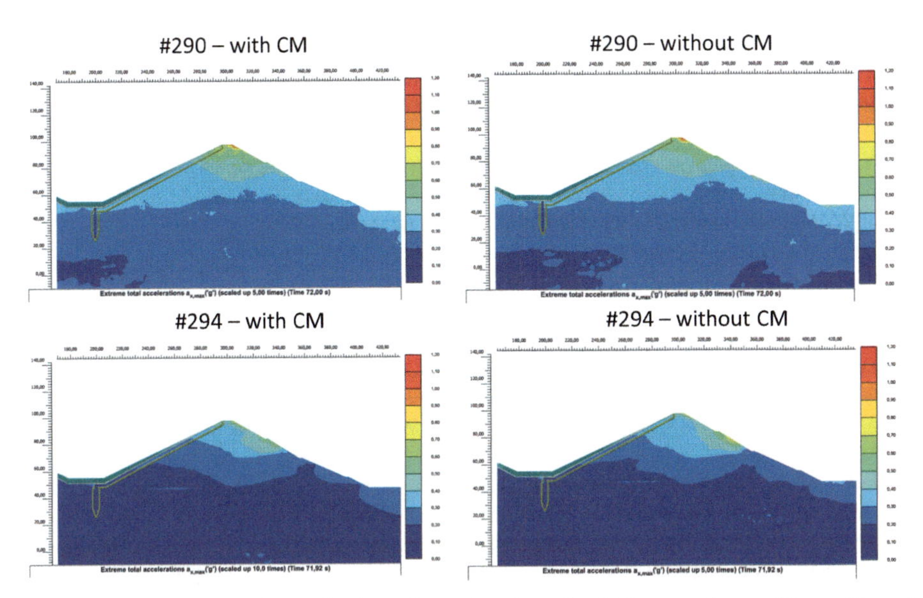

Fig. 11. Contour of maximum horizontal accelerations for full reservoir condition with and without the limestone block CM (input signals #290 and #294).

4 Conclusion

The paper illustrated the numerical procedure developed to assess the seismic response of an Italian bituminous-faced rockfill dam by means of performance-based criteria.

The peculiarity of this case-history is the presence of a strong heterogeneity (stiff limestone block) in the soil deposit so that only half of the embankment cross-section is founded on it. The presence of this stiffer block, which has been disregarded in the original design of the dam, was found to exert a major role on the static response of the dam and, in particular, on the foundation settlements at the end of the construction stage. Conversely, no relevant effects were found on the seismic performance of the dam, especially on the magnitude of crest settlement, which is the main indicator of dam performance to assess safety against freeboard loss and other damage mechanisms. As future perspective of the study, further numerical analyses will be carried out by a 3D f.e. model to better reproduce the limestone block geometry inside the foundation deposit and its effects on the overall dam performance.

References

1. Seed, H.B., Seed, R.B., Harder, L.F., Jong, H.L.: Re-evaluation for the lower San Fernando dam, Report 2, Examination of the post-earthquake slide of February 9, 1971. H. B. Seed Inc, Department of the Army, ASCE (1989)
2. Benz, T., Vermeer, P.A., Schwab, R.: A small-strain overlay model. Int. J. Num. Anal. Meth. Geomech. **33**, 25–44 (2009)
3. Benz, T.: Small-Strain Stiffness of Soils and its numerical consequences. Stuttgart: Phd Thesis, University of Stuttgart (2006)

4. Schanz, T., Vermeer, P.A., Bonnier, P.G.: The hardening soil model: Formulation and Verification. Beyond 2000 in Computational Geotechnics (1999)
5. Bentley Plaxis 2D Manual – Engineering Department License - University of Sannio, Benevento (2020)
6. Vakili, K.N., Barciaga, T., Lavasan, A.A., Schanz, T.: A practical approach to constitutive models for the analysis of geotechnical problems. Soil and Rock Mechanics, Ruhr-Universität at Bochum, Germany, Institute for Foundation Engineering (2016)
7. Lysmer, J., Kuhlemeyer, R.L.: Finite dynamic model for infinite media. J. Eng. Mech. Div. **95**, 859–878 (1969)
8. Pells, S., Fell, R.: Damage and cracking of embankment dams by earthquake and the implications for internal erosion and piping. In: Proceedings 21st International Congress on Large Dams, Montreal. ICOLD, Paris Q83–R17, Paris (2003)
9. Hynes-Griffin, M.E., Franklin, A.G.: Rationalizing the seismic coefficient method. Miscellaneous Paper GL-84–13, Department of the Army, Waterways Experiment Station, Vicksburg, MS (1984)
10. Iervolino, J., Galasso, C., Cosenz,a E.: REXEL: computer aided record selection for code-based seismic structural analysis. Bull. Earthquake Eng. **8**, 339–362 (2010)
11. Luzi, L., et al.: The Engineering Strong-Motion Database: a platform to access Pan-European Accelerometric data (2016)

Liquefaction-Induced River Levee Failure at Miwa of Inba Along Tone River During 2011 Tohoku Earthquake

Yoshimichi Tsukamoto[1](✉), Naoki Kurosaka[2], Shohei Noda[1], and Hiroaki Katayama[3]

[1] Department of Civil Engineering, Tokyo University of Science, Chiba, Japan
ytsoil@rs.tus.ac.jp
[2] Central Nippon Expressway Company, Nagoya, Japan
[3] Toho Drilling Equipment, Tokyo, Japan

Abstract. A numerous number of river levee failures were observed in Kanto region far from the epicentre during 2011 Tohoku earthquake. In this study, the earthquake reconnaissance investigations were carried out on the liquefaction-induced sliding failure of the river levee located at Miwa of Inba along Tone river. A series of weight sounding tests were conducted at a dozen of locations near the failed river levee. Soil sampling was also carried out at each location of weight sounding, and laboratory sieve and sedimentation analyses were performed to determine the grain size distributions of the soil samples. At this particular site, the outer face of the river levee slid laterally, and the evidence of sand boils was clearly found at the foot of the river levee. The liquefaction resistance of soils was estimated based on the results of weight sounding and soil sampling, by using the empirical formula developed by the authors, which explicitly incorporates the effect of fines. The values of liquefaction resistance of foundation soils at which the river levee failed and did not fail are estimated, and the cause of the sliding failure is discussed in detail.

Keywords: Liquefaction · River levee · Weight sounding

1 Introduction

Following the devastating effects of soil liquefaction observed during the recent earthquakes in Japan, the authors' group has continued to conduct the earthquake reconnaissance investigations at various locations [1–7]. Especially in the event of 2011 Tohoku earthquake, wide-spread occurrences of soil liquefaction were observed even in Kanto region far from the epicentre, where metropolitan Tokyo is located, and greatly affected the infrastructures and private houses. Many river levee failures were also observed in Kanto region. It was reported that there were river levee failures at as many as 940 locations along 10 rivers in Kanto region [8]. Soil liquefaction was apparently one of the main causes of so many river levee failures during 2011 Tohoku earthquake.

In this study, the earthquake reconnaissance investigation was carried out at Miwa of Inba along Tone river, where a tall river levee failed and slid laterally due to liquefaction

© The Author(s), under exclusive license to Springer Nature Switzerland AG 2022
L. Wang et al. (Eds.): PBD-IV 2022, GGEE 52, pp. 1995–2003, 2022.
https://doi.org/10.1007/978-3-031-11898-2_183

of foundation soils. A series of weight sounding tests and soil sampling were carried out, and the test results are discussed in detail in what follows.

2 River Levee Failure at Miwa of Inba

Soil liquefaction was observed widely in the downstream areas of Tone river, including the cities of Katori, Itako, Kamisu, Kashima and Asahi, as shown in Fig. 1. The location of Miwa of Inba is also shown in Fig. 1, where the liquefaction-induced sliding failure of a tall river levee was observed. The south face of the tall river levee slid laterally, and the sliding surface was exposed at the top of the river levee, as shown in Fig. 2. There were clear traces of sand boils at the foot of the failed river levee, and it was clear that the foundation soils had liquefied.

3 Weight Sounding and Soil Sampling

In order to clarify the effects of liquefaction of foundation soils on the river levee failure, a series of Swedish screw weight sounding (SWS) tests were carried out in this study. The details of the testing equipment and testing procedure are shown in Fig. 3(a) and described in [9]. This testing equipment is easy to carry around, and it is easy and simple to conduct field tests. The use of SWS is therefore recommended for earthquake reconnaissance investigations [10]. The penetration resistance of SWS represented by the values of N_{sw} and W_{sw} is known to be correlated with SPT N-value as well as CPT tip resistance [9]. The empirical procedure to evaluate the liquefaction resistance of soils from SWS was also proposed recently [11, 12].

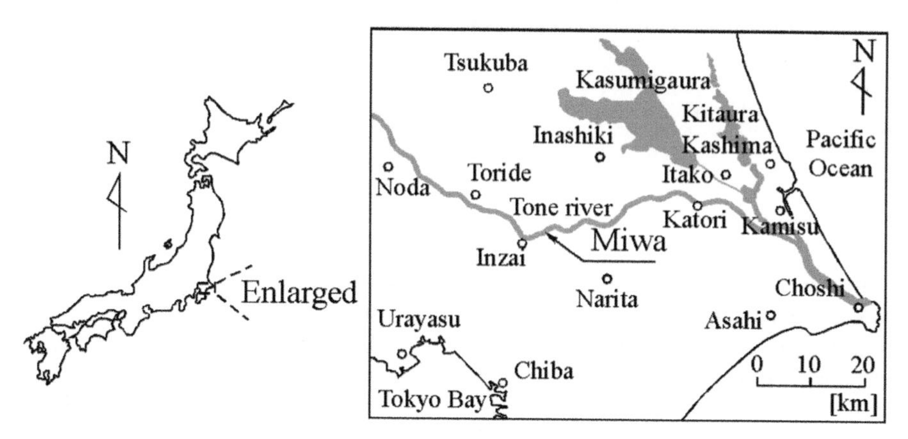

Fig. 1. Location of Miwa of Inba

In this study, soil sampling was also carried out at exactly the same locations of SWS. At each location, SWS test was first conducted, and soil samples were retrieved from various depths at the location one metre apart from the bore hole of SWS. The details of soil sampling are shown in Fig. 3(b), and are described by [12]. It was found

Fig. 2. Failed river levee (a) that slid laterally and (b) that exposed the sliding cliff (courtesy to Tonegawa Downstream Office)

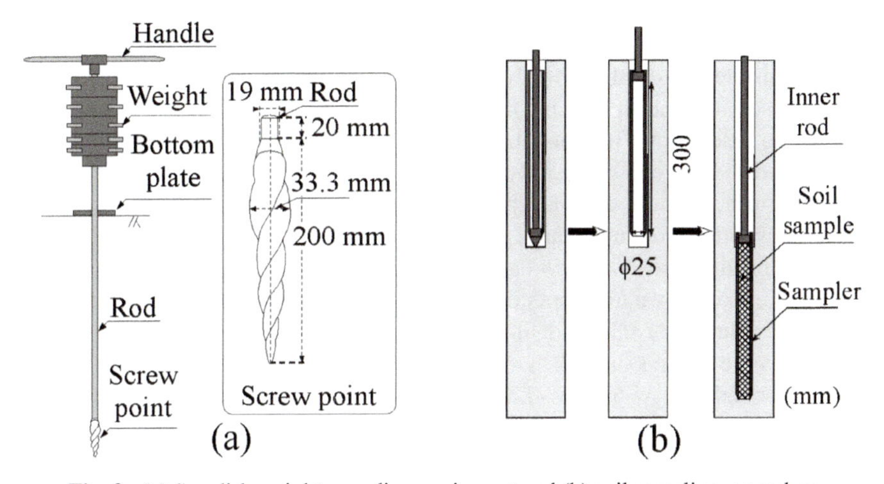

Fig. 3. (a) Swedish weight sounding equipment and (b) soil sampling procedure

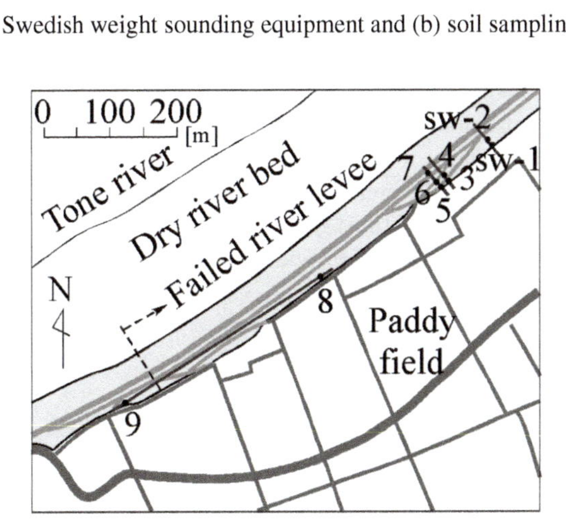

Fig. 4. Locations of weight sounding tests

in the previous study [12] that this soil sampling method is most relevant to supplement SWS tests. The sampler is 30 cm long and 2.5 cm in inner diameter. The sampler is first pushed down. The inner rod encased in the sampler is then pulled up and fixed to the sampler. The sampler fixed to the inner rod is again slowly pushed down into a soil deposit, and a soil sample is captured into the sampler. The sampler containing a soil sample is eventually pulled up to the ground surface.

A series of SWS tests and soil sampling were carried out at 9 different locations, as shown in Fig. 4. The locations of sw-1 to sw-8 correspond to the area that the river levee failed, and the location of sw-9 corresponds to the area that the river levee did not fail. The locations of sw-1 to sw-7 were determined to make it possible to draw several cross sectional profiles of the foundation soils along the failed river levee. On the other hand, the locations of sw-8 and sw-9 were determined to compare the depth-wise profiles of the foundation soils at the failed and non-failed river levees.

The test results at sw-1 and sw-2 are shown in Fig. 5. The valus of N_{sw} and W_{sw} are plotted against depth, together with the values of fines content F_c. The estimated sliding surface is also indicated. It is found that there are compacted fills with larger values of N_{sw} down to a depth of 1 to 2 m below the ground surface. There are then thick loose layers of sand and silt, exhibiting lower values of N_{sw}, down to a depth of 7 to 8 m below the ground surface. These thick layers of sand and silt should be the natural sediments that had transported from upstream and had accumulated along the old river floor.

The test results at sw-3 and sw-4 are shown in Fig. 6. The compacted fills with larger values of N_{sw} exist down to a depth of 2 to 3 m below the ground surface, underlain by the thick loose layers of sand and silt, exhibiting lower values of N_{sw}, down to a depth of 7 m below the ground surface.

The test results at sw-5 to sw-7 are shown in Fig. 7. The results at sw-5 and sw-6 indicated that the compacted fills with larger values of N_{sw} exist down to a depth of 1 m below the ground surface, underlain by the thick loose layers of sand and silt, exhibiting lower values of N_{sw}, down to a depth of 7 m below the ground surface. The results at sw-7 indicated that the body of the river levee itself exhibited lower values of N_{sw}.

The comparison of the test results at failed sw-8 and non-failed sw-9 is shown in Fig. 8. The grain size distributions of the soil samples retrieved from the bore holes at sw-8 and sw-9 are also shown in Fig. 9. There are no obvious differences between the results of SWS, though some differences in the grain compositions of the soil samples. Therefore, it would be useful to estimate the liquefaction resistance R_l of soils from the results of SWS by taking due account of fines content F_c, and to compare the values of R_l at these two locations.

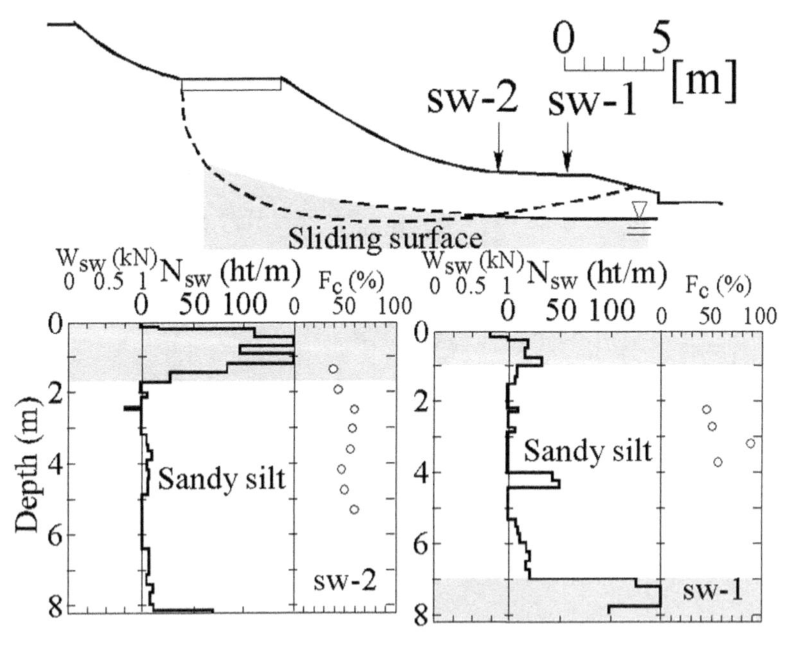

Fig. 5. Test results at sw-1 and sw-2

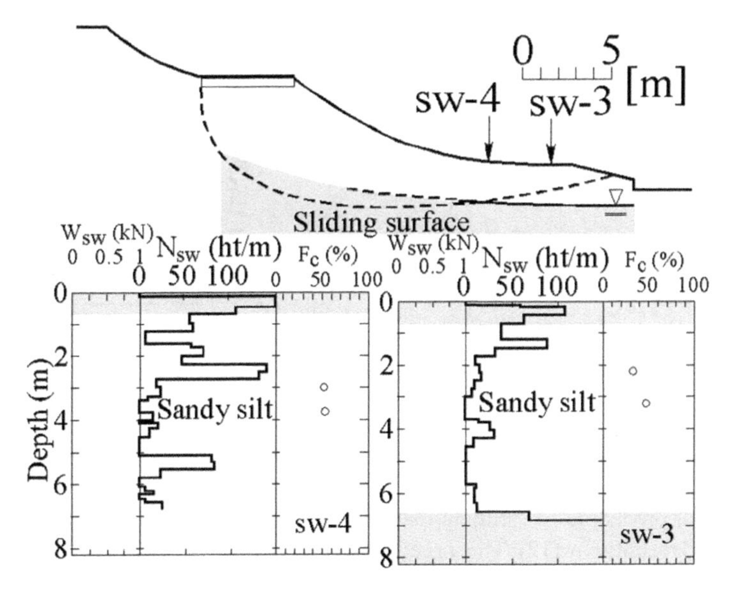

Fig. 6. Test results at sw-3 and sw-4

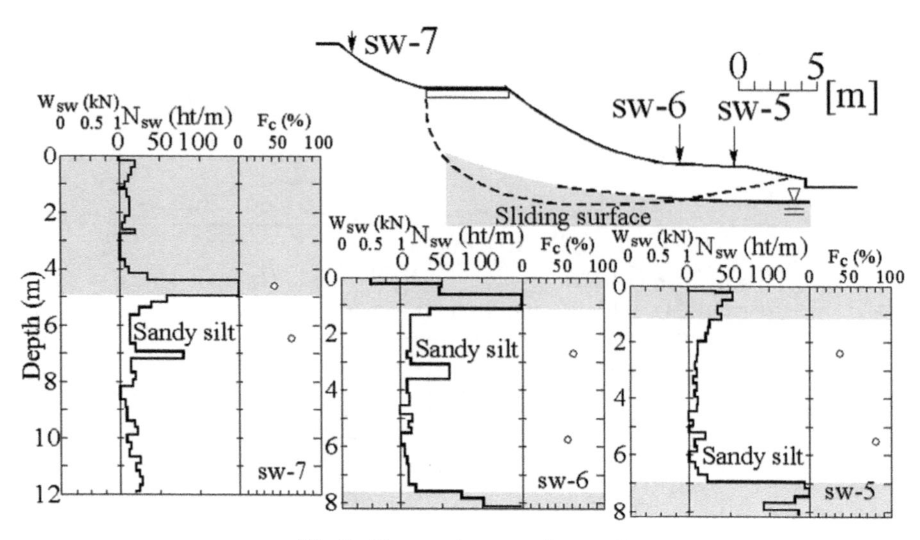

Fig. 7. Test results at sw-5 to sw-7

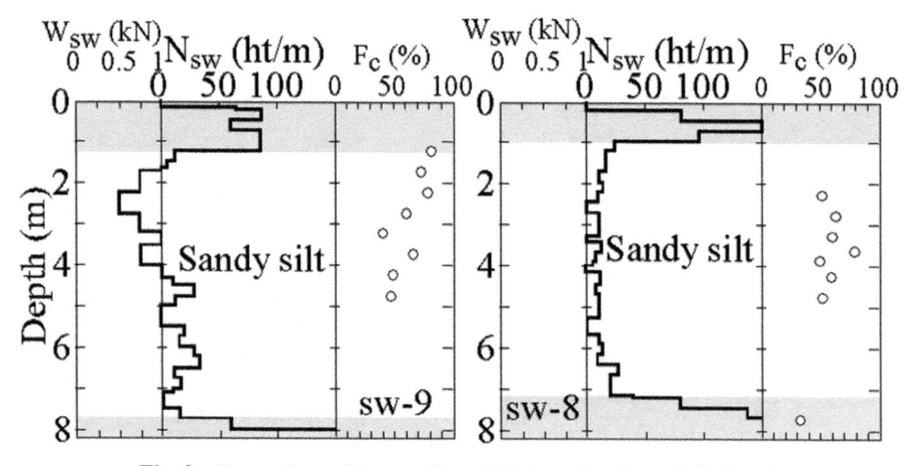

Fig. 8. Comparison of test results at failed sw-8 and non-failed sw-9

4 Resistance of Foundation Soils to Liquefaction

The empirical procedure to estimate the liquefaction resistance R_l of soils from SWS was proposed recently by [12]. This procedure explicitly incorporates the effects of fines, and is expressed as follows,

$$R_l = \alpha_{sw}\sqrt{N'_{sw1}} + \beta_{sw}\sqrt{F_c} \tag{1}$$

Herein, F_c is in percentage, α_{sw} and β_{sw} are the parameters changing with F_c as follows; $\alpha_{sw} = 0.016$ and $\beta_{sw} = 0$ for clean sand ($F_c < 5\%$), $\alpha_{sw} = 0.02$ and $\beta_{sw} = 0$ for sand with fines ($5\% \leqq F_c < 15\%$), $\alpha_{sw} = 0.02$ and $\beta_{sw} = 0.016$ for silty sand and silt ($F_c \geqq 15\%$).

It is noted that N'_{sw1} is defined as follows,

$$N'_{sw1} = (N_{sw} + 40 \times W_{sw})\alpha_{sw}\sqrt{\frac{98}{\sigma'_v}} = N'_{sw}\sqrt{\frac{98}{\sigma'_v}} \tag{2}$$

Herein, W_{sw} is in kN, and σ_v^2 is in kPa.

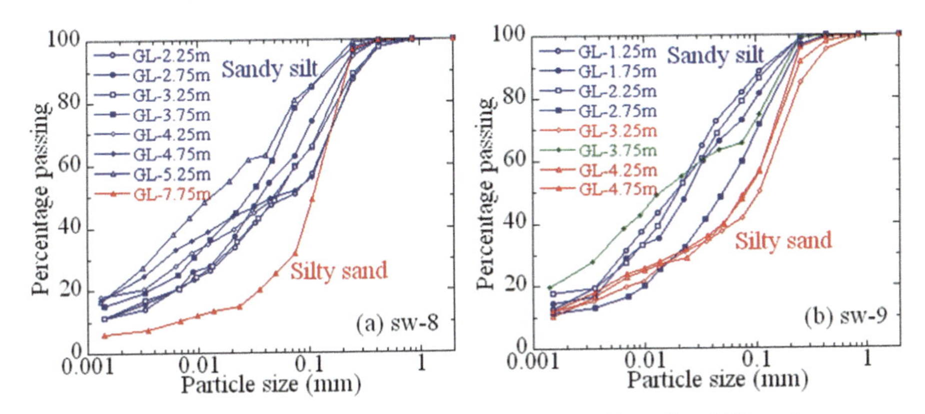

Fig. 9. Grain size distributions of soil samples at (a) sw-8 and (b) sw-9

The values of R_l are then calculated by using the above equations, for the locations of failed sw-8 and non-failed sw-9, and are plotted against depth, as shown in Fig. 10. It is assumed that the ground water level is located at 2 m deep from the ground surface. It is found that the values of R_l for non-failed sw-9 are even lower than those for failed

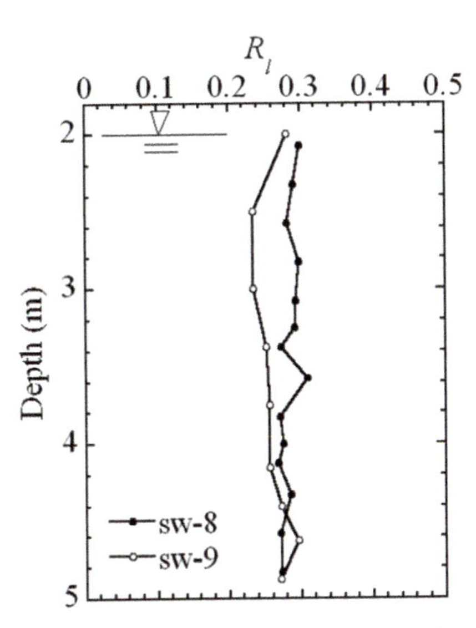

Fig. 10. Plots of liquefaction resistance R_l against depth

sw-8. It implies that there may be some specific factors that led to the triggering of soil liquefaction at sw-8, other than the susceptibility of foundation soils to liquefaction itself. In this study, the ground water level was estimated as the boundary between the wet and dry conditions of the rod surfaces of SWS. However, it was difficult to locate a ground water level correctly. Therefore, even though the susceptibilities of foundation soils to liquefaction are the same at sw-8 and sw-9, the difference in the depths of ground water levels may have attributed to the occurrence of sliding failure of the overlying river levee.

5 Conclusions

The earthquake reconnaissance investigations were conducted on the liquefaction-induced river levee failure at Miwa of Inba observed during 2011 Tohoku earthquake. A series of SWS and soil sampling were carried out to investigate the occurrence of sliding failure of the river levee. The liquefaction resistance R_l of foundations soils was successfully estimated by the empirical procedure using SWS and soil sampling. The comparisons of R_l suggested that the susceptibilities of foundation soils to liquefaction are the same at the locations where the river levee failed and did not fail. Therefore, there may be some specific factors that led to the sliding failure of the river levee. The difference in the ground water levels may be one of the possible reasons behind the occurrence of sliding failure of the overlying river levee.

Acknowledgements. The authors acknowledge that Tonegawa Downstream Office of MLIT gave permissions to conduct field tests and provided some photos of the failed river levee. Thanks are also extended to the past students of Tokyo University of Science for assisting in the field tests and laboratory tests described in this study.

References

1. Tsukamoto, Y., Ishihara, K., Kokusho, T., Hara, T., Tsutsumi, Y.: Fluidisation and subsidence of gently sloped farming fields reclaimed with volcanic soils during 2003 Tokachi-oki earthquake in Japan. In: Kokusho, T. (ed.) Earthquake Geotechnical Case Histories for Performance-based Design, pp 109–118. CRC Press (2009)
2. Tsukamoto, Y., Ishihara, K.: Use of field penetration tests in evaluating occurrence of soil liquefaction at reclaimed deposits during 2000 Tottori-ken Seibu Earthquake. In: Proceedings of 5th International Conference on Earthquake Geotechnical Engineering, 10–13 January 2011. Santiago, Chile (2011)
3. Tsukamoto, Y., Kawabe, S., Kokusho, T.: Soil liquefaction observed at the lower stream of Tonegawa river during the 2011 off the Pacific Coast of Tohoku Earthquake. Soils Found. **52**(5), Special Issue, 987–999 (2012)
4. Tsukamoto, Y., Ishihara, K., Kawabe, S., Kanemitsu, S.: Liquefaction-induced road embankment failures. Forensic Eng. Proc. Inst. Civil Eng. **166**(2), 64–71 (2013). https://doi.org/10.1680/feng.12.00020
5. Tsukamoto, Y., Hyodo, T., Fuki, N.: Some recent applications of Swedish weight sounding tests to earthquake reconnaissance investigations. In: Proceedings of 6th International Conference on Earthquake Geotechnical Engineering, 1–4 November 2015. Christchurch, New Zealand (2015)

6. Tsukamoto, Y., Kawabe, S., Kanemitsu, S.: Use of Swedish weight sounding tests for detecting liquefiable backfills reclaimed in iron sand mining pits in Asahi city of Chiba in Japan. In: Hazarika, H., Kazama, M., Lee, W.F. (eds.) Geotechnical Hazards from Large Earthquakes and Heavy Rainfalls, pp. 101–110. Springer, Tokyo (2017). https://doi.org/10.1007/978-4-431-56205-4_9
7. Hyodo, T., Tsukamoto, Y., Katayama, H.: Liquefaction-induced river levee failure during 2011 Great East Japan Earthquake: case history with Swedish weight sounding tests. In: Proceedings of International Conference on Performance-based Design in Earthquake Geotechnical Engineering, 16–20 July 2017. Vancouver, Canada (2017)
8. Kazama, M., Noda, T.: Damage statistics (Summary of the 2011 off the Pacific coast of Tohoku earthquake damage). Soils Found. **52**(5), 780–792 (2012). (Special Issue)
9. Tsukamoto, Y., Ishihara, K., Sawada, S.: Correlation between penetration resistance of Swedish weight sounding tests and SPT blow counts in sandy soils. Soils Found. **44**(3), 13–24 (2004)
10. Tsukamoto, Y.: Integrating use of Swedish weight sounding tests for earthquake reconnaissance investigations. In: Ansal, A., Sakr, M. (eds.) Perspectives on Earthquake Geotechnical Engineering. GGEE, vol. 37, pp. 467–479. Springer, Cham (2015). https://doi.org/10.1007/978-3-319-10786-8_18
11. Tsukamoto, Y., Hyodo, T., Hashimoto, K.: Evaluation of liquefaction resistance of soils from Swedish weight sounding tests. Soils Found. **56**(1), 104–114 (2016). https://doi.org/10.1016/j.sandf.2016.01.008
12. Tsukamoto, Y., Noda, S., Nakatsukasa, M., Katayama, H.: Use of weight sounding for examining the liquefaction-induced river levee failures. Proc. Inst. Civ. Eng. Geotech. Eng. (2021). https://doi.org/10.1680/jgeen.20.00062

S4: Special Session on Earthquake Disaster Risk of Special Soil Sites and Engineering Seismic Design

Effect of Rice Husk Ash and Sisal Fibre on Strength Behaviour of Soil

Benard Obbo[1(✉)], Gideon Okurut[1], Pragnesh J. Patel[1], Nisha. P. Soni[1], and Darshil V. Shah[2]

[1] Civil Engineering Department, Sitarambhai Naranjibhai Institute of Technology and Research Centre, Vidyabharti Campus, At and Po Baben, Bardoli, Surat, Gujarat 394601, India
obbobenard25@gmail.com

[2] Vidyabharti Trust Polytechnic, Umrakh, Civil Material Testing and Consultancy Cell, Bardoli, India

Abstract. Due to an increase in population, sustainable development, climate change, construction, waste management, and production of agriculture being the major challenges to be solved by researchers and scientists since the depletion of non-renewable resources has rapidly also increased. In this study the use of rice husk ash and treated sisal fibre on the strength behaviour of soil will help in the reduction of effects caused by rice husk, also increasing space for agricultural production. Rice husk ash (RHA) and single length sisal fibre (SF) (10–15 mm) at 2.5, 5.0, and 7.5% of RHA and 0.3, 0.6, and 0.9% of SF were used for this research investigation, and the optimum value of 7.5% RHA and 0.3% SF by weight of soil sample resulting to cohesion value which was over fifty (50) times the value natural soil (from 0.13 kg/cm^2 to 6.71 kg/cm^2) and also the angle of internal friction increased three times the natural soil value (from 26° to 72°).

This research also helps to obtain SDG (1, 2, 6, 9, 12, 13, 14, and 15).

Keywords: Sisal fibre · Rice husk · Environment · Geotechnical · Climate change · Ground improvement

1 Introduction

Ground Improvement is the geotechnical engineering practice of upgrading the soil properties so that it can withstand the loading of the designated structure on it [1]. Ground improvement is done for engineering structures such as buildings, highways, runways, bridges, railways, hydraulic structures, slope stability, retaining wall embankment [2]. The main objective is to increase shear strength, reduce settlement of structures, reduce shrinkage and swelling of soil, reduce permeability and reduce compressibility. Different ground improvement techniques have been adopted such as mechanical compaction, dynamic compaction, vibro-flotation, preloading, sand and stone columns, admixture method, grouting method, dewatering fine-grained soil method, inclusion, confinement, and reinforcement methods [3, 4]. In this research, the admixture method of soil improvement is explored mainly by use of rice husk ash and treated sisal fibre as an admixture to

© The Author(s), under exclusive license to Springer Nature Switzerland AG 2022
L. Wang et al. (Eds.): PBD-IV 2022, GGEE 52, pp. 2007–2023, 2022.
https://doi.org/10.1007/978-3-031-11898-2_184

investigate the variation of shear strength soil with the added proportion of admixture. Rice husk is the agricultural by-product that must be strongly managed in order to reduce waste accumulation in agriculture which affects the production of agricultural produce [5]. Asia is world-leading production of rice husk producing 770 million tons annually [6], Bangladesh producing estimation of 10 million tons [7] annually in which only less percentage is used as animal feed and fuel leading to generation of a large volume of wastes. Collection and management of rice husk have always been big problems to farmers hence hindering the production rate [8].

Admixture methods of ground improvement has been studied by different researchers. The effect of rice husk and treated sisal Fibre on lateritic soil was studied at a random inclusion of 0, 2, 4, 6, and 8% RHA and 0, 0.25, 0.5, 0.75, 1.0% SF by weight of soil sample [9] and recommended 6% RHA and 0.75% SF content by dry weight of soil sample as optimum content to receive positive results of unconfined compressive strength (UCS). Stabilization of expansive soil by using rice husk ash and calcium carbide residue. Fly ash and hydrated lime have also been widely studied [10, 11], the behaviour of hydrated lime with bagasse fibres was also studied and proven to help in soil improvement [12] where bagasse fibre was added at 0.5%, 1.0%, and 2.0% and hydrated lime was added to expansive soil to investigate its behaviour. Other soil reinforcements by admixture carried out include steel fibre admixed with fly ash [13], glass powder, plastic, and e-waste [14], random inclusion of sisal fibre on strength behaviours of soil was studied [15] which was added in four different lengths (10, 15, 20, and 25 mm) and by using four different proportions (0.25, 0.5, 0.75, and 1.0%) by dry weight of soil sample.

However, there are still research about the shear parameters (cohesion and angle of internal friction), hence in this research, shear strength parameters investigation has been carried out by use of the direct box shear test (IS:2720 (Part-13):1986) using rice husk ash burnt at 600 °C to 800 °C in an open atmosphere and treated sisal fibre of single length between 10 mm to 15 mm long. It was studied that sisal mixture with soil to form homogenous mixture is between 10 mm and 25 mm long by length of sisal fibre, difficulties in binding of soil and sisal fibres were observed at length longer than 25 mm due to sticking effects of organic fibres [15]. Sisal fibre being biodegradable material, it is treated with Sodium Borohydride (NaBH4) (1% wt/Vol) as suggested by [16]. Here weighed soil sample is mixed with different proportions of rice husk ash (2.5, 5.0, and 7.5%) and treated sisal fibre (0.3, 0.6, & 0.9%) by weight of soil sample. Additionally, this research helps in maximizing the use of agricultural wastes (Rice husk) due to its large quantity availability globally and thus helps to reduce effects of environment impact. According to this research, there is merging of engineering majors such as agricultural engineering which will deal with the agriculture sector, that is to say production, processing and management of agricultural products leading to achievement of (SDG 2). Environmental engineering will deal with the assessment, collection, management and treatment of agricultural waste hence helping in achieving (SDG 6, 9, 11, 12 and 13). Geotechnical will deal with the use of collected wastes into improvement of engineering properties of soil at low cost.

2 Methodology and Material

2.1 Materials

2.1.1 Soil Sample

The yellow soil sample was collected from Bartd village, Surat district, Gujarat state, India. Yellow soil (Fig. 1) is widely used for the surface improvement of different construction projects such as highways, railways, retaining walls, foundations, embankments, slope stability applications.

2.1.2 Sisal Fibre

Sisal fibre (Fig. 1) was procured from https://www.Fibreregion.in/natural-Fibres.html located in Chennai city, India. Fibres were ready for use hence it was only sun dried and then cut in a single-length of 10mm to 15mm [17] by using hand scissors. Sisal fibre being biodegradable material is always advisable to be treated with alkali solution [18].

2.1.3 Rice Husk

Rich Husk (Fig. 1) was procured from Amidhara Rice Mill, Navsari-Bardoli road, Tajpore Bujrang, Surat district, Gujarat state, India. For this study rice husk were burnt at a temperature of about 600 °C to 800 °C in an open atmosphere [19] to obtain ash (amorphous silica). The use of incinerators is recommended to reduce the emission of high carbon content to the atmosphere which later causes environmental pollution.

Fig. 1. Shows materials used for this study

2.2 Preparation of Samples

Earlier researcher used to investigate different engineering properties of soil by the percentage inclusion of rice husk ash and sisal Fibre at different length of sisal fibres. [20], investigated the strength behaviour of soil by use of sisal fibre at different proportions of (0.25, 0.5, 0.75, 1, and 1.25%) and four different sisal fibre lengths of (0.5, 1 and 1.5 cm). [15], studied about effect of random inclusion of sisal fib. In this study, rice husk ash proportion of 2.5, 5.0, and 7.5% by weight of soil sample was selected. Additionally, sisal fibre of single-length approximately from 10–15 mm, and proportion of 0.3, 0.6, and 0.9% by weight of soil sample was selected for this study. The mixture of composite materials of soil, RHA and SF was then bonded together by using moisture content of natural soil at maximum dry density to maintain homogeneity mixture between the composite materials. The mixing of soil-rice husk ash- sisal fibre was felt very difficult at proportion more than 10% RHA and 1% of SF, as the fibres are sticking together to form lumps. This also caused pockets of low density. Due to the complexity and variation of soil properties, it is always advisable to first determine the chemical composition of composite materials especially the oxide elements, then after the proportion of admixture can be decided as results to maintain the homogeneity and consistency of different models of soil sample. So the recommended rice husk ash & sisal fibres content shouldn't exceed 10% & 1% by weight of soil sample respectively. This investigation reports the effect of rice husk ash and sisal fibre on the strength parameters (angle of internal friction and cohesion) of soil. The comparisons of different results are evaluated and then final conclusion is based on best outcome.

2.3 Methods

2.3.1 Index Properties of Natural Soil

2.3.1.1 Grain Size Analysis

Grain size distribution of natural soil were carried out using IS: 2720 (Part-4): 1985.

2.3.1.2 ATTERBERg's (Consistency) Limit

Liquid limit, plastic limit and plastic index value were carried out using IS: 2720 (Part 5): 1985.

2.3.1.3 Compaction Test

Compaction test for natural soil and reinforced soil were carried out using standard proctor test and **IS: 2720 (Part VII)** [21] and main aim were to determine maximum dry density and optimum moisture content. Apparatus was arranged as follow by R. R. Proctor, Standard proctor mould, Standard proctor rammer of weight 2.5 kg having a fall of 30.5 cm, Tray, 2.5 kg oven dry soil sample (passing 4.75 mm sieve), fibre hammer, knife, Triple beam balance, Thermostatic controlled oven, oil, measuring cylinder, non-corrodible water content container [21]. Volume of standard proctor mould was calculated from measured diameter and height of mould. Weight of mould without base plate and collar also weighed and recorded. 3 kg of oven dried soil sample, retained 4.75 mm sieve is taken and 10% of water by volume is added by measuring cylinder. The soil sample was then mixed thoroughly and added into oiled mould in three layers. Each layer being compacted by 25 blows of full height with rammer scratching being used for every addition of each layer. Remove the collar and base plate and weight the weight of mould and compacted soil sample, by using compressor, compacted soil is removed from mould and sample are collected from centre of compacted for determination of water content. Similar procedure is repeated for soil sample by an increment of 3% of water by volume. Table 1 and Table 4 shows variation of MDD & OMC with different proportions of RHA admixed with SF. Figure 3 shows graph of dry density and water content for determining MDD & OMC (Fig. 2).

Sample Mixing

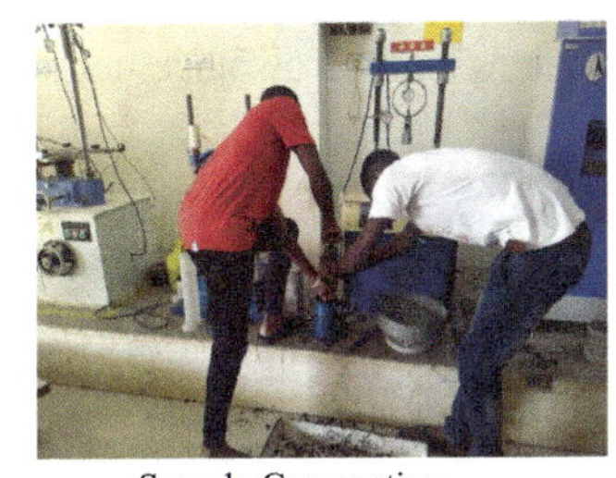

Sample Compaction

Compacted Solid

Fig. 2. Shows preparation of compaction test sample

2.3.1.4 Shear Strength Parameters

Shear strength parameters (cohesion and angle of internal friction) were determined using an undrained direct shear box test (IS:2720 (part-13):1986) [21]. Direct shear box apparatus has the following parts, Shear Box assembly including gripper plates, porous plates & holding screws, pressure pad, Direct Shear Box machine including loading yoke, driving assembly, proving ring & weights, cam type compactor, stop clock, balance (Fig. 3). Non-cohesive soils should be tamped in the shear box (6 cm × 6 cm) and (2 cm thickness of soil sample in box) itself with the base plate and grid plate or porous stone as required in place at the top and bottom of the box, using 75 blows of the cam-type compactor. The shear box containing the soil sample, plain grid plate over the base at the bottom of the soil sample, all are fitted into position. The grid plates are placed at a right angle to the direction of shear, then the loading pad was placed on the top of the plain grid plate. Both the pans were tightened together by fixing the screw. Water was added inside the water jacket to avoid drying of the sample during testing. The Shear box assembly was then fixed on the load frame. The lower part of the shear box was lowered to bear against the load jack and the upper part of the box to bear against the proving ring, then set the dial of the proving ring to zero. Loading yoke was placed on the top of the loading pad, and the dial gauge was adjusted to zero to measure the vertical displacement in the soil sample. 0.5 kg, 1.0 kg, and 1.5 kg weight was placed on the hanger of the loading yoke, this weight plus the weight of the hanger will be equivalent to the required Normal Stress. The locking screws were removed so that the pans are free to move against each other. By turning the spacing screw, the upper pan was raised slightly above the lower pans by about 1 mm. The test was carried out by applying horizontal shear to failure, reading of horizontal shear and horizontal displacement were recorded on the calibrate meter for every 30 s until the value of horizontal loading just begins to decrease. A similar procedure was repeated for all the number of samples carried out [21].

2.3.1.4.1 Computation of Results

The area of shear box (A_o) was calculated from length and width of the box, and the corrected area was calculated from the following equations.
Corrected area

$$A_c = A_o\left(1 - \left(\delta/3\right)\right);\ \text{where } \delta \text{ is the displacement obtained at maximum load kg/cm}^2$$

Shear stress in kg/cm^2

$$\tau = \frac{F}{A_c} \times 10;\ \text{where F} = \text{maximum load at failure},\ \text{A}_\text{C} = \text{corrected area},\ \tau = \text{shear stress}$$

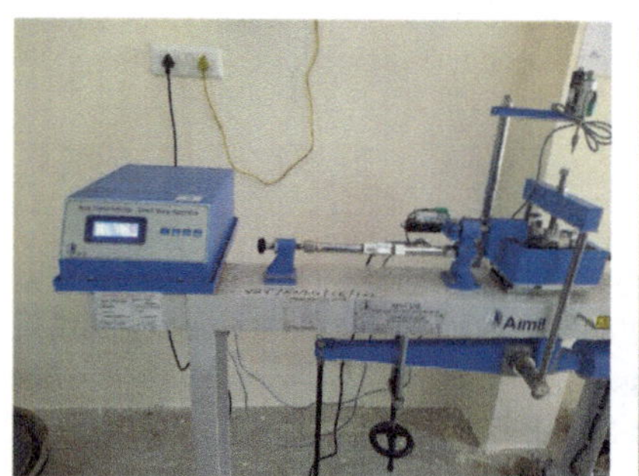

Apparatus Setup

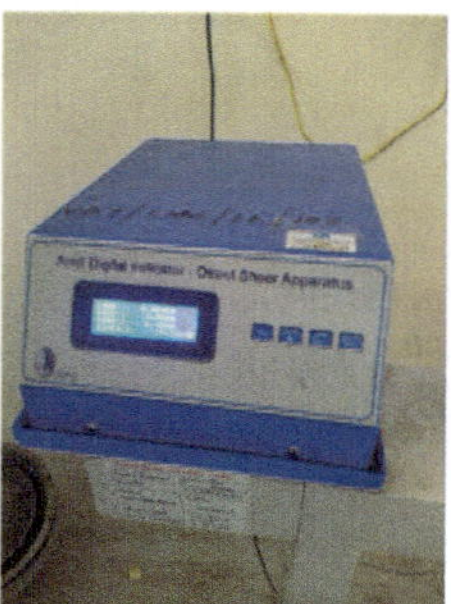

Calibration Meter

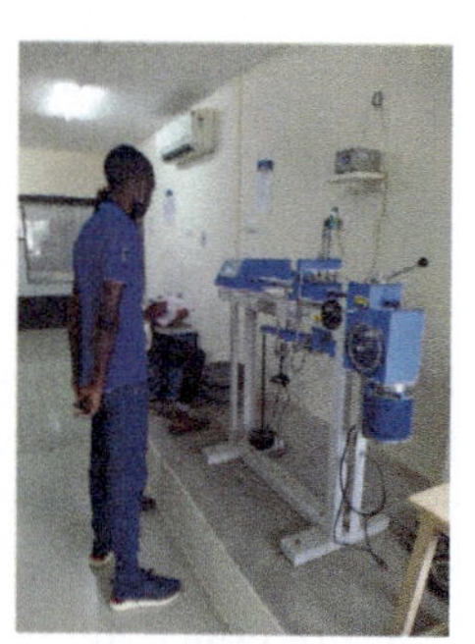

Readings

Fig. 3. Shows direct box shear apparatus

3 Results and Discussions

3.1 Index Properties of Natural Soil

First phase of this research involved geotechnical investigations of natural soil and the following were the findings. The soil had 41.2% Moisture content, 32% free swell value, 2.66% specific gravity, 24.23% plastic value, 17.00% plastic Index, 1.49 gm/cc dry density. The soil being classified as CI in IS classification (Table 2, 3 and Fig. 4).

Table 1. Shows summary of properties of natural soil before additional of admixture

Sr. No.	Properties	Values
1	Grain size analysis	
	• Gravel	3.20%
	• Sand	6.55%
	• Silt & Clay	90.25%
2	IS classification	CI
3	Specific gravity (G)	2.66%
4	Consistency limit	
	• Water limit	41.20%
	• Plastic limit	24.23%
	• Plasticity Index	17.00%
5	Free swell	30%
6	Compaction	
	• Maximum Dry Density (MDD)	1.80 gm/cc
	• Optimum Moisture Content (OMC)	16.78%
7	Shear parameters	
	• Cohesion (c)	0.13kg/cm^2
	• Angle of Internal Friction (Φ)	26^0

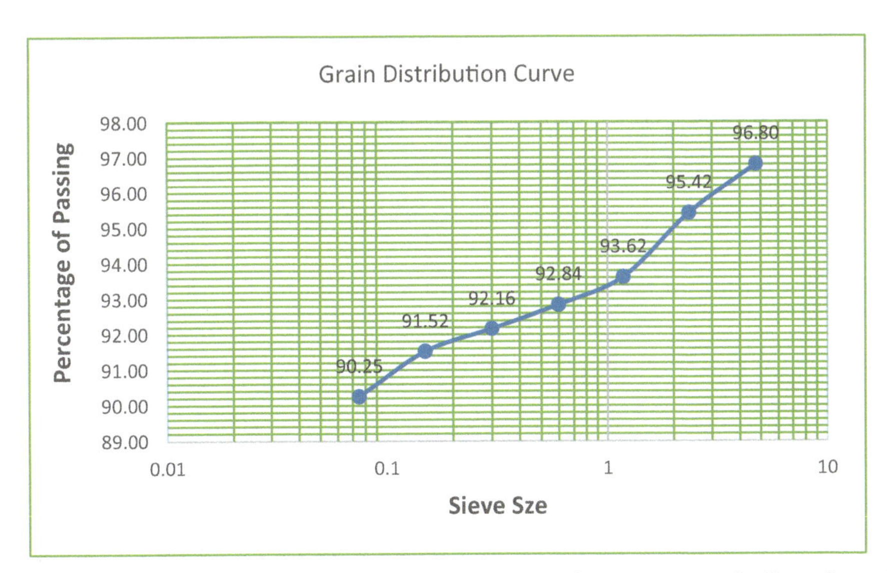

Fig. 4. Shows grain size distribution curve prepared by plotting percentage of soil passing and sieve size

Table 2. Shows results from grain distribution curve

Gravel		3.20
Sand		6.55
Silt+Clay		90.25
D_{10}		−4.66
D_{30}		−3.48
D_{60}		−1.71
C_u	D_{60}/D_{10}	0.367
C_c	$D_{30}^2/D_{60} * D_{10}$	1.5198

3.2 Compaction Characteristics

3.2.1 Maximum Dry Density (MDD)

The graph of MDD of soil-RHA mixed with soil fibre at different proportions is shown in Fig. 5. MDD values decreases with increase in RHA-SF content because of pore compaction due to SF properties. MDD values decreased from 1.8 gm/cc for the natural soil to 1.6 gm/cc at 5% RHA/0.6% SF content by dry weight of soil. Other proportions such as 2.5, and 7.5 RHA content by dry weight of soil, MDD reduced from 1.76 gm/cm^3 to 1.62 gm/cm^3, and 1.69 to 1.58 gm/cc at 0.6% SF content by dry weight of soil respectively.

The decrease in the MDD is due to the low density in RHA compared to that of soil and thus reducing the average unit weight of the compacted soil solids in the mixture. Additionally, as the SF content increased, the compaction becomes difficult because the SF occupies more spaces than soil was naturally supposed to, thereby creating some voids in the mixtures. As the fibre content increased from 0 to 0.3% by dry weight of

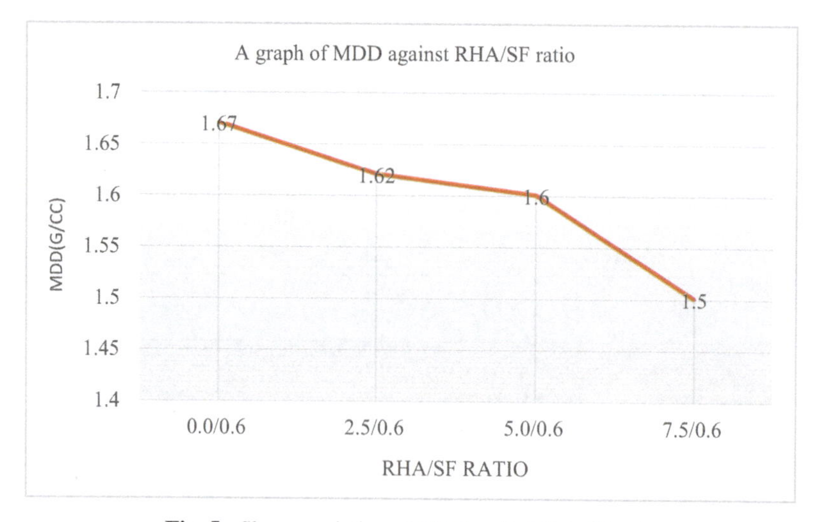

Fig. 5. Shows variation of MDD with RHA/SF ratio

soil, a decrease in MDD was recorded which further decreased as it increased from 0.3 to 0.5, 0.6 and 0.9% by dry weight of soil. These trends recorded agree with the findings of [14, 15].

Table 3. Variation of MDD at different RHA/SF proportion

Density					
Sr.No.	Particulars	RHA/SF @ 0.0/0.6	RHA/SF @ 2.5/0.6%	RHA/SF @ 5/0.6%	RHA/SF @ 7.5/0.6%
1	Dry density (d) gm/cc	1.73–1.67	1.76–1.62	1.80–1.60	1.69–1.58

3.2.2 Optimum Moisture Content (OMC)

It was noticed that an increase in RHA and SF content caused an increase in the OMC values from 17.5% to 25.2% at 5% RHA% 0.6% SF proportion by dry weight of soil. Upon further increase in the RHA and SF content, the OMC decreased. It is due to high absorption nature of SF, caused an initial increase in OMC from 17.5% to 25.2% content of the RHA by dry weight of soil and thereafter reduced the OMC with increasing aspect ratio and percentage content. This trend of reduction in the OMC after initial increase is principally associated with the water absorption property of the SF. Similar trends were observed by [14], who used laterite soils and black cotton soils respectively. On further treatment of the soil with higher RHA content (i.e. 2.5, 5.0 and 7.5% RHA) the OMC generally increased was observed by [9] (Fig. 6).

Table 4. Variation of OMC with RHA/SF proportion

Moisture content					
Sr.No.	Particulars	RHA/SF @ 0.0/0.6	RHA/SF @ 2.5/0.6%	RHA/SF @ 5.0/0.6%	RHA/SF @ 7.5/0.6%
1	Moisture Content (m) %	12.7–17.8	13.6–19.4	17.5–25.2	19.2–27.4

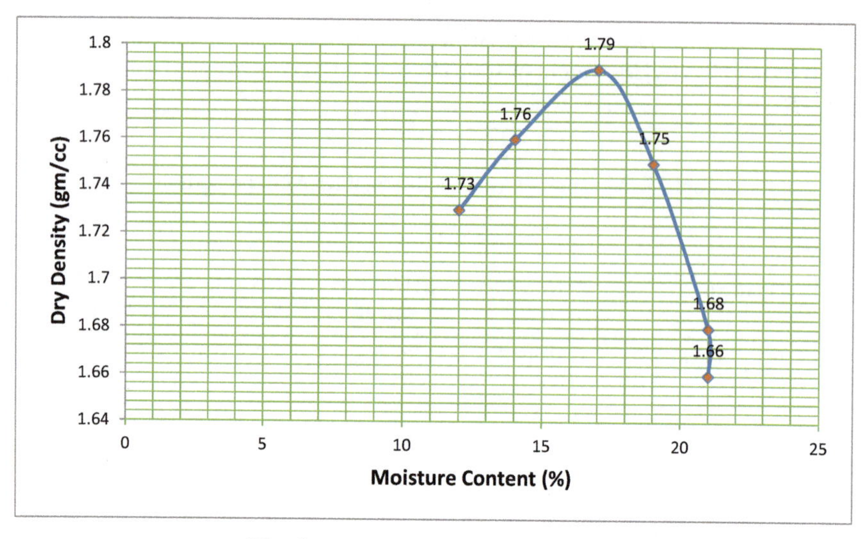

Fig. 6. A graph of MDD against OMC

3.3 Shear Strength Parameters Test

The shear strength of soil is the resistance to deformation by continuous shear displacement of soil particles or masses upon the action of force IS:2720 (part-13):1986 [21]. The shear strength of soil mainly depends upon the following factors such as the structural resistance to displacement of the soil [1, 2] because of the interlocking of the particles, the frictional resistance to the translocation between the individual soil particles at the contact points, and Cohesion or adhesion between the surfaces of soil particles. The direct Shear box assembly including gripper plates, porous plates & holding screws, pressure pad, direct shear box machine including loading yoke, driving assembly, proving ring & weights, cam type compactor, stop clock, balance. The direct box shear test was carried out in order to investigate the behaviour of soil sample under normal stress and shear stress loading. 130 gm oven dried natural soil sample passing 425 micro sieve was weighed for each sample, different proportions of RHA & SF by weight of soil sample were weighed and added to soil sample, finally moisture content at MDD was added to the mixture of soil, RHA & SF, admixed mould formed was tamped in the shear box itself with the base plate and grid plate or porous stone as required in place at the top and bottom of the box, using 75 blows of the cam type compactor. The shear strength parameters of natural soil, RHA and SF mixture was conducted according to IS:2720 (part-13):1986 [21]. The variation of shear strength parameters at different proportions of soil and RHA/SF admixture is shown in Table 5. The shear strength parameters (cohesion 'c' and angle of internal friction Φ) increased from $c = 0.13$ kPa, and $\Phi = 26°$ for natural soil to 0.50 kPa and 50° respectively at 0.3%SF and 5% RHA content. This increase trend continued to 6.7 kPa and 72° (Table 5) at optimum of 0.3% SF and 7.5% RHA (Fig. 3) content by dry weight of soil. When the content of SF was increased, there was decrease in cohesion c and increase in angle of internal friction Φ (Fig. 10). Due to cementation nature of RHA and reinforcement nature of SF, this led to increase of cohesion and angle of internal friction hence increase in shear strength (Fig. 7).

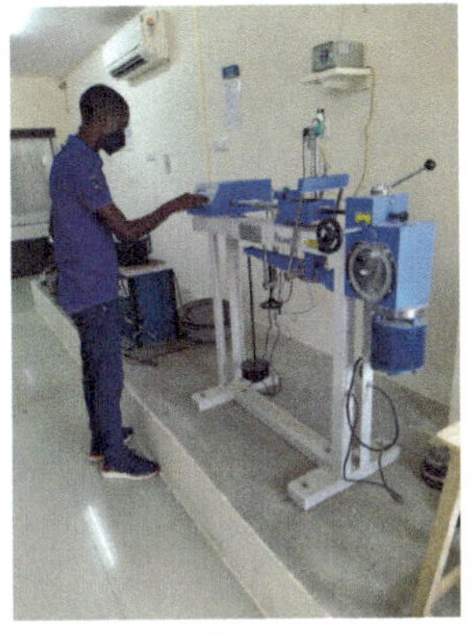

Direct Box Shear Apparatus

RHA/SF Admixed

Compressed sample

Fig. 7. Showing apparatus arrangement

Table 5. Summarises the results obtained from this test

RHA/SF (%)	Angle of internal friction (Φ^o)	Cohesion (c) (kg/cm^2)	Shear Strength (τ) (kg/cm^2) $\tau = c + 3.171 \tan \phi$
0.0/0.0	18°	0.13	0.560
2.5/0.3	49°	−0.04	3.608
5.0/0.3	50°	0.50	4.279
7.5/0.3	72°	6.71	16.469
2.5/0.6	75°	−0.76	11.074
5.0/0.6	70°	−0.34	8.372
7.5/0.6	68°	−0.26	7.588
2.5/0.9	69°	−1.02	2.030
5.0/0.9	71°	−0.95	2.450
7.5/0.9	73°	−0.82	3.010

It was observed that shear strength increased with an increased in RHA/SF ratio (Table 5), maximum value of shear strength was observed at 7.5/0.3% of RHA/SF due to an increase in shear strength parameters. Investigation also show that natural soil strength was increased from 0.56 kg/cm^2 to 16.47 kg/cm^2 (closely 30 times the original value of natural soil shear strength) at 7.5/ = 0.3% of RHA/SF proportion by weight of soil sample (Fig. 8). Since shear strength (τ) is directly proportional to cohesion (c) and angle of internal friction (Φ), it was observed that for decrease in cohesion and increase in angle of internal friction still led to increase in shear strength.

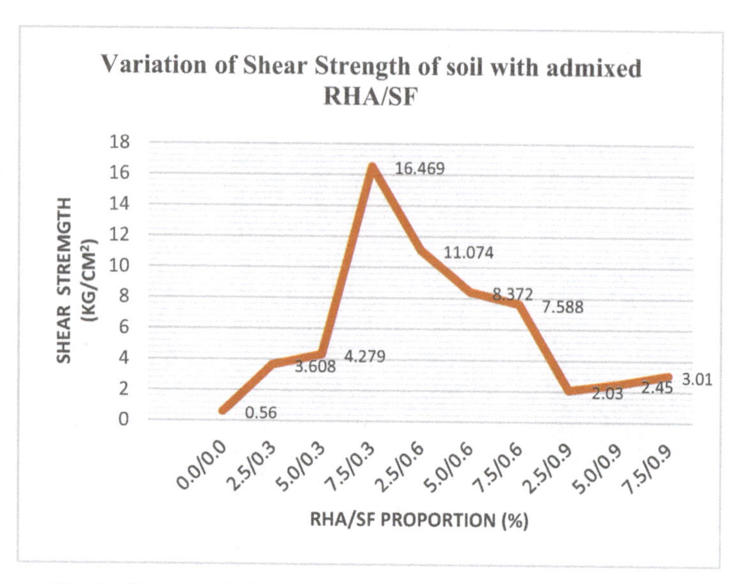

Fig. 8. Shows variation of shear strength with RHA/SF proportions

Angle of internal friction was observed increasing with all RHA/SF proportions (Table 5) and the maximum increase was observed at 2.5% RHA and 0.6% SF of about 75° (Fig. 9). Due to cementation nature of RHA and reinforcement behaviour of SF this contributed to great increase of angle of internal friction. Friction was further created due to roughness of soil-RHA-SF admixture there by leading to increase in shear strength of soil since shear strength is directly proportional to shear strength parameters.

Figure 10 shows the variation of cohesion and angle of internal friction of soil at different proportions of admixed RHA & SF, it was observed that for every increase of RHA and SF there was and an increase in the angle of internal friction due to friction created between soil, RHA and SF. The cementation nature of RHA led to binding and creating friction between the soil particles while the reinforcement nature of SF led to strength. Additionally, cohesion of soil had different variations with admixed RHA and SF. It was observed that only 5/0.3% of RHA/SF and 7.5/0.3% of RHA/SF by weight of soil gave positive results. Maximum value of cohesion was achieved at 7.5//0.3% of RHA/SF by weight of soil sample at same proportion high angle of internal friction was recorded (Table 6 and Fig. 11).

$$\text{SLOPE}(\varphi) \quad 72.5^\circ$$
$$\text{Y - Intercept(c)} \ 6.71\,\text{kg/cm}^2$$

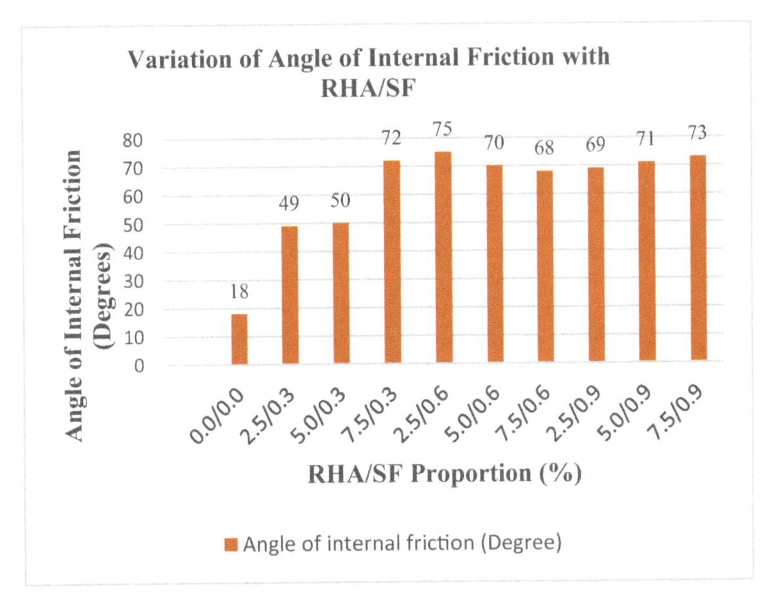

Fig. 9. Shows Variation of Angle of Internal Friction with RHA/SF proportions

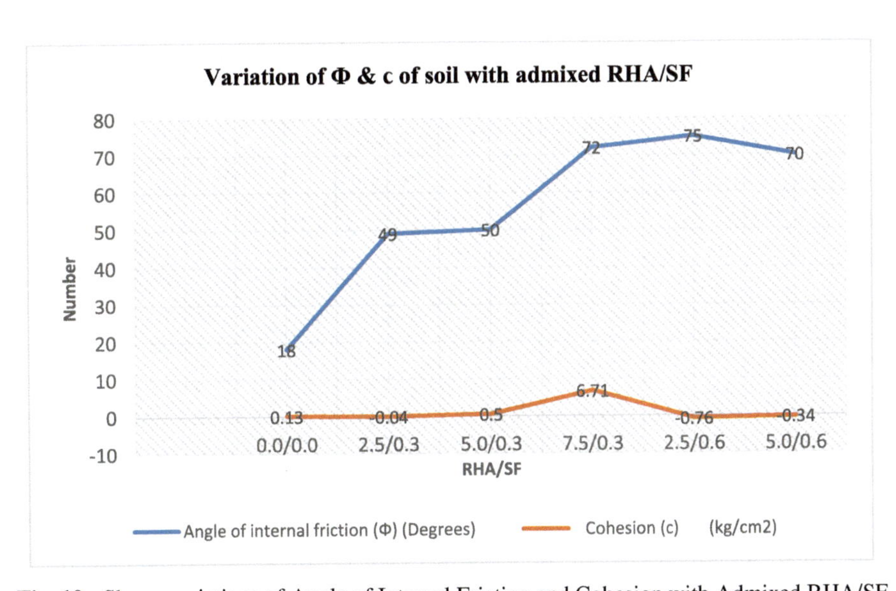

Fig. 10. Shows variations of Angle of Internal Friction and Cohesion with Admixed RHA/SF

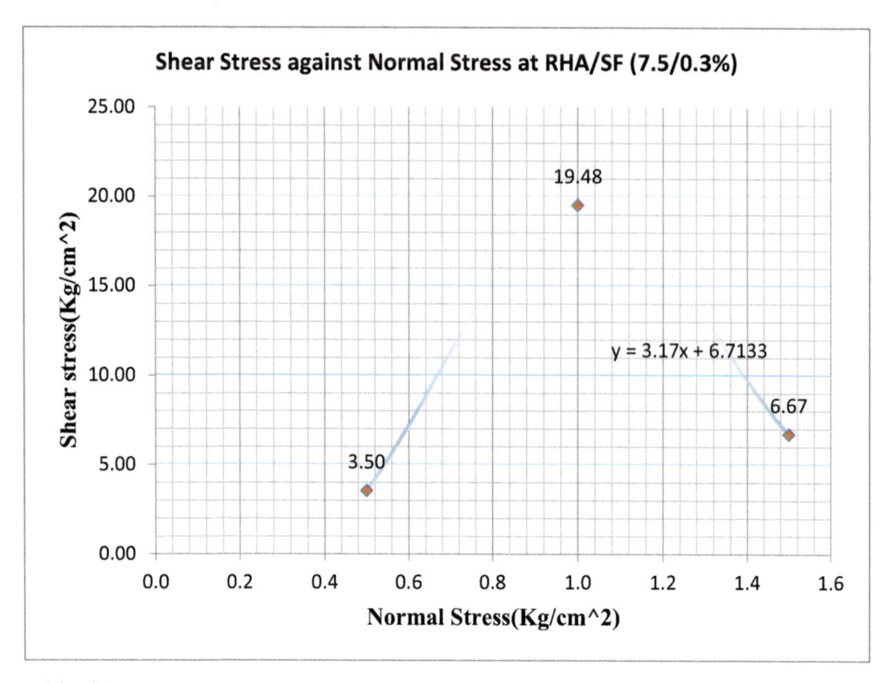

Fig. 11. Shows Variation of Shear Stress against Normal Stress at RHA/SF (7.5/0.3%)

Table 6. RHA/SF at 7.5/0.3%

Normal stress	Shear stress
0.5	3.50
1.0	19.48
1.5	6.67

Practically during road construction in Surat village, Gujarat state, India, the Government of India instructed Gandhi Road Builders Pvt Ltd to stabilise sub grade by addition of 6% of hydrated lime and 15% of fly ash. For every 1 km long, 7 m wide road and thickness of compaction is 300 mm, soil of dry density 1800 kg/m^3. Consider cost comparison between two different admixtures as seen in the calculation below;

$$\begin{aligned}\text{Total weight of Soil to be compacted (tons)} &= \text{length} \times \text{width} \times \text{thickness} \times \text{dry density}\\ &= 1000 \times 7 \times 0.03 \times 1800/1000\\ &= 378\,\text{t}\end{aligned}$$

Admixtures:

- Hydrated lime (6% by weight of total soil compacted) and fly ash (15% by weight of total soil compacted); this method is commonly used in various soil stabilization

- Rice husk ash (7.5% by weight of total soil compacted) and sisal fibre (0.3% by weight of total soil compacted), recommended soil stabilization method as recommended in this project

  Calculation of admixture required;

i. $$\text{Hydrated lime } (6\%) = 6\% \times 378 \\ = 22.68\,\text{t}$$

ii. $$\text{Fly Ash}(15\%) = 15\% \times 378 \\ = 56.70\,\text{t}$$

iii. $$\text{Rice husk ash}(7.5\%) = 7.5\% \times 378 \\ = 8.35\ \text{t}$$

iv. $$\text{Sisal fibre}(0.3\%) = 0.3\% \times 378 \\ = 1.13\,\text{t}$$

Table 7. Shows cost comparison between (hydrated lime & fly ash) and (rice husk ash & sisal fibre)

Item	Unit	Qty.	Rate (USD)	Amount (USD)
Lime (6%)	ton	22.68	$120	$2,722
Fly Ash (15%)	ton	56.70	$20	$1,134
Rice Husk Ash (7.5%)	ton	28.35	$14	$394
Sisal Fibre (0.3%)	ton	1.13	$500	$567
Sum of Lime & Fly Ash(FA)				**$3,856**
Sum of RHA & Sisal Fibre				**$961**
Difference of (Lime & FA) and (RHA&SF)				**$2,895**

4 Conclusion

An increased in cohesion (c) and angle of internal friction (Φ) was observed in RHA/SF of 5.0/0.3% and 7.5/0.3% at single length of SF 10 mm–15 mm from natural value of 18° and 0.13 kg/cm^2 to 50° and 0.67 kg/cm^2 at RHA/SF 5.0/0.3% and further increased was observed at RHA/SF 7.5/0.3% of 72° and 6.71 kg/cm^2 Shear strength being proportion to shear parameters hence there was increase in shear strength as well. From the optimum value of RHA and SF, we can conclude that maximum utilisation of agricultural wastes (RHA) for construction development can be attained due to the availability of these residues at free of cost, in return also creates space for more agricultural production and reduction of environment effect from unmanaged agricultural wastes. The following conclusion was made from this research;

i. There was an increase of angle of internal friction from 18° of natural soil to 72° at optimum proportion of 7.5% RHA and 0.3% SF by weight of soil sample.
ii. Additionally, cohesion was increased from 0.13 kg/cm^2 of natural soil to 6.71 kg/cm^2 at optimum proportion of 7.5% RHA and 0.3% SF by weight of soil sample.
iii. Shear strength also increased from 0.56 kg/cm^2 of natural soil to 16.469 kg/cm^2 at optimum proportion of 7.5% RHA and 0.3% SF by weight of soil sample.
iv. Agricultural waste management is achieved by using 7.5% RHA for soil improvement hence creating more space for agricultural production.
v. Cost of construction was also reduced due to use of cheap and available rice husk for ground stabilization, for every 1 km long of 7 m wide road embankment and 3mm thickness of compaction, approximately 3000$ USD is saved when using rice husk ash and sisal fibre for sub grade stabilization instead of using tradition admixture method of hydrated lime powder and fly ash (Table 7).
vi. Finally, environmental protection and sustainability was improved by maximum management of rice husk residue that would cause environmental pollution.

Acknowledgement. This research was supported by the Design Innovation Centre(DIC)-Gujarat Technological University Ahmedabad, and thanks goes to Vidyabharti Trust Polytechnic, Umrakh, Civil Material Testing and Consultancy Cell for providing testing equipment and tools, Dr. Chetankumar Patel for reviewing this research paper, and finally thanks to Sitarambhai Naranjibhai Institute of Technology and Research Centre for giving us conducive environment for completion of this project.

References

1. ASCE: Soil improvement (1978)
2. Murthy, V.: Geotechnical Engineering: Principles and Practices of Soil Mechanics and Foundation Engineering (2018)
3. Hausmann, M.R.: Engineering Principles for Ground Modification (2017)
4. Das, B.: Principle of Foundation Engineering, 7th edn. (2020)
5. Alhassan, M.: Potentials of rice husk ash for soil stabilization. Assumption (2008)
6. Eco-Business: Second life rice husk (2013)
7. Mustafi: Statistical Yearbook of Bangladesh Bureau of Statistics (BBS) (2005)
8. FAO: The Future of Food and Agriculture: Trends and Challenges (2017)
9. Sani, J.E., Yohanna, P., Chukwujama, I.A.: Effect of Rice husk ash admixed with treated sisal fibre on properties of lateritic soil as a road construction material. J. King Saud Univ. Eng. Sci. **32**, 11–18 (2020)
10. Andavan, S., Pagadala, V.K.: Experimental study on addition of lime and fly ash for the soil stabilization. Mater. Today Proc. **22**, 1065–1069 (2020)
11. Indiramma, P., Sudharani, C., Needhidasan, S.: Utilization of lime and fly ash to stabilize expansive soil and to sustain pollution free environment – an experimental study. Mater. Today Proc. **22**, 694–700 (2020)
12. Dang, L.C., Fatahi, B., Khabbaz, H.: Behaviour of expansive soils stabilized with hydrated lime and bagasse fibres. Proc. Eng. **143**, 658–665 (2016)

13. Praveen, G.V., Kurre, P., Chandrabai, T.: Improvement of California Bearing Ratio (CBR) value of Steel Fiber reinforced Cement modified Marginal Soil for pavement subgrade admixed with Fly Ash. Mater. Today Proc. **39**, 639–642 (2021)
14. Rai, A.K., Singh, G., Tiwari, A.K.: Comparative study of soil stabilization with glass powder, plastic and e-waste. Mater. Today Proc. **32**, 771–776 (2020)
15. Prabakar, J., Sridhar, R.S.: Effect of random inclusion of sisal fibre on strength behaviour of soil. Const. Build. Mater. **16**(2), 123–131 (2002)
16. Moraes, A.G.O., Sierakowski, M.-R., Abreu, T.M., Amico, S.C.: Sodium borohydride as a protective agent for the alkaline treatment of sisal fibers for polymer composites. Compos. Interfaces **18**(5), 407–418 (2011)
17. Tanko, A., Ijimdiya, T.S., Osinubi, K.J.: Effect of inclusion of randomly oriented sisal fibre on some geotechnical properties of lateritic soil. Geotech. Geol. Eng. **36**, 3203–3209 (2018). https://doi.org/10.1007/s10706-018-0530-y
18. FAO: International Year of Natural Fibres (2009). International Year of Natural Fibres Coordinating Unit, Trade and Markets Division (EST) (2011)
19. Shackley, S., Carter, S., Knowles, T., Middelink, E., Haefele, S., Haszeldine, S.: Sustainable gasification–biochar systems? A case-study of rice-husk gasification in Cambodia, part II: field trial results, carbon abatement, economic assessment and conclusions. Energy Policy **41**, 618–623 (2012)
20. Ramkrishnan, R., Sruthy, M.R., Sharma, A., Karthik, V.: Effect of random inclusion of sisal fibres on strength behaviour and slope stability of fine grained soils. Mater. Today Proc. **11**, 25313–25322 (2017)
21. 1. Grain Size Analysis of Soil by Dry Sieving Method, IS: 2720 (Part-4): 1985; 2. Determination of liquid limit, Plastic Limit and plastic index of Soil, IS: 2720 (Part 5): 1985; 3. Standard proctor test of soil, IS:2720(Part 7):1980; 4. Direct shear test, IS:2720(Part 13):1986; 5. SNPIT&RC Umrakh, Soil Mechanics Laboratory Manual

Cyclic Behaviour of Brumadinho-Like Reconstituted Samples of Iron Ore Tailings

Paulo A. L. F. Coelho[1(✉)], David D. Camacho[1], Felipe Gobbi[2], and Luís M. Araújo Santos[3]

[1] Department of Civil Engineering, University of Coimbra, CITTA, Coimbra, Portugal
pac@dec.uc.pt

[2] Unisinos- Federal University of Sinos River Valley- and FGS Geotecnia, São Leopoldo, Brazil

[3] Polytechnic Institute of Coimbra, CITTA, Coimbra Institute of Engineering, Coimbra, Portugal

Abstract. These days, there is an urgent call for a transition to a new low-carbon energy paradigm enabling sustainable development. This is expected to increase the demand for minerals and metals in the next decades and intensify the already massive and problematic problem of mining waste, which represents a huge proportion of the excavated material and is often accumulated in large tailings dams that have caused serious disasters around the world. To promote future sustainable mining and ensure the safety of tailings deposits, namely in seismically active regions, performance-based design needs to be applied in practice based on suitable geotechnical-based characterization of mining waste materials. This paper focus on the assessment of the undrained behaviour of iron ore tailings, which has been observed as a major factor in historic disasters in the recent past, including Brumadinho's failure in Brazil. Samples reconstituted through a slurry-based method and tested in undrained triaxial monotonic and cyclic conditions are used to investigate the response of Brumadinho-like reconstituted iron ore tailings. The results show that the samples have physical and identification properties similar to those in situ, the mechanical behaviour obeying to the principles observed in granular soils tested in similar conditions. In fact, phase transformation and critical state lines are suitably defined in undrained monotonic and cyclic loading, particular features being justified by the high angularity of crushed particles.

Keywords: Iron ore tailings · Slurry-based reconstitution · Monotonic undrained behaviour · Cyclic undrained behaviour

1 Introduction

The desired transition to a so-called low-carbon energy paradigm will greatly intensify the demand for minerals and metals, which will significantly expand the production of mining waste. Unfortunately, mining waste, which usually represents a very large proportion of the excavated material, is often accumulated in large tailings dams that have been responsible for serious disasters occurring around the world. In order to promote future sustainable mining and ensure the safety of tailings deposits, namely in seismically active regions, performance-based design (PBD) needs to be applied in practice.

© The Author(s), under exclusive license to Springer Nature Switzerland AG 2022
L. Wang et al. (Eds.): PBD-IV 2022, GGEE 52, pp. 2024–2032, 2022.
https://doi.org/10.1007/978-3-031-11898-2_185

1.1 Historic Failures of Tailings Dams

The relative recent failure of Córrego do Feijão iron ore tailings dam, which occurred in Brumadinho, Minas Gerais, Brazil, in 2019, caused almost 300 hundred deaths and massive environmental, social and economic impacts [1]. The main cause of the failure is still a matter of controversy, two major expert reports suggesting different major causes for the failure [2, 3]. Even if seismic loading is not considered a trigger of the failure, both reports recognize that the geotechnical data available before the failure occurred was insufficient. Therefore, additional and more detailed geotechnical characterization was required to assist in the numerical modelling carried out after the accident in order to identify the reasons possibly leading to the failure.

The dramatic failure of Brumadinho's tailings dam was preceded by another massive failure of an iron ore tailings dam in the same state of Minas Gerais, Brazil. The 2015 failure of Fundão tailings dam, located near Mariana, released more than 50 million m^3 of toxic mud that reached the Doce River and later the Atlantic Ocean, after flowing for 600 km during two weeks. This caused a colossal environmental disaster, seriously affecting the ecosystems and communities existing in the areas affected in different perspectives and degrees, an increased health risk to the population living in the surrounding areas of the Fundão tailings dam failure still remaining [4]. The report on the causes of the failure [5] identified several issues that contributed for the final disaster, which was the consequence of a chain of events and conditions. These included a small additional increment of loading induced in the dam as a result of some minor earthquakes that occurred in the region and accelerated the failure process that was probably already almost inevitable.

In seismically active regions, however, earthquakes pose an additional and significant risk to the stability of old and new tailings dams, various failures occurring around the world as a result of seismically-induced liquefaction. Amongst some relatively recent occurrences, the 2011 East-Japan earthquake caused significant damage in several abandoned tailings dams, including the development of surface cracks or local movement near the banks, which was particularly clear in the case of the tailings embankment at Kayakari, in Ohya mine. This dam suffered complete breach due to the seismic loading, with about 41,000 m^3 of liquefied tailings being released [6]. Also, five medium-sized mine tailings dams failed due to liquefaction induced by the 27 February 2010 Maule, Chile, earthquake, even if the earthquake occurred during the driest period of the year [7]. In the worst case, a tailings volume in excess of 100,000 m^3 flowed a distance of about 0.5 km and caused severe consequences in the region affected.

1.2 Reasons and Ways to Promote PBD Application in Practice

The repeated and tragic accidents occurring in tailings dams around the world suggest that the design of these structures, namely but not only older ones, may require an enhanced approach to ensure suitable safety levels under monotonic and seismic loading. In addition, this approach must be based on a proper characterization and modelling of tailings undrained behaviour under monotonic and cyclic loading and also safeguard economic sustainability of mining activity, which is expected to flourish in the near future. As a result, application of PBD in tailings dams' stability analysis is greatly required so that safe and sustainable mining can be promoted in the real world.

Technical and research reports of historic tailings failures have shown that the chacterization of tailings is often limited, at least from a geotechnical point of view, namely with respect to the behaviour of these mining waste materials under undrained monotonic and cyclic conditions, the latter ones being particularly relevant in seismically active regions where earthquake-induced dynamic loading is a major cause of failure. Therefore, a critical limitation to a broader implementation of PBD in design practice is an unsuitable characterization and, as a result, modelling of the behaviour of tailings under generalized, but realistic, loading conditions. This paper aims at contributing to mitigate this weakness by discussing preliminary data obtained through undrained triaxial testing carried out on Brumadinho-like reconstituted samples of iron ore tailings.

2 Preparation of Brumadinho-Like Reconstituted Samples

Considering the significance of iron ore tailings in mining activities and the major failures occurring in these types of structures in the last few years, this paper focuses on the undrained behaviour of Brumadinho-like reconstituted samples of iron ore tailings.

The reconstituted samples were prepared with iron ore obtained from the old Moncorvo Iron Mine, a recently reactivated mine near Torre de Moncorvo, in the Northeastern region of Portugal (Fig. 1). Because the grain size distribution of the tailings was finer than the grain size distribution of Brumadinho tailings, some crushed sand was mixed in suitable proportions with the natural iron tailings to achieve the desired grain size distribution. Considering the large segregation potential of a mixture of grains with very different densities and sizes and the fact that the material exhibits some plasticity, a slurry-based preparation method was used to produce uniform reconstituted samples. Figure 2 illustrates the main steps involved in the preparation of the Brumadinho-like iron ore reconstituted samples, which included: deposition of the fluid and uniformized slurry in a mould (a), consolidation of the sample under a vertical load (b) and cutting of the large reconstituted sample into smaller pieces for testing (c).

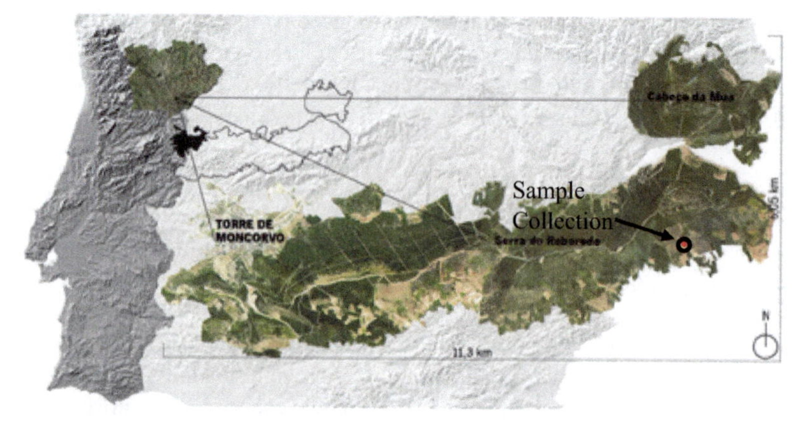

Fig. 1. Location of sample collection at Portugal's Moncorvo Iron Mine (adapted from [8])

a) slurry deposition b) slurry consolidation c) division of large sample

Fig. 2. Slurry-based preparation of Brumadinho-like reconstituted samples of iron ore tailings

3 Experimental Results

3.1 Physical and Identification Properties

Brumadinho-like reconstituted iron ore samples were prepared with similar mineral and grain size compositions to those in situ. Thus, the physical and identification properties of Brumadinho-like reconstituted samples replicate the typical field properties of Brumadinho's iron ore tailings (Table 1), being mostly composed of sand and silt and showing some plasticity (PI under 10%), with water contents slightly below the plastic limit and a quite high value of the density of solid particles, due to the presence of iron.

3.2 Undrained Behaviour Under Monotonic and Cyclic Triaxial Loading

Considering the mechanisms involved in the historic failures observed in tailings dams around the word and the importance of understanding the monotonic response of similar granular materials to explain their behaviour under cyclic loading, the Brumadinho-like reconstituted iron ore samples were firstly tested under undrained monotonic triaxial compression (UTC). All the samples were previously saturated until B values of 0.98–1, indicative of near full to full saturation. The shearing stage was implemented under constant strain rate through an increase of the vertical stress, with constant horizontal stress. Figure 3 illustrates the stress-path (a) and the variation of deviatoric stress, principal stress ratio and excess-pore-pressure with axial strain (b) in an undrained monotonic triaxial compression test carried out on a Brumadinho-like reconstituted sample that was isotopically consolidated to an effective confining stress of 200 kPa.

Table 1. Physical and identification properties of Brumadinho iron ore tailings: in situ data and Brumadinho-like reconstituted samples.

Tailings Tested	w (%)	wP (%)	w_L (%)	G	Sand (%)	Silt (%)	Clay (%)
In situ [9]	16–20	18–23	18 –33	3.8–4.9	16–85	14–67	1–7
Reconstituted	13.4	14.7	17.7	3.7	≈50	≈48	≈2

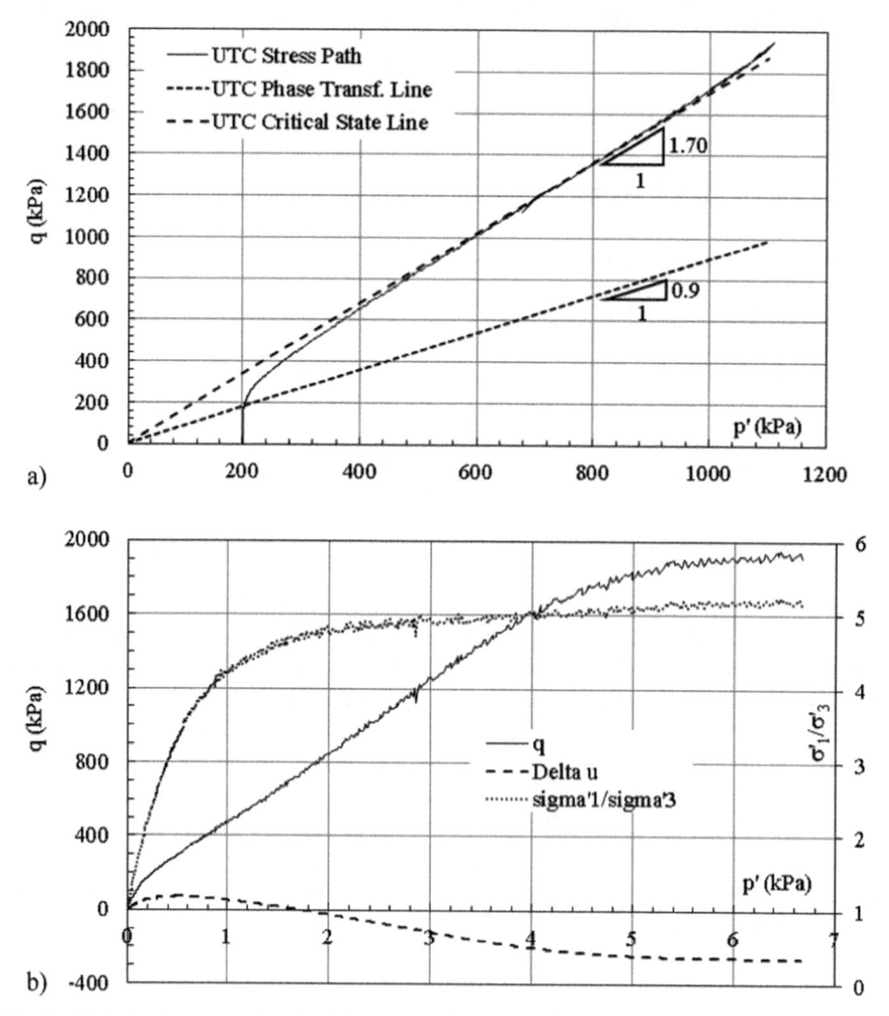

Fig. 3. Undrained monotonic triaxial compression test with $p'_0 = 200$ kPa: a) stress-path; b) variation of deviatoric stress, principal stress ratio and excess-pore-pressure with axial strain

The stress-path observed (Fig. 2a) shows continuous strain-hardening behaviour, an initial period of limited contraction being followed by dilative behaviour up to the end of the test. Two major notable stages can be defined in the stress-path observed, which adheres to the typical behaviour of granular soils in undrained monotonic compression: a) firstly, there is a point where the variation of p' drastically changes, defining the phase-transformation (PT) line [10] for the reconstituted tailings; b) secondly, the stress-path follows a relatively straight line in p'-q plane up to the end of the test where the critical state (CS) is achieved. In fact, the stable values of u, q and σ'_1/σ'_3 at the end of the test (Fig. 2b) suggest that the CS may have been achieved and that the ratio σ'_1/σ'_3 at that state is slightly above 5. Based o/n the experimental data, the slopes of the CS and PT lines in the p'-q plane can be estimated as 1.7 and 0.9, respectively. The corresponding angle of effective friction angle for a purely cohesionless material at the CS ($\approx 42^o$) is close to the values obtained for similar testing conditions in Swedish tailings ($40^o - 43^o$), the higher effective friction angle (3–5°) measured in comparison with natural granular sands being justified by the high angularity of crushed particles [11].

The undrained cyclic triaxial behaviour of the reconstituted sample also fits into the commonly observed behaviour of granular materials under similar testing conditions (Fig. 4). Firstly, the early stages of the monotonic UTC test, described in Fig. 3 and replotted in red in Fig. 4, are perfectly reproduced by the loading stage of the first cycle of the cyclic test, with respect to the variation of q (Fig. 4b) and Δu (Fig. 4c) with the axial strain, the stress-path being also fairly accurately replicated (Fig. 4a). More importantly, the slopes of the CS and PT lines determined in the monotonic UTC test are also valid to assess the behaviour of the reconstituted samples in undrained cyclic loading. In fact, the stress-path shows a sudden transition from contractive to dilative behaviour, in every cycle, every time the PT line is crossed (Fig. 4a). Even if this phenomenon is observed in compression and in extension, the slope of the PT line observed in the UTC test is only valid for the compression phase of the loading cycle. The stress-strain response in undrained cyclic loading (Fig.b) shows a progressive increment of the peak cyclic strains after the PT line is crossed for the first time, strain accumulation occurring mostly in extension. The variation of the normalized excess-pore-pressure with strain during the test (Fig. 4b) shows that the average and the peak normalized excess-pore pressure increase in every cycle, the latter one approaching 1 twice in each loading cycle. Still, significant but temporary dilation is observed in each of the loading cycles, peak negative excess-pore-pressures occurring with the peak load in extension.

Based on the data, the number of cycles required for liquefaction to initiate in the reconstituted iron tailings consolidated under an isotropic effective stress of 200 kPa and subjected to cyclic triaxial loading with Δq peaks equal to $\pm$ 200 kPa, is about 3.7, considering a double axial strain amplitude of 5%, or around 4.5, using the criterion of normalized excess-pore pressure close to 1, the two criteria yielding similar results.

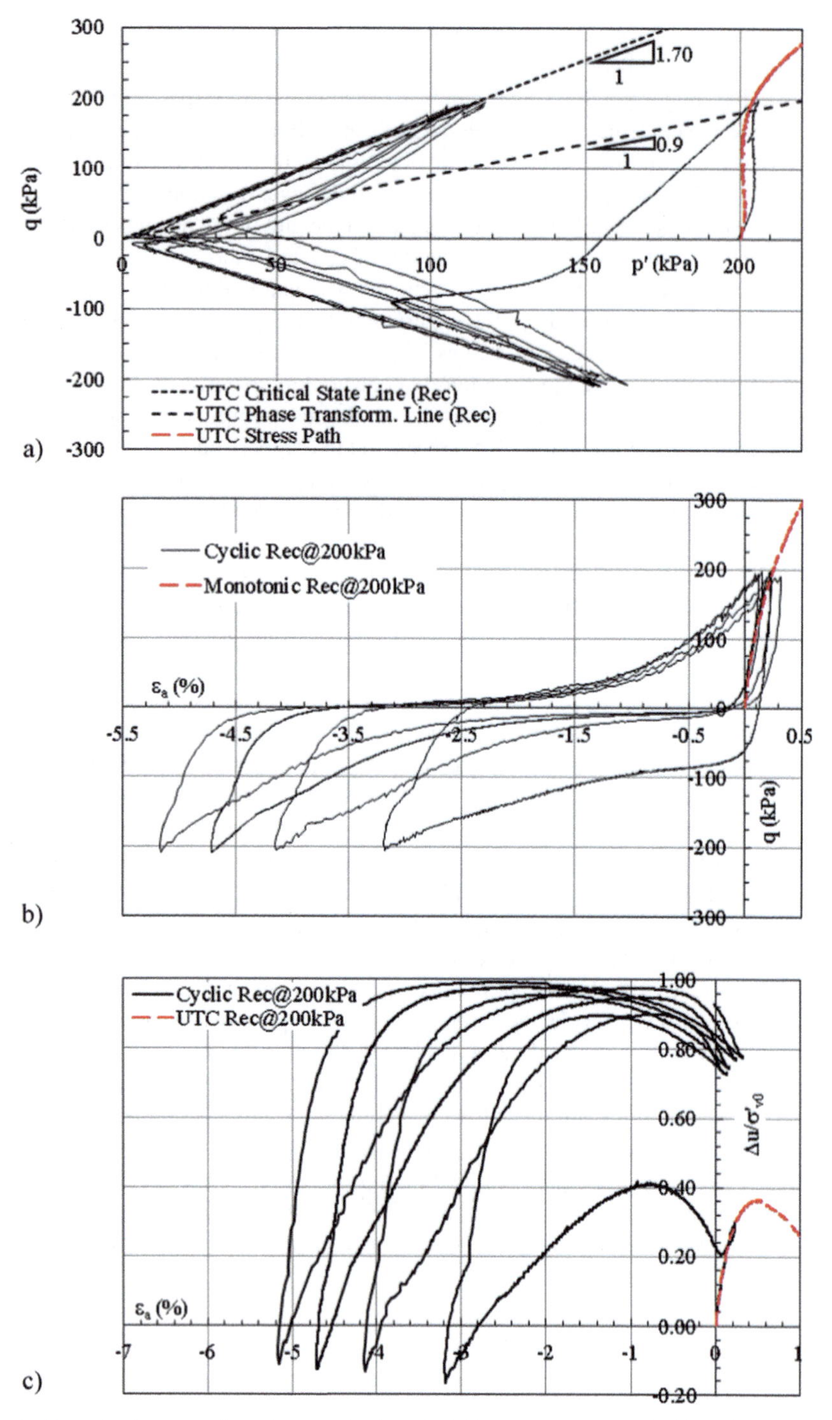

Fig. 4. Undrained cyclic triaxial test with $p'_0 = 200$ kPa and $\Delta q = \pm 200$ kPa: a) cyclic stress-path; b) cyclic stress-strain response; c) cyclic normalized excess-pore-pressure generation

4 Conclusions

This paper discusses the behaviour of reconstituted iron ore tailings samples, prepared through a slurry-based reconstitution method, in order to promote a geotechnical approach in the characterization of tailings, which is required if performance-based design is to be implemented in current design practice. In fact, this is a pressing need, as the mining sector is booming and there is a growing societal demand for an enhanced design of tailings dams around the world, namely but not only in seismic regions.

The results obtained suggest that the described slurry-based reconstitution method is a promising method to prepare representative samples for assessing the geotechnical behaviour of iron ore tailings. The undrained monotonic and cyclic triaxial tests presented illustrate the major features of the undrained behaviour of the Brumadinho-like reconstituted samples, highlighting the similarities with typical granular soils under similar undrained loading conditions. In particular, the influence of the PT and CS lines on the undrained triaxial monotonic and cyclic responses of the samples is confirmed.

Acknowledgments. This research is funded by FCT- Foundation for Science and Technology, through the research grant PTDC/ECI-EGC/4147/2021. The authors would like to acknowledge the contribution to this research by the company responsible for Torre de Moncorvo's Iron Mine, Aethel Mining, Ltd (previously MTI Ferro de Moncorvo SA), and express their gratitude to Mr. António Frazão and Arq. Carlos Guerra, who organized the iron ore tailings sample collection in the field.

References

1. Rotta, L.H.S., et al.: The 2019 Brumadinho tailings dam collapse: possible cause and impacts of the worst human and environmental disaster in Brazil. Int. J. Appl. Earth Obs. Geoinf. **90**(1), 102119 (2020)
2. Robertson, P., de Melo, L., Williams, D.J., Wilson, G.W.: Report of the expert panel on the technical causes for the failure of Feijão Dam (2019). http://www.b1technicalinvestigation.com/
3. CIMNE: Computational analyses of Dam I failure at the Córrego de Feijão mine in Brumadinho Final Report, The International Centre for Numerical Methods in Engineering, Barcelona, Spain (2021)
4. Paulelli, A.C.C., et al.: Fundão tailings dam failure in Brazil: Evidence of a population exposed to high levels of Al, As, Hg, and Ni after a human biomonitoring study. Environ. Res. **205**, 112524 (2022). https://doi.org/10.1016/j.envres.2021.112524, (ISSN 0013–9351)
5. Morgenstern N.R., Vick, S.G., Viotti, C.B., Watts, B.D.: Report on the Immediate Causes of the Failure of the Fundão Dam, Fundão Tailings Dam Review Panel (2016)
6. Ishihara, K., Ueno, K., Yamada, S., Yasuda, S., Yoneoka, T.: Breach of a tailings dam in the 2011 earthquake in Japan. Soil Dyn. Earthq. Eng. **68**, 3–22 (2015). https://doi.org/10.1016/j.soildyn.2014.10.010,ISSN0267-7261
7. Verdugo, R., et al.: Seismic Performance of Earth Structures during the February 2010 Maule, Chile, Earthquake: Dams, Levees, Tailings Dams, and Retaining Walls. Earthq. Spectra **28**, 75–96 (2012)
8. Fevereiro, C.M.P.: The iron synclinal of Torre de Moncorvo: a mineral resource as a catalyser of a transient period, MSc thesis, University of Minho, Portugal (2015) (in Portuguese)

9. Silva, W.P.: Study of the static liquefaction potential of an upstrem tailings dam using the Olson methodology, MSc thesis, UFOP, Ouro Preto, MG, Brasil (2010) (in Portuguese)
10. Ishihara, K., Tatsuoka, F., Yasuda, S.: Undrained deformation and liquefaction of sand under cyclic stresses. Soils Found. **15**(1), 29–44 (1975)
11. Bhanbhro, R.: Mechanical Behavior of Tailings- Laboratory Tests from a Swedish Tailings Dam, PhD Thesis, Luleå University of Technology, Sweden (2017)

Dynamic Properties of Organic Soils

Vincenzo d'Oriano[1] and Stavroula Kontoe[2(✉)]

[1] Graduate Geotechnical Engineer, Mott Macdonald, London, UK
[2] Reader in Soil Dynamics, Imperial College London, London, UK
stavroula.kontoe@imperial.ac.uk

Abstract. A database of laboratory stiffness degradation data of 30 samples, cering a wide range of soil types and percentage of organic content, was assembled to allow a thorough investigation of the dynamic properties of organic soils. The collated datasets are first compared with existing, commonly used empirical, hyperbolic stiffness degradation and damping curves, which are expressed mainly as a function of plasticity index and effective confining pressure. The identified discrepancies between the collated data and the empirical curves of the literature were ascribed to the presence of organic matter, leading to the introduction of a new shape variable through nonlinear regression. Based on statistical analysis of the collated information a new empirical model for organic soils is proposed, describing the reduction of normalised shear modulus and variation of damping ratio with cyclic shear strain as function of organic content in addition to other key parameters (mean effective confining pressure, plasticity index, number of cycles and loading frequency). The proposed empirical model can be used in engineering practice to infer the dynamic properties of organic soils in the absence of site-specific data.

Keywords: Organic soils · Stiffness degradation · Damping · Peat

1 Introduction

The dynamic characterisation of organic soils has received relatively little attention in the literature. The early work of Seed and Idriss [11] on peats from the Union Bay in Washington State, was followed by resonant column and torsional shear tests by [12] on soils with high organic content from the Queensboro Bridge site in New York. Further site-specific studies were carried out by [1], while recently [10, 15] performed extensive laboratory studies on highly-organic peats. The dynamic characterisation in these studies aimed to facilitate subsequent seismic site response analyses, focusing mostly on soils with very high organic content. Consequently, it is difficult to draw general conclusions on the impact of organic content on dynamic behaviour. Kishida et al. [5] performed an extensive programme of cyclic tests for a wide range of organic contents showing that the linear range of the normalized secant shear modulus G/G_{max} increases with increases of effective confining pressure and organic content, whereas under the same conditions, the damping ratio ξ decreases. In parallel studies, Kallioglou et al. [4] and Tika et al. [13] investigated extensively the effect of organic content on

© The Author(s), under exclusive license to Springer Nature Switzerland AG 2022
L. Wang et al. (Eds.): PBD-IV 2022, GGEE 52, pp. 2033–2040, 2022.
https://doi.org/10.1007/978-3-031-11898-2_186

dynamic soil properties, indicating significant impact for organic content values higher than 25%. Pagliaroli and Lanzo [8] and Pagliaroli et al. [9] also confirmed the wider linear range and the corresponding lower damping exhibited by organic soils. In the present

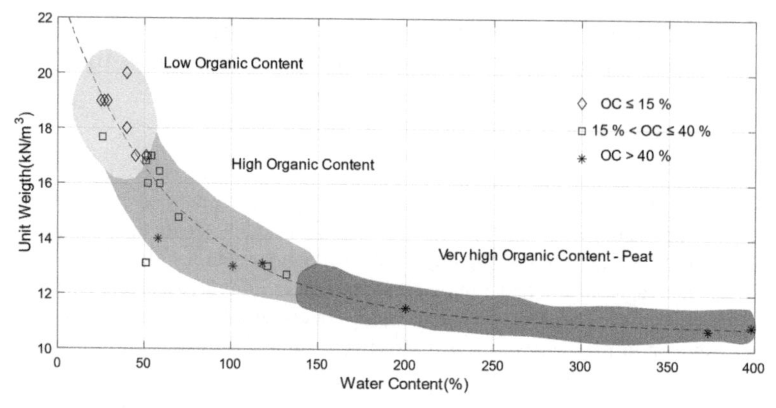

Fig. 1. Variation of unit weight with water content for the entire database. Three different zones have been identified on the basis of the organic content (OC) level.

Table 1. Summary of database sources, test types and main sample characteristics

Sample code	Source	Test type	Soil type	γ (kN/m³)	wc (%)	PI (%)	p' (kPa)	OC (%)
5D-P4	Boulanger et al. 1998	CT	Peat	11.5	200	/	200	56
S1 Cl1	Geotech. Lab Trento	RC	Silty sand	14.8	70	6	150	35
S2 C2	Geotech. Lab Trento	RC	Silty sand	17.7	26	10	150	15
S2N-C2	Pagliaroli et al. 2009	DSDSS	Silty clay	13.1	118	75	135	40
S2N-C2	Pagliaroli et al. 2009	DSDSS	Silty clay	13.1	118	75	400	40
C1	Pagliaroli et al. 2009	DSDSS	Silty clay	16.8	51	50.5	400	25
C1	Pagliaroli et al. 2009	DSDSS	Silty clay	16.8	51	50.5	800	25
26 U1	Pagliaroli et al. 2014	DSDSS	Clayey Silty	17.0	45	33	240	9.5
S1 Ost2	Socotec Italia Srl	RC	Silty clay	16.4	58.8	29.3	149	21
5D P4(1)	Stokoe et al. 1994	CT	Peat	11.5	285	/	114	65
C1	Kallioglou et al. 2009	RC	Sandy clay	19.0	25	10	130	13
C6	Kallioglou et al. 2009	RC	Sandy clay	17.0	51	30	110	6
C7s	Kallioglou et al. 2009	RC	Clay	16.0	52	40	400	33
C8s	Kallioglou et al. 2009	RC	Sandy clay	17.0	54	40	100	25
P1	Kallioglou et al. 2009	RC	Peat	13.0	101	/	370	48
P2s	Kallioglou et al. 2009	RC	Peat	14.0	58	/	400	62
S1	Kallioglou et al. 2009	RC	Silty sand	19.0	27	/	120	12
S2s	Kallioglou et al. 2009	RC	Silty clayey sand	18.0	40	/	40	13
S3s	Kallioglou et al. 2009	RC	Silty clayey sand	20.0	40	/	50	13
S4	Kallioglou et al. 2009	RC	Silty sand	19.0	29	/	50	8
S5s	Kallioglou et al. 2009	RC	Silty clayey sand	17.0	52	/	90	25
S6	Kallioglou et al. 2009	RC	Silty sand	16.0	59	/	180	25
S5	Kishida et al. 2009	CT	Clay	12.7	132	116	51	23
S6	Kishida et al. 2009	CT	Peat	10.8	398	/	17	42
S7	Kishida et al. 2009	CT	Clay	13.0	121	79	67	14
S8	Kishida et al. 2009	CTT	Peat	10.7	373	/	35	44
OC14s	Tika et al. 2010	RC	Clay	17.0	54	40	100	25
OS2s	Tika et al. 2010	RC	Silty clayey sand	18.0	40	/	40	13
OS4s	Tika et al. 2010	RC	Silty sand	19.0	29	/	50	8
OS5s	Tika et al. 2010	RC	Silty clayey sand	17.0	52	/	90	25

study experimental data from various cyclic tests were collated to explore the dynamic characteristics of organic soils covering a wide range of soil types and percentage of organic content, leading to a new empirical model for organic soils.

2 Database

A database of dynamic tests available in the literature and in engineering practice was compiled with the original sources, basic soil properties and tests conditions listed in Table 1. It includes a variety of soil types and covers a wide range of organic content levels (6%–62%). The specimens were sampled and tested in laboratories of various countries and on a variety of shear-testing devices: Resonant Column (RC), Cyclic Triaxial (CT), Double Specimen Direct Simple Shear (DSDSS). Figure 1 shows the variation of unit weight (γ) with water content (wc) for the collated soil samples. Data appears to follow closely an exponential trend with unit weight varying from 20 to 11 kN/m^3 and water content from 25% to 400%. Samples with higher organic content are characterized by higher water content and lower unit weight. Three zones can be depicted in Fig. 1 on the basis of Organic Content (OC) level: low organic content (OC $\leq$ 15%), medium-high organic content (15% $<$ OC $\leq$ 40%) and peat, for which the organic material exceeds the inorganic mineral fraction. The chart can be used for a preliminary, indirect assessment of the organic content level in a soil based on measurements of unit weight and water content.

3 Empirical Model Description

Several studies have developed laboratory-based empirical relations describing the variation of normalised shear modulus (G/G_{max}) and damping ratio (ξ) with cyclic shear strain (γ) for most common soil types, e.g. [2, 14]. Several parameters have been identified to affect the cyclic shear stress-strain behavior and these are, in increasing order of significance for common geotechnical applications, number of cycles N, loading frequency f, overconsolidation ratio OCR, mean effective confining pressure p' and plasticity index PI [3]. Despite the plethora of empirical relations, none of the existing models considers the organic content as an independent variable, possibly because organic soils have not been studied as systematically in that respect. The proposed model was developed using as a basis the empirical model of [2]. In the first instance each G/G_{max} laboratory curve collated in the database was compared with the corresponding hyperbolic curve defined by [2] for the same properties (i.e. PI, p', OCR, N and f) adopted in the laboratory test. Any resulting differences were interpreted as a consequence of the organic content presence. Subsequently the hyperbolic equation $G/G_{max}(\gamma)$ of [2] was adjusted introducing a new parameter β, as expressed by Eq. 1, denoted as $G/G_{max}(\gamma)^*$.

$$\left(\frac{G}{G_{max}}\right)^* = \frac{1}{1 + \left(\frac{\gamma}{\beta\gamma_r}\right)^{\alpha}} \tag{1}$$

where γ is the cyclic shear strain, γ_r is the reference strain defined as $\gamma_r = \tau_{max}/G_{max}$, α is a curvature coefficient, which was set equal to a mean value of 0.92 based on the statistical analysis of [2], and $\boldsymbol{\beta}$ is the new variable, exploited to account for the organic content. As in the original Darendeli formulation the damping curve, $\xi(\gamma)$, is expressed as a function of G/Gmax, the introduction of the $\boldsymbol{\beta}$ variable in Eq. 1 influences in turn the damping curve ξ^*, as shown in Eq. 2.

$$\xi^* = b\left(\left(\frac{G}{G_{max}}\right)^*\right)^{0.1} \xi_{masing} + \xi_{min} \tag{2}$$

where b is a scaling coefficient which depends on the number of cycles, ξ_{masing} is defined as a function of strain amplitude and ξ_{min} is based mainly on PI, OCR and loading frequency. These three variables are identical to the ones introduced in [2]. Finally, nonlinear regression analysis for each laboratory G/G_{max} data point and Eq. 1 was carried out to obtain the optimum β values. The procedure was carried out iteratively for all the specimens, obtaining 30 β fitting values in total. The β values are plotted against the corresponding organic content values in Fig. 2 showing an exponential variation.

$$\beta(OC) = e^{0.052*OC} \tag{3}$$

where both β and OC are expressed in percentage. The proposed shear stiffness decay curve G/G*max can now be expressed as:

$$\left(\frac{G}{G_{max}}\right)^* = \frac{1}{1 + \left(\frac{\gamma}{e^{0.052(OC)}\gamma_r}\right)^{\alpha}} \tag{4}$$

where OC is the organic content expressed in percentage. Figure 3 plots residuals, defined as the difference between the natural logarithm of the β value proposed by the exponential model and the natural logarithm of the β values obtained from the experimental data points, to assess the goodness of the fit proposed in Fig. 2. The residuals range mostly from −0.5 to 0.5 with some larger absolute values for lower organic contents, indicating an overall good fit. The choice of an exponential function reflects recognised features of the general behaviour. When the OC tends to zero, β tends to 1, so that no adjustment is applied to Eq. 1 (i.e. reverting to the original Darendeli expression). As the organic content increases, the parameter β has an increasing influence on the hyperbolic model, which is particularly strong when OC exceeds 20–25%. This OC threshold marks a significant difference in the dynamic response, as recognised in the related laboratory studies (e.g. [4]).

Figure 4 plots the proposed model for three levels of organic content (50, 25 and 10%) against data from 11 samples characterised by comparable values of organic content for each level (see Table 1). Each group of samples also refers to comparable p' and PI values (see Table 1). Samples 5D-P4, P1, P2s and 5D-P4(1) are peats characterised by OC ranging from 48% to 62%, effective confining pressure between 200 kPa (1.97 atm) and 400 kPa (3.95 atm). A zero PI was assumed in the empirical model for this set of samples; these specimens are described as highly fibrous and in such cases PI is not usually assessed. The black dash line is the model outcome for OC = 50%, p' = 2 atm

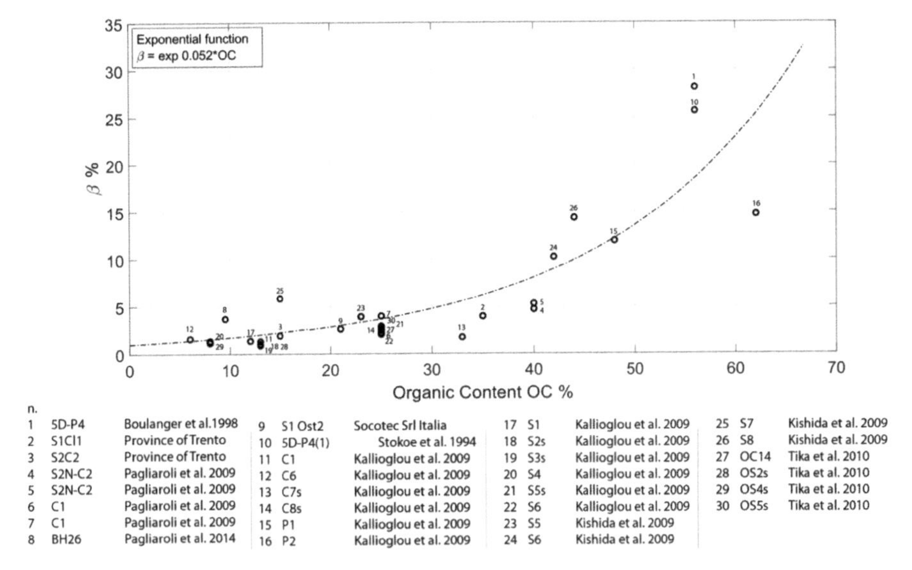

Fig. 2. Proposed exponential correlation between the parameter β and OC values.

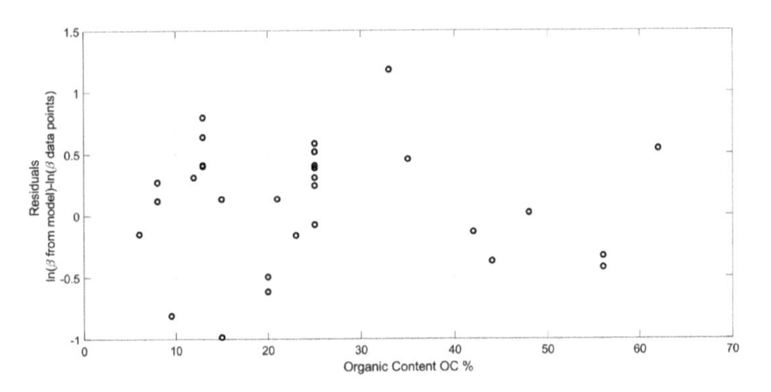

Fig. 3. Residuals for the proposed model as a function of organic content

and PI = 0. Samples S5s, S6 and OS5s are calcareous, organic sands and silty sands with OC varying from 20 to 30%, effective confining pressure ranges between 90 kPa (0.89 atm) to 180 kPa (1.78 atm) and PI = 0. For this group (grey dash line) the model corresponds to OC = 25%, p' = 1 atm (101.32 kPa) and PI = 0. Samples S2s, S3s, S4 and OS4s have OC ranging from 8 to 12%, effective confining pressure varies from 40 kPa (0.39 atm) to 50 kPa (0.49 atm) and PI = 0. The light grey dash line corresponds to OC = 8%, effective confining pressure p' = 0.5 atm (101.32 kPa) and plastic index, PI = 0. Finally, the model is plotted for a high plasticity index (PI = 79%) and low organic content (OC = 14%) case (red dotted curve), which is comparable with Sample 7 in Kishida et al. [7] (see Table 1).

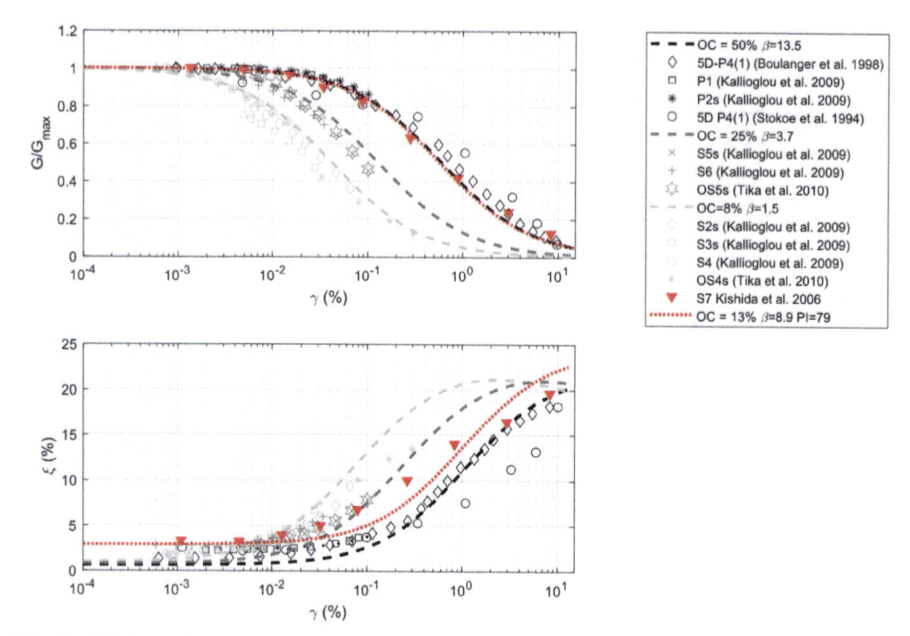

Fig. 4. OC dependent model curves (dashed lines, OC = 50, 25 and 8%) and laboratory data (points) for 11 samples corresponding to the different OC levels.

Overall, the proposed model captures well the experimental data both in terms of stiffness degradation and damping response, showing that as the organic content increases, the linear range of the G/Gmax response widens with a corresponding decrease in terms of damping. A less satisfactory performance of the damping model is observed for samples S5s, S6 and OS5s (which have been classified as calcareous organic sand) which could be due to the presence of plastic fines, resulting in an increased hysteretic damping response. Finally, the model captures well the interplay of plasticity and organic content as the proposed curve compares well the S7 sample from Kishida et al. which is characterised by relatively low organic content and high plasticity.

4 Empirical Model Performance

The model performance has been further investigated by an independent validation using additional data from Zwanenburg et al. [15], which were not included in the proposed model's database. These describe the dynamic behaviour of a peat deposit in the north of the Netherlands obtained by resonant column and cyclic direct simple shear tests. Shear modulus data have been normalised by the largest measured shear modulus value during RC testing, typically at shear strains of $\gamma < 0.003\%$. Table 2 summarises the properties of the considered dataset. LOI is defined as the percentage of material lost when heating the sample and OC is the percentage of organic content related to LOI. The proposed model is plotted in Fig. 5 assuming an average OC of 80%, a zero PI, σv' = 10.1 kPa and OCR = 2 to represent the in-situ conditions (see Table 2), as well as a frequency f = 100 Hz, and N = 10.

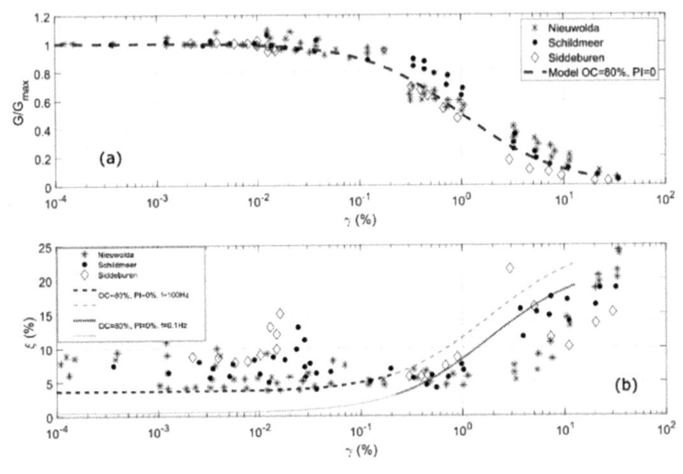

Fig. 5. Assessment of the model performance in terms of (a) G/Gmax and (b) damping curves in comparison to peat data from Zwanenburg et al. [15].

Table 2. Summary of datasets considered from Zwanenburg et al. [15]

Site	Number of tests	Wc (%)	LOI (%)	OC (%)	OCR (average)	σ'_v (kPa)
Nieuwolda	14	556 (360–748)	85.4 (70–94)	84.8 (68.8–93.8)	1.4	25–30
Siddeburen	2				2.6	5–7
Schildmeer	4				2.6	6.5–10

The proposed model predicts very well the normalised shear modulus data for the entire strain range. The damping curves are plotted for two frequencies of 0.1 Hz and 100 Hz to reflect values relevant for cyclic direct simple shear and resonant column tests respectively. Overall, the prediction of damping is not as satisfactory as the shear modulus reduction one, although the inherent scatter in the damping data is also significant.

5 Conclusions

A database of dynamic tests available in the literature and in engineering practice was compiled covering a wide range of soil types and levels of organic content. Their statistical analysis enabled the development of a new empirical model for organic soils which describes the variation of normalised shear modulus and damping ratio with cyclic shear strain as function of organic content in addition to other key parameters. The new empirical model can provide useful input to site response analyses for stratigraphies with organic materials in the absence of site-specific laboratory testing.

Acknowledgements. We would like to express our gratitude to Dr. Polyxeni Kallioglou, Prof. Alessandro Pagliaroli, Fabio Fedrizzi and Massimo Romagnoli for generously sharing their data.

References

1. Boulanger, R.W., Arulnathan, R., Harder, L.F., Torres, R.A., Driller, M.W.: Dynamic properties of Sherman Island Peat. J. Geotech. Geoenv. Eng. **124**(1), 12–20 (1998)
2. Darendeli, M.B.: Development of a new family of normalized modulus reduction and material damping curves. PhD thesis, The University of Texas at Austin (2001)
3. Guerreiro, P., Kontoe, S., Taborda, D.: Comparative study of stiffness reduction and damping curves. In: 15th World Conference on Earthquake Engineering, Portugal, Lisbon (3012)
4. Kallioglou, P., Tika, T., Koninis, G., Papadopoulos, S., Pitilakis, K.: Shear Modulus and damping ratio of organic soils. Geotech. Geol. Eng. **27** (2009)
5. Kishida, T., Boulanger, R.W., Abrahamson, N.A., Wehling, T.M., Driller, M.W.: Regression models for dynamic properties of highly organic soils. J. Geotech. Geoenv. Eng. **135**(4), 533–543 (2009)
6. Kishida, T., Boulanger, R. M., Wehling, T.W., Driller, M.: Variation of small strain stiffness for peat and organic soil. In: 8th U.S. National Conference on Earthquake Engineering (2006)
7. Kishida, T., Wehling, T.M., Boulanger, R.W., Driller, M.W., Stokoe, K.H., II.: Dynamic properties of highly organic soils from montezuma slough and Clifton court. J. Geotech. Geoenv. Eng. **135**(4), 525–532 (2009)
8. Pagliaroli, A., Lanzo, G.: Comportamento ciclico di terreni argillosi della città di Roma (2009)
9. Pagliaroli, A., Lanzo, G., Tommasi, P., Di Fiore, V.: Dynamic characterization of soils and soft rocks of the central archeological area of Rome. Bull. Earthq. Eng. **12**(3), 1365–1381 (2013). https://doi.org/10.1007/s10518-013-9452-5
10. Sarkar, G., Sadrekarimi A.: Cyclic shearing behavior and dynamic characteristics of a fibrous peat. Can. Geotech. J. in-press (2021). https://doi.org/10.1139/cgj-2020-0516
11. Seed, H., Idriss, I.: Analyses of ground motions at Union Day, Seattle during earthquakes and distant nuclear blasts. BSSA. **60**(1), 125–136 (1970)
12. Stokoe, K.H.I., Bay, J.A., Rosenblad, B.L., Hwang, S.K., Twede, M.R.: In situ seismic and dynamic laboratory measurements of geotechnical material at Queensboro Bridge and Roosevelt Island. Geotech. Eng. Rep. **GR94–5** (1994)
13. Tika, T., Kallioglou, P., Koninis, G., Michaelidis, P., Efthimiou, M., Pitilakis, K.: Dynamic properties of cemented soils from Cyprus. Bull. Eng. Geol. Environ. **69**(2) (2010)
14. Vucetic, M., Dobry, R.: Effect of soil plasticity on cyclic response. J. Geotech. Geoenv. Eng. **117**(1), 89–107 (1991). https://doi.org/10.1061/(ASCE)0733-9410(1991)117:1(89)
15. Zwanenburg, C., et al.: Assessment of the dynamic properties of holocene peat. J. Geotech. Geoenv. Eng. **146** (7)

Experimental Study on the Correlation Between Dynamic Properties and Microstructure of Lacustrine Soft Clay

Yurun Li[1(✉)], Zhongchen Yang[1], Jingjuan Zhang[1(✉)], Yingtao Zhao[2], and Chuang Du[3]

[1] Hebei University of Technology, Tianjin 300401, China
{iemlyr7888,abc123}@hebut.edu.cn
[2] Xingtai Road and Bridge Construction Group Co., Ltd, Xingtai 054001, China
[3] Henan Key Laboratory of Special Protective Materials, Luoyang 471023, China

Abstract. Due to the influential mesostructure of lacustrine soft clay on the dynamic characteristic parameters, the dynamic shear modulus and damping ratio of the lacustrine soft clay are obtained by selecting different consolidation confining pressures for indoor resonance column test and scanning electron microscope (SEM) test, and the microstructure parameters of the lacustrine soft clay are acquired by analyzing the SEM image through the IPP (Image-Pro Plus) image processing software as well. By analyzing the relationship between some dynamic characteristic parameters and microstructure parameters is analyzed to explain the dynamic shear modulus and damping ratio from the perspective of soil microstructure characteristics. It is shown that the mean maximum dynamic shear modulus has a strong correlation with the maximum radius of the particles and the shape of the soil pores (abundance, circularity, fractal dimension), indicating that the size of the soil particles and the shape of the pores have a strong influence on dynamic shear modulus of soil; the correlation between the maximum damping ratio and the microstructure characteristics is weak, and the influence law is more complicated.

Keywords: Lacustrine soft clay · Resonance column test · Scanning electron microscope · Dynamic characteristics · Mircostructure

1 Introduction

The lacustrine soft clay of the plain wetland is distributed in the lake wetlands of the North China Plain or the surrounding areas of ancient lakes, such as Hengshui Lake and Baiyangdian. Natural lacustrine soft clay has poor basic physical properties such as high water content and high void ratio, and high shear strength. Therefore, the good mechanical properties and poor physical properties of lacustrine soft clay itself are special characteristics of soft soil. Therefore, under the action of dynamic load during construction or use, soft soil may constitute an unfavorable site. Therefore, the dynamic characteristics of soft soils have attracted the attention of many researchers and engineers.

© The Author(s), under exclusive license to Springer Nature Switzerland AG 2022
L. Wang et al. (Eds.): PBD-IV 2022, GGEE 52, pp. 2041–2053, 2022.
https://doi.org/10.1007/978-3-031-11898-2_187

The dynamic shear modulus and the damping ratio are two key parameters in the dynamic characteristics of the soil. Among them, the dynamic shear modulus represents the ability of the soil to resist shear deformation; the damping ratio represents the scale of the energy absorption of the soil, that is the speed ofthe amplitude attenuation. At this stage, there have been many research results using laboratory tests to determine dynamic characteristics of soft soil or clay such as dynamic shear modulus and damping ratio. Among them, Sun et al. (1988) [1] earlier gave the dynamic characteristics of cohesive soil under cyclic loading; Kawaga et al. (1992) [2] used the resonance column test to study the dynamic shear modulus and damping ratio of marine soft soil; Teachavorasinskun et al. (2002) [3], Lanzo et al. (2009) [4], Sas et al. (2015) [5] measured the dynamic shear modulus and damping ratio of soft clay from different regions, and studied the influence of different testing conditions on the dynamic characteristics; Jallu et al. (2017) [6] used the resonance column test to study the influence of saturation on the dynamic characteristics of compacted clay; Lin et al. (2019) [7] studied the influence of mineral content in clay on the dynamic shear modulus and damping ratio under small strain. Comparing these research results, it can be shown that soft clay has certain regional characteristics, and the dynamic characteristics of soft clay or clay from different regions are different. Therefore, the relevant research results cannot be directly applied to the engineering test.

Most sedimentary clays are structural. From the perspective of microstructure, soil structure refers to the properties and arrangement of soil particles, pores, and the interaction between particles. In the early studies of soil microstructure, due to many limitations such as observation technology and image processing and analysis technology, most of the research was limited to qualitative analysis, and many conclusions were inferred from the macroscopic characteristics of soil. With the emergence of new microscopic observation equipment such as scanning electron microscope (SEM) and computerized X-ray tomography (CT) and the improvement of electronic computer technology, it provides technical support for qualitative observation and quantitative research and analysis of soil microstructure, and Scholars quickly applied it to the study of soil mechanics. In recent years, the research on the characteristics of microstructure has become more and more deeper. For example, Moore et al. (1995) [8] use fractal theory to quantitatively analyze the microstructure of soil; Liu et al. (2011) [9] use image processing Technology to quantitatively analyze the apparent porosity of clay microscopic images; Ndèye et al. (2012) [10] used generalized cylinders to model the microstructure of soil to analyze the laws of pore space characteristics. In addition, many studies have shown that there is a relationship between the macroscopic properties of soil and its microstructure, including thixotropy, consolidation and compression properties, and mechanical properties. (Osipov et al. (1984) [11]; Pusch and Weston (2003) [12]; Sivakumar et al. (2002) [13]; Cetin and Gökoğlu (2013) [14]; Jiang et al. (2014) [15]). At present, there are few research results on the correlation between the dynamic characteristics of special soils and the microstructure characteristics. Therefore, this paper is based on the resonance column test, combined with the scanning electron microscope test and image processing software, to analyze the dynamic characteristics and microstructure of the wetland lake clay. The characteristics are studied, and the correlation analysis between the two is carried out. It is explored to explain the change law of the macroscopic dynamic characteristics of the soil from the perspective of the microstructure characteristics, which provides a reference for the study of the dynamic characteristics of the soft soil.

2 Experimental Program

The correlation test of lacustrine soft clay mainly includes two parts according to the sequence of the test. The first is to use the resonance column test to measure the dynamic characteristics of the soil. The second part is to perform a scanning electron microscope test on the soil sample after the resonance column test is completed. To study the microstructure characteristics of the soil. After completing the two parts of the experiment, then comprehensively analyze the correlation between the two.

2.1 Basic Physical Properties of Lacustrine Soft Clay

The lacustrine soft clay sample used in the test was taken from Hengshui Lake in the North China Plain. The test soil was remolded soil. Before studying the dynamic characteristics of lacustrine soft clay, the specific gravity test and the limit moisture content were firstly completed. And compaction test and other indoor basic physical and mechanical property tests. The basic physical and mechanical parameters of soil are shown in Table 1.

Table 1. Basic physical properties of lacustrine soft clay.

ω (%)	γ (kN/m^3)	e	ω_L(%)	ω_p(%)	I_p	$\rho_{d\max}$(g/cm^3)
39	18.33	1.059	47.3	26.1	21.2	1.73

2.2 Resonance Column Test

The dynamic characteristics test use the GZZ-50B resonance column testing ma-chine, as shown in Fig. 1a. The soft lacustrine clay used in the resonance column test is remolded soil. The specimens were prepared by drying, crushing and sieving the lacus-trine soft clay according to the test conditions. In order to restore the condition of the road site where the soil is located, the soil samples are given a certain compaction state during sample preparation, and different confining pressures are used as working conditions. Therefore, when preparing the sample, this article controls the degree of compaction $K = 0.85$, and the formula of the degree of compaction is shown in (1). Therefore, the dry density of the sample is 1.47 g/cm3, and the moisture content is 27% (the excel-lent moisture content obtained by the compaction test). The size of the test piece is Φ39.1 mm × 80 mm. The confining pressure is 50, 100, 150 and 200 kPa for resonance column test.

$$K = \frac{\rho_d}{\rho_{d\max}} \tag{1}$$

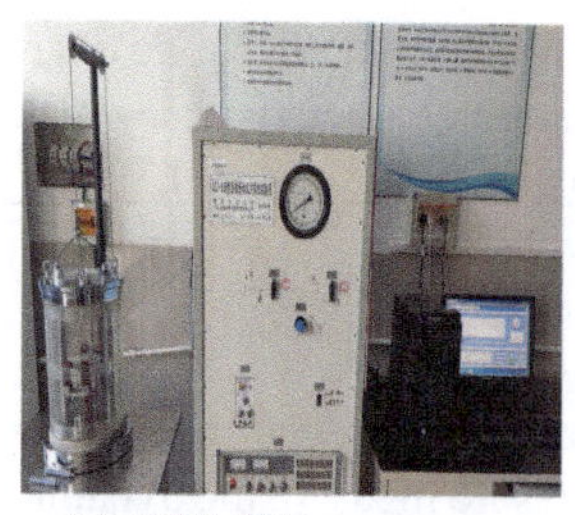

(a) GZZ-50B resonance column test machine

(b) QUANTA FEG 450 Scanning Electron Microscope

(c) Centre resentative sample of soil used in SEM

Fig. 1. Test equipment used in correlation research

2.3 Scanning Electron Microscope Test

The microstructure characteristic test use the QUANTA FEG450 field emission environmental scanning electron microscope, as shown in Fig. 1b. After the resonance column test is completed, the four specimens with consolidated confining pressures of 50, 100, 150, and 200 kPa are taken from the representative part of the center and cut into soil samples of 1.0 cm × 2.0 cm × 1.0 cm. And air-dried indoors under constant temperature and humidity. Before the scanning electron microscope test, the soil sample was broken, and the fresh broken section was used as the observation section of the scanning electron microscope, shown in Fig. 1c. Observe the relatively flat position of the interface. The magnification of shooting is 10000 times. After the SEM image is obtained, the SEM image is measured and analyzed by Image-Pro Plus 6.0(IPP 6.0) image processing software. The steps include binary processing, frame selection of the measurement area, Calibrate the size, select the parameters to be measured, complete the measurement and export the data, and perform processing and analysis.

3 Experimental Results and Discussion

3.1 Dynamic Characteristics

Dynamic Shear Modulus and Damping Ratio. The dynamic shear modulus and damping ratio can be directly measured through the resonance column test. The test results of the relationship curve between dynamic shear modulus and damping ratio and dynamic shear strain are shown in Fig. 2. The following rules can be seen from Fig. 2: (1) The dynamic shear modulus G decreases with the increase of the dynamic shear strain γ. Damping ratio D starts to rise significantly with the increase of the dynamic shear strain γ, and show nonlinearity. (2) Under the same strain condition, the dynamic shear modulus G increases with the increase of confining pressure. The confining pressure change has little effect on the damping ratio D.

Maximum Dynamic Shear Modulus. The results of the maximum dynamic shear modulus G_{max} are shown in Table 2, and the relationship curve with the change of confining pressure is shown in Fig. 2. It is not difficult to find that the maximum dynamic shear modulus Gmax increases with the increase of consolidation confining pressure (Fig. 3).

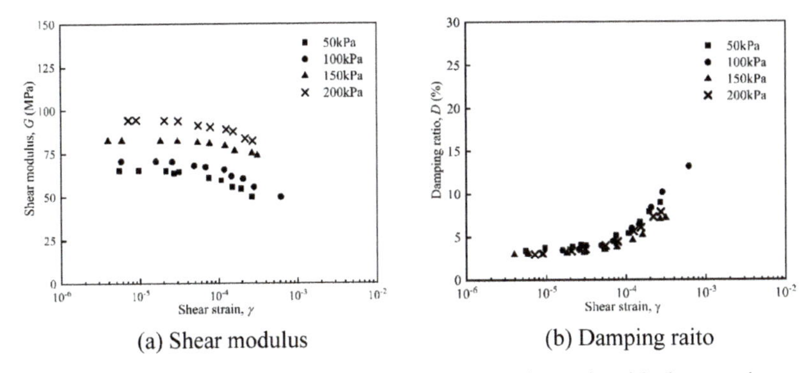

(a) Shear modulus (b) Damping raito

Fig. 2. Variation of dynamic shear modulus, damping ratio with shear strain.

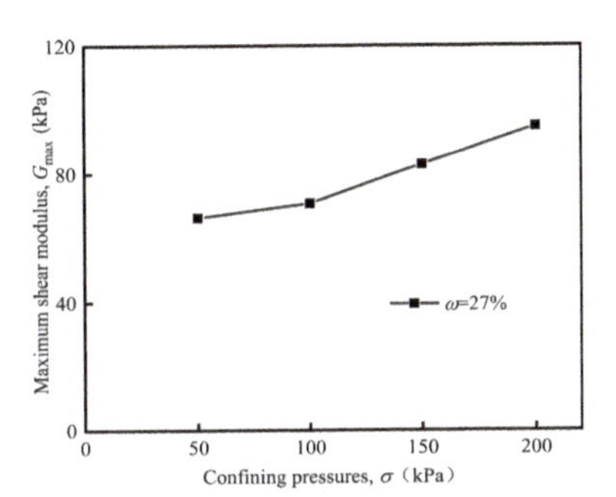

Fig. 3. The relationship curve between maximum dynamic shear modulus G_{max} and consolidation confining σ pressure.

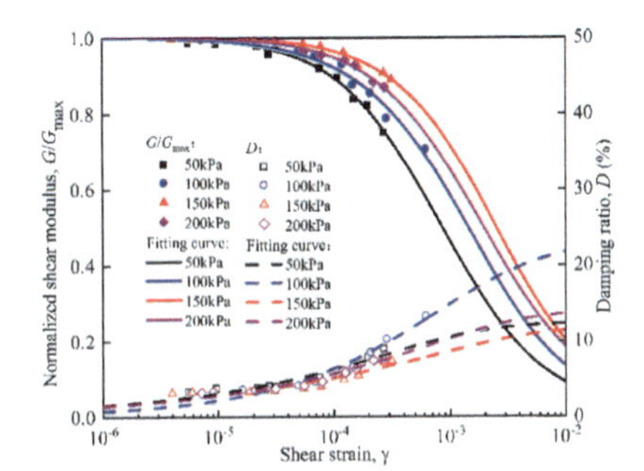

Fig. 4. G/G_{max}-γ and λ-γ fitting curves of lacustrine soft clay.

Dynamic Shear Modulus Ratio and Damping Ratio. At present, in site seismic response analysis, soil-structure dynamic interaction response analysis, and seismic stability evaluation, most of them use equivalent nonlinear viscoelastic models to consider soil nonlinearity and hysteresis.

The Davidenkov model proposed by Martin et al. (1982) [16] is used here. The expression of the model is:

$$\frac{G}{G_{max}} = 1 - \left[\frac{(\gamma/\gamma_0)^{2B}}{1 + (\gamma/\gamma_0)^{2B}}\right]^A \tag{2}$$

where A, B and γ_0 are test parameters.

For the damping ratio λ, the M-D model does not fit the test results ideally, so the following empirical formula is generally used:

$$D = D_{max}(1 - \frac{G}{G_{max}})^n \tag{3}$$

where n is the test parameter, which is related to the nature of the soil.

The Fig. 4 is the relationship curve of dynamic shear modulus ratio and damping ratio with dynamic shear strain of lacustrine soft clay obtained by Davidenkov model. The fitting parameters are shown in Table 2.

By observing Fig. 4 and Table 2, it is not difficult to find: (1) There is an obvious demarcation point between the shear strain $10^{-5}-10^{-4}$.There is a big difference in the change law of the curve before and after the cut-off point.Before the shear strain γ, the dynamic shear modulus ratio G/G_{max} decreases slowly, and shows an obvious linear change. After the shear strain is greater than the boundary point, the dynamic shear modulus ratio G/G_{max} begins to decrease significantly with the increase of the shear strain. (2) Before the boundary point, the damping ratio D increases linearly and slowly with the increase of the shear strain γ. When the shear strain exceeds the cut-off point, the rate of D rise increases, and the test curves of each group begin to show a certain discrete type. This is due to the characteristics of the clay itself and the limitation of the test method. The accuracy and stability of the damping ratio of lacustrine soft clay in predicting large strains are not ideal. Therefore, further research is needed for the damping ratio of lacustrine soft clay.

Table 2. Fitting parameters of Davidenkov model for lacustrine soft clay.

σ (kPa)	Fitting parameters					
	G_{max}(MPa)	D_{max}(%)	A	B	γ_0	n
50	66.41	12.57	0.93	0.55	0.00085	0.31
100	70.88	23.89	1.12	0.42	0.00128	0.54
150	82.93	12.02	1.01	0.50	0.00249	0.27
200	94.93	14.30	1.02	0.49	0.00174	0.33

3.2 Microstructure Characteristics

Binarization. The microstructure of lacustrine soft clay observed by scanning electron microscope (SEM) is shown in the Fig. 5. From the figure, the structure of lacustrine soft clay can be clearly observed. The lacustrine soft clay is dominated by a granular structure. The particles gather around the large particles to form soil aggregates, which constitute the basic structural unit of the soil. Fine particles and debris form a cemented connection between soil particles and soil aggregates. The structural unit body is mainly in contact with the edge and the edge surface, and the soil skeleton is arranged chaotically and has no obvious orientation. Pores are mainly the intergranular pores of soil particles and the pores inside soil aggregates. As the consolidation pressure increases, the connection between the particles is broken. Large particles and soil aggregates are crushed and broken into smaller and smaller particles. At the same time, the broken particles move and roll, and the large and medium pores are divided and compressed.

Use the IPP image processing software to binarize the SEM electron microscope image, as shown in the figure. The purpose of binarization is to distinguish the particles and pores in the electron microscope image in a black-and-white image (the white is the particle and the black is the pore), the method steps of binarization refer to other research results [17–19].

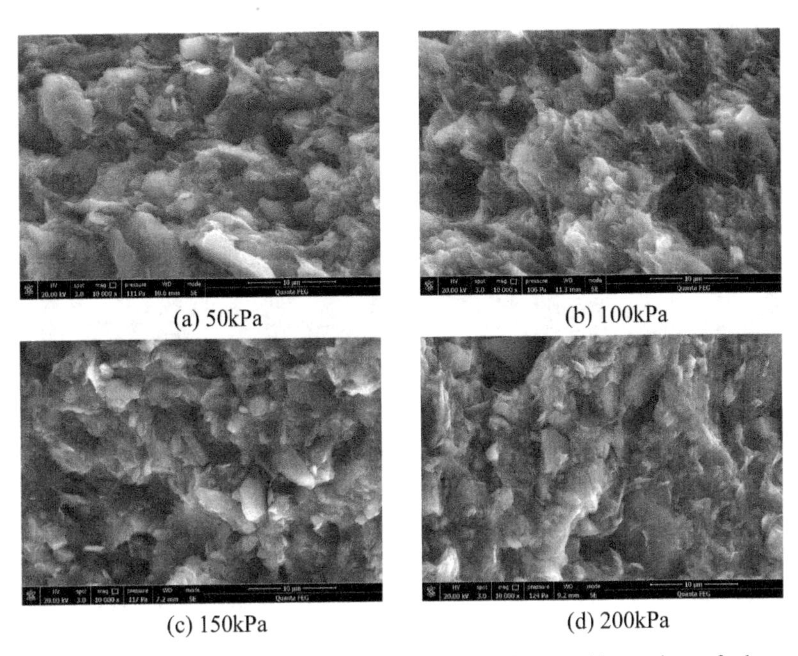

(a) 50kPa (b) 100kPa (c) 150kPa (d) 200kPa

Fig. 5. Microscopic scanning electron microscope image of lacustrine soft clay.

Definition of Various Microstructure Characteristic Parameters. The main research contents of the microstructure test include:

(1) Area: The area of the object to be measured in the image.
(2) Maximum radius: The farthest distance from the contour line of the equivalent ellipse center of the measured object in the image.
(3) Apparent void ratio: the ratio of the pore area to the particle area in the image.
(4) Abundance: The definition is the ratio of the short axis to the long axis of the equivalent ellipse of the measured object, ranging from 0 to 1, which is used to represent the geometry of particles and pores.
(5) Roundness: used to analyze the shape of soil particles and pores. Its definition is:

$$\text{Roundness} = \frac{[\text{Perimeter}]^2}{4\pi * [\text{Area}]} \tag{4}$$

A circularity equal to 1 indicates a circular shape, and greater than 1 indicates a non-circular shape (Fig. 6). The circularity of a square is about 1.27, and the circularity of an equilateral triangle is 1.67.

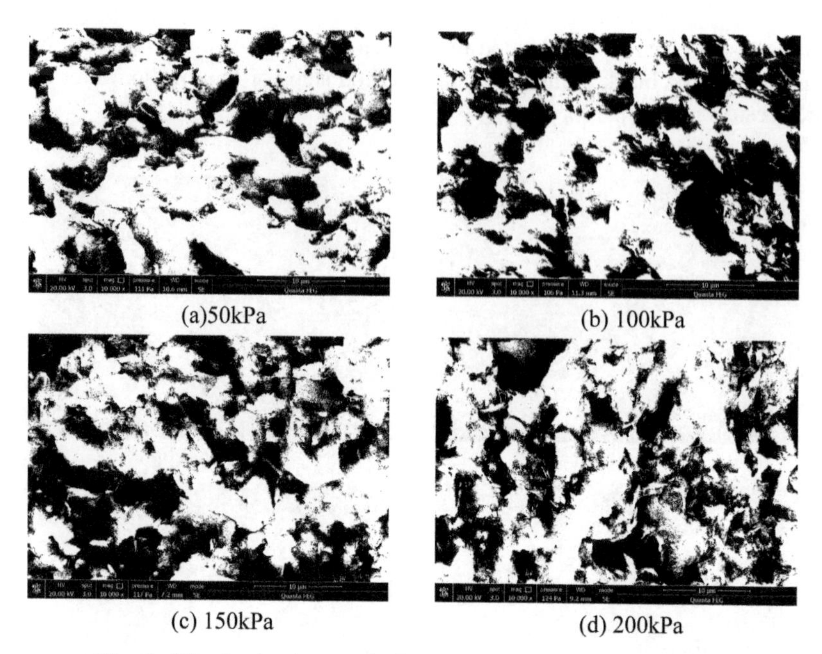
(a)50kPa (b) 100kPa (c) 150kPa (d) 200kPa

Fig. 6. Binarization image of microstructure of lacustrine soft clay.

(6) Fractal Dimension: It is used to express the complexity of particle or pore morphology. Fractal theory can be used to solve similar complex configuration problems. Some scholars believe that the microstructure of soil has similarities and the configuration is complex, so they use the fractal dimension to express the complexity of the particles and pores. Here, the area-perimeter method is used to define the fractal dimension D_f [20]:

$$\lg(C) = D_f/2 \cdot \lg(S) + c_1 \tag{5}$$

where C is the circumference of the measured object; S is the area of the measured object; D_f is the fractal dimension; c_1 is the fitting parameter.

Processing and Analysis of Microstructure Characteristics. After processing and analyzing the microstructure data measured by the IPP software, the above-mentioned applied parameters are summarized in Table 3 and Fig. 7.The parameters measured here are the average values of particles or pores

(1) The maximum radius of the particles decreases with the increase of the consolidation confining pressure. This is consistent with the conclusion obtained by direct observation and analysis of SEM images, while the other parameters are not It does not show a certain rule with the increase of the consolidation confining pressure, which is mainly affected by the randomness of the sample.
(2) The increase of the consolidation confining pressure is not obvious to the regularity of the pore area and maximum radius of the soil, but the shape parameters

of the microstructure of the pores, including the abundance, circularity and fractal dimension, have obvious effects. The degree and circularity decrease with the increase of the consolidation confining pressure, indicating that with the increase of the consolidation confining pressure, the pore shape of the soil becomes more and more elongated rather than circular, while the fractal dimension increases with the increase of the consolidated confining pressure. Here, the fractal dimension is mainly used to describe the complexity of the pore profile,it is shown that the pore profile becomes more complex, as the confining pressure increases.

(3) The apparent void ratio is the ratio of the total area of the particles in the SEM image to the total area of the pores, where the void ratio does not show regularity with the increase of the consolidation confining pressure.

(4) It can be found that the microstructure characteristics of the soil after the completion of the dynamic characteristics test have a certain regularity under different confining pressures. But if we want to explain whether the microstructure and dynamic characteristics are related, we still need to study the correlation between the two.

Table 3. Microstructure characteristic parameters of lacustrine soft clay particles and pores

Σ (kPa)	Particle				
	Area (μm²)	Max radius (μm)	Abundance	Roundness	Fractal dimension
50	17.302	1.104	0.393	2.449	1.117
100	6.963	1.149	0.364	2.679	1.114
150	7.022	0.929	0.450	2.304	1.133
200	8.646	0.906	0.404	3.148	1.123
Apparent void ratio	Pore				
0.470	2.985	1.107	0.469	2.615	1.129
0.810	6.200	1.199	0.449	2.580	1.123
0.858	5.620	1.236	0.440	2.911	1.134
0.663	3.708	1.059	0.411	3.158	1.141

3.3 Correlation Analysis of Macro and Mirco Characteristics

Regarding the relationship between the dynamic characteristic parameters and microstructure parameters of lacustrine soft clay, this paper studies the dynamic characteristics and influence laws of lacustrine soft clay under different confining pressures, but if you want to explain the reasons for these changes In fact, it can be analyzed from the microstructure. Therefore, after studying the microstructure characteristics of lacustrine soft clay, it provides a way of thinking, that trying to reveal the essence of the macroscopic dynamic characteristics from the perspective of the microstructure.

Since the regularity of the various dynamic and microstructure parameters of the soft clay measured in this experiment is not obvious, the Pearson correlation value formula will be used. The Pearson correlation value formula is shown in (6). The range of the formula is [−1–1]. When the result is negative, it means that the correlation is negative correlation. The absolute value of the Pearson correlation value in different intervals indicates that the two groups of data have different degrees. Correlation of: |0.8–1.0| very strong correlation, |0.4–0.8| moderate correlation, |0.2–0.4| weak correlation, |0.0–0.2| very weak or no correlation, so use Pearson correlation value to study Correlation between dynamic characteristic parameters and microstructural characteristic parameters of lacustrine soft clay.

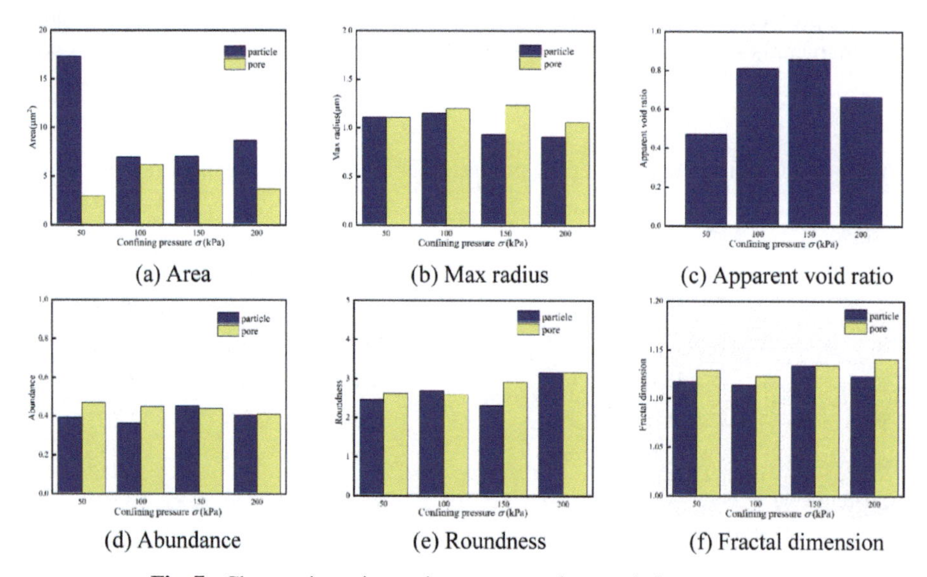

(a) Area (b) Max radius (c) Apparent void ratio

(d) Abundance (e) Roundness (f) Fractal dimension

Fig. 7. Changes in various microstructure characteristic parameters

$$\rho_{X,Y} = \frac{\mathrm{cov}(X, Y)}{\sigma_X \sigma_Y} = \frac{E[(X - \mu_X)(Y - \mu_Y)]}{\sigma_X \sigma_Y} \tag{6}$$

where cov(X,Y) is covariance, σ_x is standard deviation of the sample; μ_x is $E(x)$.

Due to the large number of dynamic characteristic parameters, the two more critical parameters G_{max} and D_{max} are selected for correlation analysis. The Pearson correlation value is shown in Table 4 and Table 5. This article will focus on the analysis of the parameters with extremely strong correlation ($|\rho_{X,Y}| \geq 0.8$).

It is not difficult to find that the maximum radius of the soil particles has an extremely significant influence on the maximum dynamic shear modulus, which is a strong negative correlation. The shape characteristics of the pores have an extremely obvious influence

Table 4. Pearson correlation value of dynamic characteristic and particle microstructure

Micro	Particles				
Dynamic	Area	Max radius	Abundance	Roundness	Fractal dimension
G_{max}	−0.529	−0.909	0.476	0.624	0.560
D_{max}	−0.424	0.631	−0.778	0.234	−0.663

Table 5. Pearson correlation value of dynamic characteristic and pore microstructure

Micro	Pore					
Dynamic	Area	Max radius	Abundance	Roundness	Fractal Dimension	Apparent void ratio
G_{max}	−0.048	−0.310	−0.965	0.981	0.901	0.305
D_{max}	0.616	0.254	0.044	−0.471	−0.662	0.374

on the maximum dynamic shear modulus, for instance, the abundance, circularity and fractal dimension of the pores have a strong correlation with the maximum dynamic shear modulus. Various microstructure parameters have a certain influence on the maximum damping ratio, but they do not constitute a strong correlation. This may be due to the unsatisfactory accuracy and stability of the lacustrine soft clay damping ratio measured by the resonance column test. It may be that the damping ratio is more susceptible to influences outside the microstructure.

4 Conclusion

This paper conducted a series of resonance column tests to study the dynamic shear modulus and damping ratio of lacustrine soft clay under different confining pressure conditions. After the resonance column test is completed, the representative part of the soil sample center is taken for scanning electron microscopy test to study the microstructure characteristics of lacustrine soft clay, and the correlation analysis of the dynamic characteristics and microstructure characteristics of the lacustrine soft clay is carried out. The conclusions of this study are summarized as follows:

(1) The dynamic shear modulus of lacustrine soft clay increases with the increase of confining pressure. However, the results of the damping ratio under different confining pressures are relatively concentrated, indicating that the influence of the confining pressure is small.
(2) Using the Davidenkov model, the relationship model between the dynamic shear modulus ratio and the damping ratio of the lacustrine soft clay is established, and the model curves and simulations of G/G_{max}-γ and λ-γ of the lacustrine soft clay under different confining pressures are given. The parameters are combined, and the characteristics and influence laws of the curve are analyzed.

(3) The average maximum radius of the particles and the microscopic shape characteristics of the pores (abundance, circularity, fractal dimension) in the microstructure characteristics of lacustrine soft clay show certain regularity with the changes of the consolidation confining pressure, that is, with the consolidation surrounding pressure. As the pressure increases, the average maximum radius of soil particles decreases, and the shape of the pores becomes more and more striped, non-circular, and the fractal of the outline becomes more and more complex. This is consistent with the conclusion obtained when directly observing the soil microstructure image.
(4) The maximum radius of the particles and the shape of the soil pores (abundance, circularity, fractal dimension) have a strong correlation with the maximum dynamic shear modulus in the dynamic characteristic parameters of the soil. In terms of damping ratio, the correlation with various parameters of the microstructure is not strong.
(5) Finally, it should be emphasized: This paper uses the Pearson correlation coefficient to study the correlation between the dynamic characteristics of lacustrine soft clay and the microstructure, and analyzes the law of influence between the two. There is no problem with this method itself, but the limitation is that the Pearson correlation value used in the research process is suitable for expressing the linear correlation between the two sets of data. However, if the dynamic characteristic parameters in the sample and the microstructure parameters have a curvilinear or non-linear correlation, it is not well represented here. Due to the limited ability of the author, this article only focuses on the analysis of the linear relationship between the two, and whether there is a more complex correlation or influence law between them, further research and discussion are needed.

Funding Information. This work was supported by the National Natural Science Foundation of China (Grant no. 51778207) and the Henan Key Laboratory of Special Protective Materials (Grant no. SZKFJJ202005).

References

1. Sun, J.I., Golesorkhi, R., Seed, H.B.: Dynamic Moduli and Damping Ratios for Cohesive Soils (1988)
2. Kagawa, T.: Moduli and damping factors of soft marine clays. J. Geotech. Eng. **118**(9), 1360–1375 (1992). https://doi.org/10.1061/(ASCE)0733-9410(1992)118:9(1360)
3. Teachavorasinskun, S., Thongchim, P., Lukkunaprasit, P.: Shear modulus and damping of soft Bangkok Clays. Can. Geotech. J. **39**(5), 1201–1208 (2002). https://doi.org/10.1139/t02-048
4. Lanzo, G., Pagliaroli, A., Tommasi, P., et al.: Simple shear testing of sensitive, very soft offshore clay for wide strain range. Can. Geotech. J. **46**(11), 1277–1288 (2009). https://doi.org/10.1139/T09-059#.Ud8JDcxH5Ag
5. Sas, W., Gabryś, K., Szymański, A.: Effect of time on dynamic shear modulus of selected cohesive soil of one section of express way no. S2 in Warsaw. Acta Geophys. **63**(2), 398–413 (2015). https://doi.org/10.2478/s11600-014-0256-z
6. Jallu, M., Saride, S., Dutta, T.T.: Effect of saturation on dynamic properties of compacted clay in a resonant column test. Geomech. Geoeng. Int. J. **12**(3), 1 (2017). https://doi.org/10.1080/17486025.2016.1208849

7. Lin, P., Ni, J., Garg, A., et al.: Effects of clay minerals on small-strain shear modulus and damping ratio of saturated clay. Soil Mech. Found. Eng. **57**(1), (2019). https://doi.org/10.1007/s11204-020-09644-5
8. Moore, C.A., Donaldson, C.F.: Quantifying soil microstructure using fractals. Géotechnique **45**(1), 105–116 (1995). https://doi.org/10.1016/0148-9062(95)99096-G
9. Liu, C., Shi, B., Zhou, J., et al.: Quantification and characterization of microporosity by image processing, geometric measurement and statistical methods: application on SEM images of clay materials. Appl. Clay Sci. **54**(1), 97–106 (2011). https://doi.org/10.1016/j.clay.2011.07.022
10. Ndèye, F.N., Monga, O., Mohamed, M.M.O., et al.: 3D Shape extraction segmentation and representation of soil microstructures using generalized cylinders. Comput. Geosci. **39**(2), 50–63 (2012). https://doi.org/10.1016/j.cageo.2011.06.010
11. Osipov, V.I., Nikolaeva, S.K., Sokolov, V.N.: Microstructural changes associated with thixotropic phenomena in clay soils. Géotechnique **34**(3), 293–303 (1984). https://doi.org/10.1680/geot.1984.34.3.293
12. Pusch, R., Weston, R.: Microstructural Stability controls the hydraulic conductivity of smectitic buffer clay. Appl. Clay Sci. **23**(1), 35–41 (2003). https://doi.org/10.1016/S0169-1317(03)00084-X
13. Sivakumar, V., Doran, I.G., Graham, J.: Particle orientation and its influence on the mechanical behaviour of isotropically consolidated reconstituted clay. Eng. Geol. **66**(3), 197–209 (2002). https://doi.org/10.1016/S0013-7952(02)00040-6
14. Cetin, H., Goekoglu, A.: Soil structure changes during drained and undrained Triaxial shear of a clayey soil. Soils Found. **53**(5), 628–638 (2013). https://doi.org/10.1016/j.sandf.2013.08.002
15. Jiang, M., Zhang, F., Hu, H., et al.: Structural characterization of natural loess and remolded loess under Triaxial Tests. Eng. Geol. **181**, 249–260 (2014). https://doi.org/10.1016/j.enggeo.2014.07.021
16. Philippe, P., Martin, H., et al.: One-dimensional dynamic ground response analyses. J. Geotech. Geoenviron. Eng. **108**(7), 935–952 (1982). https://doi.org/10.1016/0022-1694(82)90165-2
17. Dathe, A., Eins, S., Niemeyer, J., et al.: The surface fractal dimension of the soil–pore interface as measured by image analysis. Geoderma **103**(1), 203–229 (2001). https://doi.org/10.1016/S0016-7061(01)00077-5
18. Obara, B., Kozusnikova, A.: Utilisation of the image analysis method for the detection of the morphological anisotropy of calcite grains in marble. Comput. Geosci. **11**(4), 275–281 (2007). https://doi.org/10.1007/s10596-007-9051-0
19. Prakongkep, N., Suddhiprakarn, A., Kheoruenromne, I., et al.: SEM image analysis for characterization of sand grains in Thai Paddy soils. Geoderma **156**(1), 20–31 (2010). https://doi.org/10.1016/j.geoderma.2010.01.003
20. Cox, M.R., Budhu, M.: A practical approach to grain shape quantification. Eng. Geol. **96**(1), 1–16 (2008). https://doi.org/10.1016/j.enggeo.2007.05.005

Small-Strain Shear Modulus of Coral Sand with Various Particle Size Distribution Curves

Ke Liang[1(✉)], Guoxing Chen[2], and Qing Dong[3]

[1] The Key Laboratory of Urban Security and Disaster Engineering of Ministry of Education, Beijing University of Technology, Beijing, China
Liangk91@163.com
[2] Institute of Geotechnical Engineering, Nanjing Tech University, Nanjing, China
[3] College of Architecture and Civil Engineering, Beijing University of Technology, Beijing, China

Abstract. Particle size distribution plays an important role in the small-strain shear modulus G_0 of granular soils. Numerous expressions were proposed for predicting the G_0 of siliceous sands in the literature. There are significant differences between the mechanic characteristics of coral sand and siliceous sand. In this paper, a series of resonant column tests were conducted on the coral sand with various particle size distribution curves. The influences of uniformity coefficient C_u, mean-grain size $d_{50,}$ and fines content F_c on G_0 of coral sand are investigated. For the given void ratio and confining stress, the G_0 of coral sand is smaller than that of siliceous sand. The G_0 of coral sand decreases with the increase of C_u, increases with the increase of d_{50}, and first decreases then increases with the increase of F_c. The minimum void ratio e_{min}, instead of the C_u, d_{50}, and F_c, is adopted as a unified index for capturing the characteristics of the particle size distribution of granular soils. The relationships between the parameters in Hardin's equation and e_{min} are established and a minimum void ratio-based G_0 expression for coral sand is proposed. The validity of the new expression for the other types of granular soils is confirmed using the G_0 data in the literature.

Keywords: Small-strain shear modulus · Coral sand · Particle size distribution curve · Resonant column test

1 Introduction

For $\gamma_a <$ linear elastic threshold shear strain γ_{tL} (10^{-6}–10^{-5}, depending on the soil types), soil exhibits quasi-elastic behavior, at which the shear modulus is practically constant, termed as small-strain shear modulus G_0. It is the key factor in liquefaction evaluation or site seismic response analysis. Extensive researches have been carried out to study the G_0 characteristics of granular soils through the bender element (BE) or resonant column (RC) tests (Youn et al. 2008; Van Impe et al. 2015; Xu et al. 2020). Comparisons between the G_0 values tested by BE and RC tests have been analyzed by several researchers (Yang and Gu 2013; Gu et al. 2015; Hoyos et al. 2015; Xu et al.

© The Author(s), under exclusive license to Springer Nature Switzerland AG 2022
L. Wang et al. (Eds.): PBD-IV 2022, GGEE 52, pp. 2054–2072, 2022.
https://doi.org/10.1007/978-3-031-11898-2_188

2020). They found the G_0 values obtained from the BE tests are slightly larger than those from the RC tests by 10% or lower, between which the difference can be ignored.

The G_0 of granular soils is primarily affected by the void ratio e and the confining stress σ_c. For a given e, the G_0 increases with the increase of σ_c as a power law; for a given σ_c, the G_0 decreases with the increase of e. Hardin's equation is the widely used empirical formula for predicting the G_0 of granular soils (Hardin and Drnevich 1972; Hardin and Black 1966), which is given in its general dimensionless form as

$$G_0 = A \cdot F(e) \cdot \left(\frac{\sigma_c}{P_a}\right)^n \tag{1}$$

where A and n are the best-fit parameters related to soil types, most researchers have proposed the mean value of n equal to 0.5 (Hardin and Drnevich 1972; Kokusho 1980; Bui 2009; Clayton 2011); P_a is the atmospheric pressure (= 100 kPa); $F(e)$ is the function of void ratio with the typical forms of

$$F(e) = \frac{(a-e)^2}{1+e} \text{ by Hardin and Black (1966)} \tag{2}$$

$$F(e) = \frac{1}{0.3+0.7e^2} \text{ by Hardin (1978)} \tag{3}$$

$$F(e) = e^b \text{ by Bellotti } et\ al. \text{ (1996) and Menq (2003)} \tag{4}$$

where a and b are dimensionless parameters.

Particle size distribution, particle shape, and mineral composition also have important influences on the G_0 of granular soils (Menq 2003; Wichtmann and Triantafyllidis 2009; Senetakis et al. 2012; Yang and Gu 2013; Payan et al. 2016; Liu and Yang 2018, Lashin et al. 2021). Uniformity coefficient C_u, mean-grain size d_{50}, and fines content F_c are the major characteristic indexes of the particle size distribution curve. Menq (2003) investigated the influence of C_u and d_{50} on G_0 of granular soils by RC test. He found that for the given σ_c and e, the G_0 slightly increases with the increase of d_{50}, but significantly decreases with the increase of C_u. Most of the studies on G_0 of granular soils in the present literature ignored the influence of d_{50} (Wichtmann and Triantafyllidis 2009; Senetakis et al. 2012; Ha Giang et al. 2017). Wichtmann and Triantafyllidis (2009) performed 163 RC tests on quartz sand with 25 different particle size distribution curves. Their tests indicate that for the given σ_c and e, the G_0 decreases with the increases of C_u but does not depend on the d_{50}. They calibrated the parameters A, a, and n in Hardin's equation as a function of C_u. Yang and Gu (2013) conducted a series of RC and BE tests on uniformly graded types of glass beads of different mean sizes. They concluded that the G_0 of sand is independent of particle size, and confirmed the conclusion by the micromechanics-based analysis using the Hertz-Mindlin contact law. Natural sand is usually not clean but contains a certain amount of fines. The influence of F_c is usually studied separated from the other characteristic parameters of the particle size distribution curve (Liu and Yang 2018; Shi et al. 2020; Ruan et al. 2021). Liu and Yang (2018) conducted a series of RC tests on sand-fines mixtures. They found the parameter A in Hardin's equation decreases with the increase of F_c, but the parameter n increases accordingly. The effect of fines content decreases as particle size disparity (ratio of the mean size of base sand to the mean size of fines) increases from 4 to 7. They proposed the combined size disparity to quantitatively capture the coupled effect of fines content and particle size disparity on the G_0 of sand-fines mixtures and proposed an empirical equation for estimating G_0 values. Payan et al. (2016) evaluated the validity of four G_0 equations based on the concept of state parameter and found that none of the formulas can sufficiently capture the effects of void ratio and confining stress on G_0 due to the exclusion of the effect of particle shape. It proofs that the particle shape has a significant influence on the G_0 of granular soils. Senetakis et al. (2012) compared the G_0 values of quartz and volcanic sands with the same particle size distribution curve. The G_0 of volcanic sand is lower than that of quartz sand attributed to the difference in particle morphology and mineralogy. Recently, the correlations between the G_0 and other soil parameters (e.g. equivalent skeleton void ratio, constrained modulus, ect) are investigated (Chen et al. 2020; Lashin et al. 2021).

Carbonate sediments that contain a high content of calcium carbonate ($CaCO_3$) in the forms of calcite and aragonite are conventionally referred to as calcareous sand (Lopez et al. 2018). Calcareous sand with $CaCO_3 \geq 90\%$, from coral reefs in the tropical and subtropical regions, is referred to as coral sand hereafter. Compared with siliceous sand, coral sand has several significant features, such as angular and irregular particles with numerous intraparticle pores and lower particle hardness (Brandes 2011; Sharma and Ismail 2006; Rui et al. 2020; Salem et al. 2013). It is proved that there are significant differences between the shear modulus of coral sand and siliceous sand (Ha Giang et al. 2017; Morsy et al. 2019; Liu et al. 2020). The G_0 of coral sand is higher than that of siliceous sand sine the less sphericity and more angularity particle shape of coral sand produce a better fabric for shear wave propagation (Ha Giang et al. 2017). Morsy et al. (2019) investigated the G_0 of coral sand from the North Coast of Egypt by RC test. They found the stress exponent n in Hardin's equation of coral sand is higher than that of siliceous sand due to the more irregular particle shape.

There are very few systematic studies on the G_0 characteristics of coral sand. This paper presents a series of RC tests on the coral sand to investigate the G_0 characteristics. The influence of particle size size distribution, i.e., the C_u, d_{50}, and F_c, on the G_0 of coral sand is analyzed, and a new minimum void ratio-based empirical equation was established for estimating the G_0 of granular soils with various particle size distribution curves.

2 Testing Material and Testing Program

2.1 Testing Material

The tested coral sand, sampled from a coral reef in the Nansha Islands, South China Sea, is composed of aragonite, high magnesian calcite, and calcite, with mass contents of 55.5%, 41.5%, and 3.0%, respectively. The $CaCO_3$ content is higher than 95%. The specific gravity G_s of the tested coral sand particle is 2.77. The scanning electron microscope images of the coral sand particles are presented in Fig. 1. The particles have rough surfaces and numerous intraparticle pores, and their shapes are very irregular, mostly flaky and angular. Particles were mixed according to the particle size distribution curves plotted in Fig. 2. The IDs and index properties of the tested coral sands are listed in

Table 1. S and S0 (grain size between 0.075 and 2 mm) and the fines (grain size smaller than 0.075 mm) in Fig. 2 are sieved from the in-situ soil sample. The sands are divided into three groups according to the research purposes:

(1) Group CU, the particle size distribution curves have similar d_{50} (about 0.53 mm) and constant F_c (0%), but variable C_u (ranges from 2.10 to 11.20) to investigate the influence of C_u on the G_0 of coral sand;
(2) Group D, the particle size distribution curves have similar C_u (about 2.95) and constant F_c (0%), but variable d_{50} (ranges from 0.21 to 2.00 mm) to investigate the influence of d_{50} on the G_0 of coral sand;
(3) Group FC, the fines are mixed with sand S0 by weight of 5%, 10%, 15%, 20%, 30%, and 40% to investigate the influence of F_c on the G_0 of coral sand.

As shown in Fig. 3, for coral sands in group FC, the C_u first increases and then slowly decreases as the increase of F_c. The C_u gets the maximum value when $F_c \approx 30\%$. The d_{50} of coral sands in group FC decreases with the increase of F_c.

Fig. 1. Scanning electron microscope images of the coral sand particles

The values of the maximum void ratio e_{max} and minimum void ratio e_{min} are closely related to the particle size distribution curve of the sand. The variations of e_{max} and e_{min} of the tested coral sand in each group are plotted in Fig. 4. For sands in group CU

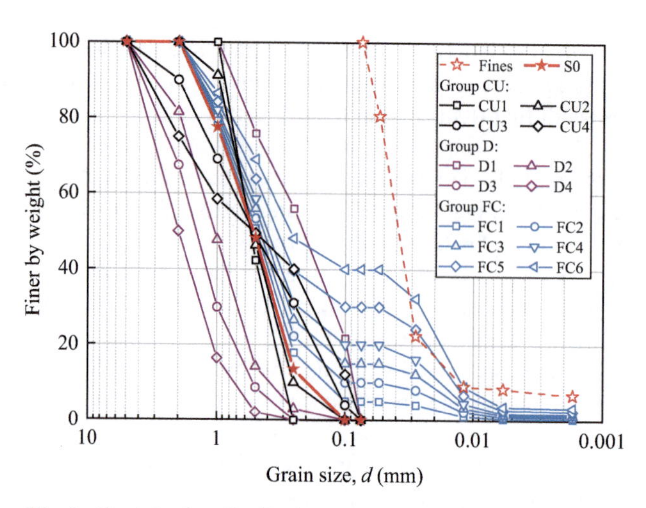

Fig. 2. Particle size distribution curves of the tested coral sands

Table 1. The index properties of the tested coral sands and the test programs

Sand group	Sand ID	Index properties					e_0 of the specimens
		C_u	d_{50} (mm)	F_c (%)	e_{max}	e_{min}	
CU	CU1	2.10	0.55	0	1.360	0.751	1.124、1.029、0.942、0.877
	CU2	2.47	0.53	0	1.182	0.646	1.047、0.947、0.866、0.781
	S0*	3.27	0.52	0	1.162	0.631	1.007、0.910、0.799、0.775
	CU3	5.99	0.52	0	0.958	0.516	0.852、0.772、0.694、0.643
	CU4	11.20	0.52	0	0.890	0.453	0.816、0.657、0.617、0.602
D	D1	3.05	0.21	0	1.165	0.582	0.968、0.820、0.680
	S0*	3.27	0.52	0	1.162	0.631	1.007、0.910、0.799、0.775
	D2	3.35	1.05	0	1.162	0.646	1.036、0.945、0.826、0.754
	D3	2.99	1.45	0	1.148	0.650	1.081、0.920、0.852
	D4	3.26	2.00	0	1.183	0.733	1.053、0.946、0.864
FC	S0*	3.27	0.52	0	1.162	0.631	1.007、0.910、0.799、0.775

(*continued*)

Table 1. (*continued*)

Sand group	Sand ID	Index properties					e_0 of the specimens
		C_u	d_{50} (mm)	F_c (%)	e_{max}	e_{min}	
	FC1	4.40	0.49	5	1.101	0.548	0.851、0.765、0.673
	FC2	5.95	0.46	10	1.019	0.450	0.818、0.726、0.561
	FC3	13.00	0.43	15	1.009	0.412	0.768、0.695、0.584
	FC4	22.19	0.40	20	0.999	0.375	0.728、0.653、0.563
	FC5	26.86	0.34	30	0.958	0.369	0.658、0.588、0.512
	FC6	26.32	0.27	40	1.025	0.401	0.663、0.573、0.487

Notes: (1) $C_u = d_{60}/d_{10}, d_{10}, d_{50}$, and d_{60} are the particle sizes corresponding to 10%, 30%, 50%, and 60% finer on the cumulative particle size distribution curve, respectively; (2) "*" represents the duplicate sand in each sand group; (3) e_0 is the initial void ratio after specimen preparation.

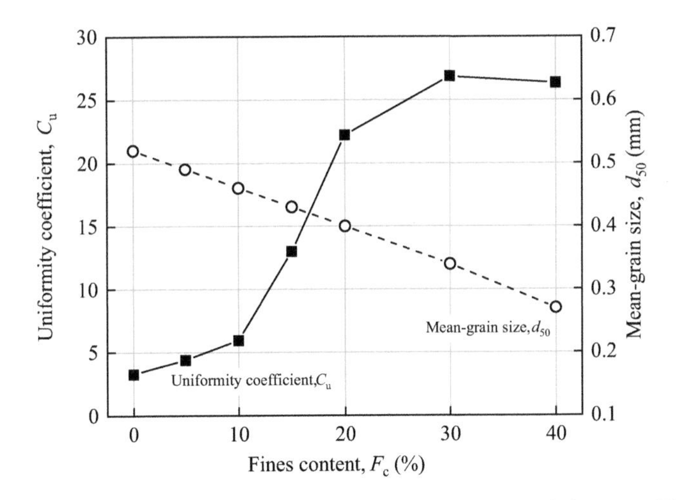

Fig. 3. Variations of the C_u and d_{50} with F_c for coral sands in group FC

(Fig. 4(a)), Both e_{max} and e_{min} decrease with the increase of C_u. For sands in group D (Fig. 4(b)), both e_{max} and e_{min} increase with the increase of d_{50}. The increasing rate as d_{50} for e_{min} is faster than that for e_{mix}. For sands in group FC (Fig. 4(c)), both e_{max} and e_{min} first decrease and then increase with the increase of F_c and get the minimum values at $F_c \approx 30\%$.

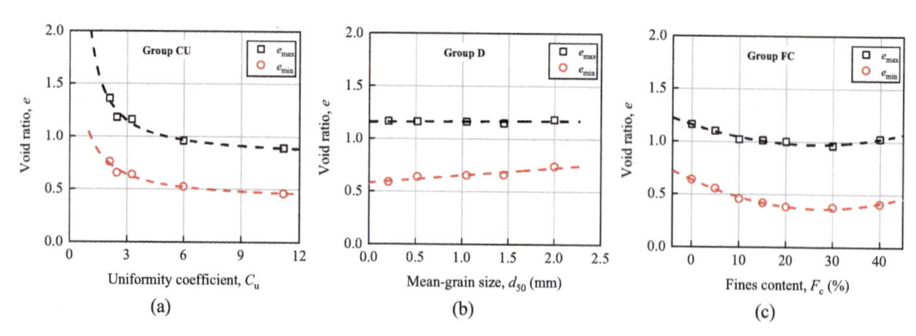

Fig. 4. Variations of e_{max} and e_{min} for the tested coral sand: (a) group CU; (b) group D; (c) group FC

2.2 Test Apparatus, Specimen Preparation, and Testing Procedure

The TSH-100 resonant column test apparatus manufactured by GCTS Instruments (Tempe, Arizona) was used in this research. It is of "fixed-free" type as shown in Fig. 5. The capacity, deviation, precision, and additional details of the test apparatuses are provided in Chen et al. (2019). The dry specimen is fixed on the pedestal, and sinusoidal excitations are applied by the motor attached to the top end of the specimen. The shear strain is derived from the acceleration measured by the accelerometer by integration. The frequency of the excitations with constant stress amplitude varied to determine the resonant frequency f_1 at which the shear strain amplitude γ_a of the specimen gets the maximum value. The secant shear modulus G of the specimen can be derived as follows,

$$G = \rho\left(\frac{2\pi hf_1}{\varphi}\right)^2 \tag{5}$$

where ρ and h are the density and height of the specimen, respectively; φ is a parameter related to the polar mass moments of inertia of the specimen (I_s) and top mass (I_t). The φ can be obtained from the function $\varphi\tan(\varphi) = I_s/I_t$. The value of I_t is calibrated using the calibration aluminum specimens of known mechanical properties ($= 2.10 \times 10^{-4}$ kg·m^2 for the RC apparatus in this research).

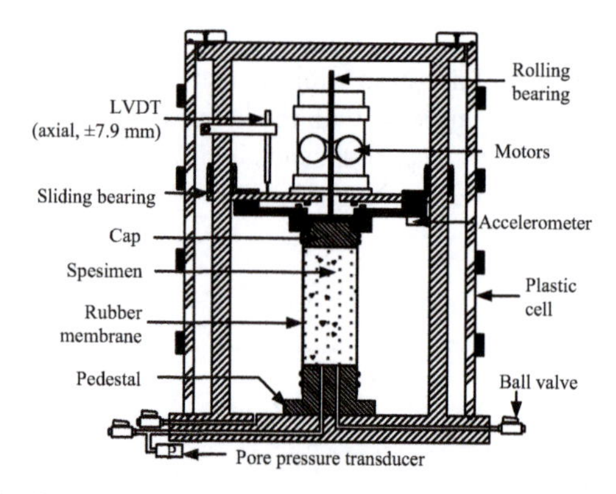

Fig. 5. Schematic of the resonant column test apparatuses

All solid cylinder specimens were 50 mm in diameter and 100 mm in height. According to ASTM D3999-11, the dry vibration method was used to prepare the specimens in this study. The oven-dried coral sand was placed in four layers into the membrane-lined split mold with a funnel. Considering the compaction of each succeeding layer would densify the sand in layers below it, the layer masses were slightly decreased from top to bottom to obtain uniform density. The values of the initial void ratio e_0 of the tested specimens after preparation are listed in Table 1.

Each specimen was isotropically consolidated in six stages with the sequence of σ_c = 20, 50, 100, 150, 200, and 300 kPa. After each stage of consolidation, the compaction of the soil was derived from the axial deformation to calculate the exact void ratio *e*. Then, the RC test was conducted to measure the G_0 (G at $\gamma_a < 10^{-5}$). The typical test outputs of the RC test are shown in Fig. 6. Figure 6(a) exemplifies the variation of the amplitudes of the strain responses during the frequency sweep excitation process. The value of f_1 was identified as the frequency of the peak of the response curve.

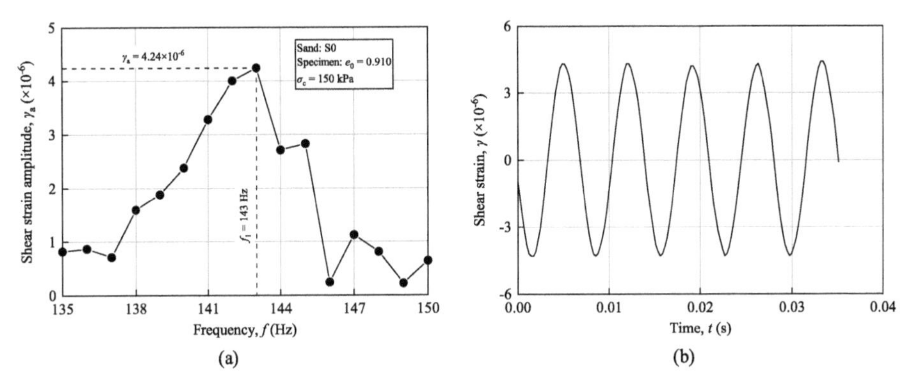

Fig. 6. Typical resonant column test output: (a) shear strain amplitude for frequency sweep excitation; (b) shear strain time history of resonance

3 Test Results

3.1 Effects of Void Ratio and Confining Stress on G_0

Figure 7 plotted the test data of G_0 against void ratio *e* for all the tested specimens. It can be observed that the G_0 decreases with *e*, and increases with σ_c. As shown in Fig. 8, the relationships between G_0 and *e* for coral sand are investigated in the log-log scale. It is found that the lg(G_0) decreases linearly with the increase of lg(*e*) at constant σ_c. Equation (4) is adopted to evaluate the influence of the void ratio on G_0, in which the parimeter *b* represents the salope of the lg(G_0)-lg(*e*) trend line. The best-fit values of parimeter *b* for coral sand S0 at σ_c of 20 – 300 kPa varies between −1.02 and −1.07 as shown in Fig. 8(a). The values of *b* vary between −1.06 and −1.19 for the coral sands with different particle size distribution curves shown in Fig. 8(b). It can be concluded that the influence of confining stress and gradation curve on the value of *b* is neglectable. The average value of *b* for all the tested coral sands is −1.09, and the G_0 of the tested coral sand is proportional to $e^{-1.09}$.

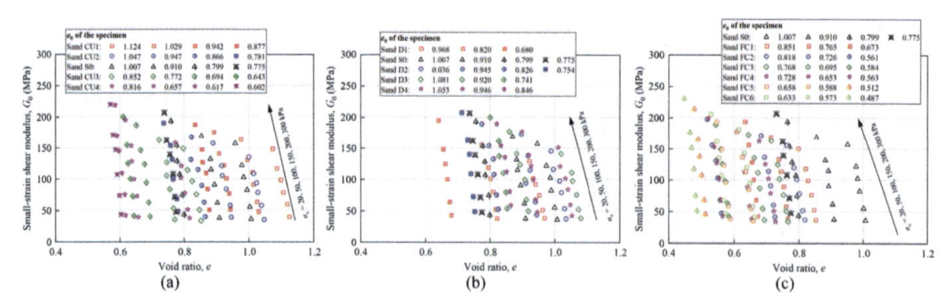

Fig. 7. Test results of G_0 as a function of e for all the coral sand specimens: (a) graduation group CU; (b) gradation group D; (c) gradation group FC

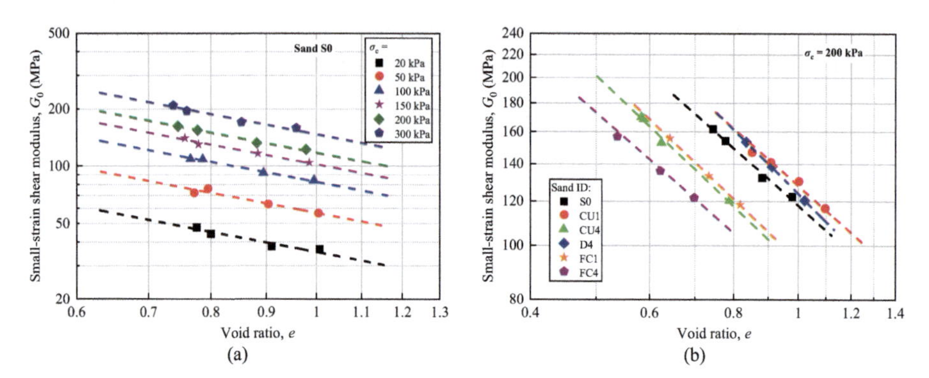

Fig. 8. The variation of G_0 as a function of e in log-log scale: (a) sand S0 ($\sigma_c = 20 \sim 300$ kPa); (b) coral sands with different gradation curves ($\sigma_c = 200$ kPa)

The volume of each specimen, along with the void ratio, decreases with the increase of the confining stress. In order to investigate the effect of σ_c, the effect of void ratio on G_0 of coral sands is eliminated by dividing $e^{-1.09}$. Figure 9 gives the relationship between the void ratio-corrected small-strain sear modulus $G_0/(e^{-1.09})$ and σ_c in the log-log scale. The lg[$G_0/(e^{-1.09})$]-lg(σ_c) relationship is also linear. Thereafter, G_0 of the tested coral sand is proportional to $\sigma_c{}^n$, Eqs. (1) and (4) are adopted to evaluate the relationship among G_0, e, and σ_c, which is expressed as follows:

$$G_0 = A \cdot e^{-1.09} \cdot \left(\frac{\sigma_c}{P_a}\right)^n \tag{6}$$

The best-fit values of A and n for all the tested coral sands are summarised in Table 2. The coefficients of determination R^2 are higher than 0.96, suggesting that the empirical Eq. (6) works reasonably well for coral sands with different particle size distribution curves.

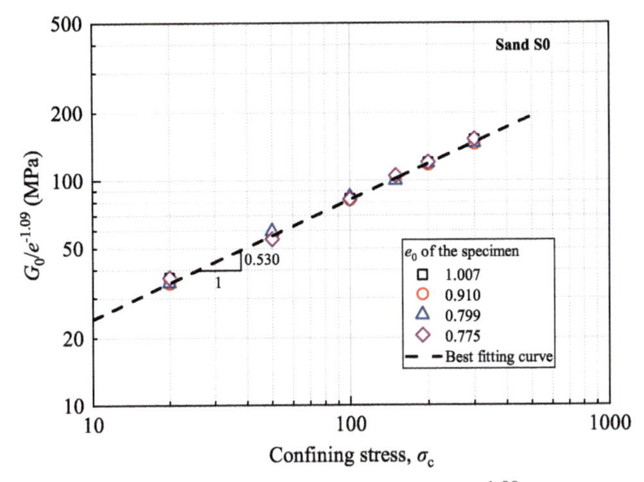

Fig. 9. The void ratio-corrected small-strain sear modulus $G_0/e^{-1.09}$ of sand S0 as a function of σ_c in log-log scale

Table 2. The best-fit parameters A and n of the G_0 prediction equation for the tested coral sand

Sand ID	Paremeters in Eq. (6)		R^2
	A (MPa)	n	
S0	81.929	0.530	0.998
CU1	93.117	0.457	0.993
CU2	77.541	0.513	0.995
CU3	62.134	0.562	0.997
CU4	61.365	0.570	0.997
D1	66.672	0.539	0.988
D2	83.865	0.516	0.991
D3	86.157	0.473	0.984
D4	89.706	0.461	0.992
FC1	64.554	0.538	0.998
FC2	71.204	0.551	0.964
FC3	67.848	0.585	0.982
FC4	60.705	0.583	0.994
FC5	61.396	0.601	0.994
FC6	69.818	0.566	0.985

3.2 Effect of Gradation Curves on G_0

The C_u, d_{50}, and F_c are the main characteristic parameters of the particle size distribution curve. Wichtmann and Triantafyllidis (2013) and Ha Giang et al. (2017) took C_u as the representative index characterizing the feature of the particle size distribution of soils. They investigated the variations of parameters in the G_0 expression of Eq. (1) against C_u and proposed the empirical equations for predicting G_0 (referred to as WT and HG model hereafter) which are listed in Table 3. The G_0 values of the tested coral sands are predicted by the WT and HG models. The comparisons between the predicted and measured G_0 values are plotted in Fig. 10. The WT expression was calibrated by the test results of quartz sands with C_u ranging from 1.5 to 8.0. As a result, the dispersion of the predicted G_0 values for the coral sands with C_u higher than 8.0 are much larger than those for coral sands within the C_u range of 1.5 to 8.0 (Fig. 10(a)). The HG expression was calibrated by the test results of coral sands with C_u ranging from 1.44 to 5.43. Also, the dispersion of the predicted G_0 values for the coral sands with C_u higher than 5.43 are much larger than those for coral sands within the C_u range of 1.44 to 5.43 (Fig. 10(b)). This indicates that the reliability of the WT and HG expressions cannot be guaranteed for the sands out of the C_u range involved in the model calibration, and they haven't fully captured the effect of the particle size distribution on the G_0 of coral sand. Other factors, such as d_{50} and F_c, may also have significant influences on the G_0 of granular soils. On the other hand, WT expression was established for predicting the G_0 of quartz sand. The G_0 values of the coral sands with $1.5 < C_u < 8.0$ are significantly underestimated by the WT expression, which means that the G_0 of coral sand is larger than that of quartz sand at the same void ratio and confining stress.

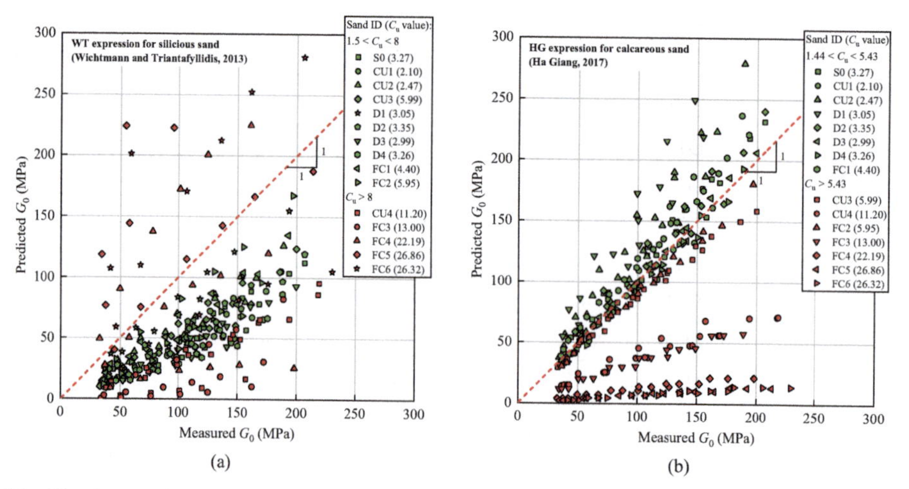

Fig. 10. Comparison between the predicted and measured G_0 for the tested coral sands: (a) WT expression; (b) HG expression

The variations of $G_0/e^{-1.09}$ with C_u, d_{50}, and F_c are plotted in Fig. 11. The coral sands in group CU have similar d_{50} and the same F_c and the $G_0/e^{-1.09}$-C_u relationships are shown in Fig. 11(a). For constant σ_c and e, the G_0 of coral sand decreases with the increase of C_u, and the decreasing rate decreases as the increasing C_u. Figure 11(b) plots the relationship between $G_0/e^{-1.09}$ and d_{50} for coral sands in group D whose C_u is similar and F_c is the same. For constant σ_c and e, the G_0 of coral sand increases with the increase of d_{50}. The increasing rate decreases as the increasing d_{50}. For coral sands in group FC, since the C_u and d_{50} vary synchronously as F_c, the variation of G_0 with F_c is much more complex. As shown in Fig. 11(c), for constant σ_c and e, the G_0 of coral sand in group FC first decreases and then increases with the increase of F_c. The G_0 gets the minimum value at $F_c \approx 25\%$.

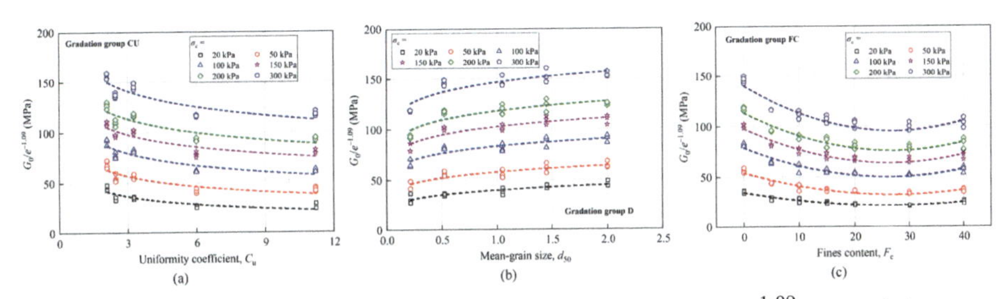

Fig. 11. Variation of void ratio-corrected small-strain sear modulus $G_0/e^{-1.09}$: (a) gradation group CU; (b) gradation group D; (c) gradation group FC

4 Expression for G_0

As shown in Fig. 12, the calibrated parameters A and n (Table 2) are plotted against C_u, d_{50}, and F_c for the tested coral sands in groups CU, D, and FC, respectively. The parameter A decreases with increase of C_u, increases with the increase of d_{50}, and first decreases then increases with the increase of F_c. Conversely, the parameter n increases with the increase of C_u, decreases with the increase of d_{50}, and first increases then decreases with the increase of F_c. It also indicates that all of C_u, d_{50}, and F_c have significant influences on the G_0 of coral sand.

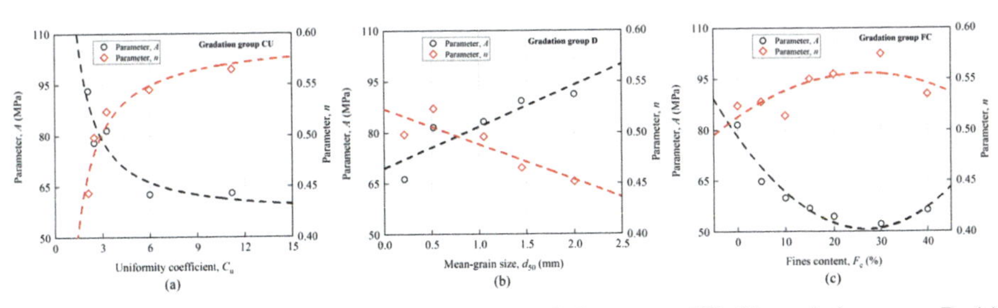

Fig. 12. Variation of the parameters A and n: (a) gradation group CU; (b) gradation group D; (c) gradation group FC

Table 3. The proposed expressions for G_0 of sand in the literature

Reference	Model name	Expression	Parameters	Tested sands for calibration
Wichtmann and Triantafyllidis (2013)	WT	$G_0 = A \cdot \frac{(a-e)^2}{1+e} \cdot \left(\frac{\sigma_c}{P_a}\right)^n$	A (MPa) = 156.3+0.313·$C_u^{2.98}$; a = 1.94·exp(-0.066·C_u); n = 0.40·$C_u^{0.18}$	Quartz sand $1.5 < C_u < 8.0$
Ha Giang *et al.* (2017)	HG	$G_0 = A \cdot e^b \cdot \left(\frac{\sigma_c}{P_a}\right)^n$	A (MPa) = 115.371·exp(-0.107·C_u); b = -4.416·exp(-0.249·C_u); n = 0.421·$C_u^{0.125}$	Coral sand $1.44 < C_u < 5.43$

The packing structure of soil, which is strongly dependent on the particle size distribution, is an important factor governing the mechanical properties. Both the extreme void ratios (e_{max} and e_{min}) and the small-strain shear modulus (or shear wave velocity) are closely related to the particle size distribution of soils (Aberg 1992, 1996; Cubrinovski and Ishihara 2002; Menq 2003; Cho et al. 2006; Chang et al. 2017). It is difficult to capture the coupled effect of C_u, d_{50}, and F_c on the G_0 of coral sand, i.e., on the parameters A and n. Encouragingly, according to Fig. 4 and Fig. 12, the variation patterns of e_{max} or e_{min} against C_u, d_{50}, and F_c are similar to those of parameter A, but opposite to those of parameter n. It is not surprising that the minimum or maximum void ratio is feasible for comprehensively mirroring the packing characteristics of granular soils with different particle size distribution curves. Since the measurement error of e_{min} is smaller than that of e_{max} in practice, the minimum void ratio is a better choice for estimating G_0 of granular soils.

The variation of parameters A and n with e_{min} for the tested coral sand are presented in Fig. 13. As expected, parameter A increases with the increase of e_{min}, parameter n decreases with the increase of e_{min}. Both A and n can be expressed as the function of e_{min} by the following empirical equations,

$$A = A_1 \exp(A_2 e_{min}) \tag{7}$$

$$n = n_1 e_{min} + n_2 \tag{8}$$

where A is in MPa. Combined with Hardin's equation, the G_0 of sands with various particle size distribution curves can be uniformly predicted by the following equation,

$$A = A_1 \exp(A_2 e_{min}) e^b \left(\frac{\sigma_c}{P_a}\right)^{n_2 e_{min} + n_2} \tag{9}$$

where b is a constant. The best-fit coefficients of A_1, A_2, n_1, and n_2 for the tested coral sands are listed in Table 4.

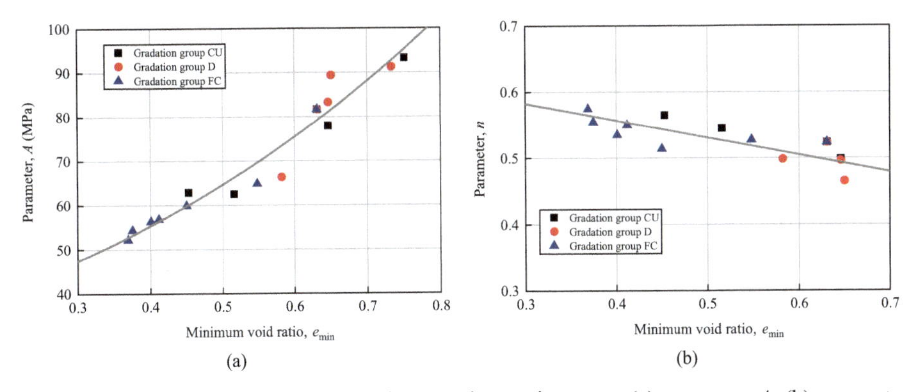

Fig. 13. The best-fit parameters in Hardin' equation against e_{min}: (a) parameter A; (b) parameter n

Table 4. Best-fit parameters in Eq. (9) for dfferent sands

Material	A_1	A_2	n_1	n_2	b
Nansha coral sand (this paper)	29.79	1.55	−0.26	0.66	−1.09
Dorsten quartz sand (Wichtmann and Triantafyllidis 2013)	10.9	2.94	−0.62	0.78	−1.54
Abu Dhabi calcareous sand (Ha Giang et al. 2017)	31.17	1.03	−0.09	0.55	−1.53

The comparison between the predicted G_0 by Eq. (9) and the measured G_0 for all the tested Nansha coral sands is presented in Fig. 14(a). The deviation of the predicted G_0 is mostly less than 10%. The good prediction of the proposed G_0 expression is confirmed. It also confirmed that e_{min} is very efficient at uniformly representing the complex effect of particle size distribution on G_0 of sands.

In order to investigate the general applicability of the G_0 expression in capturing the effect of particle size distribution. The test data of G_0 of quartz sands and calcareous sands with various gradations are obtained from the previous literature. Wichtmann and Triantafyllidis (2013) tested the G_0 of Dorsten quartz sands with 26 grain-size distribution curves by RC test. The detailed G_0 values of 16 of the sands can be obtained from the paper. Ha Giang et al. (2017) investigate the G_0 of Abu Dhabi calcareous sands with 5 grain-size distribution curves by BE test. The detailed information of the 16 quartz sands and 5 calcareous sands is summarized in Table 5. The G_0-e relationship for each sand was reanalyzed, the constant b in Eq. (9) is -1.54 and -1.53 for Dorsten quartz sand and Abu Dhabi calcareous sand, respectively. The relationships of the best-fit parameter A and n against e_{min} are investigated for the sands in literature. They can also be expressed by Eqs. (7) and (8) with the coefficients A_1, A_2, n_1, and n_2 listed in Table 4. The G_0 of Dorsten quartz sands and Abu Dhabi calcareous sands are also predicted by Eq. (9) with the parameters in Table 4. The comparisons between the predicted and measured G_0 values for quartz sand calcareous sands are shown in Figs.14(b) and (c), respectively. The predicted G_0 values show very good agreement with the measured values with the maximum difference of 15%, indicating the reliability of the proposed minimum void ratio-based G_0 expression for different types of sandy soils.

Note that the empirical relationships of $A-e_{min}$ (Eq. 7) and $n-e_{min}$ (Eq. 8), and the values of b are quite different for different types of soils according to Table 4. This might be due to (a) the test error from different researchers and different test methods; (b) the effect of mineral composition and particle shape. Consequently, further research on the minimum void ratio-dependent G_0 expression considering the effect of mineral composition and particle shape is necessary.

Table 5. Summary of the sand in literature

Reference	Test method	Sand	G_s	Sand ID	C_u	d_{50}/mm	F_c/%	e_{max}	e_{min}
Wichtmann and Triantafyllidis (2013)	RC	Dorsten quartz sand	2.65	L2	1.50	0.20	0	0.994	0.595
				L4	1.50	0.60	0	0.892	0.571
				L6	1.50	2.00	0	0.877	0.591
				L10	2.00	0.60	0	0.865	0.542
				L11	2.50	0.60	0	0.856	0.495
				L12	3.00	0.60	0	0.829	0.474
				L14	5.00	0.60	0	0.748	0.395
				L16	8.00	0.60	0	0.673	0.356
				L17	2.00	2.00	0	0.826	0.554
				L18	2.50	2.00	0	0.810	0.513
				L19	3.00	2.00	0	0.783	0.491
				L21	5.00	2.00	0	0.703	0.401
				L23	8.00	2.00	0	0.520	0.398
				L24	2.00	0.20	0	0.959	0.559
				L25	2.50	0.20	0	0.937	0.545
				L26	3.00	0.20	0	0.920	0.541
Ha Giang et al. (2017)	BE	Abu Dhabi calcareous sand	2.79	S	3.46	0.73	-	1.330	0.903
				VS	5.43	0.43	-	0.956	0.508
				S1	1.86	0.23	-	1.471	0.933
				SVS	5.43	0.43	-	1.129	0.652
				SMOL	1.44	0.17	-	1.340	0.843

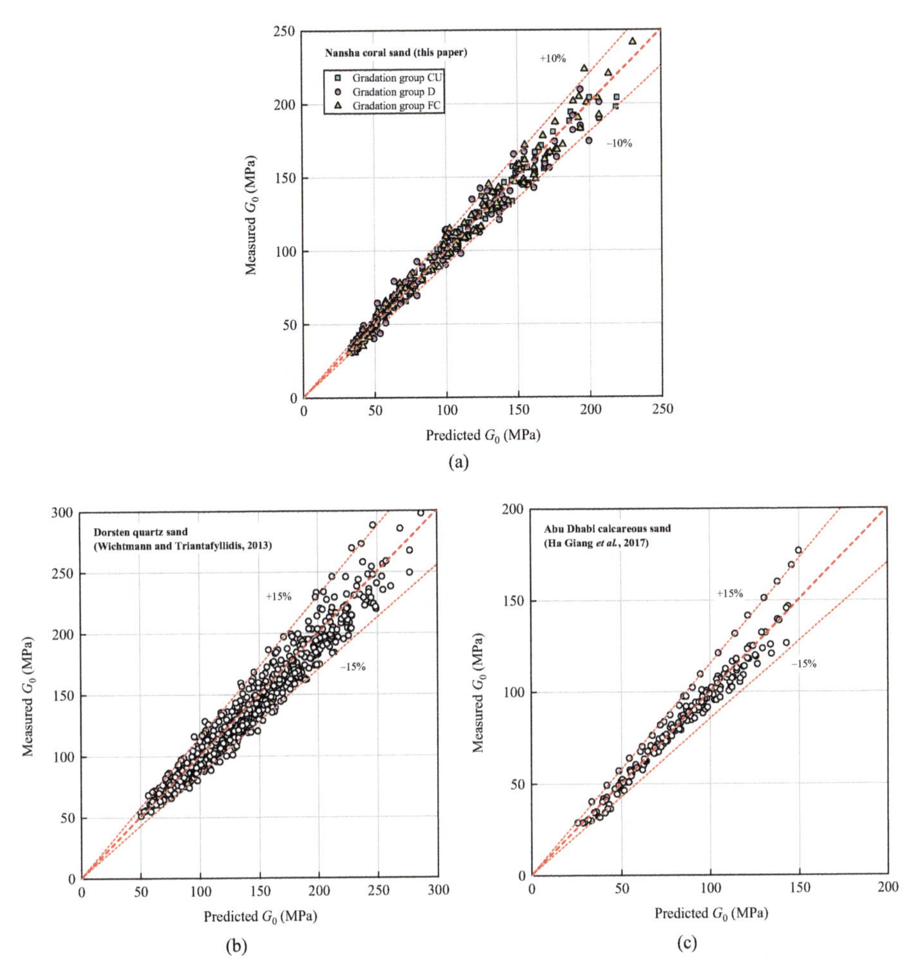

Fig. 14. Comparison between the predicted G_0 by Eq. (9) and the measured G_0 for various sands: (a) Nansha coral sand tested in this paper; (b) Dorsten quartz sands by Wichtmann and Triantafyllidis (2013); (c) Abu Dhabi calcareous sand by Ha Giang et al. (2017)

5 Conclusion

A series of RC tests were conducted on coral sands, from the Nansha Islands, South China Sea, with 15 different particle size distribution curves. This paper presents the research on the characteristics of the small-strain shear modulus of coral sand considering the effect of confining stress, void ratio, and particle size distribution. A minimum void ratio-based modified Hardin's equation for predicting G_0 of granular soils was proposed. The main conclusions are summarized as follows.

(1) The G_0 of coral sand decreases with the increase of e and increases with the increase of σ_c, which is proportional to $e^{-1.09}$ and $\sigma_c{}^n$. The particle size distribution of soil

has a significant influence on G_0 of coral sand. At constant σ_c, the void ratio-corrected small-strain sear modulus $G_0/(e^{-1.09})$ decreases with the increase of C_u, increases with the increase of d_{50}. With the increase of F_c, the $G_0/(e^{-1.09})$ first decreases for $F_c < 25\%$, and then increases slightly for $25\% < F_c < 40\%$.

(2) The general applicability of the proposed G_0 expressions, considering the effect of particle size distribution, in the literature is mostly limited by the range of the particle size distribution curves of the tested soils for calibration. The G_0 of coral sand is higher than that of siliceous sand at constant e and σ_c. The previous G_0 expressions for siliceous sand will underestimate the G_0 of coral sand.

(3) Extreme void ratios (e_{max} and e_{min}) are the comprehensive indexes reflecting the characteristics of the particle size distribution of granular soils. Both the parameters A and n in Hardin's equation are closely related to e_{min} for granular soils with various particle size distribution curves. Specifically, parameter A increases with the increase of e_{min}, and parameter n decreases with the increase of e_{min}. A minimum void ratio-based modified Hardin's equation was proposed for predicting the G_0 of granular soils considering the effect of particle size distribution. Noting that the A-e_{min} and n-e_{min} relationships and the values of b are different for various types of soils, further systematic study on the calibration of the proposed G_0 expression considering the effect of mineral composition and particle shape is needed.

Acknowledgments. The study in this paper was supported by the project funded by China Post-doctoral Science Foundation (2021M700309). This financial support is gratefully acknowledged.

References

Åberg, B.: Void ratio of noncohesive soils and similar materials. J. Geotechn. Eng. **118**(9), 1315–1334 (1992). https://doi.org/10.1061/(ASCE)0733-9410(1992)118:9(1315)

Åberg, B.: Grain-size distribution for smallest possible void ratio. J. Geotechn. Eng. **122**(1), 74–77 (1996). https://doi.org/10.1061/(ASCE)0733-9410(1996)122:1(74)

Bellotti, R., Jamiolkowski, M., LoPresti, D.C.F., O'Neill, D.A.: Anisotropy of small strain stiffness in ticino sand. Géotechnique **46**(1), 115–131 (1996). https://doi.org/10.1680/geot.1996.46.1.115

Brandes, H.G.: Simple shear behavior of calcareous and quartz sands. Geotech. Geol. Eng. **29**(1), 113–126 (2011). https://doi.org/10.1007/s10706-010-9357-x

Bui, M.T.: Influence of some particle characteristics on the small strain response of granular materials. Southampton: School of Civil Engineering and the Environment, University of Southampton (2009)

Chang, C.S., Deng, Y., Yang, Z.: Modeling of minimum void ratio for granular soil with effect of particle size distribution. J. Eng. Mech. **143**(9), 04017060 (2017). https://doi.org/10.1061/(ASCE)EM.1943-7889.0001270

Chen, G.X., Wu, Q., Zhao, K., Shen, Z.F., Yang, J.: A binary packing material-based procedure for evaluating soil liquefaction triggering during earthquakes. J. Geotechn. Geoenviron. Eng. **146**(6), 04020040 (2020). https://doi.org/10.1061/(ASCE)GT.1943-5606.0002263

Chen, G.X., Zhao, D.F., Chen, W.Y., Juang, C.H.: Excess pore-water pressure generation in cyclic undrained testing. J. Geotechn. Geoenviron. Eng. **145**(7), 04019022 (2019). https://doi.org/10.1061/(asce)gt.1943-5606.0002057

Cho, G.C., Dodds, J., Santamarina, J.C.: Particle shape effects on packing density, stiffness, and strength: natural and crushed sands. J. Geotechn. Geoenviron. Eng. **132**(5), 591–602 (2006). https://doi.org/10.1061/(ASCE)1090-0241(2006)132:5(591)

Clayton, C.R.I.: Stiffness at small strain: research and practice. Géotechnique **61**(1), 5–37 (2011). https://doi.org/10.1680/geot.2011.61.1.5

Cubrinovski, M., Ishihara, K.: Maximum and minimum void ratio characteristics of sand. Soils Found. **42**(6), 65–78 (2002). https://doi.org/10.3208/sandf.42.6_65

Gu, X., Yang, J., Huang, M., Gao, G.: Bender element tests in dry and saturated sand: signal interpretation and result comparison. Soils Found. **55**(5), 951–962 (2015). https://doi.org/10.1016/j.sandf.2015.09.002

Ha Giang, P.H., Van Impe, P.O., Van Impe, W.F., Menge, P., Haegeman, W.: Small-strain shear modulus of calcareous sand and its dependence on particle characteristics and gradation. Soil Dyn. Earthq. Eng. **100**, 371–379 (2017). https://doi.org/10.1016/j.soildyn.2017.06.016

Hardin, B.O., Drnevich, V.P.: Shear modulus and damping in soils: design equations and curves. J. Soil Mech. Found. Div. **98**(SM7), 667–692 (1972). https://doi.org/10.1061/JSFEAQ.0001760

Hardin, B.O., Black, W.L.: Sand stiffness under various triaxial stresses. J. Soil Mech. Found. Div. **92**(2), 27–42 (1966). https://doi.org/10.1061/JSFEAQ.0000865

Hardin, B.O.: The nature of stress-strain behavior of soils. In: Proceedings of Conference on Earthquake Engineering and Soil Dynamics, pp. 3–90. ASCE, New York (1978)

Hoyos, L.R., Suescún-Florez, E.A., Puppala, A.J.: Stiffness of intermediate unsaturated soil from simultaneous suction-controlled resonant column and bender element testing. Eng. Geol. **188**, 10–28 (2015). https://doi.org/10.1016/j.enggeo.2015.01.014

Kokusho, T.: Cyclic triaxial test of dynamic soil properties for wide strain range. Soils Found. **20**(2), 45–60 (1980). https://doi.org/10.3208/sandf1972.20.2_45

Lashin, I., Ghali, M., Hussien, M.N., Chekired, M., Karray, M.: Investigation of small- to large-strain moduli correlations of normally consolidated granular soils. Can. Geotech. J. **58**(1), 1–22 (2021). https://doi.org/10.1139/cgj-2019-0741

Liu, X., Yang, J.: Influence of size disparity on small-strain shear modulus of sand-fines mixtures. Soil Dyn. Earthq. Eng. **115**, 217–224 (2018). https://doi.org/10.1016/j.soildyn.2018.08.011

Liu, X., Li, S., Sun, L.: The study of dynamic properties of carbonate sand through a laboratory database. Bull. Eng. Geol. Env. **79**(7), 3843–3855 (2020). https://doi.org/10.1007/s10064-020-01785-z

Lopez, F.A.F., Taboada, V.M., Ramirez, Z.X.G., Roque, D.C., Nabor, P.B., Dantal, V.S.: Normalized modulus reduction and damping ratio curves for bay of campeche carbonate sand. In: Proceedings of the Offshore Technology Conference (2018). https://doi.org/10.4043/29010-ms

Menq, F.: Dynamic Properties of Sandy and Gravelly Soils. The University of Texas, Austin (2003)

Morsy, A.M., Salem, M.A., Elmamlouk, H.H.: Evaluation of dynamic properties of calcareous sands in Egypt at small and medium shear strain ranges. Soil Dyn. Earthq. Eng. **116**, 692–708 (2019). https://doi.org/10.1016/j.soildyn.2018.09.030

Payan, M., Khoshghalb, A., Senetakis, K., Khalili, N.: Effect of particle shape and validity of $G_{\max}$ models for sand: a critical review and a new expression. Comput. Geotech. **72**, 28–41 (2016). https://doi.org/10.1016/j.compgeo.2015.11.003

Rui, S., Guo, Z., Si, T., Li, Y.: Effect of particle shape on the liquefaction resistance of calcareous sands. Soil Dyn. Earthq. Eng. **137**, 106302 (2020). https://doi.org/10.1016/j.soildyn.2020.106302

Ruan, B., Miao, Y., Cheng, K., Yao, E.L.: Study on the small strain shear modulus of saturated sand-fines mixtures by bender element test. Eur. J. Environ. Civ. Eng. **25**(1), 28–38 (2021). https://doi.org/10.1080/19648189.2018.1513870

Salem, M., Elmamlouk, H., Agaiby, S.: Static and cyclic behavior of North Coast calcareous sand in Egypt. Soil Dyn. Earthq. Eng. **55**(12), 83–91 (2013). https://doi.org/10.1016/j.soildyn.2013.09.001

Senetakis, K., Anastasiadis, A., Pitilakis, K.: The small-strain shear modulus and damping ratio of quartz and volcanic sands. Geotech. Test. J. **35**(6), 20120073 (2012). https://doi.org/10.1520/GTJ20120073

Sharma, S.S., Ismail, M.A.: Monotonic and cyclic behavior of two calcareous soils of different origins. J. Geotechn. Geoenviron. Eng. **132**(12), 1581–1591 (2006). https://doi.org/10.1061/(ASCE)1090-0241(2006)132:12(1581)

Shi, J., Haegeman, W., Xu, T.: Effect of non-plastic fines on the anisotropic small strain stiffness of a calcareous sand. Soil Dyn. Earthq. Eng. **139**, 106381 (2020). https://doi.org/10.1016/j.soildyn.2020.106381

Van Impe, P.O., Van Impe, W.F., Manzotti, A., Mengé, P., Van den Broeck, M., Vinck, K.: Compaction control and related stress-strain behaviour of off-shore land reclamations with calcareous sands. Soils Found. **55**(6), 1474–1486 (2015). https://doi.org/10.1016/j.sandf.2015.10.012

Wichtmann, T., Triantafyllidis, T.: Influence of the grain-size distribution curve of quartz sand on the small strain shear modulus G_{max}. J. Geotechn. Geoenviron. Eng. **135**(10), 1404–1418 (2009). https://doi.org/10.1061/(ASCE)GT.1943-5606.0000096

Xu, K., Gu, X., Hu, C., Lu, L.: Comparison of small-strain shear modulus and Young's modulus of dry sand measured by resonant column and bender-extender element. Can. Geotech. J. **57**(11), 1745–1753 (2020). https://doi.org/10.1139/cgj-2018-0823

Yang, J., Gu, X.Q.: Shear stiffness of granular material at small strains: does it depend on grain size? Géotechnique **63**(2), 165–179 (2013). https://doi.org/10.1680/geot.11.P.083

Youn, J.U., Choo, Y.W., Kim, D.S.: Measurement of small-strain shear modulus G_{max} of dry and saturated sands by bender element, resonant column, and torsional shear tests. Can. Geotech. J. **45**(10), 1426–1438 (2008). https://doi.org/10.1139/T08-069

Flowability of Saturated Calcareous Sand

Lu Liu[1], Jun Guo[1], Armin W. Stuedlein[2], Xin-Lei Zhang[1], Hong-Mei Gao[1(✉)], Zhi-Hua Wang[1], and Zhi-Fu Shen[1]

[1] Nanjing Tech University, Nanjing, China
hongmei54@163.com

[2] Oregon State University, Corvallis, USA

Abstract. The potential for damaging flow liquefaction failure of calcareous sands subject to seismic, wave, or other dynamic loading is of increasing importance owing to its predominance along coastal areas and increased use as a fill material. Undrained cyclic triaxial tests were performed on saturated calcareous sand specimens prepared with different relative densities and subjected to various effective confining pressures to study the flowability of viscous, liquefied calcareous sand. The cyclic shear stress-strain rate relationship for the saturated calcareous sand specimens transitioned from an elliptical shape to a dumbbell shape as excess pore pressures accumulated under cyclic loading. The dumbbell-shaped relationship demonstrates that the saturated calcareous sand exhibited low shearing resistance and high fluidity under elevated excess pore pressures for certain conditions evaluated. The average flow coefficient, defined as the maximum shear strain rate triggered by the unit average cyclic shear stress, and the flow curve describing the flowability, of the saturated calcareous sand are used to quantify the cyclic failure potential of the calcareous sand. The relative density and cyclic stress ratio has a significant influence on the average flow coefficient: the smaller the relative density and larger the cyclic stress ratio, the smaller the number of cycles to failure. In contrast, the effective confining pressure has little effect on the magnitude of flow potential or number of cycles to triggering flow-like behavior.

Keywords: Saturated calcareous sand · Flowability · Average flow coefficient · Flow curve

1 Introduction

Calcareous sand is widely distributed along coastal areas spanning the globe. In ocean engineering, calcareous sands are commonly used as the foundation material for land reclamations, breakwaters, and as the backfill material for road embankments or airport runways [1]. The potential for damaging flow liquefaction failure of calcareous sands subject to seismic, wave, or other dynamic loading is of increasing importance. It is recognized that very loose to loose, saturated sand may be subject to the loss of shear strength which can cause large flow deformations following cyclic loading. Yasushi et al. [2] showed that such sands developed a considerable deformation under low levels of shear stress and took the form of a very soft solid material or liquid in shaking

© The Author(s), under exclusive license to Springer Nature Switzerland AG 2022
L. Wang et al. (Eds.): PBD-IV 2022, GGEE 52, pp. 2073–2079, 2022.
https://doi.org/10.1007/978-3-031-11898-2_189

table tests. Hwang et al. [3] found that the viscosity of Jumoonjin sand decreased with increasing shear strain rate by sinking ball tests and pulling bar tests. Mele [4] processed a dataset of undrained cyclic triaxial tests on five different sands within the principles of viscous fluids, and the results showed that the relationship between the apparent viscosity and shear rate is consistent with the characteristics of non-Newtonian fluids. Previous research has demonstrated that liquefied silica sand is a shear-thinning non-Newtonian fluid. However, little research work from fluid mechanical viewpoint has been done to investigate the characteristics of calcareous sand. This study presents the results of undrained cyclic triaxial tests performed on saturated calcareous sand specimens prepared with different relative densities and subjected to various effective confining pressures to investigate the flowability of viscous, liquefied calcareous sand.

2 Experimental Program

A series of undrained cyclic triaxial tests were carried out on calcareous sands using the GDS dynamic triaxial apparatus. The specific gravity G_s of the calcareous sand is 2.82. The maximum, e_{max}, and minimum e_{min}, void ratios of calcareous sand is 1.45 and 0.81, respectively. The grain size distribution is presented in Fig. 1. The specimens with 50 mm diameter and 100 mm height are prepared using the water sedimentation method. The 400 kPa back-pressure was used to saturation. The detail testing conditions of undrained cyclic triaxial tests in this paper were shown in Table 1. All specimens were subjected to sinusoidal deviator stresses with a frequency of 0.1 Hz.

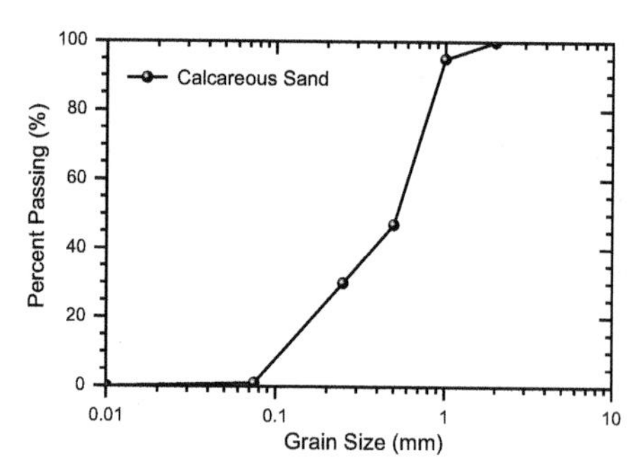

Fig. 1. Grain size distribution curve of the calcareous sand

Table 1. Summary of cyclic triaxial tests

Test No	D_r (%)	σ'_c (kPa)	*CSR*	N_L
CS1	30	100	0.20	25
CS2	30	200	0.15	250
CS3	30	200	0.20	29
CS4	30	200	0.25	8
CS5	30	300	0.20	30
CS6	45	200	0.20	58
CS7	60	200	0.20	151

Note: D_r = Initial relative density; σ'_c= Effective confining pressure; *CSR* = Cyclic stress ratio; N_L = The number of cycles to liquefaction.

3 Flowability Characteristics of Saturated Calcareous Sand

3.1 Characteristics of the Shear Stress-Strain Rate Behavior of Liquefiable Calcareous Sand

Liquefiable soil subjected to cyclic soil can be represented as a non-Newtonian fluid, with shear stress varying as function of the shear strain rate [5]. The shear strain rate, $\dot{\gamma}_i$, at time t_i can be expressed as follows:

$$\dot{\gamma}_i = \frac{1}{2}\left(\frac{\gamma_{i+1} - \gamma_i}{t_{i+1} - t_i} + \frac{\gamma_i - \gamma_{i-1}}{t_i - t_{i-1}}\right) \tag{1}$$

where γ_{i+1}, γ_i and γ_{i-1} are the shear strains at the time t_{i+1}, t_i and t_{i-1}, respectively. Typical shear stress-strain rate curves of the calcareous sand are presented in Fig. 2. When the excess pore pressure ratio, r_u, defined as the ratio of excess pore pressure, u_e, and σ'_c, was lower than 0.8, the shape of shear stress-strain rate curves was elliptical and the responses of shear stress and shear strain rate were basically in phase. As the number of cycles, N, and r_u increased, the shear stress-strain rate curves gradually altered from a symmetrical elliptical shape to an asymmetrical dumbbell shape. The dumbbell-shaped relationship demonstrates that the saturated calcareous sand exhibited transient periods of very low shearing resistance and high fluidity.

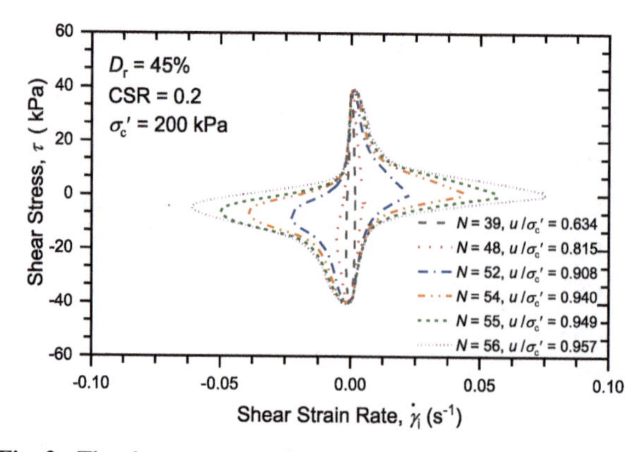

Fig. 2. The shear stress-strain rate curves of the calcareous sand

3.2 Introduction of the Average Flow Coefficient

To describe the flowability characteristics of saturated calcareous sand under cyclic loading and improve the definition of the threshold between relatively stable and unstable magnitudes of cyclic excess pore pressure, a new term designated the average flow coefficient, is proposed in this paper. The average flow coefficient refers to the maximum shear strain rate triggered by unit average shear stress in single cycle of loading of saturated calcareous sand. The average flow coefficient, $\overline{\kappa}$ (s^{-1}/kPa) is defined as follows:

$$\overline{\kappa}=\frac{\left(\dot{\gamma}_{max}-\dot{\gamma}_{min}\right)\bullet\left(\dot{\gamma}_{max}-\dot{\gamma}_{min}\right)}{A} \tag{2}$$

where

$$A=\int\int d\tau d\dot{\gamma} \tag{3}$$

is the area enclosed by the saturated calcareous sand shear stress-strain rate curve under any cycle of loading cycle (kPa•s^{-1}), τ is the average shear stress, $\dot{\gamma}_{max}$ is the maximum shear strain rate (s^{-1}), and $\dot{\gamma}_{min}$ is the minimum shear strain rate (s^{-1}), in a single cyclic loading.

Equation (2) can be transformed by:

$$\overline{\kappa}=\frac{(\tau_{max}-\tau_{min})\left(\dot{\gamma}_{max}-\dot{\gamma}_{min}\right)}{A}\bullet\frac{\left(\dot{\gamma}_{max}-\dot{\gamma}_{min}\right)}{(\tau_{max}-\tau_{min})} \tag{4}$$

where τ_{max} and τ_{min} is the peak and trough shear stress in a single cyclic loading, respectively. Then, considering an outer bounding rectangle enclosing the shear stress-strain rate curve, β_a, can be described as ratio of the bounding rectangle and the area enclosed by the shear stress-strain rate curve in any cycle can be introduced as:

$$\beta_a=\frac{(\tau_{max}-\tau_{min})\left(\dot{\gamma}_{max}-\dot{\gamma}_{min}\right)}{A} \tag{5}$$

If one considers the accepted definition of the apparent viscosity, η, of a fluid:

$$\eta = \frac{\tau_{\max} - \tau_{\min}}{\dot{\gamma}_{\max} - \dot{\gamma}_{\min}} \tag{6}$$

which can be interpreted as the instantaneous diagonal slope of the outer bounding rectangle of the shear stress-strain rate curve for a given cycle, then the average flow coefficient can be concisely expressed as:

$$\overline{\kappa} = \frac{\beta_a}{\eta} \tag{7}$$

The average flow coefficient therefore describes both the flowability of a given soil in a given cycle, but also the apparent viscosity characteristics of the saturated calcareous sand as a fluid. It has a clear physical meaning and is suitable for describing the flowability of calcareous sand under cyclic loadings.

3.3 Flow Curves of Saturated Calcareous Sand

The flow curve, which describes the variation of average flow coefficient, $\overline{\kappa}$, with number of cycles, N, may be used to easily compare the flow potential of soils under various conditions including relative density, effective initial confining pressure, particle shape, and mineralogy. Furthermore, the flow curve can be used to identify the number of cycles, N_f, upon which a threshold in relatively stable cyclic behavior transitions to that of a transiently viscous fluid. The average flow coefficient corresponding to N_f is termed the viscous flow failure coefficient, $\overline{\kappa}_f$.

Figure 3a presents the comparison of the flow curves for the saturated calcareous sand specimens prepared with relative densities of 30, 45 and 60% under $\sigma'_c = 200$ kPa and $CSR = 0.2$. These flow curves initiate at low magnitudes of $\overline{\kappa}$ ranging from 2.2×10^{-5} to 3.1×10^{-5} (s^{-1}/kPa) and are relatively insensitive to D_r. As the cyclic test continues, little variation in $\overline{\kappa}$ is observed up until a cycle, N_f, upon which $\overline{\kappa}$ increases dramatically with each additional cycle, rapidly increasing by up to three orders of magnitude. This dramatic change in the average flow index marks the threshold between relatively stable cyclic behavior and viscous flow behavior up on which strain can accumulate rapidly, and corresponds to $\overline{\kappa}_f$. Figure 3b shows the variation in N_f and r_u corresponding to $\overline{\kappa}_f$, $r_{u,f}$, with relative density; clearly, D_r controls N_f similar to numerous previous observations of cyclic resistance. However, for the same CSR and σ'_c, there is little variation in $r_{u,f}$ with D_r, suggesting that the excess pore pressure corresponding to threshold or viscous flow failure coefficient may be somewhat independent of the number of initial particle contacts.

Figure 3c and 3d present the flow curves and variation of N_f and r_u corresponding to $\overline{\kappa}_f$, respectively, for specimens of $D_r = 30\%$ and subjected to a CSR of 0.2 and under effective confining pressures of 100, 200, and 300 kPa. These flow curves initiate at low magnitudes of $\overline{\kappa}$ ranging from 1.8×10^{-5} to 5.9×10^{-5} (s^{-1}/kPa) and appear to vary linearly with the logarithm of σ'_c. Similar to the flow curves in Fig. 3a, $\overline{\kappa}$ varies little with N until N_f is reached, upon which the average flow coefficient increases dramatically.

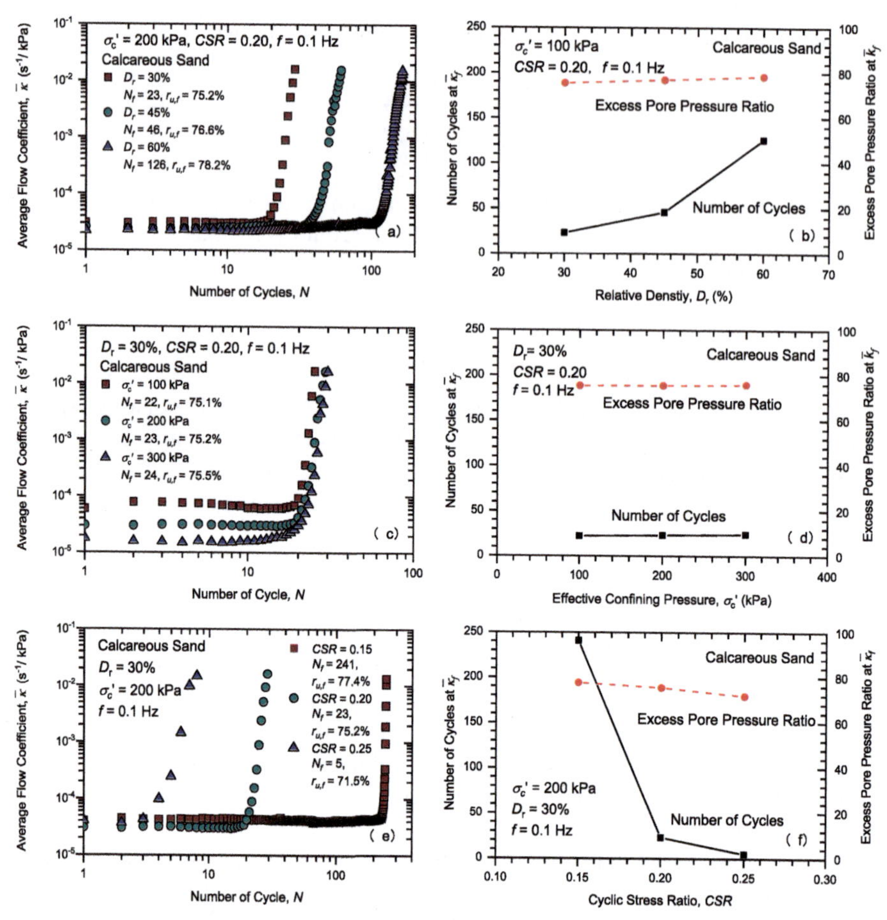

Fig. 3. Flow representation of the cyclic triaxial test results and the: (a) effect of D_r including (b) variation of N_f and $r_{u,f}$ with D_r, (c) effect of σ'_c including (d) variation of N_f and $r_{u,f}$ with σ'_c and (e) effect of *CSR* including (f) variation of N_f and $r_{u,f}$ with *CSR*.

However, these results show that N_f is relatively insensitive to σ'_c, suggesting that the effects of stress dilatancy are somewhat muted when interpreted within the viscous fluid framework.

Figures 3e and 3f present the flow curves and variation of N_f and $r_{u,f}$, respectively, for specimens of $D_r = 30\%$ under effective confining pressures of 200 kPa and subjected to a *CSR*s of 0.15, 0.20 and 0.25. The effect of loading intensity on N_f and $r_{u,f}$ is apparent. Similar to the cyclic stress framework, N_f decreases rapidly with increases in *CSR*. However, $r_{u,f}$ appears likewise sensitive to *CSR*, with viscous flow failure initiating with smaller excess pore pressures as the loading intensity increases.

4 Conclusion

In this paper, the flowability characteristics of saturated calcareous sand under cyclic loading is studied using cyclic triaxial tests. The main conclusions are summarized as follows:

(1) The cyclic shear stress-strain rate relationship for the saturated calcareous sand specimens transitioned from an elliptical shape to a dumbbell shape as excess pore pressures accumulated under cyclic loading. The dumbbell shaped relationship demonstrates that the saturated calcareous sand exhibited transient, low shearing resistance and high fluidity.

(2) The average flow coefficient, defined as the maximum shear strain rate triggered by the unit average cyclic shear stress, is suitable for describing the flowability characteristics of saturated calcareous sand.

(3) The flow curve is used to describe the viscous flow failure of saturated calcareous sand, defined as the number of cycles upon which the average flow coefficient increases dramatically corresponding to rapid accumulation of cyclic strain. Threshold average flow index is reached prior to the development of large strains and excess pore pressures, and therefore can be used as a more stringent measure of cyclic failure. The number of cycles to failure under the viscous flow failure criterion increases with relative density, reduces with cyclic shear stress amplitude, but appears insensitive to effective confining pressure.

Acknowledgements. We would like to express our thanks to the financial support for this study from the Project of the National Natural Science Foundation of China (Grant No. 52008207, 51908284 and 51678300), and Middle-aged & Young Science Leaders of Qinglan Project of Universities in Jiangsu Province, China (Grant No. QL20200203 and QL20210210).

References

1. Wang, X.Z., Jiao, Y.Y., Wang, R., Hu, M.J., Meng, Q.S., Tan, F.Y.: Engineering characteristics of the calcareous sand in Nansha Islands South China Sea. Eng. Geo. **120**, 40–47 (2011)
2. Yasushi, S., et al.: Mechanism of permanent displacement of ground caused by seismic liquefaction. Soils Found. **32**(3), 79–96 (1992)
3. Hwang, J.I., Kim, C.Y., Chung, C.K., Kim, M.: Viscous fluid characteristics of liquefied soils and behavior of piles subjected to flow of liquefied soils. Soil Dyn. Earthq. Eng. **26**, 313–323 (2006)
4. Mele L.: An experimental study on the apparent viscosity of sandy soils: From liquefaction triggering to pseudo-plastic behavior of liquefied sands. Acta Geotechnica 1–19 (2021)
5. Wang, Z.H., Ma, J.L., Gao, H.M., Stuedlein, A.W., He, J., Wang, B.H.: Unified thixotropic fluid model for soil liquefaction. Géotechnique **70**(10), 1–45 (2019)

Liquefaction Characteristics of Saturated Coral Sand Under Anisotropic Consolidations

Weijia Ma[1] and Guoxing Chen[2,3](✉)

[1] School of Mechanical Engineering, Nanjing University of Science and Technology, Nanjing, China
[2] Institute of Geotechnical Engineering, Nanjing Technology University, Nanjing, China
[3] Civil Engineering and Earthquake Disaster Prevention Center of Jiangsu Province, Nanjing, China
gxc6307@163.com

Abstract. A series of undrained cyclic triaxial tests were conducted on saturated coral sand by using GDS dynamic traixial apparatus. This research mianly investigates the characteristics of excess pore water pressure (u_e), axial strain (ε_a), stress-strain relationship and liquefaction resistance of coral sand under different consolidation ratios (*R*) and cyclic stress ratios (CSR). The test results indicate that for the specimens under anisotropic consolidations, the development of pore water pressure always show the trend from fast to slow, and eventually tends to be stable. Two development modes of axial strain can be found in saturated coral sand: cyclic mobility and plastic cumulative deformation. The stress-strain relationships of saturated coral sand under isotropic and anisotropic consolidations are quite different, and the difference could be associated with stress reversal. The relation of static deviatoric stress (q_s) and cyclic deviatoric stress (q_d) plays an important role in liquefaction characteristics. Compared with the Fujian sand, the saturated coral sand has higher liquefaction resistance due to the special physical properties. The liquefaction resistance of saturated coral sand increased with the increasing of consolidation ratio.

Keywords: Coral sand · Consolidation ratios · Excess pore water pressure · Axial strain · Liquefaction resistance

1 Introduction

Coral sand is a special soil of marine biogenesis, and its calcium carbonate ($CaCO_3$) content is more than 90%. Due to the special formation process of coral sand, the particle has the properties of high friction angle, low strength, crushable, irregular shape and inner pores (Coop 1990). Compared with terrigenous sand, coral sand has significantly different liquefaction characteristics. The complex marine environment of Nansha Islands will have a significant impact on the consolidation condition of coral sand. Moreover, the Nansha Island reefs have been suffered in seismic risk, which could have great impact on coastal structures. In order to ensure the safety of Nansha Island Reef structures, it is

© The Author(s), under exclusive license to Springer Nature Switzerland AG 2022
L. Wang et al. (Eds.): PBD-IV 2022, GGEE 52, pp. 2080–2087, 2022.
https://doi.org/10.1007/978-3-031-11898-2_190

necessary to make an experimental study on the liquefaction characteristics of Nansha coral sand.

Although the effect of consolidation conditions, usually referred to as static shear stress, on liquefaction resistance of sandy soils has been studied for several decades, the undrained anisotropy liquefaction resistance of sandy soils involving in the various cyclic laboratory tests is always a concern (e.g., Oda et al. 1985; Yang and Sze 2011; Georgiannou and Konstadinou 2014; Sivathayalan et al. 2015; Chen et al. 2020), and significantly different correction factors relating the cyclic resistance of anisotropically consolidated saturated sandy soils have been suggested to consider the effect of static shear stress levels (e.g., Seed and Harder 1990; Harder and Boulanger 1997; Boulanger 2003; Wei and Yang 2019). Furthermore, Sharma and Ismail (2006) found that for Goodwyn and Ledge Point coral sand, the static shear stress could affect the axial strain development modes. Rasouil et al. (2020) carried out cyclic undrained bidirectional simple shear tests, mainly investigated the influence of initial static shear ratios (α) on the liquefaction resistance of Hormuz Island calcareous sand. They claimed that the increase of α has the opposite effect on the liquefaction resistance of dense and loose Hormuz coral sand. However, only a little study is focus on the liquefaction characteristics of Nansha coral sand under anisotropic consolidation.

This study is to explore the liquefaction characteristics of Nansha coral sand under various anisotropic consolidation conditions. The experimental data presented in this paper not only provide a basis for developing a better and generalized understanding of the undrained anisotropic behavior of coral sand under various consolidation conditions, but also serve as a worthful reference for costal engineering construction.

2 Testing Programme

2.1 Material

The tested coral sand was taken from the Nansha Islands. The particle distribution of coral sand is shown in Fig. 1, and the physical properties are illustrated in Table 1. The mineral composition of coral sand is aragonite, calcite, and high magnesium calcite.

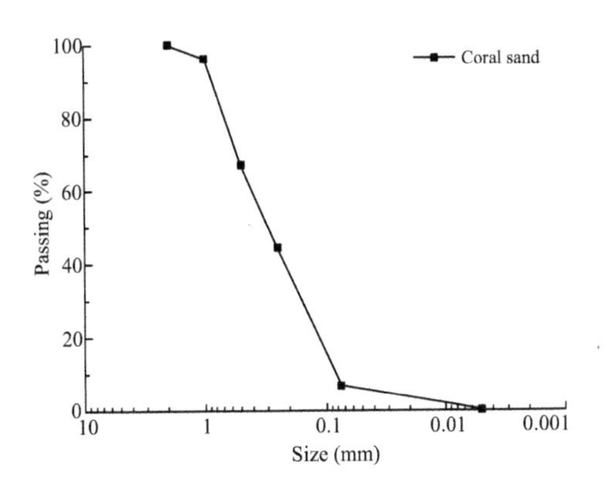

Fig. 1. Particle distribution of coral sand

Figure 2 shows the scanning electron microscope picture of coral sand. It can be seen in this picture that the coral sand particle are mostly irregular shapes.

Fig. 2. Scanning electron microscope picture of coral sand

2.2 Testing Apparatus

The advanced GDS dynamic triaxial apparatus was used to carry out the tests (Fig. 3). This apparatus can accurately control the application of axial load, and the accuracy of the results is guaranteed.

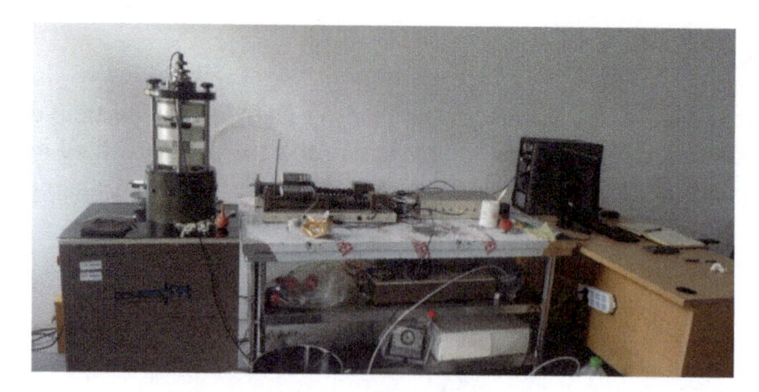

Fig. 3. GDS dynamic triaxial apparatus

2.3 Specimen Preparation

The height and diameter of the cylinder specimen is 100 mm and 50 mm, respectively. The dry deposition method is adopted to prepare the specimens. The specimens were prepared by pouring the oven-dried coral sand into the mold coverd by rubber membrane in five layers. A funnel was used for ensuring the maintenance of zero falling head, and a compaction mold was applied to ensure that each layer could achieve the target relative density (D_r). The initial relative density for all specimens in this study is 45%.

The process of saturation and consolidation for all specimens followed the ASTM D5311-92 standard. The mature saturation method of flushing CO_2 and de-aired water into the specimen, applying back pressure to 400 kPa was used. The specimens were considered to be saturated as the Skempton's *B*-value reached up to 0.97 or higher. After saturation, the specimens were isotropically or anisotropically consolidated to an effective confining pressure (p'_0) of 100 kPa with different consolidation ratios (R = 1.0, 1.5, 2.0). The cyclic axial loading was applied on the consolidated specimens with different cyclic stress ratio (CSR).

Stress reversal is an important problem in anisotropic consolidation, which has a great influence on the liquefaction characteristics. When the static deviator stress q_s is smaller than the amplitude of cyclic deviator stress q_d, the stress reversal will occur. However, when q_s is larger than q_d, the stress reversal will not occur. Figure 4 shows the schematic diagram of loads on triaxial specimen. The angle of the shear band and the shear stress and normal stress applied on the shear band can be obtained (Zhang 1984). This is benefit to analysis the stress reversal on the shear band. The stress state on the shear band is similar in direct simple shear test. Table 1 shows the specific test scheme for this test as well as the stress reversal condition for each specimen.

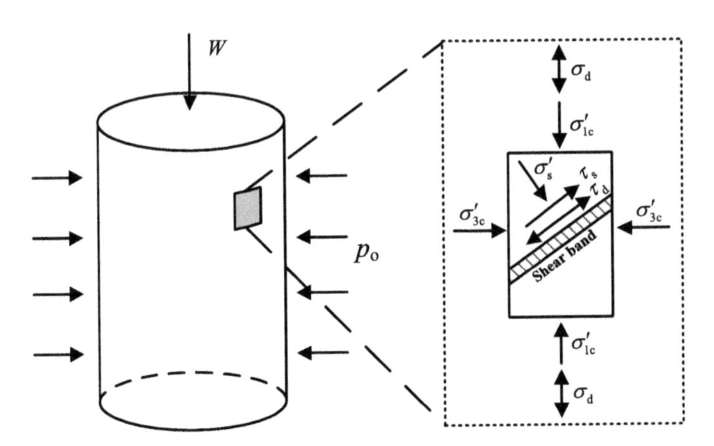

Fig. 4. The schematic diagram of loads on triaxial specimen

Table 1. Test scheme

Test no.	Consolidation condition			Cyclic loading	Stress reversal
	D_r (%)	p_0'(kPa)	R	CSR	
1	45	100	1.0	0.15	Yes
2			1.0	0.20	Yes
3			1.0	0.25	Yes
4			1.5	0.20	No
5			1.5	0.25	Yes
6			1.5	0.30	Yes
7			2.0	0.35	No
8			2.0	0.40	No
9			2.0	0.50	Yes

3 Result Analysis

3.1 Excess Pore Water Pressure

Figure 4 illustrates the developments of excess pore water pressure (u_e) for coral sand under isotropic and anisotropic consolidations. For isotropic consolidation, u_e can eventually reaches p_0', which means the complete loss of effective stress (see Fig. 5a). However, the fluctuation of u_e is very large, and with the increase of cyclic number, this fluctuation will become larger and larger. For anisotropic consolidation, the development of u_e has a rapid rising process and then becomes stable, but u_e cannot reaches p_0'. This indicates that the structure of coral sand specimens is not completely destroyed under anisotropic consolidation (see Fig. 5b, 5c).

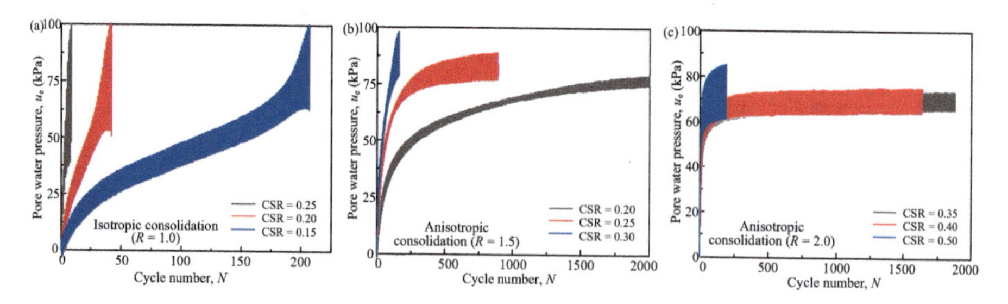

Fig. 5. The developments of excess pore water pressure for coral sand

3.2 Axial Strain

The developments of axial strain (ε_a) for various consolidation conditions are quite different in Fig. 6. During anisotropic consolidation, the static deviatoric stress (q_s)

would be produced, and the coupling of static deviatoric stress (q_S) and cyclic deviatoric stress (q_d) will deeply affect the liquefaction characteristics, especially in the feature of ε_a. Specifically, when $q_s > q_d$, the axial strain shows cumulative plastic strain, and the development rate of ε_a become slowly with the increasing of cycle number (N). When $q_s = q_d$, the development rate of ε_a seems more stable along with N. When $q_s < q_d$, the development rate of ε_a shows the mode of plastic cumulative deformation from slow to fast. When $q_s = 0$, the specimen is under isotropic consolidation, the development mode of cyclic mobility can be observed.

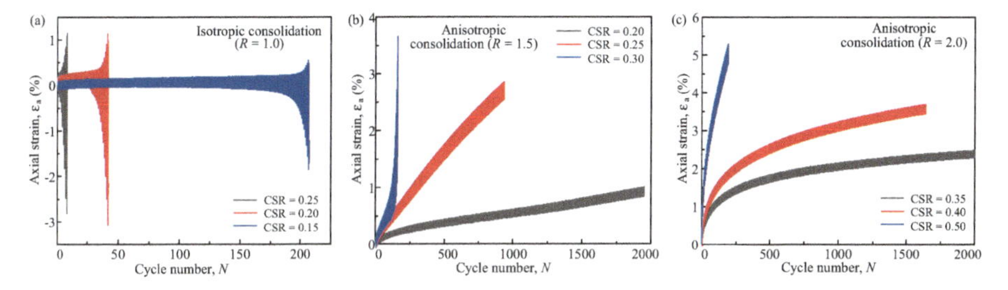

Fig. 6. The developments of axial strain for coral sand

3.3 Stress - Strain Relationship

The consolidation condition can also influence the stress - strain relationship. Figure 7 exhibits the relation of deviatoric stress and axial strain. For isotropic consolidation, with the increasing of ε_a, the area of hysteresis loop is becoming larger, and the nonlinearity of coral sand is becoming stronger. However, for anisotropic consolidation, with the increasing of ε_a, the development of hysteresis loop is not obvious.

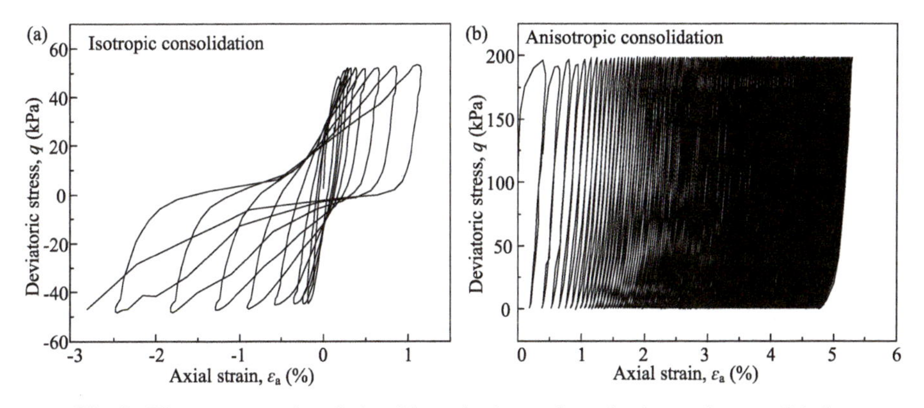

Fig. 7. The stress - strain relationship under isotropic and anisotropic consolidation

3.4 Liquefaction Resistance

The liquefaction resistance of saturated coral sand is also an important part of liquefaction characteristics. The liquefaction assessment is usually considered from two aspects: effective stress and deformation. In this study, the failure criteria for coral sand are: (I) excess pore water pressure reaching effective confining pressure; (II) the axial strain reaching 2.5%. Based on the above analysis, Fig. 8 shows the liquefaction resistance curves for various consolidation ratios. It is clear that the liquefaction resistance curves become higher with the increase of *R*. It should be pointed out that the axial strain of test No. 4 and No. 7 cannot reach 2.5% after 2000 cycles, which means they did not liquefy. A conclusion can be drawn that the smaller the CSR, the greater the consolidation ratio, and the less likely the coral sand is to liquefy, which is probably due to the stress reversal. The liquefaction resistance curve of saturated Fujian sand is also plotted in this figure for comparison. Obviously, the saturated coral sand has higher liquefaction resistance due to the special physical properties.

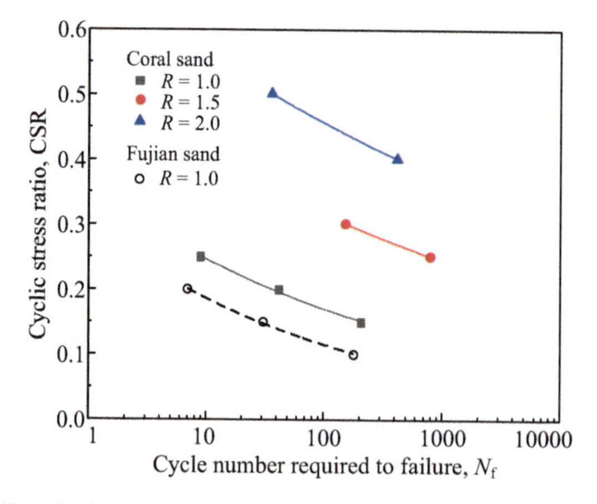

Fig. 8. The liquefaction resistance of coral sand under various consolidation ratios

4 Conclusion

1. The excess pore water pressure increases with cyclic loading. For isotropic consolidation, u_e can reach p_0'; for anisotropic consolidation, u_e cannot reach p_0', and the development of u_e will be stable at last.
2. The development modes of axial strain under various consolidation conditions are different. The cyclic mobility and cumulative plastic strain can be observed for coral sand.
3. With the increase of consolidation ratio (*R*), the liquefaction resistance of coral sand becomes larger. Specimen applied on small CSR and large *R* is hard to liquefy.

Acknowledgement. The financial support provided by the National Natural Science Foundation of China (51678299) is gratefully acknowledged.

References

Coop, M.R.: The mechanics of uncemented carbonate sands. Géotechnique **40**(4), 607–626 (1990). https://doi.org/10.1680/geot.1990.40.4.607

Oda, M., Nasser, S.N., Konishi, J.: Stress-induced anisotropy in granular masses. Soils Found. **25**(3), 85–97 (1985)

Yang, J., Sze, H.Y.: Cyclic behaviour and resistance of saturated sand under nonsymmetrical loading conditions. Géotechnique **61**(1), 59–73 (2011). https://doi.org/10.1680/geot.9.P.019

Georgiannou, V.N., Konstadinou, M.: Effects of density on cyclic behaviour of anisotropically consolidated Ottawa sand under undrained torsional loading. Géotechnique **64**(4), 287–302 (2014). https://doi.org/10.1680/geot.13.P.090

Sivathayalan, S., Logeswaran, P., Manmatharajan, V.: Cyclic resistance of a loose sand subjected to rotation of principal stresses. J. Geotechn. Geoenviron. Eng. **141**(3), 04014113 (2015). https://doi.org/10.1061/(ASCE)GT.1943-5606.0001250

Chen, G.X., Wu, Q., Zhao, K., Shen, Z.F., Yang, J.: A binary packing material-based procedure for evaluating soil liquefaction triggering during earthquakes. J. Geotechn. Geoenviron. Eng. ASCE **146**(6), 04020040 (2020). https://doi.org/10.1061/(ASCE)GT.1943-5606.0002263

Seed, H.B., Harder, L.F.: SPT-based analysis of cyclic pore pressure generation and undrained residual strength. In: Proceedings of H. Bolton Seed Memorial Symposium, pp. 351–376. BiTech Publishers, Ltd., Vancouver, British Columbia, Berkeley, CA (1990)

Harder, L.F., Boulanger, R.W.: Application of Kσ and Kα correction factors. In: Proceedings of the NCEER Workshop on Evaluation of Liquefaction Resistance of Soils, National Center for Earthquake Engineering Research (NCEER), pp. 21819195–2181997. State University of New York at Buffalo (1997)

Boulanger, R.W.: Relating Kα to relative state parameter index. J. Geotechn. Geoenviron. Eng. ASCE **129**(8), 770–773 (2003)

Wei, X., Yang, J.: Cyclic behavior and liquefaction resistance of silty sands with presence of initial static shear stress. Soil Dyn. Earthquake Eng. **122**, 274–289 (2019). https://doi.org/10.1016/j.soildyn.2018.11.029

Sharma, S.S., Ismail, M.A.: Monotonic and cyclic behavior of two calcareous soils of different origins. J. Geotechn. Geoenviron. Eng. **132**(12), 1581–1591 (2006). https://doi.org/10.1061/(ASCE)1090-0241(2006)132:12(1581)

Rasouli, M.R., Moradi, M., Ghalandarzadeh, A.: Effects of initial static shear stress orientation on cyclic behavior of calcareous sand. Marine Georesources Geotechnol. (2020). https://doi.org/10.1080/1064119X.2020.1726535

Zhang, K.X.: Stress condition inducing liquefaction of saturated sand. [In Chinese.] Earthquake Eng. Eng. Vibration **4**(1), 99–109 (1984)

Microscopic Test Analysis of Liquefaction Flow of Saturated Loess

Xingyu Ma[1,2], Lanmin Wang[2,3](✉), Qian Wang[2,3], Ping Wang[2,3], Xiumei Zhong[2,3], Xiaowu Pu[2,3], and Fuqiang Liu[4]

[1] China Earthquake Administration, Institute of Engineering Mechanics, Harbin 150080, China
[2] China Earthquake Administration, Key Laboratory of Loess Earthquake Engineering, Lanzhou 730000, China
wanglm@gsdzj.gov.cn
[3] China Earthquake Administration, Lanzhou Institute of Seismology, Lanzhou 730000, China
[4] Department of Civil Engineering and Mechanics, Lanzhou University, Lanzhou 730000, China

Abstract. On the basis of existing research, triaxial shear tests were carried out on the remolded samples after liquefaction. The research shows that the apparent viscosity of the remolded sample has been increasing, indicating that the flow characteristics have been "shear thickening". The samples in different test stages were tested by scanning electron microscope, and the microstructure parameters of the two samples in different test stages were obtained. The analysis shows that the difference of particle and pore size between undisturbed loess and remolded loess leads to the rising mode of pore water pressure in the process of liquefaction test, and then affects the change of apparent viscosity in the process of shear.

Keywords: Saturated loess · Liquefaction · Apparent viscosity · Microstructure

1 Introduction

Large soil liquefaction-sliding disaster is one of the important secondary disasters induced by earthquake [1–3]. In terms of soil liquefaction deformation, relevant personnel have done a lot of research work. Chen Yumin et al. [4] analyzed the mechanical properties and deformation characteristics of saturated sand after liquefaction by fluid mechanics method. Zhuang Haiyang et al. [5, 6] and Takashi Kiyota et al. [7]. Used cyclic torsional shear test to study the changes of several physical parameters of saturated soil under large liquefaction deformation.

The change of soil microstructure has a great influence on its macroscopic mechanical behavior. Jiang Mingjing et al. [8] analyzed the microstructure evolution process of saturated undisturbed loess along two stress paths in the triaxial test of Jingyang South Tableland in China. Sun Hong et al. [9] studied the microstructure evolution of Shanghai soft clay in triaxial tests along different stress paths. Wen Baoping et al. [10] analyzed the microstructure changes of unsaturated loess in Lanzhou, China under shear. Xu Panpan et al. [11] studied the changes of microstructure parameters in the infiltration process of loess samples with the optimum moisture content and different dry densities.

© The Author(s), under exclusive license to Springer Nature Switzerland AG 2022
L. Wang et al. (Eds.): PBD-IV 2022, GGEE 52, pp. 2088–2097, 2022.
https://doi.org/10.1007/978-3-031-11898-2_191

In summary, although studies on large deformation of soil liquefaction have been carried out at home and abroad, such studies are mainly carried out for sand and silt, and less for loess. Based on previous research work, In this paper, the undisturbed site of the trailing edge of the liquefaction slip zone of the Shibei tableland in Ningxia during the 1920 Haiyuan earthquake was selected for sampling, and the dynamic triaxial liquefaction test of saturated undisturbed loess and triaxial shear test of soil after liquefaction were carried out in the laboratory. The microstructure characteristics of the samples under different test status were analyzed by scanning electron microscopy. Study on liquefaction process of saturated loess and microstructure change in undrained shear test after liquefaction. The research results can provide a reference basis for the understanding of liquefaction slip disaster mechanism and disaster prevention in loess area.

2 Liquefaction Deformation Test

2.1 Sample Preparation

The sampling site is located in the non-sliding loess strata at the rear edge in the south of Shibei tableland, Ningxia, China. The physical properties of the sample are shown in Table 1.

Table 1. Physical properties of loess samples

Sampling horizon	Initia void ratio	Natural water content /%	Plastic index	Granulometric composition%		
				Sand particle	Silt particle	Clay particle
Shibei tableland	7.0	9.1	7.6	53.8	33.8	12.4

The undisturbed cube loess blocks were uniformly cut into cylindrical undisturbed samples with a diameter of 50 mm × a height of 100 mm. The preparation steps of remolded loess samples are as follows. Firstly, after the undisturbed loess samples were fully dried and ground, the scattered loess with water content of 9.1% was prepared. After curing for 24 h, the loose was put into a cylindrical mold with a diameter of 50 mm, and the sample was prepared by static pressure at both ends. The height of the remolded sample is 100 mm. By controlling the mass of soil during sample preparation, the density of remolded sample is equal to the undisturbed sample.

2.2 Test Equipment and Test Method

The laboratorial dynamic and static triaxial test are conducted on the WF-12440 dynamic triaxial-torsion shear test system in the Key Laboratory of Loess Seismic Engineering, China Earthquake Administration. The saturability of the samples is determined by detecting the pore water pressure coefficient b-value. The test adopts the dynamic load with a frequency of 1 Hz. The liquefaction failure criteria of the sample is set as axial dynamic strain ≥ 3% and pore water pressure ratio $U_d/\sigma_0^{'}\geq0.2$ [12].Triaxial shear test is carried out immediately after the sample reaches liquefaction. The confining pressure is 200 kPa and the shear rate is 0.4 mm/min [13] .

2.3 Analysis of Liquefaction Deformation Test Results

The saturated loess has a certain fluidity after liquefaction, the liquefied loess can be assumed as fluid for research [14]

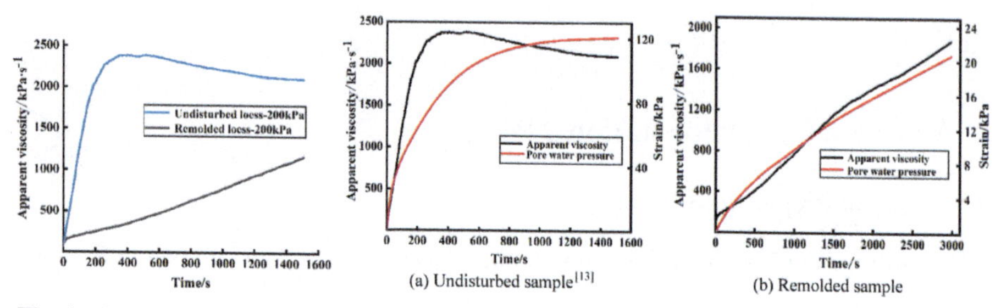

(a) Undisturbed sample[13]

(b) Remolded sample

Fig. 1. Curves of Apparent viscosity versus time

Fig. 2. Curve of apparent viscosity – pore water pressure – time relationship

In order to comprehensively analyze the influence of the change of pore water pressure on the apparent viscosity in the shear process, the curves of the apparent viscosity and pore water pressure with shear time are plotted, as shown in Fig. 2. As can be seen from Fig. 2 (a) (The data source of undisturbed loess pattern is reference 13), the pore water pressure of the undisturbed sample increases rapidly in the process of shear. When the pore water pressure reaches 80 kPa, the apparent viscosity reaches the peak, the change of apparent viscosity is stable when the pore pressure is stable. As can be seen from Fig. 2 (b), the pore water pressure of the remolded sample increases slowly in the process of shearing. At the end of the test, the pore water pressure is only up to 22 kPa. During this period, its apparent viscosity has been increasing. Figure 2 shows that the change of apparent viscosity of loess samples is greatly related to the change of pore water pressure in the shearing process. In order to study the relationship between them, SEM scanning electron microscope experiments were carried out on the samples at different test stages.

3 Microstructure Test

3.1 Test Instrument and Sample Preparation

The scanning electron microscope test was performed by KYKY-2800B scanning electron microscope in the Key Laboratory of Loess Seismic Engineering, China Earthquake Administration (Fig. 3). The working principle is that in vacuum and voltage environment, the electron beam emitted by the electron gun is irradiated on the sample, and the secondary electrons are excited to form an image reflecting the surface morphology of the sample.

Natural air drying of samples under different test status. At least 3 samples were prepared at each sampling point. The fresh section was broken at the maximum deformation of the sample, and the back was milled and flattened to prepare a square sheet. The square sheet was put into the ion sputtering equipment to spray gold so that the surface can reflect secondary electron imaging (Fig. 4). The shooting times include 100, 200, 400, 500 and 800 times. Each multiple takes 4–6 shots. In order to obtain enough information and make the photo processing results more accurate, the magnification is selected as 400 times. 4 to 6 representative images are selected from pictures with a magnification of 400 t for follow-up analysis.

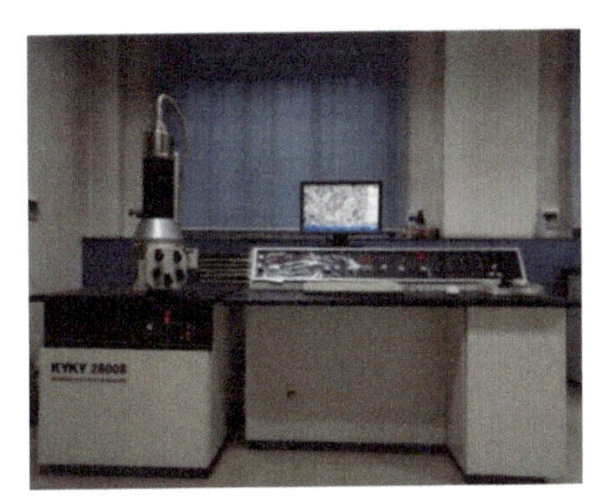

Fig. 3. KYKY-2800B scanning electron microscope

Fig. 4. Ion sputtering equipment

3.2 Microstructure Analyses

The PCAS system developed by Liu Chun of Nanjing University is used to process the pictures obtained from the experiment [15]. The PCAS system can accurately extract statistical calculation parameters such as apparent void ratio, fractal dimension of porosity, probability entropy of pore, average shape factor and fractal dimension from SEM images.

Apparent void ratio refers to the ratio of pore area to particle area in SEM images, which can indirectly reflect the change of three-dimensional void ratio. Figure 5 is the statistical results of apparent void ratio of undisturbed and remolded samples under different test status. It can be seen from Fig. 5 that the apparent porosity of the undisturbed sample and the remolded sample increased before and after the liquefaction test. This is because in the liquefaction process, the pore water pressure of the sample increases due to the external load, which makes the original large, medium and small pores in the loess continuously fuse into large pores, and improves the pore connectivity in the loess. After the shear test, the apparent porosity of undisturbed and remolded samples decreased. This is because in the shear process, the loess particles squeeze pores under external load, so that the large pores formed by liquefaction are reduced to medium and small pores to a certain extent, which reduces the pore connectivity in the soil and reduces the apparent pore area of the sample.

Statistical analysis of pore shape characteristics by average shape factor. The larger the average shape factor is, the smoother the pore shape is and the closer the spatial arrangement is. Based on fractal theory, the fractal dimension of porosity is a quantitative evaluation standard to describe the pore size distribution of soil. It represents the degree of homogenization of pores. The larger the fractal dimension of porosity is, the greater difference of pore size is and the worse the uniformity is.

It can be seen from Figs. 6 and 7 that before and after liquefaction, the average shape factor of the undisturbed sample decreased by 0.0382, indicating that the pore became narrow and long, and the fractal dimension of pore porosity increased by 0.745, showing that the pore size gap was significantly increased. The average shape factor of the remolded sample increases by 0.0086, the porosity tends to be smooth. The fractal dimension of porosity increases slightly, only 0.1346, showing that the pore size changes little. This is because the remolded loess sample is artificially made from the undisturbed sample after drying, crushing and screening. Compared with the undisturbed sample, the remolded sample lost the natural structure of the soil, and the particles and pores were relatively uniform (Fig. 8(a) and (c)). The remolded sample is hard to form long and narrow pores in the liquefaction process, and it is easy to form more smooth pores (Fig. 8(d)). The undisturbed sample in the liquefaction process due to the particle size compared with the remolded sample in terms of a large difference, making it in the liquefaction process formed more long and narrow pores (Fig. 8(b)).

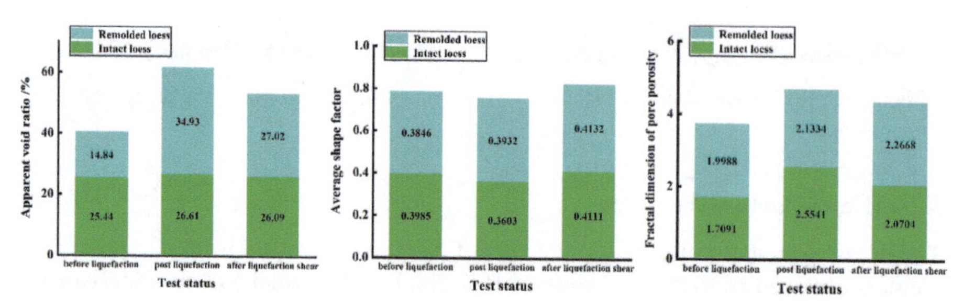

Fig. 5. Apparent void ratio under different test status

Fig. 6. Average shape factor under different test status

Fig. 7. Fractal dimension of pore porosity under different test status

Long and narrow pores are more conducive to the increase of pore water pressure than smooth pores, which makes the rising rate and final pore pressure of undisturbed samples are higher than those of remolded samples in the process of liquefaction test. Because of the difference of pore morphology, in the shear test after liquefaction, the pores of both are in the process of continuous smoothness, but the difference between the two is that the pore change of the undisturbed sample is from long and narrow voids to smooth pores, while the pores of the remolded samples are smaller round pores merged into larger round pores. It is shown that the fractal dimension of porosity of the undisturbed sample decreases and the uniformity of pores becomes better. The fractal dimension of porosity of the remolded samples increases and the uniformity of pores becomes worse. And because of the difference between the two pore morphology changes, in the process of shear test, the pore shape adjustment rate of the undisturbed sample is higher than that of the remolded sample, and the rising rate of pore water pressure and the maximum pore water pressure of the remolded sample are higher than that of the remolded sample. Therefore, in this process, compared with the response of the remolded specimen to the external load, the original sample adjusts the internal structure more quickly.

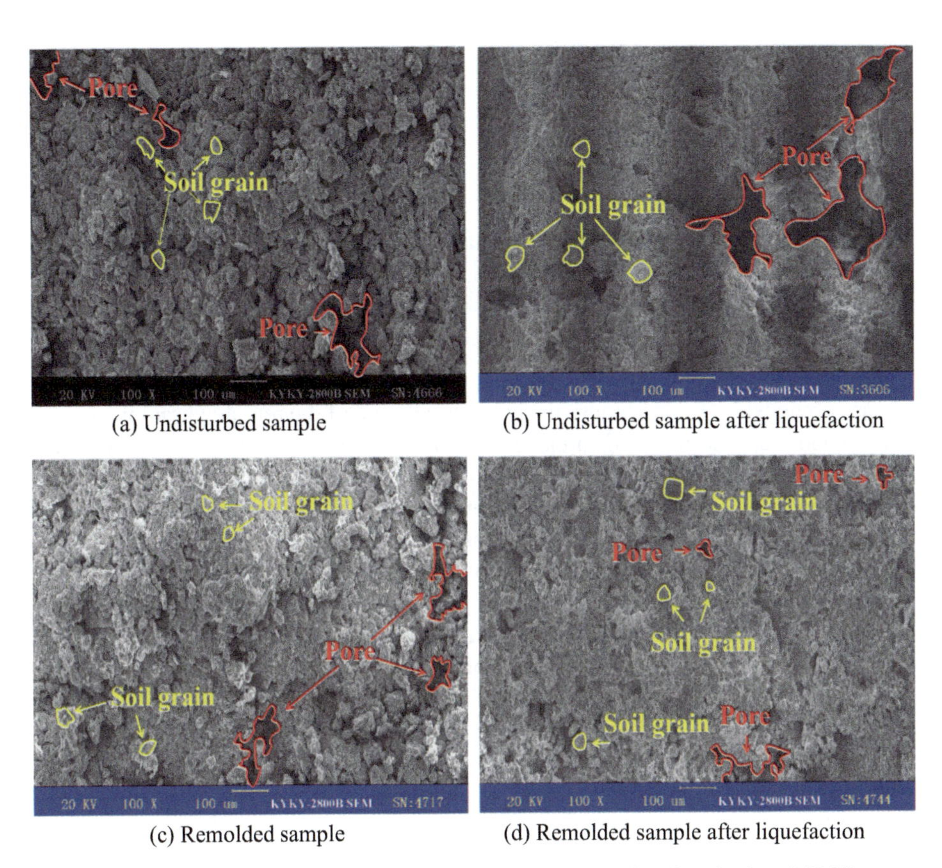

(a) Undisturbed sample (b) Undisturbed sample after liquefaction

(c) Remolded sample (d) Remolded sample after liquefaction

Fig. 8. Scanning electron microscope photo before and after liquefaction (100 X)

Probability entropy can describe the overall arrangement of loess pores. The smaller the probability entropy is, the more regular the pore arrangement is, the stronger the order and orientation are, and the more stable the structure is. The Complexity of pore structure reflected by pore fractal dimension. The larger the fractal dimension of pores, the more complex the pore structure is, the farther the spatial morphology of pores deviates from the smooth surface.

Figure 9 and Fig. 10 show the statistical results of pore probability entropy and fractal dimension of samples under different states. It can be seen from Fig. 9 and Fig. 10 that the probability entropy and fractal dimension of undisturbed samples increase obviously before and after liquefaction. This is because the original orderly arrangement of pores is disrupted in the liquefaction process of loess. Due to the rapid development of the failure process, the pores do not have sufficient time to rearrange the rules, making

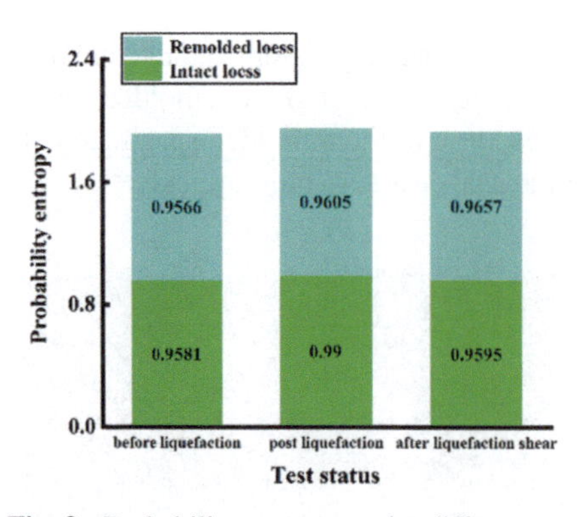

Fig. 9. Probability entropy under different test status

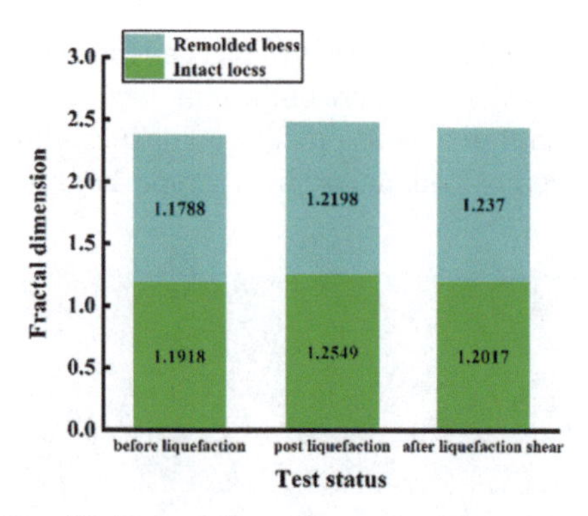

Fig. 10. Fractal dimension under different test status

their order worse and the pore structure more complex. Compared with the undisturbed sample, the probability entropy and fractal dimension of the remolded sample do not increase obviously. This is because the pore distribution of the remolded sample is more uniform and mainly round, which makes the pore arrangement phase more orderly and the structure relatively simple after liquefaction.

After the shear test, the probability entropy and fractal dimension of the undisturbed sample decrease significantly, the pore arrangement is orderly, and the structure tends to be simple. This is because the loading rate of the static triaxial shear specimen is slower than that of the dynamic triaxial liquefaction test, which makes the soil have time to arrange pores. According to the average shape coefficient and the fractal dimension of porosity, the pores of the undisturbed samples tend to be smooth in the shear process, and the difference of pore size decreases. On the other hand, the probability entropy and fractal dimension of the remolded samples increase to a certain extent, the pore arrangement is disordered, and the pore structure tends to be complex. From the changes of the average shape coefficient and the fractal dimension of porosity, it can be seen that the gap of pore size in the smooth process of the remolded sample is further enlarged in the shearing process. The reason is that the rise of pore water pressure needs a channel, but the remolded sample is not easy to form a long and narrow channel, so the size of the circular pore is increasing. Therefore, the probability entropy further increases, the order of pore arrangement becomes worse, and the pore structure tends to be complex.

In summary, the change of pore shape affects the rising rate of pore pressure in the process of liquefaction test. Due to the great difference of particle size and pore size, the undisturbed sample is easy to form long and narrow pores and penetrating pores in the process of liquefaction, which is beneficial to the increase of pore water pressure. However, because the small difference in particle size, the remolded sample is not easy to form narrow and through pores, which leads to the obstruction of the rising channel of pore water pressure and affects the increase of pore pressure. In the process of static triaxial test after liquefaction, the pore changes of the two samples tend to be smooth, but due to the difference of their approaching starting points, the original samples change from long and narrow pores to smooth, while the remolded samples are continuous smooth. The order and complexity of pore arrangement affect the change of apparent viscosity during shear. The more regular the arrangement of pores, the simpler the structure and the smaller the apparent viscosity.

4 Relationship Between Microstructure and Macroscopic Mechanical Behavior

The change of microstructure of soil directly affects its macroscopic mechanical behavior. Because the structural particles and pores of undisturbed loess are formed in the process of deposition of undisturbed loess, and the structure of undisturbed loess is strong, the loess sample has strong resistance at the initial stage of external load. In the liquefaction test, the difference of particle and pore size makes it easy to form narrow, long and penetrating pores. These channels are beneficial to the increase of pore water pressure. In the macroscopic mechanical behavior, the growth rate of pore pressure accelerated, and the dynamic stress and dynamic strain began to change obviously. In the process of sample preparation, the remolded sample loses the structure of the original soil, and the difference between the size of particles and pores is small. Due to the loss of structure, the remolded specimen can't resist the external cyclic load effectively. Because of the good uniformity of particle and pore size, the narrow and long and through pores can't be formed effectively in the process of liquefaction. In the macroscopic mechanical behavior, the pore water pressure increases slowly and can't reach the pore pressure of the undisturbed sample.

In the process of undrained shear after liquefaction, from the point of view of microstructure, the pores of both samples show the continuous smoothness of pore shape. However, due to the different starting points of the two changes, the undisturbed sample changes from long and narrow pores to smooth pores, while the remolded samples continue to be smooth. This makes the macroscopic mechanical behavior show that the growth rate of pore water pressure of the undisturbed sample decreases continuously during the shearing process, and the pore pressure growth rate of the remolded sample is also decreasing, but the amplitude is smaller than that of the undisturbed sample. The final pore pressure value is also quite different. The change rate of pore pressure is different, so there is a great difference in the change of microstructure between them. Due to the rapid growth rate of pore pressure of the undisturbed sample, the arrangement of loess particles is faster in the microstructure, so that the pores can be arranged quickly and orderly, and the pore size can achieve the effect of uniformity. In terms of macroscopic mechanical properties, the apparent viscosity increases at first and then decreases, that is, its flow characteristics change from shear thickening to shear thinning. On the other hand, the continuous smooth pores of the remolded samples are not conducive to the increase of pore pressure, which leads to the slow distribution rate of particles and the deterioration of uniformity and sequence. In terms of macroscopic mechanical properties, its apparent viscosity continues to increase, indicating that its flow characteristics continue to weaken.

5 Concluding Remarks

Based on the existing research, the movement characteristics of loess after liquid were analyzed and studied by using the method of fluid mechanics, and the samples of different test status were tested by SEM scanning electron microscope test in this paper. The main conclusions are as follows:

(1) In the process of undrained shear test, under the same confining pressure, different from the undisturbed pattern, the apparent viscosity of the remolded sample increases linearly, and its flow characteristics show a state of "shear thickening", indicating that its flow performance is getting worse, and its difference is closely related to the rising mode of pore water pressure.

(2) The pore morphology changes of the undisturbed and remolded samples before and after liquefaction affect the rising mode of pore water pressure in the liquefaction test. In the process, due to the great difference of particle and pore size, the undisturbed sample is easy to form narrow and through pores, which is beneficial to the increase of pore water pressure. However, the pore water pressure of the remolded sample increases slowly because of the small difference of particle and pore size, which is not easy to form the above pores.

(3) The change of apparent viscosity in the process of undrained shear test after liquefaction is closely related to the characteristics of pore structure and arrangement. The more regular the arrangement of pores is, the simpler the structure is, and the smaller the apparent viscosity is.

In this paper, due to the limitations of the test method, only the SEM images of the samples after different test stages were analyzed, which failed to dynamically analyze the pore changes of the samples during the whole test process, and the study on the pore changes of the "shear thickening" section of the original samples was insufficient. Further research is needed in the future.

Acknowledgements. This project is sponsored by the National Natural Science Foundation of China (U1939209,51778590).

References

1. Bai, M., Zhang. S.: Landslide induced by liquefaction of loessial soil during earthquake of high intensity. Geotech. Invest. Surv. (6), 1–5 (1990) (in Chinese with English abstract)
2. Cao, Z., Yuan, X., Chen, L., Sun, R., Meng, F.: Summary of liquefaction macrophenomena in Wenchuan Earthquake. Chin. J. Geotech. Eng. **32**(04), 645–650 (2010) (in Chinese with English abstract)
3. Watkinsonim, H.R.: Impact of communal irrigation on the, Palu earthquake-triggered landslides. Nat. Geosci. **2019**(12), 940–945 (2018)
4. Chen, Y., Liu, H., Wu, H.: Laboratory study on flow characteristics of liquefied and post-liquefied sand. Eur. J. Environ. Civ. Eng. **17**(S1), S23–S32 (2013)
5. Zhuang, H., Wang, R., Chen, G., Miao, Y., Zhao, K.: Shear modulus reduction of saturated sand under large liquefaction-induced deformation in cyclic torsional shear tests. Eng. Geo. **240**, 110–122 (2018)
6. Zhuang, H., Yang, J., Chen, S., Li, H., Zhao, K., Chen, G.: Liquefaction performance and deformation of slightly sloping site in floodplains of the lower reaches of Yangtze River. Ocean Eng **217**, 107869 (2020)
7. Kiyota, T., Koseki, J., Sato, T.: Relationship between limiting shear strain and reduction of shear moduli due to liquefaction in large strain torsional shear tests. Soil Dyn. Earthqu. Eng. **49**, 122–134 (2013)

8. Jiang, M., Zhang, F., Hu, H., Cui, Y., Peng, J.: Structural characterization of natural loess and remolded loess under triaxial tests. Eng. Geo. **181**, 249–260 (2014)
9. Sun, H., Hou, M., Chen, C., Ge, X.: Microstructure investigation of soft clay subjected to triaxial loading. Eng. Geo. **274**, 105735 (2020)
10. Wen, B., Yan, Y.: Influence of structure on shear characteristics of the unsaturated loess in Lanzhou, China. Eng. Geo. **168**, 46–58 (2014)
11. Xu, P., Zhang, Q., Qian, H., Li, M., Yang, F.: An investigation into the relationship between saturated permeability and microstructure of remolded loess: a case study from Chinese Loess Plateau. Geoderma **382**, 114774 (2021)
12. Wang, L., Liu, H., Li, L., Sun, C.: Laboratory study on the mechanism and behaviors of saturated loess liquefaction. Chin. J. Geotech. Eng. **22**(1), 89–94 (2000) (in Chinese with English abstract)
13. Ma, X., et al.: Flow characteristics of large-scale liquefaction-slip of the Loess Strata in Shibei Table land, Guyuan City, induced by the 1920 Haiyuan M8(1/2) Earthquake. Earthquake Res. China **34**(04), 469–481 (2020). https://doi.org/10.19743/j.cnki.0891-4176.202004007
14. Pei, X., Zhang, X., Guo, B., Wang, G., Zhang, F.: Experimental case study of seismically induced loess liquefaction and landslide. Eng. Geo. **223**, 23–30 (2017)
15. Liu, C., Shi, B., Zhou, J.: Quantification and characterization of microporosity by image processing,geometric measurement and statistical methods: application on SEM images of clay materials. Applied Clay Sci. **54**(1), 97–106 (2011)

Geotechnical Behavior of the Valley Bottom Plain with Highly Organic Soil During an Earthquake

Motomu Matsuhashi[1(✉)], Keisuke Ishikawa[2], and Susumu Yasuda[2]

[1] Graduate School of Tokyo Denki University, Hatoyama, Saitama, Japan
21rmg05@ms.dendai.ac.jp
[2] Tokyo Denki University, Hatoyama, Saitama, Japan

Abstract. The valley bottom plains on the plateau around the center of Tokyo form a dendritic pattern. During the Great Kanto Earthquake of 1923, destruction of houses and damage to buried pipes were concentrated along the Kanda River and the valley bottom plain of Tameike. This damage might have been caused by the amplification of seismic motions due to the presence of soil with abundant organic matter and soft clay near the ground surface in the valley bottom plain and the unconformity of the valley bottom. However, the geological structure of the valley (including its width and depth) varies greatly depending on the formation process and location, and it is unknown how these factors affect the seismic response. The purpose of this study was to investigate the effect of the ground structure on the seismic response in a valley bottom plain using the two-dimensional finite element method for seismic response analysis (FLUSH). The study area consisted of three cross-sections with different valley widths, specifically, upstream, midstream, and downstream sections in the valley bottom plain of Magome, Ota-ku, where the thickest organic-rich soil layer (9 m) in Tokyo was deposited. The results showed that the different valley widths affected the position and magnitude of the maximum velocity response at the valley floor and the magnitude of the horizontal ground strain. In the midstream section with medium valley width, the maximum amplitude of the ground strain within the ground was approximately 0.2–0.47%, which could cause damage to buried pipes.

Keywords: Valley plain · Seismic response analysis · High organic soil

1 Introduction

The valley bottom plains on the plateau around the center of Tokyo form a dendritic pattern and are characterized by an unconformable base attributable to the deposition of highly organic soil and soft clay during the formation process of the plains. These features are thought to amplify seismic motion in the valley floor during earthquakes, causing significant damage [1]. As shown in Fig. 1, following the Great Kanto Earthquake of 1923, collapsed houses and buried pipes were concentrated along the Kanda River and along the valley bottom plain of Tameike [1]. Figure 1 shows the topographic map of

© The Author(s), under exclusive license to Springer Nature Switzerland AG 2022
L. Wang et al. (Eds.): PBD-IV 2022, GGEE 52, pp. 2098–2105, 2022.
https://doi.org/10.1007/978-3-031-11898-2_192

Tokyo created by Kubo [2], who overlain a depiction of the damage points in central Tokyo caused by the Great Kanto Earthquake over that of Yasuda et al. [1]. In these valley bottom plains, the Great East Japan Earthquake of 2011 caused the ceiling panels of buildings in Kudanshita to fall.

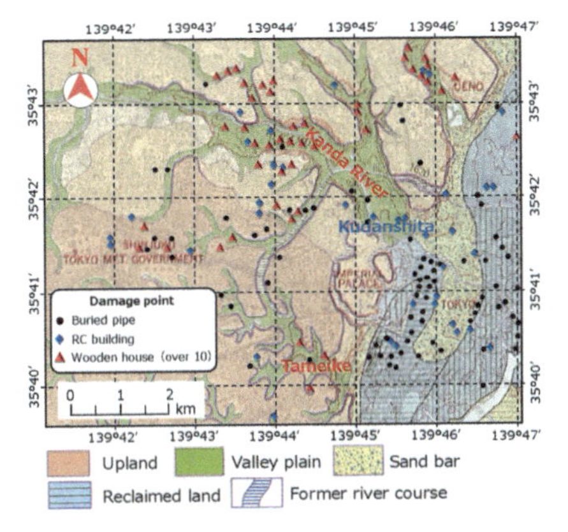

Fig. 1. Damage in central Tokyo from the 1923 Great Kanto Earthquake [1, 2]

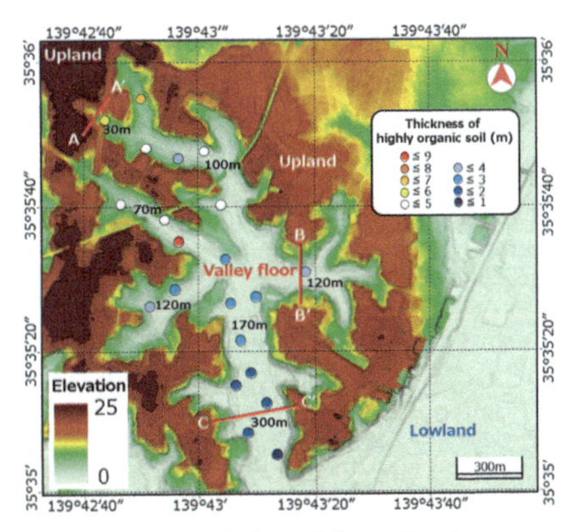

Fig. 2. Characteristics of the valley bottom plain and position of the analytical cross section at Magome Higashitani, Ota-ku, Tokyo [4–7]

There are many structures important to society and the lives of people, such as government corporate offices, in the valley bottom plains of central Tokyo. In the event of a major earthquake, such as the Tokyo metropolitan area epicentral earthquake, these structures could suffer significant damage. However, it is difficult to evaluate the danger points quantitatively because the ground structure of each valley (width and shape of the valley, thickness of the soft layer) varies depending on the formation process and the location of the valley. Therefore, in this study, to investigate the effect of different ground structures on the ground behavior during an earthquake, a two-dimensional finite element seismic response analysis (FLUSH) was performed on a valley bottom plain with different ground structures.

2 Selection of Study Sites and Overview of Seismic Response Analysis

2.1 Selection of Study Sites

In this study, the valley bottom plain of Higashitani in Magome, Ohta-ku, Tokyo, was selected as the study site. At the southern end of the Musashino Plateau, the Magome Higashitani, Magome Nishitani, and Ikegamitani are carved from the east, and the highly organic soil layer of the Higashitani to be examined is not only the thickest layer in Tokyo but also one of the thickest in Japan [3]. In addition, the ground structure of Higashitani, such as the thickness of the highly organic soil layer and the width of the valley, differs greatly depending on the position in the valley bottom. Figure 2 shows a map of the Higashitani. It also indicates the distribution and thickness of the highly organic soil at each location, as obtained from the existing geotechnical information [4-6] on the elevation map of the Geospatial Information Authority of Japan [7] and shows the original valley width as confirmed from old maps. The highly organic soil in Higashitani is distributed over the entire valley. The layer thickness is thicker upstream, but it becomes thinner downstream from the confluence of the middle branch valleys. In addition, the layer thickness of the highly organic soil is thicker in the tributary valleys than in the main valley, with a maximum thickness of 9.0 m. The valley width is narrowest in the upper reaches of the valley, where it ranges from 30 to 100 m. It exceeds 100 m in the middle reaches and is greatest in the lower reaches, where it ranges from 150 to 300 m.

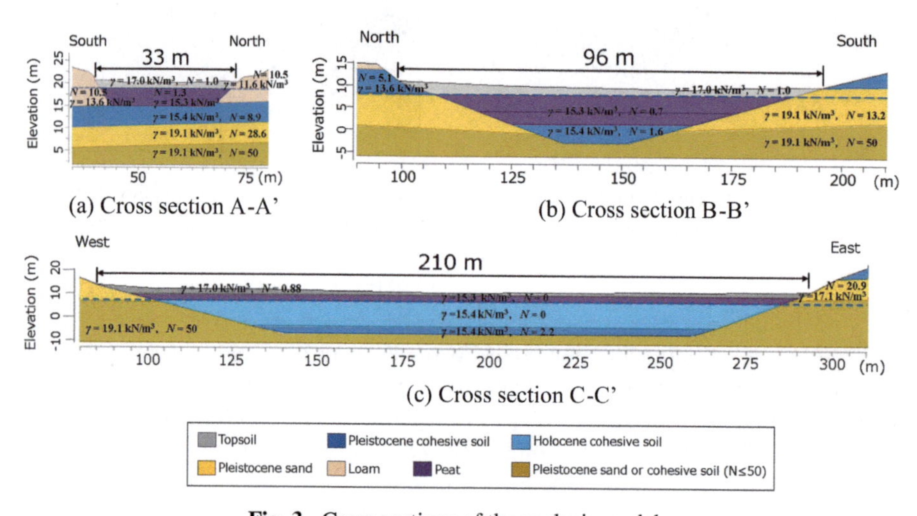

Fig. 3. Cross sections of the analysis models

2.2 Outline of Earthquake Response Analysis

As shown in Fig. 2, three cross-sections with different valley widths were selected for the analysis. The analysis cross-sections were prepared according to the following procedure.

(1) A 5-m digital elevation model (DEM) [7] was used for the ground surface geometry of the analysis model.
(2) The slope of the plateau was assumed to be the lateral slope of the valley bottom.
(3) The stratigraphic configuration of the analysis section was determined from the existing geotechnical information available from the local government.

Figure 3 shows the prepared analytical cross-sections. The middle part of the valley is approximately three times wider than the upstream part, and the downstream part is approximately two times wider than the middle part. The slope of the base of the valley is steep in the upstream part, but it is gentle in the midstream and downstream parts. The thickness of the highly organic soil in each analysis section was 4 m in the upstream section, 6.7 m in the midstream section, and 2 m in the downstream section, with the thickest accumulation in the midstream section. FLUSH was used for the seismic response analysis. The shear wave velocity and shear modulus were calculated from the soil properties obtained from the existing geotechnical information and the N values of the standard penetration test using the following estimation equations:

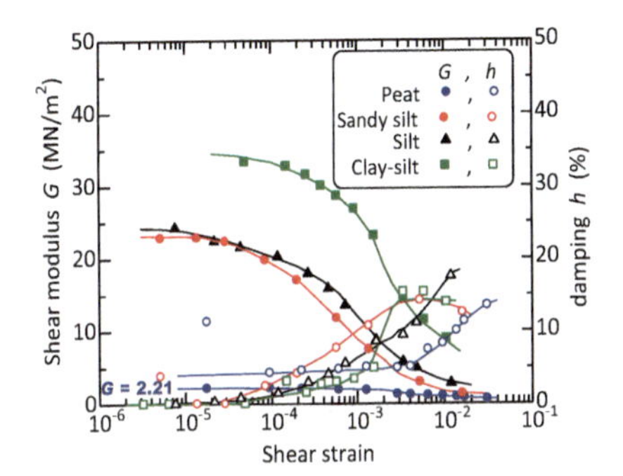

Fig. 4. Dynamic deformation characteristic test result [8]

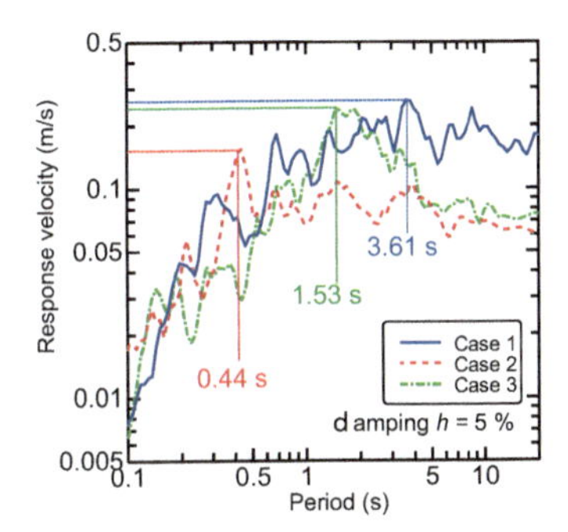

Fig. 5. Velocity response spectra for case 1 to 3

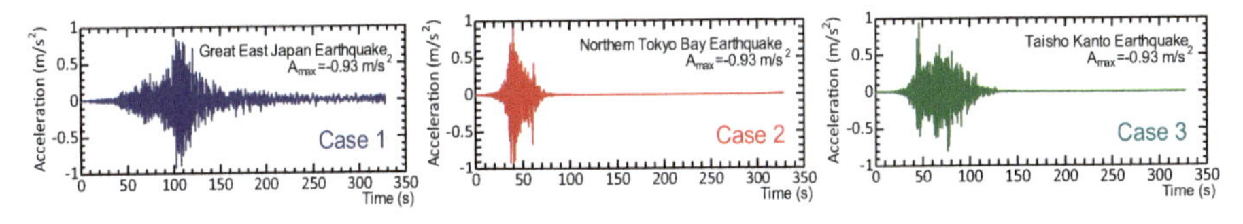

Fig. 6. Time history of input seismic motion (Case 1, Case 2, and Case 3) [11]

$$\text{Coarse grained soil}: V_S = 80 \times N^{\left(1/3\right)} (\text{m/s}) \tag{1}$$

$$\text{Fine grained soil}: V_S = 100 \times N^{\left(1/3\right)} (\text{m/s}) \tag{2}$$

$$\text{Shear modulus} : G_0 = \rho \times V_S^2 (\text{kN/m}^2) \tag{3}$$

The shear modulus of the highly organic soil was set as $G_0 = 2210$ kN/m^2 using the results of a previous dynamic deformation characteristic test, shown in Fig. 4. In previous studies, the natural water content ratio of the highly organic soil was determined as 345%, which is very high, and the porosity ratio was 4.15, which is also a very large value, indicating that the soil has very unique characteristics [8]. The deformation characteristics (G/G_0–γ, h–γ) were based on the chart proposed by Yasuda and Yamaguchi [9], which shows the deformation characteristics as a function of the average grain size and confining pressure. The average grain size, D_{50} (mm), was set from the Tokyo Metropolitan Government liquefaction prediction chart [10]. However, for highly organic soils, the test results shown in Fig. 4 were used. The bottom layer of the analytical cross-section was assumed to correspond to the supporting layer with an N value greater than 50, and the shear wave velocity was assumed to be 300 m/s.

Three waveforms with different period characteristics (Case 1, Case 2, and Case 3) were used for the input earthquake motion. The acceleration waveforms for each case are presented in Fig. 6. Case 1 is the engineering basic observation record of the Great East Japan Earthquake observed at the Shinagawa Observatory of the Bureau of Port and Harbor, Tokyo Metropolitan Government (the maximum acceleration is 0.93 m/s^2) [11]. Case 2 (Northern Tokyo Bay Earthquake) and Case 3 (Taisho Kanto Earthquake) are simulated seismic waves published by the Bureau of Port and Harbor, Tokyo Metropolitan Government [11]. The amplitudes of Case 2 and Case 3 were adjusted to the same maximum acceleration rate as in Case 1. Figure 5 shows the velocity response spectra for each of the cases. Case 1 was widely dominant around 0.5 to 5 s, Case 2 was dominant in the short period of 0.44 s, and Case 3 was dominant in the long period of 1.53 s, revealing a difference in the period characteristics of each seismic motion. The boundary condition was determined to be an energy transfer boundary on the sides and an elastic base at the bottom ($G = 188900$ kN/m^2, damping = 0.5%).

3 Results of Two-Dimensional Seismic Response Analysis Using a Cross-Sectional Model

Figures 7, 8 and 9(a) show the distribution of the maximum velocity at the ground surface, and parts (b)–(d) show the maximum velocity contours. In the maximum velocity contour results, the seismic waves input for each analysis result were (b) Case 1, (c) Case 2, and (d) Case 3. In all of the analysis sections, the maximum velocity was significantly larger for the Case 1 seismic wave. In the distribution of the velocity response at the ground surface, the response was different between the slope and the valley bottom in all analysis sections, and the difference in responses was especially pronounced in the midstream region. This indicates that the difference in valley width affects the maximum velocity at the valley bottom.

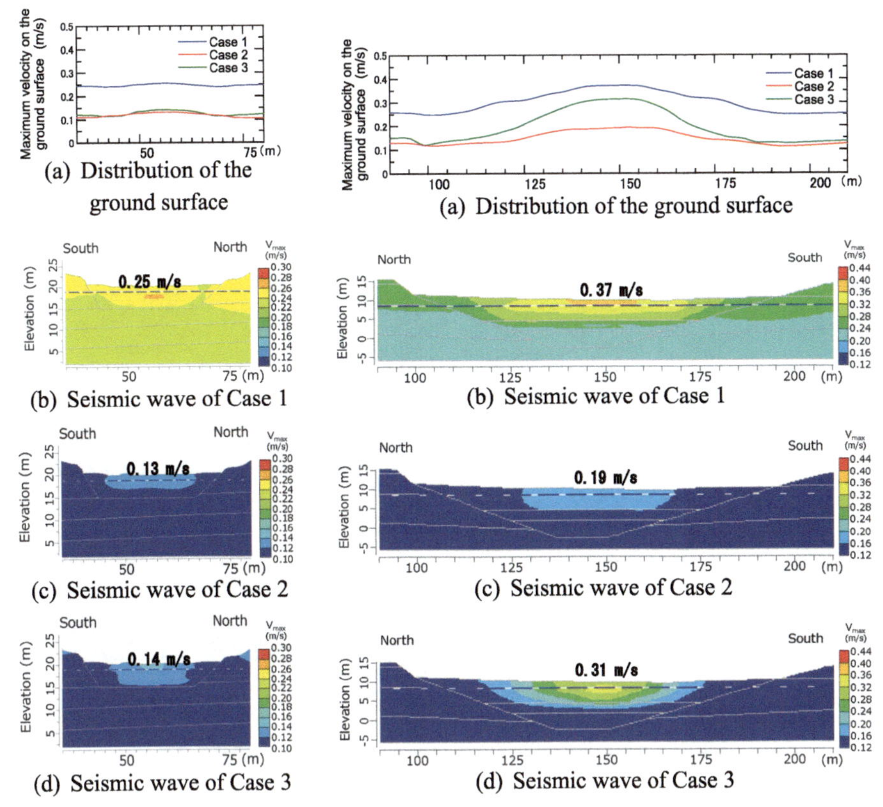

(a) Distribution of the ground surface
(b) Seismic wave of Case 1
(c) Seismic wave of Case 2
(d) Seismic wave of Case 3

Fig. 7. Maximum velocity distribution in cross section A–A

(a) Distribution of the ground surface
(b) Seismic wave of Case 1
(c) Seismic wave of Case 2
(d) Seismic wave of Case 3

Fig. 8. Maximum velocity distribution in cross section B–B′

Figure 10 shows the response distribution of the horizontal ground strain in the ground as the (a) upstream, (b) midstream, and (c) downstream section. "In-ground" refers to the depth of the buried pipe, around G.L − 1.5 m. The " + " sign indicates compressive strain, and the " − " sign indicates tensile strain. Horizontal ground axial strain was determined by dividing the displacement difference between the nodes by the distance between the nodes based on the time history of horizontal displacement at each node. In the upstream area, where the base of the valley is steep, the ground strain occurs near the boundary between the valley bottom and the plateau; however, in the midstream and downstream areas, where the slope is gentle, the ground strain increases on the slope of the base of the valley. This indicates that the shape of the valley base influences the position of horizontal ground strain generation. The magnitude of the ground strain was approximately 2–3 times greater in the midstream section than in the upstream section and half or approximately the same in the downstream section as in the upstream section. This can be attributed to the fact that the velocity response difference within the valley bottom plain increases where the valley width is approximately the same as the width in the midstream section, resulting in a ground displacement difference and a larger ground strain. The significantly lower ground strain in the downstream section might be due to

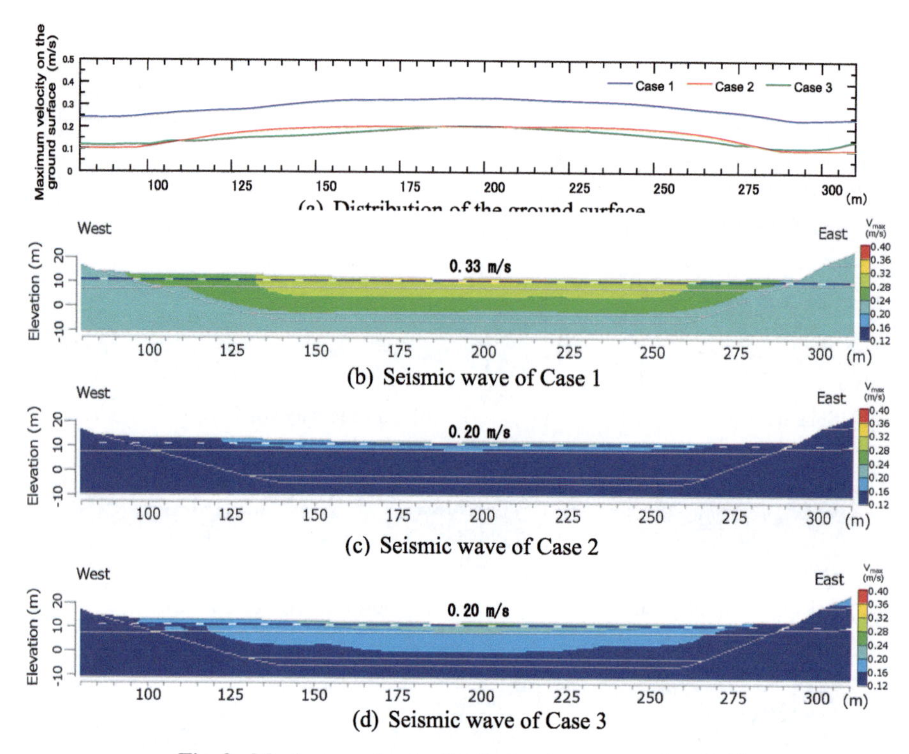

Fig. 9. Maximum velocity distribution in cross section C–C′

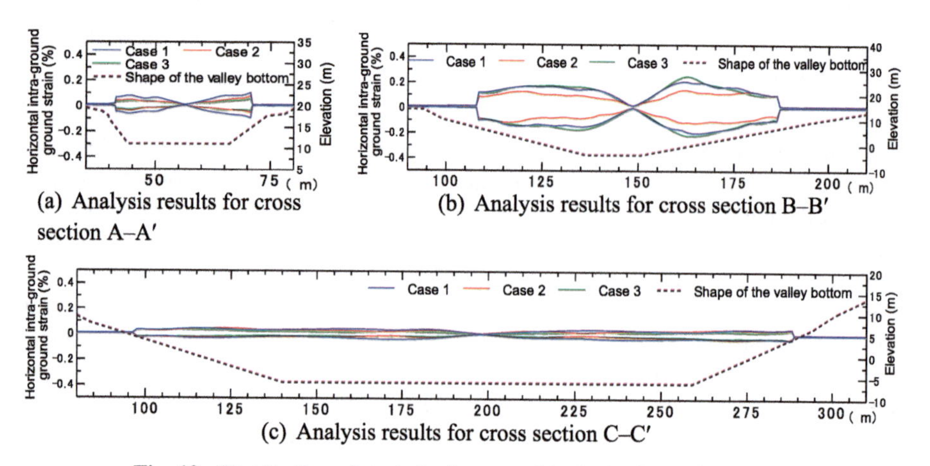

Fig. 10. Distribution of strain in the ground in the horizontal direction

the thinner layer of highly organic soil than in the upstream and midstream sections. In the midstream section, where the valley width is moderate, the maximum amplitude of the ground strain was approximately 0.2–0.47%, which could cause damage to the buried pipe.

4 Conclusion

In this study, we conducted a seismic response analysis (FLUSH) using the two-dimensional finite element method for valley bottom plains with different ground structures and investigated the influence of the different ground structures on the seismic response. The main conclusions of the study follow.

(1) The response of the maximum velocity distribution at the surface in the valley bottom plain differed between the slope and bottom portions of the valley bottom, and the difference in response was especially pronounced in the midstream region, where the valley width was approximately 100 m.
(2) Horizontal ground strain occurred on the valley slope at the base of the valley when the slope was gentle but near the boundary between the valley bottom and the plateau when the slope was steep.
(3) The most severe ground strain for the buried pipe occurred in the plain at the bottom of the valley where the highly organic soil was thick and the valley width was approximately 100 m.

References

1. Yasuda, S., Yoshikawa, Y., Ushijima, K.: Earthquake damage and layer structure in a valley-bottom plain of Tokyo. In: Proceedings of the 48th Japan National Conference JSCE, III-184, pp.422–423 (1993) [in Japanese]
2. Kubo, S.: Geomorphological map of the Tokyo lowland showing environmental changes on water space. In: Proceedings of the General Meeting of the Association of Japanese Geographers, vol.45, pp.114–115 (1994)
3. Sakaguchi, Y.: Geology of Peatlands. University of Tokyo Press, pp. 5–29, Komaba 4-chome, Meguro-ku, Tokyo 153-0041, Japan (1974) [in Japanese]
4. Metropolitan Tokyo HP: The Ground of Tokyo: http://www.kensetsu.metro.tokyo.jp/jigyo/tech/start/03-jyouhou/geo-web/00-index.html. Accessed 10 Oct 2021
5. The Inst. of Civil Engineering of the Tokyo Metropolitan Government: Tokyo Total Subgrade Map (I). GIHODO SHUPPAN Co, Hatoya Wakuri Building, 1-2-5 Kanda Jimbocho, Chiyoda-ku, Tokyo (1977) [in Japanese]
6. Ota Ward Office.: Ota Ward Geotechnical Data Browsing System. https://www.city.ota.tokyo.jp/seikatsu/sumaimachinami/kenchiku/jiban-shiryou/index.html. Accessed 10 Oct 2021
7. Geospatial Information Authority of Japan: GSI Maps. 16 Oct 2021
8. Yasuda, S., Aizu, K., Fujita, Y., Tochio, T., Naoi, K., Matsumoto, S.: Dynamic shear modulus of the peat deposited in the valley plain in Tokyo. In: 45th Japan National Conference on Geotechnical Engineering, pp. 737–738, Japan, 2010 [in Japanese]
9. Yasuda, S., Yamaguchi, I.: Dynamic soil properties of undisturbed samples. In: Proceedings of the 20th Japan National Conference on SMFE, pp.539–542 (1985) [in Japanese]
10. Metropolitan Tokyo: http://doboku.metro.tokyo.jp/start/03-jyouhou/ekijyouka/pdf/00_zenbun_.pdf. Accessed 20 Oct 2021
11. Tokyo Metropolitan Government Bureau of Port and Harbor: https://www.kouwan.metro.tokyo.lg.jp/business/keiyaku/kisojoho/index.html. Accessed 20 Oct 2021

Development of Constitutive Model Describing Unsaturated Liquefaction Characteristics of Volcanic Ash Soil

Takaki Matsumaru[1(✉)] and Toshiyasu Unno[2]

[1] Railway Technical Research Institute, Kokubunji, Tokyo 1858540, Japan
takaki.matsumaru.35@rtri.or.jp
[2] Utsunomiya University, Utsunomiya, Tochigi 3218585, Japan

Abstract. In this paper, the numerical simulations were conducted for reproducing the tendency obtained in the unsaturated liquefaction tests (cyclic tri-axial tests) using volcanic soils damaged in the 2018 Hokkaido Iburi eastern earthquake. In the numerical simulations, the three-phase porous media theory was applied. However, the simulated results underestimated the unsaturated liquefaction resistance in unsaturated conditions because the constitutive models used in the simulations were developed for describing the tendency of liquefaction characteristics for saturated soils. Therefore, the evolution rules focused on the change of void ratio were proposed for modifying the dilatancy characteristic of unsaturated volcanic ash soils. The simulated results using proposed method could successfully reproduce the tendency obtained in the tests.

Keywords: Liquefaction · Unsaturated soil · Volcanic ash soil · Constitutive model

1 Introduction

In Japan, a lot of earthquakes occurred and caused severe damages to natural slopes and embankments, especially in mountainous regions. In the 2003 Minami-sanriku earthquake, 2016 Kumamoto earthquake, and 2018 Hokkaido Eastern Iburi Earthquake, a lot of natural slopes collapsed [1]. The slopes were composed from volcanic ash soils, whose densities were very loose. Though the water levels in the slopes were very low or not observed, the failure would be caused from liquefaction in unsaturated conditions.

For these reasons, the unsaturated liquefaction tests have been conducted in order to evaluate the seismic behavior of unsaturated volcanic soils. The test results showed the increase of liquefaction resistance in unsaturated condition. However, the occurrence of liquefaction was also observed if the water content is large even in unsaturated condition. Compared to other soils, this tendency is remarkable for volcanic ash soils. Furthermore, the constitutive model for numerical simulations were also developed for describing the behaviors of unsaturated soils and were applied in boundary problems of unsaturated slopes and embankments. However, a lot of constitutive models still have problems of difficulty to describe the behavior of unsaturated liquefaction of volcanic ash soils [2].

© The Author(s), under exclusive license to Springer Nature Switzerland AG 2022
L. Wang et al. (Eds.): PBD-IV 2022, GGEE 52, pp. 2106–2113, 2022.
https://doi.org/10.1007/978-3-031-11898-2_193

This is because the volcanic ash soils are very loose, and according to the change of water content, the resistance of liquefaction changes more largely than other soils.

In this paper, the authors tried the numerical simulations of the unsaturated liquefaction tests of volcanic ash soils and developed the new technique of consideration of change of stiffness according to void ratio for the constitutive model of soil skeleton. These methods were employed in governing equations for describing cyclic unsaturated triaxial tests, considering soil skeleton, pore water and pore air. The validation of the developed methods was confirmed by the simulations of the series of unsaturated triaxial tests in different conditions.

2 Numerical Technique

2.1 Governing Equations for Simulation of Unsaturated Cyclic Triaxial Test

In the simulations of the triaxial tests, the finite element formulation is not used. Assuming homogeneous variables in the specimen, only local equilibrium is considered. Equilibrium in the triaxial tests under the unsaturated condition is obtained using the mass balance equations of the soil skeleton, and the pore water and air under constant total stress and undrained air and water conditions [3].

A right-handed coordinate system is adopted and the vertical axial direction of triaxial specimen is adopted as the z-direction. Taking material time derivative of the under the boundary condition that the lateral total stresses are constant, and under the undrained conditions for pore water and air, the governing equation is obtained as [4]

$$\begin{bmatrix} ns^w/K^w - nc & nc & 2s^w & 0 \\ nc & ns^w/K^w - nc & 2s^a & 0 \\ s^w - cp^w + cp^a & s^w + cp^w - cp^a & -D_{11} - D_{12} & 0 \\ s^w - cp^w + cp^a & s^w + cp^w - cp^a & -D_{31} - D_{32} & 1 \end{bmatrix} \begin{Bmatrix} \dot{p}^w \\ \dot{p}^a \\ \dot{\varepsilon}_x \\ \dot{\sigma}_z \end{Bmatrix} = \begin{Bmatrix} -s^w \dot{\varepsilon}_z \\ -s^a \dot{\varepsilon}_z \\ D_{13}\dot{\varepsilon}_z \\ D_{33}\dot{\varepsilon}_z \end{Bmatrix} \quad (1)$$

where n is the porosity, K^w is the bulk modulus of the pore water, s^w and s^a is the degree of water and air saturation, c is the specific water capacity and D_{ij} is the tangential module. The unknown variables, namely the lateral strain rate $\dot{\varepsilon}_x$ the vertical total stress rate $\dot{\sigma}_z$, the pore water pressure rate $\dot{p}^w$ and the pore air pressure rate $\dot{p}^a$ can be calculated from the input axial strain rate $\dot{\varepsilon}_z$.

2.2 Soil Water Characteristic Curve

The SWCC is assumed as

$$s_e^w = \frac{s^w - s_r^w}{s_s^w - s_r^w} \qquad s_e^w = \left\{1 + \left(\frac{p^c}{\alpha_{vg}}\right)^{m_{vg}}\right\}^{-n_{vg}} \quad (2)$$

where s_s^w is the saturated (maximum) degree of saturation, s_r^w is the residual (minimum) degree of saturation and s_e^w is the effective water saturation. The relationship between s_e^w and suction p^c is assumed as van Genuchten (VG) model [5] with material parameters α_{vg}, m_{vg} and n_{vg}.

The parameters α_{vg}, $\mathrm{m_{vg}}$ and $\mathrm{n_{vg}}$ are determined independently for the main drying and wetting curve of the SWCC. The relationship between the suction and the degree of saturation is located in the region covered by the main drying and wetting curve. However, in the case where the mean effective stress decreases due to the cyclic loading, the state will become close to the main wetting curve with the increase of the degree of saturation and the decrease of the suction. For describing this behavior in the SWCC, the scanning curve proposed by Zhou et al. [6] is employed. In the method, the partial differential of the effective degree of saturation to the suction is assumed as

$$\frac{\partial s_{es}^{w}}{\partial p^{c}} = \left(\frac{p_{w}^{c}}{p^{c}}\right)^{b^{wa}} \frac{\partial s_{e}^{w}}{\partial p^{c}} \tag{3}$$

where ${p^c}_w$ is the value of the suction on the main wetting curve corresponding to the current effective degree of the saturation ${s^w}_{es}$ and b^{wa} is the parameter which controls the shape of the scanning curve.

2.3 Constitutive Equation for Soil Skeleton

A simplified constitutive model for saturated sandy soil is used for the unsaturated soil with skeleton stress in place of the effective stress of saturated soil [7]. Assuming that plastic deformation occurs only when the deviatoric stress ratio changes, the yield function is assumed as

$$f = \sqrt{\frac{3}{2}}\|\boldsymbol{\eta} - \boldsymbol{\alpha}\| - k = 0 \tag{4}$$

where η is the stress ratio and k is the material parameter that defines the elastic region. The kinematic hardening parameter (back stress) $\boldsymbol{\alpha}$ and its nonlinear evolution rule are assumed as

$$\dot{\alpha} = a\left(\frac{2}{3}b\dot{\mathbf{e}}^{p} - \boldsymbol{\alpha}\dot{\varepsilon}_{d}^{p}\right) \tag{5}$$

where a and b are the hardening parameters and $\dot{\varepsilon}_{d}^{p}$ is the plastic deviatoric strain rate tensor.

In order to describe the plastic strain rate precisely, the non-associated flow rule was adopted. The plastic deviatoric strain rate and the plastic volumetric strain are derived as

$$\dot{\mathbf{e}}^{p} = \dot{\gamma}\frac{\partial g}{\partial \mathbf{s}} \quad \dot{\varepsilon}_{v}^{p} = D\dot{\gamma}\frac{\partial g}{\partial p'} \tag{6}$$

where $\dot{\gamma}$ is the hardening coefficient and D is the coefficient of dilatancy.

Based on the non-associated flow rule, the Cam-clay-type plastic potential function is assumed as

$$g = \sqrt{\frac{3}{2}}\|\boldsymbol{\eta} - \boldsymbol{\alpha}\| + M_{m}\ln\left(\frac{p'}{p'_{a}}\right) \tag{7}$$

where M_m is the critical state stress ratio, and p'_a is when $||\boldsymbol{\eta}\text{-}\boldsymbol{\alpha}|| = 0$.

In this paper, for considering the effect of the change of the void ratio on the constitutive model, the coefficient of dilatancy D is assumed to be controlled by the void ratio. D is modified during cyclic loading as following equation.

$$D^* = D \cdot (e/e_0)^{\psi^{sd}} \tag{8}$$

where ψ^{sd} is the model parameter which shows positive value. By using this equation, it is possible to adjust the amount of the value of decrease of the skeleton stress due to the compression, as shown in the liquefaction tests of unsaturated soils.

3 Experimental and Numerical Conditions

3.1 Experimental Conditions

The material used in the unsaturated liquefaction tests (cyclic triaxial tests) was volcanic ash soil (Spfa-1) damaged in the 2018 Hokkaido Iburi eastern earthquake, whose particle density G_s was 2.283; and the fine fraction content F_c was 5.2%. Unsaturated cyclic triaxial tests were conducted in different initial conditions (degree of saturation and suction). Table 1. shows the test conditions. The unsaturated triaxial tests were conducted in stress control condition. The net stress before starting cyclic loadings was about 20 kPa.

Table 1. Properties of materials used in unsaturated liquefaction tests

Dry density [g/cm^3]	Void ratio	Degree of saturation [%]
0.575	2.97	Saturated
0.507	3.50	72.6
0.495	3.62	81.9

3.2 Analytical Conditions

For the constitutive model, the bulk modulus, shear modulus, and critical state stress parameter were determined from the monotonic and cyclic triaxial tests for saturated soils. The other parameters were determined by try and error.

Figure 1 shows the SWCC. The experimental results contained the ones obtained from the water retention tests and the state of the specimens for the unsaturated triaxial tests before and after conducting cyclic loading.

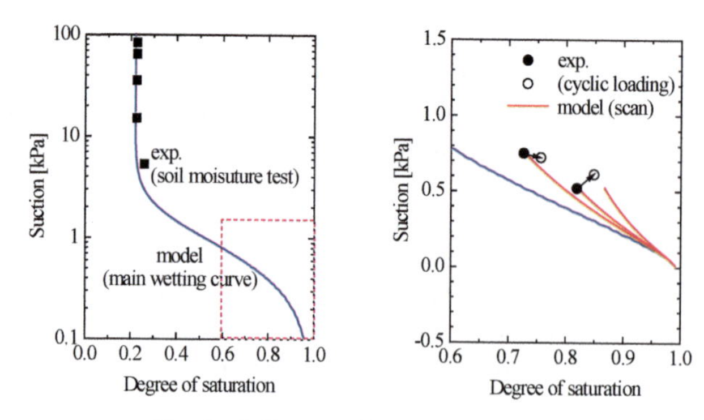

Fig. 1. Soil water characteristic curves

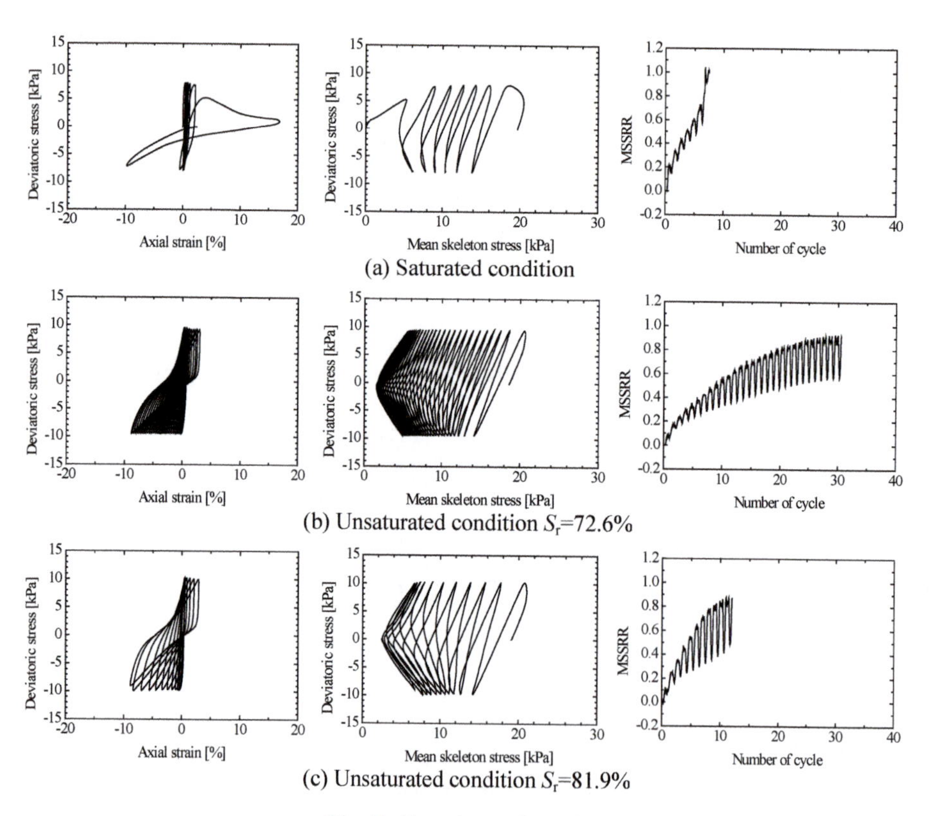

Fig. 2. Experimental results

4 Results and Discussions of Experiments and Simulations

Figure 2 shows the experimental results of all cases. These figures contain stress-strain relationships, stress pass (relationship between mean skeleton stress and deviatoric stress) and relationship between number of cycle and mean skeleton stress reduction ratio (MSSRR). In the saturated condition, the skeleton stress decreased rapidly and MSSRR reached 1.0. Compared to the saturated condition, MSSRR did not reach 1.0 and the mean skeleton stress maintained after cyclic loading in the unsaturated condition. The axial strain increased only to the direction of extension in the unsaturated conditions.

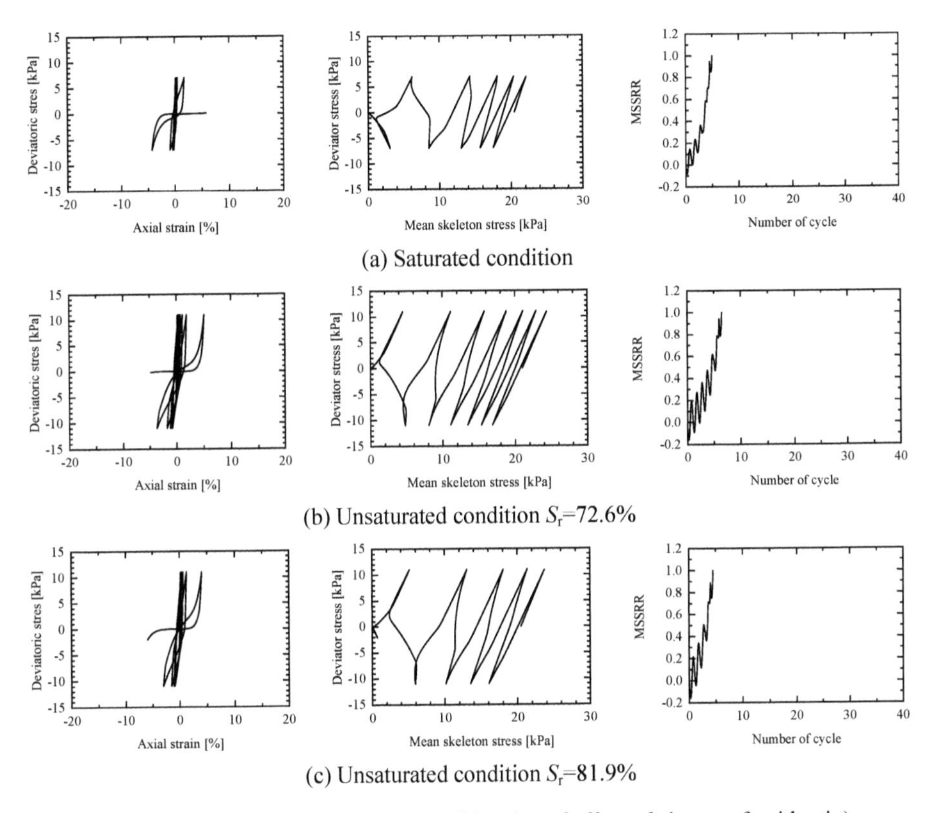

Fig. 3. Simulated results (without consideration of effect of change of void ratio)

Figure 3 shows the results obtained by the numerical simulations without consideration of the effect of change of void ratio in constitutive equations. The tendency of the experiment in the saturated condition was reproduced well. However, the simulated results in unsaturated conditions showed that MSSRR reached 1.0 and the mean skeleton stress became zero. This means that the liquefaction occurred even in the unsaturated conditions, so the simulation in the unsaturated condition could not reproduce the tendency obtained in the experiments.

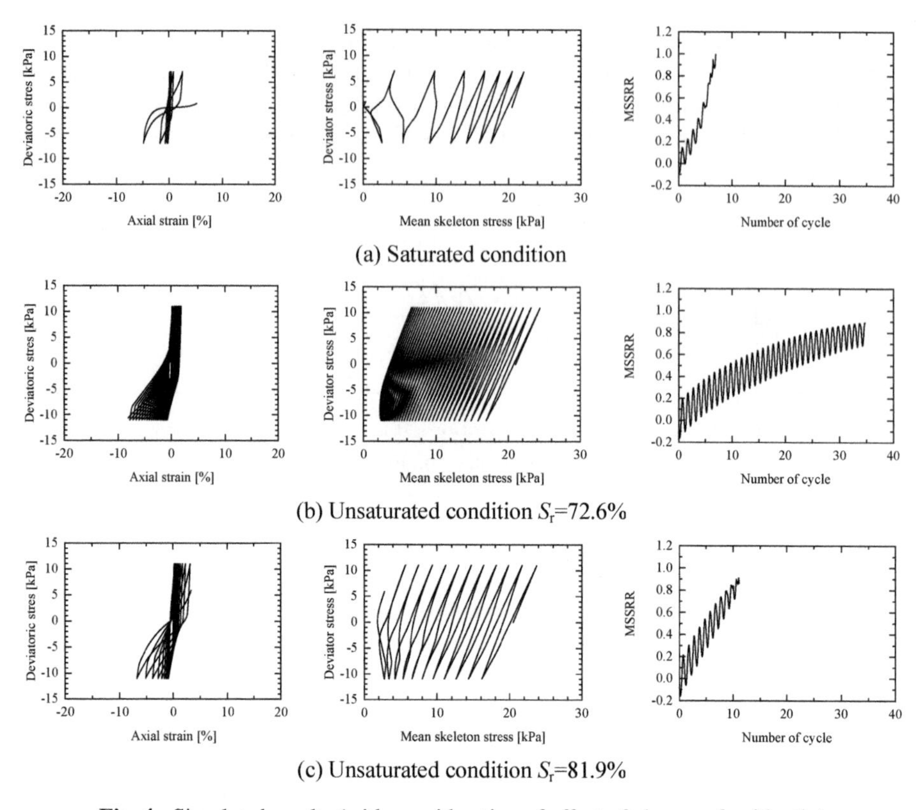

(a) Saturated condition

(b) Unsaturated condition S_r=72.6%

(c) Unsaturated condition S_r=81.9%

Fig. 4. Simulated results (with consideration of effect of change of void ratio)

Figure 4 shows the simulated results with consideration of the effect of change of void ratio in the constitutive model. In the saturated condition, liquefaction occurred without a large change compared to the simulation shown in Fig. 2. In the unsaturated conditions, the mean skeleton stress decreased, but the value did not reach zero. So, MSSRR did not became 1.0. The axial strain increased to the direction of the extension in the unsaturated conditions. Therefore, the simulated results could successfully reproduce the tendency obtained by the experiments.

5 Conclusions

In order to increase the accuracy of the prediction of the liquefaction behavior of the unsaturated volcanic ash soil, a method to considering the effect of change of void ratio for the constitutive equation of the soil skeleton was developed in the simulation of the unsaturated cyclic triaxial tests. The validation of the developed method was examined by the simulations of the unsaturated triaxial tests aimed to the volcanic ash soil damaged in the 2018 Hokkaido Iburi eastern earthquake. It was revealed that the proposed method contributed to increase of the reproduction of the tendency obtained by the experiments.

Acknowledgement. This works was financially supported by JSPS KAKENHI Grant Numbers JP21H01423 and JP21K18741. The authors special thanks to Mr. Midorikawa, former student of Utsunomiya University, for his contributions of conducting the cyclic triaxial tests.

References

1. Uzuoka, R., Sento, N., Kazama, M., Unno, T.: Landslides during the earthquakes on May 26 and July 26, 2003 in Miyagi. Japan, Soils Found. **45**(4), 149–163 (2005)
2. Uzuoka, R., Unno, T., Sento, N., Kazama, M.: Effect of pore air pressure on cyclic behavior of unsaturated sandy soil. Unsaturated Soils: Res. Appl. **1**, 783–789 (2014)
3. Uzuoka, R., Borja, R.I.: Dynamics of unsaturated poroelastic solids at finite strain. Int. J. Numer. Anal. Meth. Geomech. **36**, 1535–1573 (2012)
4. Matsumaru, T., Uzuoka, R.: Numerical simulation of unsaturated cyclic triaxial test considering effect of void change and scanning soil water characteristic curve, In: Proceedings on 3rd International Conference on Performance-based Design in Earthquake Geotechnical Engineering, vol. 353, pp. 1–6, Vancouver (2017)
5. van Genuchten, M.: A closed-form equation for predicting the hydraulic conductivity of unsaturated soils. Soil Sci. Soc. Am. J. **44**(5), 892–898 (1980)
6. Zhou, A.Z., Sheng, D., Sloan, S.W., Gens, A.: Interpretation of unsaturated soil behavior in the stress - saturation space, I: Volume change and water retention behavior. Comput. Geotech. **43**, 178–187 (2012)
7. Matsumaru, T., Uzuoka, R.: Three-phase seepage-deformation coupled analysis about unsaturated embankment damaged by earthquake, Int. J. Geomech. **16**(5), C4016006 (2016)

Field Observations and Direct Shear Tests on the Volcanic Soils Responsible for Shallow Landslides During 2018 Hokkaido Eastern Iburi Earthquake

Shohei Noda[1], Hiroya Tanaka[2], and Yoshimichi Tsukamoto[1(✉)]

[1] Department of Civil Engineering, Tokyo University of Science, Chiba, Japan
ytsoil@rs.tus.ac.jp

[2] Japan Railway Construction, Transport and Technology Agency, Kanagawa, Japan

Abstract. A numerous number of landslides were observed in the local town of Atsuma during 2018 Hokkaido Eastern Iburi Earthquake. The area of landslides was found the largest ever recorded in Japan during the last 150 years. It was noteworthy that there were many landslides observed on relatively gentle slopes of less than 20°, and most of the landslides were shallow-seated. This area is known to be covered by several layers of volcanic pyroclastic fall deposits, intervened with their weathered clayey and loam soils. The effects of rainfall recorded one day before the earthquake may not be negligible for triggering of so many shallow slips of the local volcanic soil deposits. In this study, field investigations were conducted to scrutiny the layers of soil deposits responsible for shallow landslides, and two different soil samples of Shikotsu and Tarumae-d pumice fall deposits were retrieved. Multiple series of volume-constant direct shear tests were then conducted on these two volcanic soils, and the effects of degree of saturation were especially examined. In order to mechanically characterise the crushability of particles of these pumice fall deposits, a series of particle crushing tests were also conducted.

Keywords: Volcanic soil · Field observation · Direct shear test

1 Introduction

The intensive seismic shaking measuring the magnitude of M = 6.7 and the two large aftershocks hit the eastern Iburi region in Hokkaido on September 6, 2018. A numerous number of landslides occurred in the local town of Atsuma located close to the epicentre, as shown in Fig. 1. The seismic intensity of 7 was measured at the town of Atsuma, which is the largest intensity defined in the scale of the meteorological agency in Japan. This is even the first time that it was recorded in Hokkaido. The area of landslides was also estimated to be the largest ever recorded in Japan during the last 150 years. Most of the landslides were shallow-seated with clear sliding surfaces and long run-out distances. The effects of rainfall recorded one day before the earthquake may not be negligible for triggering of so many shallow landslides.

© The Author(s), under exclusive license to Springer Nature Switzerland AG 2022
L. Wang et al. (Eds.): PBD-IV 2022, GGEE 52, pp. 2114–2122, 2022.
https://doi.org/10.1007/978-3-031-11898-2_194

This area is known to be covered by several layers of volcanic pyroclastic fall deposits, associated with the eruptions of the volcanic mountains indicated in Fig. 1. In a typical depth-wise sequence observed from a ground surface, there are several layers of Tarumae pumice fall deposits, associated with several events of volcanic eruptions of Tarumae Mt. During the last 10 thousand years. Located below these layers is a relatively thin layer of Eniwa pumice fall deposit, associated with the event of volcanic eruption of Eniwa-dake Mt., that occurred approximately 20 thousand years ago. Below this layer is a thick layer of Shikotsu pumice fall deposit, associated with the event of eruption of a volcano now existing as Shikotsu caldera lake, that occurred approximately 40 to 45 thousand years ago.

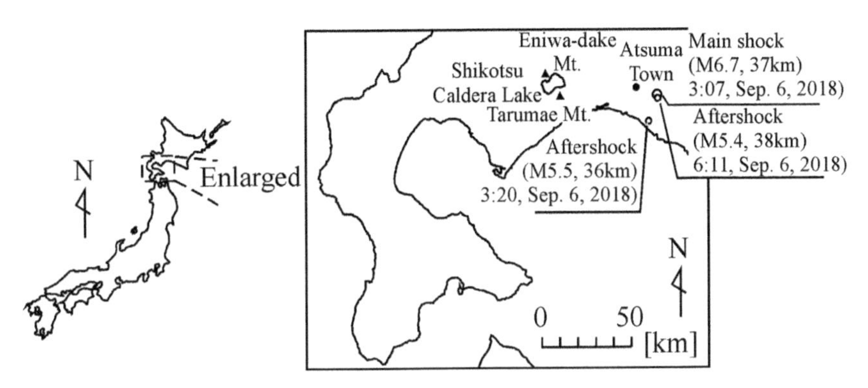

Fig. 1. Location of Atsuma town

In this study, the earthquake reconnaissance investigations were carried out at two sites of shallow landslides, denoted as site A and site B. A landslide was observed on a steep slope at site A, and the soil layer responsible for sliding was investigated. On the other hand, a landslide occurred on a gentle slope at site B, and the soil layer responsible for sliding was also investigated. The soil samples were retrieved from these two sites, and a series of laboratory direct shear tests were conducted to investigate the effects of saturation on the mobilisation of shear strength of the pumice fall deposits that exhibited long run-out sliding.

2 Field Observations

The shallow landslide was observed on a steep slope at site A, as shown in Fig. 2. The sliding debris covered the entire road surface at the foot of the slope and spread over the paddy fields. The close scrutiny of the stratification of the shallow soil deposits as indicated in Fig. 3 revealed that there are relatively thin layers of Tarumae fall deposits on this steep slope, underlain by Eniwa fall deposits, below which Shikotsu fall deposits exist. The exposed sliding surface is therefore located within Shikotsu fall deposits, and seems to consist of a weathered clayey deposit. It is noteworthy that there are lenses of weathered clayey strata within Shikotsu fall deposits, and those clayey strata may have exposed as sliding surfaces during the earthquake.

The shallow landslide was also observed on a gentle slope at site B, as shown in Fig. 4. The slope angle at the downstream area was about 15°. The close scrutiny of the stratification of the shallow soil deposits as indicated in Fig. 5 revealed that there are distinct and thick layers of Tarumae fall deposits on this gentle slope, which are classified as Tarumae-a, b, d1 and d2, each of which is associated with the event of volcanic eruption of Tarumae Mt. The exposed sliding surface is therefore located within Tarumae fall deposits, and seems to consist of weathered loam deposits.

Fig. 2. Steeply inclined shallow landslide at site A

Fig. 3. Stratification of volcanic pumice fall deposits at site A

A series of portable dynamic cone penetration tests were conducted on this gentle slope at site B, immediately next to the sliding area, as shown in Fig. 4. The portable dynamic cone penetration test shown in Fig. 6 is the portable in situ penetration test developed by PWRI in Japan, and easy to carry around and conduct field tests. It is especially suitable for field testing on slopes. A field test proceeds as follows. The cone attached to the head of the rod is placed on a ground surface to make the rod stand vertically. The cone is then penetrated vertically by repeating a free fall of the 5kg hammer onto the anvil from a height of 50 cm through the guiding rod. The number of free falls necessary to penetrate the cone for a depth of 10 cm is counted and denoted as N_d. The value of N_d is known to be correlated with SPT N-value, tip resistance of CPT as well as N_{sw} & W_{sw} values of Swedish weight sounding [1]. The results of a series of the penetration tests are shown in Fig. 7. The values of N_d are less than 5 and very low, until the cone hit the layer of a loam deposit at a depth of 2 m from the ground surface. This loam layer exhibiting the value of N_d larger than 10 is found to serve as a sliding surface.

Fig. 4. Gently inclined shallow landslide at site B with locations of penetration tests

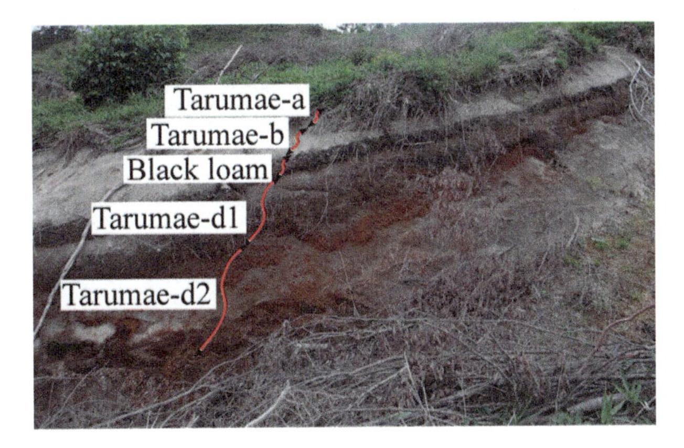

Fig. 5. Stratification of volcanic pumice fall deposits at site B

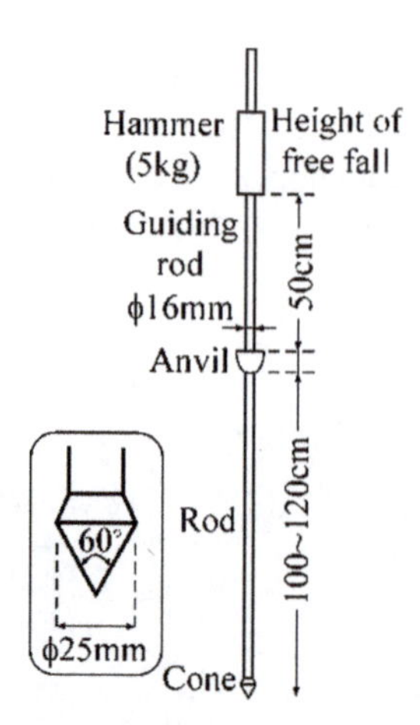

Fig. 6. Portable dynamic cone penetration test (after [1])

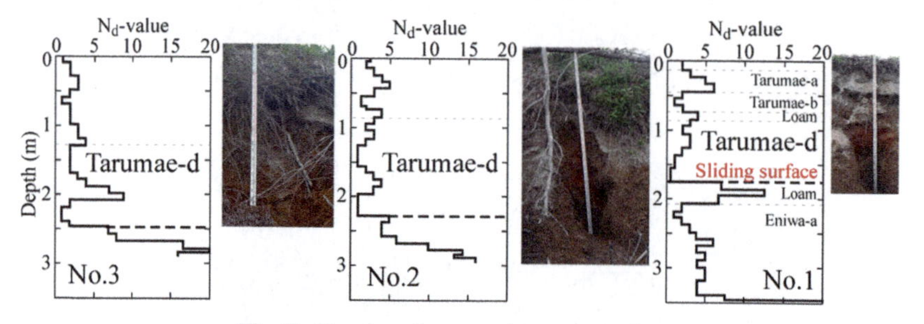

Fig. 7. Results of penetration tests at site B

3 Laboratory Test Results

The soil samples of Shikotsu and Tarumae-d fall deposits were retrieved from sites A and B, respectively. These soil samples were taken from the exposed side faces of the non-sliding soil masses shown in Figs. 3 and 5. These soil samples contain large amounts of gravels. However, the soil particles larger than 2 mm are removed for the purpose of laboratory testing. The physical properties and grain size distributions of the soil samples are shown in Table 1 and Fig. 8. These soil samples consist of pumice fall deposits, and are expected to be extremely crushable. The crushability of soil particles is expected to affect the fragmentation process of a landslide, leading to a long run-out distance. On this vein, a series of laboratory single-particle crushing tests were conducted, as shown in Fig. 9. In this diagram, the plots of the single-particle crushing strength against particle size are shown for the soil samples of Shikotsu and Tarumae-d fall deposits. The younger deposit of Tarumae-d consistently exhibits lower strength than the old deposit of Shikotsu. For comparison purposes, the plots for the samples of local Futtsu and Kakegawa mudstones are also shown in Fig. 9. The soil samples of the fall deposits tested in this study exhibited lower strength than those of the mudstones.

In this study, a series of laboratory volume-constant direct shear tests were conducted on the unsaturated specimens of Shikotsu and Tarumae-d fall deposits. It makes good reasons to use volume-constant direct shear tests for testing unsaturated soil specimens [2, 3]. The laboratory testing methods can be characterised by the stress and volume conditions imposed on soil specimens, as summarised in Table 2, where σ is total stress, u_a is pore air pressure and u_w is pore water pressure. In saturated undrained triaxial tests, the effective stress defined as $\sigma' = \sigma - u_w$ is allowed to change in such a manner that the total confining stress σ is kept constant and u_w may change. On the other hand, in volume-constant direct shear tests, the effective stress is also allowed to change in such a manner that the total confining stress σ may change and u_w is kept constant. In addition, in both of the testing methods, the volume of a specimen is kept constant.

Table 1. Physical properties of soil samples

	Shikotsu	Tarumae-d
ρ_s (g/cm^3)	2.67	2.54
$\rho_{d\,min}$ (g/cm^3)	0.920	0.814
$\rho_{d\,min}$ (g/cm^3)	0.584	0.624
F_c (%)	45	52

Table 2. Stress and volume conditions imposed in the laboratory tests (after [2, 3])

Laboratory testing method		σ	u_a	u_w	Volume
Saturated	Undrained TX	constant	----	change	constant
	u_a drained & u_w undrained TX	constant	constant	change	change
Unsaturated	u_a undrained & u_w undrained TX	constant	change	change	change
	Volume-constant DS	change	constant	constant	constant

N.B.) TX: triaxial test, DS: direct shear test.

The rectangular shear box with 12 cm long, 12 cm wide and 12 cm high was used in this study. All the soil specimens were prepared by moist placement. The soil sample with a pre-determined value of water content was placed in the shear box, and tamped equally to achieve one-fifth of the full height of the soil specimen. This procedure was repeated five times, and the soil specimens with various values of S_r and ρ_d were prepared. They were then consolidated to the vertical overburden stress of 100 kPa. The test results for the specimens of Shikotsu fall deposits are shown in Figs. 10 and 11. Based on the data shown in Figs. 10 and 11, the value of shear stress corresponding to the state of phase

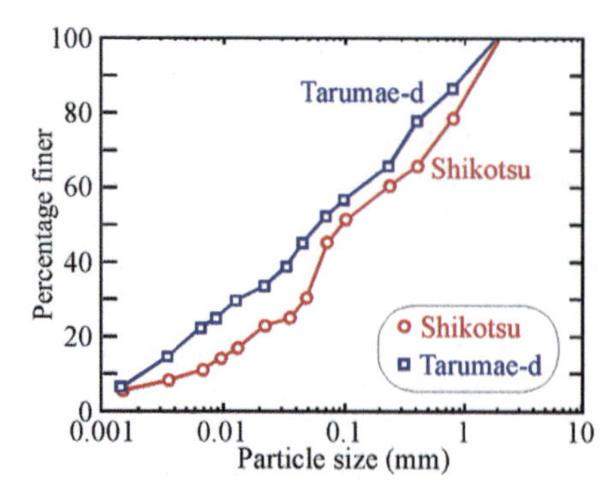

Fig. 8. Grain size distributions of soils of sliding mass retrieved from sites A and B

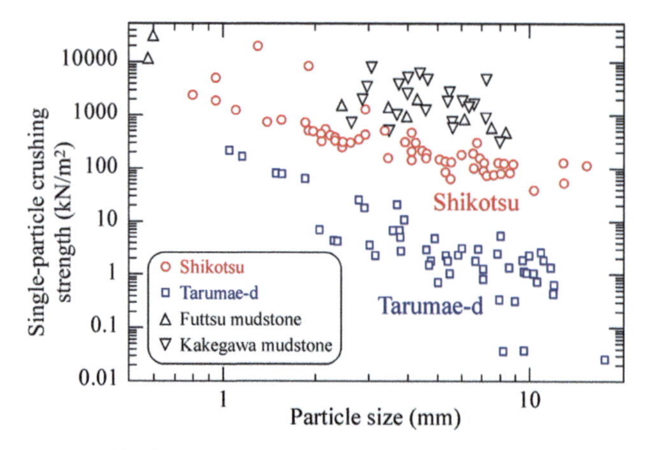

Fig. 9. Results of single-particle crushing tests

transformation is determined as the direct shear strength τ_u, and the direct shear strength ratio τ_u/σ_{vo} is determined as τ_u divided by the initial vertical overburden stress σ_{vo}. The plots of the direct shear strength ratio τ_u/σ_{vo} against S_r are shown for the specimens of Shikotsu and Tarumae-d fall deposits in Figs. 12(a) and (b). It is found that the direct shear strength ratio significantly reduces as the degree of saturation reaches $S_r = 80\%$.

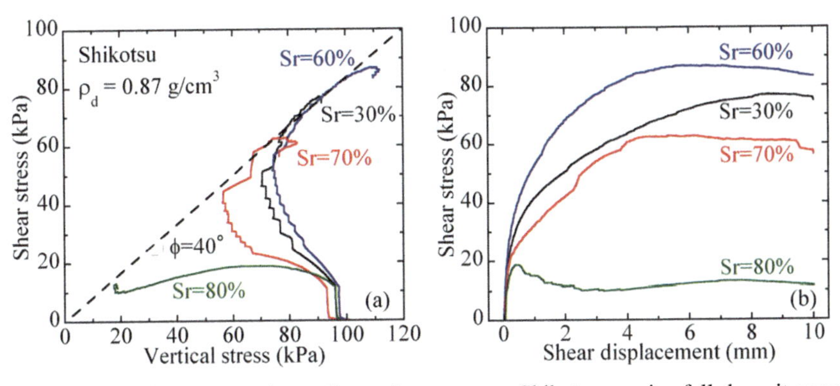

Fig. 10. Results of constant-volume direct shear tests on Shikotsu pumice fall deposit samples with different values of S_r

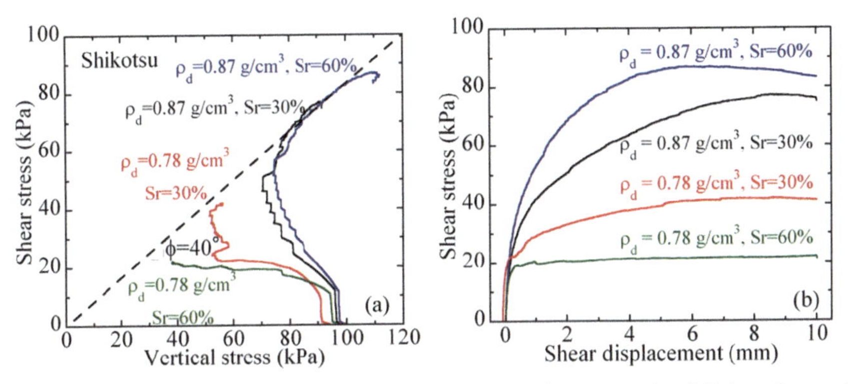

Fig. 11. Results of constant-volume direct shear tests on Shikotsu pumice fall deposit samples with different values of ρ_d

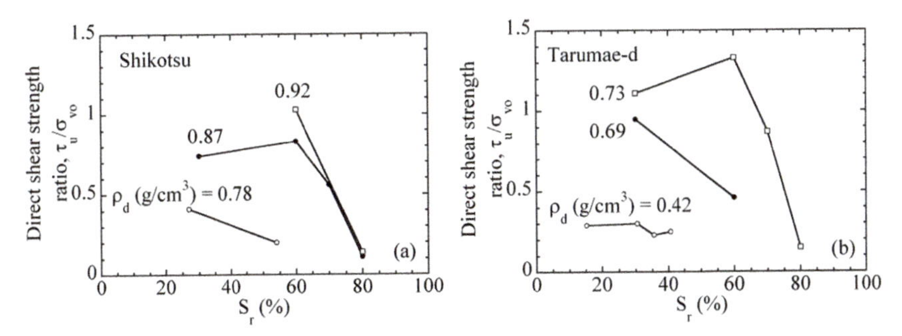

Fig. 12. Plots of direct shear strength ratio against S_r for (a) Shikotsu and (b) Tarumae-d pumice fall deposit samples

4 Conclusions

The field observations were conducted on the stratifications of the volcanic fall deposits at two sites of shallow landslides in the town of Atsuma. The soil layers responsible for landslides were detected, and the sliding surfaces were found to consist of weathered clayey and loam deposits. The results of the laboratory tests disclosed that the particles of the volcanic soils responsible for landslides are extremely crushable, and the direct shear strengths of these soils reduce significantly as the degree of saturation reaches $S_r = 80\%$.

Acknowledgements. The authors acknowledge the kind cooperation of Dr. K. Hashimoto, T. Abe and M. Ikeda of Chuo Kaihatsu Corporation. Dr. M. Ikeda provided the useful information on the origins and stratifications of the volcanic fall deposits in the town of Atsuma.

References

1. Japanese geotechnical society: Standards for geotechnical and geoenvironmental investigation methods – standards and explanations (2004)
2. Uemura, K., Onishi, K., Tsukamoto, Y.: Use of volume-constant direct shear tests in determining cyclic and residual strength of unsaturated soils. In: Proceedings of 3rd Taiwan – Japan Joint Workshop on Geotechnical Hazards from Large Earthquakes and Heavy Rainfall, Keelung, Taiwan, Oct 31 – Nov 3, pp. 97–102 (2008)
3. Tsukamoto, Y.: Influence of degree of saturation on liquefaction resistance and undrained shear resistance of silty sands. In: Proceedings of International Conference on Performance-based Design in Earthquake Geotechnical Engineering, Vancouver, Canada, Jul 16–20 (2017)

Investigation on Saturation State of Loess Using P-wave Velocity

Zehua Qin, Yuchuan Wang, and Xin Liu(✉)

College of Geological Engineering and Geomatics, Chang'an University, Xi'an, China
xliu67@chd.edu.cn

Abstract. Pore pressure parameter, also known as B-value, is used to characterize the saturation state of soils. Recent advances have established a relationship between B-value and P-wave velocity based on the poroelasticity theory, hence it sheds a light to evaluate the degree of saturation of soils using P-wave velocity. Previous studies in this regard often focus on granular materials, the feasibility of using the relationship for loess was less extensively studied. This paper presents experimental results on the P-wave velocity of loess using a pair of piezoceramic extender elements at a sequence of saturation states. It was found that, the P-wave velocity do not change with excitation frequency. At an excitation frequency of 5 kHz, the wave signal is most clear. Besides, the effects of initial water content and sand content on P-wave velocity were both examined. Though the P-wave velocity increases with the B-value, that is approximately in line with theoretical predictions, departures from predictions were observable at higher B-values. The outcome of this study suggests a potentially good use of P-wave velocity as an indicator of saturation in loess.

Keywords: P-wave velocity · B-value · Saturation

1 Introduction

Among a number of factors that affect liquefaction in loess (i.e., soil density, fines content, etc.), the degree of saturation is crucial (Sherif et al. 1977; Yoshimi et al. 1989). For instance, from the research performed by Karam et al. (2009) the cyclic resistance ratio of intact loess varied from 0.4 to 1.9 that was strongly dependent on the initial water content. Evaluation of the saturation condition of loess is of great significance in the geotechnical discipline at both academic and practical perspectives.

It is pertinent to note that a conventional method to characterize the saturation of soils uses the B-value in laboratorial experiments as follow (Skempton 1954).

$$B = \frac{1}{1 + n\frac{K_b}{K_f}} = \frac{1}{1 + n\frac{K_b}{K_w} + n\frac{K_b}{P_a}(1 - S_r)} \tag{1}$$

where n is the porosity; K_b, K_f, and K_w are the bulk modulus of the soil skeleton, the pore fluid and the deaired water, respectively; P_a is denoted as the atmospheric pressure;

© The Author(s), under exclusive license to Springer Nature Switzerland AG 2022
L. Wang et al. (Eds.): PBD-IV 2022, GGEE 52, pp. 2123–2130, 2022.
https://doi.org/10.1007/978-3-031-11898-2_195

and S_r is denoted as the degree of saturation. In a triaxial test, by applying a small increment of confining pressure ($\Delta\sigma$) on a soil specimen, the B-value is determined as the ratio between $\Delta\sigma$ and the corresponding change in pore pressure (Δu). It is clear that a specimen becomes fully saturated ($S_r = 1$) as the B-value approaches to a unit. Though the B-value is readily measured in the laboratory, the above method is not applicable in the field condition.

To this end, recent advances have proposed an approach to evaluate the saturation of granular soils using P-wave velocity (V_p) (Tsukamoto et al. 2002; Yang 2002), and it specifies the relationship as follows.

$$V_p = \left[\frac{4G/3 + K_b/(1-B)}{\rho}\right]^{1/2} \tag{2}$$

where G is the shear modulus of the soil skeleton; ρ is the bulk density; K_b is the bulk modulus of skeleton.

The advantage of using the P-wave velocity is that it can be applied both in the field and in the laboratory. Hence, it correlates field observations and various types of soil properties that are determined from the laboratory. Yet, previous studies using the above correlation focus on granular materials (Tamura et al. 2002; Naesgaard et al. 2007; Gu et al. 2013). Given a more complex nature of loess as compared with sand, it is still unclear as to whether the above mothed is applicable in loess. In fact, a number of factors including soil fabric, initial water content, sand content may lead to notable variations of measured B-values during the saturation process (Sugiyama et al. 2016). In the above context, concerns about the feasibility of using V_p as an indicator of saturation were raised over the observed discrepancies in loess. In current literatures, however, experimental studies addressing these issues were rather limited.

This paper aims to examine the feasibility of using V_p as an indicator of saturation of loess. Moist tamping method was adopted to prepare loess specimens of a similar void ratio. By comparing measured B-values and V_p in experiments and in the prediction, discussions were made on the effect of initial water content and the sand content, along with a hypothesis to account for observed discrepancies.

2 Experimentation

2.1 Materials and Test Procedures

Figure 1 shows the particle size distribution curves of test materials measured by sieving and sedimentation tests. The average sand content of the original loess used in this experiment was determined as 11% (i.e., greater than 63 μm). To produce a sequence of sand-loess mixtures, a uniformly graded quartz sand was adopted as an addictive, and the quantity of the add-in quartz sand varied from 0% to 10% by mass. In each test, a loess sample of 50 mm in diameter and 100 mm in height was used in a triaxial apparatus that is equipped with a pair of piezoceramic extender elements. The extender elements were used to generate the P-wave that propagates in phase with the direction of vibration (Liu et al. 2020). In each test, the bottom pedestal was used as a transmitter and the top cap was used as a receiver. A total number of 10 received signals were stacked per measurement to reduce the influence of uncorrelated noises. A schematic diagram in Fig. 2 shows the working principle of extender element test. In this study, an initial void ratio, e = 0.88, was chosen as the target. By using the moist tamping (MT) method, loess specimens were prepared at water contents of 5% and 14%, and sand contents of 0% and 10%, respectively. Note that the 14% of water content is close to the optimal water content of the test material determined from the compaction test. To obtain a similar void ratio via the MT method, a pre-determined mass of wet soil was deposited into a split mold and then it was subjected to continuous tamping. The process was repeated five times to create successive compacted soil layers.

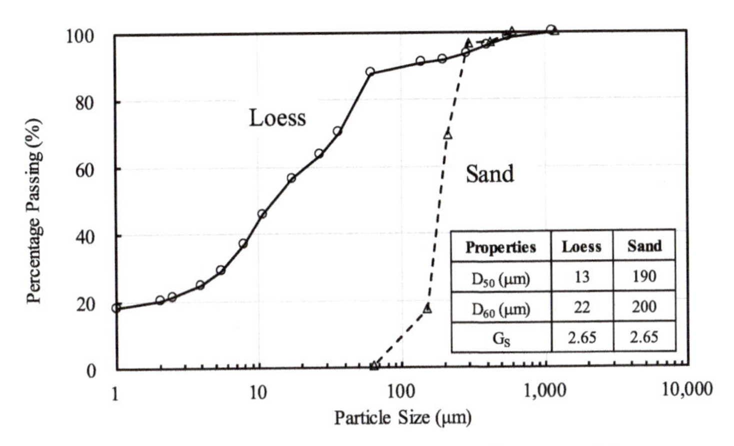

Fig. 1. Particle size distribution curves of test materials.

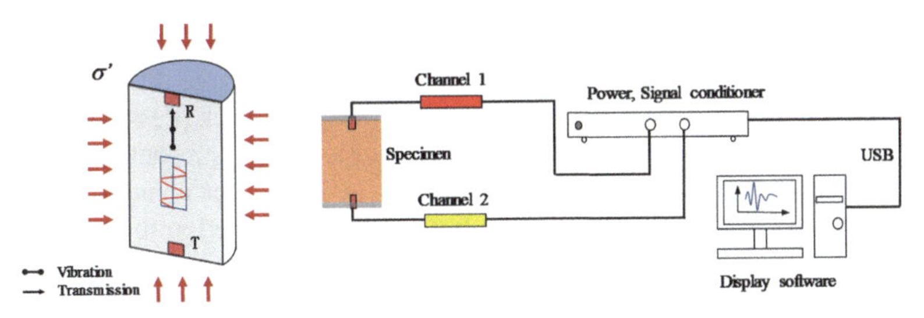

Fig. 2. Working principles of the extender element.

2.2 Sample Saturation

To achieve a series of saturation states, a stepwise procedure was implemented for all specimens. Firstly, with a small suction (i.e., around −5kPa) applied at the drainage outlet, the tap water was able to circulate consistently through the loess specimen with enough time to fill the void space inside of the sample. Upon the end of water flushing, no air bubbles at the drainage outlet can be observed, indicating a continuous water phase in the specimen, and the leftover air is assumed only in bubble-like forms. During the saturation, the circulation of carbon dioxide as often required in a triaxial test was not applied in this experiment. As a matter of fact, there is no such feasibility of using the carbon dioxide in the field condition. To further increase the degree of saturation, a stepwise increment of the back pressure (i.e., 100 kPa) was applied on each specimen, whilst the effective stress maintains constant (σ' = 30 kPa). By following above procedures, an equilibrium of inflow was finally reached at each saturation stage, in the meantime, it enables to measure the P-wave velocity and the associated B-value of the specimen.

3 Experimentation

3.1 Signal Interpretations

Figure 3(a) presented the received P-wave signals in a stepwise-saturated loess specimen having 5% of initial water content. Here an excitation signal of 5 kHz was used, because it provided the most clear waveforms under otherwise the same conditions. It was revealed that the B-value increases with the back pressure, indicating an enhanced degree of saturation. Given that the P-wave component travels in phase with the excitation signal, the first arrival time of the P-wave was identified at the first bump in the received signals, denoted as upward triangles. As the back pressure (or B-value) increases, the travel time becomes shorter. For example, at BP = 800 kPa, the arrival time of the P-wave is about 62 μs, which is significantly less than that at BP = 200 kPa. Besides, wave signals of the loess specimen prepared at 14% of water content were displayed in Fig. 3(b). At a given back pressure little difference in terms of the B-value was observed. To facilitate comparisons, the arrival time in Fig. 3(a) was also included in this plot, denoted as downward arrows. It is evident that as the water content increases, the arrival time of the P-wave becomes longer at the first several saturation stages. Bearing in mind that the P-wave travels much faster in water than in solid grains, the above observation indicates that the loess specimen prepared with 14% of water content is more difficult to saturate than the one with 5% of water content under a relatively low back pressure, though the changes of B-value are quite similar in both cases. Figure 3(c) presented the received P-wave signals of the loess specimen prepared at 14% of water content and 10% of sand content. Likewise, the arrival time in Fig. 3(b) was included in this plot, denoted as downward triangles. It is evident that as the sand content increases, the arrival time of the P-wave becomes shorter at the first several saturation stages. It implies that the loess sample with sand addition is easier to saturate than the original loess specimen.

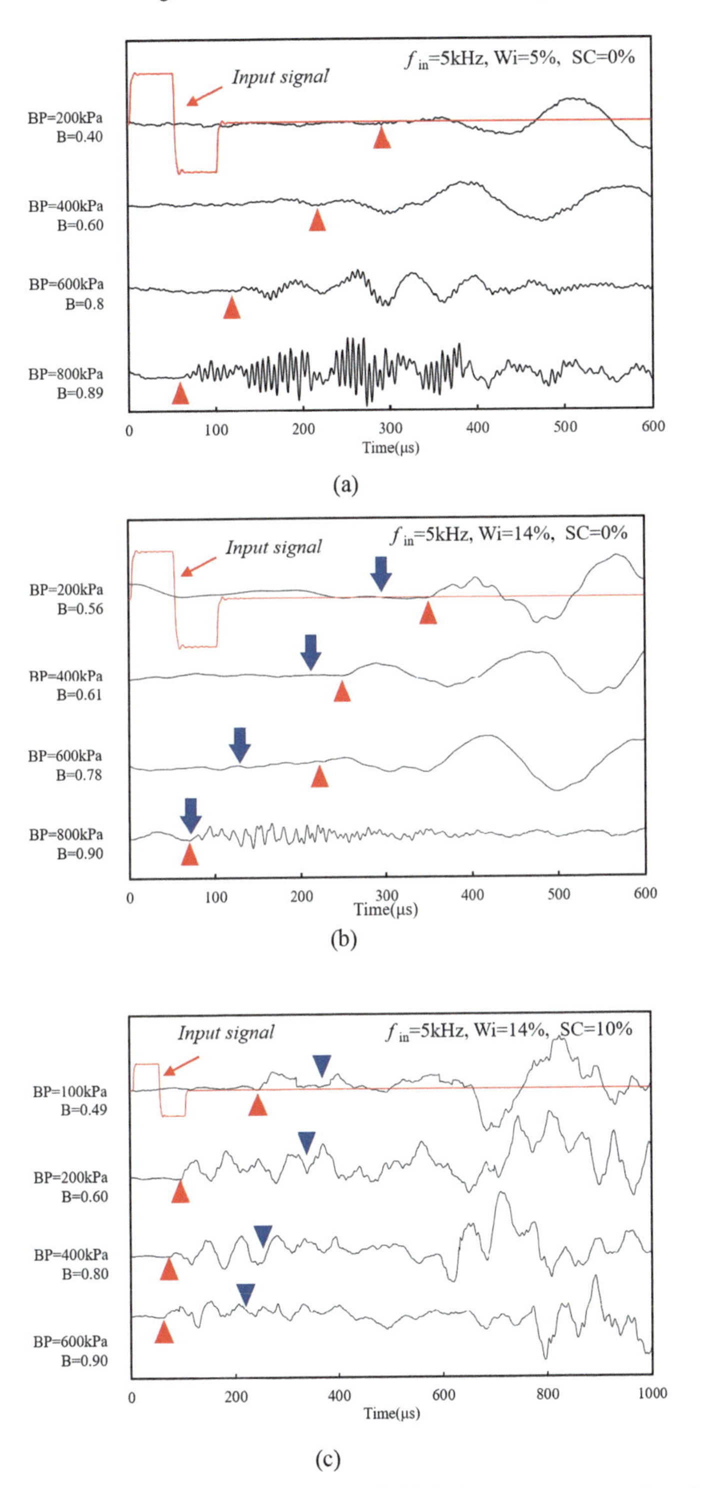

Fig. 3. Evolution of received signals in loess at varied initial water contents and sand content (a) Wi = 5%, SC = 0%; (b) Wi = 14%, SC = 0%; (c)Wi = 14%, SC = 10%.

3.2 P-wave Velocity and Saturation

By performing a careful interpretation on the arrival time of received signals, the P-wave velocity (V_p) was calculated by measuring the travel distance and the travel time. It has been generally confirmed that the travel distance is the length between two tips of extender elements (Liu and Yang 2018).

In Figs. 4 and 5, the measured V_p and the associated B-values were plotted. Besides, a prediction using Eq. 2 was also included in this plot. Note that the stiffness of loess ($K_b = 50$ MPa) in the prediction was determined in an isotropic consolidation test under an effective stress σ' = 30 kPa. In Fig. 4, several features were observed very interesting in this plot. Firstly, V_p increases with the B-value, that agrees with the theoretical prediction. At the final saturation stage, the measured P-wave velocity is about 1600 m/s, which is very close to that in water, indicating a fully saturated state (i.e., $S_r = 1$); Secondly, the above relationship depends on the initial water content. For a loess specimen reconstituted at 14% of water content, the measured V_p increases with the B-value that in general agrees with the prediction. On the other hand, for a specimen reconstituted at a 5% of water content, the measured V_p increases slightly with increasing B-value when the B-value is smaller than 0.4. Yet, when the B-value is larger than 0.4, discrepancies between the measured V_p and the prediction are observable, and it is quite remarkable at higher B-values.

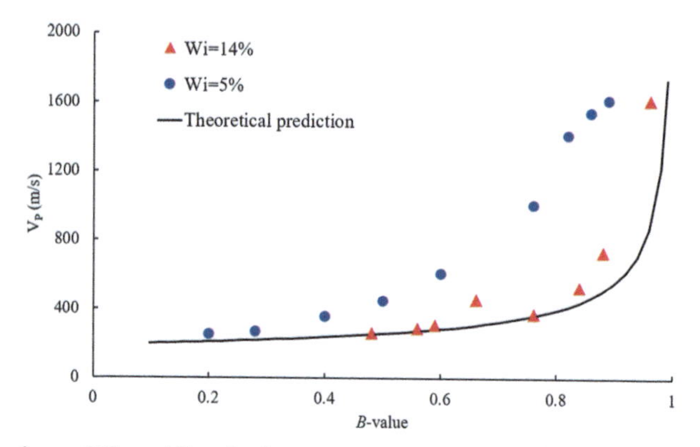

Fig. 4. Comparison of V_p and B-value between measurements and the prediction for loess with different water content.

Furthermore, in Fig. 5, effect of sand content on V_p was also investigated. It is clear that, at similar conditions (i.e., stress level, packing density), the measured V_p is greater for a loess specimen reconstituted with 10% of sand content. The rational explanation to this observation is due to an increased permeability of loess with the add-in sand.

To gain a better understanding, magnetic resonance imaging (MRI) tests were carried out for saturated loess specimens that were initially prepared at different water contents, and the results were shown in Fig. 6. It is evident from this figure that small pores in terms of the pore throat diameter (He et al. 2020) are dominant in loess samples. Of more interest, the loess specimen at 5% of initial water content contains more small size pores.

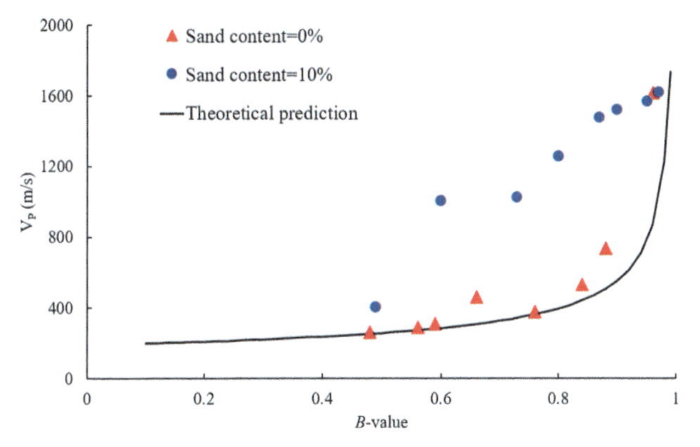

Fig. 5. Comparison of V_P and B-value between measurements and the prediction for loess with different sand content.

Thereby, it was hypothesized that less air bubbles were likely trapped within the small size pores (i.e., $W_i = 5\%$) under seepage conditions. In other words, the connected pores are prevailing, and more air bubbles accumulate preferentially to the top layer through the fluid channel owning to a smaller density ($\rho_{air} = 1.2$ kg/m^3, $\rho_{water} = 1000$ kg/m^3). With that being the case, though the B-value increases with back pressure that reflects an overall saturation of a specimen, the compression wave has to transmit through the spatially ordered "air-filled" local pores of varying compressibility, resulting different V_P.

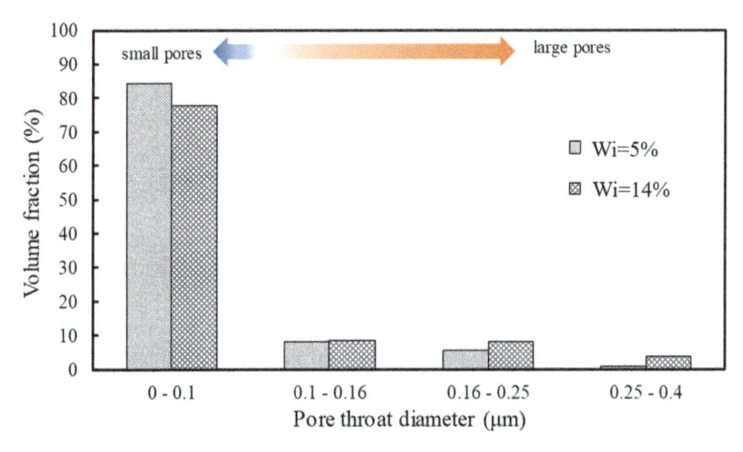

Fig. 6. MRI results of loess specimens at different initial water contents.

4 Conclusions

In this study, discussions were highlighted on the feasibility of using the P-wave velocity to evaluate the saturation state of loess. Several main findings were summarized as follows. (a) At each saturation stage, the arrival time of the compression wave is shorter in the loess specimen with 5% of initial water content as compared with that of a higher water content (i.e., 14%), under the same water content, the arrival time of the compression wave is shorter in the loess specimen with 10% of sand content as compared with the loess specimen with 0% of sand content; (b) The measured V_p increases with the B-value that is in line with the prediction based on the poroelasticity theory. Yet, departures from the prediction were observable at higher B-values in particular for the specimen prepared with 5% of water content and the specimen prepared with 10% of sand content; (c) It was hypothesized that the distinct trends between V_p and the B-value as compared with the theoretical prediction were attributed to different characteristics of the pore size and the associated distribution of air bubbles. The outcomes of this study provide useful guidance to use the P-wave velocity as an indicator of saturation in loess materials.

References

1. Gu, X., Yang, J., Huang, M.: Laboratory investigation on relationship between degree of saturation, B-value and P-wave velocity. J. Cent. South Univ. **20**, 2001–2007 (2013)
2. He, P., Xu, Q., Liu, J., Li, P., Pu, C., Chen, D.: Effect of pore size distribution on percolation characteristics of loess based on nuclear magnetic resonance technique. Sci. Tech. Eng. **20**(30), 12355–12360 (2020)
3. Karam, J.P., Cui, Y.J., Tang, A.M., Terpereau, J.M., Marchadier, G.: Experimental study on the cyclic resistance of a natural loess from northern France. Soils Found. **49**(3), 421–429 (2009)
4. Liu, X., Yang, J.: Shear wave velocity in sand: effect of grain shape. Géotechnique **68**(8), 742–748 (2018)
5. Liu, X., Qin, H., Lan, H.X.: On the relationship between soil strength and wave velocities of sandy loess subjected to freeze-thaw cycling. Soil Dyn. Earthq. Eng. **136**, 106216 (2020)
6. Naesgaard, E., Byrne, P.M., Wijewickreme, D.: Is P-wave velocity an indicator of saturation in sand with viscous pore fluid? Int. J. Geomech. **7**(6), 437–443 (2007)
7. Sherif, M.A., Tsuchiya, C., Ishibashi, I.: Saturation effects on initial soil liquefaction. Am. Soc. Civil Eng. **103**(8), 914–917 (1977)
8. Skempton, A.W.: The pore-pressure coefficients A and B. Géotechnique **4**(4), 143–147 (1954)
9. Sugiyama, Y., Kawai, K., Iizukac, A.: Effects of stress conditions on B-value measurement. Soils Found. **56**(5), 848–860 (2016)
10. Tamura, S., Tokimatsu, K., Abe, A., Sato, M.: Effects of air bubbles on B-value and P-wave velocity of a partly saturated sand. Soils Found. **42**(1), 121–129 (2002)
11. Tsukamoto, Y., Ishihara, K., Nakazawa, H., Kamada, K., Huang, Y.: Resistance of partly saturated sand to liquefaction with reference to longitudinal and shear wave velocities. Soils Found. **42**(6), 93–104 (2002)
12. Yang, J.: Liquefaction resistance of sand in relation to P-wave velocity. Géotechnique **52**(4), 295–298 (2002)
13. Yoshimi, Y., Tanaka, K., Tokimatsu, K.: Liquefaction resistance of a partially saturated sand. Soils Found. **29**(3), 157–162 (1989)

Relationship Between Shear Wave Velocity and Liquefaction Resistance in Silty Sand and Volcanic Sand

Masataka Shiga(✉) and Takashi Kiyota

Institute of Industrial Science, The University of Tokyo, Bunkyo, Japan
{shiga815,kiyota}@iis.u-tokyo.ac.jp

Abstract. Quantification of influencing factors to liquefaction is essential for accurate prediction of in-situ liquefaction resistance. Previous studies have shown that there is a significant relationship between shear wave velocity and liquefaction resistance for a limited number of samples under the same density as in-situ with different soil fabric. However, samples such as silty sand or volcanic sand have not been investigated. In this study, reconstituted specimens were prepared using in-situ samples collected from two locations, and shear wave velocity measurements and undrained cyclic loading tests were conducted. The results show that shear wave velocity and liquefaction resistance rise with increasing over-consolidation ratio, and when they are normalized to specimens without over-consolidation history, the results are in good agreement with previous trends.

Keywords: Liquefaction · Shear wave velocity · Cyclic resistance ratio

1 Introduction

It has been recognized since the 1960s that liquefaction can cause severe damage to structures[1, 2]. This recognition has led to more precise methods for assessing the possibility of liquefaction of the present ground in response to expected earthquake ground motions. A typical example is a stress-based method by Seed et al. [3], which calculates the factor of safety using the ratio of seismic shear stress to liquefaction resistance of the ground. Although this concept is now widely used [4, 5], how calculated seismic shear stress and liquefaction resistance differs from one guideline.

Most liquefaction assessment methods in Japan calculate liquefaction resistance based on N-values from standard penetration tests. In the 2011 off the Pacific coast of Tohoku earthquake (referred to as the Tohoku earthquake), the current N-value-based method underestimated liquefaction resistance in 76 out of 199 surveyed sites [6]. Previous studies [7, 8] reported that the N-value has a specific correlation with the density of the ground. Several other studies [9, 10] have also examined the positive correlation between in-situ density and liquefaction resistance. Therefore, the underestimation in the Tohoku earthquake can be attributed to in-situ ground factors other than density.

© The Author(s), under exclusive license to Springer Nature Switzerland AG 2022
L. Wang et al. (Eds.): PBD-IV 2022, GGEE 52, pp. 2131–2140, 2022.
https://doi.org/10.1007/978-3-031-11898-2_196

Seed et al.[3] identified the following soil factors influencing liquefaction resistance: density, soil types, stress history, initial depositional environment, aging effect and stress state. In subsequent studies, the correlation of liquefaction resitance with above factors has been discussed in detail. Ye et al.[11] or Suzuki and Toki [12] conducted a series of undrained cyclic loading tests after pre-shearing. It was found that small and large pre-shear play different roles in changing liquefaction resistance. Clough [13] or Porcino et al. [14] used sandy material with a cementing agent to study the cementation effect, and the treated specimens exhibited an improved cyclic liquefaction resistance. Although many types of research were conducted for those factors with laboratory experiments, it had still been challenging to quantify the extent to which those factors affect liquefaction resistance.

To estimate the in-situ liquefaction resistance depending on these multiple factors, Kiyota et al. [15] compared shear wave velocity and liquefaction resistance of reconstituted specimens with different soil fabric and the same density as in-situ ground. This study relies on the hypothesis that the current soil fabric reflects stress history and initial depositional environment and can be quantified by shear wave velocity. In conclusion, Kiyota et al. showed that OCR1 specimen normalized shear wave velocity and liquefaction resistance show a notable trend regardless of sand type. However, since this relationship was established for a limited number of samples, such as the Toyoura sand and samples collected at the liquefaction site, it was not investigated whether the relationship is valid for a broader range of soil types.

The above previous studies can be summarized as follows.

1. The estimation of in-situ liquefaction resistance depends on many factors.
2. The relationship between shear wave velocity and liquefaction resistance is established for a limited number of samples.

In this paper, the author conducted a series of undrained cyclic loading tests with shear wave velocity measurements to investigate the relationship between shear wave velocity and liquefaction resistance at the same density for silty sand and volcanic sand and compare the results with the past studies.

2 In-Situ Survey and Specifications of Laboratory Testing

In this study, two locations in Japan were chosen to take in-situ samples, as shown in Fig. 1. The selection of these sites was based on the fact that FL value calculated by the liquefaction estimation method was below 1. However, liquefaction did not occur in the vicinity during a particular earthquake in the past.

One sample is silty sand at the site in Ukishima-Town, Kawasaki-City, Kanagawa-Pref. ($\rho_s = 2.702\ \text{g/cm}^3, D_{50} = 0.110\ \text{mm}, F_c = 34.4\%$). Although a maximum horizontal acceleration of 128.1 cm/s^2 was recorded at the seismic station located 1 km west-southwest of this point during the Tohoku earthquake, there was no ground deformation induced by liquefaction. However, the current liquefaction estimation method showed that FL value was below one from five meters to 11.5 m depth using standard penetration test (JIS A1219), as illustrated in Fig. 2.

The other sample was collected at the site in Azahoutoku, Bihoro-Town, Hokkaido. The collected sample was volcanic Sand ($\rho_s = 2.547$ g/cm^3, $D_{50} = 0.443$ mm, $F_c = 1.3\%$), geologically classified as Kutcharo pumice flow deposit. The Tokachi-oki Earthquake in 2003 caused severe liquefaction damage in the southern part of Hokkaido. Although the maximum horizontal acceleration of 85.3 cm/s^2 was recorded at the site's seismic station located 3 km southwest, no liquefaction damage was confirmed in the vicinity. Borehole investigation and SPT test showed that loose volcanic sand was deposited from 2 m to 4 m depth, and the FL value at the shallow region was lower one, as shown in Fig. 3.

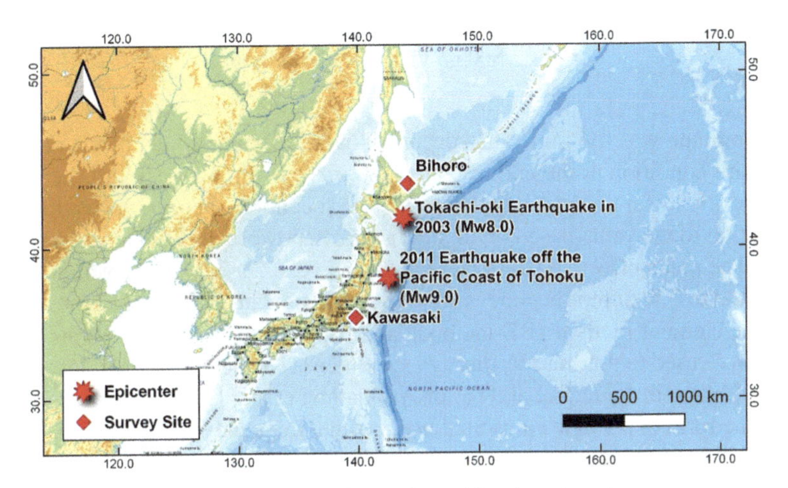

Fig. 1. Survey location and considered earthquakes

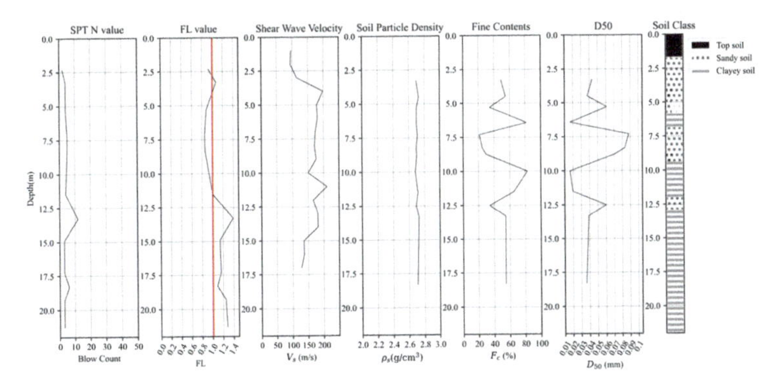

Fig. 2. Soil profile at the site in Kawasaki

A series of undrained cyclic loading tests of those two samples were conducted with a triaxial apparatus to obtain stress and strain parameters and shear wave velocity (V_s). The silty sand and the volcanic sand specimens have a height of 100 mm and a diameter of 50 mm, a height of 150 mm, and a diameter of 75 mm, respectively. The thickness

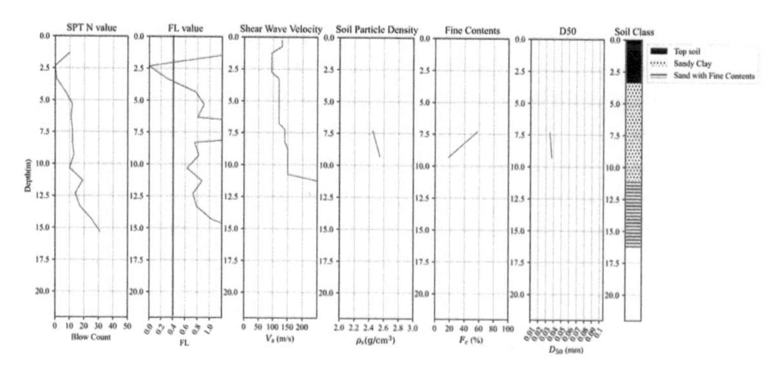

Fig. 3. Soil profile at the site in Bihoro

of the membrane was 0.3 mm. The specimens were prepared to achieve a dry-density at the in-situ condition at the end of consolidation. The initial density of the specimen was set slightly lower than the target density. The specimens were subjected to isotropic consolidation to the value based on in-situ mean effective stress with a back pressure of 200 kPa. The mean effective stress was calculated based on effective overburden stress and K_0 equal to 0.5. Some specimens were over-consolidated to change their soil fabric.

Bender element method (BE method) and trigger and accelerometer method (TA method) were used to measure V_s through specimens. The detailed methodology for BE method is shown in Jovičić et al. [16] or Arulnathan et al. [17]. In TA method, a shear wave, in the form of a single sinusoidal wave with a frequency of 1 kHz, was generated by a pair of wave sources attached to the top cap that were simultaneously excited in the torsional direction [18]. A pair of accelerometers was employed to measure the arrival time of the shear wave at two different heights of the specimen.

After the measurement of V_s, , undrained cyclic loading tests were conducted with a constant amplitude of cyclic deviator stress until double amplitude of axial strain ($\varepsilon_{a(DA)}$) reached 5%. The detailed specification of the test for each specimen is given in Table 1..

Table 1. Detailed specification of undrained cyclic loading tests

Name	Apparatus	Specimen Dimension	Target Dry Density ρ_d (g/cm^3)	Mean Effective Stress σ_c' (kPa)	Cyclic Stress Ratio CSR	Over-consolidation Ratio OCR	Obtained Dry Density $\rho_{d(ob)}$ (g/cm^3)
Kawa_01	Stress-controled	D50 mm × H100 mm	1.45	55	0.08	1	1.517
Kawa_02					0.1		1.448
Kawa_11					0.12		1.432
Kawa_12					0.12		1.527
Kawa_14					0.08		1.444

(*continued*)

Table 1. (*continued*)

Name	Apparatus	Specimen Dimension	Target Dry Density ρ_d (g/cm^3)	Mean Effective Stress σ_c' (kPa)	Cyclic Stress Ratio CSR	Over-consolidation Ratio OCR	Obtained Dry Density $\rho_{d(ob)}$ (g/cm^3)
Kawa_16					0.1		1.585
Kawa_04					0.25	5	1.502
Kawa_17					0.3		1.543
Kawa_18					0.4		1.525
Kawa_19					0.25	3	1.556
Kawa_20					0.15		1.52
Kawa_21					0.2		1.521
Biho_01	Strain-controled	D75mm × H150mm	1.47	30	0.4	5	1.492
Biho_05					0.25	1	1.503
Biho_08					0.2	1	1.45
Biho_11					0.3	5	1.465
Biho_12					0.35	5	1.478

3 Undrained Cyclic Loading Tests

3.1 Silty Sand Obtained in Kawasaki

Figure 4 and Fig. 5 show the undrained cyclic loading test results with the silty sand obtained in Kawasaki-City. The loading was stopped when $\varepsilon_{a(DA)}$ exceeded 5%. However, since excess pore water pressure ratio ($\Delta u/\sigma_c'$) did not reach 95% in some tests, the maximum value of $\Delta u/\sigma_c'$ is written next to the number of cycles in the legend box.

The OCR1 specimen shows positive excess pore water pressure during the first loading cycle, while the OCR5 specimen shows less excess pore water pressure in the early stage of loading. The stress-strain relationship of both specimens shows that the incremental axial strain for one cyclic loading became large after the axial strain exceeded 1%. This is the reason why the difference between $N_{c\left(\Delta u/\sigma_c'=0.95\right)}$ and $N_{c(DA=5\%)}$ is less than one for all tests.

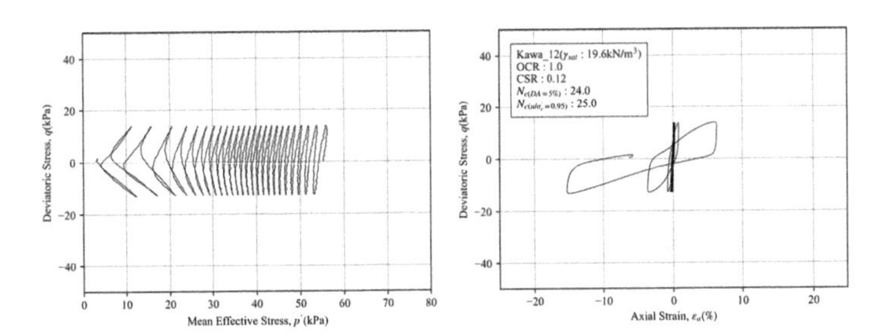

Fig. 4. Stress path and stress-strain curve of OCR1 specimens of the silty sand

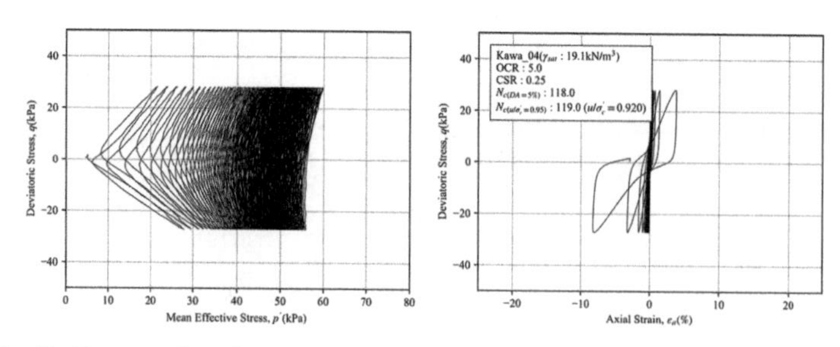

Fig. 5. Stress path and stress-strain curve of OCR5 specimens of the silty sand

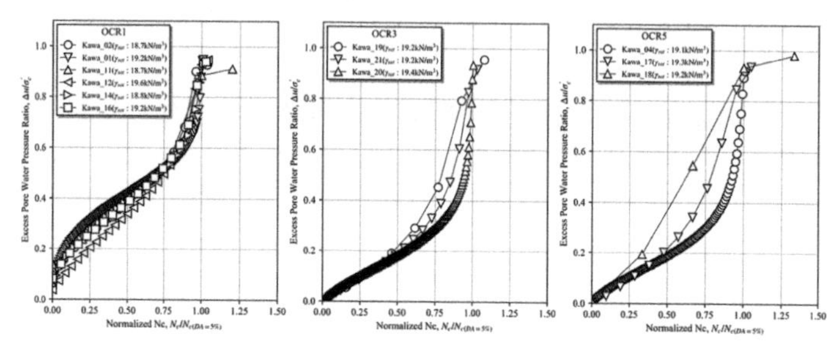

Fig. 6. The correlation between normalized N_c and $\Delta u/\sigma_c'$

Figure 6. Shows the difference between N_c normalized by $N_{c(\Delta u/\sigma_c'=0.95)}$. And the maximum excess pore water pressure generated during each cycle. In the case of OCR1, it can be seen that the excess pore water pressure increases sharply as the normalized N_c approaches 1. In the cases of OCR3 and OCR5, the specimens with higher CSR have higher excess pore water pressure values for the same number of cycles. This can be

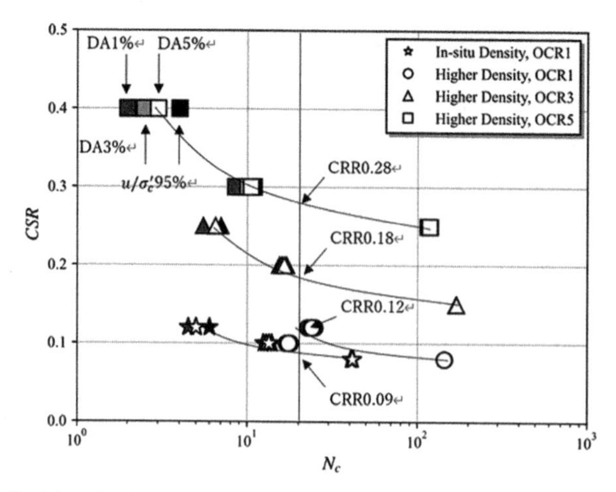

Fig. 7. Liquefaction curve with different OCR of silty sand in Kawasaki

attributed to the fact that higher shear stresses cause greater negative dilatancy, further generating excess pore water pressure.

Figure 7. Shows the plot of N_c and CSR with different OCR. The different colors of the plots represent N_c at $\varepsilon_{a(DA)} = 1\%$, $\varepsilon_{a(DA)} = 3\%$, $\varepsilon_{a(DA)} = 5\%$ and $\Delta u/\sigma_c' = 95\%$, respectively. The curve in the figure is the liquefaction resistance curve of each OCR, based on $N_{c(DA=5\%)}$. The CRR of the OCR1 specimen with the same density as in-situ was 0.09. For specimens with equal density after over-consolidation, the values of CRR were 0.12, 0.18, and 0.28 for specimens with OCR 1, 3, and 5, respectively.

3.2 Volcanic Sand Obtained in Bihoro

Figure 8. And Fig. 9. Show the stress-strain relationship and the effective stress path of OCR 1 and 5. As the same tendency can be seen in the previous section, positive excess pore water pressure is sharply accumulated in the OCR 1 specimen at the first cycle, while the OCR 5 specimen less increment of excess pore water pressure immediately after the initial loading.

Figure 10. Shows the relation between CSR and N_c, which is the number of cyclic when $\varepsilon_{a(DA)}$ reaches 5%. It is found that as the OCR increases, the liquefaction curve moves upward on the figure. The liquefaction resistance for OCR 1 is 0.22 and for OCR 5 was 0.33.

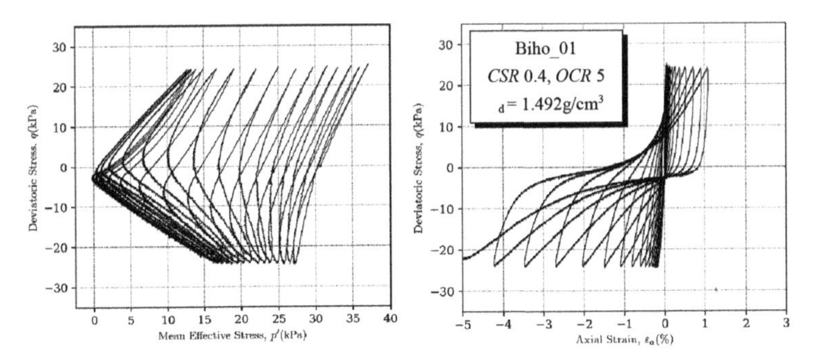

Fig. 8. Stress path and stress-strain curve of OCR5 specimens of the volcanic sand

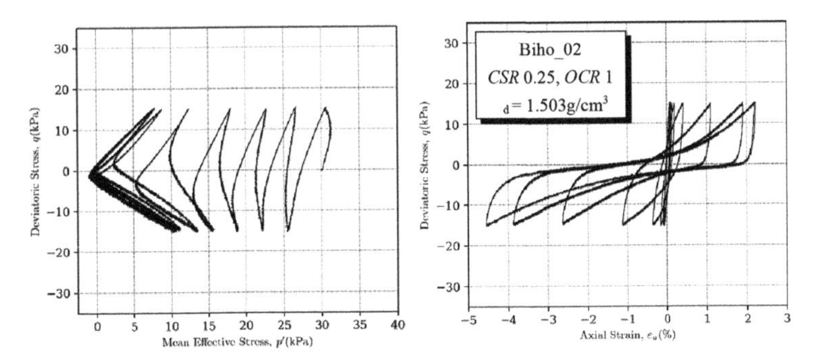

Fig. 9. Stress path and stress-strain curve of OCR1 specimens of the volcanic sand

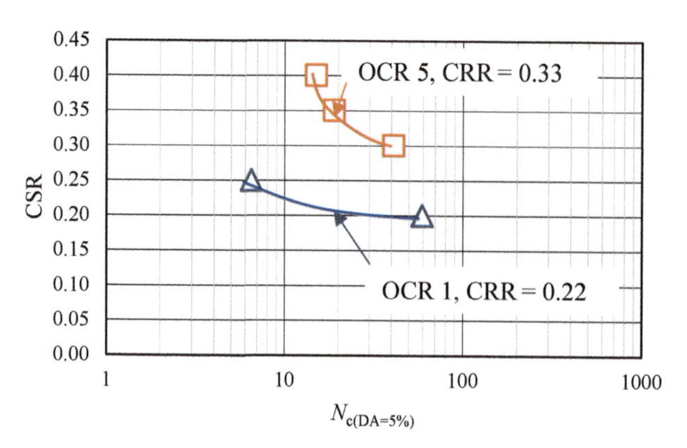

Fig. 10. Liquefaction curve with different OCR of volcanic sand in Bihoro

3.3 Relationship Between Shear Wave Velocity and CRR

Figure 11 combines the relationship between normalized V_S (V_S/V_S^*) and normalized CRR (CRR/CRR*) of the silty and volcanic sands compared with the result obtained by Kiyota et al.[15]. The denominators are V_S and CRR of OCR 1 specimen of each sand, respectively. The original paper stated that they could obtain the normalized V_S less than one in case that they conducted undrained cyclic loading test with liquefied and re-consolidated specimens.

Although many previous pieces of research indicated that the relationship between V_S and CRR depends on many parameters such as density, soil type, soil fabric, aging effect, and stress state, Kiyota et al. reported that the specimen composed of the same material with the same density and stress state as the in-situ condition has a unique relationship between V_S/V_S^* and CRR/CRR* regardless of the type of sample and specimen density. Based on the compiled data plot, Kiyota et al. proposed a regression equation of the relationship as;

$$\frac{CRR}{CRR*} = \left(\frac{V_S}{V_S^*}\right)^{5.02} \tag{1}$$

Figure 11 was drawn to confirm whether the relationship is also valid for the silty sand and volcanic sand. It can be seen that the relationship between V_S and CRR in the two series of undrained cyclic loading tests follows the existing relationship. This implies a unique relationship between shear wave velocity and liquefaction resistance of specimens with different soil fabric in various types of sandy soils, such as silty sand and volcanic sand without crushable properties.

As Kiyota et al. described, the unique relationship can be utilized to obtain in-situ liquefaction resistance. Gamma-lay density logging and PS logging are quite common non-destructive method to get in-situ density and shear wave velocity, respectively, which is equivalent to V_S in Eq. (1). In addition, sampling of the liquefied layer can be conducted even by tube sampling known to have some disturbance during sampling [19]. The disturbed samples are subjected to cyclic undrained loading tests with shear wave velocity measurement to obtain V_S^* and CRR* under the same density as in-situ one.

Although the method of falling or layering apparently varies V_s^* and CRR*, the unique relationship, Eq. (1) can calculate in situ CRR with involving the soil fabric effect caused by different preparation methods.

In future studies, we are going to estimate in-situ liquefaction resistance and safety factor of liquefaction at several depths by utilizing the uniform relationship, and compare the estimation accuracy with some exiting liquefaction assessment method.

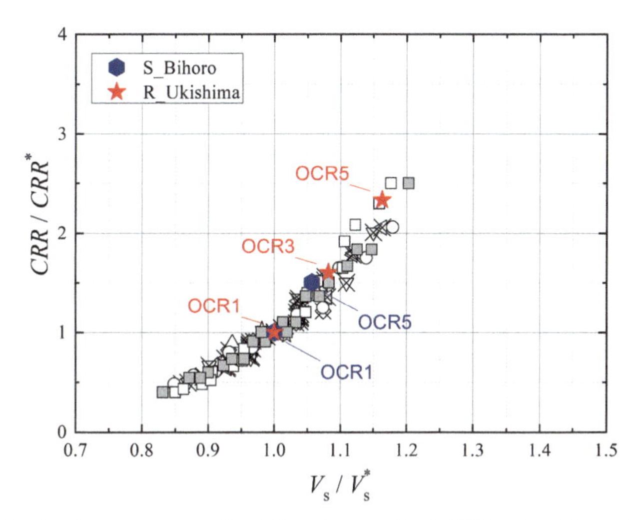

Fig. 11. Relationship between normalized V_s and normalized CRR of the silty and volcanic sands (after Kiyota et al. [15])

4 Conclusion

A series of undrained cyclic loading tests of two samples were conducted with a triaxial apparatus to investigate the effect of soil fabric on liquefaction resistance by using shear wave velocity. The specimens were prepared using in-situ samples collected from two locations and adjusted to the same density as the in-situ condition. In addition, some specimens are subjected to over-consolidation to change their soil fabric. The current conclusions at this point are as follows.

1. Both two in-situ samples showed a decrease in excess pore water pressure generation in the first cycle and an increase in liquefaction resistance as the over-consolidation ratio increased.
2. Shear wave velocity rises as the over-consolidation ratio increases. In terms of liquefaction resistance and shear wave velocity normalized by the values of the virgin specimens that were not subjected to over-consolidation, the trend was consistent with the previous articles.

Acknowledgments. This research was funded by Grant-in-Aid for JSPS Fellows Number 19J12349. The author is grateful to Mr. Jun Kawamura and Mr. Makoto Kitayama in Chemical Grout Co., Ltd., and Dr. Takemine Yamada in Kajima Corporation for providing the insightful in-situ data and samples. I also thank Dr. Takuya Egawa in Civil Engineering Research Institute for Cold Region for sharing the data and giving technical comments to our papers.

References

1. Ishihara, K.: Simple method of analysis for liquefaction of sand deposits during earthquakes. Soils Found. **17**(3), 1–17 (1977)
2. Seed, H.B., Idriss, I.M.: Analysis of soil liquefaction: niigata earthquake. J. Soil Mech. Found. Div. **93**(3), 83–108 (1967)
3. Seed, H.B., Idriss, I.M.: Simplified procedure for evaluating soil liquefaction potential. J. Soil Mech. Found. Div. **97**(9), 1249–1273 (1971)
4. Architectural Institute of Japan, Recommendations for Design of Building Foundation (2019)
5. Japan Road Association, Specifications for Highway Bridges -Part 5Seismic Design (2017)
6. Soil Dynamics Team, Geotechnical and Geotechnical Research Group, Public Works Research Institute, A Study on Liquefaction Determination Method Based on Liquefaction in the 2011 off the Pacific coast of Tohoku Earthquake (2014)
7. Cubrinovski, M., Ishihara, K.: Empirical correlation between SPT N-value and relative density for sandy soils. Soils Found. **39**(5), 61–71 (1999)
8. Marcuson, W.F., Bieganousky, W.A.: SPT and relative density in coarse sands. J. Geotech. Eng. Div. **103**(11), 1295–1309 (1977)
9. Tokimatsu, K., Yoshimi, Y.: Empirical correlation of soil liquefaction based on Spt N-value and fines content. Soils Found. **23**(4), 56–74 (1983)
10. Yoshimi, Y., Tokimatsu, T., Ohara, J.: In situ liquefaction resistance of clean sands over a wide density range. Geotechnique **44**(3), 479–494 (2015). https://doi.org/10.1680/geot.1994.44.3.479
11. Ye, B., Lu, J., Ye, G.: Pre-shear effect on liquefaction resistance of a Fujian sand. Soil Dyn. Earthq. Eng. **77**, 15–23 (2015)
12. Suzuki, T., Toki, S.: Effects of preshearing on liquefaction characteristics of saturated sand subjected to cyclic loading. Soils Found. **24**(2), 16–28 (1984)
13. Clough, G.W., Iwabuchi, J., Rad, N.S., Kuppusamy, T.: Influence of cementation on liquefaction of sands. J. Geotech. Eng. **115**(8), 1102–1117 (1989)
14. Porcino, D., Marcianò, V., Granata, R.: Cyclic liquefaction behaviour of a moderately cemented grouted sand under repeated loading. Soil Dyn. Earthq. Eng. **79**, 36–46 (2015)
15. Kiyota, T., Maekawa, Y., Wu, C.: Using in-situ and laboratory-measured shear wave velocities to evaluate the influence of soil fabric on in-situ liquefaction resistance. Soil Dyn. Earthq. Eng. **117**, 164–173 (2019)
16. Jovičić, V., Coop, M.R., Simić, M.: Objective criteria for determining Gmax from bender element tests, Geotechnique **46**(2), 357-362, (2015). https://doi.org/10.1680/geot.1996.46.2.357
17. Arulnathan, R., Boulanger, R.W., Riemer, M.F.: Analysis of bender element tests. Geotech. Test. J. **21**(2), 120–131 (1998)
18. AnhDan, L.Q., Koseki, J., Sato, T.: Comparison of young's moduli of dense sand and gravel measured by dynamic and static methods. Geotech. Test. J. **25**(4), 349–368 (2002)
19. Tokimatsu, K., Hosaka, Y.: Effects of sample disturbance on dynamic properties of sand. Soils Found. **26**(1), 53–64 (1985)

Small to Medium Strain Dynamic Properties of Lanzhou Loess

Binghui Song[1](✉), Angelos Tsinaris[2], Anastasios Anastasiadis[2], Kyriazis Pitilakis[2], and Wenwu Chen[3]

[1] First Institute of Oceanography, MNR, Qingdao 266061, People's Republic of China
bhsong@fio.org.cn
[2] Aristotle University of Thessaloniki, 54124 Thessaloniki, Greece
[3] Lanzhou University, Lanzhou 730000, People's Republic of China

Abstract. Loess is a kind of world-recognized problematic soil and, as a consequence, it is a major hazard in geotechnical engineering. This study examines the small to medium strain dynamic properties of Lanzhou intact and recompacted loess specimens from different burial depths through a set of resonant column (RC) tests. The influence of soil structure on the initial shear modulus G_0, the variations of G/G_0 and damping ratio DT (%) with shear strain γ were analyzed and discussed, including also comparisons between undisturbed and recompacted loess. The test results show that the structure effect has an important influence on the dynamic properties of dry loess at small and medium strain, whereas, with the increase of moisture, the structure effect weakens, causing the dynamic properties (i.e., G_0, G/G_0-γ and DT (%)-γ) of undisturbed loess are comparable to that of recompacted ones. Finally, it is proved that a normalized correlation between G/G_0 and $\gamma/\gamma_{0.7}$, where $\gamma_{0.7}$ is a reference soil strain is practically identical for undisturbed and recompacted loess in Lanzhou, irrespective of soil structure, which may be useful in practical applications.

Keywords: Loess · Small to medium strain · Shear modulus · Damping ratio · Structure

1 Introduction

Loess Plateau in China is an earthquake-prone area where tectonic movements are very active. Throughout history, several catastrophic disasters caused by earthquakes have been reported and documented [1, 2]. To this end, the knowledge of the dynamic properties of loess is of utmost importance and significance for earthquake protection and risk mitigation.

During the past few years, Lanzhou has carried out a large-scale land-cutting and land-filling project in order to expand the development space of the urban area, and numerous filling sites are built, destroying the inherent structure of loess. Therefore, it has important application value to evaluate the structure effect on the dynamic properties of loess through comparing the dynamic characteristics between the natural and

© The Author(s), under exclusive license to Springer Nature Switzerland AG 2022
L. Wang et al. (Eds.): PBD-IV 2022, GGEE 52, pp. 2141–2150, 2022.
https://doi.org/10.1007/978-3-031-11898-2_197

recompacted loess, which provides technical reference for the implementation of relevant engineering activities.

Structure effects on the dynamic shear modulus and damping ratio of intact loess had been particularly addressed by Wang et al., [3] where it was reported that the initial deviatoric stress ratio, which was regarded as a parameter representing the degree of structure degradation of loess, significantly influenced the dynamic properties of undisturbed loess. In general, while the significance of structure effect on the dynamic properties of loess has been acknowledged, the available findings are very limited, especially for dynamic properties of loess at small to medium shear strain levels.

The present paper aims to contribute to the understanding of the structure effect of loess found in Lanzhou, Northwest of China, on its dynamic properties at small to medium shear strain levels. Research could provide an important reference for revealing the dynamic characteristics of structural loess in the small- and medium-strain range.

2 Loess Samples and Testing Program

2.1 Loess Sample

The loess samples (both disturbed and undisturbed) were collected from the New Area of Lanzhou with the burial depths of 4 m, 6 m, and 23 m, respectively (Table 1.). In the laboratory, conventional tests were conducted to measure the basic physical properties of Lanzhou loess according to the ASTM specifications. Table 1. summarizes the index properties of Lanzhou loess tested in this program. Based on the age of deposition and physical properties, the loess in the present study could be designated as Q_4 loess [4].

The granulometric characteristics of Lanzhou loess are depicted in Fig. 1. As shown, the main part of Lanzhou loess is composed of silt (0.005 mm–0.05 mm), occupying more than 70% in total, while the clay fraction (<0.005 mm) is less than 20% accompanying by about 10% of sand (0.05 mm–2 mm). Though the clay fraction is limited, it has a significant role in supporting the open microstructure of loess as bonding agents [5].

Table 1. Index properties of natural loess in Lanzhou

Sample	Depth/m	Moisture W_0 (%)	Density ρ (g/cm^3)	Void ratio e	Specific gravity G_s	Saturation S_r (%)	LL (%)	PI (%)
LZL-4	4	14.33	1.50	1.06	2.70	36.6	24.3	7.7
LZL-6	6	13.42	1.53	1.00	2.70	36.2	24.8	8.1
LZL-23	23	14.22	1.60	0.93	2.70	41.4	25.5	9.3

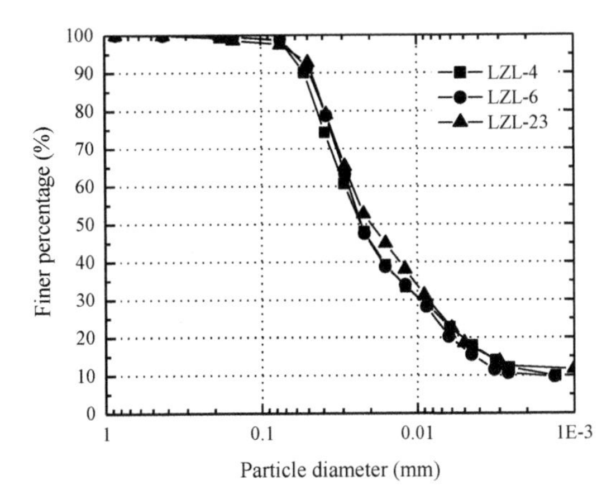

Fig. 1. Grain size distribution of Lanzhou loess specimens from different burial depth

2.2 Testing Program

In the present study, a series of resonant column tests were performed on intact and recompacted loess specimens having different water contents under various confining pressures. Table 2. summarizes the moisture condition of the tested loess specimens. W_0 (%) refers to the moisture content of loess in natural condition. Considering that water content of loess could be altered greatly due to evaporation, rain, or irrigation activities, two kinds of moisture condition were selected in the present study, i.e., the drying condition ($W_{RC}\approx7\%$) is approximated to the water content of surficial loess, and the wetting condition ($W_{RC}\approx25\%$) represents the highly saturated state of loess. It should be noted that W_{RC} (%) refers to the moisture of the specimen after resonant column test.

In terms of sample preparation, loess specimens with lower moistures were prepared using the well-documented method for drying samples illustrated by Haeri et al., [6] and Wen and Yan [7]. Back-pressure saturation method following ASTM D4767–11 regulations was used to prepare loess with a high degree of saturation (e.g. LZL-4-U). Recompacted specimens were formed at the target initial dry densities similar to the corresponding intact loess following the standard compaction method proposed by Ladd [8]. The compaction moisture is around 7%, which approaches the water content of surficial loess. The range of confining pressures applied during RC test is also listed in Table 2..

The free-fixed end resonant column device comprehensively tested and evaluated through numerous past projects has been used [9–11]. The whole operating procedure and the analysis of the RC test results were conducted following the ASTM 4015–92 specification. Detailed information about experimental equipment and testing procedure had been given in Song et al., [12].

Table 2. Moisture conditions of loess specimens and confining pressures σ'_m for RC test

Sample	Natural condition	Resonant column tests		
	W_0 (%)	W_{RC} (%)	$S_{r\text{-}RC}$ (%)	σ'_m (kPa)
LZL-4-U(1)	14.33	24.41	78.6	25, 50, 100
LZL-4-R(2)	14.33	23.99	80.6	25, 50, 100
LZL-6-U	13.42	6.83	18.9	25, 50, 100, 200
LZL-6-R	13.42	6.52	18.2	25, 50, 100, 200
LZL-23-U	14.22	24.52	79.2	50, 100, 200
LZL-23-R	14.22	25.88	78.4	50, 100, 200

Note: U(1): undisturbed specimen; R(2): recompacted specimen.

3 Experimental Results and Discussions

3.1 Shear Modulus and Damping of Lanzhou Loess at Small and Medium Strain Levels

Initial Shear Modulus G_0

Shear modulus G_0 at small shear strain levels is one of the key parameters in seismic site response analysis. Comparisons of G_0 for undisturbed and recompacted loess under different confining pressures are given in Fig. 2. As shown, the initial shear modulus G_0 of loess generally rises monotonically with effective confining pressure σ'_m, but different trends are observed between dry and wet loess. For dry loess (LZL-6), the rate of increase between G_0 and σ'_m is more pronounced than for wet loess (LZL-4/23). Additionally, the G_0 of undisturbed specimen is much bigger than that of recompacted one, and the difference increases significantly with effective confining pressures. When loess becomes wet, and in particular highly saturated (LZL-4/23), the values of G_0 between undisturbed and recompacted loess approach each other quite closely, the relationships between G_0 and σ'_m for such loess are almost identical. This phenomenon could be mainly related to the structural characteristics of loess. When loess is relatively dry, there is a large structural strength difference between the undisturbed and the recompacted loess, which increases the shear modulus of undisturbed loess compared to that of recompacted one under the same confining pressure [7, 13]. As the saturation of loess increases, the inter-particle cementation inside loess is softened gradually, so the corresponding structural strength of undisturbed loess reduces too, resulting in a closer and closer shear modulus between undisturbed and recompacted loess [7].

It should be noted that the loess samples used in the present study are unsaturated, in that case, the effect of matric suction on G_0 cannot be ignored according to unsaturated soil mechanics [14]. As previously described, the recompacted loess was prepared with a similar void ratio and degree of saturation as intact loess in the present study, and the confining pressure adopted in RC tests kept the same too. In addition, as stated by Cheng et al., [15] the microstructure of intact and remolded loess compacted under the same moisture ($w = 10.9\%$) is similar based on scanning electron microscope (SEM) measurements, resulting in a similar soil-water retention behavior between intact and recompacted loess. However, compared with the recompacted specimen, more clay aggregates are accumulating at inter-particle contacts in intact loess, which reduces slippages at silt contacts and stiffens the intact soil skeleton [16]. Because of this, it may be concluded that the difference in G_0 between intact and recompacted loess could be mainly due to the soil structure effect in the present study, and matric suction function is relatively limited [7, 13].

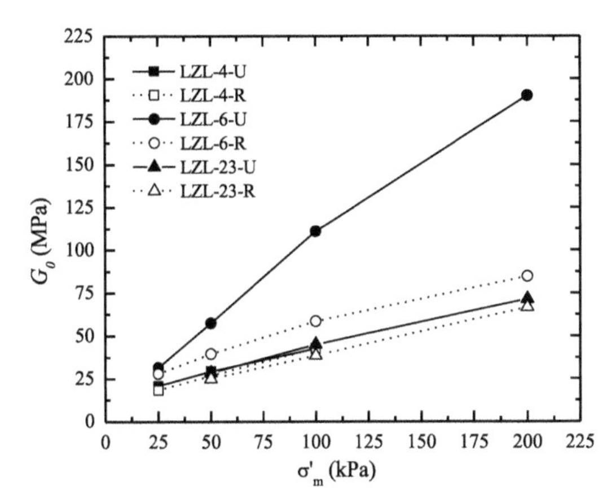

Fig. 2. Relationships between small-strain shear modulus G_0 and effective confining pressure σ'm for undisturbed (U) and recompacted (R) loess specimens tested

Degradation of Shear Modulus G/G$_0$-γ

Figure 3 illustrates the measured normalized shear modulus G/G_0 versus shear strain γ for undisturbed and recompacted loess specimens under similar moistures and confining pressures. One can notice that the normalized shear modulus G/G_0 of undisturbed loess degrades more quickly with shear strain than recompacted ones, indicating the ductility is enhanced during dynamic loading after remolding for loess, which is similar to the case of structural Augusta clay [13].

When loess is highly saturated, (e.g., W≈24%), the differences in G/G_0-γ curves between undisturbed and recompacted specimens are practically negligible. It appears that although there is a difference in structure composition between undisturbed and recompacted loess [17], just like in the G_0 case, the structure effect becomes insignificant and even negligible on G/G_0-γ curves at high moisture levels.

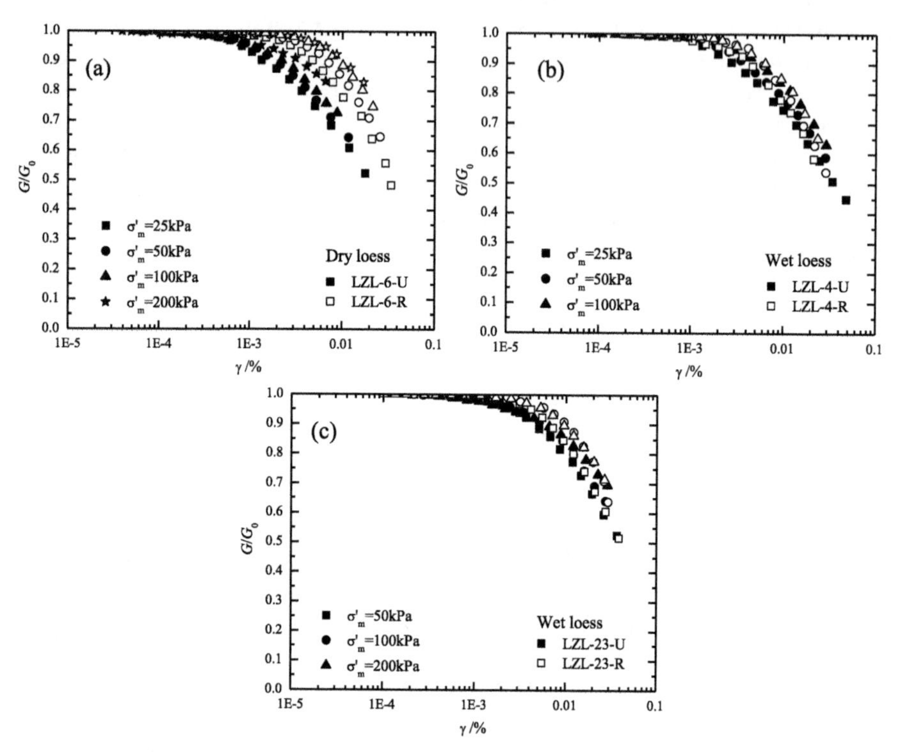

Fig. 3. Effect of cementation on the normalized shear modulus (G/G_0) vs. shear strain γ (%) of loess: (a) specimens in dry condition, (b, c) specimens highly saturated

Development of Damping Ratio DT(%)-γ

Similar behaviour is measured and observed in the damping (Fig. 4). For dry loess (Fig. 4a), the damping ratios DT (%) of undisturbed loess are relatively larger than recompacted ones, while for wet loess (Fig. 4b and Fig. 4c), the difference about DT (%) between undisturbed and recompacted loess is small. A possible reason for the observed discrepancy of damping ratio between intact and recompacted loess in dry conditions may be attributed to the strong inter-particle bonding of undisturbed loess [18].

3.2 Normalization Characteristics of G/G_0-γ of Loess

According to Santos and Correia [19], a reference shear strain γ_r at $G/G_0 = 0.7$ was proposed to depict the normalization characteristics of G/G_0-γ curves, based on the

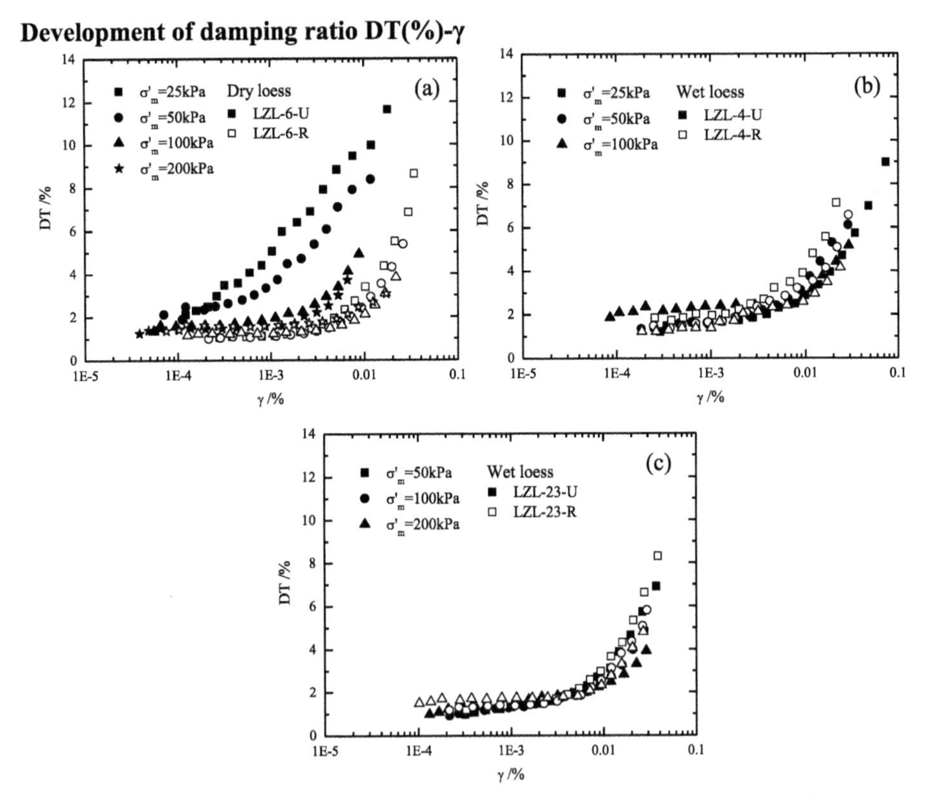

Fig. 4. Effect of cementation on damping ratio DT (%) vs. shear strain γ (%) of loess: (a) specimens in dry condition, (b, c) specimens highly saturated

conventional stress-strain hyperbolic model given by Hardin and Drnevich [20]:

$$G/G_0 = 1/[1 + a(\gamma/\gamma_r)] \tag{1}$$

where G is the secant shear modulus of soil; G_0 is the initial shear modulus; a is a fitting parameter; γ is shear strain; γ_r is reference shear strain (hereinafter called $\gamma_{0.7}$) according to $G/G_0 = 0.7$.

Figure 5 illustrates the normalized relationships between G/G_0 and $\gamma/\gamma_{0.7}$ at small and medium strain levels for Lanzhou loess. As shown, an almost perfect coincidence could be observed for all the curves of G/G_0 versus $\gamma/\gamma_{0.7}$ for Lanzhou loess, irrespective of moisture, confining pressure and structure characteristics, which provides a piece of useful information to predict the degradation curve of G/G_0-γ for structural loess.

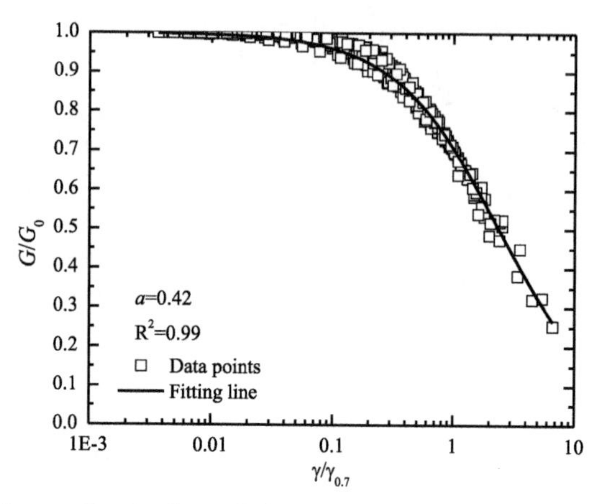

Fig. 5. Variation of normalized values of shear modulus G/G_0 and $\gamma/\gamma_{0.7}$ for Lanzhou loess specimens tested and relationship based on the hyperbolic model given by Hardin and Drnevich [20]

4 Conclusions

In the present study, the influence of soil structure on small and medium strain dynamic properties of Lanzhou loess was examined under isotropic torsional resonant column testing, and the corresponding mechanism analysis was defined in terms of structure composition of loess. The main conclusions derived from this research are as follows:

(1) Structure effect has an important influence on the dynamic properties of loess at small and medium strain, increasing the initial shear modulus G_0 and damping ratio DT (%), and reducing its ductility in dynamic response. However the structure effect on the dynamic properties becomes weak and even negligible for saturated loess.

(2) Under the same confining pressure, the small-strain shear modulus G_0 of undisturbed loess is much bigger than that of recompacted one at dry conditions, while for saturated loess, no apparent disparity is found in G_0 between undisturbed and recompacted loess.

(3) Under the same confining pressure, the normalized shear modulus G/G_0 of undisturbed loess degrades more quickly with shear strain than for recompacted ones, i.e., the ductility is enhanced for loess after remolding, while for saturated loess, G/G_0-γ curves approach each other closely.

(4) For dry undisturbed loess, the damping ratio DT (%) is much higher compared to recompacted specimens, whereas DT (%)-γ curves of undisturbed loess are getting close to that of recompacted ones at saturated state.

(5) The correlations between G/G_0 and $\gamma/\gamma_{0.7}$ could be normalized for loess, irrespective of soil structure.

References

1. Luo, W.B.: Seismic problems of cave dwellings on China's loess Plateau. Tunn. Undergr. Space Technol. **2**(2), 203–208 (1987)
2. Zhang, Z., Wang, L.: Geological Disasters in loess areas during the 1920 Haiyuan earthquake. China. GeoJournal **36**(2), 269–274 (1995)
3. Wang, Z.J., Luo, Y.S., et al.: Effects of initial deviatoric stress ratios on dynamic shear modulus and damping ratio of undisturbed loess in China. Eng. Geol. **143–144**, 43–50 (2012)
4. Liu, D.S., Zhang, Z.H.: Chinese loess. Acta Geol. Sin. **42**(1), 1–14 (1962). (in Chinese)
5. Liu, Z., Liu, F., et al.: Collapsibility, composition, and microstructure of loess in China. Can. Geotech. J. **53**(4), 673–686 (2015)
6. Haeri, S.M., Zamani, A., et al.: Collapse potential and permeability of undisturbed and remolded loessial soil samples Research and applications, Unsaturated soils, pp. 301–308. Springer, New York (2012). https://doi.org/10.1007/978-3-642-31116-1_41
7. Wen, B.P., Yan, Y.J.: Influence of structure on shear characteristics of the unsaturated loess in Lanzhou, China. Eng. Geol. **168**, 46–58 (2014)
8. Ladd, R.S.: Preparing Test Specimens Using Undercompaction. Geotech. Test. J. **1**(1), 16–23 (1978)
9. Drnevich, V.P., Hardin, B.O., et al. (1978). Modulus and Damping of soils by the Resonant-Column Method. Dynamic Geotechnical Testing, ASTM STP 654, American Society for Testing and Materials, 91–125
10. Pitilakis, K., Raptakis, D., et al.: Geotechnical and Geophysical Description of Euro-Seistest, Using Field, Laboratory Tests and Moderate Strong Motion Recordings. J. Earthquake Eng. **3**(3), 381–409 (1999)
11. Anastasiadis, A., Senetakis, K., et al.: Small-Strain Shear Modulus and Damping Ratio of Sand-Rubber and Gravel-Rubber Mixtures. Geotech. Geol. Eng. **30**(2), 363–382 (2011)
12. Song, B., Tsinaris, A., et al.: Small-strain stiffness and damping of Lanzhou loess. Soil Dyn. Earthq. Eng. **95**, 96–105 (2017)
13. Lanzo, G., Pagliaroli, A.: Stiffness of natural and reconstituted Augusta clay at small to medium strains. In: Ling, H.I., Callisto, L., Leshchinsky, D., Koseki, J. (eds.) Soil Stress-Strain Behavior: Measurement, Modeling and Analysis. Solid Mechanics and Its Applications, vol. 146, pp. 323–331. Springer, Dordrecht (2007). https://doi.org/10.1007/978-1-4020-6146-2_16
14. Ning, L., Godt, J.W., et al.: A closed-form equation for effective stress in unsaturated soil. Water Resour. Res. **46**(5), 567–573 (2010)
15. Cheng, Q., Zhou, C., Ng, C.W.W., Tang, C.S.: Effects of soil structure on thermal softening of yield stress. Eng. Geol. (2020). https://doi.org/10.1016/j.enggeo.2020.105544
16. Ng, C.W.W., Cheng, Q., Zhou, C.: Thermal effects on yielding and wetting-induced collapse of recompacted and intact loess. Can. Geotech. J. **55**, 1095–1103 (2018)
17. Kruse, G.A.M., Dijkstra, T.A., et al.: Effects of soil structure on soil behaviour: Illustrated with loess, glacially loaded clay and simulated flaser bedding examples. Eng. Geol. **91**(1), 34–45 (2007)
18. Ashmawy, A.K., Salgado, R., et al. (1995). Soil damping and its use in dynamic analyses. Proceeding of the 3rd International Conferences on Recent Advances in Geotechnical Earthquake Engineering and Soil Dynamics, St. Louis, Missouri, Vol. I: 35–41

19. Santos, J.A. and Correia, A.G. (2000). Shear modulus of soils under cyclic loading at small and medium strain level. 12th World Conference on Earthquake Engineering, paper ID 0530. Auckland, New Zealand
20. Hardin, B.O., Drnevich, V.P.: Shear modulus and damping in soil: design equations and curves. Journal of the Soil mechanics and Foundation Engineering Division **98**(7), 667–692 (1972)

Cyclic Simple Shear Tests of Calcareous Sand

Kai-Feng Zeng and Hua-Bei Liu(✉)

Huazhong University of Science and Technology, Wuhan 430074, China
hbliu@hust.edu.cn

Abstract. A series of simple shear tests under different test conditions were carried out on calcareous sand from the South China Sea. Notably, constant stress and volume tests under monotonic and cyclic testing conditions were conducted. This study was focused on the simple shear behavior of calcareous sand and the relationship between the relative breakage B_r and the input energy E. The test results showed that the calcareous sand has obvious particle breakage during the simple shear test procedure, and there is a unique relationship between the relative breakage B_r and the input energy E at the end of the tests, regardless of the testing conditions. In addition, the study also found that a unique critical state line (CSL) exists for both constant stress and volume monotonic and cyclic tests, and there is a unique cyclic phase transformation line (CPTL) for constant volume cyclic tests. The volumetric strain accumulation increases with cycle number, vertical stress, and cyclic shear strain during constant stress cyclic tests.

Keywords: Calcareous sand · Particle breakage · Input energy · Simple shear test · Monotonic and cyclic

1 Introduction

Calcareous sand has become one type of important geotechnical materials in off-shore and ocean engineering. However, it has been reported in many studies that the mechanical properties of calcareous sand are very different from those of quartz sand [1, 2]. Most existing studies on the mechanical properties of calcareous sand are usually concerned with the triaxial stress states [3–5].

For simple shear test condition [6–8], Mao and Fahey [6] conducted a series of undrained simple shear tests on calcareous sand and found that the calcareous sands have unique monotonic phase transformation line and cyclic phase transformation line, and the backbone curve is close to the monotonic loading curve. Ji et al. [7] conducted a comparative study on the drained cycle simple shear behavior of calcareous sand and siliceous sand, it showed that there are obvious differences between the hysteresis curve shape and particle breakage of siliceous sand and calcareous sand. In general, there exist few studies on the cyclic simple shear behavior of calcareous sand, while the simple shear stress path involved is not adequately extensive. Therefore, in this study a series of simple shear tests was carried out on calcareous sand to study the mechanical behavior and particle breakage of calcareous sand under the simple shear condition.

© The Author(s), under exclusive license to Springer Nature Switzerland AG 2022
L. Wang et al. (Eds.): PBD-IV 2022, GGEE 52, pp. 2151–2158, 2022.
https://doi.org/10.1007/978-3-031-11898-2_198

2 Testing Program

2.1 Test Material and Apparatus

The calcareous sand used in this study was obtained from the South China Sea, and their particle shapes are very irregular and angular. Its representative particle shapes, initial grain-size distribution curve and physical properties are shown in Fig. 1(a).

The apparatus used in the simple shear tests was the large cyclic direct simple shear machine (CDSS). As shown in Fig. 1(b), the apparatus is composed of the horizontal and vertical loading devices, horizontal and vertical displacement sensors, shear box, shear ring, and data acquisition systems. Two sizes of specimens were used in this article, and the influence of specimen size also was discussed in the end. For the large diameter specimen (LDS), the thickness of the shear ring is 6.7 mm, the diameter and height of the specimen were 305 and 101 mm, respectively. Those of small diameter specimen (SDS) are 3, 101 and 50 mm, respectively. The surface of the shear ring is polished and the friction between the rings is very small and can be ignored.

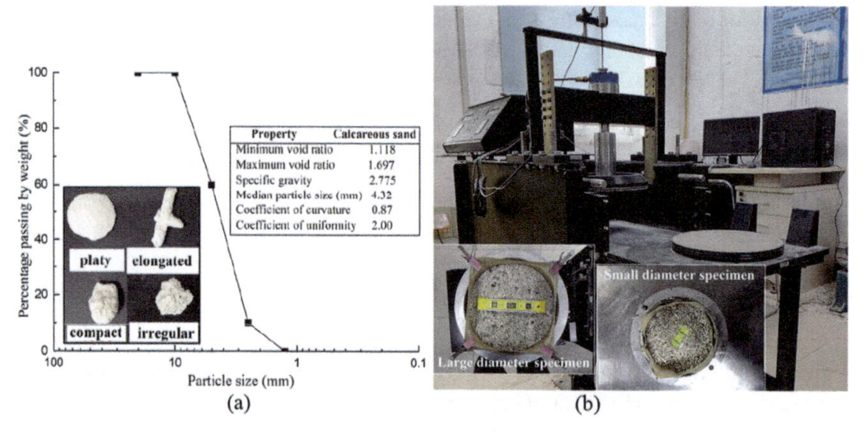

Fig. 1. Initial grain-size distribution curve of calcareous sand and simple shear test apparatus: (a) test material; (b) test apparatus.

2.2 Testing Methods

The specimen preparation is mainly divided into three steps. Firstly, the calcareous sand sample was immersed in distilled water and placed in a vacuum environment to saturate for 24 h. Secondly, the saturated calcareous sand particles were divided into four parts, and the specimen was prepared by a moist tamping method. By dividing the sample into layers and using light tamping, the targeted density can be achieved with negligible particle breakage, the prepared sample is shown in Fig. 1(b). Finally, the specimen is fixed in the shear box by bolts, and the subsequent tests were carried out according to the test protocols (see Table 1).

The loading method of monotonic and cyclic tests are ramp and triangular wave, respectively. The loading rate of displacement and force are 0.02 mm/s and 10 N/s, respectively. The cyclic tests are two-way symmetrical cyclic test, and the interval of repeated loading-unloading tests are 1% or 2%. The LDS is the large diameter specimen and the SDS is small diameter specimen.

Table 1. Test parameters of simple shear test.

Test name	Relative density *Dr*	Initial vertical stress σ_{v0} (kPa)	Cycle number	Cyclic shear strain/stress	Specimen size
Constant stress monotonic	25%	150	-	-	LDS
	85%	50;100;150	-	-	LDS
Constant volume monotonic	25%	150	-	-	LDS
	85%	50;100;150	-	-	LDS
Constant stress/volume repeated loading-unloading	85%	150	-	-	LDS
Constant stress cyclic	85%	150	16	1%; 2%; 5%; 10%	LDS
Constant volume cyclic	85%	150	16	$0.1\sigma_{v0}$; $0.15\sigma_{v0}$; $0.2\sigma_{v0}$	LDS
Constant stress monotonic	25%	150; 300; 600; 1000	-	-	SDS
Constant stress cyclic	25%	300; 600; 1000	16	5%	SDS

3 Test Results and Analyses

3.1 Monotonic Test Results

Figure 2 and Fig. 3 show the constant stress and volume monotonic test results of large diameter specimen, respectively. Calcareous sand has a higher compressibility and lower strength than quartz sand, and the relative density *Dr* has obvious influence on the monotonic simple shear behavior of calcareous sand. In the constant stress monotonic tests, the dense sand (*Dr* = 85%) has a higher shear strength and dilatancy, and dilatancy decreases as the vertical stress increases. In addition, the loose sand has basically reached the critical state at the end of test, and the critical state friction angle φ_{CSL} is approximately 34.5°, but not for dense sand. In the constant volume monotonic tests, the shear stress of dense sand (*Dr* = 85%) will increase significantly after the phase transformation point, but that of loose sand (*Dr* = 25%) will remain stable. However, there is a unique monotonic phase transformation line (MPTL) regardless of the particle breakage, vertical stress, and relative density (see Fig. 3(b)).

Results of repeated loading-unloading tests are also shown in Fig. 2 and Fig. 3. The results show that the unloading-reloading modulus E_{ur} decreases with shear strain increases. The contraction occurs when unloading, and the volumetric strain is larger than that of the monotonic test in the constant stress tests. In the constant volume tests, the phase transformation line of repeated loading-unloading test is different from MPTL, but it is close to the cyclic phase transformation line (CPTL, see Fig. 5(b)). Further, all simple shear tests have the same critical state line (CSL) regardless of the test conditions (see Fig. 2, Fig. 3, and Fig. 5) and particle breakage.

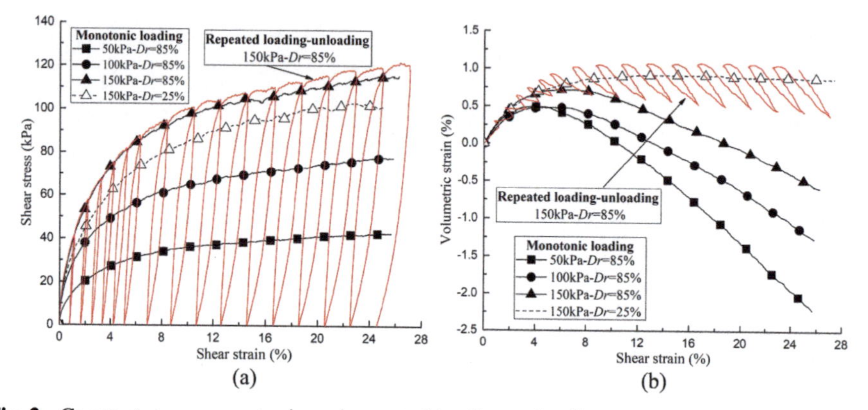

Fig. 2. Constant stress monotonic and repeated loading-unloading test results of large diameter specimen: (a) shear stress-shear strain; (b) volumetric strain-shear strain.

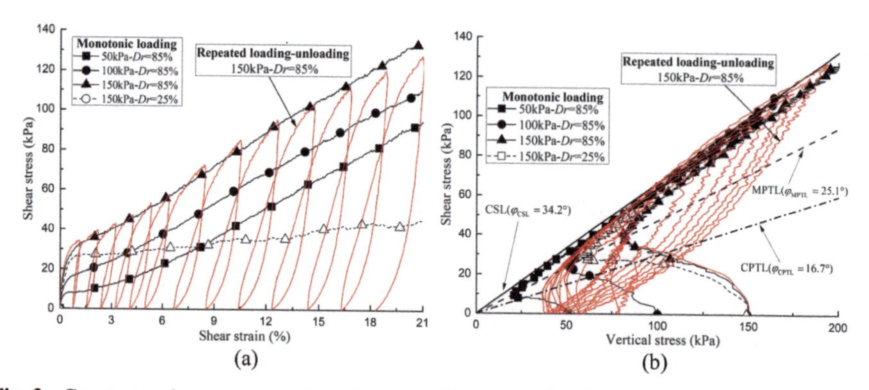

Fig. 3. Constant volume monotonic and repeated loading-unloading test results of large diameter specimen: (a) shear stress-shear strain; (b) stress path.

3.2 Cyclic Test Results

Figure 4 shows the constant stress cyclic test results of large diameter specimen, the cyclic tests are the strain control two-way symmetrical cyclic (average shear strain $\gamma_{ave} = 0\%$). The shape of the hysteresis loop is ellipse. As the cyclic shear strain γ_{cyc} increases, the dynamic shear modulus decreases and the area of hysteresis loop increases. There is dilatancy and contraction during every cycle, but the volumetric strain accumulation ε_{ac} increases with cycle number. And under the same cycle number, ε_{ac} increases with γ_{cyc}.

Figure 5 shows the constant volume cyclic test results of large diameter specimen, the cyclic tests are the stress control two-way symmetrical cyclic (average shear stress $\tau_{ave} = 0$ kPa). It can be found that the shape of the hysteresis loop is butterfly-shaped, and this butterfly shape becomes more and more obvious as the stress path moves towards the critical state line. Figure 5(c) shows that the shear strain remains stable at the beginning of the cycle, but it increases sharply after reaching the critical value (N_{liq}), and the N_{liq} decreases with the cycle shear stress τ_{cyc} increases. This failure mode is 'cyclic liquefaction' [8]. Further, Fig. 5(d) shows that the backbone curve is still basically consistent with the monotonic loading curve despite the presence of particle breakage.

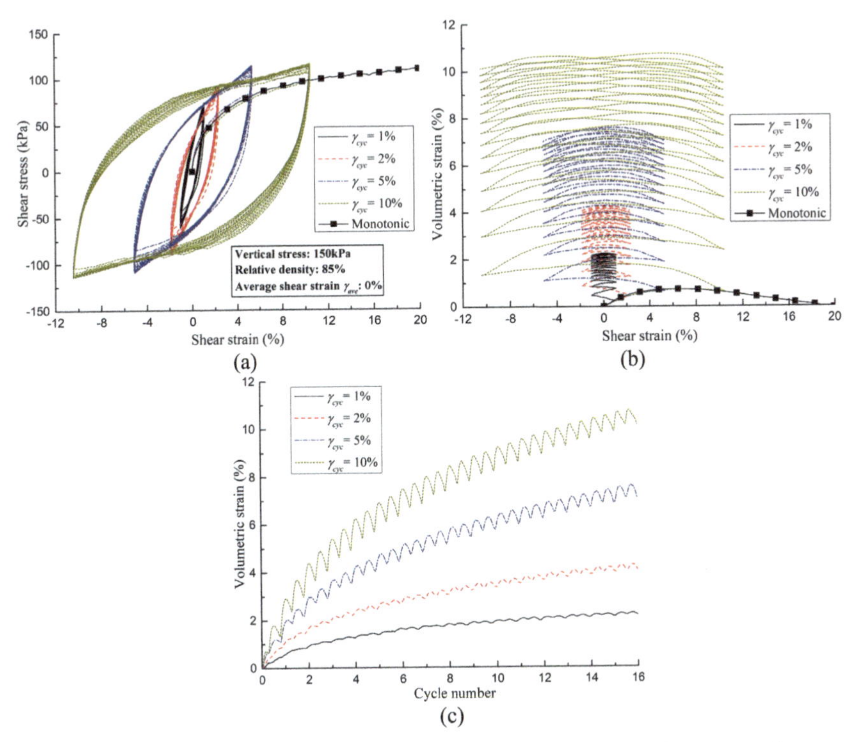

Fig. 4. Constant stress cyclic test results of large diameter specimen: (a) shear stress-shear strain; (b) volumetric strain-shear strain; (c) volumetric strain-cycle number.

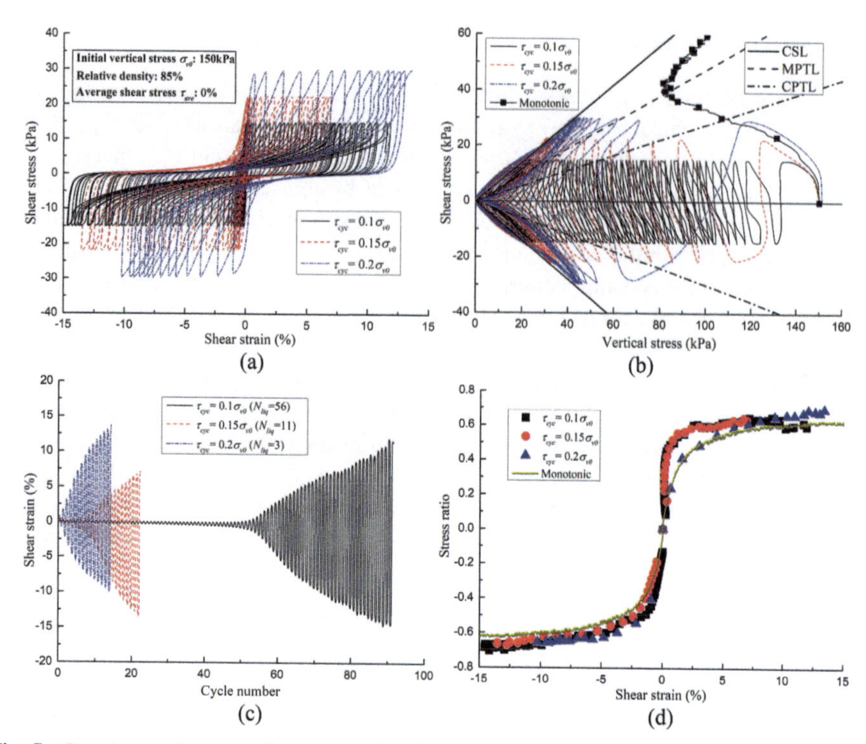

Fig. 5. Constant volume cyclic test results of large diameter specimen: (a) shear stress-shear strain; (b) stress path; (c) shear strain-cycle number; (d) backbone curve.

3.3 Influence of Specimen Size

Due to apparatus limitations, the maximum vertical stress of large diameter specimen is only 150 kPa. In order to study particle breakage under high stress levels, a series of small diameter specimen tests were carried out. Figure 6 shows the comparison of large and small diameter specimen. The specimen size has a small effect on the simple shear characteristics of calcareous sand, and the small diameter specimen can be employed to study the particle breakage under high stress levels. Other test phenomena are similar to those of large diameter specimen, only the constant stress cyclic test results of small diameter specimen show that the volumetric strain accumulation ε_{ac} increases with vertical stress.

4 Particle Breakage Analyses

Figure 7(a) shows the grain-size distribution curves before and after tests. It clearly indicates that the calcareous sand has obvious particle breakage during the simple shear test procedure, and the particle breakage increases with the vertical stress and cyclic strain.

As commonly used in the literature [9, 10], the relative breakage B_r proposed by Hardin [11] is used to quantify the particle breakage, and the input energy E is used to describe the evolution of particle breakage, and the input energy per unit volume E in the simple shear test can be expressed as follows:

$$E = \sum_{SOT}^{EOT} (\sigma_v d\varepsilon_v + \tau d\gamma) \quad (1)$$

where σ_v and τ are the vertical and shear stress; $d\varepsilon_v$ and $d\gamma$ are the volumetric and shear strain increment; and SOT and EOT are the start and end of tests, respectively. Figure 7(b) shows the relationship between the relative breakage B_r and the input energy E. Although the relative density, vertical stress, cyclic shear strain/stress, specimen size, loading method of the tests are different, there is a unique relationship between the relative breakage B_r and the input energy E at the end of simple shear tests, and the B_r increases with E.

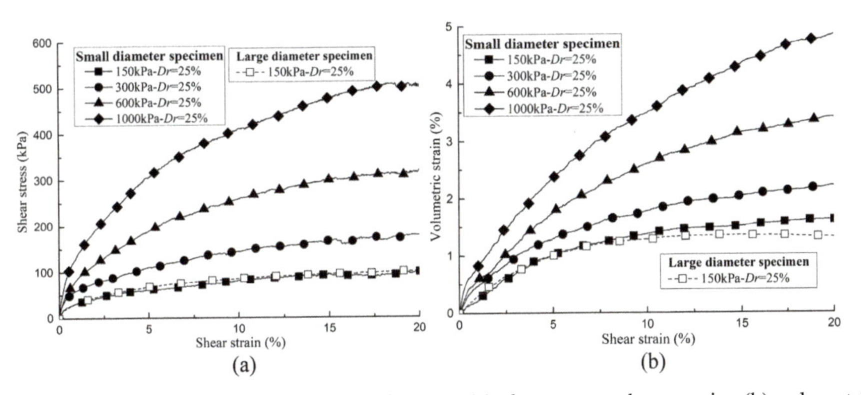

Fig. 6. Results of constant stress monotonic tests: (a) shear stress-shear strain; (b) volumetric strain-shear strain.

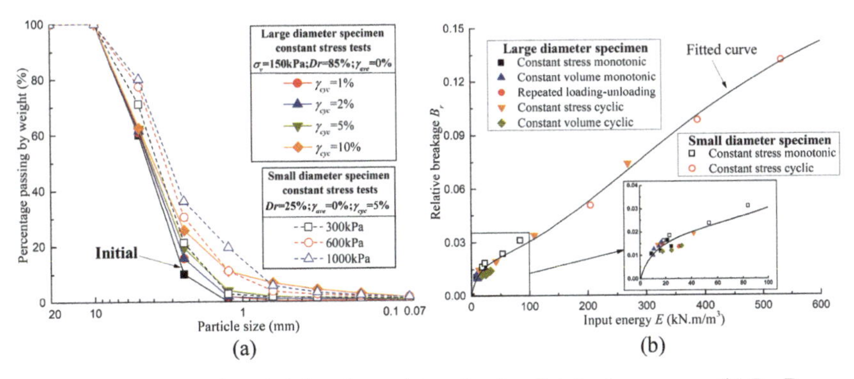

Fig. 7. Results of particle breakage: (a) grain-size distribution curves; (b) B_r-E.

5 Conclusions

The major findings are summarized as follows:

(1) The calcareous sand has obvious particle breakage during the simple shear test procedure, which leads to an increase in compression and a decrease in strength.
(2) Although the particle breakage is obvious, there still exist unique critical state line (CSL), monotonic phase transformation line (MPTL) and cyclic phase transformation line (CPTL), which are independent of particle breakage.
(3) The particle breakage increases with the vertical stress and cyclic strain. In addition, although the simple shear test conditions are different, there exists a unique relationship between the relative breakage and the input energy at the end of simple shear tests.

References

1. Liu, C.Q., Yang, Z.Q., Wang, R.: The present condition and development in studies of mechanical properties of calcareous soils. Rock Soil Mech. **16**(4), 74–84 (1995)
2. Wu, J.P., Lou, Z.G.: Research on the behavior of calcareous soils in offshore pile foundations. Ocean Eng. **14**(3), 75–83 (1996)
3. Wu, J.P., Chu, Y., Lou, Z.G.: Influence of particle breakage on deformation and strength properties of calcareous sands. Chin. J. Geotech. Eng. **19**(5), 51–57 (1997)
4. Shahnazari, H., Rezvani, R.: Effective parameters for the particle breakage of calcareous sands: an experimental study. Eng. Geol. **159**(9), 98–105 (2013)
5. Yu, F.W.: Particle breakage and the drained shear behavior of sands. Int. J. Geomech. **17**(8), 4017041 (2017)
6. Mao, X., Fahey, M.: Behaviour of calcareous soils in undrained cyclic simple shear. Geotechnique **53**(8), 715–727 (2003)
7. Ji, W.D., Zhang, Y.T., Wang, Y., et al.: Comparative study of shear performance between coral sand and siliceous sand in cycles simple shear test. Rock Soil Mech. **39**(S1), 282–288 (2018)
8. Porcino, D., Caridi, G., Ghionna, V.N.: Undrained monotonic and cyclic simple shear behaviour of carbonate sand. Geotechnique **58**(8), 635–644 (2008)
9. Lade, P.V., Yamamuro, J.A., Bopp, P.A.: Significance of particle crushing in granular materials. J. Geotech. Eng. **2**(5), 99–110 (2016)
10. Liu, H.B., Zeng, K.F., Zou, Y.: Particle breakage of calcareous sand and its correlation with input energy. Int. J. Geomech. **20**(2), 04019151 (2020)
11. Hardin, B.O.: Crushing of soil particles. J. Geotech. Eng. ASCE **111**(10), 1177–1192 (1985)

S5: Special Session on Soil Dynamic Properties at Micro-scale: From Small Strain Wave Propagation to Large Strain Liquefaction

Effect of Fabric Anisotropy on Reliquefaction Resistance of Toyoura Sand: An Experimental Study

Pedram Fardad Amini and Gang Wang(✉)

Hong Kong University of Science and Technology, Kowloon, Hong Kong
gwang@ust.hk

Abstract. Recent earthquakes in New Zealand and Japan demonstrated that preshaking histories can significantly influence the liquefaction resistance of sandy soils, which has not been fully understood yet. It has also been reported that soils prepared with different depositional methods referred to as initial or inherent anisotropy show different liquefaction behavior. Therefore, a comprehensive experimental study was carried out for the first time to explore the combined effects of inherent anisotropy and induced anisotropy due to cyclic shearing on the reliquefaction resistance of loose ($D_r = 45\%$) and dense ($D_r = 70\%$) Toyoura sands subjected to various cyclic stress ratios (CSR). A hollow cylinder torsional shear apparatus (HCTSA) was used in this study since it can mimic the field stress conditions more realistically during earthquake events. In order to investigate the influence of inherent anisotropy on liquefaction and reliquefaction resistance of Toyoura sand, the hollow cylindrical specimens were reconstituted with different methods of dry deposition (DD) and moist tamping (MT), representing naturally and artificially deposited soils, respectively. Furthermore, sand specimens were cyclicly sheared up to medium and large shear strain levels, then reconsolidated at different states to evaluate the effect of induced anisotropy with different shear histories on reliquefaction resistance in the sand. A large preshearing reduces the reliquefaction resistance, while the reliquefaction resistance increases when the sandy soil is moderately presheared, irrespective of relative density and reconsolidation state. It was also found that the effect of reconsolidation state is more significant in the dense sand than the loose sand. Another key finding of this study is that the effects of initial fabric acquired by different reconstitution methods and reconsolidation states are lost in the reliquefaction tests once the loose Toyoura sands experienced large preshaking.

Keywords: Liquefaction · Reliquefaction · Fabric anisotropy

1 Introduction

The field evidence of multiple liquefaction or reliquefaction phenomenon was witnessed during the recent earthquakes in Japan (2011) and New Zealand (2010–2011), where several aftershocks followed the mainshock. Upon consecutive earthquakes, it was found

© The Author(s), under exclusive license to Springer Nature Switzerland AG 2022
L. Wang et al. (Eds.): PBD-IV 2022, GGEE 52, pp. 2161–2170, 2022.
https://doi.org/10.1007/978-3-031-11898-2_199

that the liquefaction-induced damage caused by the second stage was the most severe. The reduced reliquefaction resistance owing to the previous liquefaction was observed while, as it is known, soil density increases because of the postliquefaction reconsolidation. These contrary observations imply that the interaction between a mainshock with and without liquefaction manifestation and aftershock is very complex and a great matter of concern, especially for practical engineers, as preshaking affects the design of structures after an earthquake. It also indicates that there exists another more dominant factor than density in controlling reliquefaction resistance. Therefore, to safeguard structures, a fundamental understanding of the reliquefaction behavior of sandy soils is needed, as in some cases, successive earthquakes can be detrimental.

Soil fabric describes the spatial arrangement of particles and accompanied voids distribution [1]. In general, soil fabric anisotropy can be categorized into two sources of inherent anisotropy (or initial fabric) resulting from soil sedimentation and induced anisotropy created during different stress histories. There have been few studies investigating the effect of inherent anisotropy and induced anisotropy on the cyclic behavior of granular soils separately. For example, it has been indicated that specimens with different initial fabrics of dry deposition (DD) and moist tamping (MT), the two most commonly used specimen preparation methods (SPM) in the laboratory corresponding to naturally and artificially deposited soils, respectively, can have totally different liquefaction behavior of sandy soils [2–4]. In this regard, further studies are essential as the initial fabric is a key factor controlling soil's mechanical response and its undrained cyclic response. Furthermore, sandy soils exhibit anisotropic behavior under cyclic loading, significantly influencing the reliquefaction resistance [5, 6]. However, most prior studies were conducted under cyclic triaxial stress path, which can not accurately represent field seismic stress conditions. In a recent preliminary study, Fardad Amini et al. [7, 8], with the aid of hollow cylinder torsional shear apparatus (HCTSA), investigated the impact of stress-induced anisotropy on the reliquefaction resistance of dense Toyoura sand at different strain histories. They concluded that the reliquefaction resistance of dense Toyoura sand is greatly influenced by the different strain history levels and reconsolidation states created before and during the first liquefaction stage.

Nevertheless, no research has been conducted before to examine the coupled effects of inherent and induced anisotropies on reliquefaction resistance of sandy soil at different strain histories. For this reason, extensive multi-stage experiments were carried out on Toyoura sand in this study by employing an HCTSA, which can better mimic the field stress conditions during earthquakes. Different SPMs of DD and MT were used to create distinct initial fabrics. The specimens reconstituted with either method were imposed different strain histories and stopped at different reconsolidation states. Additionally, tests were performed on the loose and dense DD specimens to explore the effect of soil density on the reliquefaction resistance of Toyoura sand. This paper reports a fundamental experimental study to elucidate the effects of initial fabric and fabric change under cyclic shearing on the reliquefaction response of sand, which can also be considered a valuable source for numerical and theoretical analyses.

2 Experimental Program

All the multi-stage experiments were performed in a hollow cylinder torsional shear apparatus (HCTSA), built at The Hong Kong University of Science and Technology (HKUST) and described by Fardad Amini et al. [7] and Fardad Amini and Wang [9]. Standard research sand, the Toyoura sand, uniform quartz sand with subangular particles and mean particle size (D_{50}) of 0.22 mm, was used in this study. This sand has minimum (e_{min}) and maximum (e_{max}) void ratios of 0.988 and 0.639, respectively, and specific gravity $G_s = 2.65$.

The hollow cylindrical specimens with outer and inner diameters of 200 mm and 150 mm, respectively, and a height of approximately 310 mm, were prepared by dry deposition (DD) and moist tamping (MT) techniques. In the DD technique, the oven-dried Toyoura sand was funneled into the hollow space between the inner and outer molds in ten layers at the constant falling height [10]. The DD specimens were prepared at nominal relative densities (D_r) of 45% (loose) and 70% (dense) to investigate the effect of density state on the reliquefaction resistance of DD Toyoura sands. In the MT technique, the Toyoura sand with a water content of 5% was placed into the hollow space between the inner and outer molds in ten layers and tamped layer by layer using a tamper. Undercompaction procedure proposed by Ladd [11] was adopted to create homogeneous specimens. The MT specimens were formed at a target relative density of 45% (loose). After reconstitution, all the specimens' full saturation was corroborated by obtaining a Skempton B-value of 0.97 or higher. All the specimens were subsequently consolidated isotropically to reach an initial effective mean principal stress (EMPS) of $P_0^{'} = 100$ kPa.

Several reliquefaction tests (two-stage shearing) were performed, accounting for the influence of fabric anisotropy involving initial fabric and stress-induced fabric change on the subsequent cyclic characteristic of Toyoura sand. The experimental procedure is schematically demonstrated in Fig. 1(a). In the first stage, the isotropically consolidated specimens were sheared cyclically under undrained conditions upon reaching the desired residual shear strain levels (i.e., γ_{res}= 0.4, 2.0, and 3.7/5.0%) and stopped at different states A and B called reconsolidation states (see Fig. 1(b) for details). After isotropic reconsolidation, the preshaken specimens were cyclicly sheared again under undrained conditions until reaching a double-amplitude shear strain (γ_{DA}) of at least 7.5% indicating liquefaction/reliquefaction. In all experiments, the fixed single-amplitude cyclic torsional shear stress (τ_{SA}) was applied at a constant shear strain rate of 0.5%/min, and varying cyclic stress ratios (CSRs) in the range of 0.10–0.25. CSR is defined as the ratio between τ_{SA} and $P_0^{'}$. Cyclic liquefaction tests (one-stage shearing) without previous shearing history (virgin soil) were also performed under the prescribed CSRs as benchmark tests to evaluate the effect of preshearing history. It should also be noted that during undrained shearing, the height of hollow cylinderical specimens was kept unchanged to simulate quasi-simple shear conditions. The method suggested by Tatsuoka et al. [12] was used to correct the shear stress for membrane force. In addition, the effect of membrane penetration was negligible and was not taken into account. Details of the experimental program are summarized in Table 1.

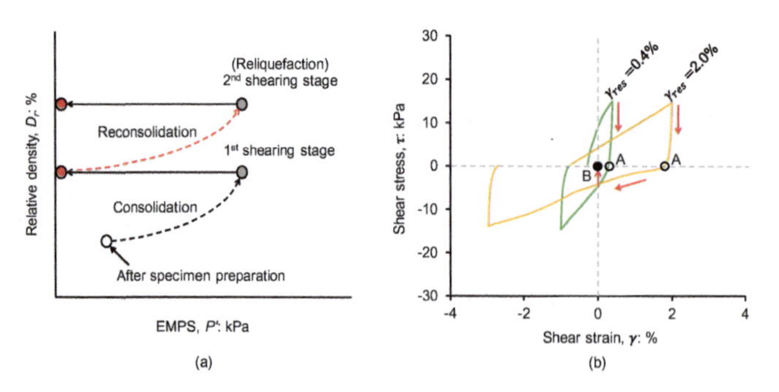

Fig. 1. Schematic illustration of reliquefaction tests and reconsolidation states

Table 1. Details of the experimental program.

Test pattern	SPM	D_r (%)	CSR	γ_{res}	Recon. States
Liquefaction	DD and MT	45 and 70%	0.10, 0.15, 0.20 and 0.25	-	-
Reliquefaction				0.4, 2.0, 3.7% (or 5.0%)	A and B

Note: Recon. = Reconsolidation

3 Experimental Results

Figure 2 shows the cyclic liquefaction behavior of loose Toyoura sand ($D_r = 45\%$) with different DD and MT initial fabrics. For the DD specimen, as the cyclic shear stress proceeds, the EMPS decreases gradually except the first cycle because of excess pore water pressure (EPWP) generation. During the last cycle of loading in the 9th cycle, a limited flow response is observed, followed by initial liquefaction, which is defined as the zero EMPS. Limited flow is regarded as sudden but limited development of deformation due to the rapid rise in EPWP. It can be seen that initial liquefaction and liquefaction, defined as the state when the double-amplitude shear strain becomes at least 7.5%, happen together in the last loading cycle. On the other hand, the MT specimen shows a cyclic mobility response, in which the stress path moves recurrently through the zero EMPS and follows a typical "butterfly loops" (Castro [13]). EPWP and shear strain evolution in the MT specimen is more progressive than in the DD specimen. The MT specimen liquefied in almost 33rd cycles, more than three times in the DD specimen.

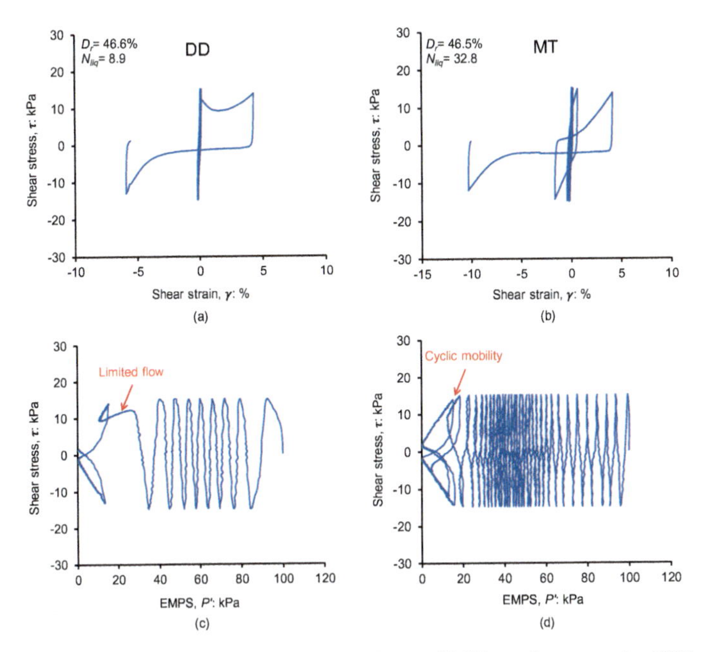

Fig. 2. Cyclic liquefaction response of loose DD and MT specimens under CSR = 0.15

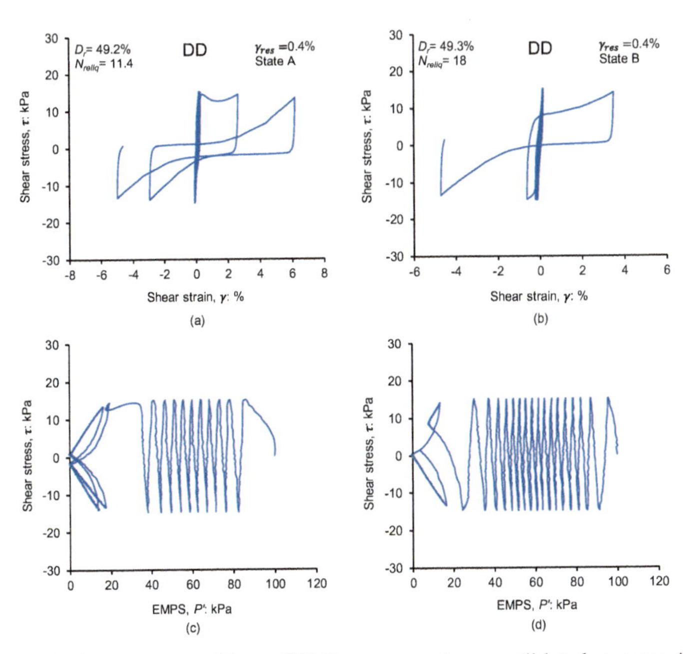

Fig. 3. Reliquefaction response of loose DD Toyoura sand reconsolidated at states A and B with $\gamma res = 0.4\%$ and CSR = 0.15

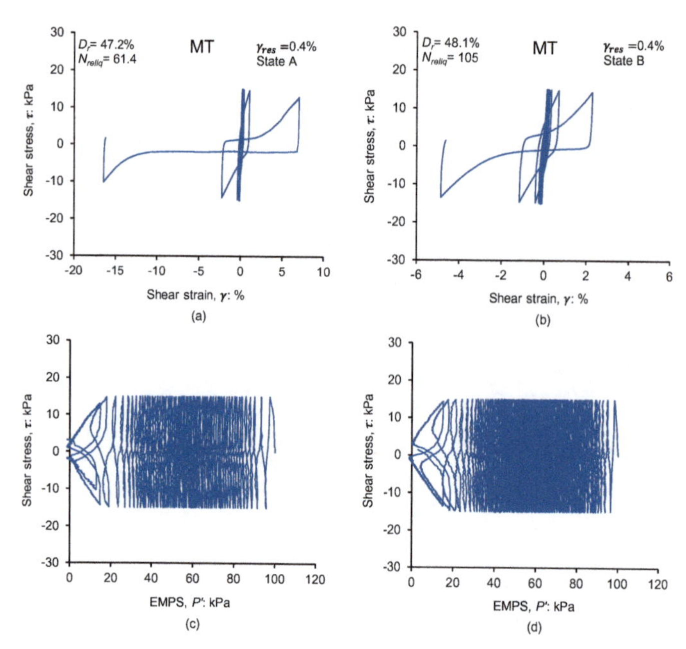

Fig. 4. Reliquefaction response of loose MT Toyoura sand reconsolidated at states A and B with $\gamma res = 0.4\%$ and CSR = 0.15

Figures 3 and 4 illustrate the reliquefaction response of two pairs of DD and MT specimens, in which both pairs experienced medium γ_{res} of 0.4%, at CSR = 0.15, reconsolidated at different states of A and B. The first pair of specimens were prepared using the DD method, while the second pairs were formed by the MT method. Both pairs of medium preshaken DD and MT specimens depict an increase in reliquefaction resistance irrespective of reconsolidation states, although relative density did not increase significantly. DD specimens reconsolidated at states A and B reliquefy during the 12th and 18th cycle, respectively. While MT specimens reconsolidated at states A and B reliquefy with much more number of cycles during 62nd and 105th cycle, respectively. Reliqufaction resistance of both DD and MT specimens increases because a moderately preshaking creates a stronger structure that persists even after reconsolidation. However, the fabric created in both DD and MT specimens reconsolidated at state A is more anisotropic than those reconsolidated at state B. This anisotropic structure leads to lower reliquefaction resistance, evident from a more asymmetric stress path during the first cycle.

Figures 5 and 6 illustrate the reliquefaction response of two pairs of DD and MT specimens, in which both pairs experienced large γ_{res} of 2.0%, at CSR = 0.15, reconsolidated at different states of A and B. Both pairs of large preshaken DD and MT specimens show a reduction in reliquefaction resistance irrespective of reconsolidation states, although relative density increases significantly. The DD specimen reconsolidated at state A reliquefies during the 4th cycle, and the one reconsolidated at state B reliquefies during the 3rd cycle. On the other hand, MT specimens reconsolidated either at states A or B need almost three cycles to reliquefaction. Decrease in reliquefaction resistance of large preshaken specimens is due to the high degree of induced anisotropy. Regardless of slight differences in the number of cycles to reliquefaction in both pairs of DD and MT specimens, it seems that once the loose soil is largely preshaken, there is no effect of initial fabric and reconsolidation state. The initial anisotropy is completely lost when the EMPS gets zero. Moreover, the effect of the reconsolidation state is canceled may be due to the high compressibility of loose soils.

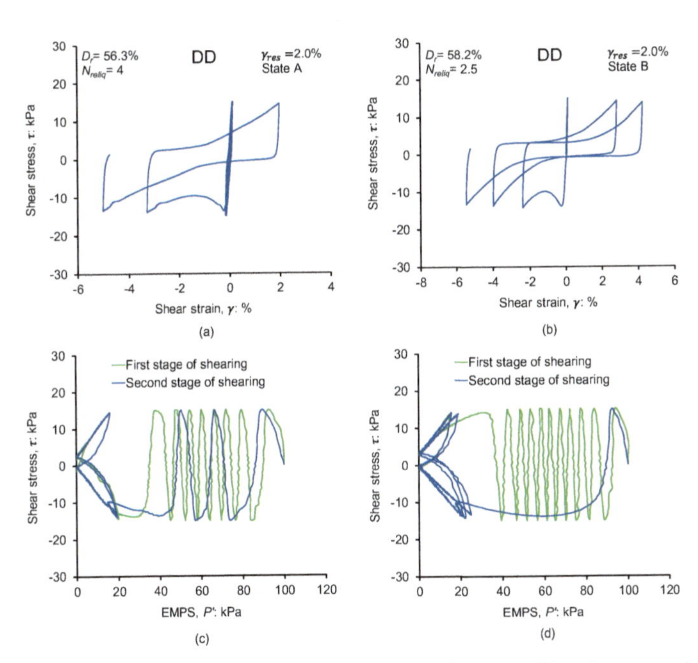

Fig. 5. Reliquefaction response of loose DD Toyoura sand reconsolidated at states A and B with $\gamma res = 2.0\%$ and CSR = 0.15

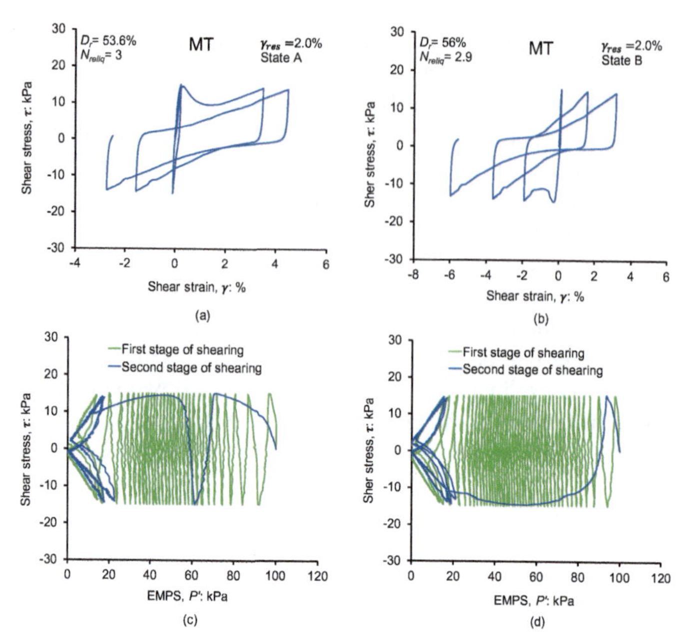

Fig. 6. Reliquefaction response of loose MT Toyoura sand reconsolidated at states A and B with $\gamma res = 2.0\%$ and CSR = 0.15

Figure 7 displays the relationship between CSR and cycle number N_{liq} or N_{reliq} to liquefaction or reliquefaction for all the tests conducted in this study. For all DD (loose and dense) and MT (loose) specimens, the strongest reliquefaction resistance occurs once the specimens have experienced a medium γ_{res}= 0.4%. All medium preshaken specimens reconsolidated at state B depict greater resistance than those reconsolidated at state A. On the other hand, a large γ_{res}= 2.0–5.0% reduces the reliquefaction resistance, irrespective of initial fabrics or reconsolidation states. For dense DD large preshaken sands, specimens reconsolidated at state B show greater reliquefaction resistance than those reconsolidated at state A. However, loose DD and MT specimens represent almost similar reliquefaction resistance once they are largely preshaken, irrespective of reconsolidation state. This similar resistance can be attributed to the significant disturbance of loose DD and MT soil structures.

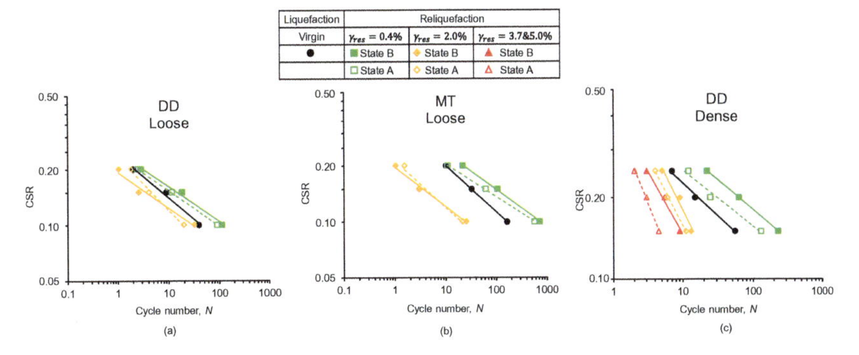

Fig. 7. CSR versus cycle number, *Nliq* or *Nreliq* for all tests conducted in this study

4 Conclusions

A comprehensive experimental research program was carried out using an HCTSA to study the coupled effects of inherent and induced anisotropies on the reliquefaction resistance of Toyoura sand ($D_r = 45\%$) at different strain histories. It was shown that for loose Toyoura sands, moist tamped (MT) specimens with cyclic mobility failure mechanism show higher liquefaction resistance than dry deposited (DD) specimens, which showed limited flow failure mode. Furthermore, it was found that preshearing history significantly affects the reliquefaction resistance of eighter loose (DD and MT) or dense (DD) Toyoura sand. For both loose and dense sands, a medium preshearing (i.e., γ_{res}= 0.4%) increases reliquefaction resistance, while the resistance reduces when the Toyoura sand experiences a large preshearing (i.e., γ_{res}= 2.0–5.0%), irrespective of initial fabrics. Regardless of the initial fabric and density, medium preshaken specimens reconsolidated at state A show a lower reliquefaction resistance than those at state B. The same behavior was observed for dense DD large preshaken sands. However, the effects of the initial fabric and reconsolidation state are lost once the loose Toyoura sand is largely persheared. Therefore, reliquefaction resistance is almost similar for loose large preshaken specimens irrespective of reconstitution method and reconsolidation state.

Acknowledgements. The authors acknowledge support from Hong Kong Research Grants Council (Grant No. 16214220).

References

1. Mitchell, J., Soga, K.: Fundamentals of Soil Behavior, 3rd edn. Wiley, New York (2005)
2. Yang, Z.X., Li, X.S., Yang, J.: Quantifying and modelling fabric anisotropy of granular soils. Géotechnique **58**(4), 237–248 (2008)
3. Sze, H.Y., Yang, J.: Failure modes of sand in undrained cyclic loading: impact of sample preparation. J. Geotech. Geoenviron. Eng. **140**(1), 152–169 (2014)
4. Ni, X., Ye, B., Zhang, F., Feng, X.: Influence of specimen preparation on the liquefaction behaviors of sand and its mesoscopic explanation. J. Geotech. Geoenviron. Eng. **147**(2), 04020161 (2021)

5. Yamada, S., Takamori, T., Sato, K.: Effects on reliquefaction resistance produced by changes in anisotropy during liquefaction. Soils Found. **50**(1), 9–25 (2010)
6. Oda, M., Kawamoto, K., Suzuki, K., Fujimori, H., Sato, M.: Microstructural interpretation on reliquefaction of saturated granular soils under cyclic loading. J. Geotech. Geoenviron. Eng. **127**(5), 416–423 (2001)
7. Fardad Amini, P., Huang, D., Wang, G., Jin, F.: Effects of strain history and induced anisotropy on reliquefaction resistance of Toyoura sand. J. Geotech. Geoenviron. Eng. **147**(9), 04021094 (2021)
8. Fardad Amini, P., Huang, D., Wang, G.: Dynamic properties of Toyoura sand in reliquefaction tests. Géotechnique Lett. **11**(4), 1–8 (2021)
9. Fardad Amini, P., Wang, G.: Effect of induced anisotropy on postliquefaction and reliquefaction resistance of sand: experimental study. In: Proceedings of 17th World Conference on Earthquake Engineering (17WCEE), Sendai, Japan (2020)
10. Ishihara, K.: Liquefaction and flow failure during earthquakes. Géotechnique **43**(3), 351–415 (1993)
11. Ladd, R.S.: Preparing test specimens using undercompaction. Geotech. Test. J. **1**(1), 16–23 (1978)
12. Tatsuoka, F., Sonoda, S., Hara, K., Fukushima, S., Pradhan, T.B.S.: Failure and deformation of sand in torsional shear. Soils Found. **26**(4), 79–97 (1986)
13. Castro, G.: Liquefaction and cyclic mobility of saturated sands. J. Geotech. Eng. Div. **101**(6), 551–569 (1975)

Variation of Elastic Stiffness of Saturated Sand Under Cyclic Torsional Shear

Yutang Chen and Jun Yang(✉)

The University of Hong Kong, Hong Kong, China
junyang@hku.hk

Abstract. For soil elements surrounding the foundations of offshore wind turbines or oil platforms, very many loading cycles of varying amplitudes and frequencies are quite common. The influence of such cyclic loadings on the soil stiffness is a critical concern in the evaluation of the long-term performance of the foundation-structure system during its service life. Current understanding of this influence is however inadequate due to the scarcity of experimental data. This study investigates the variation of small-strain shear modulus (G_0) of saturated Toyoura sand subjected to cyclic torsional shear stress cycles with a small amplitude by using a RC/TS apparatus. Sand specimens of different densities have been tested. It is found that the G_0 value tends to decrease with loading cycles and the decrease is mainly associated with the initial loading cycles. When the number of cycles is beyond around 3500, the G_0 value tends to become stable. The degree of reduction of G_0 is not sensitive to the density of sand specimen and is generally within 10% for the range of densities investigated.

Keywords: Dynamic properties · Elastic stiffness · Saturated sand · Long-term cyclic loadings

1 Introduction

The characterization of small strain shear modulus (G_0) of granular soils is a subject of both practical and theoretical interest since the pioneering work of Hardin and Richart (1963). This key parameter is required in a wide range of geotechnical applications such as earthquake ground response, structural vibration and liquefaction evaluation [2–4]. Based on the early work of laboratory testing of clean, uniform quartz sands, some empirical equations have been developed to predict G_0 value in which the confining stress and void ratio are two main factors [1, 5]. Further work focusing on the effects of grain characteristics was reported by some researchers and several empirical equations have been suggested that account for the influence of coefficient of uniformity, particle shape and fines content [6–9].

In practical applications such as offshore wind turbines (OWTs), the soil surrounding the foundation is often subjected to a large number of loading cycles with varying amplitudes and frequencies during their service time. The repeated loading may cause

© The Author(s), under exclusive license to Springer Nature Switzerland AG 2022
L. Wang et al. (Eds.): PBD-IV 2022, GGEE 52, pp. 2171–2179, 2022.
https://doi.org/10.1007/978-3-031-11898-2_200

changes in the properties of the surrounding soil and the foundation stiffness and, subsequently, changes in the natural frequency of the system [11, 12]. It is well recognized that if the natural frequency of OWTs is close to the loading frequencies from wind, waves, rotational frequency (1P) and blade passing frequency (3P), the devastating resonance or drastic fatigue damage is possible to occur. There is a concern about the impact of cyclic loading histories on the stiffness of the soil surrounding these foundations.

In this study, a series of sand specimens at different densities and confining stresses were subjected to a large number of torsional shear stress cycles and the G_0 values of these specimens were measured by the resonant column apparatus (RC) after each package of loading cycles. All the specimens were tested at the saturated conditions. Selected results are presented together with analysis and discussion.

2 Experimental Program

2.1 Testing Apparatus

The apparatus used in this study, as shown in Fig. 1(a), combines the functions of resonant column (RC) and cyclic torsional shear (TS). The specimen, measured 100 mm in height and 50 mm in diameter, can be confined with water or air for saturated and dry conditions, respectively. The TS function controlled by input voltage was used to apply cyclic shear stress to the specimen. The small strain shear modulus G_0 was measured by the RC function (see Fig. 1(b)).

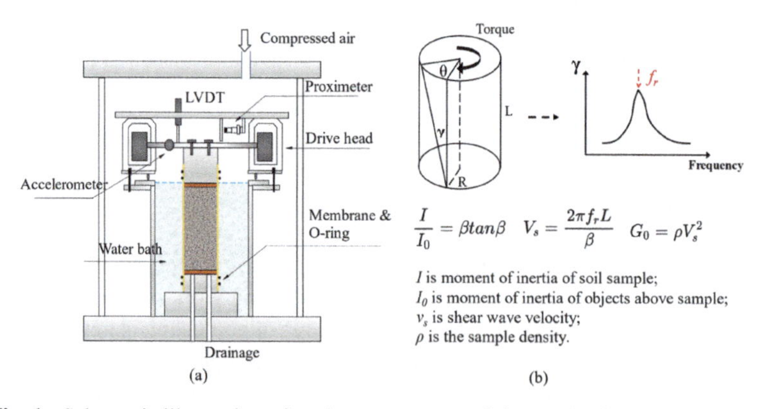

Fig. 1. Schematic illustration of testing apparatus and the mechanism to measure *G0*

2.2 Testing Materials and Procedures

Toyoura sand, a clean and uniform quartz sand, was adopted as the testing material. The basic physical properties and the particle size distribution curve of Toyoura sand along with its SEM image are provided in Fig. 2. The sample preparation method adopted was dry tamping method (DT) and the procedure of this method is illustrated in Fig. 3.

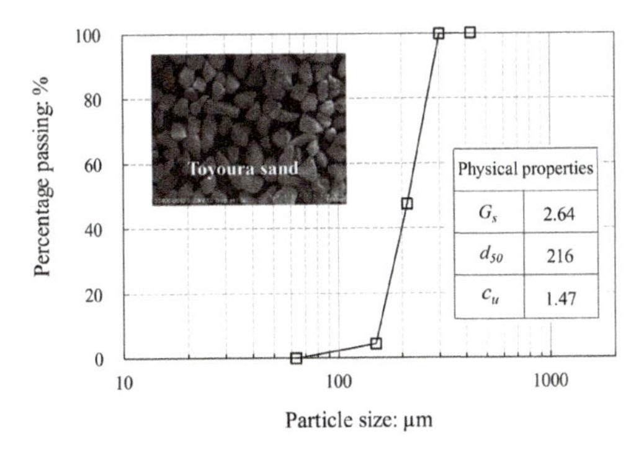

Fig. 2. Particle size distribution curve of Toyoura sand

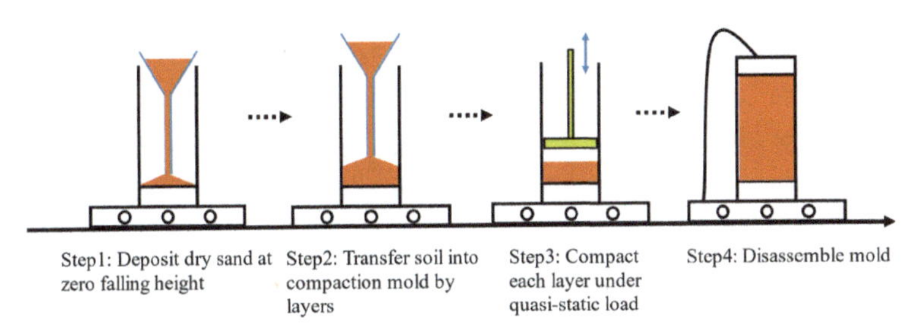

Fig. 3. Schematic illustration of dry tamping method

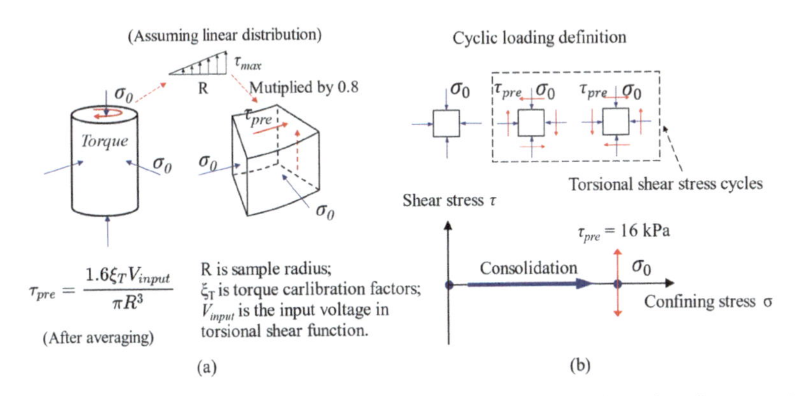

Fig. 4. Schematic illustration of soil element in torsional shear mode and testing procedure

As shown in Fig. 4(b), torsional shear stress cycles τ_{pre} were applied after the specimens were isotropically consolidated to σ_0. The shear stress amplitude was controlled by the input voltage to be 16 kPa for all specimens. The induced shear strains of specimens were between 0.01%–0.02% for confining stresses of 150 kPa and 300 kPa and were around 0.035% for the confining stress of 50 kPa. The small strain shear modulus G_0 was measured after the specified loading cycles (14000 in all tests reported here). In total, 12 saturated sand specimens of different densities were tested as summarized in Table 1.

Table 1. Test series and results

Testing No	Confining stress p' (kPa)	e_c	e_p	$[G_0]_{14000}/[G_0]_0$	$[G_0/f(e)]_{14000}/[G_0/f(e)]_0$
1	150	0.832	0.826	0.904	0.893
2	150	0.833	0.828	0.937	0.927
3	150	0.790	0.788	0.934	0.929
4	150	0.792	0.790	0.928	0.924
5	150	0.795	0.792	0.911	0.906
6	150	0.750	0.749	0.926	0.924
7	150	0.742	0.741	0.913	0.911
8	150	0.675	0.675	0.938	0.938
9	300	0.802	0.802	0.926	0.925
10	300	0.707	0.707	0.954	0.953
11	50	0.768	0.747	0.955	0.917
12	50	0.790	0.766	0.953	0.907

e_c is the post-consolidation void ratio and e_p is the void ratio after 14000 cycles;
$[G_0]_{14000}/[G_0]_0$ is the ratio of G_0 at 14000 cycles and initial state;
$[G_0/f(e)]_{14000}/[G_0/f(e)]_0$ is the ratio of $G_0/f(e)$ at 14000 cycles and initial state

3 Results and Discussions

3.1 Evolution of Void Ratio, Volumetric, Axial and Radial Strain

The volume of the specimen can be altered under long-term torsional shear cycles, and quantifying this volume change is important for analysing the loading history effects on G_0. Since all tests were carried out under saturated conditions, the volume change of each specimen was measured more reliably as compared with the experiments under dry conditions [13–15]. Figure 5(a) shows the evolution of void ratio of two specimens with different post-consolidation void ratios. It is clear that the change of void ratio is quite small, and the most apparent variation occurs at first several thousand cycles. Even for the relatively loose specimen ($e_c = 0.832$), the void ratio eventually became 0.826 after 14000 loading cycles. For the case of post-consolidation void ratio of 0.742, the loading cycles almost had no influence. The evolutions of axial, volumetric and radial strains of these two specimens are given in Fig. 5(b), (c) and (d) respectively. It is noted that the radial strain is always much larger than the axial strain and this observation is consistent with that reported in [17]. This indicates that the reduction of sample radius is much larger than its length when subjected to cyclic torsional shearing. To quantify the densification effects on stiffness, the void ratio function $f(e) = (2.17 - e)^2/(1 + e)$ that has been frequently used in the literature [9, 16] is adopted here. When the void ratio becomes smaller due to cyclic loadings, the value of $f(e)$ becomes larger. The value of $[f(e)]_N/[f(e)]_0$ ($[f(e)]_N$ is the value of $f(e)$ at loading cycle N) versus loading cycles for several specimens tested at different confining stress is plotted in Fig. 6(a). It is found that the contribution of densification on stiffness is approximately 5% when the confining stress is 50 kPa (corresponding to the cyclic stress ratio CSR of 0.32). However, such densification effects can generally be ignored when the confining stress is increased to 150 and 300 kPa (CSR is 0.11 and 0.05, respectively). Such phenomenon can be clearly observed in Fig. 6(b).

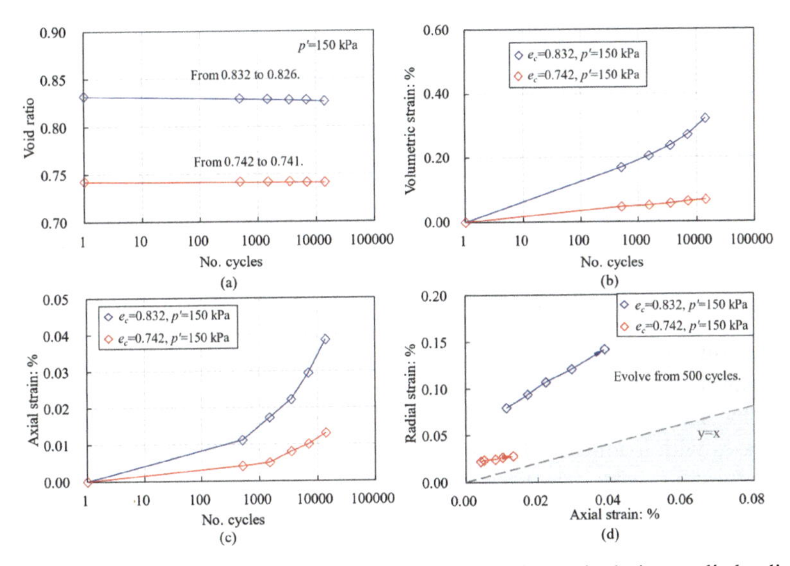

Fig. 5. Variations of void ratio, volumetric, axial and radial strain during cyclic loadings

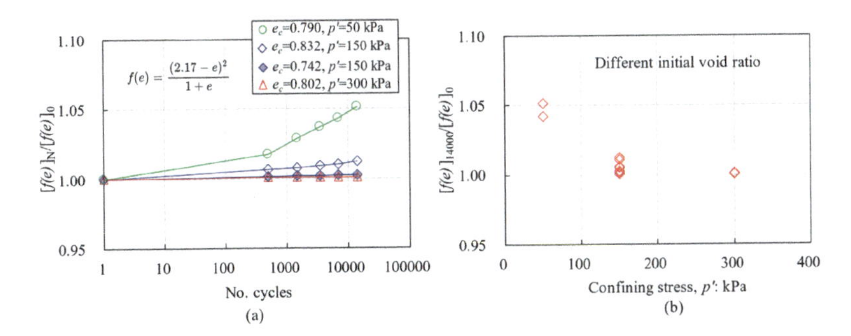

Fig. 6. The contributions of densification effects on G_0

3.2 Variation of Stiffness

Figure 7 shows an example of RC signals of a saturated specimen subjected to different numbers of torsional shear stress cycles. Its post-consolidation void ratio is 0.675 and the confining stress is 150 kPa. For clarity, only results at 0, 500 and 14000 cycles are given. It is clear that the resonant frequency decreases after the application of stress cycles, suggesting that there is degradation of shear wave velocity as well as the small strain shear modulus G_0.

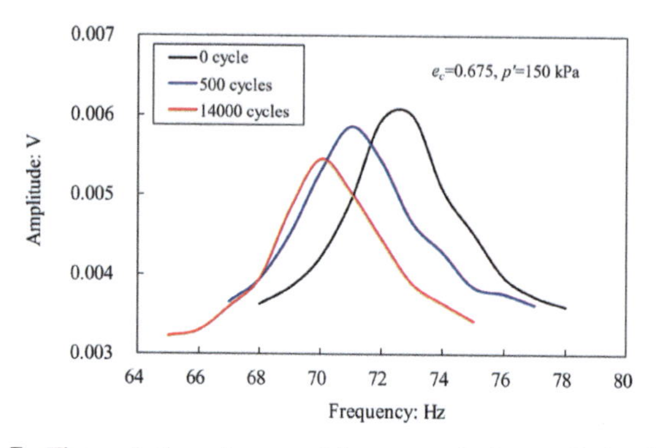

Fig. 7. The evolution of resonant frequency during cyclic loadings

The measured G_0 values of the three specimens confined at 150 kPa but with different densities are shown as a function of the number of cyclic loading in Fig. 8(a). The figure with the logarithmic axis of cycle number is provided in Fig. 8(b). It is noted that G_0 decreases with loading cycles for all these three specimens and an apparent decay emerges after 500 cycles. The G_0 becomes quite stable, however, as the loading cycles go beyond 3500, exhibiting only a very small decreasing tendency up to 14000 cycles. The initial G_0 of the specimen ($e_c = 0.742$) in Fig. 8 (a) is 125.3 MPa and it decreases to around 115.8 MPa after 3500 cycles. Further loading did not bring about further notable degradation and the final value of G_0 is 114.4 MPa. To consider the influence of void ratio, the data of $G_0/f(e)$ is displayed in Fig. 8(c). The initial values of $G_0/f(e)$ are located in a narrow range, which indicates that $f(e)$ does a good job to quantify the effects of void ratio. The values of $[G_0]_N/[G_0]_0$ and $[G_0/f(e)]_N/[G_0/f(e)]_0$ are plotted versus loading cycles as shown in Fig. 8(d), (e) and (f), where $[G_0]_N$, $[G_0/f(e)]_N$ are the values of G_0 and $G_0/f(e)$ at loading cycles of N, respectively. It is noted that the reduction of G_0 of all specimens after 14000 cycles is generally within 10%. The measured values of stiffness and void ratio before and after 14000 loading cycles are given in Table 1.

3.3 Effects of Post-consolidation Void Ratio and Confining Stress

As the G_0 value becomes stable after 14000 cycles, the ratio of G_0 values at 0 and 14000 cycles is used here to characterize the stiffness loss induced by cyclic loadings. It is interesting to examine whether the post-consolidation void ratio can cause any differences. To answer this question, the data of $[G_0]_{14000}/[G_0]_0$, obtained for tests

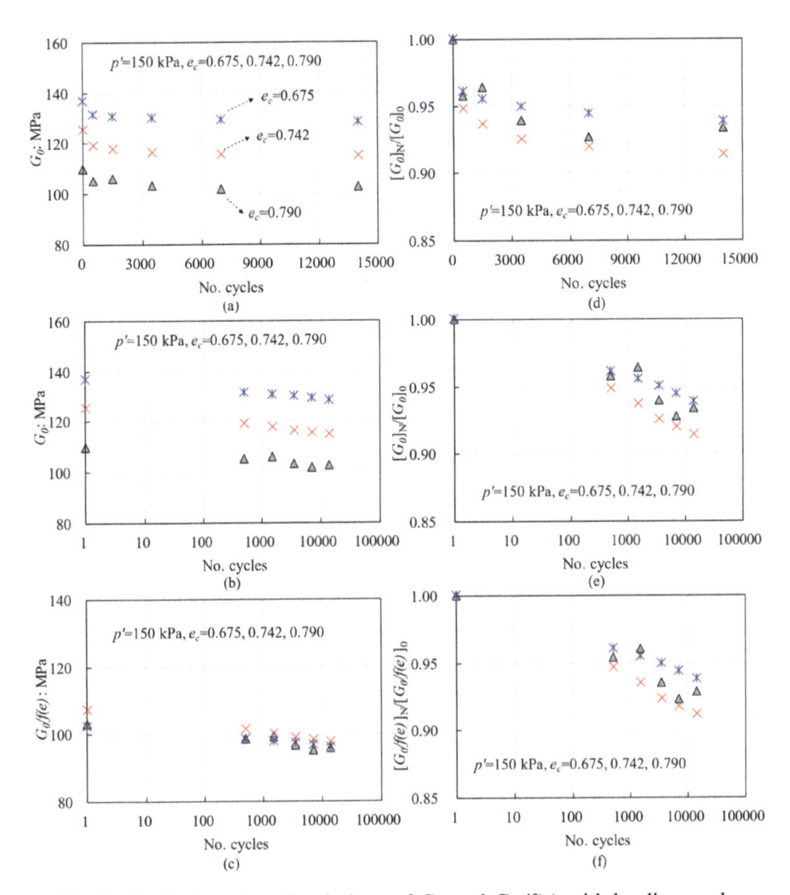

Fig. 8. Typical results of variations of G_0 and $G_0/f(e)$ with loading cycles

conducted at 150 kPa confining stress, are plotted as a function of e_c in Fig. 9(a). It appears that a looser initial state causes a slightly larger loss of G_0, but the difference is not very significant. The degree of degradation stays in a narrow range for different initial densities. When the effect of void ratio is removed by using the void ratio function, a similar trend is observed. It is considered reasonable because the void ratio variation due to cyclic loading is quite small according to the results in Sect. 3.1.

Several tests were also carried out at a lower confining stress 50 kPa and a higher confining stress of 300 kPa. The data of $[G_0]_{14000}/[G_0]_0$ and $[G_0/f(e)]_{14000}/[G_0/f(e)]_0$ is shown in Fig. 10. The degradation of G_0 of all the specimens is within 10%. Two specimens conducted at 50 kPa show a smaller stiffness loss and this is probably due to the larger densification effect. When the void ratio function is incorporated to remove the effect of void ratio, there is a trend that a higher confining stress (lower CSR) brings about a smaller stiffness loss as show in Fig. 10(b).

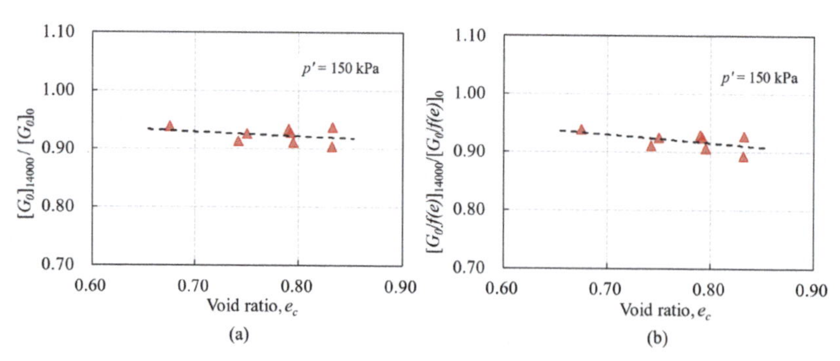

Fig. 9. Effects of post-consolidation void ratio e_c on G_0 and $G_0/f(e)$ degradation after 14000 cycles

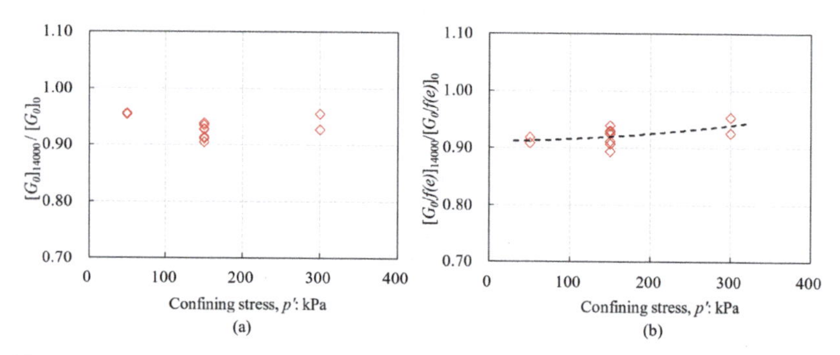

Fig. 10. Degradation of G_0, $G_0/f(e)$ after 14000 cycles for all the specimens at different confining stress

4 Conclusions

In this study, the variation of small strain shear modulus G_0 of sand due to large numbers of loading cycles at small amplitude was studied. Specimens of Toyoura sand with different initial densities and confining stress were tested using an RC/TS apparatus. The main findings of this study are summarized as:

(a) The G_0 value tends to decrease with loading cycles and the decrease is mainly associated with the initial loading cycles. When the number of cycles goes beyond around 3500, the G_0 value tends to become stable.

(b) The reduction of G_0 is not obviously affected by the initial density of sand specimen and is generally within 10% for the range of densities considered. A higher confining stress tends to bring about a little smaller stiffness loss when the densification effect is removed.

(c) The void ratio change due to the applied cyclic loading is small even after 14000 cycles. The contribution of densification to G_0 is only within 5% for specimens at the confining stress of 50 kPa and is even lower for specimens under higher confining stresses (150 and 300 kPa).

Acknowledgement. The work was supported by the Research Grants Council of Hong Kong (No. C7038-20G and 17206119). This support is gratefully acknowledged.

References

1. Hardin, B.O., Richart, F.E., Jr.: Elastic wave velocities in granular soils. J Soil Mech. Found. Div. **89**, 33–65 (1963)
2. Yang, J., Yan, X.R.: Factors affecting site response to multi-directional earthquake loading. Eng. Geol. **107**, 77–87 (2009)
3. Clayton, C.R.I.: Stiffness at small-strain: research and practice. Géotechnique **61**, 5–38 (2011)
4. Yang, J., Liu, X., Guo, Y., Liang, L.B.: A unified framework for evaluating in situ state of sand with varying fines content. Géotechnique **68**, 177–183 (2018)
5. Hardin, B.O., Drnevich, V.P.: Shear modulus and damping in soils: design equations and curves. J. Soil Mech. Found. Div. **98**, 667–692 (1972)
6. Liu, X., Yang, J.: Shear wave velocity in sand: effect of grain shape. Géotechnique **68**, 742–748 (2018)
7. Shin, H., Santamarina, J.C.: Role of particle angularity on the mechanical behavior of granular mixtures. J. Geotech. Geoenviron. Eng. **139**, 353–355 (2013)
8. Wichtmann, T., Triantafyllidis, T.: Influence of the grain-size distribution curve of quartz sand on the small strain shear modulus G_{max}. J. Geotech. Geoenviron. Eng. **135**, 1404–1418 (2009)
9. Yang, J., Gu, X.Q.: Shear stiffness of granular material at small strains: does it depend on grain size? Géotechnique **63**, 165–179 (2013)
10. Nikitas, G., Arany, L., Aingaran, S., et al.: Predicting long term performance of offshore wind turbines using cyclic simple shear apparatus. Soil Dyn. Earthq. Eng. **92**, 678–683 (2017)
11. Abadie, C.N., Byrne, B.W., Houlsby, G.T.: Rigid pile response to cyclic lateral loading: laboratory tests. Géotechnique **69**, 863–876 (2019)
12. Ma, H.W., Yang, J., Chen, L.Z.: An investigation into the effect of long-term cyclic loading on the natural frequency of offshore wind turbines. Int. J. Offshore Polar Eng. **30**, 266–274 (2020)
13. Drnevich, V.P., Richart, F.E.: Dynamic prestraining of dry sand. J. Soil Mech. Found. Div. **96**, 453–468 (1970)
14. Wichtmann, T., Triantafyllidis, T.: Influence of a cyclic and dynamic loading history on dynamic properties of dry sand, part I: cyclic and dynamic torsional prestraining. Soil Dyn. Earthq. Eng. **24**, 127–147 (2004)
15. Li, X.S., Yang, W.L.: Effects of vibration history on modulus and damping of dry sand. J. Geotech. Geoenviron. Eng. **124**, 1071–1081 (1998)
16. Gu, X.Q., Yang, J., Huang, M.S., Gao, G.Y.: Bender element tests in dry and saturated sand: signal interpretation and result comparison. Soils Found. **55**, 951–962 (2015)
17. Bai, L.D.: Preloading effects on dynamic sand behavior by resonant column tests. PhD thesis, Technical University of Berlin, Berlin, German (2011)

Liquefaction Resistance and Small Strain Shear Modulus of Saturated Silty Sand with Low Plastic Fines

Xiaoqiang Gu, Kangle Zuo, Chao Hu, and Jing Hu(✉)

Department of Geotechnical Engineering and Key Laboratory of Geotechnical and Underground Engineering of Ministry of Education, Tongji University, Shanghai 200092, China
jhu@tongji.edu.cn

Abstract. The liquefaction resistance and the small strain shear modulus G_0 of saturated Fujian sand with low plastic fines were evaluated with cyclic triaxial and bender element tests. The testing program encompasses a wide range of initial void ratios, confining stresses, and fines content from 0% to 30%. The results show that the influence of fines content on the liquefaction resistance and G_0 can be characterized predominantly by the equivalent granular void ratio (e*). The liquefaction resistance of Fujian sand with Shanghai silt decreases as e* increases, which resembles the results of sands containing non-plastic fines. Furthermore, a notable relationship between the liquefaction resistance and the G_0 was presented, which can potentially be used to evaluate the liquefaction resistance of silty sand using G_0.

Keywords: Silty sand · Liquefaction resistance · Small strain shear modulus · Equivalent granular void ratio

1 Introduction

Sand liquefaction is a common earthquake disaster. It often leads to foundation instability, collapse of buildings and destruction of urban lifeline projects. Both clean sand and silty sand are susceptible to liquefaction as reported, and the mechanism of the former has been studied for decades with a firm understanding achieved [1]. While liquefaction potential of silty sand has been investigated by many researchers, contradictory results have been reported on the effect of fines content on liquefaction resistance. Several studies have reported that with the increase of fines content, the liquefaction resistance of soil could either increase [2], or decrease [3]. In addition, some studies showed that liquefaction resistance first decreases as the fines content increases up to a limit fines content, and then increases as the fines content continues to increase [4]. The conflicting trend may be due to the choice of the index describing the behaviors of the soil mixture (e.g. relative density, global void ratio, skeleton void ratio and equivalent skeleton void ratio). The small strain shear modulus G_0 is a common parameter used in liquefaction evaluation. The notable Hardin's formula is usually used to calculate G_0 accounting for the influences of confining stress and void ratio. Nevertheless, previous studies show that

© The Author(s), under exclusive license to Springer Nature Switzerland AG 2022
L. Wang et al. (Eds.): PBD-IV 2022, GGEE 52, pp. 2180–2187, 2022.
https://doi.org/10.1007/978-3-031-11898-2_201

the small strain shear modulus of silty sand decreases with the increase of fines content at a constant global void ratio, where Hardin's formula is not applicable [5]. Therefore, it is worthwhile to explore the suitable index to characterize the liquefaction resistance and the small strain shear modulus of silty sand as well as their relation.

Different state indexes are defined to describe the properties of soils, in which the global void ratio (e) and relative density (D_r) are widely used for clean sand. However, they are not effective to capture the influence of fines content on the liquefaction resistance of silty sands. For the binary system, three states exist [6]: (1) when the fines content is very small, the sand particles are in direct contact with each other to form the sand skeleton structure, and the fines particles only exist in the pores between coarse particles and do not take part in the force chain. The skeleton void ratio (e_{sk}) was later proposed to characterize the state, which is defined as

$$e_{sk} = \frac{e + FC}{1 - FC} \tag{1}$$

where FC is the fines content. (2) When the fines content continues to increase, some fines particles begin to participate in the formation of soil skeleton, while some still stays in the pores between coarse particles. While the properties of the mixture are different than that with low fines content, these two are both sand-dominated. (3) When the fines content is high, the fines particles are in direct contact with sand particles to form the main skeleton of soil, and the sand particles are suspended between the fines particles. The behavior of the mixture is fines-dominated. Rahman et al. [7] proposed the limit fines content (TFC) to distinguish the "sand-dominated" and "fines-dominated" behaviors, which is defined as

$$TFC = 0.40 \times \left(\frac{1}{1 + \exp(\alpha - \beta \cdot \chi)} + \frac{1}{\chi} \right) \tag{2}$$

where $\alpha = 0.5$ and $\beta = 0.13$ are fitting constants, $\chi = D_{10}/d_{50}$ is the particle size ratio, D_{10} is the effective grain size of coarse sand and d_{50} is the mean grain size of fines particle. To delineate the participation of fines particles in the force chain of sand-dominated soil (FC < TFC), the equivalent skeleton void ratio (e^*) was proposed by Thevanayagam et al. [6] as

$$e^* = \frac{e + (1 - b) \cdot FC}{1 - (1 - b) \cdot FC} \tag{3}$$

where b ($0 \leq b < 1$) reflects the active fraction of fines particle engaging in the force chain. The fines particles do not engage in the force chain when b is zero. The value of b can be determined by [8]

$$b = \left\{ 1 - \exp\left[-\frac{0.3}{k} \right] \right\} \left(r \times \frac{FC}{TFC} \right)^r \tag{4}$$

where $k = 1 - r^{0.25}$ and $r = 1/\chi$.

Common methods for evaluating the liquefaction potential of saturated soils are standard penetration testing (SPT), cone penetration testing (CPT) and shear wave velocity

(V_s), in conjunction with the liquefaction triggering charts [1, 9]. The advantages of the in situ tests are obvious. However, most of the measured field data are collected after earthquakes, and the uncertainties embedding in the adjustment factors for determining the liquefaction resistance are difficult to quantify. In comparison, laboratory tests (e.g. cyclic triaxial tests or cyclic simple shear tests) have merits including simple operation, strong repeatability and low cost. It could be convenient to evaluate the liquefaction potential of silty sands based on laboratory tests.

In this study, a series of cyclic triaxial and bender element tests were conducted to investigate the effect of fines content on the liquefaction resistance and small strain shear modulus of Fujian sand with low plastic fines. The relationships among the liquefaction resistance, small strain shear modulus and different indexes were studied. The result lays down a foundation of a potential empirical method for predicting liquefaction potential of silty sands with the small strain shear modulus.

2 Experimental Setup

2.1 Test Materials

Fujian sand was used as the host sand in this study, which is a common sand with round particle shapes in China, and a low plastic silt from Shanghai was used as the fines particle, which has a liquid limit of 24.7 and a plasticity index of 7.2. The mass of fines content (FC) of the mixture ranges from 0% to 30%. Figure 1 shows the grain size distribution of the test materials. The maximum and minimum void ratios, e_{max} and e_{min}, of the mixtures were determined according to the geotechnical testing method in the Chinese standard GB/T 50123–2019. Basic physical property indexes are listed in Table 1.

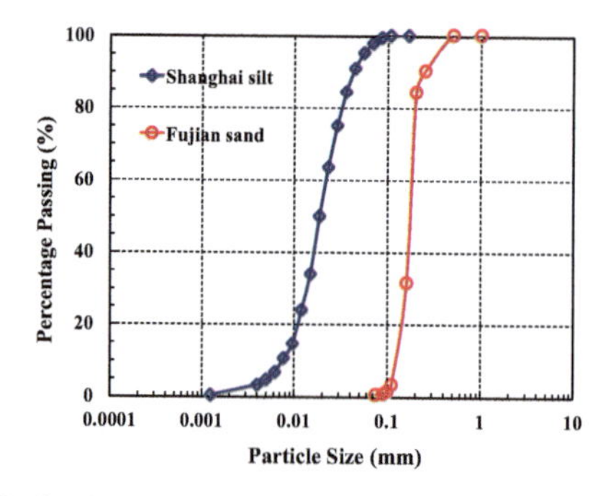

Fig. 1. Grain size distribution of Fujian sand and Shanghai silt.

2.2 Experimental Procedure

The effects of low plastic fines on the liquefaction resistance and the small strain shear strain of silty sand were investigated by conducting undrained cyclic triaxial and bender

Table 1. Physical properties of the sand-silt mixtures.

FC	G_s	$\rho_{d,max}$	$\rho_{d,min}$	e_{max}	e_{min}
0%	2.638	1.595	1.308	1.017	0.654
10%	2.647	1.660	1.315	1.013	0.594
20%	2.655	1.761	1.320	1.011	0.508
30%	2.664	1.825	1.322	1.015	0.460

element tests. All specimens with a height of 140 mm and a diameter of 70 mm were prepared by the moist tamping method using the undercompaction procedure.

The specimen was prepared in 7 layers and each layer with a 5% moisture content was compacted to the desired relative density with a tamper. In order to reach a high degree of saturation, specimens were first flushed with carbon dioxide (CO_2) and de-aired water, then back pressure was applied. The sample is deemed to be fully saturated when the B-value is larger than 0.98. After saturation, all specimens were isotropically consolidated under an effective confining stress of 100 kPa. All the void ratios reported in this manuscript are postconsolidation void ratios except as noted.

Bender element tests were conducted after consolidation, the input signal were sinusoidal waves with different excitation frequencies to better identify the arrival time [10]. The frequencies were 2, 5, 10, 20, and 50 kHz for S-waves. The output signal of the 20 kHz excitation frequency was chosen to confirm the travel time with the first arrival method [11]. The small strain shear modulus G_0 is determined by

$$G_0 = \rho(V_s)^2 \tag{5}$$

where V_s is the velocity of the shear wave and ρ is the soil mass density.

Specimens were subjected to sinusoidal double-amplitude uniform cyclic loadings with frequency of 0.2 Hz at various cyclic stress ratios, while the cell pressure was kept constant. Cyclic stress ratio (CSR) is the ratio of half the cyclic deviator stress σ_d to the initial effective confining pressure σ'_{3c}. The test was terminated until the specimen was liquefied. Excess pore water ratio (R_u) reaching unity is adopted as the criteria for initial liquefaction, where R_u is defined as the ratio of excess pore water pressure Δu to the initial effective confining pressure σ'_{3c}.

Specimens with two initial relative densities (D_r = 44.7 and 70.5%), four fines contents (FC = 0, 10, 20 and 30%) and various CSRs were selected for investigating the effect of various attributes on the liquefaction resistance and G_0 of the silty sand.

3 Test Results and Discussions

3.1 Cyclic Triaxial Tests

The results of a typical cyclic triaxial test are presented in Fig. 2, which is performed on a sample with 10% silt with an initial relative density of 70.5% subjected to CSR of 0.220. As shown in Fig. 2a, CSR remains constant until R_u reaches unity. Figure 2b shows the development of the axial strain under cyclic loadings. The axial strain remains small in the beginning and increases dramatically when excess pore pressure reaches the initial confining pressure of 100 kPa. As shown in Fig. 2c, R_u ramps up gradually to reach 100% at the 22th cycle. Figure 2d shows the stress path in the q–p' space. The specimen loses its strength as the mean effective stress becomes zero due to the generation of excess pore water pressure. Similar phenomena were observed for other samples.

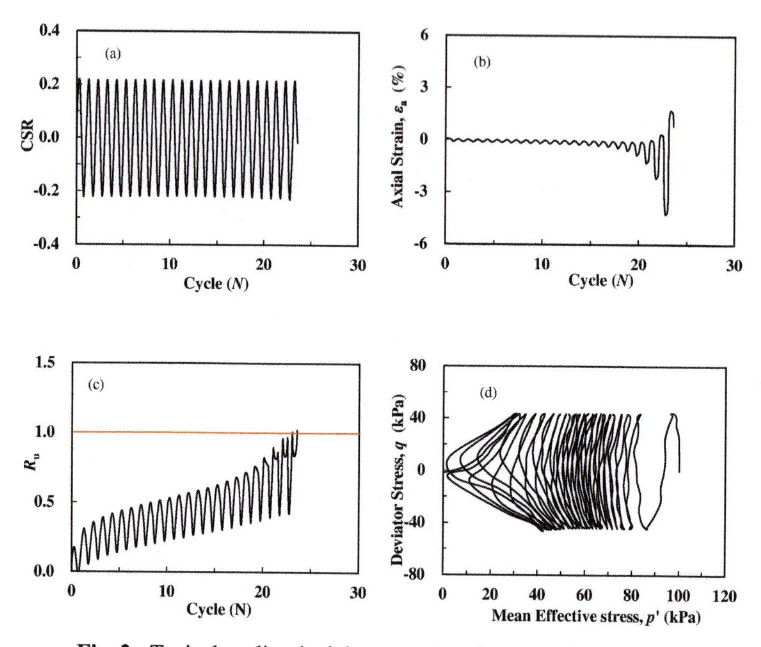

Fig. 2. Typical cyclic triaxial test results of sand with 10% fines.

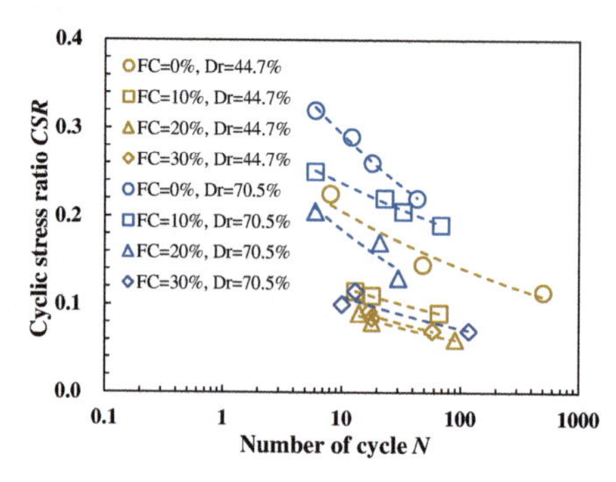

Fig. 3. The number of cycles of initial liquefaction versus the CSR for all the tests.

All the curves of the number of cycles reaching initial liquefaction versus CSR are presented in Fig. 3. As expected, the number of cycles to reach initial liquefaction decreases as CSR increases, and the cyclic resistance increases dramatically as the relative density increases. In addition, for the denser specimens (i.e. $D_r = 70.5\%$), the liquefaction resistance decreases continuously as the fines content increases to 30%, which resonates with the observations in Wichtmann et al. [12]. While for the looser specimens (i.e. $D_r = 44.7\%$), the liquefaction potential first increases with the increase of fines content up to 20%, then it remains the same with further increases of fines content, which agrees well with Karim and Alam [13].

3.2 Suitable Index for Characterizing Liquefaction Resistance and G_0

Consensus has yet been achieved on the effects of fines content on liquefaction resistance and small strain shear modulus of silty sand [14, 15]. To discover the suitable index dictating the properties of silty sands, we investigated the relationships among liquefaction resistance, small strain shear modulus and widely used indexes including relative density (D_r), global void ratio (e), skeleton void ratio (e_{sk}) and equivalent skeleton void ratio (e^*).

In the following, CRR_{15} is adopted to represent the liquefaction resistance of specimens, which is defined as the CSR giving rise to initial liquefaction in the 15th cycle of loading. It is found that CRR_{15} increases with the increase of D_r, while it decreases with the increase of e, e_{sk} and e^*. The correlation between liquefaction resistance and D_r, e or e_{sk} is greatly influenced by the fines content. However, it is of interest to note that CRR_{15} can be uniquely determined by e^* irrespective of fines content, as shown in Fig. 4a. Therefore, e^* serves better as an index for liquefaction evaluation compared with D_r, e and e_{sk}.

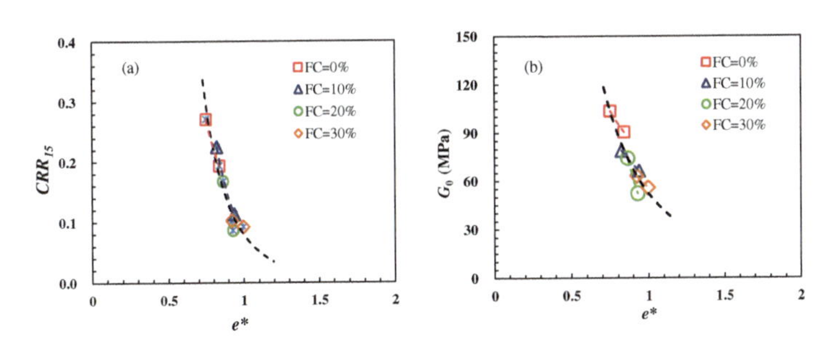

Fig. 4. The relationships between liquefaction resistance and G_0 with e^*.

The relationships between G_0 and the four indexes were also obtained, and it is found that G_0 increases with the increase of D_r, while it decreases with the increase of e, e_{sk} and e^*. It is remarkable that G_0 cannot be uniquely determined by D_r, e or e_{sk}, and e^* also serves best as the index to characterize the small strain shear modulus, as shown in Fig. 4b. It is deduced that e^* could be an intrinsic quality connecting CRR_{15} and G_0. As shown in Fig. 5, CRR_{15} can be captured well with G_0 irrespective of fines content, which indicates a potential empirical method for predicting liquefaction resistance of silty sands with its small strain shear modulus.

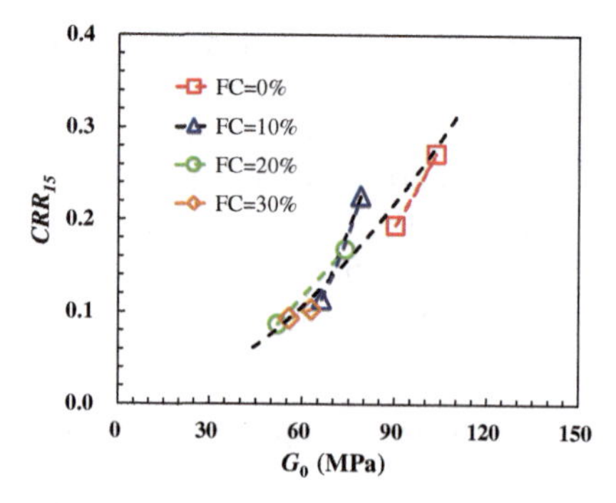

Fig. 5. The relationship between liquefaction resistance and G_0.

4 Conclusions

In this study, a series of cyclic triaxial and bender element tests has been performed on sand-dominated specimens with low plastic fines contents smaller than the limit fines content. The relationships among CRR_{15}, G_0 and e^* are carefully investigated. The main findings of the study are summarized as follows:

(a) Liquefaction resistance and small strain shear modulus of silty sand with low plastic fines and their relations with four basic indexes, i.e. D_r, e, e_{sk} and e^*, are presented. It is found that both liquefaction resistance and small strain shear modulus of the silty sand increase with the increase of D_r and decrease with the increase of e, e_{sk} and e^*.

(b) The liquefaction resistance and small strain shear modulus of silty sand with low plastic fines can be captured well by the equivalent skeleton void ratio e^* irrespective of fines content. A unique relationship can be established between CRR_{15} and G_0 through e^*, which suggests a potential empirical method to evaluate liquefaction potential of silty sand using G_0.

Acknowledgements. The work presented in this paper is supported by National Natural Science Foundation of China (Grant Nos. 41772283, 51822809). These supports are gratefully acknowledged.

References

1. Youd, T.L., Idriss, I.M., Andrus, R.D., et al.: Liquefaction resistance of soils: summary report from the 1996 NCEER and 1998 NCEER/NSF workshops on evaluation. J. Geotech. Geoenviron. Eng. **127**(4), 297–313 (2001)
2. Amini, F., Qi, G.Z.: Liquefaction testing of stratified silty sands. J. Geotech. Geoenviron. Eng. **126**, 208–217 (2000)
3. Lade, P.V., Yamanmuro, J.A.: Effects of nonplastic fines on static liquefaction sands. Can. Geotech. J. **34**(6), 918–928 (1997)

4. Polito, C.P., Martin, J.R.: Effects of nonplastic fines on the liquefaction resistance of sands. J. Geotech. Geoenviron. Eng. **127**(5), 408–415 (2001)
5. Goudarzy, M., Rahman, M.M., Koenig, D., et al.: Influence of non-plastic fines content on maximum shear modulus of granular materials. Soils Found. **56**(6), 973–983 (2016)
6. Thevanayagam, S.: Effect of fines and confining stress on undrained shear strength of silty sands. J. Geotech. Geoenviron. Eng. **124**(6), 479–491 (1998)
7. Rahman, M.M., Lo, S.R., Gnanendran, C.T.: On equivalent granular void ratio and steady state behaviour of loose sand with fines. Can. Geotech. J. **45**(10), 1439–1455 (2008)
8. Mohammadi, A., Qadimi, A.: A simple critical state approach to predicting the cyclic and monotonic response of sands with different fines contents using the equivalent intergranular void ratio. Acta Geotech. **10**(5), 587–606 (2014). https://doi.org/10.1007/s11440-014-0318-z
9. Robertson, P.K., Wride, C.E.: Evaluating cyclic liquefaction potential using the cone penetration test. Can. Geotech. J. **35**, 442–459 (1998)
10. Yang, J., Gu, X.Q.: Shear stiffness of granular material at small strain: does it depend on grain size? Géotechnique **63**(2), 165–179 (2013)
11. Baxter, C.D.P., Bradshaw, A.S., Green, R.A., Wang, J.H.: Correlation between cyclic resistance and shear-wave velocity for providence silts. J. Geotech. Geoenviron. Eng. **134**(1), 37–46 (2008)
12. Wichtmann, T., Kimmig, I., Steller, K., Triantafyllidis, T., et al.: Correlations of the liquefaction resistance of sands in spreader dumps of lignite opencast mines with CPT tip resistance and shear wave velocity. Soil Dyn. Earthq. Eng. **124**(9), 184–196 (2019)
13. Karim, M.E., Alam, M.J.: Effect of non-plastic silt content on the liquefaction behavior of sand–silt mixture. Soil Dyn. Earthq. Eng. **65**, 142–150 (2014)
14. Yang, J., Wei, L.M., Dai, B.B.: State variables for silty sands: Global void ratio or skeleton void ratio? Soils Found. **55**(1), 99–111 (2015)
15. Chen, G., Wu, Q., Zhao, K., et al.: A binary packing material-based procedure for evaluating soil liquefaction triggering during earthquakes. J. Geotech. Geoenviron. Eng. **146**(6), 04020040 (2020)

Volumetric Strains After Undrained Cyclic Shear Governed by Residual Mean Effective Stress: Numerical Studies Based on 3D DEM

Mingjin Jiang(✉), Akiyoshi Kamura, and Motoki Kazama

Tohoku University, Sendai, Japan
jiang.mingjin.r1@dc.tohoku.ac.jp

Abstract. A series of numerical tests is performed using the three-dimensional discrete element method to study the volume contraction characteristics of granular material during the reconsolidation process following undrained cyclic shear. Results show that post-liquefaction reconsolidation can be categorized into a liquefied and a solidified portion. The decrease in the void ratio in the liquefied portion is not accompanied by an increase in the mean effective stress, while the opposite is true for the solidified part. The residual mean effective stress affects the volumetric strain significantly during reconsolidation. The decrease in void ratio during the reconsolidation beginning from effective stress reduction ratios of 0.1, 0.5, and 0.9 is 86%, 37%, and 7% of that beginning from the initial liquefaction state, respectively. In addition, the volumetric strain during reconsolidation is associated with the change in pore uniformity.

Keywords: Volumetric strain · Undrained cyclic shear · Residual mean effective stress · 3D DEM

1 Introduction

Post-liquefaction settlement of surface ground is one of the causes of severe damage to buildings and infrastructures after an earthquake. It is primarily attributed to soil volume change as a result of drained pore water (reconsolidation), accompanied by excess pore water pressure dissipation. In terms of performance-based design, the amount of volumetric strain during or after liquefaction must be predicted accurately such that necessary countermeasures can be implemented.

The volume contraction characteristics of saturated sand have been investigated via various cyclic undrained tests, followed by drained reconsolidation. Lee and Albaisa [1] discovered that the volumetric strain during reconsolidation was affected by the particle size, relative density, and excess pore water pressure after cyclic shear. Nagase and Ishihara [2] and Shamoto et al. [3] reported that the reconsolidation volumetric strain was significantly associated with the maximum shear strain during cyclic shear. Tokimatsu and Seed [4] as well as Ishihara and Yoshimine [5] proposed simplified prediction models for post-liquefaction settlement based on experiment results. Sento et al. [6] discovered that the reconsolidation volumetric strain demonstrated a higher correlation with the

© The Author(s), under exclusive license to Springer Nature Switzerland AG 2022
L. Wang et al. (Eds.): PBD-IV 2022, GGEE 52, pp. 2188–2195, 2022.
https://doi.org/10.1007/978-3-031-11898-2_202

accumulated shear strain than the maximum shear strain generated during cyclic shear. Uzuoka et al. [7] proposed a prediction model for liquefaction and post-liquefaction settlement based on the minimum effective stress. Zhou et al. [8] discovered that the compression index during reconsolidation was 1.3–1.5 times as great as that during consolidation and proposed a model for post-liquefaction settlement estimation based on an assumed initial stress.

The reconsolidation process after liquefaction can be categorized into liquified and solidified portions [9]. The liquified portion is known as re-sedimentation [8]. A consensus was achieved, i.e., the volume contraction in the liquefied portion occupies a significant proportion of the total volume change during post-liquefaction reconsolidation. Therefore, understanding the re-sedimentation process is vital to the prediction of the total volume strain. However, owing to the limited measurement range in experiments, typically 10^{-1}–10^{0} kPa, the nonlinear relationship between the void ratio e and mean effective stress $\sigma_{\mathrm{m}}^{'}$ during re-sedimentation is yet to be clarified.

The discrete element method (DEM), developed by Cundall and Strack [10], is a mesh-free method. It allows the direct observation of specimen response at the particle scale without the limitations of measurement. A series of three-dimensional (3D) DEM simulations was conducted to investigate the reconsolidation characteristics of granular materials after cyclic undrained shear, and the results were analyzed based on the macroscopic and microscopic responses.

2 Simulation

2.1 DEM Model

Commercial code Rocky 4 [11] was used in this study. A DEM assembly was prepared using 16830 single-sized particles measuring 2 mm with a density of 2.667 g/cm^3. As shown in Fig. 1, it was composed of two loading layers and one interior layer, as well as enclosed with periodic boundaries in the x-direction and stiff platens in the other directions. The interaction model between particles or between particles and stiff platens was consistent with those used by Jiang et al. [12], i.e., a linear spring dash model in the normal direction, a linear spring Coulomb limit model in the tangential direction, and a rolling resistance model. The main parameters used were the same: The friction coefficient between particles and stiff lateral platens was zero; the elastic modulus of the particles and stiff platens was 1.0×10^{8} and 1.0×10^{11} N/m^2, respectively; the coefficient of restitution was 0.3; the friction coefficient between particles was 0.5; the rolling resistance coefficient was 0.35. In particular, the particles in the top and bottom layers were not allowed to rotate, and they were prevented from sliding on the stiff vertical platens by applying an extremely high friction coefficient. The specimen was compressed gradually under the K_0-condition by fixing stiff lateral platens and periodic boundaries and moving the top platen downward simultaneously without applying gravity. The consolidated specimen measured 48.23 mm (z-direction) $\times$ 50 mm $\times$ 50 mm.

In the center of the specimen, a measuring cube with a side length of 25 mm was set to obtain the stress information. The mean effective stress of the consolidated specimen (initial mean effective stress $\sigma_{\mathrm{m0}}^{'}$) was 100.2 kPa. In addition, without considering the

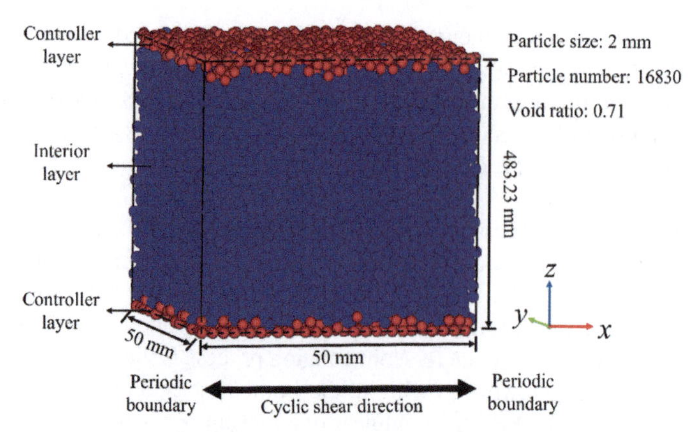

Fig. 1. Particle assembly in DEM after K_0-consolidation. Boundaries in x-direction (cyclic shear direction) are periodic, whereas those in other directions are stiff. Blue particles represent interior particles (which have no contact with vertical boundaries); red particle layers represent controllers for transferring displacement to interior particles.

overlap volume between particles, the overall void ratio of the entire specimen was 0.71 with ignoring the overlap volume between particles.

2.2 Undrained Cyclic Shear and Reconsolidation

After consolidation, cyclic shear strain γ with an amplitude of 1% was applied to the specimen via the horizontal movement (x-direction) of the bottom vertical platen. The top vertical platen and the periodic boundaries were maintained stationary. During cyclic shear, the specimen volume was maintained constant without applying gravity to simulate the undrained condition. The cyclic loading frequency was set to 2 Hz, and a quasi-static response was ensured by satisfying the $I = \dot{\varepsilon} d\sqrt{\rho / p'_{PT}} < 2.5 \times 10^{-3}$ criterion [13], where I is the inertial number, $\dot{\varepsilon}$ the strain rate, d the diameter of the particles, ρ the solid density, and p'_{PT} the mean effective stress at the phase transformation.

Figure 2 shows the shear stress–strain relationship and effective stress path during the undrained cyclic shear, where τ is the shear stress. The macroscopic response obtained using the 3D DEM was consistent with the pattern obtained from laboratory element tests. In a loading cycle, shear stress and strain formed the hysteresis loop, and negative and positive dilatancy behaviors appeared alternately. As cyclic loading continued, the stiffness and mean effective stress of the specimen decreased, and ultimately it liquefied and could not bear any stress.

As shown in Fig. 3, reconsolidations were commenced at the zero shear strain state to avoid the dilatancy induced by the residual shear strain. Cases 1–5 corresponded to reconsolidations beginning from a residual mean effective stress of 91.3, 50.7, 9.5 and kPa, the initial liquefaction, and a full liquefaction, respectively. Initial liquefaction refers to the first time at which the mean effective stress was below 10^{-3} kPa, and full liquefaction refers to the state where the stiffness of the specimen would not recover as shear strain was applied. The effect of changes in the pore water drainage rate was

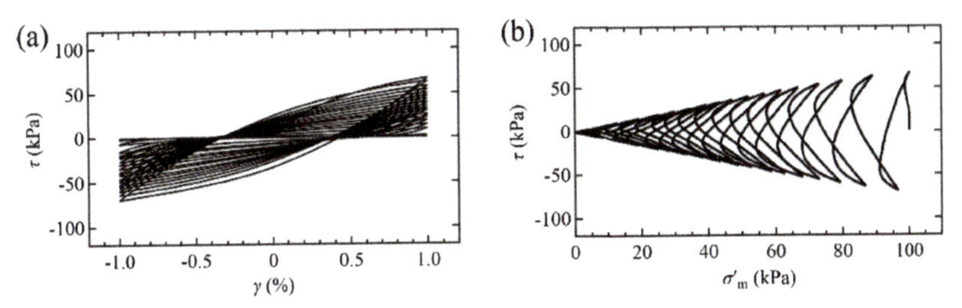

Fig. 2. Dynamic responses during undrained cyclic shear until full liquefaction: (a) Shear stress–strain relationship; (b) effective stress path.

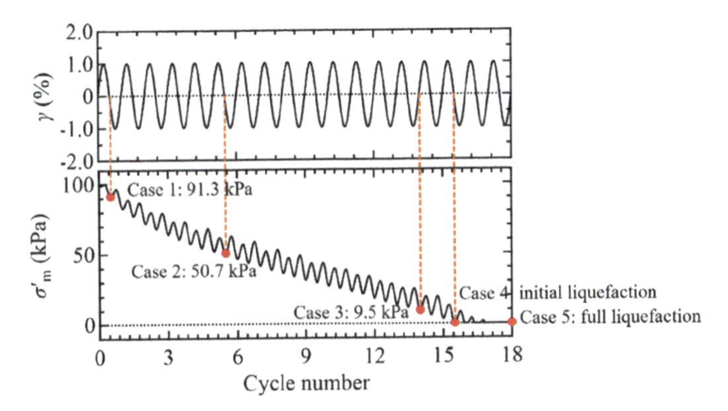

Fig. 3. Reconsolidation beginning at different cycles.

not considered in this study. The top platen moved downward at a constant velocity of 0.01 mm/s under the assumption that the volumetric strain rate was extremely low during reconsolidation.

3 Result

The coordination number Z is defined as the average number of interparticle contacts and is expressed as $Z = (N_{cb} + 2N_{cp})/N_p$, where N_{cb} and N_{cp} represent the total number of contacts between the particle and platen, and between particles, respectively; N_p is the total number of particles. Z is highly associated with the stability of the specimen. Jiang et al. [12] reported that Z reduced to less than 2 when the specimen liquefied. Figures 4 and 5 show the evolutions of Z and void ratio e from consolidation to reconsolidation, respectively. During consolidation, Z increased from 2.2 to 4.7 as the mean effective stress reached 100 kPa. Subsequently, during the undrained cyclic shear, it decreased gradually until initial liquefaction and fluctuated until full liquefaction. During reconsolidation, the interparticle contacts that vanished during undrained cyclic shear recovered; however, they remained at a slightly lower level compared with the level before undrained cyclic loading. In the liquefied cases (Cases 4 and 5), Z and the mean effective stress did not increase as e decreased when Z is less than 2. It was regarded as the liquefied portion of reconsolidation in this study, which indicates that the contact between particles was primarily in the form of impact, and a stable structure had not been formed. In the solidified portion, both Z and the mean effective stress increased as e decreased as a stable structure was formed. In addition, owing to the measurement limitations, the boundary between the liquefied and solidified portion observed in previous laboratory tests [7, 8] was larger than that observed in this study (between 10^{-1} and 10^{-2} kPa), and the process where e decreased extremely slowly at a stress level of less than 10^{0} kPa in the solidified portion of reconsolidation had not been observed.

The volumetric strain after undrained cyclic shear was associated with the residual mean effective stress, as shown in Fig. 5. A larger residual mean effective stress corresponded to a smaller decrease in e during reconsolidation. The decrease in e during reconsolidation in Cases 1, 2, and 3, corresponding to an effective stress reduction ratio of 0.1, 0.5, and 0.9, respectively, was 86%, 37%, and 7% of that in Case 4 (initial liquefaction), respectively. The effective stress reduction ratio is defined as $1 - \sigma'_{m}/\sigma'_{m0}$. The slope of the reconsolidation curve was similar to that of the normal consolidation curve in the stable zone. In particular, although the changes in e were different in the liquefied portion of the initial and full liquefaction cases (Case 4 and 5), they were similar after reconsolidation.

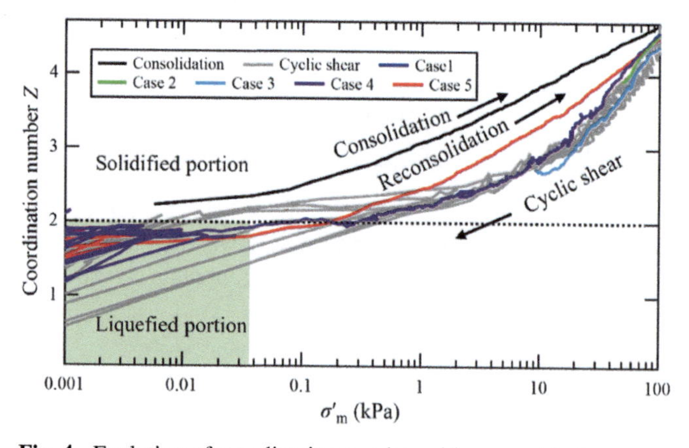

Fig. 4. Evolution of coordination number with mean effective stress.

Wei et al. [14] indicated that the degree of pore uniformity measured by the standard deviation of porosity was closely associated with the volume contraction characteristics of sands. In this study, as shown in Fig. 6, a measuring cube with a side length of 30 mm in the center of the specimen was divided into 27 sub-cubes of the same size. Considering that the porosity of specimen changes during reconsolidation process, the coefficient of variation (CV), defined as the ratio of the standard deviation to the mean, would be more representative of the degree of pore uniformity from cyclic shear to reconsolidation. It also should be noted that the sub-cube size (sub-cube number) within a reasonable range would affect the value of CV of porosity, however, its evolution trend was unchanged. The mean and maximum/minimum value and CV of porosity before cyclic shear (BCS), before and after reconsolidation (BR and AR) were obtained and shown in Fig. 7. During

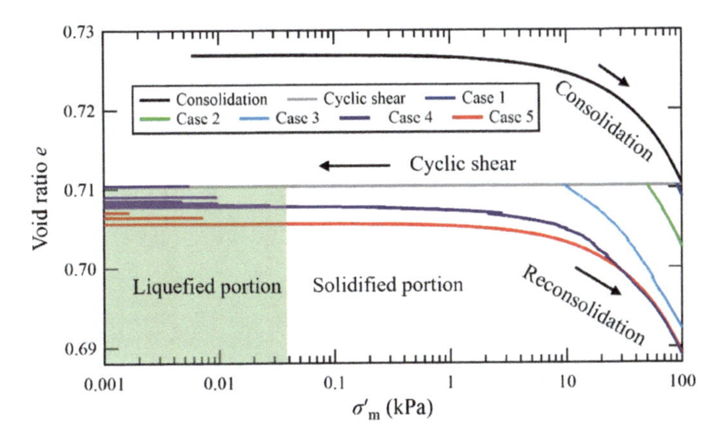

Fig. 5. Evolution of void ratio with mean effective stress.

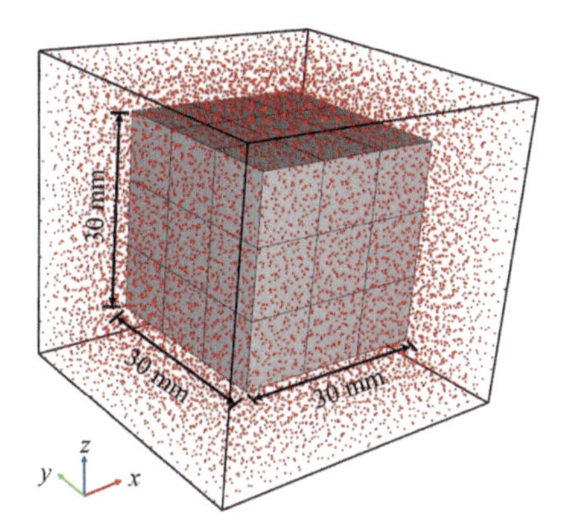

Fig. 6. Measuring cube and sub-cubes (Red points imply the centroid of particles).

the undrained cyclic shear, the CV of porosity decreased until the initial liquefaction, indicating that the pore in the specimen became increasingly more uniform. The higher CV of porosity during full liquefaction compared with that during the initial liquefaction may be attributed to the suspended particle movement, which is easily disturbed by the continued application of shear strain. In addition, during reconsolidation, the CV of porosity increased, which indicates that reconsolidation resulted in an uneven distribution of pores. Specifically, during reconsolidation processes began before liquefaction, the increase in the CV of porosity was associated with the volumetric strain. A larger decrease in e (Fig. 5) or mean value of porosity indicated a larger increase in the CV of porosity standard deviation.

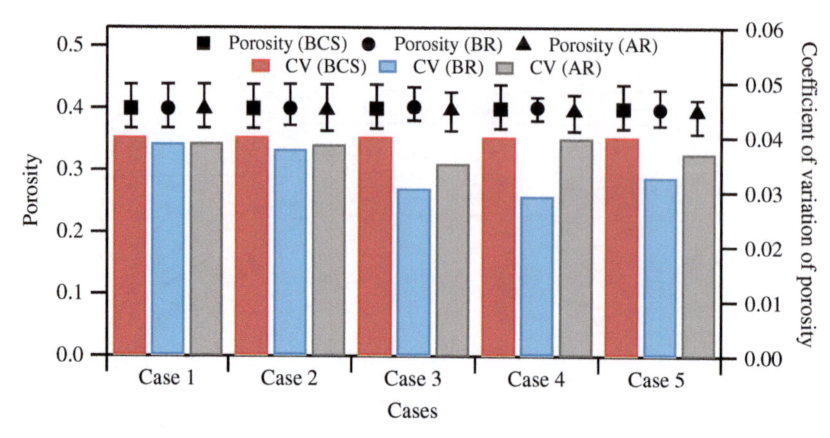

Fig. 7. Pore uniformity before cyclic shear (BCS), before reconsolidation (BR), and after reconsolidation (AR) in different cases. The marker and error bars denote the mean and maximum/minimum values of porosity, respectively. The bar chart represents the coefficient of variance (CV) of porosity.

4 Conclusion

To avoid the measuring limitation of laboratory tests, a series of 3D DEM simulations was performed to investigate the reconsolidation characteristics of K_0-consolidated materials. The drainage rate effect was disregarded by applying volumetric strain at a constant and low rate without gravity. The main conclusions are as follows:

1. The post-liquefaction reconsolidation process can be categorized into a liquefied and a solidified portion. In the liquefied portion, the void ratio *e* decreased without an increase in the mean effective stress. In the solidified portion, *e* decreased as the mean effective stress increased; however, *e* decreased extremely slowly at a stress level less than 10^0 kPa.
2. The residual mean effective stress significantly affected the volumetric strain during reconsolidation after an undrained cyclic shear. The higher the residual mean effective stress, the smaller was the volumetric strain during reconsolidation. The decrease in *e* during reconsolidation that began from effective stress reduction ratios of 0.1, 0.5, and 0.9 was 86%, 37%, and 7% of that beginning from the initial liquefaction state, respectively. In addition, although the total volumetric strains were similar for reconsolidation beginning from the initial and full liquefaction states, the volumetric strains in the liquefied portion were different.
3. The pore uniformity in the specimen increased during undrained cyclic shear, whereas it decreased during reconsolidation. Specifically, during reconsolidation processes began before liquefaction, the change in the CV of porosity during reconsolidation was positively correlated with the volumetric strain.

References

1. Lee, K.L., Albaisa, A.: Earthquake induced settlements in saturated sands. J. Geotech. Eng. Div. **100**(4), 387–406 (1974)
2. Nagase, H., Ishihara, K.: Liquefaction-induced compaction and settlement of sand during earthquakes. Soils Found. **28**(1), 65–76 (1988)
3. Shamoto, Y., Sato, M., Zhang, J.: Simplified estimation of earthquake-induced settlements in saturated sand deposits. Soils Found. **36**(1), 39–50 (1996)
4. Tokimatsu, K., Seed, H.B.: Evaluation of settlements in sands due to earthquake shaking. J. Geotech. Eng. **113**(8), 861–5878 (1987)
5. Ishihara, K., Yoshimine, M.: Evaluation of settlements in sand deposits following liquefaction during earthquakes. Soils Found. **32**(1), 173–188 (1992)
6. Sento, N., Kazama, M., Uzuoka, R.: Experiment and idealization of the volumetric compression characteristics of clean sand after undrained cyclic shear. Doboku Gakkai Ronbunshu **2004**(764), 307–317 (2004)
7. Uzuoka, R., Shimizu, Y., Kamura, A., Sento, N., Kazama, M.: A unified prediction for liquefaction and settlement of saturated sandy ground. In: 4th International Conference on Earthquake Geotechnical Engineering, Thessaloniki, Greece (2010)
8. Zhou, Y., et al.: Characterization of reconsolidation volumetric strain of liquefied sand and validation by centrifuge model tests. Chinese J. Geotech. Eng. **36**(10), 1838–1845 (2014)
9. Florin, V.A., Ivanov, P.L.: Liquefaction of saturated sandy soils. In: Proceedings of the 5th International Conference on Soil Mechanics and Foundation Engineering, pp. 107–111, Paris (1961)
10. Cundall, P.A., Strack, O.D.: A discrete numerical model for granular assemblies. Geotechnique **29**(1), 47–65 (1979)
11. ESSS. ROCKY: user's manual, version 4.4 (2019)
12. Jiang, M., Kamura, A., Kazama, M.: Comparison of liquefaction behavior of granular material under SH-and Love-wave strain conditions by 3D DEM. Soils Found. **61**(5), 1235–1250 (2021)
13. Perez, J.L., Kwok, C.Y., Huang, X., Hanley, K.J.: Assessing the quasi-static conditions for shearing in granular media within the critical state soil mechanics framework. Soils Found. **56**(1), 152–159 (2016)
14. Wei, X., Zhang, Z., Wang, G., Zhang, J.: DEM study of mechanism of large post-liquefaction deformation of saturated sand. Rock Soil Mech. **40**(4), 1596–1602 (2019)

One-Dimensional Wave Propagation and Liquefaction in a Soil Column with a Multi-scale Finite-Difference/DEM Method

Matthew R. Kuhn(✉)

University of Portland, Portland, OR 97203, USA
kuhn@up.edu

Abstract. The paper describes a multi-phase, multi-scale rational method for modeling and predicting the free-field wave propagation and liquefaction of soils. The one-dimensional time-domain model of a soil column uses the discrete element method (DEM) to track stress and strain within a series of representative volume elements (RVEs), driven by seismic rock displacements at the column base. The RVE interactions are unified with a time-stepping finite-difference algorithm. The Darcy's principle is applied to resolve the momentum transfer between a soil's solid matrix and its interstitial pore fluid. The method can analyze numerous conditions and phenomena, including site-specific amplification, down-slope movement of sloping ground, dissolution or cavitation of air in the pore fluid, and drainage that is concurrent with shaking. Several refinements of the DEM are necessary for realistically simulating soil behavior and for solving a range of propagation and liquefaction factors: most importantly, the poromechanic stiffness of the pore fluid and the pressure-dependent stiffness of the grain matrix. The model is verified with successful modeling-of-models simulations of several well-document centrifuge tests. The open-source DEMPLA code is available on the GitHub repository.

Keywords: Liquefaction · Wave propagation · Poromechanics · DEM

1 Introduction

During the past two decades, great progress has been made in understanding the mechanics of seismic propagation in near-surface soils and their weakening, lateral spreading, and liquefaction during ground shaking. This understanding has been aided by experimental studies, including the use of down-hole instrumentation arrays and of scaled centrifuge tests. In spite of this progress, geotechnical practitioners usually address wave propagation and liquefaction separately: site-specific ground motions are analyzed with one-dimensional linear models, and the likelihoods of lateral spreading and liquefaction are assessed with empirically-based estimates of resistance and seismic demand. Geotechnical researchers and practitioners now need unified approaches that realistically

© The Author(s), under exclusive license to Springer Nature Switzerland AG 2022
L. Wang et al. (Eds.): PBD-IV 2022, GGEE 52, pp. 2196–2203, 2022.
https://doi.org/10.1007/978-3-031-11898-2_203

include the interplay of soil weakening and wave propagation, and the effects of one on the other. Although finite-element methods hold promise, they require a macro-scale constitutive model that is faithful to a soil's complex behavior, and upon a deft choice of the model's parameters. Discrete element methods (DEM) have the advantage of relying solely on micro-scale information, and this information is largely derived from well-established physical tests (grain stiffness, water bulk modulus, etc.).

During the past ten years, the DEM has been shown to simulate the seismic behavior of granular soils in small assemblies of a few thousands of particles, capturing their small-strain stiffness and damping, but also tracking the large-strain pre- and post-liquefaction behaviors [8]. Some DEM studies have focused on the underlying micro-scale fabric and its changes during cyclic loading, and for these studies, a close agreement with the behavior of a target sand is not required [9]. However, a DEM model with close fidelity to real sand is necessary for simulating the complex problem of wave propagation and liquefaction in a soil column. The DEM code must model the pressure-dependence of stiffness (i.e., depth-dependent wave speed), the interplay of the pore fluid and the granular matrix as water is redistributed during shaking, and the complex progression and ratcheting of the normal and shearing forces at the contacts between particles. A resolution of these difficult issues is presented in [4, 5, 7, 8], with some of the details given in Sect. 2.3.

In Sect. 2.1, a summary of the field equations of wave propagation are presented; and the numerical implementation of these equations is briefly described in Sect. 2.2. An example is given in Sect. 3. Full details of the algorithm are given in [5], and the open-source DEMPLA program code is available on GitHub in the repository mrkuhn53/dempla, which includes source code, documentation, and examples [6].

2 Field Equations and Implementation

2.1 Field Equations

A one-dimensional column of soil extends from a rock base to the ground surface, with the two surfaces parallel (Fig. 1a). A rotated frame is used, in which the upward coordinate x_3 of a point $\mathbf{x}$ is measured perpendicular to the rock and ground surfaces (Fig. 1b). Lateral x_1 and x_2 directions are parallel to the two surfaces. The unit normal of the rock and ground surfaces is $\mathbf{n}$. Water can be present, either below or above the ground surface. When the water surface is below the ground surface, the water table can have a different slope than the ground surface; when the ground surface is submerged, the water surface is assumed horizontal. The upward normal of the water surface has unit direction $\mathbf{n}_{\mathrm{w}}$. The soil can be comprised of multiple stratigraphic layers with interfaces that are parallel to the ground surface and rock.

Displacements of the column's soil matrix $\mathbf{u}(\mathbf{x})$ have components u_1, u_2, and u_3 in the coordinate directions, and these movements are induced by a three-dimensional displacement history at the rock base. Vector field $\mathbf{w}(\mathbf{x})$ is the displacement of the pore fluid relative to the soil matrix. The spatial gradient $\nabla\mathbf{u}$ is the strain of the soil matrix, and these strains are imparted at representative volume elements (RVEs) located at material points $\mathbf{x}$. In the numerical implementation, an RVE is a DEM assembly, and the DEM algorithm is used to compute the effective stress $\boldsymbol{\sigma}'(\mathbf{x})$ within an RVE that results from the history of strain at $\mathbf{x}$. . The DEM algorithm allows nearly arbitrary sequences (and control) of the six components of stress or strain, allows stress rotations, allows control of either total or effective stresses, and permits the modeling of conditions that are drained, undrained, or partially drained [5, 7].

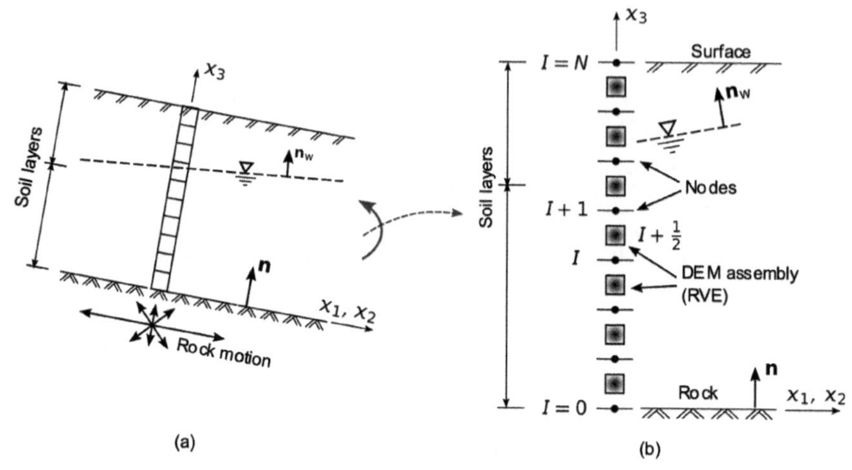

Fig. 1. Multi-scale model of a soil slope: (a) sloping ground with a sloping water table and multiple soil layers; (b) rotated system showing the locations of the displacement nodes and the inter-nodal RVEs, composed of DEM assemblies.

The field equations are those of Biot [1, 13], which are the momentum equations of the interacting solid and fluid phases,

$$\rho \mathbf{g} + \nabla \cdot \boldsymbol{\sigma} = \rho \ddot{\mathbf{u}} + n \rho_{\mathrm{f}} \ddot{\mathbf{w}} \tag{1}$$

$$\rho_f \mathbf{g} - \nabla p - (n/k) \dot{\mathbf{w}} = \rho_{\mathrm{f}} \ddot{\mathbf{u}} + \rho_{\mathrm{f}} \ddot{\mathbf{w}} \tag{2}$$

$$\boldsymbol{\sigma} = \boldsymbol{\sigma}^{'} - p\mathbf{I} \tag{3}$$

where $\rho_{\mathrm{f}}(\mathbf{x})$ is the pore fluid density; $\rho(\mathbf{x})$ is the bulk soil density, including the pore fluid; $n(\mathbf{x})$ is the soil porosity; $\mathbf{g}$ is the gravity acceleration vector, with magnitude g; $k(\mathbf{x})$ is the soil permeability in the x_3 direction, with units m^2 s^{-1} Pa^{-1}; $\boldsymbol{\sigma}(\mathbf{x})$ is the total stress (tension positive), given by Eq. (3); $\boldsymbol{\sigma}^{'}(\mathbf{x})$ is the effective stress; $\mathbf{I}$ is the identity tensor; and $\dot{\mathbf{w}}(\mathbf{x})$ is the fluid's seepage velocity relative to the granular matrix. Note that tensile stress is considered positive. Gravity $\mathbf{g}$ is the 3-vector $[g_1, g_2, g_3]$, where g_i is the inner product, $\mathbf{n}_i = \mathbf{g} \cdot \mathbf{n}_i$.

Because the system is one-dimensional, the stress gradients, $\sigma_{i1,1}$ and $\sigma_{i2,2}$, are zero; the relative fluid accelerations $\ddot{w}_1$ and $\ddot{w}_2$, are zero; the pressure gradients, $p_{,1}$ and $p_{,2}$, are stationary at their initial values; and the initial vertical gradient of fluid pressure is hydrostatic, $\left(p_{,3}\right)_{t=0} = -\rho_f g(\mathbf{n}_{\mathrm{w}} \cdot \mathbf{n}_3)$. These conditions establish the initial geostatic and hydrostatic conditions at the RVEs. Combined with Eqs. (1)–(3), the conditions give the four coupled equations of the one-dimensional problem,

$$\sigma'_{13,3} + \rho g_1 + \rho_{\mathrm{f}} g(\mathbf{n}_{\mathrm{w}} \cdot \mathbf{n}_1) = \rho \ddot{u}_1 \tag{4}$$

$$\sigma'_{23,3} + \rho g_2 + \rho_{\mathrm{f}} g(\mathbf{n}_{\mathrm{w}} \cdot \mathbf{n}_2) = \rho \ddot{u}_2 \tag{5}$$

$$\sigma'_{33,3} + \rho g_3 - p_{,3} = \rho \ddot{u}_3 + n\rho_{\mathrm{f}} \ddot{w}_3 \tag{6}$$

$$-p_{,3} + \rho_{\mathrm{f}} g_3 - (n/k)\dot{w}_3 = \rho_{\mathrm{f}} \ddot{u}_3 + \rho_{\mathrm{f}} \ddot{w}_3 \tag{7}$$

in which the four kinematic variables, u_1, u_2 u_3, and w_3, must be solved as functions of time t and position x_3.

2.2 Solution Algorithm

Equations (4)–(7) are solved with a leap-frog finite difference algorithm. At each node, the displacements are solved in a series of time steps Δt, and at a given time t, the nodal displacements are used for computing the displacement gradients (strains) at the RVEs. For step Δt, the increment of strain is computed for each RVEs, and this increment is imposed on the corresponding DEM assembly, each of which contains thousands of particles. (The strain increment that is imposed on a DEM assembly is divided into a number smaller increments, ensuring that the assembly is deformed in a quasi-static manner.) After the DEM assemblies are deformed at time t, the effective stress $\boldsymbol{\sigma}'$ in each assembly is computed using the Love–Weber formula, and the water pressure p is computed using the assembly's poromechanic model. The simultaneous Eqs. (4)–(7) are then solved to compute the accelerations $\ddot{u}_i$ and $\ddot{w}_3$, which are used for updating the velocities and for finding the nodal displacements of the soil matrix and fluid at $t + \Delta t$. The full algorithm is presented in [5].

For points below the water table, the DEM model includes a poromechanic component that computes the fluid pressure in the water, and the model accommodates either pure water or water that is quasi-saturated with isolated air bubbles. With soil above the water table, the air in the dry soil is also treated as a fluid, with its own density, compressibility, and permeability (see [7] and the appendix in [5]).

Although the field equations are quite general, some computational speedup is realized for seismic simulations by distinguishing two regimes of behavior [13]: the dynamic period of ground shaking and the post-shaking period of consolidation. During shaking, we must consider inertial effects and apply the full Biot equations. The time step Δt for the dynamic regime is limited by the Courant–Friedrichs–Lewy (CFL) condition, which prolongs the time-driven computations.

For the post-shaking period, slow consolidation is the dominant process, and accelerations are insignificant: $\ddot{\mathbf{u}} \to 0$, $\ddot{\mathbf{w}} \to 0$. By assuming that the accelerations are zero, Eq. (7) is simplified as follows [13], abrogating the CFL limitation:

$$\dot{w}_3 = \frac{k}{n}\left(-p_{,3} + \rho_{\mathrm{f}} g_3\right) \tag{8}$$

with constant stresses, $\sigma_{i3} = 0$, imposed during the consolidations (see [5]).

2.3 DEM Algorithm

The DEM model presupposes an RVE response that is quasi-static and rate-independent, since a significant computational speed-up can be exploited when this assumption applies. This assumption is verified by monitoring and controlling the dimensionless inertial number I, defined as $I = \dot{\gamma} d \sqrt{\rho_s / \overline{p}}$, where d is the particle size, $\dot{\gamma}$ is the strain rate, ρ_s is the density of the particles, and $\overline{p}$ is the mean effective stress [2]. Knowing that the problem is quasi-static, one can reduce the DEM computation time, by choosing a DEM time-step and/or a DEM particle density ρ_s that gives a smaller value of I, thus maintaining numerical stability with a speed-up of simulation time [11], a technique that is commonly used in quasi-static DEM simulations [12]. In the author's simulations, I was maintained well below 0.001, while the DEM algorithm advanced the strain in increments of between 1×10^{-7} and 1×10^{-6} per DEM time-step.

Realistic simulations of wave propagation and liquefaction require four DEM features, features that are not included in most DEM codes:

1. Because the simulations model the migration and redistribution of the pore fluid (displacement $\mathbf{w}$), the DEM code must include a poromechanic component for all pore fluids that are present in the soil column. The model must account for the coupling of the inter-granular effective stress, pore fluid pressure, strain of the soil matrix, and strain of the pore fluid. The solution in [7] applies to water-saturated and quasi-saturated soils, accounting for bubble size, surface tension, vapor pressure, and air solubility. The model for dry soil is derived in the appendix of [5].
2. The DEM code must include a robust servo-algorithm that can maintain target stresses and strains for different control types (strain-control and multiple types of stress-control) while adjusting to abrupt changes in loading during seismic motion. The code uses an adaptive control algorithm that maintains close fidelity to the target stresses and strains (see appendix in [7]).
3. Seismic wave speed depends upon the mean effective stress: seismic waves slow as they approach the ground surface. A simple linear contact model in DEM yields no such pressure-dependence of the wave speed. A sphere–sphere Hertzian model also underestimates the dependence; whereas, a cone–cone Hertzian model overestimates the dependence. For sands, the relationship between shear modulus G and mean effective stress $\overline{p}$ is of the proportionality $G \propto \overline{p}^{\beta}$, where exponent β is usually about 0.50. The code uses a contact model of blunted cones, which gives the correct β [8].
4. The DEM contact model must also find tangential force. The two most common tangential models—the linear model and the Modified Mindlin (MM) model with a sphere–sphere Hertzian normal force—are inadequate for realistic (predictive) simulations. With an MM model, a closed loading path that lies below the friction limit in the normal–tangential force-space can produce a spurious creation or loss of elastic energy. This deficiency must be disallowed when modeling cyclic loading, since contacts will follow erratic loading–unloading paths but must not produce energetically infeasible outcomes. The author uses an implementation of the full Mindlin–Deresiewicz solution, developed by Jaeger [3, 4], and this model correctly accounts for annular slip and disallows energy creation.

3 Example: Comparison with Centrifuge Experiments

The example is a simulation of a centrifuge test that was conducted at Rensselaer Polytechnic Institute (RPI) [10]. The centrifuge test modeled a prototype 10 m submerged deposit of medium-dense sand at relative density $D_r = 65\%$, with a 5° slope that liquefied during dip-parallel near-sinusoid shaking. The centrifuge model was 20 cm deep and spun at 50 g, replicating a prototype 10m depth. Similar to centrifuge tests, the simulations were conducted at model-scale, avoiding the problem of matching the scales of multiple quantities in the prototype system (particle size, viscosity, stress, depth, surface tension, permeability, etc.). The soil density was multiplied by 1.294 to account for the mass of the centrifuge box's laminar rings. The centrifuge water was treated with additives to produce a viscosity 50 times greater than that of pure water, and these conditions were also included in the simulation model.

A series of DEM assemblies, each with 10,648 particles was prepared to simulate the Nevada Sand of the centrifuge model. DEM particles were created with a distribution of sizes similar to that of Nevada Sand, and a non-spherical particle shape was adopted. The author chose a non-convex cluster of seven spheres, allowing interlocking between particles, and thus inhibiting inter-particle rolling, without using any rolling stiffness or rolling friction. In three different simulations, the DEM assemblies had void ratios equal to those of Nevada Sand at relative densities D_r of 45%, 65%, and 75%, as in the RPI tests. A contact model was adopted that very closely fit the small-strain stiffness of Nevada Sand, including the stiffness' dependence on mean stress and void ratio (see Sect. 2.3). An inter-particle coefficient of 0.40 was found to give results similar to cyclic simple-shear tests of Nevada Sand.

The input base motion of the RPI centrifuge tests was a near-sinusoid application of 22 cycles of acceleration amplitude ≈0.20 g at 2 Hz, with a peak acceleration of 0.23 g. The simulation requires the base displacements $\mathbf{u}(t)$, so accelerations were double-integrated and, when necessary, adjusted to end with zero velocity.

Results are shown in Fig. 2, which compares simulations with the RPI centrifuge experiments. Figure 2a shows the down-slope movement at the ground surface and at the mid-depth of 5 m with a sand of density $D_r = 65\%$. The simulation closely matches the centrifuge results until the final few cycles of acceleration, when the simulation ended with about 25% more lateral movement than the centrifuge model. The soil column liquefied during both the simulation and centrifuge test. Figure 2b shows the final accumulated downslope movements for sands with $D_r = 45\%$, 65%, and 75%. Again, the simulations give similar movements as the RPI centrifuge tests, with decreasing lateral movement as density is increased. Finally, Fig. 2c compares the post-shaking consolidation settlement versus time for simulations and centrifuges test with $D_r = 45\%$, showing similar consolidation behaviors for the simulation and centrifuge test.

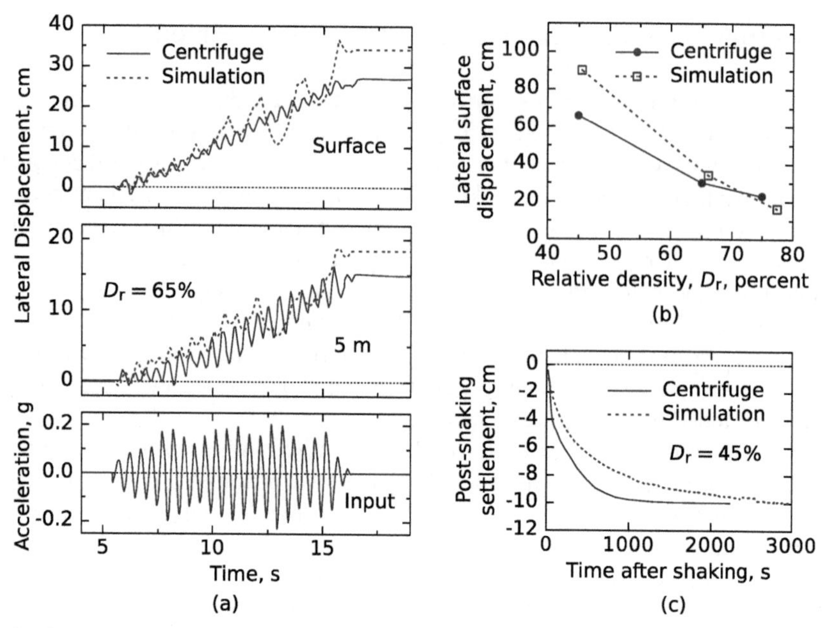

Fig. 2. Comparison of RPI centrifuge tests and simulations of a 10 m prototype: (a) lateral down-slope displacements ($D_r = 65\%$) at the surface, at depth 5 m, and at the base; (b) final down-slope displacements for three densities; and (c) post-shaking settlement ($D_r = 45\%$). See [5, 10].

4 Example and Conclusions

In the paper, a multi-phase multi-scale method is developed for modeling and predicting the wave propagation and the weakening and liquefaction of near-surface soils. Verification results of a simple example are promising, with down-slope lateral movements correctly predicted with reasonable accuracy. The full paper [5] compares other results and at multiple soil densities. The paper [5] also includes comparisons with three other published centrifuge studies: comparisons of surface response spectra, sub-surface pore water pressure rise, and post-liquefaction settlements. Besides promising better understanding of the diverse phenomena that occur during ground shaking, the method's results are sufficiently favorable to warrant use as a predictive tool for analyzing site-specific seismic response and liquefaction potential. The DEMPLA code is open-source and freely available on GitHub [6], and all examples in [5] are included on the site, along with documentation and sample assemblies.

Other than the restriction to a one-dimensional column, the method is general and allows diverse conditions and phenomena: (1) three-dimensional motions of rock and soil, propagating as both p-waves and s-waves; (2) nearly arbitrary stress and strain paths during ground shaking; (3) sloping ground surface with down-slope movement; (4) multiple stratigraphic soil layers; (5) sub-surface water table or complete submergence of the site; (6) sloping water table with down-dip seepage forces; (7) onset and depth of liquefaction; (8) saturated soil, dry soil, or quasi-saturated soil with entrained air at a specified saturation; (9) dissolution or cavitation of entrained air; (10) effect of saturation on wave propagation and liquefaction; (11) effect of drainage that is concurrent with shaking; (12) effect of pore fluid viscosity; (13) site-specific amplification of surface accelerations relative to those of the rock base; (14) pressure-dependence of wave speed and the slowing of waves as they approach the surface; (15) dilation and the coupling of s-waves and p-waves; (16) voids redistribution and development of a water film beneath less permeable layers; (17) softening of soil during shaking, with a slowing of wave speeds and shifting of frequency content due to build-up of pore fluid pressure; (18) post-shaking consolidation and settlement; post-consolidation reshaking and post-triggering behavior; and (20) relative rise of the water table during and after shaking.

References

1. Biot, M.A.: Mechanics of deformation and acoustic propagation in porous media. J. Appl. Phys. **33**(4), 1482–1498 (1962)
2. da Cruz, F., Emam, S., Prochnow, M., Roux, J.N., Chevoir, F.: Rheophysics of dense granular materials: discrete simulation of plane shear flows. Phys. Rev. E **72**(2), 021309 (2005)
3. Jäger, J.: New Solutions in Contact Mechanics. WIT Press, Southampton, UK (2005)
4. Kuhn, M.R.: Implementation of the Jager contact model for discrete element simulations. Int. J. Numer. Methods Eng. **88**(1), 66–82 (2011)
5. Kuhn, M.R.: Multi-scale simulation of wave propagation and liquefaction in a one-dimensional soil column: hybrid DEM and finite-difference procedure. Acta Geotech. **17**, 1–22 (2021). https://doi.org/10.1007/s11440-021-01402-7
6. Kuhn, M.R.: DEMPLA and OVAL: programs for analyzing particle assemblies and for simulating wave propagation and liquefaction with the discrete element method (2021). https://github.com/mrkuhn53/dempla
7. Kuhn, M.R., Daouadji, A.: Simulation of undrained quasi-saturated soil with pore pressure measurements using a discrete element (DEM) algorithm. Soils Found. **60**(5), 1097–1111 (2020)
8. Kuhn, M.R., Renken, H., Mixsell, A., Kramer, S.: Investigation of cyclic liquefaction with discrete element simulations. J. Geotech. Geoenvron. Eng. **140**(12), 04014075 (2014)
9. Nguyen, H.B.K., Rahman, M.M., Fourie, A.: The critical state behaviour of granular material in triaxial and direct simple shear condition: a DEM approach. Comput. Geotech. **138**, 104325 (2021)
10. Sharp, M.K.: Development of centrifuge based prediction charts for liquefaction and lateral spreading from cone penetration testing. Ph.d. thesis, Rensselaer Polytechnic Institute, Troy, N.Y. (1999)
11. Suzuki, K., Kuhn, M.R.: Uniqueness of discrete element simulations in monotonic biaxial shear tests. Int. J. Geomech. **14**(5), 06014010 (2014)
12. Thornton, C.: Numerical simulations of deviatoric shear deformation of granular media. Géotechnique **50**(1), 43–53 (2000)
13. Zienkiewicz, O.C., Bettess, P.: Soils and other saturated porous media under transient, dynamic conditions general formulation and the validity of various simplifying assumptions. In: G.N. Pande, O.C. Zienkiewicz (eds.) Soil Mechanics–Transient and Cyclic Loads, pp. 1–16. John Wiley & Sons, Chichester (1982)

Microscopic Insight into the Soil Fabric During Load-Unload Correlated with Stress Waves

Yang Li(✉), Masahide Otsubo, and Reiko Kuwano

Institute of Industrial Science, The University of Tokyo, Tokyo, Japan
liyang16@iis.u-tokyo.ac.jp

Abstract. Soils have a complex contact network (i.e. soil fabric), which affects the mechanical characteristics (e.g. stiffness anisotropy, shear strength) and further determines the performance in geotechnical constructions. Conventional laboratory tests on soils can give macro-scale mechanical responses as a physical relevance, while the discrete element method (DEM) has been used as a numerical tool to gain insights into micro-scale mechanical characteristics such as soil fabric. In this study, laboratory element tests are conducted using spherical glass beads subjected to several cycles of load-unload reversals at selected pre-peak strain ranges, where stress waves are continuously measured during the loading process. DEM is adopted using spherical particles to simulate the equivalent load-unload reversals and wave propagations. Based on the experimental and DEM results, the compression velocity (V_p) and shear wave velocity (V_s) are found to be approximately reversible when the load-unload cycles are at low stress levels, while a significant reduction in both velocities can be observed at higher stress levels. The variation of soil fabric anisotropy during load-unload reversals is explored using DEM; the fabric anisotropy increases markedly beyond a stress ratio threshold of 1.8 and cannot be fully recovered during the subsequent unloading process. Fabric anisotropy is found to be linearly correlated with wave velocity ratio (V_p/V_s) at the pre-peak stage, although both quantities are respectively affected by the given stress states.

Keywords: Soil fabric · Stress waves · Load-unload · DEM · Laboratory tests

1 Introduction

It is essential to explore the evolution of soil fabric under load-unload processes which take place commonly in practical geotechnical engineering such as seismic impulses or moving vehicle loads. The variation of soil fabric during cyclic loading is complicated; the fabric firstly shows significant variations and then gets stabilized as the number of loading cycles increases, indicating the effect of loading history [1]. More importantly, soil fabric cannot always be recovered when the deviator stress is unloaded to the initial stress state [2]. Specifically, the change in stress is not always accompanied by a synchronized change in soil fabric. The particle rearrangement is less significant when the loading direction is reversed to the initial stress state [3].

© The Author(s), under exclusive license to Springer Nature Switzerland AG 2022
L. Wang et al. (Eds.): PBD-IV 2022, GGEE 52, pp. 2204–2211, 2022.
https://doi.org/10.1007/978-3-031-11898-2_204

Stress wave measurements have been widely used to correlate the liquefaction resistance with shear wave velocity [4] or estimate the degradation in small strain shear modulus (G_O) during cyclic loading [5]. However, there is a lack of studies related to the continuous trace of fabric change correlated with wave velocities during the load-unload reversals. Li et al. [6] found a good correlation between wave velocity ratio (V_p/V_s) and fabric anisotropy during monotonic loading in DEM. This study investigates the variation of stress wave velocities subjected to load-unload reversals using both laboratory tests and DEM simulations. Building on [6], the fabric change is explored during load-unload reversals at the pre-peak stage, where stress waves are detected continuously during the load-unload process.

2 Laboratory Test

2.1 Tested Material and Testing Procedure

Nominally spherical and smooth glass beads (GB) were used in the laboratory tests. The material density (ρ_p), Young's modulus (E_p) and Poisson's ratio (ν_p) of GB grains are 2500 kg/m^3, 71.6 GPa and 0.23, respectively. The particle size distribution (PSD) is between 0.4 and 0.65 mm. The particle shape was measured using a QICPIC apparatus [7] with a sphericity of 0.940, convexity of 0.974 and aspect ratio of 0.975. The root mean square of surface roughness (RMS) was measured as 69 nm using the Gaussian filter method [8]. In this study, a sample was prepared with an initial void ratio (e_O) of 0.595 and a relative density (D_r) of 94% with an isotropic confining stress of $\sigma' =$ 50 kPa in a dry state. The sample had a height of 150 mm and a diameter of 75 mm. as shown in Fig. 1(a).

A stress-controlled algorithm was adopted to conduct the load-unload cycles with a constant strain rate of 6×10^{-6} [s^{-1}], which was slow enough to perform the wave measurement suitably at selected axial strains (ε_a). Three cycles were carried out with deviator stress ratios of $q/q_{max} = 0.23$, 0.57 and 0.86 to represent the first, second and third cycles, where q_{max} is the maximum deviator stress during monotonic loading for the given e_O, which was measured in a separate test. Once the desired q/q_{max} was attained, the sample was unloaded with a reversed loading direction while keeping the same strain rate until q became zero. It is noted that a creep duration was allowed to maintain the stability of the sample before each cycle. Following Dutta et al. [9], a linear variable differential transducer (LVDT) was used to measure the axial strain.

2.2 Stress Wave Measurement

Wave measurements were performed using disk transducers (DTs) installed in the triaxial apparatus [9]. A single period of sinusoidal-shaped wave pulse was excited by the transmitter DTs in both compression (P-) and shear (S-) motions with an input frequency of $f_{in} = 7$ kHz and a double amplitude of 140 V. P- and S-waves propagated in the loading direction (longitudinal side of the sample) and the velocities (V_p and V_s) were determined based on $V = L/t$, where L is the wave travel distance, equivalent to the sample height; and t is the travel time. A check was made to confirm that f_{in} ensured the suitable ratio of L to the wavelength to avoid near field effect [10].

3 DEM Simulation

3.1 Sample Preparation in DEM

A random packing sample was prepared with spherical particles using a servo-control algorithm in DEM [6], which was run in LAMMPS [11]. The particle properties were adjusted to those in the laboratory tests, except for a 10% smaller E_p (64.5 GPa) according to Cavarretta et al. [12]. The DEM sample was also consolidated using an isotropic confinement of 50 kPa, giving e_0 value of 0.597 using a friction coefficient (μ) of 0.05 (Fig. 1(b)). The sample was adjusted as 150 mm in the longitudinal direction where the wall boundaries were used, while the periodic boundaries were used in the lateral directions. The similar three load-unload cycles were carried out in the longitudinal direction with deviator stress ratios of about $q/q_{max} = 0.24$, 0.65 and 0.82/0.91 with μ of 0.4. Note that there are two cases of the third cycle considering the discrepancy in the stress-strain response between laboratory tests and DEM. The constant strain rate was 2 $\times$ 10^{-3} [s^{-1}], which aimed to ensure the quasi-static flow regime [13] and improve the computational efficiency.

3.2 Wave Propagation Procedure in DEM

The wall boundaries were excited in either longitudinal or lateral direction to generate P- or S- planar stress waves, respectively, using a sinusoidal wave pulse with f_{in} of 7 kHz and a double amplitude of 5 nm referring to [14]. Prior to the wave propagation, a creep process was allowed to suppress the particle movement and maintain the stress state. μ was increased to 0.5 to ensure the elastic response during wave propagation. Otsubo et al. [15] confirmed that the wave velocities are not influenced by a further increase in μ from the load-unload condition. The resultant changes in the stress on both wall boundaries were used as an analogue to the wave signals in the laboratory tests. The selections of damping factors consistently followed [6]. The testing program is sketched in Fig. 2.

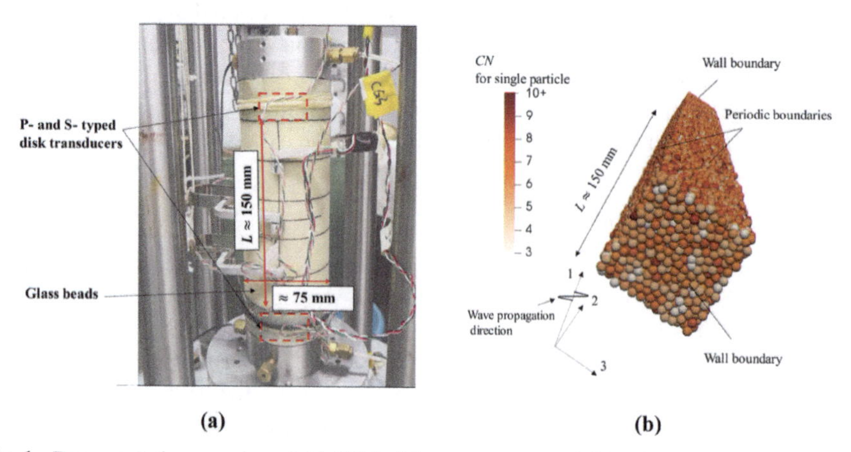

Fig. 1. Representative samples of (a) GB in laboratory tests and (b) spherical particles in DEM simulations.

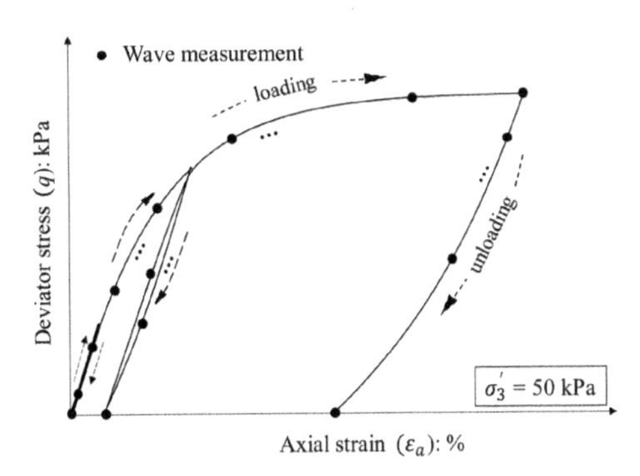

Fig. 2. Schematic of testing program of load-unload reversals and wave measurements.

4 Variation of Stress Waves During Load-Unload Reversals

V_p and V_s are plotted against the major stress (σ_1') and geometric mean stress ($\sqrt{\sigma_1'\sigma_3'}$), respectively, during load-unload for laboratory tests and DEM simulations in Fig. 3. As three cycles were performed at selected pre-peak stress states, the rightward variation refers to the loading path, while that in the pposite direction denotes the unloading path, as indicated by dash arrows. In the first two cycles, V_p is reversible as the curves along loading and unloading paths converge. In contrast, V_s displays a certain degree of reduction where the curve of unloading develops below that of loading. In the third cycle, V_p and V_s are not reversible and show markable declines along unloading, while V_p of the DEM sample has a relatively more significant reduction in the second case of the third cycle (i.e. $q/q_{max} = 0.91$) due to a larger q/q_{max}. The irreversibility of V_p and V_s reveals that the stress state is not the only governing factor of wave velocities during load-unload reversals, but also the shearing history.

The variations of normalized wave velocity ratio $(V_p/V_s)^*$ with stress ratio (σ_1'/σ_3') are given in Fig. 4, where the wave velocity ratio is normalized by its value in the initial isotropic stress state. Following the loading and unloading paths, $(V_p/V_s)^*$ increases and decreases accordingly with σ_1'/σ_3'. The variations are reversible in the first cycle since the curves converge; however, $(V_p/V_s)^*$ values for the second and third cycles are not recoverable during unloading path, leading to larger $(V_p/V_s)^*$ values at the ending point of unloading.

The variations of $(V_p/V_s)^*$ with ε_a are portrayed in Fig. 5. Upon each cycle, the wave ratio increases along axial loading while decreases along unloading. The overall trend is qualitatively comparable between those from Lab and DEM samples although $(V_p/V_s)^*$ of the Lab case is slightly smaller than that of the DEM case. From Figs. 4 and 5, it is important to notice that $(V_p/V_s)^*$ cannot return to its value shown at the beginning of each cycle and appears to be larger at the end of unloading, more likely happening at the cycle with a higher q/q_{max}.

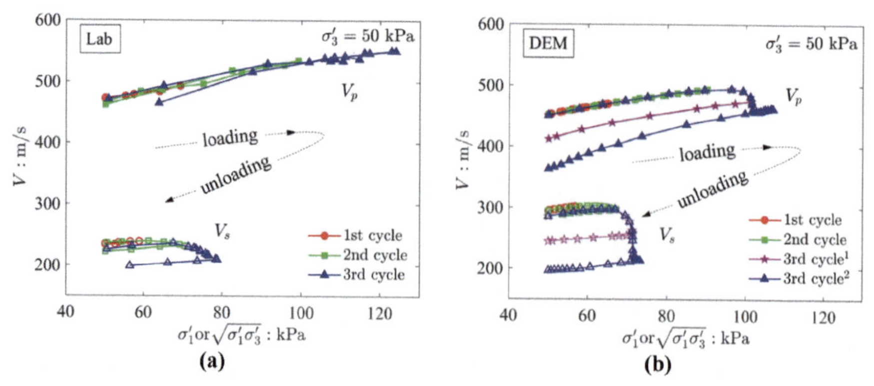

Fig. 3. Variations of V_p with major stress component (σ_1') and V_s with geometric mean stress ($\sqrt{\sigma_1'\sigma_3'}$) during load-unload for (a) laboratory tests and (b) DEM simulations.

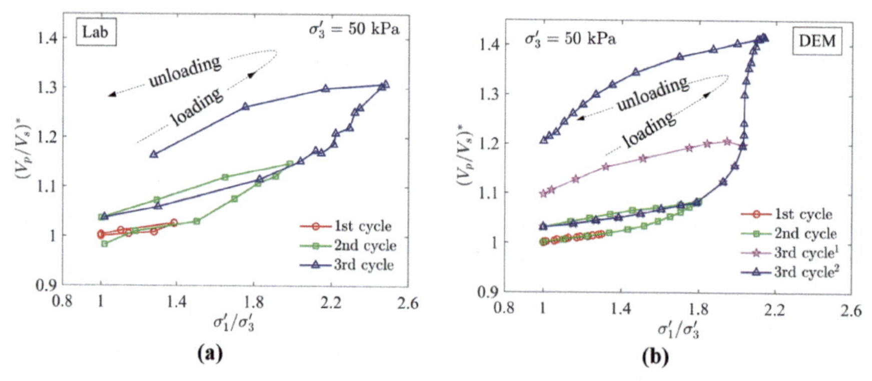

Fig. 4. Variations of normalized wave velocity ratio $(V_p/V_s)^*$ with stress ratio (σ_1'/σ_3') during load-unload for (a) laboratory tests and (b) DEM simulations.

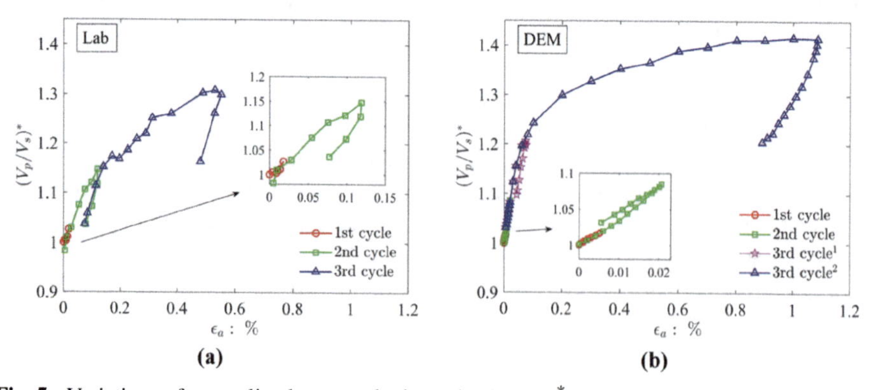

Fig. 5. Variations of normalized wave velocity ratio $(V_p/V_s)^*$ with axial strain (ε_a) during load-unload for (a) laboratory tests and (b) DEM simulations.

5 Evolution of Soil Fabric During Load-Unload Reversals

Satake [18] proposed the second order fabric tensor, which is defined as:

$$\Phi_{ij} = \frac{1}{N}\sum_{k=1}^{N} n_i^k n_j^k \tag{1}$$

where N is the number of contacts and n_i^k, n_j^k are unit orientation vectors in the i and j directions of the principal axes. Referring to the coordinate system in Fig. 1(b), Φ_{11} stands for the fabric in the vertical direction (Φ_{ver}) and the mean of Φ_{22} and Φ_{33} denotes that in the horizontal directions (Φ_{hor}). The fabric ratio is defined as Φ_{ver}/Φ_{hor} to quantify the fabric anisotropy.

As stress ratio (σ_1'/σ_3') was found to have a significant impact on the fabric change [17], Fig. 6 gives the variation of Φ_{ver}/Φ_{hor} with σ_1'/σ_3' and ε_a, respectively. Φ_{ver}/Φ_{hor} is approximately recoverable in the first two cycles, while there is a hysteresis in the later cycles. According to experimental results [18], the normalized Young's modulus (E/σ_1') does not reduce when $\sigma_1'/\sigma_3' < 2$. Regarding DEM simulation results in the literature [19], contact networks do not significantly rearrange until $\sigma_1'/\sigma_3' > 1.6$. From Fig. 6(a), $\sigma_1'/\sigma_3' = 1.8$ can be considered as the threshold beyond which fabric ratio increases markedly. This slightly larger stress ratio might be attributed to the effect of inter-particle friction [20]. Figure 6(b) provides the variations of Φ_{ver}/Φ_{hor} with ε_a during load-unload reversals. Consistent with the observation in [6], the fabric ratio increases during loading, indicating that the soil fabric is more dominant in the direction of loading than that in the horizontal plane. Subsequently, the fabric ratio decreases with unloading, in line with [3]. Similar to the wave velocity ratio, the variation of Φ_{ver}/Φ_{hor} upon unloading is not fully reversible compared with that of loading when the load is reversed at $\sigma_1'/\sigma_3' > 1.8$, ending up with a relatively larger value at the end of unloading.

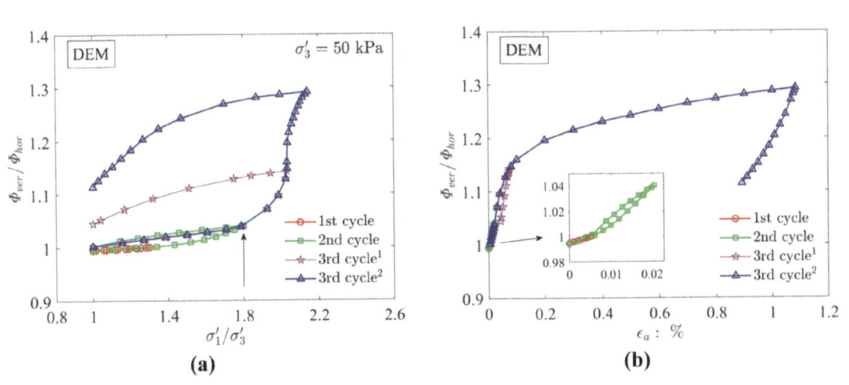

Fig. 6. Variations of fabric ratio (Φ_{ver}/Φ_{hor}) during load-unload in DEM simulations (a) with stress ratio (σ_1'/σ_3') and (b) with axial strain (ε_a).

Figure 7 provides the relationship between Φ_{ver}/Φ_{hor} and $(V_p/V_s)^*$, irrespective of the stress state or axial strain. An overall linear correlation can be observed for the results from three cycles, with a best-fit line of $\Phi_{ver}/\Phi_{hor} = 0.72(V_p/V_s)^* + 0.27$ ($R^2 = 0.995$), signifying that the change in the wave ratio is strongly affected by the evolution of fabric anisotropy. In turn, the macroscopically measured wave velocities turn out to be a useful index to infer the micromechanical soil fabric anisotropy during load-unload reversals. This observation has enhanced the finding reported in [6] by extending this correlation from monotonic loading to load-unload reversals. It should be noted that these loading cycles were investigated at the pre-peak stage, and thus loading cycles at later strain range, i.e. post-peak stage, should be checked in future work.

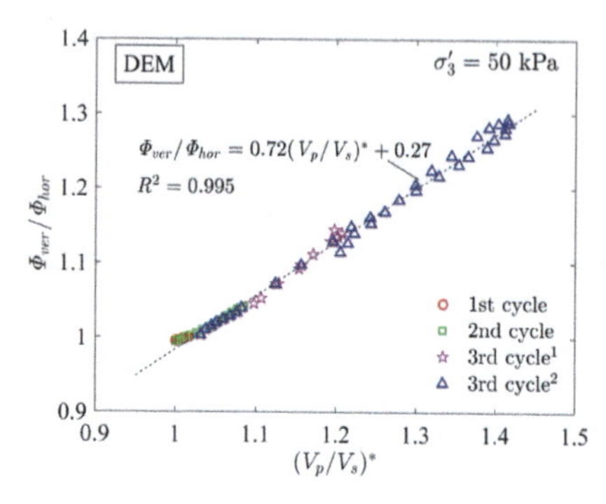

Fig. 7. Correlation between fabric ratio (Φ_{ver}/Φ_{hor}) and normalized wave velocity ratio $(V_p/V_s)^*$.

6 Conclusions

This study has conducted laboratory tests using spherical GB subjected to three load-unload reversals at the pre-peak stage, during which stress waves were continuously measured. Equivalent loading tests and stress wave propagation tests were performed using supplementary DEM simulations where micromechanical fabric anisotropy was investigated. Based on the results, the following conclusions can be drawn:

1. V_p and V_s are almost reversible during the first two cycles at lower q/q_{max} where the fabric has not significantly changed. On the other hand, V_p and V_s exhibit a more obvious reduction during the unloading path in the third cycle where q/q_{max} is larger than 0.8.
2. The stress ratio (σ_1'/σ_3') has a marked effect on the change of soil fabric, beyond which the fabric anisotropy significantly develops and cannot be recovered during the unloading path. $\sigma_1'/\sigma_3' = 1.8$ is found to be the threshold in this study. The exact value depends on the material property and testing condition.
3. The change in soil fabric during load-unload reversals is found to be well correlated with wave velocity ratio (V_p/V_s) at the pre-peak stage. This correlation gives a possibility to predict the evolution of soil fabric using macroscopically measured wave velocities.

References

1. Wei, J., Huang, D., Wang, G.: Fabric evolution of granular soils under multidirectional cyclic loading. Acta Geotech. **15**, 2529–2543 (2020)
2. Jiang, M., Zhang, A., Li, T.: Distinct element analysis of the microstructure evolution in granular soils under cyclic loading. Granular Matter **21**(2), 1–16 (2019). https://doi.org/10.1007/s10035-019-0892-8
3. O'Sullivan, C., Cui, L.: Micromechanics of granular material response during load reversals: combined DEM and experimental study. Powder Technol. **193**, 289–302 (2009)
4. Andrus, R., Stokoe, K., II.: Liquefaction resistance of soils from shear-wave velocity. J. Geotech. Geoenviron. Eng. **126**, 1015–1025 (2000)
5. Zhou, Y., Chen, Y.: Influence of seismic cyclic loading history on small strain shear modulus of saturated sands. Soil Dyn. Earthq. Eng. **25**, 341–353 (2005)
6. Li, Y., Otsubo, M., Kuwano, R.: DEM analysis on the stress wave response of spherical particle assemblies under triaxial compression. Comput. Geotech. **133**, 104043 (2021)
7. Altuhafi, F., O'Sullivan, C., Cavarretta, I.: Analysis of an image-based method to quantify the size and shape of sand particles. J. Geotech. Geoenviron. Eng. **139**, 1290–1307 (2013)
8. Li, Y., Otsubo, M., Kuwano, R., Nadimi, S.: Quantitative evaluation of surface roughness for granular materials using Gaussian filter method. Powder Technol. **388**, 251–260 (2021)
9. Dutta, T., Otsubo, M., Kuwano, R., O'Sullivan, C.: Evolution of shear wave velocity during triaxial compression. Soils Found. **60**, 1357–1370 (2020)
10. Leong, E., Yeo, S., Rahardjo, H.: Measuring shear wave velocity using bender elements. Geotech. Test. J. **28**, 12196 (2005)
11. Plimpton, S.: Fast parallel algorithms for short-range molecular dynamics. J. Comput. Phys. **117**, 1–19 (1995)
12. Cavarretta, I., O'Sullivan, C., Ibraim, E., et al.: Characterization of artificial spherical particles for DEM validation studies. Particuology **10**, 209–220 (2012)
13. Lopera Perez, J., Kwok, C., O'Sullivan, C., et al.: Assessing the quasi-static conditions for shearing in granular media within the critical state soil mechanics framework. Soils Found. **56**, 152–159 (2016)
14. Otsubo, M., O'Sullivan, C., Ackerley, S., Parker, D.: Selecting an appropriate shear plate configuration to measure elastic wave velocities. Geotech. Test. J. **43**, 20180146 (2020)
15. Otsubo, M., O'Sullivan, C., Hanley, K.J., Sim, W.W.: Influence of packing density and stress on the dynamic response of granular materials. Granular Matter **19**(3), 1–18 (2017). https://doi.org/10.1007/s10035-017-0729-2
16. Satake, M.: Fabric tensor in granular materials. Proceedings of the IUTAM conference on deformation and failure of granular materials, Delft, pp. 63–67 (1982)
17. Jang, E., Jung, Y., Chung, C.: Stress ratio–fabric relationships of granular soils under axi-symmetric stress and plane-strain loading. Comput. Geotech. **37**, 913–929 (2010)
18. Hoque, E., Tatsuoka, F.: Anisotropy in elastic deformation of granular materials. Soils Found. **38**, 163–179 (1998)
19. Gu, X., Hu, J., Huang, M.: Anisotropy of elasticity and fabric of granular soils. Granular Matter **19**(2), 1–15 (2017). https://doi.org/10.1007/s10035-017-0717-6
20. Gu, X., Yang, J., Huang, M.: DEM simulations of the small strain stiffness of granular soils: effect of stress ratio. Granular Matter **15**, 287–298 (2013)

Shear Work and Liquefaction Resistance of Crushable Pumice Sand

Rolando P. Orense[1(✉)], Jenny Ha[1], Arushi Shetty[1], and Baqer Asadi[2]

[1] University of Auckland, Auckland 1142, New Zealand
r.orense@auckland.ac.nz
[2] Jacobs, Auckland, New Zealand

Abstract. Crushable volcanic soils, such as pumice sands, are often encountered in engineering projects in the North Island of New Zealand. Due to the highly crushable nature of the pumice sand components, current empirical correlations, derived primarily from hard-grained sands, are not applicable when evaluating the liquefaction potential of pumice-rich soils. To better understand their liquefaction characteristics, cyclic undrained triaxial tests were performed on high-quality undisturbed soil samples sourced from various pumice-rich sites in the North Island. The undrained response, expressed in terms of the development of excess pore water pressure and axial strain with the number of cycles, and the shear work (or cumulative dissipated energy), defined as the energy consumed by the soil during plastic deformation until liquefaction, is examined vis-à-vis the pumice contents of the specimens. When compared to published trends for normal sands available in the literature, the results indicate that the shear work for pumice-rich sands sand is larger in specimens with higher pumice contents because some energy is spent as the particles undergo crushing. As a result, the liquefaction resistance of crushable pumice sand is higher than that of natural sand for the same level of loading applied.

Keywords: Liquefaction · Pumice sand · Cyclic test · Shear work · Particle crushing

1 Introduction

Pumice deposits are found in several areas of the North Island of New Zealand. They originated from a series of volcanic eruptions centred in the Taupo and Rotorua regions, called the "Taupo Volcanic Zone". While they exist mainly as deep sand layers in river valleys and flood plains, they are also found as coarse gravel deposits in hilly areas. As a result, they are often encountered in engineering projects being undertaken in the region.

These deposits contain pumice sands, characterised as lightweight, highly crushable, and compressible due to their vesicular nature. SEM imaging indicates that pumice sands have lots of surface and internal voids (see Fig. 1a). The single-particle crushing strength of pumice sand is one order of magnitude less than normal silica sand (see Fig. 1b); it can be crushed by fingernail pressure. Cone penetration tests on loose and dense pumice sand deposits in a CPT chamber showed that the cone resistance profile is practically the

© The Author(s), under exclusive license to Springer Nature Switzerland AG 2022
L. Wang et al. (Eds.): PBD-IV 2022, GGEE 52, pp. 2212–2219, 2022.
https://doi.org/10.1007/978-3-031-11898-2_205

same for both density states, presumably due to particle crushing when the rod penetrates the deposit (see Fig. 1c). The results indicate that cone resistance is not an appropriate index to represent the relative density of soil.

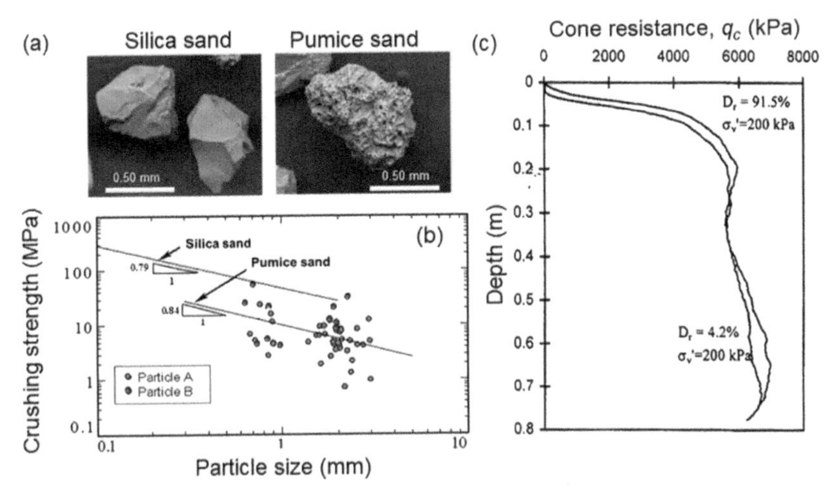

Fig. 1. Comparison between silica sand and pumice sand in terms of: (a) particle characteristics from SEM images; (b) single particle crushing strength as a function of particle size [1]); and (c) cone penetration resistance in chamber test [2]).

Because of the presence of these crushable pumice sands in the soil matrix, laboratory undrained cyclic triaxial tests conducted at the Geomechanics Laboratory (University of Auckland) showed that pumiceous sands behave differently when compared to normal (hard-grained) silica sands [3–5]. Hence, conventional methods of estimating the liquefaction resistance of soils in the field do not apply to these materials.

To better understand their liquefaction characteristics, high-quality soil samples were obtained from various pumice-rich sites in the North Island using diverse sampling techniques and then tested in the laboratory using a cyclic triaxial apparatus. Moreover, the pumice content of each sample was quantified using a recently developed method that correlates the degree of particle crushing of the samples to the amount of crushable pumice particles present. In addition to the undrained response, expressed in terms of the development of excess pore water pressure and axial strain with the number of cycles, and the shear work (or cumulative dissipated energy), defined as the energy consumed by the soil during plastic deformation until liquefaction, was also examined.

The cumulative dissipated energy, ΣW, is indicated by the area under the deviator stress and axial strain (see Fig. 2a). It represents the energy consumed by the soil during plastic deformation until liquefaction, and therefore a good indicator of the cyclic shear behaviour and liquefaction strength of the soil.

2 Materials Used and Methodology

2.1 Samples Investigated

High-quality undisturbed pumice-rich samples were obtained from six sites within the Waikato Basin and Bay of Plenty region in the central part of North Island. These samples were obtained using the Dames & Moore (DM) sampler and the Gel-push sampler (both triple-tube type, GPTR, and static type, GPS). The index properties of the samples obtained are reported by Asadi et al. [6], while details about the undisturbed soil sampling methods are discussed by Stringer [7].

2.2 Quantification of Pumice Content

The pumice-rich sands collected from various sites showed different amounts of pumice particles present in the soil mixture. To estimate the pumice content (*PC*), defined as the ratio (by weight) of the amount of the crushable pumice sand components to the total amount of the sample, the methodology developed by the authors [8, 9] was followed. According to this approach, which is based on the crushability feature of the pumice sands, the *PC* of the specimens are estimated based on the relative breakage, B_r, measured during a modified maximum dry density (MDD) test. The relation between PC and B_r is illustrated in Fig. 2b.

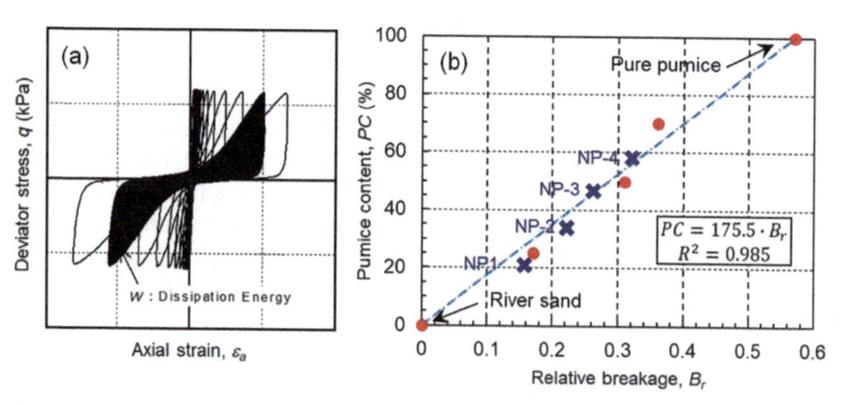

Fig. 2. (a) Schematic diagram of cumulative dissipated energy, ΣW; (b) Relation between pumice content and relative breakage from Modified MDD test [8, 9].

2.3 Testing Programme

Overall, 21 undisturbed soil samples from 6 locations were tested in the laboratory. The majority of the samples were tested three times (although some were tested 2–6 times) under different levels of cyclic shear stress ratio (*CSR*). All tests were conducted under an effective confining pressure $\sigma'_c = 100$ kPa and a loading frequency $f = 0.1$ Hz. A summary of the laboratory undrained cyclic triaxial tests performed is presented in

Table 1, including the range of pumice contents and samplers used. For each test, the development of double amplitude axial strain, ε_{DA}, and excess pore water pressure ratio, r_u ($= u/\sigma'_c$, where u is the excess pore water pressure) with the number of cycles were obtained. For each test, the shear work per cycle was calculated from the deviator stress-axial strain curves using MATLAB.

Table 1. Summary of laboratory tests conducted

Location ID	No. of samples	No. of tests/sample	Pumice content (%)	Sampler used
Tauranga 1	5	3	72, 72, 19, 19, 40	DM (×5)
Tauranga 2	3	3	0, 45, 43	GPTR (×3)
Hamilton – GS	3	2–4	35, 93, 93	DM (×1), GPTR (×2)
Hamilton – TR	2	2–3	16, 38	GPTR (×1), DM (×1)
Edgecumbe	3	3–4	36, 42, 39	GPTR (×3)
Whakatane	5	2–6	71, 35, 38, 88, 55	GPS (×1), DM (×1), GPTR (×2)

3 Results and Discussion

3.1 Development of Axial Strain and Pore Water Pressure with Cyclic Loading

To examine the undrained cyclic behavior of pumice-rich sands, the maximum values of ε_{DA} and r_u at the end of each loading cycle are plotted versus the normalised number of cycles in Figs. 3 and 4, respectively. In the figure, the number of cycles, N, was normalised by N_{liq}, defined as the number of cycles required to reach $\varepsilon_{DA} = 5\%$. The plots shown correspond to tests with roughly similar *CSR*. For the purpose of the analysis, the samples were grouped into three broad pumice content ranges such that bias in individual specimens was reduced and general trends could be observed: zero ($PC = 0\%$), low ($PC = 16$–35%) and high ($PC = 72$–93%) pumice contents.

From both figures, while there is a general scatter of results (due to the undisturbed samples having different relative densities, fines contents, fabric/structure, stress history, etc.), a general trend of the plots can be observed: (1) Specimens with low pumice content ($PC = 0\%$) show negligible deformation during the early part of the cyclic load application, although some development in pore water pressure occurred. During the middle-third of the loading, the rate of increase in r_u decreases; however, when it reaches a specific value, an immediate increase in r_u is generated, accompanied by a sudden occurrence of large deformation, leading to liquefaction in just a few cycles. (2) Specimens with high pumice content ($PC = 72$–93%) show a similar immediate increase in r_u at the initial stage of loading, but with almost twice the rate of r_u development; moreover, the initial deformation is much higher. In the next stage, the rate of increase in r_u decreases gradually until initial liquefaction (i.e., where the specimen reached $r_u > 0.95$) is reached. They also show a gradual increase in deformation, starting from the beginning of the cyclic loading and steadily increasing until liquefaction, in an almost linear fashion, due to particle crushing. Note that these specimens can undergo large deformation even under high r_u, presumably because of the formation of a more stable soil skeleton induced by crushing. (3) Specimens with low pumice content (PC = 16–35%) show deformation and pore pressure responses that are midway between the zero and high PC specimens.

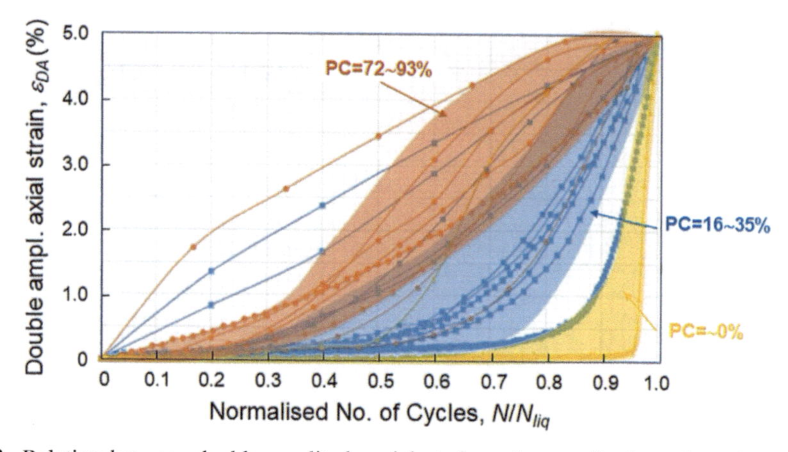

Fig. 3. Relation between double amplitude axial strain and normalised number of cycles for pumice-rich samples with different pumice contents.

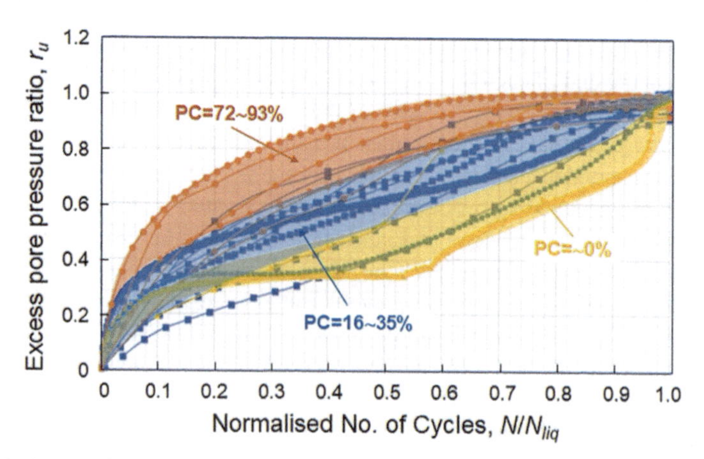

Fig. 4. Relation between excess pore water pressure ratio and normalised number of cycles for pumice-rich samples with different pumice contents.

3.2 Development of Cumulative Dissipated Energy with Cyclic Loading

As discussed earlier, the dissipated energy per cycle is calculated as the area within each hysteresis loop. Due to the increasing strain caused by an increase in the excess pore water pressure, these loops typically increase in area with an increase in the number of cycles. Figure 5 plots the development of the cumulative dissipated energy for specimens in the three *PC* ranges. For zero *PC* specimens, the initial development of ΣW is very slow because of the very minimal amounts of strain the specimens experienced. As the number of cycles increases towards liquefaction, the axial strain begins to increase in much larger increments, resulting in a significant increase in ΣW. On the other hand, specimens with high *PC* tend to have greater cumulative dissipated energy, even at the start of the cyclic load application.

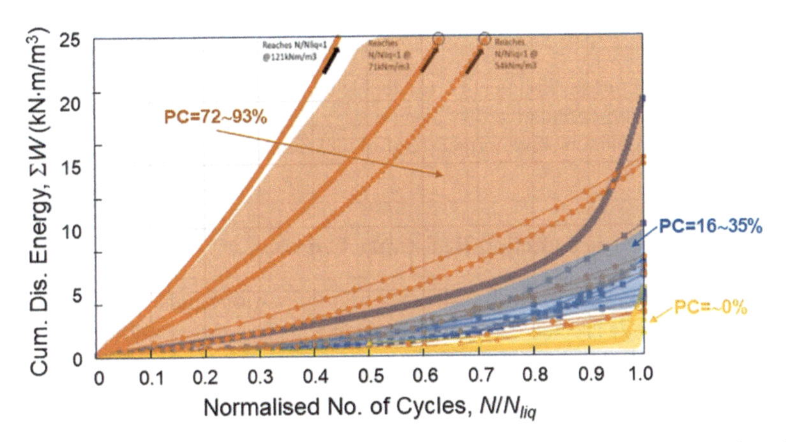

Fig. 5. Relation between cumulative dissipated energy and normalised number of cycles for pumice-rich samples with different pumice contents.

3.3 Development of Deformation and Pore Pressure with Shear Work

To observe clearly how the axial deformation and excess pore water pressure vary with the cumulative dissipated energy, the median curves of the trends of each of the above parameters with the normalised number of cycles (i.e., Figs. 3, 4 and 5) for each *PC* range are obtained and plotted with respect to each other; these are shown in Fig. 6. Also shown in the figures are trends for the natural sand materials (with relative density $D_r = 50\%$) reported by Yoshimoto et al. [10]. Based on Fig. 6a, ΣW of zero *PC* sand is similar to the other hard-grained sands, while that of medium *PC* sands is about 2–4 times higher than those of natural sands and almost the same as that of Iwakuni clay. Finally, the ΣW of high *PC* specimens is nearly one order of magnitude greater than those of normal sands. The same trends are presented in Fig. 6b in terms of the relation between ΣW and r_u. Normal sands are not resistant to liquefaction because they lack energy absorption capacity. In contrast, r_u in Iwakuni clay increases only up to about 0.6 when $\varepsilon_{DA} = 5\%$. It is also observed that natural sands and pumice-rich show similar behaviour until $r_u = 0.6$; however, in the case of high *PC* specimens, r_u does not reach 1.0 when $\varepsilon_{DA} = 5\%$.

Based on the trends, pumice-rich sands are more resistant to liquefaction than natural sands, possibly because some energy is dissipated as the particles crush due to cyclic loading.

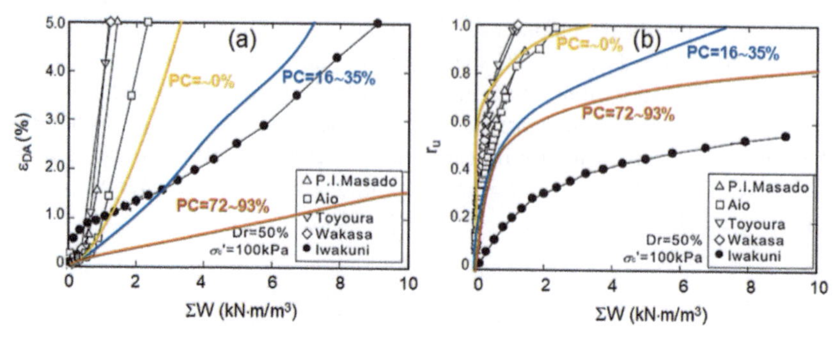

Fig. 6. Comparison of the development of: (a) double amplitude axial strain; and (b) excess pore water pressure ratio with cumulative dissipated energy for representative pumice-rich samples (with varying pumice contents) and normal (hard-grained) sands.

3.4 Comparison of Liquefaction Resistance Curves

To illustrate the higher liquefaction resistance of pumice-rich sands, representative samples for each PC range are selected and plotted in Fig. 7, together with those of natural sands [10]. Although these pumiceous sand specimens are from undisturbed samples (with different relative densities, fines contents, fabric/structure, degrees of cementation, etc.) while the natural sand specimens are reconstituted, a general trend is observed: zero PC sand has the same the liquefaction resistance as ordinary sands, while low PC and high PC sands have higher liquefaction resistance.

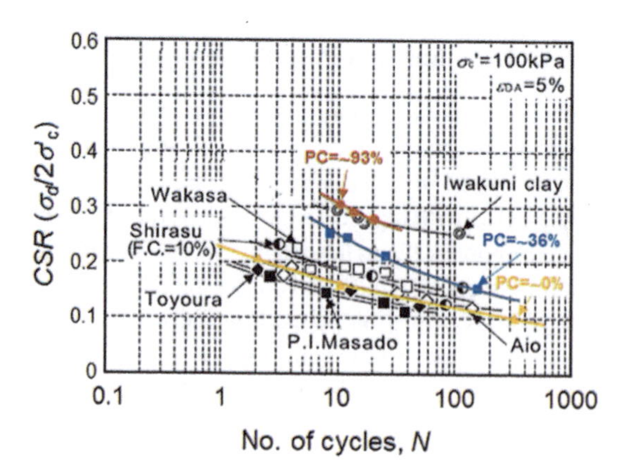

Fig. 7. Comparison of liquefaction resistance curves of representative pumice-rich samples (with varying pumice contents) and normal (hard-grained) sands.

4 Concluding Remarks

The results of undrained cyclic triaxial tests conducted on high-quality undisturbed sand samples with varying pumice contents (*PC*) were analysed. It was observed that under cyclic loading, samples with high *PC* have a greater cumulative dissipated energy response than those with low and zero *PC*. This could be attributed to particle crushing and subsequent stabilisation of pumice soil structure, manifested in a greater initial build-up of r_u and larger deformations. The presence of some amount of crushable pumice sand in the sample resulted in higher cumulative dissipated energy and greater liquefaction resistance than normal sands. Note that other factors, such as particle size, fines contents, fabric/structure, cementation and stress history, etc., may also have varying effects on the observed response.

Acknowledgment. Parts of this project were funded by the Natural Hazards Research Platform (NHRP) and QuakeCoRE, a New Zealand Tertiary Education Commission-funded Centre. This is QuakeCoRE Publication Number 0730.

References

1. Orense, R.P., Pender, M.J., Hyodo, M., Nakata, Y.: Micro-mechanical properties of crushable pumice sands. Géotechn. Lett. **3**(April–June), 67–71 (2013)
2. Wesley, L.D., Meyer, V.M., Pronjoto, S., Pender, M.J., Larkin, T.J., Duske, G.C.: Engineering properties of pumice sand. In: 8th ANZ Conference on Geomechanics, vol. 2, pp. 901–908 (1999)
3. Orense, R.P., Pender, M.J.: Liquefaction characteristics of crushable pumice sand. In: 18th International Conference on Soil Mechanics & Geotechnical Engineering, vol. 1, pp. 1559–1562 (2018)
4. Orense, R.P., Asadi, M.S., Asadi, M.B., Pender, M.J., Stringer, M.E.: Field and laboratory assessment of liquefaction potential of crushable volcanic soils. In: Theme Lecture, 7th International Conference on Earthquake Geotechnical Engineering, pp. 442–460 (2019)
5. Orense, R.P., Asadi, M.B., Stringer, M.E., Pender, M.J.: Evaluating liquefaction potential of pumiceous deposits through field testing: case study of the 1987 Edgecumbe earthquake. Bull. N. Z. Soc. Earthq. Eng. **53**(2), 101–110 (2020)
6. Asadi, M.B., Orense, R.P., Asadi, M.S., Pender, M.J.: Experimental and simplified correlations for liquefaction assessment of crushable pumiceous sand by shear wave velocity. J. Geotech. Geoenv. Eng. (2021, submitted)
7. Stringer, M., Taylor, M., Cubrinovski, M.: Advanced soil sampling of silty sands in Christchurch. Report No: 2015-06, University of Canterbury, Christchurch, NZ (2015)
8. Asadi, M.S., Orense, R.P., Asadi, M.B., Pender, M.J.: Maximum dry density test to quantify pumice content in natural soils. Soils Found. **59**(2), 532–543 (2019)
9. Asadi, M.S., Orense, R.P., Asadi, M.B., Pender, M.J.: Laboratory-based method to quantify pumice contents of volcanic deposits. In: 6th Interbational Conference on Geotechnical & Geophysical Site Characterization, Budapest, Hungary (2021)
10. Yoshimoto, N., Orense, R., Hyodo, M., Nakata, Y.: Dynamic behaviour of granulated coal ash during earthquakes. J. Geotech. Geoenv. Engg **140**(2), 0413002 (2014)

Digital Particle Size Distribution for Fabric Quantification Using X-ray μ-CT Imaging

Ana Maria Valverde Sancho and Dharma Wijewickreme(✉)

The University of British Columbia, Vancouver, Canada
dharmaw@civil.uba.ca

Abstract. Past research works on monotonic and cyclic shear behavior of natural silts at the University of British Columbia (UBC), Canada, has shown that the soil fabric and microstructure has a significant influence on the mechanical response of natural silts. With this background, a research program has been undertaken at UBC to capture non-destructive 3D images of Fraser River low-plastic silt using X-ray micro-computed tomography (X-ray μ-CT) for assessing particle fabric. In this regard, identifying individual grains to analyze the particle size distribution (PSD) of a given soil specimen forms a vital step in studying the fabric of silts.

Advancements in computing power have allowed for 3D X-ray imaging of fine-grained soil (silt) specimens. Through computer processing of these 3D images, it is possible to obtain the main dimensional and directional parameters to represent individual particles in digital form; in turn, some of this information can be used to obtain a digital PSD of a given silt matrix. A good way to assess the outcomes of the image processing approach is by comparing the PSDs from the image-based analysis to those from mechanical methods and laser technologies.

In the present work, specimens from a natural silt are analyzed using laboratory methods (i.e., mechanical sieve along with hydrometer analysis) and laser diffraction techniques to establish the benchmark PSDs, and the results are compared with those digitally derived from X-ray μ-CT imaging. In addition, the digital PSDs obtained by imaging of specimens made of pre-calibrated standard-size silica-based beads were also compared with their counterpart physical PSDs. The observed good agreement between the results obtained from physical and digital techniques support the suitability of using the X-ray μ-CT to understand the particulate fabric of silty soils.

Keywords: Fabric/microstructure of soils · Low-plastic silt · Particle size distribution · 3D image analysis · X-ray micro-computed tomography

1 Background

Observations from past earthquakes suggest that fine-grained silty soils with high levels of saturation are susceptible to earthquake-induced softening and strength reduction [1–3], giving rise to damage-causing geotechnical hazards. Past research has shown that fabric and microstructure have a marked effect on the macroscopic monotonic and cyclic behavior of soils [4–8].

© The Author(s), under exclusive license to Springer Nature Switzerland AG 2022
L. Wang et al. (Eds.): PBD-IV 2022, GGEE 52, pp. 2220–2228, 2022.
https://doi.org/10.1007/978-3-031-11898-2_206

Soil fabric is fundamentally described as the spatial arrangement of solid particles and voids [9]. There are two main components of fabric; the particle's discrete orientation and the particle's relative position with respect to adjacent particles [10]. Although significant studies have been conducted in sands and clays [11, 12], research addressing this influence specifically with respect to silt has been limited [13]; however, works on cementitious materials with similar particle sizes to silts have been performed using X-ray μ-CT technologies [14]. With this background, a research program has been undertaken to better understand the macro behavior of silts using micro particle physics; the work underway is intended to capture three-dimensional images using X-ray micro-computed tomography for assessing particle fabric.

First attempts to explore the feasibility of X-ray μ-CT in visualizing silt fabric at UBC has found to be promising [15–18]. Identifying individual grains in order to analyze the particle size distribution (PSD) of a given soil specimen can be considered as an important step in studying the fabric and microstructure of silts. In this paper, the X-ray μ-CT imaging and processing techniques are assessed with respect to the above topic using two silt sized materials - Nicomekl River silt and commercially available, pre-calibrated, standard-size silica beads. The digital particle size distributions obtained from X-ray μ-CT imaging of the standard-sized-grains are compared with those ranges from the manufacturers' calibration, laser diffraction analysis, as well as from mechanical sieving. The information related to the experimentation undertaken and the initial findings from the work are presented in the next sections.

2 Experimental Program

2.1 Material Description

Two silt-size granular materials with particle sizes ranging between 2 and 74 μm were used for this research. The properties and index parameters of the two materials are given in Table 1. Relatively undisturbed samples of silt retrieved from a site in the vicinity of Nikomekl River located in Surrey, British Columbia, Canada was chosen as one of the test materials representing natural silts. The site is underlain by swamp, and shallow lake deposits with lowland peat in part overlying silty clay and deltaic and distributary channel fills sands and silts [19]. It is recognized that since the tested material is natural silt, there is a small fraction of particles that are greater than 74 μm and as such, the material is not pure silt. The other material is a commercially available pre-calibrated, standard-size silica manufactured by Silicycle, Quebec, Canada. In this, the particle sizes ranged from 40 to 63 μm and two batches were received; one with irregular shape beads and other with spherical shape beads. PSD analysis using hydrometer, mechanical sieving, as well as laser diffraction methodologies were performed on both, the Nicomekl River Silt and the silica gel particles.

Table 1. Soil properties and index parameters of tested silts

Parameter	Standard-size silica particles		Nicomekl River silt
	I-40-63	S-45-63	
Specific gravity, G_S	2.02	1.89	2.77
Plasticity index, PI	NA	NA	7
Particle size range (μm)	40–63	45–63	2–200*
Particle shape	Irregular	Spherical	Irregular

2.2 Sample Preparation for Imaging

Fine-grained particle fabric studies require achieving a high resolution in X-ray μ-CT scans. Based on previous studies at UBC [15], it was determined that specimens of 5-mm diameter or less are required to meet the resolution requirements to characterize silty grain size material using this technology. The specimens were prepared in thin-walled plastic tubes with a nominal diameter of 5.0 mm and a wall-thickness of 0.14 mm. Four specimens were prepared. Three containing standard-size silica particles and one containing relatively undisturbed Nicomekl River silt. All the silica specimens imaged were in a loose state - two of the specimens were from irregular particles (one saturated and the other in dry conditions), while the third specimen was prepared using spherical particles in a dry state. The use of these standard particle sizes served as a way to derive supporting information to calibrate/verify the newly developed procedures and methodologies; they also provided "bench mark" information to extend the findings to study the fabric of natural silts.

2.3 Image Acquisition, Processing and Analysis

The X-ray μ-CT scanner available at the Pulp and Paper Research Center at UBC, was used to collect 3D high-resolution non-destructive images of the specimens. The scanner available at this facility is a ZEISS Xradia 520 Versa manufactured by Zeiss International, Oberkochen, Germany. The X-ray μ-CT scans produce 2D images that record the variation of X-ray attenuation within objects. The 2D images are later stalked and reconstructed into a 3D image. The steadiness of the specimen during the scan is key to achieve high-resolution images. A specimen holder was specifically designed and fabricated to provide a well secured specimen during the scan.

A number of parameters need to be controlled to achieve images with satisfactory resolution. In this case, a field of view of ~800 μm was used for the scans that resulted in a 0.8 μm resolution. Considering that the pre-calibrated silica particles imaged have sizes between 40–63 μm all particles should be visible at this resolution. Nicomekl River silt, being a natural material containing particles that are smaller than those for the above pre-calibrated silica, presented challenges in the image resolution and the resulting segmentation. The output of a μ-CT scan requires further processing in the form of quantitative and qualitative image analysis: namely, image pre-processing and image segmentation [18]. The reconstruction and analysis of the digital images yielded were performed using the commercially available Avizo 9.7 software [20] a software that has been successfully used by various researchers to study particulate geomaterials [21, 22]. After retrieving individual image particle data (i.e. length, width, breath, thickness, volume, orientation, etc.) from the image processing analysis, an in-house developed MATLAB code was used to obtain the digital PSD curves.

2.4 Particle Size Verification

As indicated earlier, characterization of individual particles is key to understanding the soil fabric. PSD is one of the well-known metrics to classify and characterize soils, and traditionally, the PSD for the coarse-grained and fine-grained soil fractions are obtained in the laboratory by mechanical sieve and hydrometer analyses, respectively, in accordance with ASTM D7928-21 and ASTM D6913/D6913M – 17 [23, 24]. Moreover, techniques such as laser diffraction can also be used to obtain grain sizes.

In order to obtain the digital PSD plots, for a given particle, the particle size for the x-axis of the PSD plot was assumed to be the smallest dimension derived from the image segmentation process. Since the y-axis of the PSD plot is based on the weight passing a given particle size, it was considered reasonable to assume a constant specific gravity and thereby using the volume of given particle as the substitution for weight. On this basis, the volume "retained" on particle sizes corresponding to each sieve size, as per ASTM D6913/D6913M – 17, was calculated; afterwards, the "retained" volume for each sieve size was divided by the total volume retained over all the sieves allowing computation of the percent passing.

3 Initial Findings

A typical section through the raw imaging data obtained by X-ray computed tomography conducted on standard-sized silica specimens and Nicomekl River silt specimens are presented in Fig. 1 – Upper Row. The greyscale intensity reflects the density detected by the imaging device at a given location, and these values formed the key input data for the image processing. The lower row of Fig. 1 presents the processed images using several filters, which were used to improve the image quality in order to apply segmentation algorithms to separate the particles, and obtain the required metrics. Avizo software was used to process the images. Through the segmentation process, each particle is separated from adjoining others with the purpose of retrieving individual particle information (i.e., length, volume, orientation, etc.). This data serves as the main input for all subsequent quantitative analyses. The processing and analysis for all images were conducted employing identical workflows - more details on this matter are presented in [18].

From visual inspection of the two raw images (Fig. 1 - Upper Row), it appears that the resolution of the silica specimens was sufficient to capture the full range of particle sizes imaged (40–63 μm); the ability to recognize their irregular and spherical shapes is also well observable. However, for the Nicomekl River silt, it is quite possible that the smaller size particles might not have been well captured. As shown in Fig. 1 – Lower Row, the segmentation algorithms seem to have been able to identify/separate most particles in all images. However, it is important to recognize that some challenges still exist in interpreting image data. Firstly, it was found that the digital image processing approach might treat the smaller particles immediately adjacent to a bigger particle as part of the latter bigger particle. This effect, in turn, could produce the virtual particle to have larger dimensions compared to the real particle. Secondly, some bigger particles could be segmented by the software as two or three smaller particles. From visual inspection, both challenges are present in the images, more noticeably in the Nicomekl River silt image; however, neither seemed to be a persistent factor.

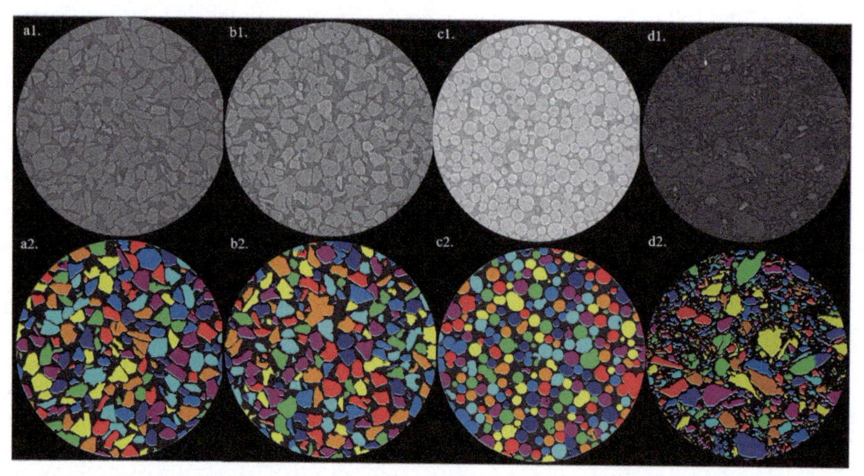

Fig. 1. Sections of tomography images: (a) dry pre-calibrated irregular silica material; (b) saturated pre-calibrated irregular silica material; (c) dry pre-calibrated spherical silica material; and (d) Nicomekl River silt. Top level – raw images and Lower level – processed images.

The PSD curve obtained from laboratory mechanical sieving for the irregular pre-calibrated silica grains is overlain in Fig. 2(a) for comparison with those obtained from the digital data analysis. Note: The range of particle sizes cited by the manufacturer for the standard soil is also marked on the same figure using a grey shaded region. Similar comparison is made in Fig. 2(b) for the spherical pre-calibrated silica grains. The PSDs for the pre-calibrated silica are well captured by the X-ray μ-CT. It can also be noted that there is good agreement between the digitally derived PSDs for the saturated and dry specimens. This suggests that the imaging process is not significantly affected by the presence or absence of water in the silt specimens, which is an important finding that confirms the suitability of the method to study the particle fabric of saturated as well as unsaturated materials. The observed mutual agreement of these results supports the suitability of the fine-grained commercial silica as a "bench mark" material for the intended research.

In addition to the data from digital processing, the PSDs generated from laser diffraction analysis and mechanical sieve analysis for the respective materials are overlain in Figs. 2(a) and 2(b). The results from laboratory analysis are somewhat deviated from those obtained from digital analysis. This difference can be attributed to experimental errors that are inherent in mechanical sieving and laser diffraction as well as segmentation procedures. All the curves are in general agreement in spite of the limitations and differences in the techniques used.

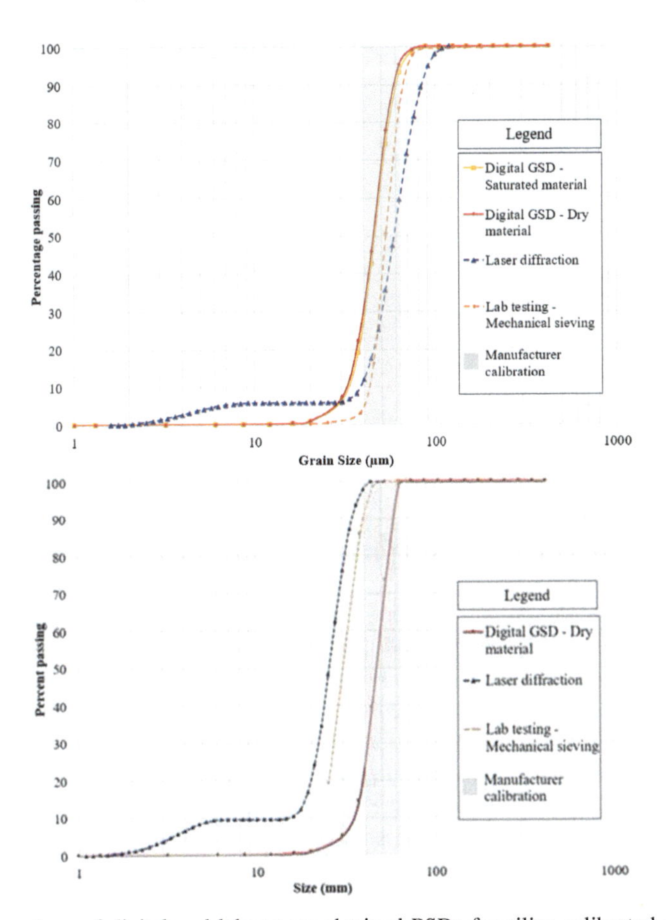

Fig. 2. Comparison of digital and laboratory obtained PSDs for silica calibrated particles. (a) Irregular shape particles (b) Spherical shape particles

Figure 3 presents the PSDs obtained for the Nicomekl River silt. The same workflows used for the silica material were extrapolated to be used in Nicomekl River silt. The results from laboratory testing including hydrometer and mechanical sieving are compared to laser diffraction analysis and digitally obtained PSD. There seems to be good agreement between the mechanical sieving and the digitally obtained GSD. However, the laser diffraction results are more in agreement with the hydrometer analysis data than those from mechanical sieving and digital processing. Although additional investigations are needed to clarify the reasoning for some of these differences, the role of X-ray micro-CT technology to study fabric on natural silt size particulate materials emerges from these findings.

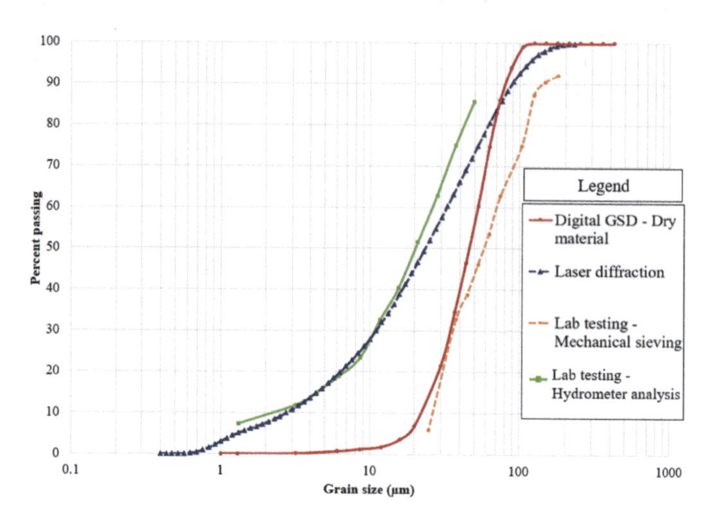

Fig. 3. Comparison of digital and laboratory obtained PSDs for Nicomekl River silt.

4 Conclusions and Future Work

The established methodology for micro-CT image processing was assessed using pre-calibrated silica grains and was found to be successful in identifying and segmenting particle ranges between 40–63 μm in both, irregular and spherical shapes. Digital calculations of PSDs in dry and saturated specimens show a very good correlation. This is of importance because most liquefaction related research would involve saturated soils, thus, increasing the potential of observations from pre-calibrated dry materials to saturated natural soils. Moreover, the methodology was expanded to include natural silty material from Nicomekl River. It is important to recognize that the segmentation of pre-calibrated silica material (representing a uniform coarse silt) is simpler in comparison to a natural material comprising the full range of silt particle sizes. Findings from this research have shown that X-ray micro-CT serves as a suitable methodology for the visualization of 3D fabric and microstructure of fine-grained material.

The ongoing work at UBC is intended to improve and refine the image analysis process in order to better account for the true 3D volume of individual particles and in turn improve the image segmentation process. Moreover, the work will also explore the particle rearrangement that occurs under different loading modes using pre-calibrated particles. It is to be noted that the imaging herein was conducted on specimens prepared in a loose state, and further studies are underway to assess imaging of specimens prepared at higher densities. The final objective of this research is to relate the microstructure to the macroscopic mechanical behavior of natural silts.

Acknowledgements. The authors would like to acknowledge the financial support of the Natural Science and Engineering Research Council of Canada (NSERC). The collaborative support provided by Dr. Mark Martinez and Mr. James Drummond at the X-ray μ-CT imaging facility at the Pulp and Paper Centre at UBC in Vancouver is deeply appreciated. Thanks are due to Carlo Corrales for his support in the sample preparation and laboratory testing performed for this research as an undergraduate research assistant.

References

1. Bray, J.D., Sancio, R.B., Riemer, M.F., Durgunoglu, T.: Liquefaction susceptibility of fine-grained soils. In: Proceedings of the 11th International Conference on Soil Dynamics and Earthquake Engineering and 3d International Conference on Earthquake Geotechnical Engineering, pp. 655–662 (2004)
2. Idriss, I.M., Boulanger, R.W.: SPT-based liquefaction triggering procedures (2010)
3. Cubrinovski, M., Rhodes, A., Ntritsos, N., Van Ballegooy, S.: System response of liquefiable deposits. Soil Dyn. Earthq. Eng. **124**(May 2018), 212–229 (2019)
4. Zlatovic, S., Ishihara, K.: Normalized behavior of very loose non-plastic soils: effects of fabric. Soils Found. **37**(4), 47–56 (1997)
5. Høeg, K., Dyvik, R., Sandbækken, G.: Strength of undisturbed versus reconstituted silt and silty sand specimens. J. Geotech. Geoenviron. Eng. **126**(7), 606–617 (2000)
6. Vaid, Y.P., Sivathayalan, S., Stedman, D.: Influence of specimen-reconstituting method on the undrained response of sand. Geotech. Test. J. **22**(3), 187–195 (1999)
7. Wijewickreme, D., Sanin, M.V.: Cyclic shear response of undisturbed and reconstituted low-plastic Fraser River silt. In: Proceedings of the Geotechnical Earthquake Engineering and Soil Dynamics IV, American Society of Civil Engineers, Sacramento, California, 10 p. (2008)
8. Wijewickreme, D., Soysa, A., Verma, P.: Response of natural fine- grained soils for seismic design practice: a collection of research findings from British Columbia, Canada. In: Proceedings of 3rd International Conference on Performance Based Design in Earthquake Engineering, Vancouver, Canada (2017)
9. Brewer, R., Sleeman, J.R.: Soil structure and fabric: their definition and description. J. Soil Sci. **11**(1), 172–185 (1960)
10. Oda, M.: Initial fabrics and their relations to mechanical properties of granular material. Soils Found. **12**(1), 17–36 (1972)
11. Hight, D.W., Leroueil, S.: Characterization of soils for engineering purposes. In: Natural Soils Conference, p. 255 (2003)
12. Fonseca, J., Nadimi, S., Reyes-Aldasoro, C.C., O'Sullivan, C., Coop, M.R.: Image-based investigation into the primary fabric of stress-transmitting particles in sand. Soils Found. **56**(5), 818–834 (2016)
13. Wijewickreme, D., Soysa, A., Verma, P.: Response of natural fine-grained soils for seismic design practice: a collection of research findings from British Columbia, Canada. Soil Dyn. Earthq. Eng. **124**(2019), 280–296 (2019)
14. Zhang, M., Jivkov, A.P.: Micromechanical modelling of deformation and fracture of hydrating cement paste using X-ray computed tomography characterization. Compos. B Eng. **88**, 64–72 (2016)

15. Wesolowski, A.: Application of computed tomography for visualizing three-dimensional fabric and microstructure of Fraser River Delta silt. M.A.Sc. thesis, Department of Civil Engineering, University of British Columbia, Vancouver, Canada (2020)
16. Valverde, A., Wesolowski, M., Wijewickreme, D.: Towards understanding the fabric and microstructure of silt – initial findings of soil fabric from X-ray u-CT. In: 17th World Conference on Earthquake Engineering, Sendai, Japan (2020)
17. Wesolowski, M., Valverde, A., Wijewickreme, D.: Towards understanding the fabric and microstructure of silt – feasibility of X-ray μ-Ct image silt structure. In: 17th World Conference on Earthquake Engineering. Sendai, Japan (2020)
18. Valverde, A., Wijewickreme, D.: Towards understanding the fabric and microstructure of silt – assessment of laboratory specimen uniformity. In: GeoNiagara, Niagara, Canada (2021)
19. Armstrong, J.E., Hicock, S.R.: Surficial geology, New Westminster, West of Sixth Meridian, British Columbia. In: Geological Survey of Canada, "A" Series Map 1484A (1980)
20. Thermo Fisher Scientific (TFS) (2019). Avizo 9.7
21. Fonseca, J.: The evolution of morphology and fabric of a sand during shearing. Ph.D. thesis, Imperial College London (2011)
22. Markussen, Ø., Dypvik, H., Hammer, E., Long, H., Hammer, Ø.: 3D characterization of porosity and authigenic cementation in Triassic conglomerates/arenites in the Edvard Grieg field using 3D micro-CT imaging. Mar. Pet. Geol. **99**, 265–281 (2019)
23. ASTM D6913/D6913M – 17: Standard test methods for particle-size distribution (gradation) of soils using sieve analysis. Annual Book of ASTM Standards, ASTM International, West Conshohocken, PA (2017)
24. ASTM D7928-21: Test method for particle-size distribution (gradation) of fine-grained soils using the sedimentation (hydrometer) analysis. Annual Book of ASTM Standards, ASTM International, West Conshohocken, PA (2021)

The Relationship Between Particle-Void Fabric and Pre-liquefaction Behaviors of Granular Soils

Jiangtao Wei(✉), Minxuan Jiang, and Yingbin Zhang

Southwest Jiaotong University, Chengdu, China
jtwei@swjtu.edu.cn

Abstract. When driven by undrained cyclic shearing, saturated granular soils will experience the increase of excess pore water pressure and the decrease of effective stress. This phenomenon is termed as "cyclic liquefaction" or "cyclic softening". Revealing the evolution of fabric in accompany with effective stress reduction provides a significant insight into the fundamental mechanism of cyclic liquefaction. In this study, numerical tests were conducted in DEM simulations to explore the cyclic liquefaction of granular packings with different void ratios. With the decrease of mean effective stress p' during cyclic liquefaction process, the decrease of particle-void descriptor E_d can be observed for all packings. From the micromechanical perspective, large size voids are redistributed and the local void distribution around particle becomes relative uniform. The change of particle-void fabric from consolidated state to initial liquefaction state is irrelevant to density of the packing. A power function is further adopted to describe the negative correlation between E_d and p'.

Keywords: Discrete-element method · Cyclic liquefaction · Particle-void fabric

1 Introduction

Cyclic liquefaction of granular soils has attracted considerable interests from researchers and engineers from both geotechnical and seismic fields since 1964 Nigaata Earthquake. After initial liquefaction, granular soils will be transformed between "fluid-like" state and "solid-like" state (Idriss and Boulanger 2008; Shamoto et al. 1997; Wei et al. 2018). Granular packing in the "solid-like" state can resist external load whereas packing in the "fluid-like" state flows under any applied loads. In terms of cyclic liquefaction of a granular system, geotechnical researchers were interested in the liquefaction potential and liquefaction assessment of granular soils (Vaid et al. 2001; Idriss and Boulanger 2008; Yang and Sze 2011), and the consequences of cyclic liquefaction such as lateral spreading and post-liquefaction settlement (Ishihara 1993; Shamoto et al. 1997).

A great number of studies have revealed that cyclic liquefaction of granular soils mainly attributes to the collapse of load bearing structure, which can be reflected by the decrease of coordination number (Wang and Wei 2016; Wang et al. 2016; Sitharam et al. 2009). This is from the contact-based fabric to explain the cyclic liquefaction. However, few studies have probed cyclic liquefaction phenomenon from void-based

© The Author(s), under exclusive license to Springer Nature Switzerland AG 2022
L. Wang et al. (Eds.): PBD-IV 2022, GGEE 52, pp. 2229–2236, 2022.
https://doi.org/10.1007/978-3-031-11898-2_207

fabric, partially due to the difficulty to define individual void (Li and Li 2009; Ghedia and O'Sullivan 2012). Wang et al. (2016) proposed a fabric entity termed as mean neighboring particle distance (MNPD) to reflect space arrangement of particles. Wei et al. (2018) proposed fabric descriptors based on Voronoi tessellation to quantify the anisotropy of local void distribution. These studies revealed that the void-based fabric was strongly correlated with large post-liquefaction deformation of granular soils. How the void-based fabric evolves with the effective stress reduction in pre-liquefaction is not well understood.

In this study, comprehensive numerical tests were performed in DEM (discrete element method) simulations to explore the cyclic liquefaction of granular soils from void-based fabric perspective. DEM has been proven to be an excellent tool to study cyclic behaviors of granular soils (Sitharam et al. 2009; O'Sullivan 2011; Huang et al. 2018). Particle-void descriptors proposed by Wei et al. (2018) were adopted to quantify void-based fabric, in terms of particle arrangement and local void distribution within the granular packing. Evolution of particle-void descriptors in accompany with the decrease of mean effective stress in cyclic liquefaction process is further explored for samples with different densities.

2 Methodology and Approach

2.1 Numerical Setups in DEM Simulation

Numerical simulations in this study are carried out in an open-source DEM platform YADE (Smilauer et al. 2015), which has been widely used to perform DEM simulations. Granular packings in the simulation are generated in a 2D square box with the periodic boundaries in all directions. The packing consists of 4,000 frictional disks, with the radii ranges from 0.3 to 0.6 mm. The simplified Hertz-Mindlin model is employed to describe the contact behaviors between contacted particles (Yimsiri and Soga 2010; Wei and Wang 2020). In the contact model, the normal contact force is given by

$$f_n = k_n \delta \tag{1}$$

$$k_n = \frac{2E\sqrt{r}}{3(1-\nu^2)}\sqrt{\delta} \tag{2}$$

where k_n is the normal stiffness and δ is the contact overlap. r is the equivalent radius determined by the radii of two contacted particles: $r = \frac{r_A r_B}{r_A + r_B}$. E and ν are Young's modulus and Poisson's ratio of particles, respectively. In the DEM modellings, $E = 67$ GPa and $\nu = 0.3$.

The tangential contact force is given by

$$df_s = k_s dU_s \tag{3}$$

$$k_s = \frac{2E\sqrt{r}}{(1+\nu)(2-\nu)}\sqrt{\delta} \tag{4}$$

where k_s is the tangential stiffness, U_s is the tangential displacement of the contact. The tangential contact force satisfies the Coulomb criterion that $f_s < \mu f_n$, and μ is the frictional coefficient of particle. After generation of particles within a 2D square box, the packing is firstly subjected to an isotropically consolidation process to reach the confining stress $p_0 = 10\,\text{kPa.m}$. Figure 1(a) demonstrates the consolidated packing. In this process, packings with different void ratios can be obtained by assigning different frictional coefficient μ to particles. In the simulation, four samples with different void ratios (i.e. e = 0.206, 0.214, 0.222, 0.231) are prepared, in order to explore the influence of density to the evolution of fabric.

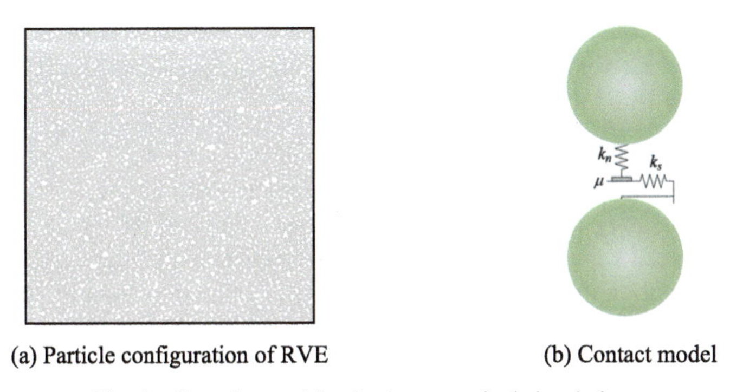

(a) Particle configuration of RVE (b) Contact model

Fig. 1. Granular packing in the numerical simulation.

2.2 Quantification of Void-Based Fabric

To quantitatively explore the void-based fabric in a granular packing, we follow the approach proposed by Wei et al. (2018). In the approach, the granular packing is firstly partitioned into microscopic particle-void cells using Voronoi tessellation. As shown in Fig. 2, each cell is polygonal shape and consists of one particle and its surrounding voids. Local void distribution in the cell is further described by the function $r(\theta)$, which is defined as the ratio between the radial dimension of the cell and the particle:

$$r(\theta) = \frac{R_c(\theta)}{R_p(\theta)} \tag{5}$$

Using Fourier descriptor analysis (Bowman et al. 2001), $r(\theta)$ can be approximated by a concise expression with three parameters:

$$r^2(\theta) \approx \frac{A_r}{\pi}[1 + e_d \cos 2(\theta - \theta_d)] \tag{6}$$

where A_r is the enclosed area of$r(\theta)$, such that $\oint r^2(\theta)d\theta/2 = A_r$. e_d is the shape factor of $r(\theta)$ and it measures the shape elongation of local void distribution. θ_d measures the principal orientation of$r(\theta)$. Figure 2 presents $r(\theta)$ of particle-void cells with different magnitude ofe_d, ranging from 0.01 to 0.30 (Wei and Zhang 2020). With the increase ofe_d, the cell becomes elongated and local void distribution around the central particle is non-uniform. Furthermore, relative large pores can be observed besides the particle with a higher value ofe_d.

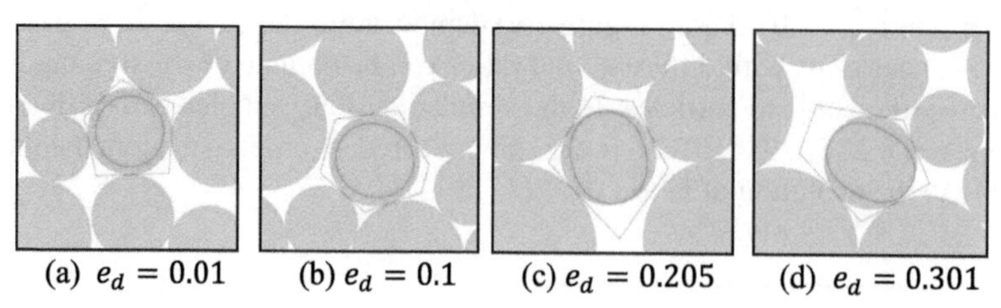

Fig. 2. $r(\theta)$ of particle-void cells with different magnitude of e_d, ranging from 0.01 to 0.3 (from Wei and Zhang 2020).

In the approach, particle-void descriptors (denoted as E_d and A_d) are derived from statistical analysis of e_d and θ_d to quantify void-based fabric of the whole packing. E_d is defined as the mean value of e_d of all cells, and it measures the shape elongation of the local void distribution throughout the packing. A_d is defined to quantify the anisotropy degree of particle-void fabric. E_d and A_d have a strong correlation with cyclic mobility and jamming transition in post-liquefaction stage of granular soils (Wei et al. 2018).

Figure 3 presents the initial value of E_d for samples with $p_0 = 10\,\text{kPa.m}$ before any shearing. A linear function can be employed to describe the relationship between E_d and void ratio e:

$$E_d = ke + T \tag{7}$$

where k is scale factor and T is intercept. The scale factor k is determined by particle size distribution of granular packing. In this study, the value of k is around 0.55.

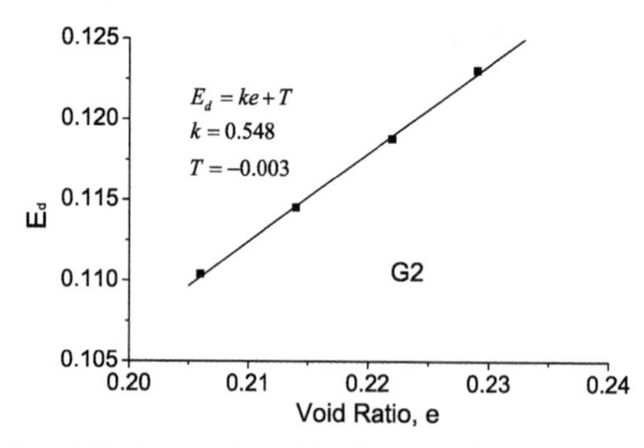

Fig. 3. The initial value of E_d for samples with different void ratio under confining stress $p_0 = 10\,\text{kPa.m}$.

3 Results and Observations

3.1 Cyclic Liquefaction

To explore the characteristics of particle-void fabric for granular packing with different confining stresses, all samples are subjected to the undrained cyclic loading tests, in which the volume of the packing is kept constant and the decrease of mean effective stress can be observed. Figure 4 demonstrates representative results from sample with e = 0.214 during the loading test with axial strain amplitude $\varepsilon_a = 0.5\%$. The simulation results are qualitatively similar to laboratory results of medium-dense sands (Figueroa et al. 1994). With the increase of cycle number, both the amplitude of deviatoric stress q and mean effective stress p' decrease. The loading is terminated when the packing reaches initial liquefaction, in which the mean effective stress p' is close to zero and coordination number Z is lower than 2. As shown in Fig. 4(c) and (d), sample reaches the initial liquefaction after 25 loading cycles, as p' decreases from initial value of 10 kPa·m to 10^{-3} kPa·m and Z decreases from initial value of 3.67 to 1.89.

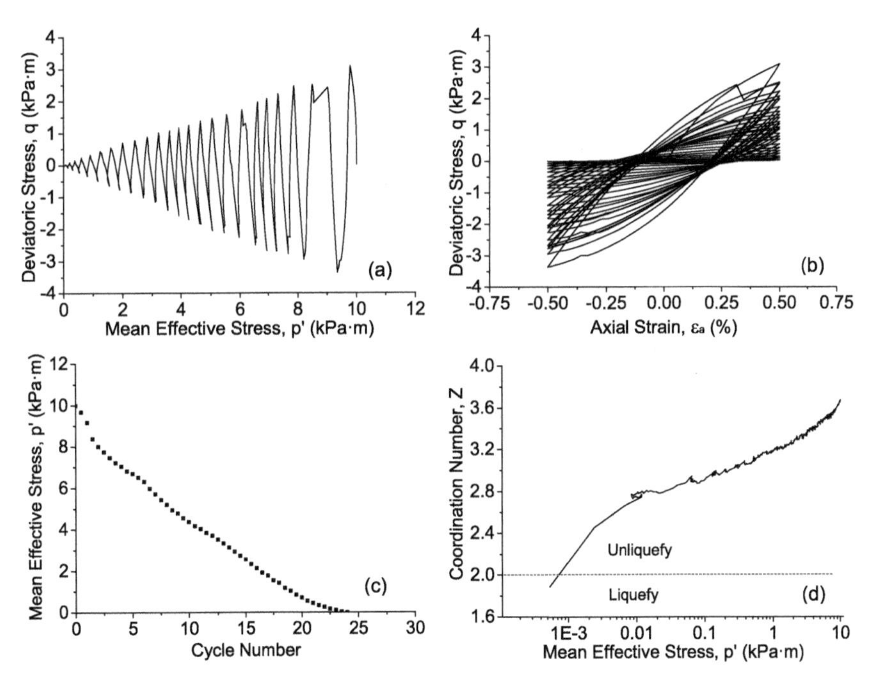

Fig. 4. Responses of sample with e = 0.214 during undrained cyclic loading test with axial strain amplitude $\varepsilon_a = 0.5\%$. (a) Effective stress path; (b) Evolution of deviatoric stress q with axial strain ε_a; (c) Decrease of mean effective stress p' with the increase of cycle number; (d) Evolution of Z with the decrease of p'.

We further check the evolution of particle-void descriptors E_d and A_d during the cyclic liquefaction process and the results are presented in Fig. 5. To get rid of the influence of shear-induced anisotropy, the data of E_d, A_d and p' in Fig. 5 are picked

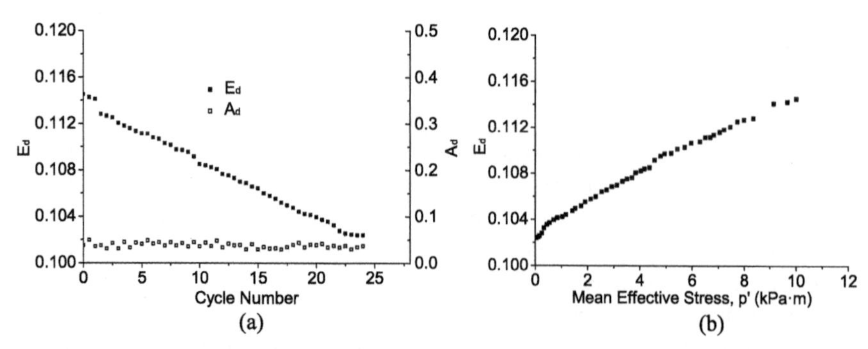

Fig. 5. Typical response of particle-void fabric evolution during cyclic liquefaction process for sample with e = 0.214. (a) Evolution of E_d and A_d with the increase of cycle number; (b) Evolution of E_d versus p'.

at zero deviatoric stress state (i.e. $q = 0$) in each loading cycle. With the increase of cycle number, the value of A_d remains at a relative low range during the whole loading process. It implies that the void-based fabric at zero deviatoric stress state can be regarded to be isotropic. On the other hand, E_d decreases from the initial value of 0.114 to 0.102. Figure 5(b) demonstrates the evolution of E_d versus p' during cyclic liquefaction process. A positive correlation between p' and E_d can be observed. Similar behaviors are also found in the monotonic loading tests (Wei and Zhang 2020).

3.2 E_d with P'

The evolution of E_d in accompany with the decrease of p' for samples with different particle size distributions and void ratios are presented in Fig. 6(a). For all samples, a positive correlation between E_d and p' can be observed, and the slope of the curve gradually increase with the decrease of p'. It is interesting to find that the curves for different samples are almost parallelled with each other. The results imply that the change of particle-void fabric during the entire cyclic liquefaction process may be irrelevant to void ratio of the sample. The insets of the figure further present ΔE_d for samples from consolidated state $p' = 10$ kPa·m to initial liquefaction state. The value of ΔE_d for samples with different void ratios are close to each other, which is −0.0122.

Considering the linear relationship between E_d and void ratio e at consolidated state with $p' = 10$ kPa·m (refers to Eq. 7), the value of E_d at initial liquefaction state (denoted as $E_{d,0}$) is also linearly related with void ratio:

$$E_{d,0} = E_d + \Delta E_d = ke + T_0 \tag{8}$$

Note that $E_{d,0}$ can be regarded as the minimum value of E_d for the packing in the pre-liquefaction stage. After initial liquefaction, E_d will keep decreasing with continued cyclic loading (Wei et al. 2018).

The relationship between E_d and p' can be well fitted by a power function:

$$E_d = a\left(p'\right)^b + E_{d,0} \tag{9}$$

where a and b are fitting parameters. Combining the linear correlation between $E_{d,0}$ and e in Eq. (8), we can get the following equation to predict E_d from p' and e of the granular packing:

$$E_d = a(p')^b + ke + T_0 \tag{10}$$

Figure 6(b) presents the curve fitting by Eq. (10). The parameters in the equation are summaries below: a = 2.44×10^{-3}, b = 0.70, k = 0.548 and $T_0 = -0.015$.

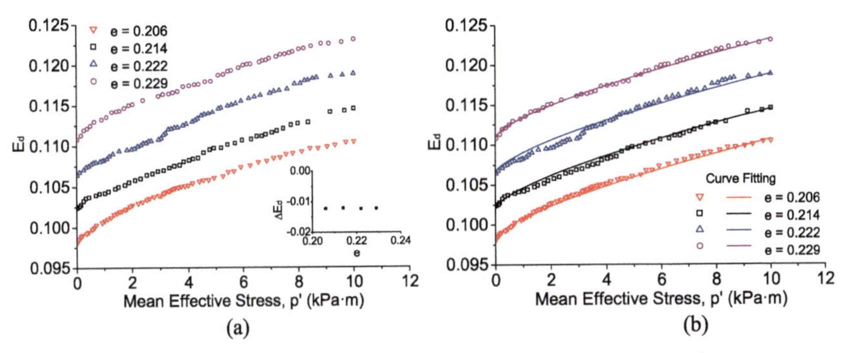

Fig. 6. Evolution of particle-void fabric indicator E_d in accompany with p' for samples with different void ratios.

4 Conclusions

In summary, the relationship between void-based fabric and mean effective stress p' during cyclic liquefaction process is explored by DEM simulations. Void-based fabric is quantified by particle-void descriptor E_d, which measures the uniformity of local void distribution around particles. Under the same confining pressure, a linear function can be employed to describe the positive correlation between E_d and void ratio e. With the decrease of p' during cyclic liquefaction process, the decrease of E_d is observed for all samples with different void ratios. From consolidated state (p' = 10 kPa·m) to initial liquefaction state, the change of E_d is almost the same for samples with different void ratios, which implies that the change of void-based fabric during cyclic liquefaction process is irrelevant to density of the packing. A power function is further adopted to describe the negative correlation between E_d and p'.

Acknowledgements. This research was supported by the National Natural Science Foundation of China (51908471). The financial supports are gratefully acknowledged.

References

Bowman, E.T., Soga, K., Drummond, W.: Particle shape characterisation using Fourier descriptor analysis. Geotechnique **51**(6), 545–554 (2001)

Christoffersen, J., Mehrabadi, M.M., Nemat-Nasser, S.: A micromechanical description of granular material behavior. J. Appl. Mech. **48**(2), 339–344 (1981)

Figueroa, J.L., Saada, A.S., Liang, L., Dahisaria, N.M.: Evaluation of soil liquefaction by energy principles. J. Geotech. Eng. **120**(9), 1554–1569 (1994)

Gu, X., Yang, J., Huang, M.: DEM simulations of the small strain stiffness of granular soils: effect of stress ratio. Granular Matter **15**(3), 287–298 (2013)

Ghedia, R., O'Sullivan, C.: Quantifying void fabric using a scan-line approach. Comput. Geotech. **41**, 1–12 (2012)

Huang, X., Kwok, C.-Y., Hanley, K.J., Zhang, Z.: DEM analysis of the onset of flow deformation of sands: linking monotonic and cyclic undrained behaviours. Acta Geotech. **13**(5), 1061–1074 (2018). https://doi.org/10.1007/s11440-018-0664-3

Idriss, I., Boulanger, R.: Soil Liquefaction During Earthquakes. Earthquake Engineering Research Institute, Oakland, California (2008)

Li, X., Li, X.S.: Micro-macro quantification of the internal structure of granular materials. J. Eng. Mech. **135**(7), 641–656 (2009)

O'Sullivan, C.: Particle-based Discrete Element Modeling: A Geomechanics Perspective. Taylor & Francis, Hoboken (2011)

Seguin, A.: Experimental study of some properties of the strong and weak force networks in a jammed granular medium. Granular Matter **22**(2), 1–8 (2020). https://doi.org/10.1007/s10035-020-01015-z

Shamoto, Y., Zhang, J., Goto, S.: Mechanism of large post-liquefaction deformation in saturated sand. Soils Found. **37**(2), 71–80 (1997)

Sitharam, T.G., Vinod, J.S., Ravishankar, B.V.: Post-liquefaction undrained monotonic behaviour of sands: experiments and DEM simulations. Géotechnique **59**(9), 739–749 (2009)

Šmilauer, V., Catalano, E., Chareyre, B.: Yade Documentation, 2nd edn. The Yade Project (2015)

Vaid, Y., Stedman, J., Sivathayalan, S.: Confining stress and static shear effects in cyclic liquefaction. Can. Geotech. J. **38**(3), 580–591 (2001)

Wang, G., Wei, J.: Microstructure evolution of granular soils in cyclic mobility and post-liquefaction process. Granular Matter **18**(3), 1–13 (2016). https://doi.org/10.1007/s10035-016-0621-5

Wang, R., Fu, P., Zhang, J.-M., Dafalias, Y.F.: DEM study of fabric features governing undrained post-liquefaction shear deformation of sand. Acta Geotech. **11**(6), 1321–1337 (2016). https://doi.org/10.1007/s11440-016-0499-8

Wei, J., Huang, D., Wang, G.: Microscale descriptors for particle-void distribution and jamming transition in pre-and post-liquefaction of granular soils. J. Eng. Mech. **144**(8), 04018067 (2018)

Wei, J., Zhang, Y.: The relationship between contact-based and void-based fabrics of granular media. Comput. Geotech. **125**, 103677 (2020)

Yang, J., Sze, H.: Cyclic behaviour and resistance of saturated sand under non-symmetrical loading conditions. Géotechnique **61**(1), 59–73 (2011)

Yimsiri, S., Soga, K.: DEM analysis of soil fabric effects on behaviour of sand. Géotechnique **60**(6), 483–495 (2010)

Zhang, J., Majmudar, T.S., Tordesillas, A., Behringer, R.P.: Statistical properties of a 2D granular material subjected to cyclic shear. Granular Matter **12**(2), 159–172 (2010)

Changes in Sand Mesostructure During Sand Reliquefaction Using Centrifuge Tests

Xiaoli Xie and Bin Ye(✉)

Tongji University, Siping Road 1239, Shanghai, China
yebin@tongji.edu.cn

Abstract. Previous studies demonstrated that the sand reliquefaction resistance may decrease though the sand deposits were densified after reconsolidation, and attributed this counterintuitive phenomenon to the changes in sand mesostructure after liquefaction. However, no previous studies visualized how sand mesostructure evolved during the whole process of a liquefaction event. In this study, a centrifuge model of the saturated sand deposit liquefied successively under the same input seismic motion. Excess pore pressure at the upper and lower deposit were measured for assessing the sand liquefaction resistance. Besides, sand mesoscopic images during the whole process of each liquefaction event were recorded for analyzing the sand particle behaviors under multiple liquefaction events. The experimental results show that large voids formed under the upward seepage effects during sand reconsolidation, and the long-axes of the sand particles neighboring the large voids rotated vertically. The sand mesostructure after reconsolidation caused the quick contraction of the sand deposit under the subsequent shaking event.

Keywords: Sand reliquefaction · Centrifuge shaking table tests · Sand mesostructure · Seepage

1 Introduction

Field reconnaissance after the 2010–2011 New Zealand earthquake and the 2011 Great East Japan Earthquake demonstrated that the liquefiable deposits could relief many times during the aftershocks, and even resulted in a larger liquefaction region or more severe damages to the infrastructures [1, 2]. Shaking table tests [3] and element tests [4] also verified that sand reliquefaction resistance could decrease with the shaking events.

Finn, et al. [5] explained that the loss of resistance was caused by the formation of a uniform metastable structure or the generation of a nonuniform structure. Mesoscopic images of the sand particles captured from 1 g shaking table tests also proved that the sand mesostructure indeed changed a lot after the previous liquefaction events [6]. However, no direct experimental results have illustrated that how the liquefaction events affected sand mesostructure and influence sand reliquefaction resistance. The evolution process of sand mesostructure during reliquefaction was important to further understand the intrinsic mechanism of the loss in the reliquefaction resistance.

© The Author(s), under exclusive license to Springer Nature Switzerland AG 2022
L. Wang et al. (Eds.): PBD-IV 2022, GGEE 52, pp. 2237–2244, 2022.
https://doi.org/10.1007/978-3-031-11898-2_208

In this study, centrifuge shaking table tests were conducted on a saturated sand deposit, with the measurement of excess pore pressure, acceleration at different depths and the ground vertical settlement. Simultaneously, a mesoscopic image acquisition system was employed to take sand particle images during the whole process of each shaking event. Based on the comprehensive analysis on the macro- and mesoscopic experimental results, the evolution process of sand mesostructure under reliquefaction was revealed, and the mesoscopic mechanism of the phenomenon of the decreasing liquefaction resistance was analyzed.

2 Experimental Method

2.1 Centrifuge Model

The TLJ 150 centrifuge at Tongji University were used in this study to conduct the shaking table tests. A rigid container with inner size at 510 × 450 × 550 mm was adopted to prepare the sand model to firmly install the micro-image acquisition system into the rigid container. The layout of the model container was displayed in Fig. 1. The container was divided into two portions which was separated by a transparent polymeric methyl methacrylate (PMMA) plate with thickness at 40 mm. One portion was used for placing the micro-image acquisition system, and another portion was used for constructing the sand deposit model. Two slices of polystyrene foam board were sticked to the two sides of the zone of sand deposit model to weaken the boundary effects during shaking.

Artificial silica sand was employed to prepare the centrifuge model. The particle size of this sand ranged from 0.075 mm to 1.0 mm. The maximum and minimum void ratio was 1.09 and 0.57 respectively. The coefficient of uniformity and curvature was 2.22 and 0.98 respectively, and the specific gravity was 2.65.

The centrifuge model of the saturated sand deposit was constructed using the wet pluviation method. Scaling factors of the dynamic time and the consolidation time were not the same if water was used as the pore fluid in the sand deposit model [7]. A methylcellulose solution was used instead to solve this disagreement, which was also adopted in the previous studies [8]. Because the centrifuge model was vibrated under 25 g centrifugal acceleration, the methylcellulose solution with 25 cSt was used in this study. The construction of the sand deposit model was completed by depositing dry sand for several layers to ensure a homogeneous model. Approximately five minutes were set between every twice sand deposition for the dissipation of some bubbles. Pore pressure transducers and accelerometer were placed at different depths during the model construction to measure the distribution of the pore pressure and the acceleration along depth. Two laser displacement transducers were fixed above the model top surface to measure the model vertical settlement.

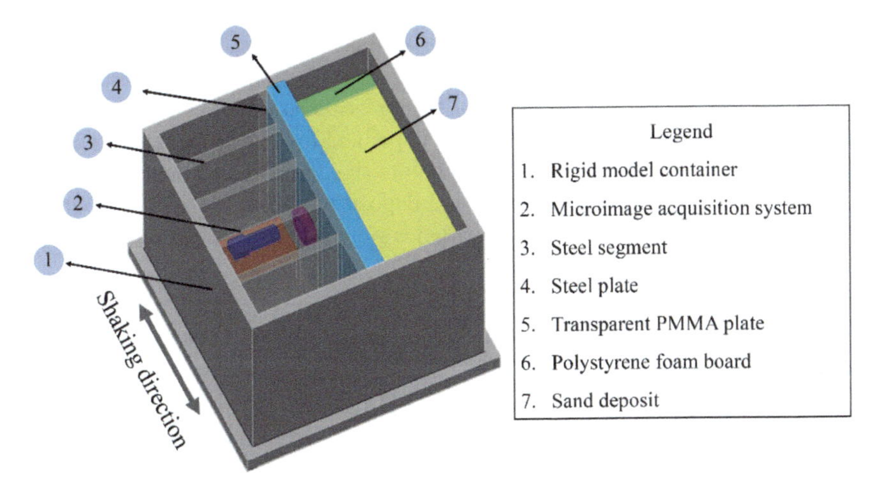

Fig. 1. A schematic diagram of the centrifuge model.

2.2 Experimental Process

The model settled and excess pore pressure was generated when the centrifugal acceleration increased to 25 g gradually. After the static consolidation was completed, the sand deposit model was shaken two times under the same earthquake wave. Enough interval time was set between the two shaking events to ensure that the excess pores pressure caused in the first shaking event was completely dissipated. For convenience, the first and the second shaking events were identified as Motions 1 and 2 respectively.

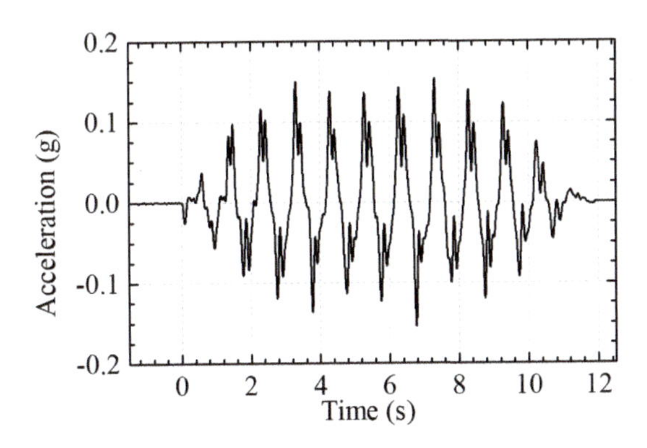

Fig. 2. The applied acceleration time history at the model bottom.

Figure 2 displays the time history of the input acceleration. The maximum acceleration and the frequency of the seismic motion was 0.15 g and 1 Hz, and the duration of the seismic motion was approximately 12 s. During each shaking event, mesoscopic sand particle images at the depth of 1. 625 m was captured with a speed at 45 images per second, together with the measurement of pore pressure, acceleration and the vertical model settlement.

3 Seismic Response of Excess Pore Pressure

Figure 3 shows the time histories of excess pore pressure at different depths. The depth of 1.625 m and 5. 625 m was referred as the upper and lower deposit respectively. The average relative density of the sand deposit before shaking was also indicated in Fig. 3. The excess pores pressure at the lower sand deposit remained higher than that at the upper sand deposit. Therefore, the sand particles were subjected to the upward seepage force during reconsolidation.

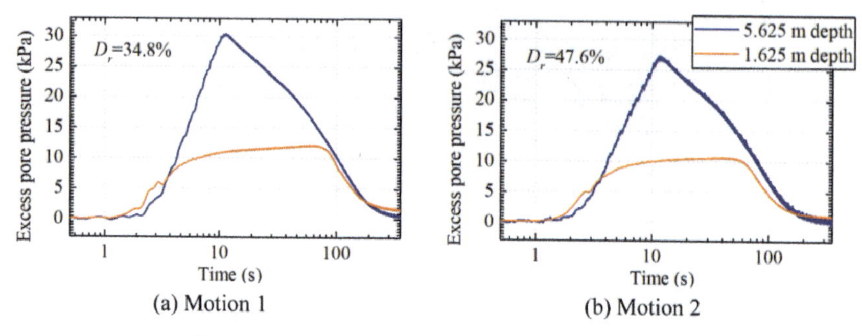

(a) Motion 1 (b) Motion 2

Fig. 3. The time histories of excess pore pressure.

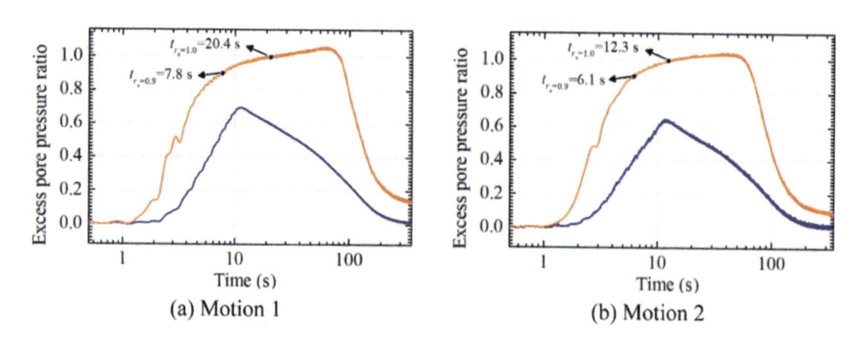

(a) Motion 1 (b) Motion 2

Fig. 4. The time histories of excess pore pressure ratio.

Figure 4 displays the time histories of excess pore pressure ratio (r_u) at the upper and lower depth. Liquefaction resistance was evaluated by comparing the moments when r_u reached 1.0 and 0.9 for liquefaction events, and the maximum r_u for non-liquefaction events. Liquefaction The maximum r_u at the lower depth decreased after Motion 1. This demonstrated that the liquefaction resistance of the lower deposit increased with the shaking events. Because the upper sand deposits liquefied in the two shaking events, the moments when r_u reached 1.0 were used to compare sand liquefaction resistance [3]. However, when r_u increased closely to 1.0, a slight fluctuation in the excess pore pressure may cause errors in determining the liquefaction triggering moments. Therefore, in this study, both the moments when r_u reached 1.0 and 0.9 were indicated in Fig. 3 for ensuring the accuracy in assessing sand liquefaction resistance. As shown in Fig. 3, the moments when r_u reached 1.0 and 0.9 in Motion 2 were earlier than those in Motion 1, indicating that the sand liquefaction resistance of the upper deposit in Motion 2 decreased.

Although the average relative density of the sand deposit increased from 34.8% to 47.6% after Motion 1, this sand densification did not enhance the reliquefaction resistance. Previous studies speculated the reliquefaction resistance was also affected by the sand mesostructure after the initial liquefaction [9]. Inspired by the previous speculations, the evolution of sand mesostructure during reliquefaction was analyzed as follows.

4 Evolution of Sand Mesostructure During Reliquefaction

Figure 5 shows the evolution of sand mesostructure during the shaking stage of Motion 1. With the gradual increase of the seismic acceleration, the sand particles moved or rotated slightly to build contacts with the neighboring particles as shown in Fig. 5(a) and (b). The continuously increased excess pore pressure caused the loss of the sand effective stress. Then sand particles moved more easily under the cyclic shear stress of the earthquake loading. Consequently, voids among some sand particles (e.g., sand particles in the Zones B and C in Fig. 5(c)) were enlarged, and the arrangement of some particles (e.g., sand particles in the Zone A in Fig. 5(d)) was redistributed.

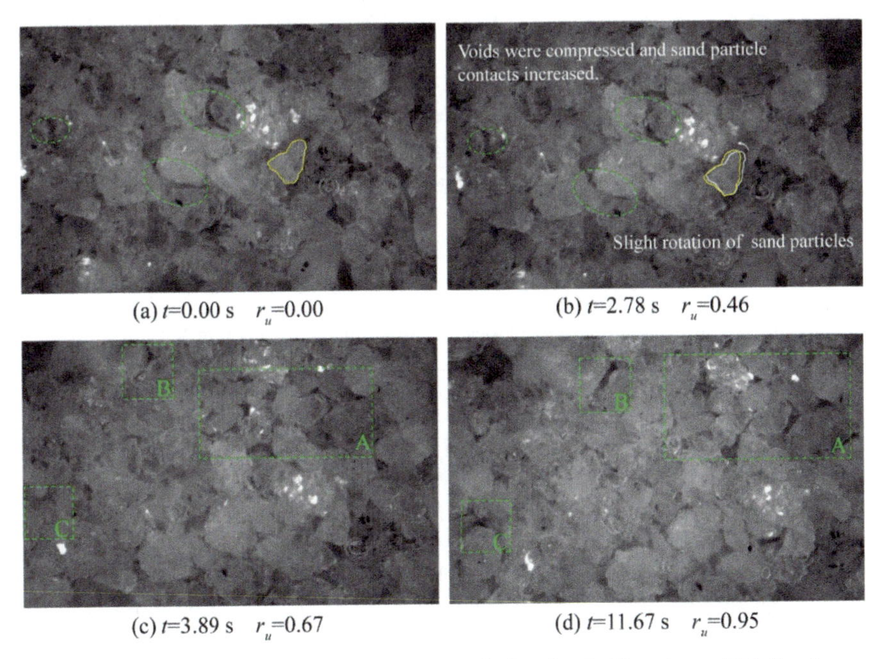

Fig. 5. The evolution of sand mesostructure during the shaking stage of Motion 1.

After the seismic motion was over, the excess pore pressure at the upper deposit continued to increase for a few seconds. This is because the excess pore pressure at the lower deposit was high than that at the upper deposit after the seismic motion was over as indicated in Fig. 3. The upper deposit liquefied approximately at 20.4 s with the excess pore pressure increasing. Firstly, small particles moved upward rapidly under the upward seepage as shown in Fig. 6(a) and (b). These small particles were previously filled into the voids among large particles. The movement of the small particles caused the voids among large particles to link gradually as shown in Fig. 6(c). These connected voids were called open voids. The large particles nearby the open voids rotated drastically under the upward seepage. Especially, the particle long-axes rotated towards the vertical direction as shown in Fig. 6(d).

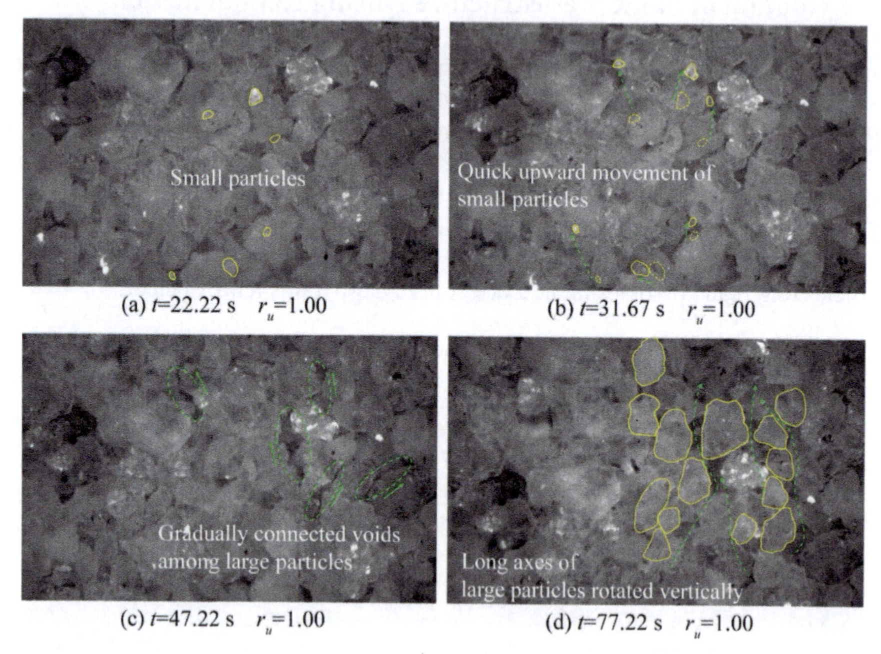

Fig. 6. The evolution of sand mesostructure during $r_u = 1.0$ in Motion 1.

With the dissipation of the excess pore pressure, the sand did not liquefy. Figure 7 displays the evolution of sand mesostructure during $r_u < 1.0$ in Motion 1. Some particles still moved upward during the initial stage of $r_u < 1.0$ as shown in Fig. 7(a). However, these particles moved downward with the increasingly weaker seepage. Consequently, thin and narrow voids called slender voids were generated along with the differentiate movement of the sand particles as shown in Fig. 7(b). It should be noted that most of the particles adjusted the contacts with the neighboring particles slightly during $r_u < 1.0$, and the open voids generated during $r_u = 1.0$ still existed after the reconsolidation.

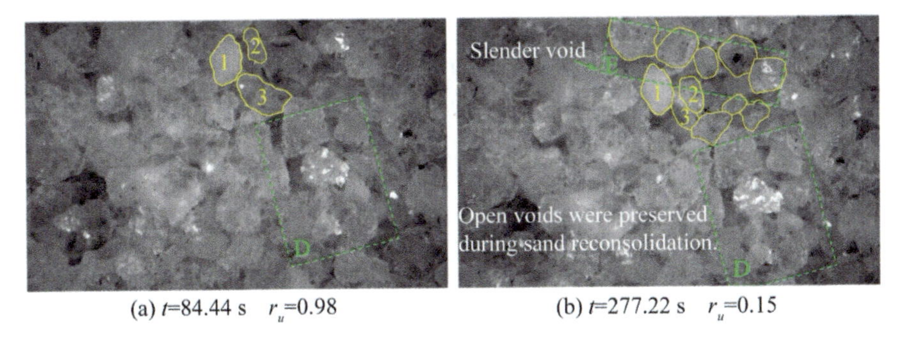

(a) t=84.44 s r_u=0.98 (b) t=277.22 s r_u=0.15

Fig. 7. The evolution of sand mesostructure during $r_u < 1.0$ in Motion 1

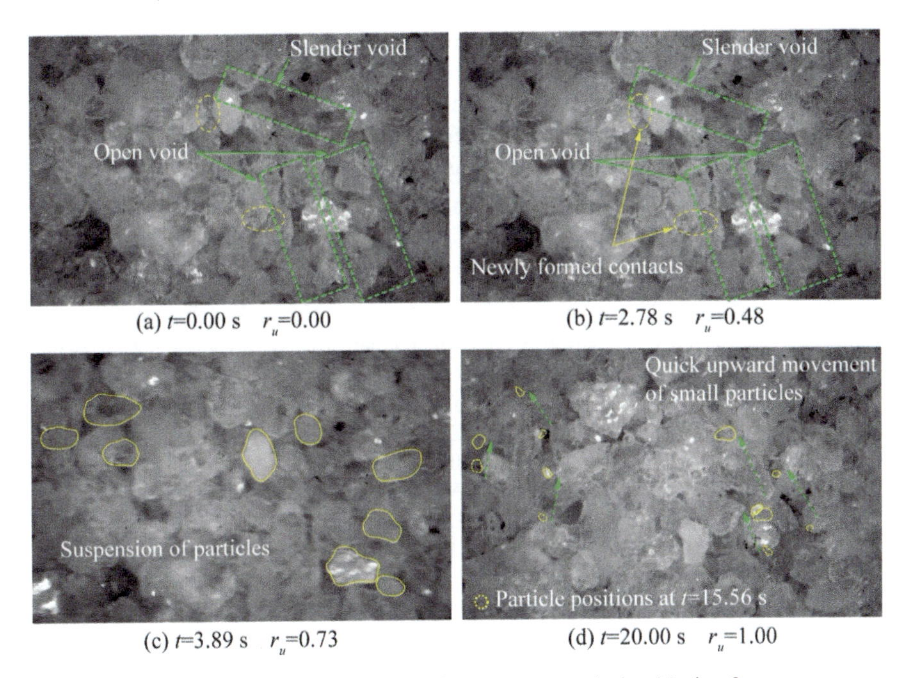

(a) t=0.00 s r_u=0.00 (b) t=2.78 s r_u=0.48

(c) t=3.89 s r_u=0.73 (d) t=20.00 s r_u=1.00

Fig. 8. The evolution of sand mesostructure during Motion 2.

Figure 8 displays the evolution of sand mesostructure during Motion 2. During the initial shaking stage, some of the particles increased contacts with the neighboring particles as shown in Fig. 8(a) and (b), which resembles the particle behaviors at the beginning of Motion 1. The large voids generated after Motion 1 were not destroyed under the initial small-amplitude shaking. With the shaking amplitude increasing, the particles nearby the large voids lost contacts and moved into the large voids immediately, as shown in Fig. 8(c). The movement of these particles further provoked the contact loss of the other particles. Therefore, the whole deposit at the photography position contracted rapidly and the excess pore pressure increased quickly in Motion 2. Then sand reliquefied with the excess pore pressure increasing. The small particles moved upward firstly, which also resembles the particle behaviors during the initial stage of liquefaction.

5 Conclusions

Centrifuge shaking table tests were conducted to explore the evolution of sand mesostructure during reliquefaction, and reveal the mesoscopic mechanism of the decrease in sand reliquefaction resistance. Conclusions are drawn as follows.

(1) The sand particles exhibited some similar mesoscopic behaviors under the seismic events. The particles tended to build new contacts with the neighboring particles under the initial small-amplitude shaking, and the small particles always firstly moved upward quickly in advance of the large particles during liquefaction.
(2) The formation of the sand mesostructure during reconsolidation induced the loss of liquefaction resistance in the subsequent shaking event. Large voids including open and slender voids were generated under the upward seepage, and the long-axes of the particles rotated to be vertical along with the generation of the open voids.

References

1. Quigley, M.C., Bastin, S., Bradley, B.A.: Recurrent liquefaction in Christchurch, New Zealand, during the Canterbury earthquake sequence. Geology **41**(4), 419–422 (2013)
2. Wakamatsu, K.: Recurrence of liquefaction at the same site induced by the 2011 Great East Japan Earthquake Compared with Previous Earthquakes. In: The 15th World Conference on Earthquake Engineering, Lisbon, Portugal (2012)
3. Ha, I.-S., Olson, S.M., Seo, M.-W., Kim, M.-M.: Evaluation of reliquefaction resistance using shaking table tests. Soil Dyn. Earthq. Eng. **31**(4), 682–691 (2011)
4. Wahyudi, S., Koseki, J., Sato, T., Chiaro, G.: Multiple-liquefaction behavior of sand in cyclic simple stacked-ring shear tests. Int. J. Geomech. **16**(5), C4015001 (2016)
5. Finn, W.D., Bransby, P.L., Pickering, D.J.: Effect of strain history on liquefaction of sand. J. Soil Mech. Found. Div. **96**(6), 1917–1934 (1970)
6. Ye, B., Hu, H., Bao, X., Lu, P.: Reliquefaction behavior of sand and its mesoscopic mechanism. Soil Dyn. Earthq. Eng. **114**, 12–21 (2018)
7. Dewoolkar, M.M.: Centrifuge modelling of models of seismic effects on saturated earth structures. Geotechnique **49**(2), 247–266 (1999)
8. Adamidis, O., Madabhushi, G.S.P.: Use of viscous pore fluids in dynamic centrifuge modelling. Int. J. Phys. Model. Geotech. **15**(3), 141–149 (2015)
9. Ye, B., Zhang, L., Wang, H., Zhang, X., Lu, P., Ren, F.: Centrifuge model testing on reliquefaction characteristics of sand. Bull. Earthq. Eng. **17**(1), 141–157 (2018)

Examining the Seismic Behavior of Rock-Fill Dams Using DEM Simulations

Zitao Zhang[2], Rui Wang[1], Jing Hu[2](✉), Xuedong Zhang[2], and Jianzheng Song[2]

[1] State Key Laboratory of Hydroscience and Engineering, Tsinghua University, Beijing 100084, China

[2] China Institute of Water Resources and Hydropower Research, Beijing 100038, China

hujing@iwhr.com

Abstract. High rock-fill dams are planned or under construction in high seismic intensity area, challenging the seismic design of those dams. This study aims to examine the seismic behavior of rock-fill dams using DEM simulations. The focus is put on the permanent deviatoric strain field and the shear stress-strain loops in rock-fills which can hardly be measured in the experiments. The acceleration response and crest settlement observed in the simulations are consistent with those monitored in centrifuge shaking table tests. The key feature of permanent deformation is surface sliding, and the size of the associated shear zone is influenced by the slope angle. Moreover, there also exist deeper shear zones with a relatively low level of shear strain compared with the surface sliding zone. The rock-fills near slope surface may experience a sharp increase in shear strain after several loading cycles, and the associated strain accounts for a large amount of the permanent shear strain. The variation rate of the shear strain can be decreased, and then the shear strain can hardly develop although the input motion is still strong, reflecting the effect of pre-shaking on the seismic behavior of dams. Moreover, the timing of sharp increase of shear strain is different for rock-fills at various locations, suggesting the complexity of the seismic behavior of rock-fill dams.

Keywords: Rock-fill dam · DEM simulation · Centrifuge test · Shear stress-strain loop · Permanent deformation

1 Introduction

Concrete-faced rock-fill dams have been widely used all over the world, especially in those places where suitable clayey core material is not available. In China, 200 m high rock-fill dams are planned or under construction in high seismic intensity area. The associated design peak ground acceleration can reach 0.4 g, challenging the seismic design of those dams.

Shaking table tests have been carried out on small-scale models to study the seismic behavior of rock-fill dams [1–3]. Due to the stress-dependency of rock-fills, the results of shaking table tests should be viewed only qualitatively. Centrifuge shaking table test is proven to be a powerful tool to examine the seismic behavior as the prototype stress

© The Author(s), under exclusive license to Springer Nature Switzerland AG 2022
L. Wang et al. (Eds.): PBD-IV 2022, GGEE 52, pp. 2245–2252, 2022.
https://doi.org/10.1007/978-3-031-11898-2_209

field can be reproduced in those tests. Many researchers carried out centrifuge shaking table tests to study the acceleration response and seismic deformation of rock-fill dams [4–6]. The authors also carried out those tests and examined the seismic stress evolution in the face slab of rock-fill dams and the failure mode of rock-fill dams subjected to strong earthquakes [7, 8]. Although many insights into the seismic behavior of rock-fill dams have been obtained, the behavior regarding the permanent deviatoric strain field and shear stress-strain loops in the rock-fills during shaking still remains unclear as stresses can hardly be measured in those tests.

In order to examine the seismic behavior of rock-fill dams, discrete element method (DEM) simulations on a centrifuge shaking table test were carried out. Details in those simulations are firstly presented, followed by experimental results and discussions.

2 Details in DEM Simulations

2.1 Contact Model

The linear rolling resistance model (Itasca Consulting Group 2019) is used to account for the interlocking between particles. The contact normal force N is given by

$$N = k^n U^n \tag{1}$$

where k^n is the secant normal stiffness and U^n is the relative normal displacement of the particles. No tension strength is allowed in the normal direction.

In the tangential direction, the incremental friction dT is given by

$$dT = -k^s dU^s \tag{2}$$

where k^s is the tangent shear stiffness and dU^s is the incremental relative sliding displacement of the particles. Moreover, due to the slip behavior between particles, the magnitude of the shear force T is restricted by the maximum allowable value, which can be described by

$$|T| \leq \mu |N| \tag{3}$$

where μ is the inter-particle friction coefficient.

The incremental rolling resistance dM_r is given by

$$dM_r = -k_r d\theta_r \tag{4}$$

where k_r is the rolling resistance stiffness and $d\theta_r$ is the relative bend-rotation increment. The rolling resistance stiffness k_r is given by

$$k_r = k_s R^2 \tag{5}$$

where R is the equivalent sphere radius of the two contacting entities. Similar to the friction force, the rolling resistance M_r is restricted by the maximum value ${M_r}^m$, which can be described by

$$|M_r| \leq |M_r^m| = \mu_r R |N| \tag{6}$$

where μ_r is the maximum rolling resistance coefficient.

Table 1 summarizes the parameters used in this study.

Table 1. Parameters used in the DEM simulations

Parameters	Value
Particle density	2650 kg/m^3
Secant normal stiffness k_n	2.0×10^5 N/m
Tangent shear stiffness k_s	2.0×10^5 N/m
Inter-particle friction coefficient μ	0.6
Maximum rolling resistance coefficient μ_r	1.2

2.2 Simulation Setup and Procedures

DEM simulations on the seismic behavior of rock-fill dams were carried out to simulate the centrifuge test denoted as G1-T1-0.23 g in [8]. In order to reduce the computation time, the model size in the simulation is half of the centrifuge model. By increasing the centrifugal acceleration from 40 g, which is the value used in the experiment, to 80 g in the simulation, the simulation results reflect the behavior of a prototype dam identical to that of the centrifuge test. Figure 1 presents the configuration of the dam with an empty reservoir at prototype scale. As face slabs, cushion layers and toe walls have minor influence on the seismic behavior of rock-fills, they are not modeled in the simulations in order to minimize the computation time. The prototype dam has a total height (H_0) of 8.4 m and a slope ratio of 1:1.6 in the upstream side and 1:1.8 in the downstream side.

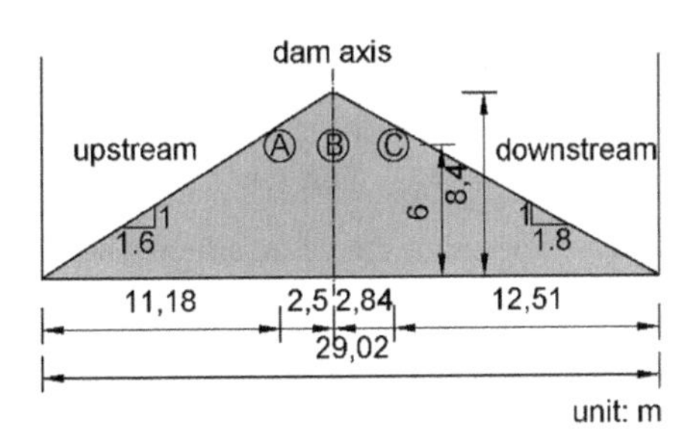

Fig. 1. Schematic drawing of the prototype dam.

The rock-fills were composed of 77% of particles with sizes of 5–10 mm and 23% of those finer than 5 mm in the simulation, which were the same as those used in the centrifuge test. A cuboid assembly of particles were firstly generated in a container composed of four side walls and one base wall, and then the particles outside the profile of the dam were deleted after consolidation at 1 g. Afterwards, the gravity was increased to 80 g and then the particles reached a new equilibrium state. The average density of the soil material was 1900 kg/m^3 before shaking, which was close to the value used in the centrifuge test. The dam model was then subjected to a dynamic excitation applied to the base wall and the side walls of the container. The prototype bedrock acceleration histories with a PGA of 0.23 g were the same as that used in the experiment (see Fig. 2). In order to minimize boundary effects, no friction and rolling resistance is allowed between particles and side walls during the whole shaking process.

3 Results and Discussions

The simulation results are presented and discussed in the following. Prototype values are used unless otherwise specified.

3.1 Acceleration Response

Figure 2 presents the horizontal acceleration records at different heights along dam axis. The input or bedrock motion is also illustrated in the figure. The rock-fills close to the crest exhibit higher acceleration, reflecting the amplification of horizontal acceleration. This resembles the behavior observed in the centrifuge test (see Fig. 5 in [8]). The peak horizontal acceleration is calculated, and furtherly, the amplification factors. As shown in Fig. 3, the amplification factor increases with height, and the factors are consistent with those observed in the experiment.

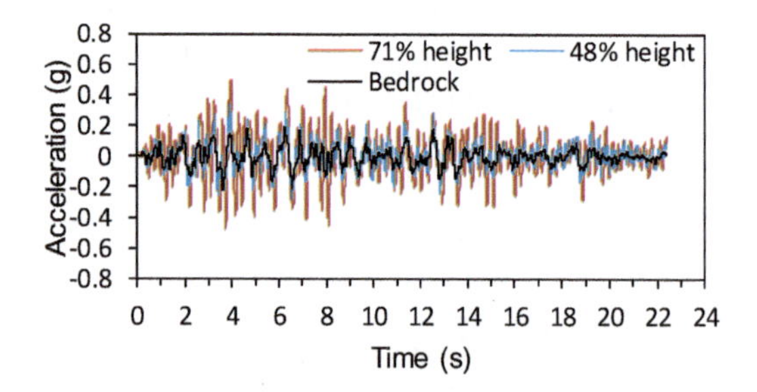

Fig. 2. Acceleration records at different heights

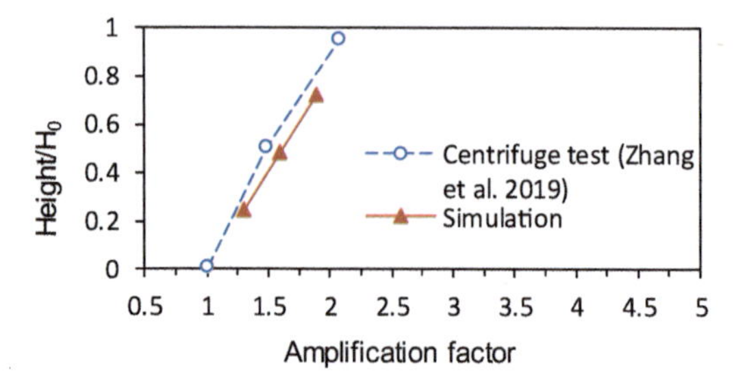

Fig. 3. Distribution of the amplification factor.

3.2 Permanent Deformation

The permanent crest settlement is about 9 mm, which accounts for 0.11% of the initial dam height. The settlement is slightly larger than that monitored in the centrifuge test. As shown in Fig. 4, driven by the gravitational force, the particles in the upstream side move towards the upstream-downward direction, while the particles in the downstream side move towards the downstream-downward direction. Particles near the slope surface demonstrates much larger displacements. This is similar to the surface sliding behavior observed in the centrifuge test (see Fig. 8 in [8]). Since the upstream slope is slightly steeper than the downstream one, most particles near crest move towards the upstream-downward direction.

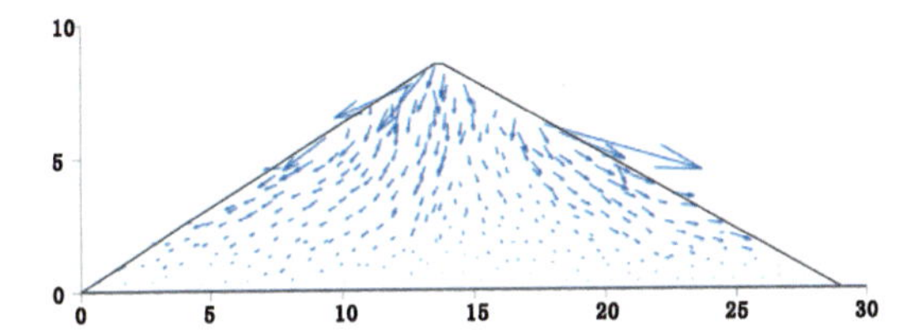

Fig. 4. Permanent displacement field of particles at the end of shaking.

Figure 5 presents the field of permanent deviatoric strain ε_q at the end of shaking, which is given by

$$\varepsilon_q = \frac{\sqrt{2}}{6}\sqrt{4[\varepsilon_x^2 + (\varepsilon_x - \varepsilon_z)^2 + \varepsilon_z^2] + 6\gamma_{xz}^2} \quad (7)$$

where ε_x, ε_z and γ_{xz} are horizontal, vertical and shear strains, respectively. The surface sliding zone can be clearly identified. The upstream surface sliding zone with a value of ε_q larger than 0.2% has a length of 9.2 m and a thickness of 1.9 m, while the downstream zone exhibits a length of 12.5 m and a thickness of 1.6 m. Due to the relatively small slope angle, the shear zone has larger area and smaller thickness in the downstream side. The deviatoric strain is as large as 10% near slope surface. Moreover, as illustrated in Fig. 5, there also exist two deeper shear zones. The deviatoric strain in those zones are 0.2–0.6%, which is much smaller than that observed near slope surface.

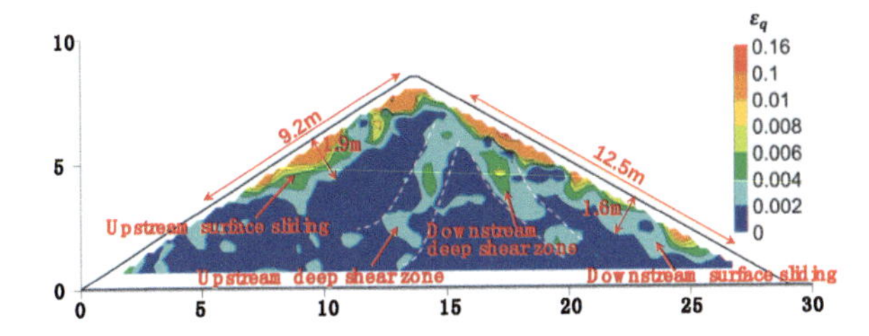

Fig. 5. Permanent deviatoric strain ε_q field at the end of shaking.

3.3 Cyclic Shear Stress-Strain Loops

As shown in Fig. 1, three measurement spheres denoted as A, B and C are used to monitor the cyclic stress (τ_{xz}) – strain (γ_{xz}) loops in rock-fills. The centroids of those spheres are located at a height of 6 m, and the radii are 0.72 m. Figures 6, 7 and 8 present the cyclic shear stress-strain loops at A, B and C, respectively. Figure 9 demonstrates the evolutions of shear strain γ_{xz} over time. As shown in Fig. 6, the shear stress τ_{xz} induced by shaking initially increases and then turns to decrease over time. This trend is consistent with the variation of input energy over time, which can be reflected by the bedrock acceleration histories (see Fig. 2). Such behavior can also be observed in other locations (see Figs. 7 and 8).

Half of the difference between the maximum and minimum values, i.e., [max(τ_{xz}) – min(τ_{xz})]/2, is used to describe the peak shear stress induced by shaking. The peak shear stresses are 14 kPa, 16 kPa and 8 kPa in the locations A, B and C, respectively. Although the peak shear stress is the largest near dam axis, the shear strain is minor (see Figs. 7 and 9). For the rock-fills near upstream slope surface, i.e., at location A, a sharp increase in shear strain γ_{xz} occurs at about 3.9–4.8 s, and the associated increment accounts for 83% of the total strain. Afterwards, the variation rate is decreased, and then the shear strain seems saturated after 8.2 s although the input seismic energy is still strong. This reflects the effects of pre-shaking on the seismic behavior of rock-fill dams. A similar trend can also be found in location C in the downstream side, however, the timing is different. The sharp increase of shear strain occurs at 1.4–1.5 s, which is much earlier than that observed in the upstream side. This suggests the complexity of the seismic behavior of rock-fill dams.

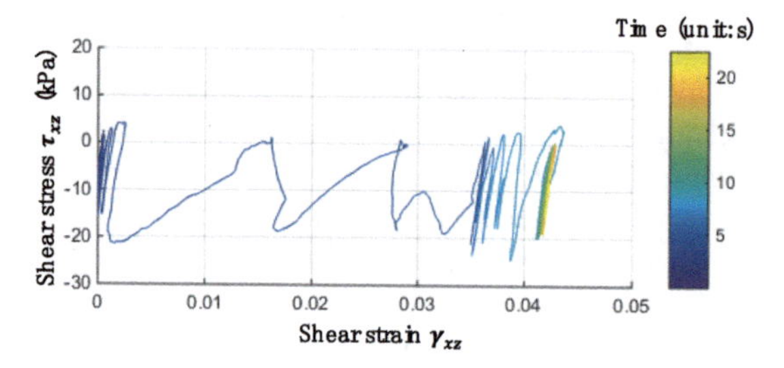

Fig. 6. Shear stress-strain behavior at location A near the upstream slope surface.

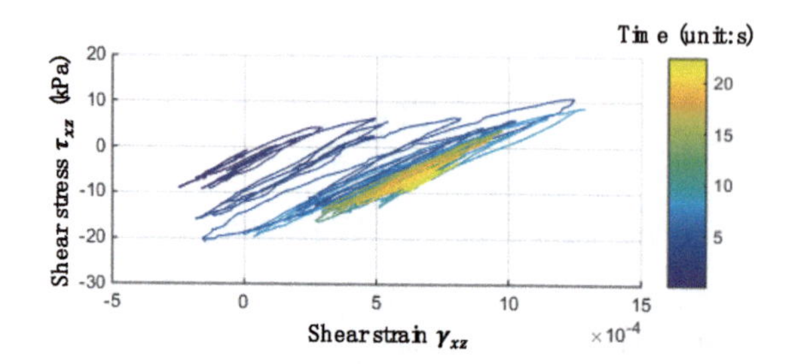

Fig. 7. Shear stress-strain behavior at location B near dam axis.

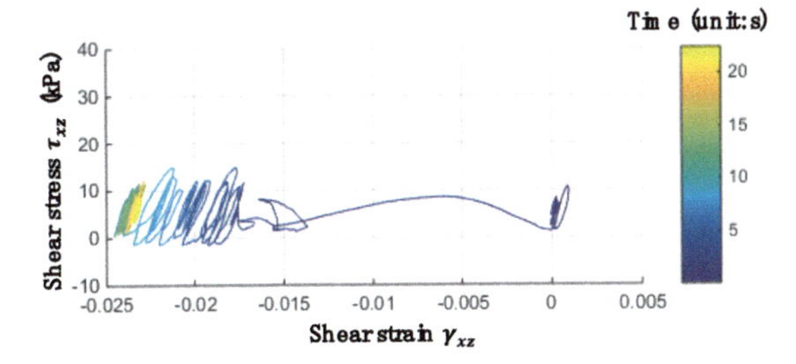

Fig. 8. Shear stress-strain behavior at location C near the downstream slope surface.

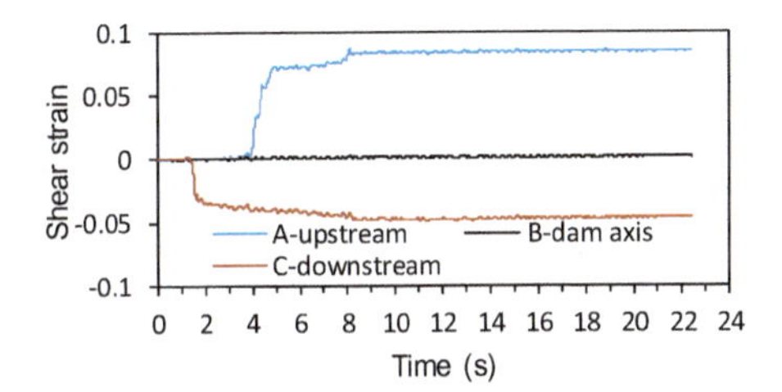

Fig. 9. Evolution of shear strain over time.

4 Concluding Remarks

In this study, the seismic behavior of rock-fill dams is examined using DEM simulations. The acceleration response and crest settlement are consistent with centrifuge test results, and the permanent deformation and shear stress-strain loops in rock-fills are further discussed. The salient findings are summarized in the following.

Due to the variation in slope angle, the sizes of surface sliding zone in the upstream and downstream sides are different. The steeper slope exhibits a shear zone with a larger thickness, while the gentler one has a shear zone with larger area. In addition to surface sliding, two deeper shear zones can be identified near dam axis, nevertheless, the deviatoric strain in those zones is much smaller than that observed near slope surface.

The peak shear stress induced by shaking near dam axis can be larger than those near slope surface, however, minor shear strain occurs near dam axis. For the rock-fills near upstream slope surface, a sharp increase in shear strain occurs after the initial several loading cycles, and the associated strain accounts for a large amount of the shear strain. Afterwards, the variation rate is decreased, and then the shear strain seems saturated although the input motion is still strong, reflecting the effect of pre-shaking on the seismic behavior of dams. Moreover, the timing of sharp increase of shear strain is different for rock-fills at various locations, suggesting the complexity of the seismic behavior of rock-fill dams.

This study suggests the potential of combining centrifuge shaking table tests and DEM simulations in the analysis of seismic behavior of rock-fill dams. Further experimental and numerical study will be carried out to fully understand the associated mechanisms.

Acknowledgements. This research was supported by Open Research Fund Program of State Key Laboratory of Hydroscience and Engineering (sklhse-2021-D-05), National Natural Science Foundation of China (51809290) and the IWHR Research and Development Support Program (GE0145B032021).

References

1. Han, G., Kong, X., Li, J.: Dynamic experiments and numerical simulations of model concrete-face rockfill dams. In: Proceedings of Ninth World Conference on Earthquake Engineering, Tokyo-Kyoto, Japan, vol. VI, pp. 331–336 (1988)
2. Liu, X., Wang, Z., Zhao, J.: Advancement of technology on shaking table model test and dynamic analysis of CFRD. J. Hydraul. Eng. 29–35 (2002). (in Chinese)
3. Zhou, G., Liu, X., Zhao, J., et al.: The shaking table model test study on acceleration seismic response of Zipingpu CFRD. Chin. Civil Eng. J. **45**, 20–24 (2012). (in Chinese)
4. Kim, M.K., Lee, S.H., Yun, W.C., Kim, D.S.: Seismic behaviors of earth-core and concrete-faced rock-fill dams by dynamic centrifuge tests. Soil Dyn. Earthq. Eng. **31**, 1579–1593 (2011)
5. Cheng, S., Zhang, J.M.: Centrifuge modeling test of dynamic response and deformation law of concrete-faced rockfill dam. Eng. Mech. **29**, 2982–2985 (2012). (in Chinese)
6. Cheng, S., Zhang, J.M.: Dynamic centrifuge model test on concrete-faced rockfill dam. J. Earthq. Eng. Eng. Vibrat. **31**, 98–102 (2011). (in Chinese)
7. Zhang, X.D., Zhang, Z., Wei, Y.Q., et al.: Examining the seismic stress evolution in the face slab of concrete-faced rock-fill dams using dynamic centrifuge tests. Soil Dyn. Earthq. Eng. **123**, 337–356 (2019)
8. Zhang, Z., Zhang, X.D., Hu, J., et al.: Dynamic centrifuge modelling of concrete-faced rock-fill dams subjected to earthquakes. In: Proceedings of the XVII ECSMGE-2019, Iceland, pp. 1–8 (2019)

DEM Simulation of Undrained Cyclic Behavior of Saturated Dense Sand Without Stress Reversals

Xin-Hui Zhou and Yan-Guo Zhou(✉)

MOE Key Laboratory of Soft Soils and Geoenvironmental Engineering, Institute of Geotechnical Engineering, Zhejiang University, Hangzhou 310058, China
qzking@zju.edu.cn

Abstract. Two-dimensional discrete element method (DEM) is used to study the undrained behavior of dense granular materials under cyclic loading without stress reversals, and to clarify the effect of initial static shear on cyclic resistance and the underlying mechanism. A series of undrained stress-controlled cyclic triaxial tests were simulated with varying values of cyclic stress ratio (CSR) and initial static shear stress ratio (α), and the type of "residual deformation accumulation" cyclic response was identified. The evolution of internal microstructure of the granular materials was quantified using a contact-normal-based fabric tensor and the coordination number. The higher α (i.e., smaller consolidation stress ratios in tests) leads to higher stress-induced initial fabric anisotropy. The cyclic resistance of dense granular materials increases with initial fabric anisotropy. During the loading process, the dense granular materials with higher initial fabric anisotropy exhibited slower reduction in coordination number. The study shed lights on the underlying mechanism that why the presence of initial static shear is beneficial to the cyclic resistance for dense sand.

Keywords: DEM · Cyclic resistance · Static shear · Fabric anisotropy · Granular materials

1 Introduction

Slope failure caused by earthquakes is one of the most serious geotechnical disasters. The soil element within a slope is different from that of a level ground because of the existence of initial static shear stress (τ_s) on the horizontal plane. During earthquake shaking, a cyclic shear stress (τ_d) will be superimposed on the horizontal plane (Zhou et al. 2020).

In the framework of traiaxial tests, the undrained cyclic triaxial tests on anisotropically consolidated samples can be used to simulate the behavior of soil elements within a slope during earthquake. According to the relative size of the static deviatoric stress (q_s) and the cyclic deviatoric stress (q_c), Hyodo et al. (1991) classified the undrained cyclic behavior into three types: "stress reversal" ($q_s < q_c$), "no reversal" ($q_s > q_c$), and "intermediate" ($q_s = q_c$). Unlike the "stress reversal" case, the excess pore water

© The Author(s), under exclusive license to Springer Nature Switzerland AG 2022
L. Wang et al. (Eds.): PBD-IV 2022, GGEE 52, pp. 2253–2261, 2022.
https://doi.org/10.1007/978-3-031-11898-2_210

pressure in the "no reversal" case can't build up to the value of σ'_3. As a result, there will be no sudden increase in strains because of liquefaction. It was widely observed that in the "no reversal" case, the residual deformation brought the samples to failure even though no liquefaction had occurred.

Most of the researchers have found that the presence of initial static shear is beneficial to the cyclic resistance for dense sand in the framework of the triaxial tests (Vaid and Chern 1983; Yang and Sze 2011). However, the underlying mechanism is still unclear. DEM (discrete element method) is a useful tool to study the macroscopic phenomenon from a microscopic perspective. For example, the onset of macroscopic liquefaction occurs when the mechanical coordination number is less than 4, which is the minimum requirement for a stable three-dimensional load-bearing structure (Edwards 1998).

In this study, a series of undrained stress-controlled cyclic triaxial tests based on two dimensional DEM with varying cyclic stress ratio CSR and initial static shear stress ratio α are simulated to investigate the undrained behavior and cyclic resistance of dense granular materials under cyclic loading without stress reversals. The induced microstructure changes are further examined, including the coordination number and fabric anisotropy, to clarify the effect of initial static shear stress on the cyclic resistance and the underlying mechanism.

2 DEM Model

2.1 DEM Program and Contact Models

In this paper, a commercial code PFC2D (Itasca 2005) was employed to perform the numerical simulations. A rectangular packing (85 mm × 40 mm) of circular particles is considered. 4069 total particles with radii ranging from 0.3 mm to 0.6 mm are randomly generated and the uniform distribution was adopted. A linear force-displacement contact law for circular disk elements was employed. All the parameters adopted in this study were summarized in Table 1. It should be noted that a calibration process was conducted

Table 1. Parameters in DEM simulation

Parameter	Value
Number of particles N_b	4069
Particle solid density ρ_s (kg/m^3)	2630
Wall-particle normal stiffness $k_{n\_w\text{-}p}$ (N/m)	4×10^{12}
Wall-particle shear stiffness $k_{s\_w\text{-}p}$ (N/m)	2×10^{12}
Wall-particle friction coefficient $\mu_{w\text{-}p}$	0
Particle-particle normal stiffness $k_{n\_p\text{-}p}$ (N/m)	2×10^8
Particle-particle shear stiffness $k_{s\_p\text{-}p}$ (N/m)	1×10^8
Particle-particle friction coefficient $\mu_{p\text{-}p}$	0.5
Local damping ratio β	0.7

to determine the parameters used in this study. The details could be found in Zhou et al. (2007).

2.2 Sample Generation

Each assembly was first prepared randomly distributed over the space, and then anisotropically consolidated under different sets of major principal stress (σ_1') and minor principal stress (σ_2') to achieve desired initial static shear stress ratio α. The relation between σ_1', σ_2' and α can be expressed as

$$\alpha = (\sigma_1' - \sigma_2')/(\sigma_1' + \sigma_2') \tag{1}$$

In order to eliminate the K_σ effect (Vaid et al. 2001), the confining stress after consolidation of all the assemblies was kept the same, $p_0' = (\sigma_1' + \sigma_2')/2 = 200\,\text{kPa}$. Different coefficients of friction were employed during consolidation while the coefficient of friction was set to 0.5 after consolidation. All the assemblies had close void ratio after consolidation, varying in a narrow range of 0.2254–0.2255, with a mean relative density $D_\text{r} = 56\%$, according to Yang et al. (2012).

2.3 Loading Method

Stress-controlled method was performed to simulate the undrained cyclic tests. A servo control scheme was adopted, in which the velocities of the boundary walls were adjusted in such a way that the sample volume remained constant while the cyclic deviator stress followed a sinusoidal cyclic stress history as follows:

$$q(t) = q_s + q_c \sin(2\pi ft) \tag{2}$$

where f is the loading frequency; q_s is the initial static deviator stress, defined as $q_s = \sigma_1' - \sigma_2'$; q_c is the magnitude of the cyclic deviator stress, defined as $q_c = \sigma_d'/2$.

The cyclic behavior is sensitive to the loading frequency. A parametric study covering various loading frequency ($f = 0.5$–5000 Hz) revealed that when f is below 5 Hz, the rates of generation of excess pore water pressure and strain were almost the same and the ratio of the maximum unbalanced force to the average contact force <0.01(Itasca 2008), which means the pseudo-static state was fulfilled. Therefore, a cyclic loading frequency f of 5 Hz was used in all the simulations.

2.4 Micromechanical Parameters

The mechanical coordination number (MCN) proposed by Thornton (2000) is adopted to examine gross fabric changes during cyclic loading. It is calculated as an average number of inter-particle contacts for each particle, but excludes particles with only one or zero contacts which are not contributing to the stable state of stress. It can be expressed as:

$$\text{MCN} = \frac{2N_c - N_{b1}}{N_b - N_{b0} - N_{b1}} \tag{3}$$

where N_b and N_c = total number of particles and contacts, respectively; N_{b1} and N_{b0} = total number of particles with only one or zero contacts, respectively.

The fabric anisotropy can be determined by the fabric tensor, as introduced by Satake (1982), which can be expressed as:

$$\phi_{ij} = \frac{1}{N}\sum_{c=1}^{N} n_i^c n_j^c = \int_{\theta} f(\theta) n_i^t n_j^t d\theta \tag{4}$$

where n_i^c is the cth unit contact normal vector. ϕ_{ij} is a symmetric tensor with the first trace $tr(\phi_{ij}) = 1$. $f(\theta)$ is the angular distribution of the unit contact normal vector. In the two-dimensional space, it fulfills:

$$\int_{\theta} f(\theta) d\theta = 1 \tag{5}$$

Expended using Fourier series and ignoring the higher-order (Sitharam et al. 2009), $f(\theta)$ can be written as:

$$f(\theta) = \frac{1}{2\pi}(1 + a_{ij} n_i n_j) \tag{6}$$

where a_{ij} is the deviatoric tensor and can be related to ϕ_{ij} by the following:

$$a_{ij} = 4\left(\phi_{ij} - \frac{1}{2}\delta_{ij}\right) \tag{7}$$

where δ_{ij} is the Kronecker delta. The norm of a_{ij} is a measure of fabric anisotropy. A larger value of the norm indicates that the assembly is more anisotropic. The norm of a_{ij} can be obtained as follows:

$$a = \sqrt{tr(a_{ij}^2)} = \frac{\sqrt{2}}{2}|a_1 - a_2| \tag{8}$$

where a_1 and a_2 are the two principal components of a_{ij} with $a_1 = -a_2$.

3 Result and Analyses

3.1 Cyclic Response

Table 2 lists the series of cyclic tests simulated with varying values of cyclic stress ratio CSR and initial static shear stress ratio α.

Figure 1 shows the cyclic response from test e2255_α0.225_c0.2, which has a target void ratio of 0.2255, initial static stress ratio of 0.225 and cyclic stress ratio of 0.2. See the note beneath Table 2 for the naming convention of the simulations. The excess pore water pressure increases during the early stages of loading and then tends to be stable at about 126 kPa (Fig. 1(a)). However, the terminal excess pore water pressure is less than the initial effective confining pressure 200 kPa, which means that no liquefaction occurs in this test. The axial strain accumulates continuously on the compression side

Table 2. Undrained cyclic biaxial test program

Test ID*	e	σ_1'/σ_2'	α	*CSR*
e2254_α0.175_*c*0.125	0.2254	235/165	0.175	0.125
e2254_α0.175_*c*0.15	0.2254	235/165	0.175	0.150
e2255_α0.2_*c*0.125	0.2255	240/160	0.200	0.125
e2255_α0.2_*c*0.15	0.2255	240/160	0.200	0.150
e2255_α0.2_*c*0.175	0.2255	240/160	0.200	0.175
e2255_α0.225_*c*0.125	0.2255	245/155	0.225	0.125
e2255_α0.225_*c*0.15	0.2255	245/155	0.225	0.150
e2255_α0.225_*c*0.175	0.2255	245/155	0.225	0.175
e2255_α0.225_*c*0.2	0.2255	245/155	0.225	0.200

* Test ID consists of the prescribed void ratio, initial static shear stress ratio α and cyclic stress ratio *CSR*.

(Fig. 1(b)), and it is interesting to note that the strain accumulation was at a more or less constant rate which was also reported by Yang and Sze (2011). The stress strain curves also develop only on the compression side (Fig. 1(c)). The stress path shifts gradually to the left and stable at last (Fig. 1(d)), and no "butterfly" shaped stress path is observed, which is a typical phenomenon in the stress reversal tests (Pang and Yang 2018).

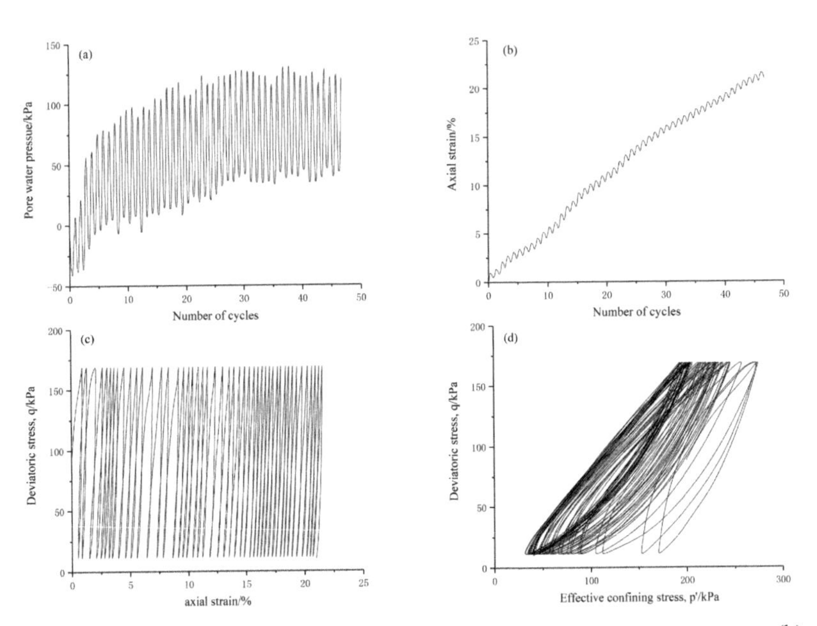

Fig. 1. Cyclic response of sand e2255_α0.225_*c*0.2: (a) excess pore water pressure; (b) axial strain; (c) stress strain curve; (d) effective stress path.

3.2 Cyclic Resistance

The usual approach is to define the cyclic resistance as the point that is accompanied by 5% double-amplitude (DA) axial strain. Sometimes, the criterions of 2.5% and 10% DA are also used. However, in the case of "residual deformation failure" response, the strain development occurs only in one direction, and 5% DA may not able to happen in this kind of situation. Therefore, 5% DA is considered to be an unreasonable criterion to define cyclic resistance. Yang and Sze (2011) proposed that the occurrence of 5% peak axial strain (PS) in compression is regarded as a reasonable criterion in the case of plastic strain accumulation response. In this study, the 5% PS criterion is considered.

As shown in Fig. 2, as α increases from 0.175 to 0.225, the presence of initial static shear increases the cyclic resistance of dense granular materials. The studies conducted by Yang and Sze (2011), Pan and Yang (2018) also revealed that the initial static shear increases the cyclic resistance of dense sands. However, the underlying mechanism is still unclear. A detailed observation of the micromechanical quantities including fabric anisotropy and mechanical coordination number will be presented in the following section, which may provide further insights into the mechanism of the macroscopic phenomenon mentioned above.

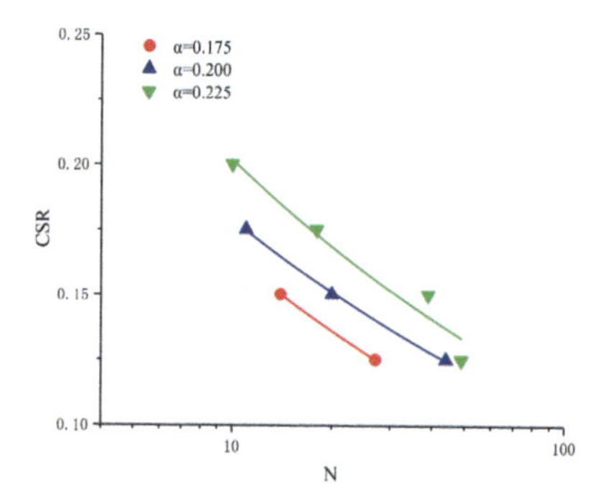

Fig. 2. Number of cycles to failure N with respect to CSR with various α

In the framework of triaxial tests, anisotropically consolidated method is used to study the initial static shear stress. The fabric anisotropy of the anisotropically consolidated samples is presented in Fig. 3(a). It can be seen that the fabric anisotropy increases from 0.092 to 0.124 as α increases from 0.175 to 0.225. For these samples, the fabric anisotropy is due to the concentration of inter-particle contacts along the compression direction, and the larger α leads to the larger stress-induced initial fabric anisotropy a. The MCN values of these samples are presented in Fig. 3(b). With the increase of α, the MCN decreases slightly from 3.429 to 3.411. So it is reasonable to assume that α has trivial effect on the contact density of the samples. In the two-dimensional space, the minimum MCN needed to support a stable load-bearing structure is 3 (Wang et al. 2016). Note that all MCN values are larger than 3, which means that all samples are stable.

To quantitatively compare the evolution of microstructure of samples under different initial static shear stress ratio α, the evolution of MCN and a under different α for the case of CSR = 0.15 is presented. As shown in Fig. 4(a), the MCN decreases sharply in the first cycle regardless of the magnitudes of α, which accounts for about 50% of the reduction. It is interesting to find that the sample with larger α exhibits slower reduction in MCN. When the MCN stabilizes, the sample with larger α will share larger terminal value of MCN, despite the fact that the initial value of MCN is smaller. With the increase of α, the fabric anisotropy *a* grows more slowly. From Fig. 3, we know that the sample with larger α has larger stress-induced initial fabric anisotropy *a*. The sample with larger α still has larger *a* at the early stages of cyclic loading process. A sudden increase of *a* is also observed in the first cycle, which is similar to the sudden change in MCN. As the loading proceeds, the sample with larger α will share smaller final fabric anisotropy instead.

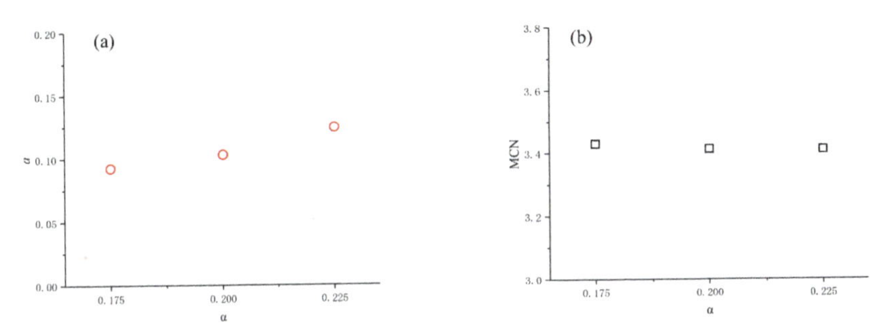

Fig. 3. The relationship between α, a and MCN: (a) a-α; (b) MCN-α.

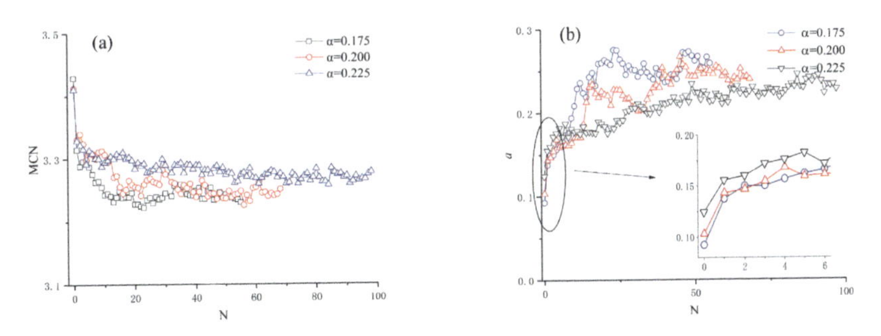

Fig. 4. The evolution of microstructure under different α for CSR = 0.150: (a) MCN; (b) *a*.

4 Conclusions

A micromechanical study has been conducted to investigate the undrained behavior of dense granular materials under cyclic loading without stress reversals. Based on the results of two-dimensional DEM simulation, the induced microstructure changes are further examined, including the mechanical coordination number and fabric anisotropy. The following conclusions are drawn from the present study:

(1) The "residual deformation accumulation" cyclic response was identified from the simulation results, whose typical characteristics are: (1) the residual strain only develops on the compression side, and there is no sudden increase in strain because of the excess pore water pressure ratio is less than 1; (2) the stress path shifts to the left first and stable at last while the hysteresis loop keeps moving to the right.
(2) Anisotropically consolidated method produces stress-induced initial fabric anisotropy a in the sample, and the larger initial static shear stress ratio α leads to the larger fabric anisotropy a. However, with the increase of α, the mechanical coordination number MCN only shows a negligible decrease, which means that the initial static shear has almost no influence on the contacts density of the samples.
(3) A sharp decrease of MCN and increase of fabric anisotropy a are observed in the first cycle of loading. The sample with larger α exhibits slower decrease in MCN and faster growth in a. With the increase of α, the sample shares larger terminal value of MCN, even though its initial MCN is smaller. Similar pattern is also found in the evolution of a, where the sample has smaller initial a shares larger terminal value instead.

Acknowledgements. The authors would like to acknowledge the National Natural Science Foundation of China (Nos. 51988101, 51978613 and 51778573) and the Chinese Program of Introducing Talents of Discipline to University (the 111 Project, No. B18047) for the funding support.

References

1. Edwards, S.: The equations of stress in a granular material. Physica A Statist. Mech. Applic. **249**(1–4), 226–231 (1998)
2. Hyodo, M., Murata, H., Yasufuku, N., Fujii, T.: Undrained cyclic shear strength and residual shear strain of saturated sand by cyclic triaxial tests. Soils Found. **31**(3), 60–76 (1991)
3. Itasca C G Inc. Manual of particle flow code in 3-dimension. Minneapolis (2008)
4. Satake, M.: Fabric tensor in granular materials. In: IUTAM Symposium on Deformation and Failure of Granular Materials, Delft, pp. 63–68 (1982)
5. Sitharam, T., Vinod, J., Ravishankar, B.: Post-liquefaction undrained monotonic behaviour of sands: experiments and DEM simulations. Géotechnique **59**(9), 739–749 (2009)
6. Thornton, C.: Numerical simulations of deviatoric shear deformation of granular media. Géotechnique **50**(1), 43–53 (2000)
7. Vaid, Y.P., Chern, J.C.: Effect of static shear on resistance to liquefaction. Soils Found. **23**(1), 47–60 (1983)
8. Vaid, Y.P., Stedman, J.D., Sivathayalan, S.: Confining stress and static shear effects in cyclic liquefaction. Can. Geotech. J. **38**(3), 580–591 (2001)
9. Wang, R., Fu, P., Zhang, J.M., DEM Dafalias, Y.F.: study of fabric features governing undrained post-liquefaction shear deformation of sand. Acta Geotech. **11**(6), 1321–1337 (2016)
10. Yang, J., Sze, H.Y.: Cyclic behaviour and resistance of saturated sand under non-symmetrical loading conditions. Géotechnique **61**(1), 59–73 (2011)
11. Yang, Z.X., Yang, J., Wang, L.Z.: On the influence of inter-particle friction and dilatancy in granular materials: a numerical analysis. Granul. Matter **14**(3), 433–447 (2012)

12. Zhou, J., Shi, D.D., Jia, M.C., Yan, D.X.: Numerical simulation of mechanical response on sand under monotonic loading by Particle Flow Code. J. Tongji Univ. (Nat. Sci.) **35**(010), 1299–1304 (2007). (in Chinese)
13. Zhou, Y.-G., Xia, P., Ling, D.-S., Chen, Y.-M.: A liquefaction case study of gently sloping gravelly soil deposits in the near-fault region of the 2008 Mw7.9 Wenchuan earthquake. Bull. Earthq. Eng. **18**(14), 6181–6201 (2020)

S6: Special Session on Underground Structures

Seismic Behaviour of Urban Underground Structures in Liquefiable Soil

Emilio Bilotta(✉)

University of Napoli Federico II, Naples, Italy
emilio.bilotta@unina.it

Abstract. Soil liquefaction has been often one of the most significant causes of damage to aboveground structures in urban areas during recent earthquakes, e.g. 2012 Emilia (northern Italy), 2011 Tohoku Oki (Japan) and particularly 2011 Canterbury- Christchurch (New Zealand), where about half of the €25 billion loss was directly caused by such a phenomenon.

In some cases, sewer pipes or open-cut tunnels in liquefied deposits have been affected by floatation and large uplift. The current and future construction of relatively shallow and light underground structures in seismic regions may involve areas that are exposed to the risk of liquefaction thus increasing possible associated damages.

This paper investigates the behaviour of a tunnel during soil liquefaction, from an experimental and numerical point of view, focusing on the combined effects of soil liquefaction in urban areas, where underground structures are likely to interfere with buildings. Such an aspect is rather unexplored, and the research in this field may contribute to the performance-based design of urban underground facilities.

Keywords: Tunnels · Liquefaction · Seismic soil-structure interaction

1 Introduction

1.1 Seismic Soil-Structure Interaction of Underground Structures in Urban Areas

Wave propagation during earthquakes in a densely urbanized environment is more complex than in ideal free-field conditions due to the presence of structures above and below the ground surface. Different aboveground structures can be close enough to influence reciprocally each other during earthquakes (seismic structure-soil-structure interaction). The seismic responses of the underground space, with tunnels and deep excavations, may affect that of buildings founded either on shallow foundations or on piles and, reciprocally, inertial effects of the buildings on the underground response can be relevant. These effects are known as "seismic site-city interaction" [1] or "city effects" [2]: Lou et al. [3] presented a detailed State-of-Art of the research on structure–soil–structure dynamic interaction.

© The Author(s), under exclusive license to Springer Nature Switzerland AG 2022
L. Wang et al. (Eds.): PBD-IV 2022, GGEE 52, pp. 2265–2276, 2022.
https://doi.org/10.1007/978-3-031-11898-2_211

Only recently the research has focused on the dynamic interaction of shallow tunnels and underground structures with the urban environment. Numerical analyses conducted by Pitilakis & Tsinidis [4] showed that the presence of the above ground structures may result in an increase of both ovalisation and lining forces in a circular tunnel during seismic shaking, being this effect more significant for shallower tunnels. Centrifuge tests carried out by Dashti et al. [5] evidenced that the presence of an adjacent high-rise building reduces racking of an underground box structure, increasing at the same time the seismic lateral earth pressures. It is worth noticing that the interaction between shallow tunnels and building foundations during an earthquake, among other seismic soil-structure interaction (SSSI) problems in urban areas, has received attention in the last decade [5–9] due to the increasing use of underground urban space.

1.2 Urban Tunnels in Liquefiable Soil

The expansion of the built environment to liquefaction-prone areas requires additional issues to be considered in performance-based design. As far as the development of the underground space is concerned, the effects of significant soil shear strength and stiffness loss due to severe excess pore pressure build-up on shallow and light structures, such as urban tunnels and pipelines, has been investigated in several works.

Significant uplift of sewer pipes and railway tunnels has been observed in past events, e.g. [10, 11], and reproduced experimentally by physical modelling. 1-g shaking table tests have been carried out on models of box structures, semi-buried roads, sewer man-holes and pipes [12–15]. Centrifuge tests on reduced-scale models of tunnels [16–18] allowed to gather data on the influence of tunnel shapes, soil density, overburden, and groundwater level. These works have shown that the tunnel uplift is noteworthy affected by the width of the underground structure and the depth of the liquefied layer beneath the tunnel invert.

The seismic soil-structure interaction between a tunnel and a building in liquefiable soil has been investigated only very recently by numerical [19–21] and experimental studies [22]. In the following, some of the results from these studies are shown, focusing on tunnels with both circular and rectangular transverse section. The effect of the building presence on the internal forces in the tunnel lining are discussed. At the same time, it is shown how the building settlement and the tunnel uplift in the liquefiable soil may be reciprocally influenced, if the distance between the tunnel and the building foundation is small. This is a common feature of shallow tunnels for urban infrastructures, particularly in densely urbanised areas.

The results are also compared to those of very recent experimental and numerical works by Zhu et al. [23–25].

2 Circular Tunnels

2.1 Internal Forces in Transverse Section

A numerical exercise has been carried out by the finite element code Plaxis 2D [26] modelling a shallow circular tunnel (0.5 < C/D < 2) in a homogeneous saturated sand layer

($D_r = 40\%$) undergoing shaking in plane strain [19]. The constitutive model UBC3D-PLM [27, 28] has been adopted for sand and the relevant mechanical parameters are shown in Table 1. The analyses aimed to compute the increment of internal forces in the tunnel lining during shaking, including the effect of soil liquefaction. A set of time histories of acceleration was used as input at the base of the model, with different characteristics and intensity (Table 2). The tunnel diameter D is equal to 6 m and the lining is made of reinforced concrete lining (EA = 10.5E^6 kN/m; EI = 78.75E^3 kNm^2/m).

Table 1. Soil model parameters [28]

φ'_{cv}	φ'_p	c'(kPa)	K^e_B	K^e_G	K^p_G	m_e	n_e	n_p	R_f	$N_{1,60}$	f_{hard}	f_{post}
32°	35.5°	0.01	300	360	180	0.5	0.5	0.4	0.93	7.4	1.6	1

Table 2. Input signals.

ID			M_w (g)	PGA (cm/s)	PGV (s)	T_p (s)	T_m (m/s)	AI (s)	$D_{5\text{-}95}$ (m/s)	CAV (g)
1	Avey	22/06/06	6.5	0.5	24.65	0.2	0.29	2.24	5.86	8.63
2	Friuli	06/05/76	6.5	0.35	23.45	0.26	0.39	0.8	4.3	5.91
3	Norcia	30/10/16	6.5	0.78	74.37	0.2	0.32	4.74	8.92	14.99
4	Northridge	17/01/94	6.7	0.68	62.19	0.26	0.54	3.93	9.06	15.51
5	South Iceland	21/06/00	6.4	0.36	55.52	0.4	0.64	1.24	3.9	5.96
6	Tirana	09/01/88	5.9	0.33	15.26	0.12	0.29	0.62	6.15	4.37

The results indicate that, as far as liquefaction occurs, a liquefied soil layer forms that acts as an isolator, reducing the amplitude of ground acceleration at surface compared to the base. Such a shallow layer of liquefied soil, that forms in the tunnel cover, may eventually reach the tunnel depth. The layer thickness increases with the intensity of the ground motion. Correspondingly, the pore pressure build-up, associated with changes in effective stresses, affects the distribution of internal forces in the tunnel lining. A larger change is induced in the hoop force when soil liquefaction approaches. The effect on bending moment depends on the position in the lining. Figure 1 shows the calculated maximum increments of normalised internal forces at the end of shaking, $\Delta N/(\tau_{eq} \cdot D)$ and $\Delta M/(\tau_{eq} \cdot D^2)$, as a function of the Arias intensity, AI (Eq. 1). Consistently with the simplified methods to compute increments of internal forces during shaking [29, 30], the equivalent shear stress, τ_e, was computed from the horizontal equilibrium of a deformable soil column from the surface to the tunnel roof, representing the effect of the soil inertia forces acting on the structure (Fig. 1).

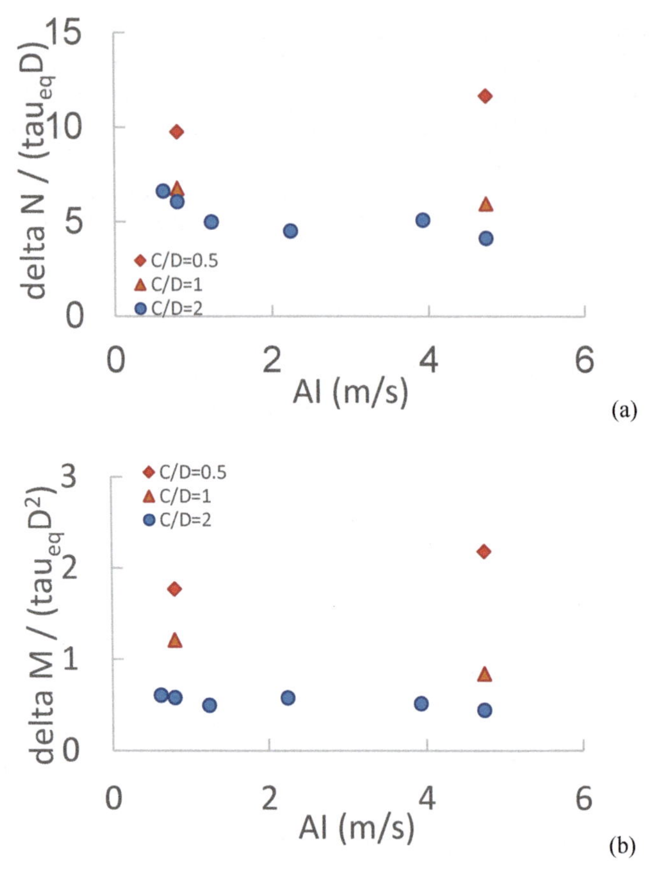

Fig. 1. Dimensionless maximum increments of hoop force (a) and bending moment (b) in the tunnel lining.

These results show that once liquefaction occurs, the dimensionless values of internal forces are generally independent on the ground motion intensity and larger for shallower cover. This indicates that the inertial shear stresses acting on the tunnel lining are not the only responsible for the increment of internal forces. Zhu et al. [24] have suggested that in liquefied ground, in addition to the dynamic increment of shear stress, the dynamic increment of total horizontal stress, caused by excess pore pressure build-up is responsible for part of the increment of internal forces in a circular tunnel lining ($C/D \approx 1$). In the analyses above, the thickness of ground that liquefies ($r_u \geq 0.95$) or reaches a significant pore pressure build-up ($r_u \geq 0.8$) increases with the ground motion intensity and is never deeper than 8 m. This implies that only the shallower tunnels ($C/D = 0.5$) are generally immersed in liquefied soil, and sometimes those at intermediate depth ($C/D = 1$). Hence, the changes of internal forces in the tunnel lining for the same ground motion intensity depend on the tunnel depth. Full interaction dynamic analyses, able to predict excess pore pressure build-up and soil liquefaction, can capture such a behavior.

2.2 Uplift

The tunnel uplift caused by soil liquefaction can be considered as an EDP (Engineering Demand Parameter) of interest. The values calculated in the abovementioned numerical analyses with C/D = 2 are plotted in Fig. 2 as a function of two IMs (Intensity Measures), that are the Arias Intensity, AI [31] and the Cumulative Absolute Velocity, CAV [32]:

$$AI = \frac{\pi}{2g} \int_0^{t_d} a(t)^2 dt \tag{1}$$

$$CAV = \int_0^{t_d} |a(t)| dt \tag{2}$$

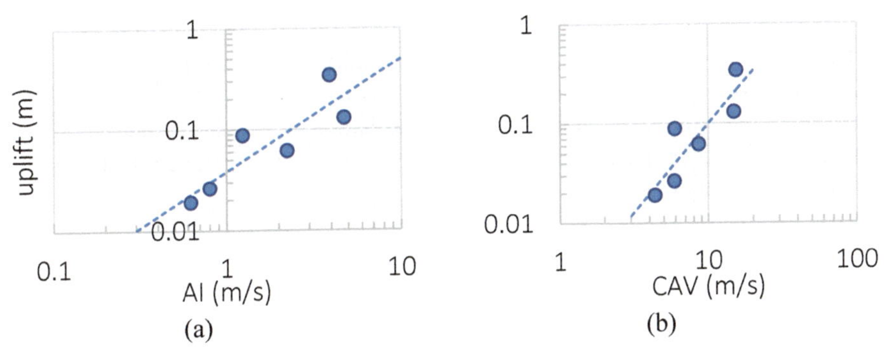

Fig. 2. Tunnel uplift as a function of Arias Intensity (a) and Cumulative Absolute Velocity (b).

As shown in the figure, a typical bi-logarithmic functional form [33] can be used to fit the data:

$$\ln(EDP) = A + B\ln(IM) \tag{3}$$

The values of the parameters A and B obviously depend on the selected IM (AI or CAV) and on the C/D ratio. Much larger uplift can be computed for very shallow tunnels, compared to the deeper ones shown in Fig. 2 (i.e. C/D = 2). The centrifuge tests carried out by Chian et al. [18] have shown that higher uplift for shallower buried structure is justified by the lower shear resistance and overburden by the overlying ground, due to the small cover between the tunnel axis and the ground surface.

2.3 Interaction with a Building

The numerical model described above was slightly modified to include a simple framed structure mimicking a two-storey building as shown in Fig. 3. It consists of a two-floors and a basement and lies to one side of the tunnel. The frame was modelled using linear elastic beam elements: EI = $1.6 \cdot 10^5$ kNm^2/m and EA = $1.2 \cdot 10^7$ kN/m for the basement; EI = $6.75 \cdot 10^4$ kNm^2/m and EA = 1.6·105 kN/m for the rest. The beam elements have a unit mass that also considers the presence of floors and walls.

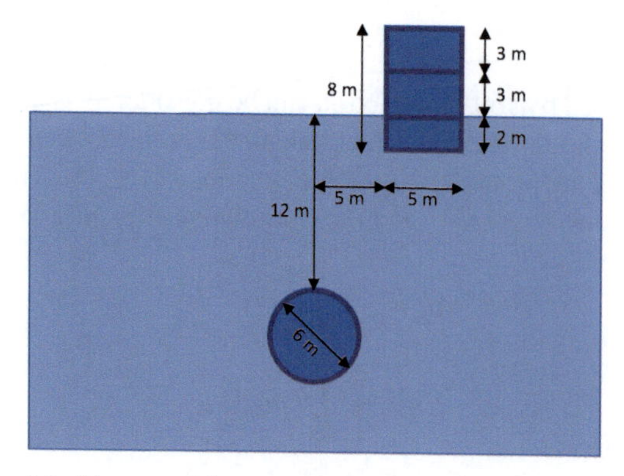

Fig. 3. Geometrical layout of the numerical model with simple framed structure.

The results have shown that the presence of the building does not change hoop forces, N, induced by pore-pressure build-up, while the bending moment, M, is generally larger than without the building, as a less uniform distribution of stresses is induced around the tunnel [19]. Furthermore, a minor effect of the presence of a building on the amount of tunnel uplift is observed.

The effect of the circular tunnel on the building movement is negligible when the cover to diameter ratio is large enough (C/D = 2). The building settlement is mainly affected by soil liquefaction occurring around its foundations. However, a shallower tunnel may interact with the building located at its side, producing larger tilt. This is shown in Fig. 4 for the case of input signal '3'. The tunnel diameter has been kept equal (D = 6 m). The plot shows a larger difference for C/D = 0.5, compared to C/D = 1 and C/D = 2, between the final values of settlements of the two side of the building (points G and I), indicating larger tilt.

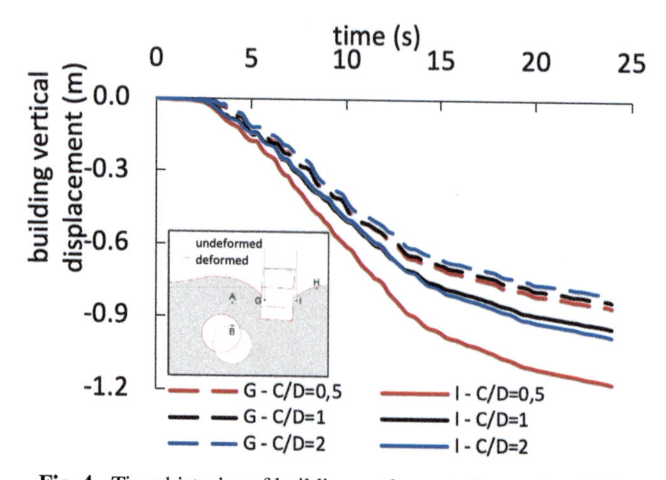

Fig. 4. Time histories of building settlements (input signal '3').

3 Rectangular Tunnels

3.1 Interaction with a Building – Numerical Evidence

In addition to the circular ones, a typical section of urban metro tunnel is that of a rectangular box that can accommodate two separate platforms. This is often the case of underground stations in urban areas, with limited cover.

As noticed in the Sect. 2, for circular tunnels, the shallower the tunnel the larger the uplift associated with the mobility of the surrounding liquefied soil. A box-type tunnel with a cover upon tunnel height ratio, $C/h_T = 1$ has been analyzed using Plaxis 2D [26] to investigate the tunnel-structure interaction in liquefiable sand [21]. The constitutive model PM4Sand [34] has been adopted for sand and the relevant mechanical parameters are shown in Table 3. The layout is shown in Fig. 5.

Table 3. Soil model parameters [34]

D_R	G_0	hp_0	p_A (kPa)	e_{max}	e_{min}	n_b	n_d	φ'_c	ν	Q	R
47%	594	0.1	101.3	0.923	0.574	0.5	0.1	33°	0.3	8	1.2

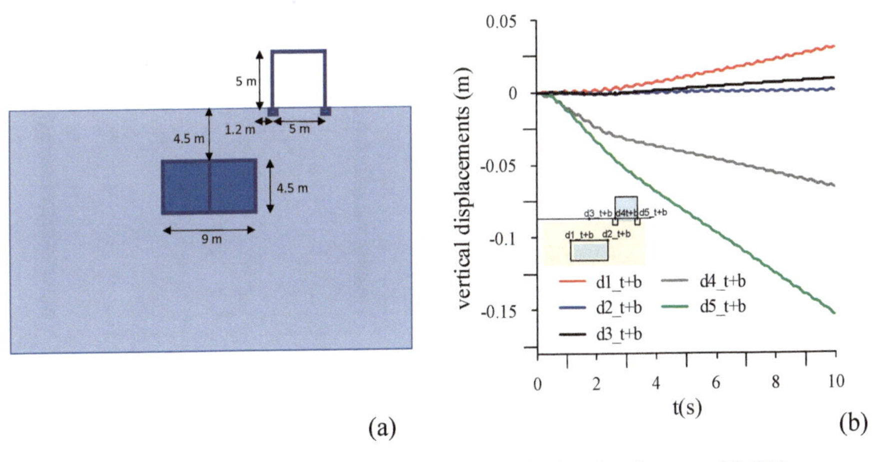

Fig. 5. Layout of numerical analyses (a) and calculated settlements (b) [21].

The analysis has been repeated without the building and without the tunnel, to show the reciprocal influence between the two structures. The results show that the presence of the tunnel reduces the average building settlement and produces tilt (d4 and d5 in Fig. 5b). On the other hand, the presence of the building creates a non-symmetrical displacement field on one side of the tunnel, that rotates consequently (d1 and d2 in Fig. 5b). A larger relative distance between the tunnel wall and the building basement upon the tunnel cover would affect the results [19].

These conclusions have been confirmed and extended in a recent study by Zhu et al. (2021), that carried out 3D FE analyses of the interaction between a box-type tunnel (rectangular section) and a single-degree-of-freedom aboveground structure. The latter has been located either directly above the tunnel or to the right, at various distances, indicating that in the first case the rotation of the tunnel is almost null, reaching a peak for at certain distance of the building from the tunnel, then reducing. The tunnel uplift is lowest when the building is directly above the tunnel. The study also highlighted the effect of the tunnel depth: the effect of the building on the tunnel movements is very small when the ratio $C/h_T > 2$.

3.2 Interaction with a Building – Experimental Evidence

Experimental evidence on the reciprocal interaction between a box-type tunnel and a model building has been gathered by Miranda et al. [22]. A series of centrifuge tests have been carried out at the Schofield Centre of the University of Cambridge (UK) on a reduced scale model of a rectangular tunnel embedded in a liquefiable layer of sand, with and without a model building founded in proximity. These tests intended to reproduce a similar layout to that described in §3.1, as a typical case of a shallow tunnel in urban environment. In the model T1 only the tunnel was modelled, while in the model T2 a frame representing a model building has been added. Figure 6a shows a photo of the centrifuge model and Fig. 6b the experimental layout of the model T2. The dimensions are shown at the prototype scale.

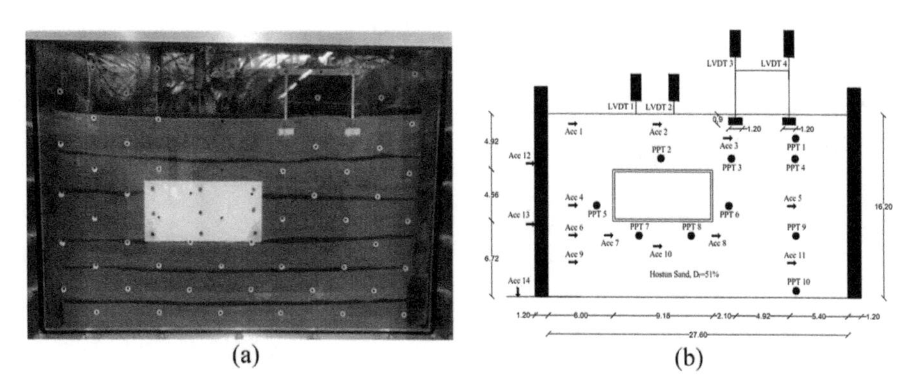

Fig. 6. Photo (a) and layout (b) of the centrifuge test #2 [22].

The model ground is a layer of saturated Hostun sand ($D_r = 45\%–50\%$). An embedment ratio $C/h_T = 1.1$ has been adopted. The model size is 60 times smaller than the actual prototype, since it has been tested at an increased g-level of 60g.

A series of four pseudo-harmonic time histories of acceleration has been applied at the model base, with PGA increasing from 0.06g to 0.34g in T1 and from 0.07g to 0.42g in T2 (corresponding to AI increasing from 0.16 cm/s to 7.8 cm/s and from 0.14 cm/s to 8.13 cm/s, respectively).

During shaking, time histories of acceleration and pore pressure have been recorded along vertical and horizontal arrays. Settlements of a few points at the ground surface have been monitored by using LVDTs (Linear Variable Differential Transformers). Particle Image Velocimetry (PIV) allowed deep ground movements to be measured from the front window of the model. Displacements of the sway frame (settlement, horizontal displacement, and tilt) have been also monitored.

During the first two ground shakings the displacements were negligible in both models and very small excess pore pressures were measured. Soil liquefaction was only achieved in the stronger shaking events, that is the third and the fourth ones. These have caused noticeable settlements of the structure and uplift of the tunnel.

The full set of results will not be shown here as they are described in [22]. Nevertheless, a comparison among the displacement field measured in the ground around the tunnel (T1) and between the tunnel and the building foundations (T2) is depicted in Fig. 7 (negative values indicate heave).

Contour maps show that a much less symmetric ground displacement field occurs in model T2 than in T1. In the model with the tunnel only (T1), the ground at the tunnel side settles, and the ground in tunnel cover is pushed upwards due to the tunnel uplift. In model T2, the ground settlements extend to part of the tunnel cover (on the right-hand side) and an interaction zone between the left building foundation and the right side of the tunnel appears.

The distribution of ground displacements in the figure shows that in T2 both the tunnel and the structure are subjected to rotation: the tunnel rotates clockwise, and the structure rotates anti-clockwise, mutually interacting.

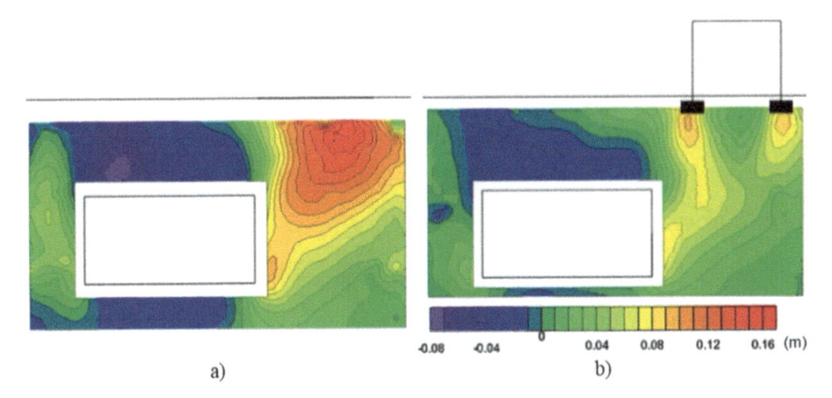

Fig. 7. Settlements at the end of shakings no. 3 in model T1 (a) and T2 (b) [22].

4 Remarks

Experimental and numerical evidence has provided a framework to identify a few issues that need to be addressed in the performance-based design of a tunnel in urban area when the subsoil is subjected to cyclic liquefaction.

Depending on the tunnel depth below the building foundations, the tunnel lining may be affected by changes of the internal forces that are driven not only by soil inertial actions, as usual occurs under seismic shaking, but also by the excess pore pressure build-up in the ground, that modifies the distribution of total stresses around the tunnel.

This behavior is not captured by the pseudo-static approach usually adopted in design; hence full dynamic analyses may be needed.

The tunnel uplift caused by large excess pore pressures may deserve attention during seismic events with high intensity and when the tunnel is shallow (up to C/D = 1). However, for circular tunnels the influence of close buildings on the tunnel movements is negligible, unless the tunnel is very shallow (C/D = 0.5) and close to the building foundation.

The liquefaction-induced movements of rectangular tunnels, in a densely built urban subsoil, may be influenced by the presence of buildings. When a building is founded at a tunnel side, the tunnel and the building rotate towards each other: in this case the tunnel uplift is larger on the side where there is no structure, thus increasing its rotation compared to a free-field condition.

It is worth noting that a tunnel is a long structure that may run across different ground conditions and it may interact with different building foundations over a short span. Spatial distribution of liquefaction phenomena in a complex underground environment can influence the tunnel performance. These aspects were not addressed in this work, as they need a proper consideration of three-dimensional effects and ground heterogeneity. More research is certainly required in this direction to address the structural effects of possible differential displacements induced by soil liquefaction along the tunnel axis.

Finally, the results of the studies suggest that if the reciprocal interference between the tunnel and the building needs to be mitigated, this might be achieved either by reducing the soil mobility around the tunnel (thus limiting the average tunnel uplift) or by disconnecting the tunnel movements within the liquefied soil from those of the building (thus limiting tunnel rotation). Different design strategies may be required in the two cases.

5 Conclusions

The paper has discussed the effect of tunnel-building interaction in liquefiable soil on the behavior of shallow tunnels in urbanized areas.

Internal forces in the structural elements of the tunnel lining are affected by the building presence, depending on several factors: tunnel depth, distance from the building, ground motion intensity, excess pore pressure build-up. Based on numerical results and experimental evidence in literature, a few indications have been provided in the paper to identify the extent at which such issue should be considered in tunnel design.

Tunnel buoyancy in liquefiable ground is relevant for very shallow depth and strong earthquakes, when the kinematic of box-type tunnels is altered by the building presence. At a small distance, such as it occurs in dense urban space, a building at the tunnel side may introduce asymmetry in the ground displacements and produce tunnel rotation. Differential movements associated with uneven distribution of buildings along the tunnels and their effect on the tunnel performance should be assessed in tunnel design. When necessary, proper mitigation actions should be undertaken.

Larger research efforts should be spent to investigate the effects of ground heterogeneity and three-dimensional layout on underground conditions, that have not been considered in this work.

References

1. Kham, M., Semblat, J.-F., Bard, P.-Y., Dangla, P.: Seismic site-city interaction: main governing phenomena through simplified numerical models. Bull. Seismol. Soc. Am. **96**(5), 1934–1951 (2006)
2. Ghergu, M., Ionescu, I.R.: Structure-soil-structure coupling in seismic excitation and "city effect." Int. Eng. Sci. **47**(3), 342–354 (2009)
3. Lou, M., Wang, H., Chen, X., Zhai, Y.: Structure–soil–structure interaction: literature review. Soil Dyn. Earthq. Eng. **31**, 1724–1731 (2011)
4. Pitilakis, K., Tsinidis, G.: Performance and seismic design of underground structures. In: Maugeri, M., Soccodato, C. (eds.) Earthquake Geotechnical Engineering Design. GGEE, vol. 28, pp. 279–340. Springer, Cham (2014). https://doi.org/10.1007/978-3-319-03182-8_11
5. Dashti, S., Hashash, Y.M.A., Gillis, K., Musgrove, M., Walker, M.: Development of dynamic centrifuge models of underground structures near tall buildings. Soil Dyn. Earthq. Eng. **86**, 89–105 (2016)
6. Tsinidis, G., Pitilakis, K., Madabhushi, G., Heron, C.: Dynamic response of flexible square tunnels: centrifuge testing and validation of existing design methodologies. Geotechnique **65**(5), 401–417 (2015)
7. Hashash, Y.M.A., Musgrove, M., Dashti, S., Cheng, P.: Seismic performance evaluation of underground structures – past practice and future trends. In: Proceedings of the PBD-III Performance Based Design in Earthquake Geotechnical Engineering, Vancouver, p. 305 (2017)
8. Abate, G., Massimino, M.R.: Numerical modelling of the seismic response of a tunnel–soil–aboveground building system in Catania (Italy). Bull. Earthq. Eng. **15**(1), 469–491 (2016). https://doi.org/10.1007/s10518-016-9973-9
9. Lončarević, D., Tsinidis, G., Pitilakis, D., Bilotta, E., Silvestri, F.: Numerical study of dynamic structure-soil-tunnel interaction for a case of Thessaloniki Metro. In: Proceedings of VII International Conference on Earthquake Geotechnical Engineering, 7ICEGE 2019, 17–20 June (2019)
10. Koseki, J., Matsuo, O., Ninomiya, Y., Yoshida, T.: Uplift of sewer manholes during the 1993 Kushiro-Oki earthquake. Soils Found. **37**(1), 109–121 (1997)
11. Yasuda, S., Kiku, H.: Uplift of sewage manholes and pipes during the 2004 Niigataken-Chuetsu earthquake. Soils Found. **6**(46), 885–894 (2006)
12. Koseki, J., Matsuo, O., Koga, Y.: Uplift behavior of underground structures caused by liquefaction of surrounding soil during earthquake. Soils Found. **37**(1), 97–108 (1997)
13. Otsubo, M., Towhata, I., Taeseri, D., Cauvin, B., Hayashida, T.: Development of structural reinforcement of existing underground lifeline for mitigation of liquefaction damage. In: Geotechnics of Roads and Railways: Proceedings of the XV Danube-European Conference on Geotechnical Engineering, Wien, vol. 1, pp. 119–125 (2014)
14. Watanabe, K., Sawada, R., Koseki, J.: Uplift mechanism of open-cut tunnel in liquefied ground and simplified method to evaluate the stability against uplifting. Soils Found. **56**(3), 412–426 (2016)
15. Castiglia, M., de Magistris, F.S., Onori, F., Koseki, J.: Response of buried pipelines to repeated shaking in liquefiable soils through model tests. Soil Dyn. Earthq. Eng. **143**, 106629 (2021)
16. Chou, J.C., Kutter, B.L., Travasarou, T., Chacko, J.M.: Centrifuge modeling of seismically induced uplift for the BART transbay tube. J. Geotech. Geoenviron. **137**(8), 754–765 (2010)
17. Chian, S.C., Madabhushi, S.P.G.: Effect of buried depth and diameter on uplift of underground structures in liquefied soils. Soil. Dyn. Earthq. Eng. **41**, 181–190 (2012)
18. Chian, S.C., Tokimatsu, K., Madabhushi, S.P.G.: Soil liquefaction- induced uplift of underground structures: physical and numerical modeling. J. Geotech. Geoenviron. Eng. **140**(10), 04014057 (2014)

19. Bilotta, E.: Modelling tunnel behaviour under seismic actions: an integrated approach. In: Physical Modelling in Geotechnics, vol. 1, pp. 3–20. CRC Press (2018)
20. Maddaluno, L., Stanzione, C., Nappa, V., Bilotta, E.: A numerical study on tunnel-building interaction in liquefiable soil. In: Earthquake Geotechnical Engineering for Protection and Development of Environment and Constructions - Proceedings of the 7th International Conference on Earthquake Geotechnical Engineering, pp. 3708–3715. CRC Press (2019)
21. Miranda, G., Nappa, V., Bilotta, E.: Preliminary numerical simulation of centrifuge tests on tunnel-building interaction in liquefiable soil. In: Calvetti, F., Cotecchia, F., Galli, A., Jommi, C. (eds.) Geotechnical Research for Land Protection and Development. CNRIG 2019. Lecture Notes in Civil Engineering, vol. 40. Springer, Cham (2020). https://doi.org/10.1007/978-3-030-21359-6_62
22. Miranda, G., Nappa, V., Bilotta, E., Haigh, S.K., Madabhushi, S.P.G.: Centrifuge tests on tunnel-building interaction in liquefiable soil. In: Geotechnical Aspects of Underground Construction in Soft Ground – Proceedings of the 10th International Symposium on Geotechnical Aspects of Underground Construction in Soft Ground, IS-CAMBRIDGE 2022, pp. 613–619 (2021)
23. Zhu, T., Hu, J., Zhang, Z., Zhang, J.M., Wang, R.: Centrifuge shaking table tests on precast underground structure–superstructure system in liquefiable ground. J. Geotech. Geoenviron. Eng. (2021). https://doi.org/10.1061/(ASCE)GT.1943-5606.0002549
24. Zhu, T., Wang, R., Zhang, J.-M.: Evaluation of various seismic response analysis methods for underground structures in saturated sand. Tunn. Undergr. Space Technol. **110**, 103803 (2021)
25. Zhu, T., Wang, R., Zhang, J.-M.: Effect of nearby ground structures on the seismic response of underground structures in saturated sand. Soil Dyn. Earthq. Eng. **146**, 106756 (2021)
26. Brinkgreve, R.B.J., Kumaeswamy, S., Swolfs, W.M.: PLAXIS2016 User's manual (2016). https://www.plaxis.com/kbtag/manuals/
27. Beaty, M., Byrne, P.: An effective stress model for predicting liquefaction behavior of sand. In: Dakoulas, P., Yegian, M., Holtz, R.D. (eds.) Geotechnical Earthquake Engineering and Soil Dynamics III, vol. 75, no. 1, pp. 766–777. ASCE Geotechnical Special Publication (1998)
28. Galavi, V., Petalas, A., Brinkgreve, R.B.J.: Finite element modelling of seismic liquefaction in soils. Geotech. Eng. J. SEAGS & AGSSEA **44**(3), 55–64 (2013)
29. Hashash, Y.M.A., Hooka, J.J., Schmidt, B., Yao, J.I.-C.: Seismic design and analysis of underground structures. Tunn. Undergr. Space Technol. **16**, 247–293 (2001)
30. Tsinidis, G., et al.: Seismic behaviour of tunnels: From experiments to analysis. Tunn. Undergr. Space Technol. **99**, 103334 (2021)
31. Arias, A.: A measure of earthquake intensity. In: Hansen, R.J. (ed.) Seismic Design for Nuclear Power Plants, pp. 438–483. MIT Press, Cambridge, Massachusetts (1970)
32. Electric Power Research Institute (EPRI): A criterion for determining exceedances of the operating basis earthquake, EPRI Report NP-5930, Electric Power Research Institute, Palo Alto, California (1988)
33. Bullock, Z., Karimi, Z., Dashti, S., Porter, K., Liel, A.B., Franke, K.W.: A physics-informed semi-empirical probabilistic model for the settlement of shallow-founded structures on liquefiable ground. Géotechnique **69**(5), 406–419 (2019)
34. Boulanger, R.W., Ziotopoulou, K.: Formulation of a sand plasticity plane-strain model for earthquake engineering applications. Soil Dyn. Earthq. Eng. **53**, 254–267 (2013)

Parametric Analyses of Urban Metro Tunnels Subject to Bedrock Dislocation of a Strike-Slip Fault

Zhanpeng Gan, Junbo Xia, Jun Du, and Yin Cheng(✉)

School of Civil Engineering, Southwest Jiaotong University, Chengdu 610031, China
yin.cheng@swjtu.edu.cn

Abstract. As an essential part of lifeline engineering, the subway tunnel is the hub of traffic and transportation. With the development of the economy, the social demand for it is increasing. Although faults should be avoided as much as possible in the process of route selection for tunnels, the tunnel, as a long linear structure, in some cities will be inevitably crossed by faults in the construction. In this paper, the seismic response of a shallow buried subway tunnel crossing an active strike-slip fault is analyzed subject to bedrock fault dislocations. Based on the 3D modeling of the tunnel using the finite element software ABAQUS, the displacement and stress of the tunnel lining structure are calculated with the change of five factors, including fault-tunnel crossing angle, bedrock overburden thickness, tunnel buried depth, lining thickness, and soil properties. The influence of the various factors on the structural seismic response of the tunnel under the bedrock dislocations is discussed. The analysis results can be used as a reference for performance-based design for tunnels crossing active faults.

Keywords: Tunnels · Crossing fault · Strike-slip fault · Bedrock fault dislocation · Structural response

1 Introduction

The past earthquake disaster showed that fault displacement could cause significant damage to tunnels that are crossed by the fault. Many studies have been carried out on the seismic performance of tunnels and have achieved meaningful results [e.g. 1–3]. The research methods mainly include experimental methods [1, 4, 5] and numerical simulations [6–8], analyzing the failure mechanism and law of the tunnel structure induced by fault displacement with the changing of fracture zone width, surrounding rock condition, fault type, dislocation type and distance, tunnel diameter, deformation joint and other factors [9–14]. However, previous studies rarely considered the structural response of urban subway shallow tunnels subject to the bedrock dislocation overburden by overlaying soil under different influencing factors.

This paper carried out a parametric analysis of urban metro tunnels subject to bedrock dislocation of a strike-slip fault to investigate the lining response of the tunnel with different factors, such as fault-tunnel crossing angle, bedrock overburden thickness,

© The Author(s), under exclusive license to Springer Nature Switzerland AG 2022
L. Wang et al. (Eds.): PBD-IV 2022, GGEE 52, pp. 2277–2284, 2022.
https://doi.org/10.1007/978-3-031-11898-2_212

tunnel buried depth, lining thickness, and soil properties. This parameter analysis can provide a reference to the seismic design and assessment of tunnel crossing faults.

2 Numerical Modeling

This study uses ABQAQUS software to establish a numerical calculation model for a subway tunnel crossing a strike-slip fault, as shown in Fig. 1. It consists of three parts: lining, soil, and bedrock. The length of the model is 400 m, and the width is 60 m. The tunnel lining has a circular cross-section, the outer diameter is 6 m, and the lining thickness is 300 mm.

The tunnel lining is modeled by shell elements; the soil and bedrock are modeled by solid elements; the contact is modeled with frictional face-to-face contact elements. The soil property is modeled using a Mohr-Coulomb constitutive model. In order to better simulate the deformation and internal force changes of the tunnel lining under fault action, the tunnel lining and bedrock are modeled using the linear elastic constitutive model. The lining is assumed to be a continuous and homogenous body; the soil layer is simplified to be homogeneous. The parameter values of soil, bedrock, and tunnel lining are shown in Table 1.

The numerical simulation is mainly divided into two steps: (1) balancing the ground stress under gravity; (2) applying a constant displacement on one side of the bottom surface of the model to simulate the fault dislocation process (Fig. 2).

Table 1. Parameter values of materials.

Name	Density (g/cm^3)	Elastic modulus (MPa)	Poisson ratio	Cohesive force (kPa)	Internal friction angle (°)
Silty clay	2	20	0.28	27	16
Sand	1.89	33	0.28	0	32
Lining	2.4	27600	0.2		
Bed rock	2.7	15000	0.3		

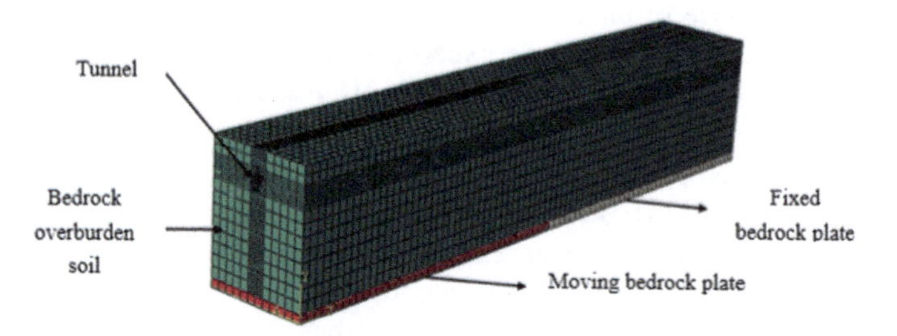

Fig. 1. Schematic diagram of simulation model

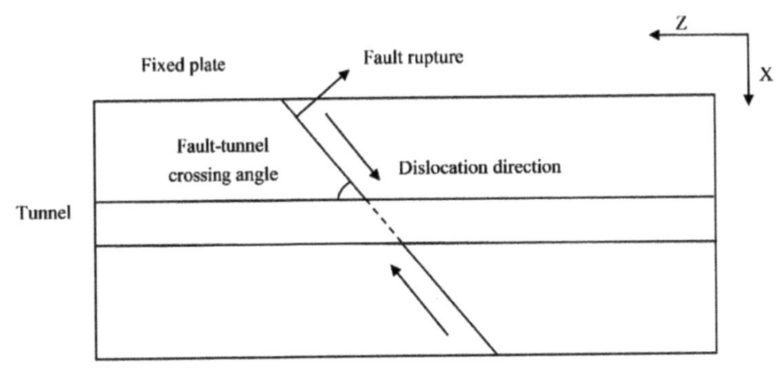

Fig. 2. Schematic diagram of strike-slip fault dislocation

In order to study the five influencing factors on the structural response of the tunnel crossing faults, five calculation cases are designed to carry out the numerical simulations, shown in Table 2.

Table 2. Parameter settings for the considered working conditions.

Working conditions	Crossing angle (°)	Tunnel buried depth (m)	Lining thickness (mm)	Bedrock overburden thickness (m)	Soil properties
1	45, 60, 90, 120	10	300	60	Silty clay
2	90	10	300	30, 50, 60, 70	Silty clay
3	90	5, 10, 15, 20	300	60	Silty clay
4	90	10	300, 325, 350, 375	60	Silty clay
5	90	10	300	60	Silty clay, sand

3 Parametric Analyses of Influencing Factors

Under the five working conditions in Table 2, the structural responses along the axial direction of the tunnel lining, including the displacement value and the maximum and minimum principal stress values, subject to the bedrock dislocation up to 3 m analyzed. The influence of five parameters on the structural responses is studied in this section.

3.1 Influence of Tunnel-Fault Crossing Angles

Figure 3 shows the structural response of the right spandrel of the tunnel lining subject to the 3 m dislocation of fault with crossing angles of 45°, 60°, 90°, and 120°. It can be observed that the horizontal displacement curve of the lining in X displacement at different crossing angles follows the same trend along the axial tunnel. The maximum principal stress and the minimum principal stress first increase and then decrease with increasing crossing angle. When the crossing angle is at 45°, the peak values of maximum and minimum principal stresses are relatively large, while their peak values are relatively small when the crossing angle is at 90°. It is demonstrated that there is an unfavorable angle for the tunnel to cross the fault for which the seismic response of the tunnel lining is more severely affected by the bedrock dislocation.

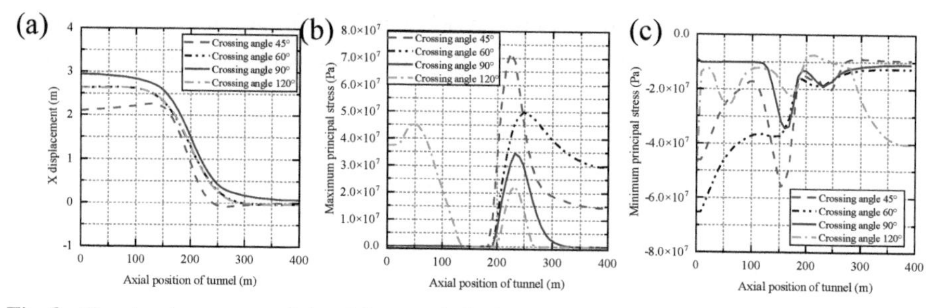

Fig. 3. Structural response of the right spandrel of the tunnel for different tunnel-fault crossing angles: (a) displacement, (b) maximum and (c) minimum principal stress

3.2 The Influence of Bedrock Overburden Thickness

Figure 4 shows the structural response of the right spandrel of the tunnel lining subject to the 3 m dislocation of fault with bedrock overburden thickness of 30 m, 50 m, 60 m, and 70 m. It is observed that the deformation of the tunnel lining can be effectively decreasing with increasing the thickness of the bedrock overburden. For different bedrock overburden thicknesses, the maximum and minimum principal stress in the area near the dislocation plane varies greatly, and they tend to be stable away from the dislocation plane. The maximum and minimum principal stress peaks show a decreasing trend as the bedrock overburden thickness increases.

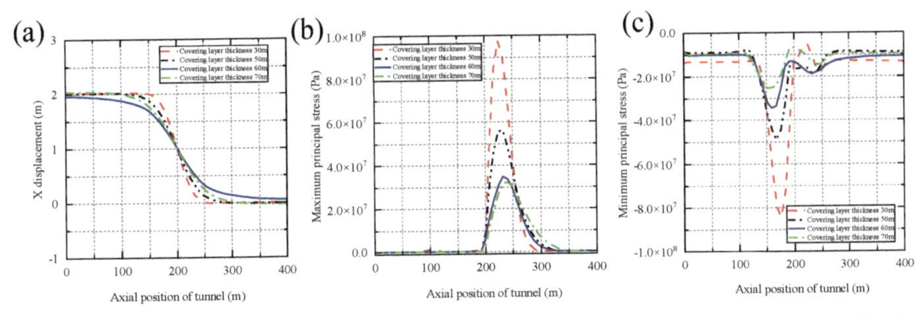

Fig. 4. Structural response of the right spandrel of the tunnel under different bedrock overburden thickness: (a) displacement, (b) maximum and (c) minimum principal stress

3.3 Influence of Tunnel Buried Depth

Figure 5 illustrates the structural response of the right spandrel of the tunnel lining subject to the 3 m dislocation of fault with tunnel buried depth of 5 m, 10 m, 15 m, and 20 m. It is observed that, given a certain bedrock overburden thickness, with the increase of the buried depth of the tunnel, the influencing range of the tunnel lining increases. Both the maximum and the minimum principal stress show an increasing trend as the tunnel buried depth increases. The maximum bending moment value increases as the buried depth of the tunnel increases.

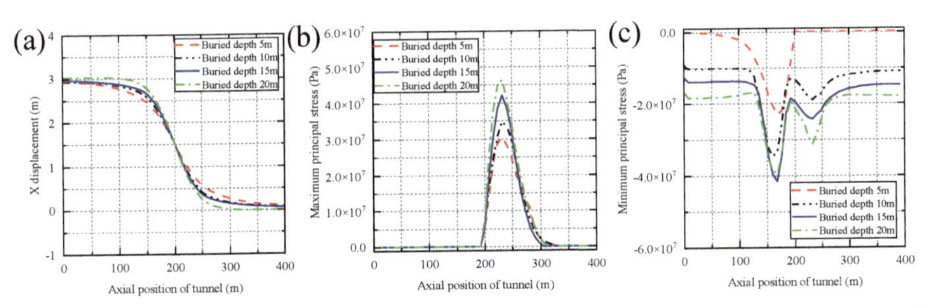

Fig. 5. Structural response of the right spandrel of the tunnel under different tunnel depths: (a) displacement, (b) maximum and (c) minimum principal stress

3.4 Influence of Lining Thickness

Figure 6 illustrates the structural response of the right spandrel of the tunnel lining subject to the 3 m bedrock dislocation with lining thickness of 300 mm, 325 mm, 350 mm, and 375 mm. When the outside diameter of the lining section is constant, the lining thickness has almost no effect on the deformation of the tunnel lining. The maximum principal stress is greater than the minimum principal stress, indicating that the tunnel lining is more susceptible to tensile failure. The peak values of maximum and minimum principal stress decrease with increasing the tunnel lining thickness.

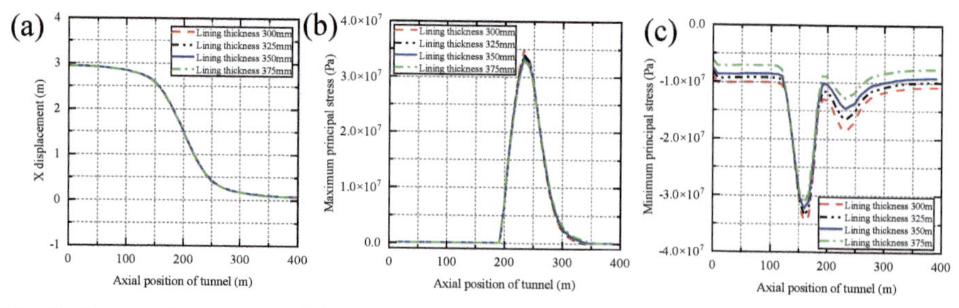

Fig. 6. Structural response of the right spandrel of tunnel subject to bedrock dislocation for different lining thicknesses: (a) displacement, (b) maximum and (c) minimum principal stress

3.5 Influence of Soil Properties

Figure 7 shows the structural response of the right spandrel of the tunnel lining subject to the 3 m bedrock dislocation with buried silty clay and sand soil. It is illustrated that the influencing range of silty clay on the tunnel lining is more extensive than that of sand. The maximum and minimum principal stresses of the tunnel lining structure decrease as the buried soil becomes softer. Based on the above analysis, it can be seen that the internal force and deformation of the tunnel lining structure buried in the soft soil layer are small under the action of the fault dislocation, so the tunnel lining structure buried in the soft soil layer can be better resist the fault dislocation.

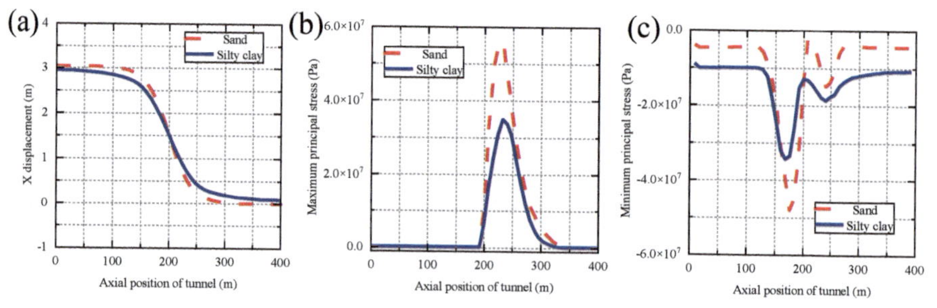

Fig. 7. Structural response of right arch waist for different soil layer properties: (a) displacement, (b) maximum and (c) minimum principal stress

4 Conclusion

In this paper, the finite element model of subway tunnel crossing strike-slip fault is established. By considering the influence of various factors, such as including fault-tunnel crossing angle, bedrock overburden thickness, tunnel buried depth, lining thickness, and soil properties, the mechanical characteristics of the tunnel under the strike-slip fault dislocation are analyzed. The following conclusions are drawn.

(1) When the fault-tunnel crossing angle is at 90°, the internal force response of the tunnel lining is relatively more minor compared to other crossing angles.
(2) For a constant tunnel burial depth, the thicker the bedrock overburden thickness is, the smaller the influence of bed dislocations on the internal force of the tunnel lining.
(3) For a constant bedrock overburden thickness, the smaller the tunnel burial depth, the less the internal force of the tunnel lining structure is affected by the bedrock dislocation.
(4) When the outside diameter of the tunnel is constant, the maximum and minimum principal stresses decrease as the lining thickness increases, but the reduction effect is not very obvious.
(5) The shallow tunnel buried in silty clay is less influenced by the fault dislocation than that buried in the sand.

References

1. Kontogianni, V.A., Stiros, S.C.: Earthquakes and seismic faulting: effects on tunnels. Turk. J. Earth Sci. **12**(1), 153–156 (2003)
2. Lin, M.L., et al.: Response of soil and a submerged tunnel during a thrust fault offset based on model experiment and numerical analysis. In: Pressure Vessels and Piping Conference, ASME 2005, pp. 313–316. American Society of Mechanical Engineers Digital Collection (2005)
3. Kiureghian, A.D., Ang, H.S.: A fault-rupture model for seismic risk analysis. Bull. Seismol. Soc. Am. **67**(4), 1173–1194 (1977)
4. Wang, S.S., Gao, B., Sui, C.Y.: Mechanism of shock absorption layer and shaking table tests on shaking absorption technology of tunnel across fault. Chin. J. Geotech. Eng. **37**(6), 1086 (2014)
5. Liu, X.Z., Li, X.F., Sang, Y.L., Lin, L.L.: Experimental study on normal fault rupture propagation in loose strata and its impact on mountain tunnels. Tunn. Undergr. Space Technol. **49**, 417–425 (2015)
6. Hu, J., Xu, L., Xu, N.W.: Numerical analysis of faults on deep-buried tunnel surrounding rock damaged zones. Adv. Civ. Eng. **1**(4), 90–93 (2011)
7. Wang, F.F., Jiang, X.L., Niu, J.Y.: Numerical simulation for dynamic response characteristics of tunnel near fault. Electron. J. Geotech. Eng. **21**(17), 5559 (2016)
8. Sabagh, M., Ghalandarzadeh, A.: Numerical modelings of continuous shallow tunnels subject to reverse faulting and its verification through a centrifuge. Comput. Geotech. **128**, 103813 (2020)
9. Zhang, Z.Q., Chen, F.F., Li, N., He, M.M.: Influence of fault on the surrounding rock stability for a mining tunnel: distance and tectonic stress. Adv. Civ. Eng. **2019**, 1–12 (2019). https://doi.org/10.1155/2019/2054938
10. Kun, M., Onargan, T.: Influence of the fault zone in shallow tunneling: a case study of Izmir Metro Tunnel. Tunn. Undergr. Space Technol. **33**, 34–45 (2013)
11. Wang, Q., Chen, G., Guo, E., Ma, Y.: Nonlinear analysis of tunnels under reversed fault. Indian Geotech. J. **47**(2), 132–136 (2017). https://doi.org/10.1007/s40098-016-0219-1
12. Su, Y.F., Zhang, M.H., Zhang, Z.X.: The influence of large fault on tunnels in underground mines. Electron. J. Geotech. Eng. **21**(9), 3535–3539 (2016)

13. Liu, G.Z., Qiao, Y.F., He, M.C., Fan, Y.: An analytical solution of longitudinal response of tunnels under dislocation of active fault. Rock Soil Mech. **41**(3), 923–932 (2020)
14. Wu, H.N., Shen, S.L., Yang, J., Zhou, A.N.: Soil-tunnel interaction modelling for shield tunnels considering shearing dislocation in longitudinal joints. Tunn. Undergr. Space Technol. **78**, 168–177 (2018)

Resilience Assessment Framework for Tunnels Exposed to Earthquake Loading

Z. K. Huang(✉), D. M. Zhang, and Y. T. Zhou

Key Laboratory of Geotechnical and Underground Engineering of Ministry of Education, Department of Geotechnical Engineering, Tongji University, P.O. Box 200092, Shanghai, China
5huangzhongkai@tongji.edu.cn

Abstract. The present paper proposes an integrated framework for the seismic resilience assessment of tunnels by using the appropriate fragility, restoration and functionality models, which consider both geotechnical and structural effects. Typical circular tunnel in Shanghai city of China is examined in this study, and the corresponding numerical model is built in ABAQUS. A couple of earthquakes are chosen to conduct non-linear dynamic analysis, so as to get the tunnel responses under increasing levels of ground shaking intensity. Fragility curves are constructed accounting for the main sources of uncertainties. Moreover, based on the proposed fragility curves and the existing empirical tunnel restoration functions, the development of resilience index (*Re*) with the peak ground acceleration (*PGA*) at the free-field ground surface for circular tunnel is evaluated and quantified. The results indicate that this type of tunnel is good of seismic resilience. The proposed framework is expected to facilitate city managers to support adaptation with preventive or retrofitting measures against seismic hazards, toward more resilient metro systems.

Keywords: Seismic resilience · Circular tunnel · Fragility curve · Numerical modelling

1 Introduction

Tunnels constitute key elements of transportation infrastructure in densely urbanized areas, thus, seismically-induced damages of tunnels may have significant consequences on the operation of the global transportation network after a major event. Hence, a careful investigation of the seismic resilience analysis of tunnels is of paramount importance in the perspective of seismic infrastructure planning and interventions. Seismic resilience describes the ability of a tunnel and the related organization or community to mitigate damage caused by seismic hazards [1], constituting the key components for the vulnerability and risk analysis of metro system. Resilient tunnels can withstand, respond and adapt to earthquake events by maintaining and even enhancing critical functionality. The seismic resilience assessment frameworks usually include fragility curves and restoration functions. Generally, fragility curves describe the probability of exceeding each damage state under an increasing level of intensity measure IM. For tunnels, the

© The Author(s), under exclusive license to Springer Nature Switzerland AG 2022
L. Wang et al. (Eds.): PBD-IV 2022, GGEE 52, pp. 2285–2294, 2022.
https://doi.org/10.1007/978-3-031-11898-2_213

existing fragility curves are usually derived based on expert judgment [2], empirical [3] or numerical [4, 5] approaches. While the restoration function expresses the rapidity of recovery of tunnel exposed to seismic damage, usually it can be obtained from the existing national codes or directly derived based on the expert elicitations approaches for specific tunnels. So far, the available seismic resilience analyses for tunnels are quite limited, as compared to buildings [6] or bridges [7].

This study focuses on the probabilistic seismic resilience analysis of circular tunnel in soft soil deposits. The results show that this type of tunnel is good of seismic resilience. The findings of this study are expected to facilitate city managers to enhance seismic risk management toward more resilient metro systems.

2 Resilience Assessment Framework

2.1 Definition of Fragility Curve

Most of the known fragility curves may be described using a lognormal probability distribution function, as illustrated in Eq. 1:

$$P(ds \geq ds_i | IM) = \Phi\left[\frac{1}{\beta_{tot}} \cdot \ln\left(\frac{IM}{IM_{mi}}\right)\right] \quad (1)$$

where $P(ds \geq ds_i|IM)$ is the exceeding probability for a particular damage state ds given a seismic intensity level, the latter defined by the seismic intensity measure IM, $\Phi(\cdot)$ represents the standard cumulative probability function, IM_{mi} stands for the median threshold value of IM required to cause the ith ds, and β_{tot} stands for the total lognormal standard deviation, which is modelled by the combination of three primary sources of uncertainty, as shown in Eq. 2:

$$\beta_{tot} = \sqrt{\beta_{ds}^2 + \beta_C^2 + \beta_D^2} \quad (2)$$

where β_{ds} is the uncertainty related to the definition of damage state, β_C is the uncertainty related to the response and resistance (capacity) of the element, and β_D is the uncertainty from the earthquake input motion (demand). The parameters β_{ds} and β_C are taken as 0.4 and 0.3, respectively [4]. A damage index is adopted to quantitatively define seismic damage state (*DS* or *ds*). This *DI* is defined as the ratio of the actual (M) over the capacity (M_{Rd}) bending moment of the tunnel cross-section. The capacity of the tunnel lining is evaluated through a section analysis based on the lining material and geometry properties, accounting for the induced axial forces (N) and bending moments (M). Five damage states are shown in Table 1.

Table 1. Definition of damages states and corresponding damage index.

Damage state (*ds*)	Range of damage index (*DI*)	Central value of *DI*
ds_0: none	$M_{sd}/M_{Rd} \le 1.0$	–
ds_1: minor	$1.0 < M_{sd}/M_{Rd} \le 1.5$	1.25
ds_2: moderate	$1.5 < M_{sd}/M_{Rd} \le 2.5$	2.00
ds_3: extensive	$2.5 < M_{sd}/M_{Rd} \le 3.5$	3.00
ds_4: complete	$M_{sd}/M_{Rd} \ge 3.5$	–

2.2 Probabilistic Resilience Assessment

Resilience *Re* is a term used to describe the ability of tunnels to recover after being subjected to seismic loads. The most widely used analytical definition is adopted in this work. The following formula can be used to calculate *Re* [7]:

$$Re = \frac{\int_{t_0}^{t_h} Q(t)dt}{t_h - t_0} \quad (3)$$

where $Q(t)$ is the functionality of the tunnel at time t under the functionality recovery function, t_0 is the occurrence time of the earthquake event (t_0 is set as 0 in this study), t_h is the investigated time point. The resilience index *Re*, as defined in Eq. 3, can be graphically illustrated as shown in Fig. 1.

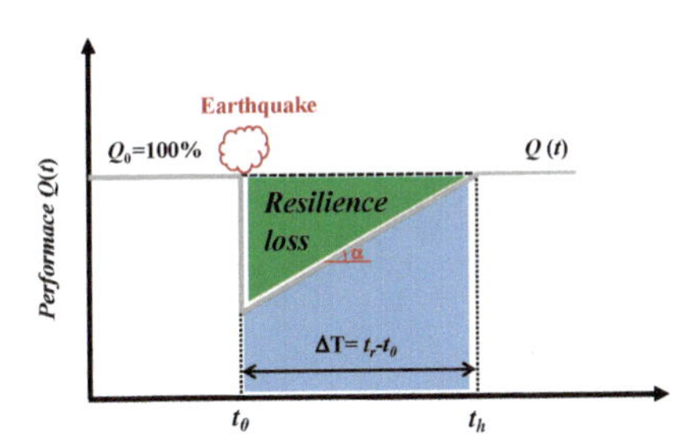

Fig. 1. Definition of resilience index *Re*.

The resilience index of a tunnel subjected to a certain ground shaking level can be generated based on the restoration functions, which describe the rapidity of functionality recovery for the different *DSs*, and the probabilities of occurrence of each *DS*. Hence, the tunnel functionality at time t can be computed as Eq. 4.

$$Q(t) = \sum\nolimits_{i=0}^{4} Q[DSi|t]P[DS_i|IM] \quad (4)$$

in which $Q[DS_i|t]$ is the functionality recovery function of the tunnel at time t with an initial damage state i, $P[DS_i|IM]$ is the conditional probability of being damage state i

for an event with a given *IM*, and can be calculated by fragility analysis, as presented in the following equations:

$$P[DS_i|IM] = P\left[ds > DS_{i+1}|IM\right] - P[ds > DS_i|IM], when\ i = 1, 2, 3 \quad (5)$$

$$P[DS_i|IM] = P[ds > DS_i|IM], when\ i = 4 \quad (6)$$

3 Examined Case

3.1 Numerical Model

A typical circular tunnel in Shanghai city of China was used to perform seismic resilience assessment. This type of tunnel has a typical outer diameter of 6.2 m, and the thickness of the lining is 0.35 m. The buried depth *h* (from tunnel crown) of the tunnel is 9 m, therefore, the overburden depth ratio *h/D* is 1.45, to consider shallow tunnel section. For the reinforced concrete tunnel, its elastic modulus E_c and Poisson's ratio v_c of is 3.55 GPa and 0.2. Three soil profiles are chosen based on the practical engineering conditions in Shanghai. They are classified into soil type D based on EC8 (2004) [8], and are denoted as soil deposits D1, D2 and D3 in this work. The detailed density ρ, cohesion *c*, friction angle φ and shear wave velocities V_s of the examined soil profiles are shown in Fig. 2 respectively.

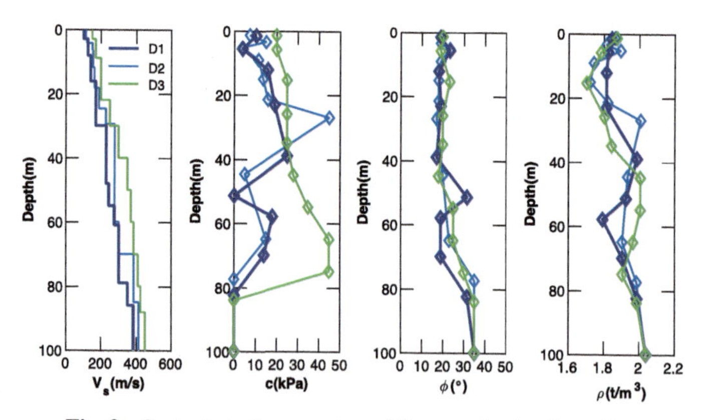

Fig. 2. Geotechnical parameters of the examined soil profiles.

Additionally, Fig. 3 shows the variations of shear modulus ratio G/G_{max} and damping ratios *D* with shear strain γ for clayey and sandy deposits in three examined soil profiles, and it is further utilized to describe the nonlinear behavior of the clayey and sand soils under seismic loadings.

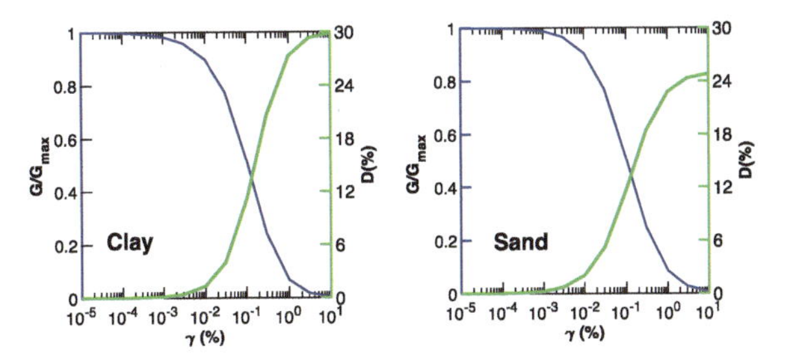

Fig. 3. G-γ-D curves for clayey and sand soil layers.

Figure 4 shows a detailed 2D numerical model of the soil-tunnel system using ABAQUS. The height of the model was set at 100 m, while the width was set as 400 m. A finite sliding hard contact algorithm, embedded in ABAQUS, was used to model the soil-tunnel interface. The tangential behavior was modelled based on the penalty friction formulation, by introducing a Coulomb frictional model with a friction coefficient μ equal to 0.6. The base boundary of the model was simulated as an 'elastic bedrock' in a way suggested by Lysmer and Kuhlemeyer [9], introducing proper dashpots, where apply the selected ground motions in terms of acceleration time series. For the side boundaries of the model, kinematic tie constraints were introduced to make the opposite vertical sides to move simultaneously.

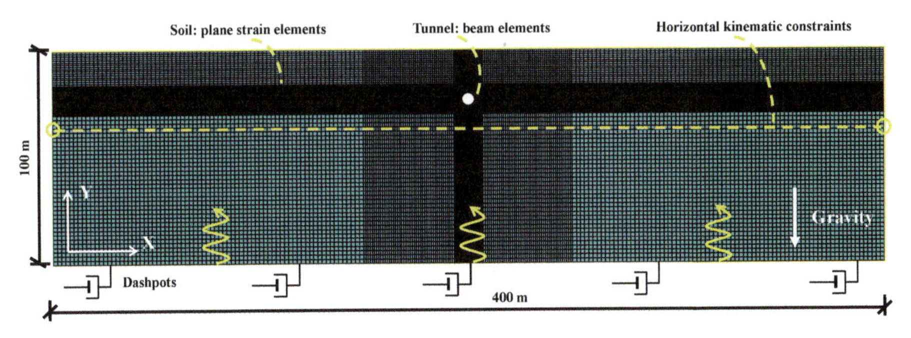

Fig. 4. Finite element model.

The response of tunnel lining was modelled using a simple linear elastic model. A uniform circular tunnel lining was adopted focusing to the computation efficiency in the numerous analysed cases. A visco-elasto-plastic model with Mohr-Coulomb yield criterion was adopted to capture the non-linear behavior of soil in the dynamic analysis. The shear moduli of soil deposits, corresponded to the strain compatible shear modulus G gradients, which were evaluated through 1D soil response analyses that preceded the 2D analyses, as discussed above. Each analysis was performed in two steps namely a geostatic step and a subsequent dynamic analysis step. The first step aimed at establishing the geostatic stress equilibrium in the model, whereas the in the second step the acceleration time history was introduced via dashpots at the model's base.

3.2 Selected Earthquakes

The seismic ground motions are associated with a high level of uncertainty, which is propagated within the seismic vulnerability analysis of any element at risk, thus, it's quite important to choose a set of ground motions, appropriately. In this study, 12 records from different earthquakes, recorded in soil conditions similar to soil type B of EC 8 (2004), were chosen as seismic ground motions for the analyses. All selected motions were provided by the Pacific Earthquake Engineering Research Strong Motion Database and their basic properties are shown in Table 2.

Table 2. Selected seismic records.

No.	Earthquake	Station name	Year	Mag. (M_w)	R (km)	PGA (g)
1	Northridge-01	LA - Hollywood Stor FF	1994	6.69	19.73	0.23
2	Parkfield	Cholame-Shandon Array	1966	6.19	12.9	0.24
3	Loma Prieta	Treasure Island	1989	6.93	77.32	0.16
4	Kern County	Taft Lincoln School	1952	7.36	38.42	0.15
5	San Fernando	Castaic - Old Ridge Route	1971	6.61	19.33	0.34
6	Imperial Valley-02	El Centro Array #9	1940	6.95	6.09	0.28
7	Superstition Hills-01	Imperial Valley W.L. Array	1987	6.22	17.59	0.13
8	Parkfield-02_CA	Parkfield-Cholame 2WA	2004	6.00	1.63	0.62
9	Imperial Valley-07	El Centro Array #11	1979	5.01	13.61	0.19
10	Tottori_Japan	TTR008	2000	6.61	6.86	0.39
11	Kobe_Japan	Port Island	1995	6.9	3.31	0.32
12	Borrego Mtn	El Centro Array #9	1968	6.63	45.12	0.16

3.3 Probabilistic Seismic Demand Analysis

According to the above considerations, two parameters are required to derive the corresponding fragility curves, i.e. IM_{mi} and β_{tot}. The median threshold values of PGA, corresponding to the prescribed damage states are determined based on a regression analysis of the nonlinear numerical analysis results. In this study, a linear regression analysis was adopted using the natural logarithm of the damage index (LnDI) and PGA (LnPGA) as the dependent and independent variable, respectively. As shown in Fig. 5, the different points (360 in totals) represent the corresponding values of damage index under increasing input intensity for three examined soil profiles, while the black solid line stands for the regression fit curve for these damage index data. It is noted that IM_{mi} is computed based on the regression fit equation with the different thresholds of damage states. β_D is computed as the dispersion of the simulated damage indices with regard to the regression fit, and it's equal to 0.18 here. Thus, the total lognormal standard deviation is equal to 0.53 calculated by Eq. 2. The detailed fragility function parameters in terms of IM_{mi} and β_{tot} are presented in Table 3.

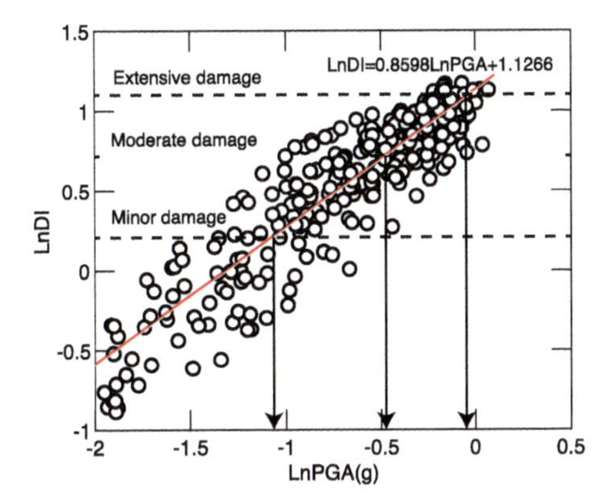

Fig. 5. Evolution of the damage index with *PGA* at free-field ground surface.

Table 3. Derived parameters of the fragility curves at free-field ground surface.

Damage state	Minor (g)	Moderate (g)	Extensive (g)	β_{tot}
Soil type D	0.35	0.60	0.97	0.53

3.4 Fragility Curves

Figure 6 presents the computed analytical fragility curves of the examined tunnel in soft soil deposits with respect to PGA at the ground free field conditions. As expected, the probability of damage increases with increasing PGA. When PGA is lower than 0.2g, the probability of extensive damage is negligible, which means that the examined shallow tunnel may withstand this level of earthquake intensity well, practically with no damages. The probabilities of minor and moderate damages are 15% and 2%, respectively. When PGA is equal to 0.4g, the probabilities of minor, moderate and extensive damage increase to 60%, 22% and 5%, respectively, while as the PGA increases to near 1.0g, there is a 97% probability the tunnel to be exposed to minor damage. The relevant probabilities for moderate or extensive damages are 83% and 52%. The latter observations highlight the significant importance of the sufficient seismic design for shield tunnel in the similar sites.

3.5 Probabilistic Resilience Assessment

The above developed fragility curves are used to calculate the seismic resilience index of the studied tunnel based on the procedure shown in Sect. 2. For the following seismic

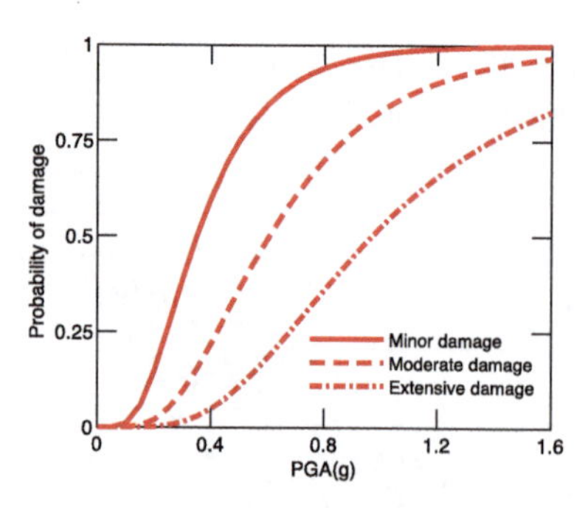

Fig. 6. Developed fragility curves.

resilience analysis, because of the lack of a specific restoration function for the studied tunnel, the empirical ones provided by [4] are finally used in this work, as presented in Fig. 7 in the term of minor, moderate and extensive damage, which are derived on based on the results from the experts' surveys. It is noted that there is a need for development of rigorous restoration models, which will account more accurately for critical parameters, such as the availability of relevant sources and funds, management approaches in constructions and repairs, maintenance strategies.

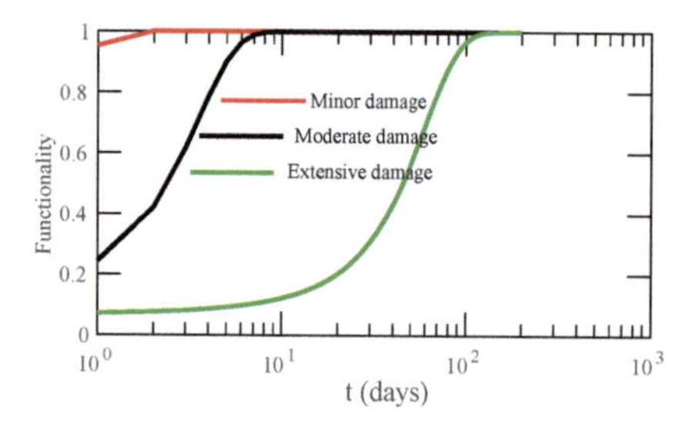

Fig. 7. Restoration functions by (FEMA H., 2003).

According to the Eq. 3, the tunnel functionality *Q(t)* at time t is calculated by the developed fragility curves and functionality restoration functions [10]. Figure 8 presents the influences of five different values of *PGA* (0.2, 0.4, 0.6, 0.8 and 1.0g), on the development of tunnel functionality *Q(t)*. As expected, the damage induced by a lower *PGA* seems easier to be recovered, indicating a higher resilience index *Re* with lower *PGA* for the studied tunnel.

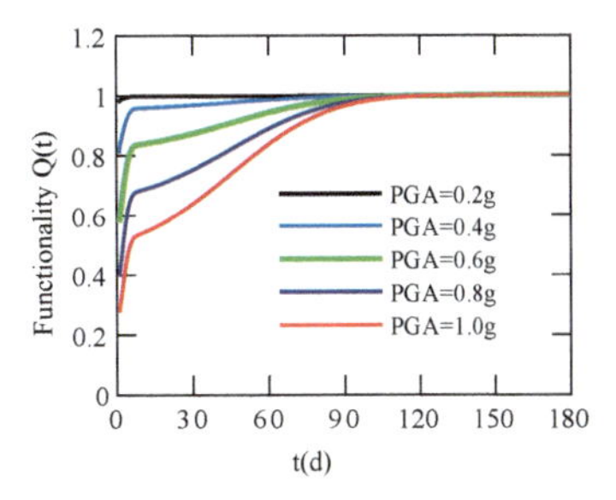

Fig. 8. Functionality recovery function of the examined tunnel.

Figure 9 shows the evolution of resilience index *Re* for the studied tunnel under the different values of *PGA*. Naturally, it is observed that the seismic resilience index of tunnels decreases slowly as the *PGA* increases. However, it indicates that the resilience of tunnels keeps a high level between 0.875 and 0.990, as the *PGA* is ranged between 0.1g and 1.0g. Therefore, the calculated results show that the examined tunnel is good of seismic resilience.

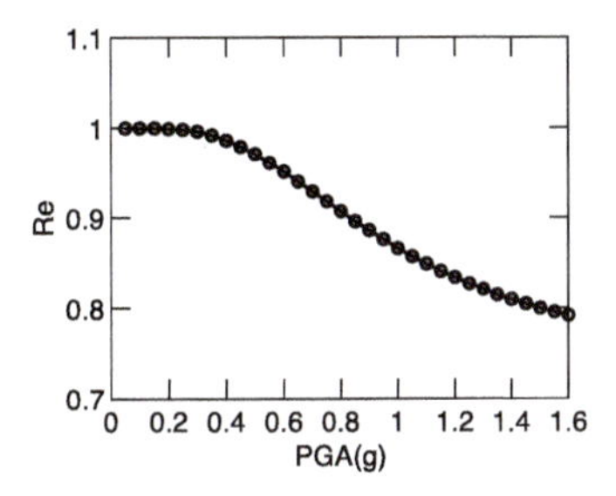

Fig. 9. Resilience index *Re* of the examined tunnel.

4 Conclusions

The probabilistic seismic resilience of a circular tunnel in soft soils is presented in this paper. A numerical simulation approach is used to construct the relevant seismic fragility curves. The resilience index Re for circular tunnel is then analyzed and quantified using the proposed fragility curves and current empirical tunnel restoration functions. The results indicate that the resilience of the examined tunnel keeps a high level between 0.875 and 0.990, as the *PGA* is ranged between 0.1g and 1.0g. The developed framework is clear and easy to be conducted. The findings of this study are likely to aid city administrators in improving seismic risk assessment and resulting in more resilient metro systems. Furthermore, it is recommended to propose appropriate thresholds for the resilience indices *Re* in the future study.

Acknowledgements. This work was supported by National Natural Science Foundation of China (Grant No. 52108381), and China Postdoctoral Science Foundation (Grant No. 2021M702491).

References

1. Woods, D.D.: Four concepts for resilience and the implications for the future of resilience engineering. Reliab. Eng. Syst. Saf. **141**, 5–9 (2015)
2. ATC-13. Earthquake Damage Evaluation Data for California, Applied Technology Council, Redwood City (1985)
3. ALA (American lifelines alliance): Seismic Fragility Formulations for Water Systems: Part 1-Guideline. American Society of Civil Engineers-FEMA (2001)
4. Argyroudis, S., Pitilakis, K.: Seismic fragility curves of shallow tunnels in alluvial deposits. Soil Dyn. Earthq. Eng. **35**, 1–12 (2012)
5. de Silva, F., Fabozzi, S., Nikitas, N., et al.: Seismic vulnerability of circular tunnels in sand. Géotechnique **71**, 1056–1070 (2020)
6. Hosseinpour, F., Abdelnaby, A.E.: Effect of different aspects of multiple earthquakes on the nonlinear behavior of RC structures. Soil Dyn. Earthq. Eng. **92**, 706–725 (2017)
7. Dong, Y., Frangopol, D.M.: Risk and resilience assessment of bridges under mainshock and aftershocks incorporating uncertainties. Eng. Struct. **83**, 198–208 (2015)
8. EC8. Eurocode 8: Design of Structures for Earthquake Resistance. The European Standard EN 1998-1, Brussels, Belgium (2004)
9. Lysmer, J., Kuhlemeyer, R.L.: Finite dynamic model for infinite media. J. Eng. Mech. D ASCE **95**(4), 859–878 (1969)
10. FEMA (Federal Emergency Management Agency): HAZUS-MH multi-hazard loss estimation methodology, earthquake model. Federal Emergency Management Agency, Washington, DC, USA (2003)

Stability Analysis of Tunnel in Yangtze Estuary Under Dynamic Load of High-Speed Railway

Liqun Li[1], Qingyu Meng[2], Leming Wang[2], and Zhiyi Chen[1](✉)

[1] Department of Geotechnical Engineering, Tongji University, Shanghai 200092, China
zhiyichen@tongji.edu.cn

[2] China Railway Design Corporation, Tianjin 300456, China

Abstract. The section from Shanghai to Hefei along the Yangtze River is located in the territory of Shanghai, Jiangsu and Anhui provinces in East China. There is one tunnel in the whole line, which is the Yangtze River tunnel along the Yangtze River. The total length of the tunnel is 14,150 m, in which the total length of the section under the Yangtze River is 10,360 m, with the maximum soil covering of about 39 m and the maximum underwater burial depth of about 67 m. The project is located at the mouth of the Yangtze River, and there are some unfavorable geology, such as saturated soft soil (silty) and fine sand formation. From the perspective of train-track-structure-stratum coupling effects, the interaction mechanism between tunnel and stratum under long-term cyclic load is explored. Combined with the actual geological parameters and engineering parameters, the dynamic response of the train structure under cyclic load is numerically simulated. Through the results of numerical simulation, the tunnel life under the train load is evaluated from the perspective of cyclic load fatigue failure, and the necessity of the tunnel lining structure is analyzed from the perspective of structural dynamic response.

Keywords: High-speed railway tunnel · Saturated soft soil · Dynamic response · Optimization of lining type

1 Introduction

When a vehicle passes through a high-speed railway ballastless track tunnel, the vibration of the train and the dynamic response of the tunnel structure and the surrounding ground are the result of the combined effect of the dynamic characteristics of the train-track-structure-stratum system and the dynamic coupling effect between the systems. Train-Track-Structure-Stratum is regarded as a unified system for dynamic analysis, which reflects the dynamic process of the train, track, and structure participating in the vibration, the basic mechanical characteristics, and the dynamic interaction regular pattern between vehicle and tunnel structure.

The dynamic analysis process of the complete train-track-structure-stratum coupling system is divided into three main steps. First, the real train, track, and tunnel structure is abstracted into a physical calculation model that can be used for calculation, and the mechanical theory is used to deform the structure under force. Carry out analysis, and

© The Author(s), under exclusive license to Springer Nature Switzerland AG 2022
L. Wang et al. (Eds.): PBD-IV 2022, GGEE 52, pp. 2295–2303, 2022.
https://doi.org/10.1007/978-3-031-11898-2_214

then use general finite element software to parameterize established physical calculation model, combine material properties and geological data of the actual construction to numerically simulate the dynamic response of the entire coupling system. To ensure the accuracy of the numerical simulation, the dynamic response data of the model was compared with the existing measured data [1] to determine the rationality of the numerical model.

2 Finite Element Modeling

The modeling size of finite element model is 200 m * 80 m * 350 m, and the surrounding area of shield uses high-dense grid, and the element type is C3D8R element. In order to avoid the influence of dynamic load reflection on the calculation results, infinite element boundary is adopted for the left and right boundaries of the model and the boundary of shield sections at both ends. The base adopts fixed constraint boundary, while the surface adopts free boundary. The finite element model is shown in Fig. 1 below.

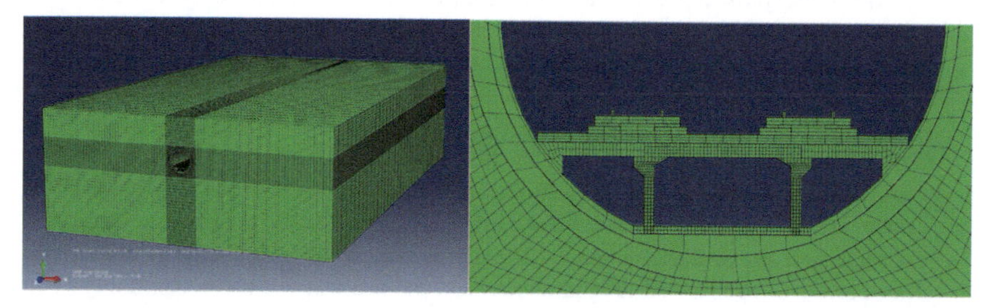

Fig. 1. Train-load time history curve.

2.1 Railway Track Model

Compared with other ordinary railway track structures, the standard of high-speed railways is much higher than ordinary railways. Ballastless tracks are commonly used in high-speed railway tracks, and most high-speed railways nowadays use CRTS series ballastless tracks. The track structure from top to bottom is: steel rails, elastic fasteners, precast concrete track slabs, CA mortar, concrete support layer.

The rail material standard rail (60 kg/m). In the actual high-speed railway, the section shape of the rail is more complicated. Because this article mainly focuses dynamic response of foundation hence rail is not the focus of research, section of the rail is simplified to a rectangular shape; Gauge of railway tracks is 1.435 m according to the actual situation. The track slab is made of precast concrete using C55 reinforced concrete. For fasteners and under-rail rubber pads, spring-damping units are used to simplify the simulation. The distance between fasteners is 0.6 m. The damping stiffness is 60 kN/mm, and the damping coefficient is 50 kN·s/m. Concrete support layer also uses C55 concrete. The track material parameters are shown in Table 1. Track slab, CA mortar cushion and concrete base all use Abaqus C3D8R solid elements; the fasteners between the rail and the track are simulated by spring damping elements, spring stiffness $k = 6E + 07$ N/m, damping coefficient $\zeta = 5E + 04$ N·s/m.

Table 1. Physical and mechanical parameters of rail track materials.

Type	E (GPa)	P (kg/m^3)	μ	H (m)
Railway track	210	7800	0.3	0.2
Track plate	35.5	2500	0.2	0.2
CA mortar	7	1800	0.2	0.03
Bearing layer	30	2500	0.2	0.3

2.2 Shield Model

For the interval design of general shield tunnel engineering in China, modified routine method are usually adopted. It is assumed that the bending stiffness of each shield segment is uniform, that is, the uneven weakening of the stiffness of the entire ring by weak parts such as joints is not considered. For the coupled dynamic analysis of train-track-structure-stratum, to calculate the efficiency, a homogeneous ring model is used to simulate the shield segment structure. Existing studies have shown that the change trend of the principal stress time history curve at the same part of the homogeneous ring model and the segmented segment model is basically the same under the action of the vehicle actuation load [2], so the homogeneous ring model can be used to simulate the shield Structure tube sheet structure.

2.3 Soil Conditions

The soil in this model is simulated by the C3D8R element, and the soil is layered according to the results of the on-site geological survey report. The Mohr-Coulomb constitutive model is used to simulate the dynamic response of the soil under the action of the train. The bottom of the soil adopts a fixed constraint boundary, and the side adopts a non-displacement boundary.

2.4 Train Load

For the simulation of train load, two methods are usually used. The first one is to consider only the axle load, that is, the force state biased towards the static force; in the actual situation, due to the track irregularity theory [3], the train acts The load on the rail is not the axle load of the train itself, so in this numerical simulation, the track irregularity theory will be used to simulate the train load for calculation. We use an exciting force function to simulate the load of the high-speed rail train, including the static load part and the vibration load superimposed by a series of sine functions. Corresponding to the high, medium, and low frequencies, they reflect track irregularities, additional dynamic loads, and The excitation force of the rail surface corrugation effect simulates the interaction force between the wheel and the rail. Its expression is:

$$F(t) = P_0 + P_1 \sin \omega_1 t + P_2 \sin \omega_2 t + P_3 \sin \omega_3 t \tag{1}$$

where P_0 is the static wheel load, P_1, P_2, and P_3 respectively correspond to the typical vibration load amplitude under the three control conditions. According to foreign high-speed railways, the axle load is generally 16–17 t, and the unilateral static wheel weight P_0 = 80 kN. The unsprung mass is taken as $M_0 = 750$ kg $= 750$ N·s^2/m. For the three control conditions of I, II and III, the typical irregular vibration wavelength and the corresponding vector height are $L_1 = 10$ m, $a_1 = 3.5$ mm; $L_2 = 2$ m, $a_2 = 0.4$ mm; $L_3 = 0.5$ m, $a_2 = 0.08$ mm. After introducing the above formula, the train load time history curve at the speed of 70 m/s (252 km/h) can be obtained as shown in Fig. 2:

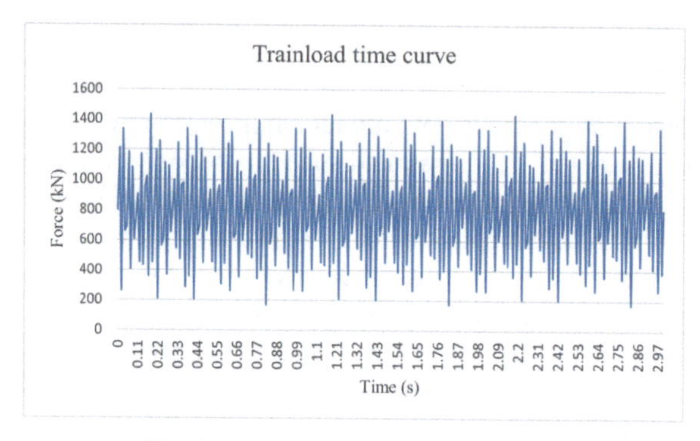

Fig. 2. Train-load time history curve.

3 Structural Safety Analysis of Shield Tunnel

3.1 Selection of Dangerous Sections

According to geological data, three representative positions were selected as typical sections to analyze the maximum principal stress amplitude of the whole structure under train dynamic load. The results are shown in Fig. 3.

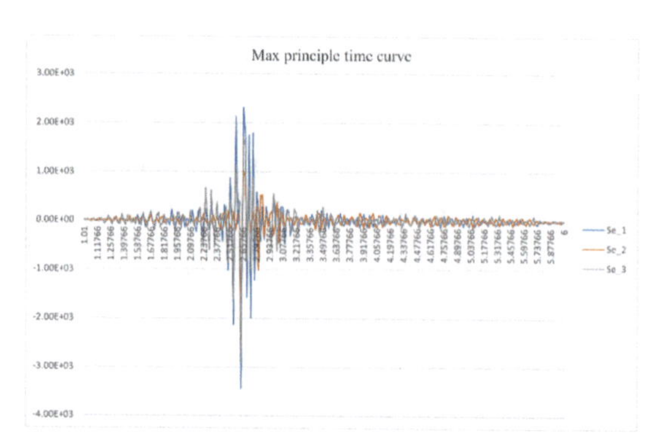

Fig. 3. Time history curve of maximum principal stress amplitude of structure.

It can be seen from the stress time history curve that the dynamic stress generated by Sect. 1 (Se_1) under the action of the train load has a greater fluctuation range, and can well reflect the effect of multiple wheelsets on the structure of the high-speed train. Therefore, the soil parameters of Sect. 1 (Se_1) are selected for modeling and subsequent data analysis.

3.2 Load Transmission Mechanism and Selection of Dangerous Parts

According to stress cloud diagram of shield tunnel (Fig. 4), the structural stress peaks mainly appear on the top, bottom and internal track structure of the shield. Therefore, select: 1. The connection between track structure and shield, 2. The bottom segment of shield, 3. The top segment of shield, 4. The top of support structure under track as the dangerous part of the dynamic analysis, the dangerous part selection diagram is shown in the Fig. 5.

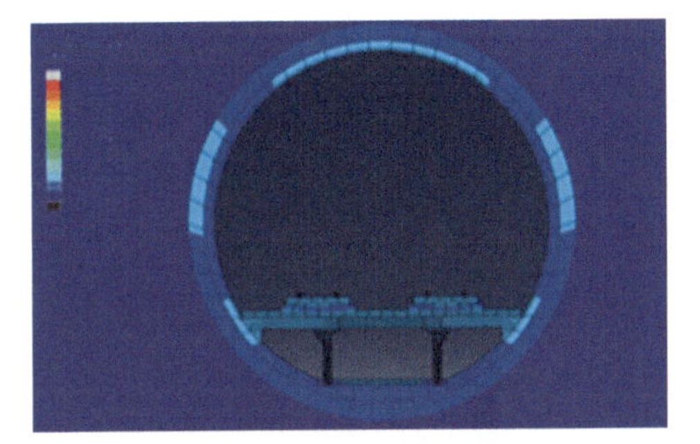

Fig. 4. Structure stress cloud map (max principal).

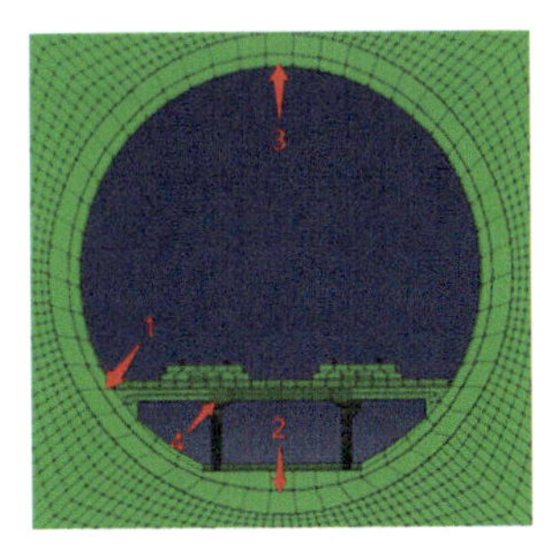

Fig. 5. Schematic diagram of dangerous parts selection.

Comparative analysis of the maximum principal stress amplitude of the above-mentioned dangerous parts, the result is shown in Fig. 6. From the comparative analysis of the maximum principal stress amplitude of the structure, it can be seen that the maximum principal stress of the structure is within a safe level, and the maximum stress position appears at the connection between the track structure layer and the shield tunnel segment (i.e., position 1 in Fig. 5). In construction process, focus should be paid to this location.

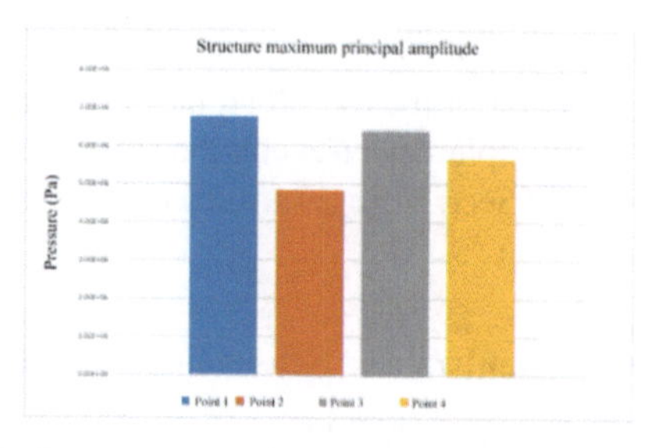

Fig. 6. Structure's maximum principal stress amplitude.

In order to further explore the spread of dynamic trainload in Stratum, the dynamic stress amplitude of the soil under the shield tunnel was compared and analyzed, and the results are shown in Fig. 7. It can be seen from the figure that as the depth increases, the dynamic stress amplitude of the soil layer has a significant decrease. The dynamic stress amplitude at 25 m below the plane of the track slab is only 25% of the soil at the position under the shield tunnel.

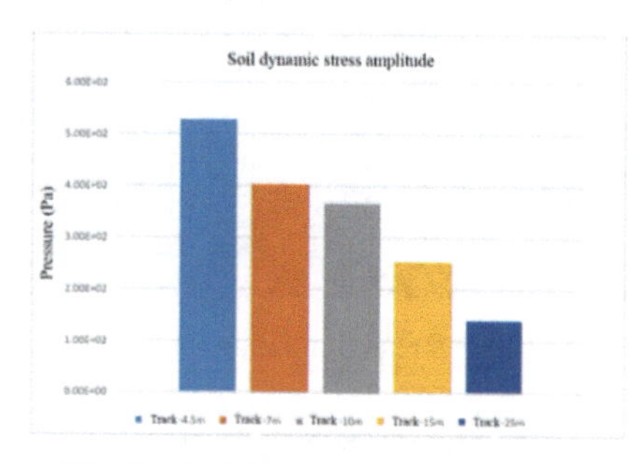

Fig. 7. Soil dynamic stress amplitude.

3.3 Analysis of Tunnel Convergence Deformation Under Trainload

According to the requirements of railway tunnel design deformation control, the horizontal and vertical relative displacement of the tunnel under the dynamic load of the train are analyzed. The schematic diagram of the model is shown in Fig. 8.

The relative displacement analysis result of the model is shown in Fig. 9. The vertical deformation of the shield tunnel is more obvious, and the peak value reaches 0.025 mm. According to the requirements of railway tunnel deformation control of 1‰D–5‰D, the deformation value of shield tunnel is much smaller than the deformation control value of railway tunnel. That is, under tunnel operation conditions, the tunnel deformation caused by the dynamic load of the train is very small, which meets the requirements of normal use.

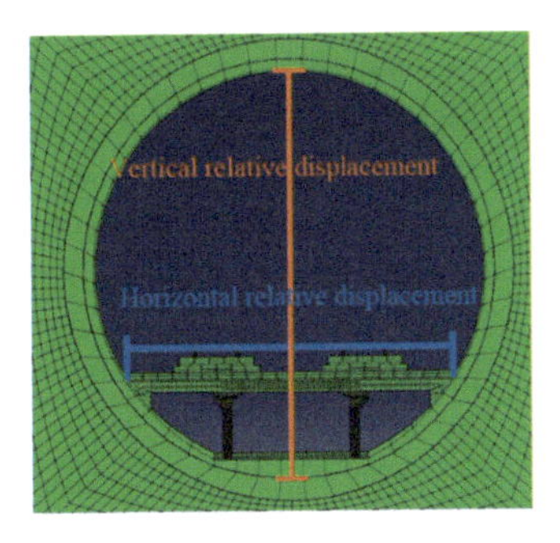

Fig. 8. Schematic diagram of deformation control analysis.

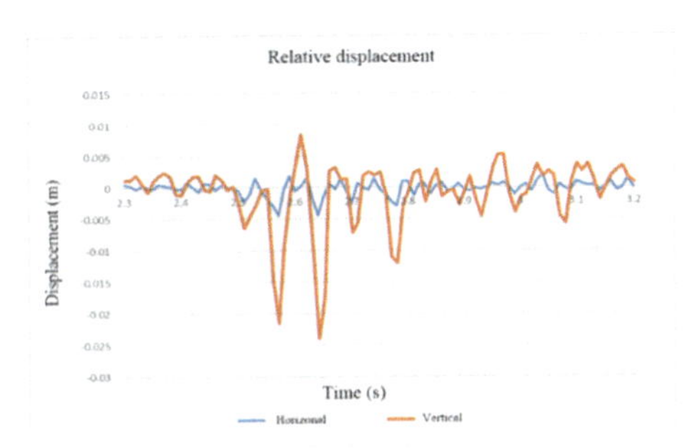

Fig. 9. Convergent deformation time history curve.

4 Optimization of Shield Tunnel Structure

In order to compare the dynamic response of the structure caused by the vehicle load under the single/double-layer lining condition, the soil parameters of Sect. 1 (Se_1) were selected, and the finite element models of the single-layer and double-layer lining were established respectively, and static considerations were applied. The train load based on the theory of track irregularity is used to simulate the most unfavorable situation. Comparing the stress cloud diagrams under the condition of single/double lining, select the following three nodes for further stress analysis. Node selection is shown in Fig. 10.

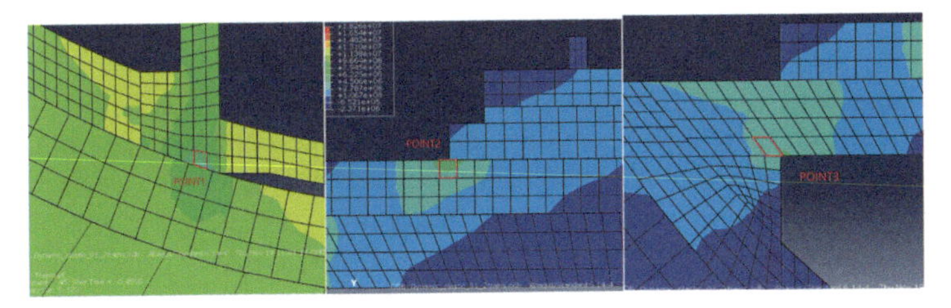

Fig. 10. Node selection.

A comparative analysis of the dynamic stress amplitude of the above three nodes (point1, point2, point3) shows that by increasing the inner lining, the dynamic stress amplitude caused by the train dynamic load on the structure can be reduced, thereby reducing the vibration amplitude of the structure. By comparing the peak value of the relative displacement between the top and bottom of the shield, it is found that the double-layer lining structure has better stability under the cyclic load of the train. The stress results are shown in Fig. 11.

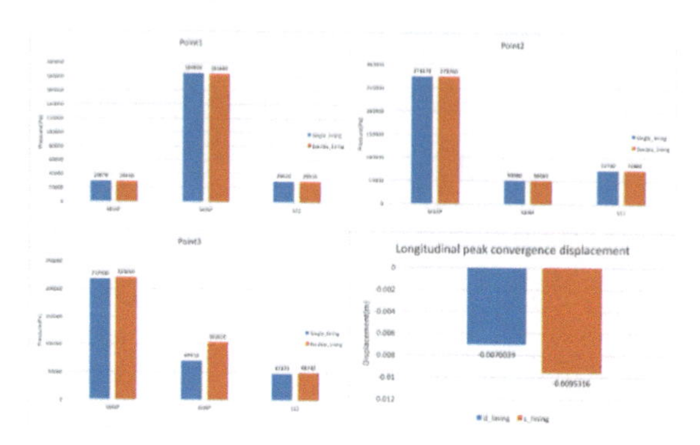

Fig. 11. Comparison of stress amplitude and convergence deformation.

5 Conclusions

(1) From the results of stress analysis, the dynamic stress amplitude of the structure caused by the train dynamic load is relatively small, which is not a controlling factor that affects the long-term operation of the tunnel.
(2) The maximum value of the top-bottom relative displacement of the shield tunnel is 9.532 mm (single-layer lining) and 7.004 mm (double-layer lining), which is less than the commonly used design deformation control value of railway tunnels 1‰D–5‰D (1‰D = 15.4 mm), The results obtained meet the design specifications.
(3) From the perspective of tunnel convergence deformation, double lining can effectively reduce the convergence deformation range of shield tunnel, which is more beneficial to improve the structural integrity.

Acknowledgments. This research was supported by the National Natural Science Foundation of China (Grant No. 51778464), State Key Laboratory of Disaster Reduction in Civil Engineering (SLDRCE19-B-38) and the Fundamental Research Funds for the Central Universities (22120210145). All supports are gratefully acknowledged.

References

1. Xu, X.: Study on field measurement and full-scale model test on dynamic response of high-speed railway subgrade. Zhejiang University, Hangzhou (2014)

2. Liu, Y.X.: Study on fatigue failure form and life prediction of shield segment under high-speed train dynamic load. Southwest Jiaotong University, Chengdu (2017)
3. Liang, B., Cai, Y.: Dynamic analysis of high-speed railway subgrade under uneven conditions. J. China Railw. Soc. **21**(2), 84–88 (1999)
4. Zhang, W.C.: Study on vibration of subgrade-ballastless track-vehicle coupled system under irregularities condition. Southwest Jiaotong University, Chengdu (2016)
5. Chen, L.Y.: Analysis on dynamic response of ballastless track subgrade of high-Speed railway. Chongqing Jiaotong University, Chongqing (2015)

Seismic Performance of an Integrated Underground-Aboveground Structure System

Wen-Ting Li(✉), Rui Wang, and Jian-Min Zhang

Tsinghua University, Beijing 100084, China
li.wenting@outlook.com

Abstract. Underground structures such as subway stations, which are operated at a high capacity and frequency, are prone to damage by large earthquakes. The seismic behavior of underground-ground structure system, which refers to underground structures that directly connect to above ground buildings, is a complex issue involving the dynamic interaction between ground building and underground structures, as well as the kinematic interaction between underground structures and surrounding soil. To study the seismic performance of integrated underground-aboveground structure systems, this paper presents a three-dimensional finite element simulation analysis, addressing the influence of aboveground high-rise buildings on the seismic response of underground structures. The results show that above ground buildings significantly increase the underground structure's racking responses and dynamic forces during earthquakes.

Keywords: Integrated underground-aboveground structure system · Seismic performance · Soil-structure interaction · Finite element simulation

1 Introduction

Integrated underground-aboveground structure is defined as underground structure that directly connect to above ground buildings. The seismic behavior of integrated underground-aboveground structure systems is a complex issue involving the dynamic interaction between above ground building and underground structures, as well as the dynamic interaction between underground structures and surrounding soil. Therefore, the integrated underground-aboveground structure systems may significantly differ from traditional single underground structures.

For a long time, the general perception has been that underground structures face minimal seismic risk unless when located on either active faults or liquefiable sites; few countries consider the seismic design of underground structures in their national design standards. The seismic risk of underground structures took center stage after the 1995 Hyogoken Nanbu Earthquake (also called the 1995 Kobe Earthquake), in which 6 of 21 subway stations in Kobe, Japan, were severely damaged. One of the six stations, the Daikai subway station, is the most heavily damaged large underground structure in recorded history. The total collapse of Daikai station resulted in the subsidence (2.5 m) of

© The Author(s), under exclusive license to Springer Nature Switzerland AG 2022
L. Wang et al. (Eds.): PBD-IV 2022, GGEE 52, pp. 2304–2312, 2022.
https://doi.org/10.1007/978-3-031-11898-2_215

the street's surface [1]. Researchers have investigated on the seismic failure performance of underground structures [2, 3] and the interaction influences of nearby ground structures on underground structure [4, 5]. The seismic performance of connected ground and underground structures needs to be further investigated.

In this paper, three-dimensional finite element analysis was conducted to investigate the seismic performance of integrated underground-aboveground structure systems, addressing the influence of ground high-rise buildings on seismic responses of underground structures. Through time history analysis, the structural inter-story racking responses and dynamic forces during earthquake were investigated.

2 Problem Definition

This study is focused on an integrated underground-aboveground structure system (Fig. 1) contains three above ground high-rise buildings and a large underground structure. The total length of the four-storied underground structure is 280 m, total width is 197 m and total height is 40 m. The three ground high-rise buildings are located at the northern side of the underground structure. The highest ground building is 120 m high with 25 stories, the middle ground building is 78 m high with 19 stories and the lowest ground building is 38 m high with 9 stories. The concrete material of ground building is primarily C50 while the core wall is C60. The concrete material of underground structure is primarily C40. The surrounding soil is simplified as 23 horizontal layers whose profile properties are listed in Table 1.

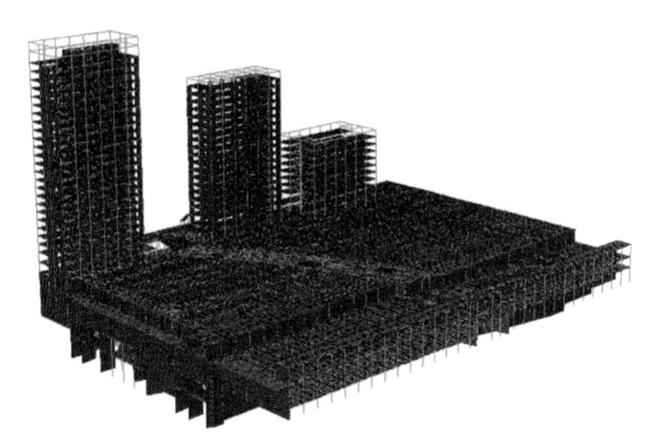

Fig. 1. Finite element model of the integrated underground-aboveground structure system

The artificial seismic wave of Tangshan record is used as incident shaking whose acceleration history and Fourier spectra are plotted in Fig. 2. The Tangshan record lasts 20 s. Its peak acceleration is scaled to 0.3g.

Table 1. Site properties.

No.	Thickness /m	Density/$\left(kg/m^3\right)$	$v_s/(m/s)$	G_{max} (MPa)	No.	Thickness /m	Density/$\left(kg/m^3\right)$	$v_s/(m/s)$	G_{max} (MPa)
1	3.7	1900	161	50.20	13	3	1990	373	282.23
2	1.8	1940	188	69.90	14	3.3	2080	397	334.18
3	1.4	1930	191	71.77	15	3.9	2010	354	256.76
4	1.3	1940	188	69.90	16	0.7	2070	383	309.53
5	10.3	2050	270	152.34	17	11.5	2100	436	406.93
6	9.8	2050	311	202.12	18	1.3	1990	383	297.56
7	2.8	1940	326	210.17	19	10.4	2100	468	468.86
8	3.4	2060	363	276.70	20	3.5	2030	402	334.41
9	1	1940	326	210.17	21	3.3	2060	422	373.96
10	0.9	2060	363	276.70	22	1.5	2100	455	443.17
11	5.2	1980	366	270.37	23	3.1	2010	419	359.71
12	6.6	2080	401	340.94					

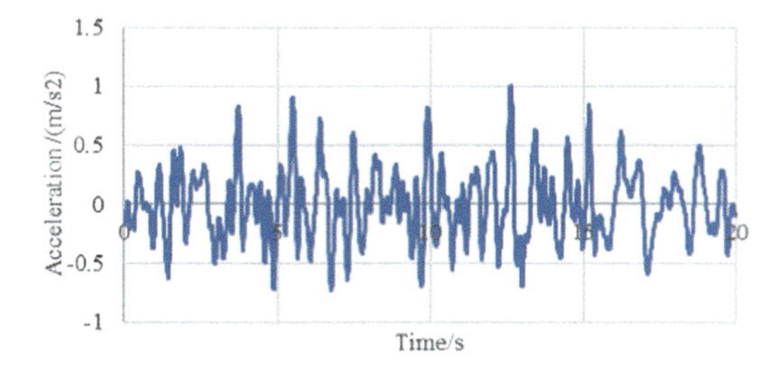

(a) Acceleration history

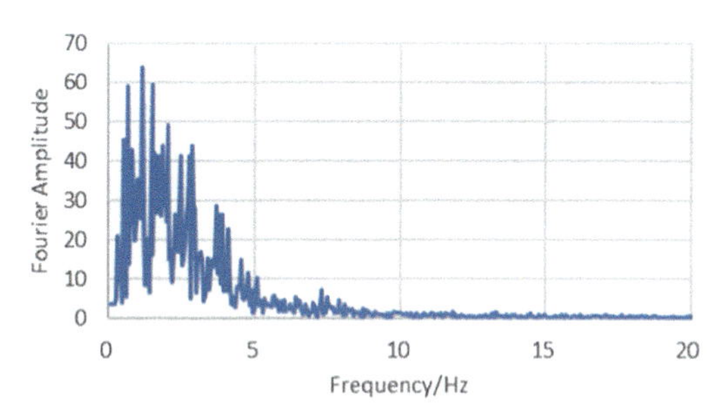

(b) Fourier spectrum

Fig. 2. Seismic records

The FE simulation was developed with procedure ABAQUS, which is widely used in the analysis of underground structures. The structures were simulated by beam element B32 and shell element S4R while the soil was simulated by solid element C3D8R. The finite element model is displayed in Fig. 3. In time history analysis, the implicit time integration method is used to calculate the transient dynamic response. The maximum increment time step is 0.02 s.

Boundary conditions are important to prevent waves generated at the model boundaries from re-entering the FE model. The model boundaries in this simulation were modeled carefully to make sure that the free-field condition is attained around the structures. The nodes on the lateral boundaries were constrained to have the same vertical and horizontal displacements. This type of boundary condition, which was suggested by Zienkiewicz et al. [6] for seismic problems, is called a repeatable boundary condition. The bottom boundary represents the bedrock and was assumed to be perfectly rigid and rough. The input shaking is imposed at the finite element model's bottom boundary, i.e., bedrock level.

To investigate the influence of above ground buildings on seismic performance of underground structures, the FE model without above ground buildings was established for comparison.

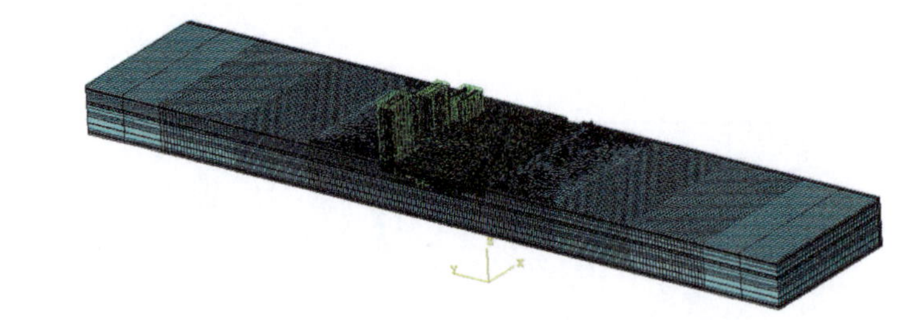

(a) Underground-ground structure-soil system

(b) Underground structure-soil system

Fig. 3. Simulation models

3 Results

3.1 Structural Inter-story Drift

The structural inter-story drift ratio is a widely used factor in structural vulnerability assessments. To be clearer, the highest above ground building is labelled as Structure A; the middle above ground building is labelled as Structure B. To evaluate the influence of above buildings quantitatively, three coordinate systems are defined as X_A-Y_A and X_B-Y_B, as shown in Fig. 4.

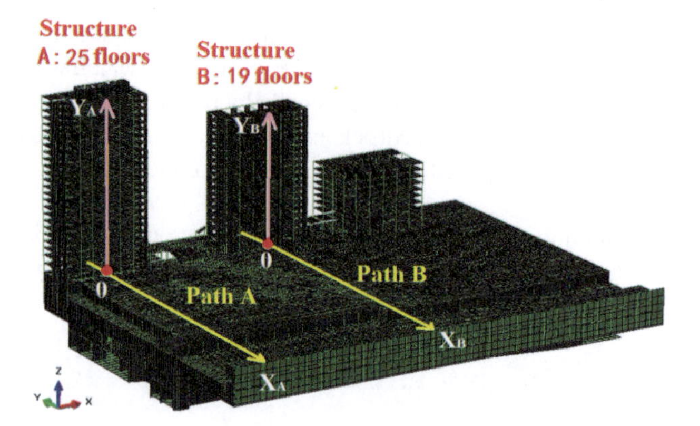

Fig. 4. Coordinate systems

The inter-story racking responses of underground structure along axis X_A and X_B are plotted in Fig. 5 and Fig. 6, respectively. In this paper, B1 refers to basement first floor, B2 refers to basement second floor, B3 refers to basement third floor and B4 refers to basement fourth floor. It is seen that the maximum inter-story racking ratio of underground structure along axis X_A is 1/827, along X_B is 1/1200. Generally, racking responses of underground structure are increased as the depth increased. The special fluctuation is caused by the structure irregularity.

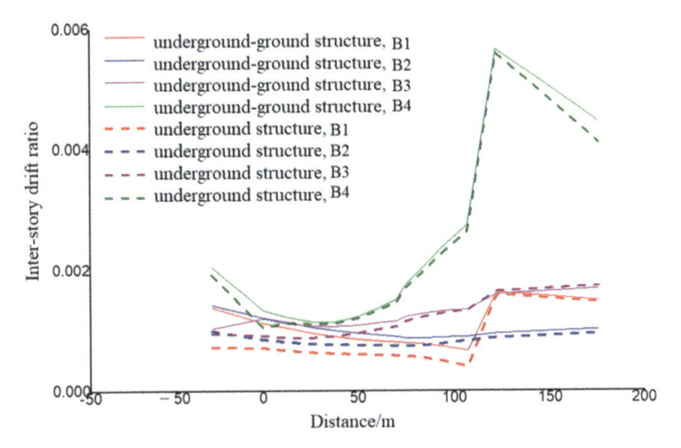

Fig. 5. Inter-story racking responses of underground structure along axis X_A

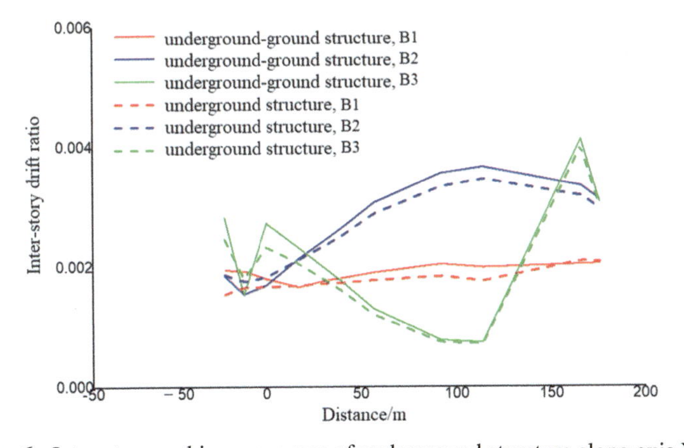

Fig. 6. Inter-story racking responses of underground structure along axis X_B

The existence of ground buildings generally increases the inter-story racking responses of underground structure. The increased ratio is listed in Table 2 and Table3, for Structure A, and Structure B, respectively. It is seen that: (1) ground buildings significantly increase inter-story racking responses of underground structure. The higher the ground buildings, the larger the influence of ground buildings on underground buildings. (2) The influence is largest at first floor of underground structure. (Increase ratio by Structure A is 93% and by Structure B is 26%). As the depth increases, the influence of ground buildings sharply decreases. (3) Horizontal influence distance of ground buildings is around 100 m (Influence distance by Structure A is 107 m and by Structure B is 114 m). As depth increases, the horizontal influence distance decreases.

Table 2. Increase ratio on underground structure's inter-story racking by Structure A/%

X_A/m	−26.85	0	12	26	38	50	60	70	75	83	94	107	122	176
First floor	93	59	58	52	47	42	40	39	40	42	50	69	2	2
Second floor	44	43	41	36	31	27	24	21	19	17	14	8	8	8
Third floor	9	32	28	23	18	14	11	9	6	5	3	1	−3	−2
Fourth floor	7	28	8	3	2	2	2	3	3	3	4	4	1	8

Table 3. Increase ratio on underground structure's inter-story racking by Structure B /%

X_B /m	−22	−12	0	17	32	40	57	92	114	166	176
First floor	26	16	8	−1	3	5	7	11	13	−3	−2
Second floor	−1	−12	−7	1	4	5	6	7	6	5	5
Third floor	15	−15	17	13	11	10	8	5	3	3	3

3.2 Structural Force

For underground structures, structural column is proved important in seismic resistance. Thus, this study focused on the structural column's axial force. Generally, the ground buildings increase the column's axial force of underground structure. The increase ratio caused by Structure A, Structure B respectively are plotted in Fig. 7 and Fig. 8. It is seen that: (1) ground buildings significantly increase the column's axial force of underground structure, especially at structural first floor. The maximum increase ratio by Structure A is 1945% and by Structure B is 2063%. (2) The influence is largest at first floor of underground structure. As the depth increases, the influence of ground buildings decreases. At the third floor of underground structure, the increase ratio is 83% caused by Structure A, 220% caused by Structure B. (3) Normally, the influence of ground building on axial force at columns' top is larger than that at columns' base. (4) Horizontal influence distance of ground buildings is around 94 m for Structure A and is 42 m for Structure B.

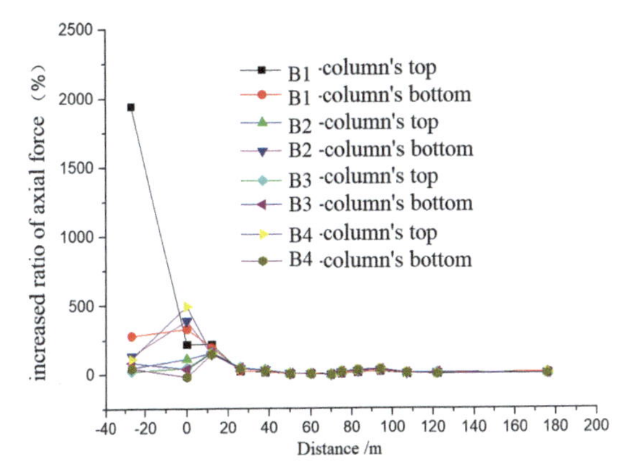

Fig. 7. Increase ratio on underground structure columns' axial force by Structure A

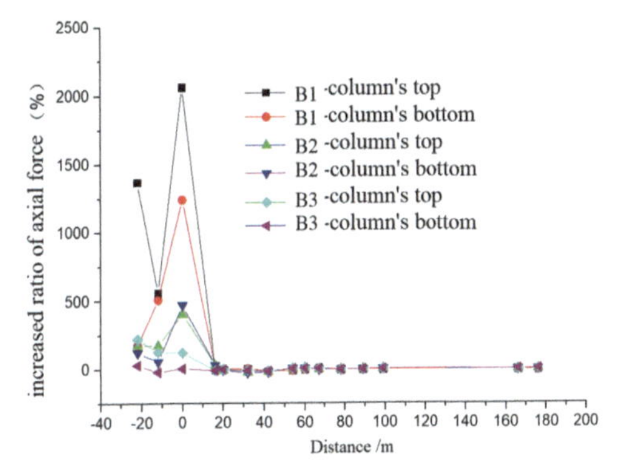

Fig. 8. Increase ratio on underground structure columns' axial force by Structure B

4 Conclusions

In this paper, a three-dimensional finite element analysis was conducted to investigate the seismic performance of integrated underground-aboveground structure systems, addressing the influence of above ground high-rise buildings on seismic responses of underground structures. The structure's inter-story racking responses and dynamic forces were investigated. The following conclusions were drawn for this specific structure system model:

(1) Ground buildings significantly increase inter-story racking responses of underground structure. The higher the ground buildings, the larger the influence of ground buildings on underground buildings. The influence on inter-story drift is largest at first floor of underground structure. As the depth increases, the influence sharply decreases.
(2) Horizontal influence distance of ground buildings on inter-story racking responses of underground structure is around 100 m. As depth increases, the horizontal influence distance decreases.

(3) Ground buildings significantly increase the columns' axial force of underground structure. The influence is largest at first floor of underground structure. As the depth increases, the influence of ground buildings decreases.
(4) Normally, the influence of ground building on structural axial force at columns' top is larger than that at columns' base. Horizontal influence distance on columns' axial force varies from 42 m to 94 m.

References

1. Pitilakis, K., Tsinidis, G.: Performance and seismic design of underground structures. In: Maugeri, M., Soccodato, C. (eds.) Earthquake Geotechnical Engineering Design. GGEE, vol. 28, pp. 279–340. Springer, Cham (2014). https://doi.org/10.1007/978-3-319-03182-8_11
2. Debiasi, E., Gajo, A., Zonta, D.: On the seismic response of shallow-buried rectangular structures. Tunn. Undergr. Sp. Technol. **38**, 99–113 (2013). https://doi.org/10.1016/j.tust.2013.04.011
3. Li, W., Chen, Q.: Effect of vertical ground motions and overburden depth on the seismic responses of large underground structures. Eng. Struct. **205**, 110073.1-110073.18 (2020). https://doi.org/10.1016/j.engstruct.2019.110073
4. Zhu, T., Wang, R., Zhang, J.M.: Effect of nearby ground structures on the seismic response of underground structures in saturated sand. Soil Dyn. Earthq. Eng. **146**, 106756 (2021). https://doi.org/10.1016/j.soildyn.2021.106756
5. Li, W.T., Chen, Q.J.: Seismic damage evaluation of an entire underground subway system in dense urban areas by 3D FE simulation. Tunn. Undergr. Space Technol. **99**, 103351 (2020). https://doi.org/10.1016/j.tust.2020.103351
6. Zienkiewicz, O.C., Bicanic, N., Shen, F.Q.: Earthquake input definition and the trasmitting boundary conditions. In: Doltsinis, I.S. (ed.) Advances in Computational Nonlinear Mechanics. ICMS, vol. 300, pp. 109–138. Springer, Vienna (1989). https://doi.org/10.1007/978-3-7091-2828-2_3

A Simplified Seismic Analysis Method for Underground Structures Considering the Effect of Adjacent Aboveground Structures

Jianqiang Liu[1], Tong Zhu[2,3](✉), Rui Wang[3], and Jian-Min Zhang[3]

[1] CCCC Second Highway Engineering Co., Ltd., Xi'An 710065, China
[2] China State Construction Engineering Co., Ltd., Beijing 100029, China
zhutongt@163.com
[3] School of Civil Engineering, Tsinghua University, Beijing 100084, China

Abstract. Adjacent aboveground structures can significantly affect the seismic response of underground structures, and the structure soil-structure dynamic interaction should be considered in the seismic design of underground structures. Three-dimensional numerical models of underground-aboveground structure systems in saturated soil are established and fluid-solid coupling elastoplastic dynamic analysis is conducted. Results show that the aboveground structure can amplify the seismic internal force and deformation response of the underground structure due to the change of the horizontal stress distribution of the near-field soil, which is an adverse seismic effect. A simplified seismic analysis method is proposed based on this mechanism. The effect of adjacent aboveground structure can be approximately considered by applying the additional loading generated by the overground structure to both sides of the underground structure in the form of pseudo-static force. The effectiveness of this simplified method is preliminarily verified by dynamic analysis.

Keywords: Underground structure · Adjacent structure system · Seismic analysis method

1 Introduction

The utilization of underground space is an important development tendency of urbanization, and the seismic design of underground structures is the key of design. The earthquake data show that the seismic damage of the underground structure located in the saturated liquefiable soil is particularly significant [1–3], which should be paid great attention in the seismic design analysis.

Due to the increased density of urban buildings, the underground structure will inevitably be adjacent to the aboveground structure, forming more and more interaction systems. This structure-soil-structure dynamic interaction have a significant impact on the seismic response of underground structures [4–6]. Most of current seismic design codes and guidelines for underground structures use the simplified analysis methods

© The Author(s), under exclusive license to Springer Nature Switzerland AG 2022
L. Wang et al. (Eds.): PBD-IV 2022, GGEE 52, pp. 2313–2321, 2022.
https://doi.org/10.1007/978-3-031-11898-2_216

based on the pseudo-static concept [7, 8], and do not consider the effect of adjacent aboveground structures, which need urgent relevant research.

In this study, three-dimensional finite element models of the underground structure-adjacent aboveground structure systems in saturated sand are conducted, and the fluid-solid coupling elastoplastic dynamic analysis is carried out to study the effect of the adjacent aboveground structure on the seismic internal force and deformation response of the underground structure. Based on the mechanism, a simplified seismic analysis method for underground structures that can consider the effect of adjacent aboveground structures is proposed, whose effectiveness is preliminarily verified.

2 Numerical Analysis Model and Method

2.1 Analysis Conditions

The research condition in this study is based on a centrifuge shaking table test of adjacent structure systems in saturated liquefiable soil [9]. The Chinese Fujian standard sand with water level located on the surface is used in the model, and the ground depth is 50 m. The underground structure is a simplified rectangular model with length of 9.8 m and height of 6.2 m. The slab and wall thickness are both 0.6 m and the structure burial depth is 5.5 m, and the structure material is C50 concrete. The aboveground structure is 8m high, simulated as a single degree of freedom system with a concentrated mass of 380 t at the top, and the natural frequency of the structure is 0.93 Hz.

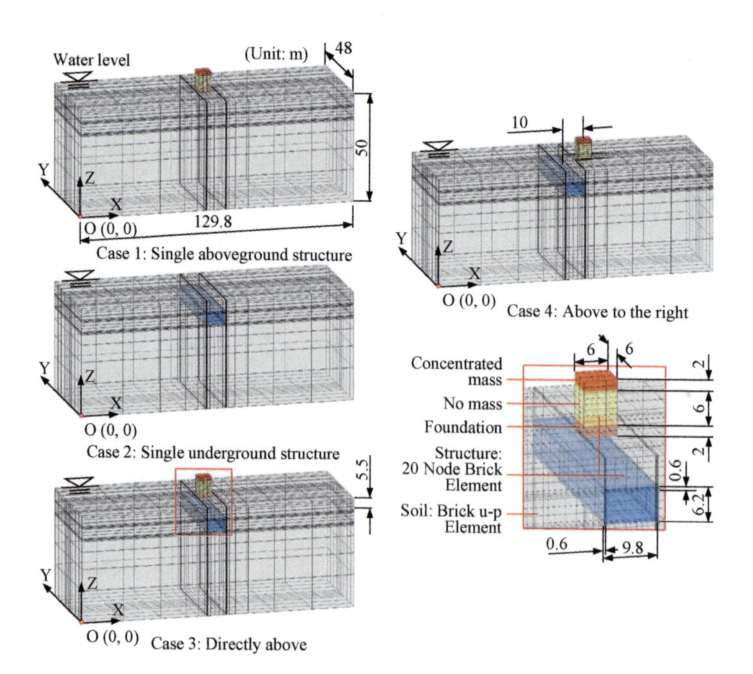

Fig. 1. Analysis model

In order to analyze the effect of adjacent aboveground structures on the seismic response of underground structures, 4 analysis cases are conducted as shown in Fig. 1: 1) a single aboveground structure without underground structure (Case 1: Single aboveground structure); 2) a single underground structure without aboveground structure (Case 2: Single underground structure); 3) a ground structure located directly above an underground structure (Case 3: Directly above); 4) a ground structure located above to the right side of an underground structure (Case 4: Above to the right).

2.2 Numerical Analysis Method

Three-dimensional numerical analysis models are established in the OpenSees finite element program [10]. The soil is simulated by the fluid-solid coupling brick element, and the saturated sand is simulated by the CycLiqCPSP model developed by Wang et al. [11]. This model is based on the physical mechanism of large deformation of sand liquefaction [12], and can accurately describe the development process of deformation before and after liquefaction of the saturated sand. The model parameters in this study are determined according to the study by Zhu et al. [13]. The underground structure and the aboveground structure are simulated by 20 node brick element, and the structure is simulated by linear elastic constitutive model, as shown in Fig. 1.

The models have free drainage surfaces, while the lateral and base boundaries are impermeable. Displacements at corresponding heights of the two lateral boundaries of the model are tied to simulate a free field boundary condition. After the initial stress field is obtained with a static analysis step, the ground motions are input from the base. Rayleigh damping with an initial damping ratio is used for the model. The ground motion is selected from Kobe earthquake data, and the ground motion amplitude is adjusted to 0.4 g and the main frequency is 0.59 Hz.

3 Effect of Adjacent Aboveground Structures on the Seismic Response of Underground Structures

3.1 Effect Laws

Seismic internal force and deformation are the key seismic response of underground structures in design. This research will study the effect laws and physical mechanism of adjacent aboveground structures located directly and laterally above on the seismic internal force and deformation response of the underground structure.

First, the effect on the seismic internal force response is analyzed. In Case 2, Case 3, and Case 4, the peak dynamic bending moment at the top of the walls on both sides of the underground structure is shown in Fig. 2(a) and (b) (the bending moment is positive when the right side is in tension). When the aboveground structure is directly above the underground structure, the peak positive bending moment at the top of the left wall of the underground structure is increased by 25.6%, and the peak negative bending moment at the top of the right wall of the underground structure is increased by 31.5%. The results show that the aboveground structure directly above amplifies the dynamic bending moment of the side walls of the underground structure, resulting in an additional bending moment in tension on the inside. However, when the aboveground structure is located above to side of the underground structure, the amplifying effect on the seismic internal force of the underground structure is weak.

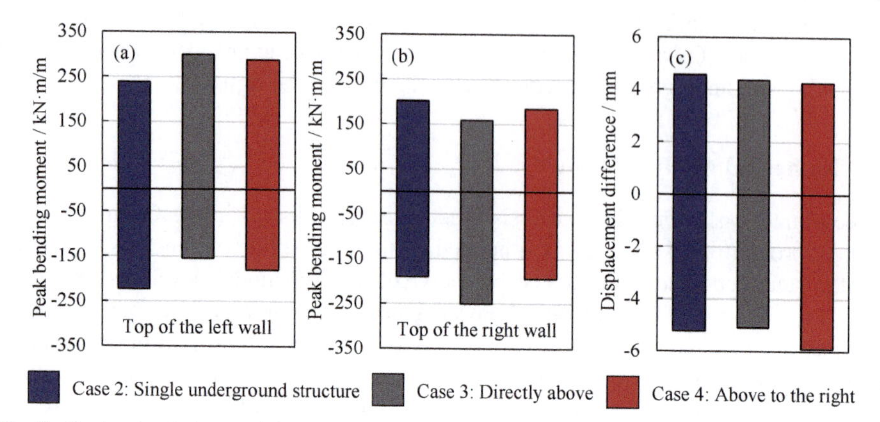

Fig. 2. Peak seismic internal force at (a) top of the left wall and (b) top of the right wall, and (c) peak deformation response of the underground structure in Case 2, Case3, and Case 4

Second, the effect on the seismic deformation response is analyzed, which can be represented by the horizontal displacement difference between the top and bottom slabs of the underground structure. In Case 2, Case 3, and Case 4, the peak dynamic deformation of the underground structure is shown in Fig. 2(c). When the aboveground structure is directly above the underground structure, the effect on the seismic deformation response of the underground structure is weak. However, when the aboveground structure is located above to side of the underground structure, it will significantly affect the seismic deformation of the underground structure, amplifying the shear deformation pointing away from the aboveground structure by 13.4%.

3.2 Mechanism Analysis

The physical mechanism of the effect of adjacent aboveground structures on the seismic response of underground structures is further analyzed. The seismic internal force response is analyzed first. At the end of the earthquake in Case 1, at a depth of 5.5 m (the depth corresponding to the top slab of the underground structure in Case 2–4), the total horizontal compressive stress increment distribution of the soil is shown in Fig. 3. This total horizontal compressive stress increment is caused by the accumulation of excess pore pressure in the saturated soil during the earthquake, which will produce lateral pressure on the underground structure. It can be seen from the figure that the gravity effect of the aboveground structure will amplify the total horizontal compressive stress increment of the soil within a certain range below. The position of the underground structure in Case 3 and Case 4 is also drawn with dotted lines in the figure. When the underground structure is located directly below the aboveground structure (Case 3), the total horizontal compressive stress increment of the soil on both sides is amplified most significantly, resulting in an obvious additional dynamic bending moment on the side walls. However, when the underground structure is located under the side of the above-ground structure (Case 4), the total horizontal compressive stress increment of the soil on both sides is relatively small, so the additional dynamic bending moment of the side walls is also small.

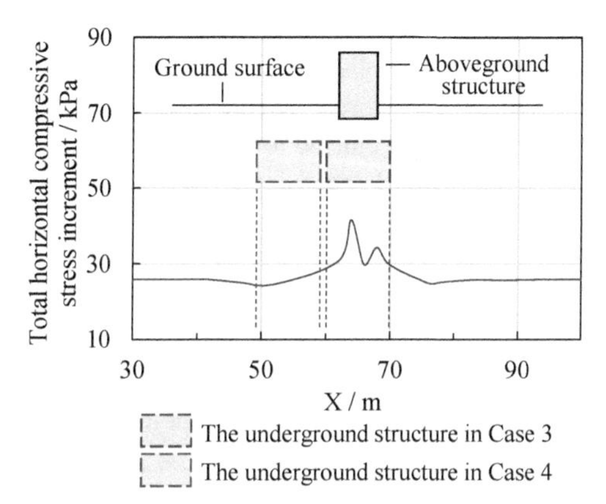

Fig. 3. Distribution of total horizontal compressive stress increment in the near-field soil at the depth of 5.5 m in the Case 1

The seismic deformation response of the underground structure is next analyzed. It can be seen from Fig. 3 that when the underground structure is located directly below the aboveground structure (Case 3), the total horizontal compressive stress increment of the soil on both sides is almost symmetrically distributed. Under the symmetrical action of this soil, the structure produces little additional shear deformation to one side. However, when the underground structure is located under the side of the aboveground structure (Case 4), the total horizontal compressive stress increment of the soil on the side close to the aboveground structure (right side) is greater than that on the side far away from the aboveground structure (left side). This asymmetric dynamic interaction results in a distinct unidirectional shear deformation of the underground structure that points away from the aboveground structure (left side).

4 Simplified Seismic Analysis Method for Underground Structures Considering the Effect of Adjacent Aboveground Structures

4.1 Principle of the Simplified Method

We have concluded in last section that the mechanism of the effect of adjacent above-ground structures on the seismic internal force and deformation response of underground structures is that the gravity increases the horizontal stress of the near-field soil. Under the action of the stress increment, the underground structure will generate additional internal force, and the asymmetry of the stress increment on both sides will cause additional deformation of the underground structure. Therefore, to establish a practical simplified analysis method for seismic design, it is necessary to reasonably simplify the description of the near-field soil compressive stress increment caused by the aboveground structure, and then apply it to the underground structure.

Boussinesq proposed an analytic formula for the stress distribution in spatial domain when the vertical force acts on the surface of the elastic semi-infinite space [14]. The distribution of the horizontal soil compressive stress increment at the corresponding position of the side walls of the underground structure under the aboveground structure can be obtained by this formula and applied to the walls on both sides of the underground structure in form of static distributed force. Then the internal force and deformation of the underground structure calculated can be approximately regarded as the additional seismic response of the underground structure caused by the action of the adjacent aboveground structure.

4.2 Validation of the Simplified Method

The effectiveness of the proposed simplified method can be demonstrated by comparing the additional response of the underground structure obtained by the simplified method and dynamic analysis.

For the additional internal force, when the aboveground structure is located directly above the underground structure, the effect on the seismic internal force response of the underground structure is the most significant, so the structure responses in Case 3 and Case 2 are compared. Under the action of gravity of the aboveground structure, the additional horizontal soil compressive stress increment distribution at the corresponding positions of the walls on both sides of the underground structure in Case 3 is the same, as shown in Fig. 4(a). The distributed force is applied to the two side walls of the underground structure, and the calculated additional bending moment distribution of the side walls is shown in Fig. 4(b). Compared with the results obtained from the dynamic analysis, it can be seen that the simplified analysis method can approximately predict the peak value of the additional dynamic bending moment.

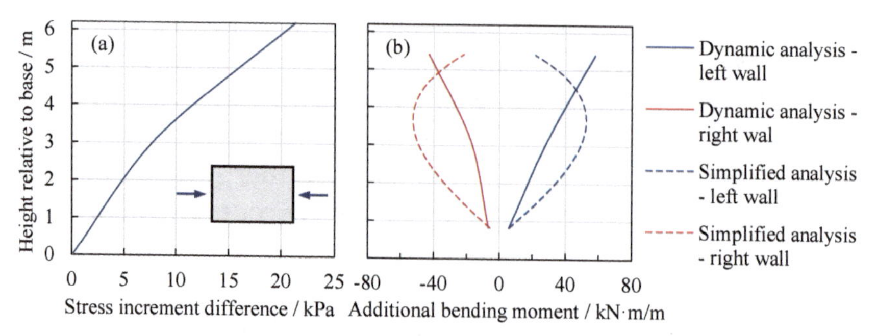

Fig. 4. (a) Distribution of the difference of the total horizontal compressive stress increment between near and far field soil under the action of the aboveground structure directly above the underground structure. (b) The additional bending moment of the underground structure calculated by dynamic analysis and simplified method

For the additional deformation, when the aboveground structure is located above to the side of the underground structure, the effect on the seismic deformation response of the underground structure is the most significant, so the structure responses in Case 4 and Case 2 are compared. Under the action of gravity of the aboveground structure, the additional horizontal soil compressive stress increment distribution at the corresponding positions of the right wall of the underground structure is larger than that of the left wall in in Case 4, as shown in Fig. 5(a). The distributed force is applied to the two side walls of the underground structure, and the calculated additional deformation of the underground structure (represented by the displacement difference relative to the bottom slab) distribution is shown in Fig. 5(b). Compared with the results obtained from the dynamic analysis, it can be seen that the simplified analysis method can approximately predict the distribution and peak value of the additional dynamic deformation.

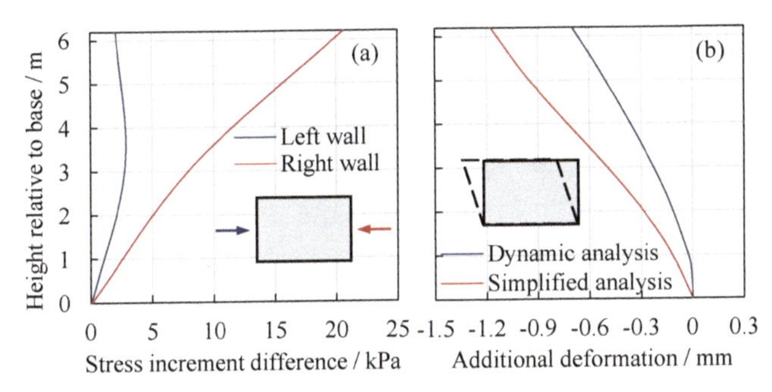

Fig. 5. (a) Distribution of the difference of the total horizontal compressive stress increment between near and far field soil under the action of the aboveground structure above to the right side of the underground structure. (b) The additional deformation of the underground structure calculated by dynamic analysis and simplified method.

5 Conclusion

In this study, three-dimensional numerical models are conducted for the underground structure-adjacent aboveground structure systems in saturated soil ground, and the fluid-solid coupling elastoplastic dynamic analysis is carried out to study the effect of the adjacent aboveground structure on the seismic internal force and deformation response of the underground structure. A practical simplified analysis method for seismic design of underground structures which can consider the effect of adjacent aboveground structures is proposed. The following conclusions are drawn:

(1) The adjacent aboveground structure has an amplified action on the seismic internal force response of the underground structure, which will cause the additional bending moment with inner tension of the side walls. The reason is that the gravity of the aboveground structure amplifies the horizontal compressive stress increment of the near-field soil on both sides of the underground structure.

(2) When the aboveground structure is located above to the side of the underground structure, it will cause additional shear deformation of the underground structure pointing away from the aboveground structure. The reason is the asymmetry of the horizontal compressive stress increment of the near-field soil on both sides of the underground structure caused by the gravity of the aboveground structure.
(3) A practical simplified analysis method for seismic design of underground structures that can consider the effect of adjacent aboveground structures is proposed based on the physical mechanism. For the underground structure-adjacent aboveground structure system, the distribution of the horizontal soil stress increment at the corresponding positions of the two side walls of the underground structure can be calculated first, and then applied to the underground structure in the form of static distribution force, and the additional internal force and deformation response of the underground structure caused by the effect of the aboveground structure can be approximately obtained.

Acknowledgments. The authors would like to thank the China Postdoctoral Science Foundation (No. 2021M703050) for funding this study.

References

1. Samata, S., Ohuchi, H., Matsuda, T.: A study of the damage of subway structures during the 1995 Hanshin-Awaji earthquake. Cement Concr. Compos. **19**(3), 223–239 (1997)
2. Yasuda, S., Kiku, H.: Uplift of sewage manholes and pipes during the 2004 Niigataken-Chuetsu earthquake. Soils Found. **46**(6), 885–894 (2006)
3. Tokimatsu, K., Tamura, S., Suzuki, H., et al.: Building damage associated with geotechnical problems in the 2011 Tohoku Pacific Earthquake. Soils Found. **52**(5), 956–974 (2012)
4. Lou, M., Wang, H., Chen, X., et al.: Structure–soil–structure interaction: literature review. Soil Dyn. Earthq. Eng. **31**(12), 1724–1731 (2011)
5. Pitilakis, K., Tsinidis, G., Leanza, A., et al.: Seismic behaviour of circular tunnels accounting for above ground structures interaction effects. Soil Dyn. Earthq. Eng. **67**, 1–15 (2014)
6. Dashti, S., Hashash, Y.M.A., Gillis, K., et al.: Development of dynamic centrifuge models of underground structures near tall buildings. Soil Dyn. Earthq. Eng. **86**, 89–105 (2016)
7. Pitilakis, K., Tsinidis, G.: Performance and seismic design of underground structures. In: Maugeri, M., Soccodato, C. (eds.) Earthquake Geotechnical Engineering Design. GGEE, vol. 28, pp. 279–340. Springer, Cham (2014). https://doi.org/10.1007/978-3-319-03182-8_11
8. Tsinidis, G.D.E., Silva, F., Anastasopoulos, I., et al.: Seismic behaviour of tunnels: from experiments to analysis. Tunn. Undergr. Space Technol. **99**(May), 103334 (2020)
9. Zhu, T., Hu, J., Zhang, Z., et al.: Centrifuge shaking table tests on precast underground structure-superstructure system in liquefiable ground. J. Geotech. Geoenviron. Eng. **147**(8), 106756 (2021)
10. Mckenna, F., Fenves, G.L.: OpenSees manual (PEER Center) (2001). http://OpenSees.berkeley.edu
11. Wang, R., Zhang, J.M., Wang, G.: A unified plasticity model for large post-liquefaction shear deformation of sand. Comput. Geotech. **59**, 54–66 (2014)
12. Zhang, J.M., Wang, G.: Large post-liquefaction deformation of sand, Part I: physical mechanism, constitutive description and numerical algorithm. Acta Geotech. **7**(2), 69–113 (2012)

13. Zhu, T., Wang, R., Zhang, J.M.: Effect of nearby ground structures on the seismic response of underground structures in saturated sand. Soil Dyn. Earthq. Eng. **2021**(146), 106756 (2021)
14. Boussinesq, J.: Application des Potentiels á l'Etude de l'Equilibre et du Mouvement des Solides Elastiques. Gauthier-Villars, Paris (1885)

Seismic Response Analysis on the Tunnel with Different Second-Lining Construction Time

Weigong Ma[1(✉)], Lanmin Wang[2], and Yuhua Jiang[3]

[1] College of Civil Engineering and Mechanics, Lanzhou University, Lanzhou 730000, China
370523730@qq.com
[2] Lanzhou Institute of Seismology, China Earthquake Administration, Lanzhou 730000, China
[3] China Railway Northwest Science Research Institute Co., Ltd., Lanzhou 730000, China

Abstract. For large deformation of soft rock tunnel, the construction time of secondary line plays a key role in the research of controlling the large deformation. In this paper, taking the Qamdo Tunnel of Sichuan-Tibet Railway, which is the greatest engineering project in construction in the 21th century, the stress states of secondary lining with different installation times of secondary lining are analyzed in the statics and dynamics. The results indicated that dynamic stress has a more remarkable decreasing tendency than static stress as the construction time for secondary lining; tunnel structure will have a lower sensitivity to dynamic excitation when the displacement ratio is up to 98% to construct secondary lining. If the secondary lining has to be constructed with a displacement ratio less than 98%, the concrete strength must be strengthened based on the real construction time of secondary lining, so that it may safely bear more partial load for controlling large deformation of the tunnel.

Keywords: Construction time · Secondary lining · Large deformation · Tunnel · Seismic response

1 Introduction

The construction time of tunnel secondary lining have always been hot issue in the underground engineering. As is known the magnitude of resistance force offered by tunnel secondary lining due to the construction time. If the secondary lining is constructed too early, there will be too larger stress state in concrete, causing concrete cracking and fissure. And if the time lags the optimal opportunity, the secondary lining will can bear no its responsibility excellently about itself, yet further the surround rock before supporting may lose stability and cause collapse accident.

The determine of optimal time for constructing secondary lining is especially important in soft rock tunnel under high geostress state, in which there will usually have been a large deformation. According the relevant tunnel code, secondary lining should be applied when the initial support deformation rate is less than 0.1–0.2 mm/d, or the prior deformation reaches 80%–90% of the total predicted deformation (ultimate deformation). In the case of shallow buried area, especially extremely shallow buried area, and

© The Author(s), under exclusive license to Springer Nature Switzerland AG 2022
L. Wang et al. (Eds.): PBD-IV 2022, GGEE 52, pp. 2322–2330, 2022.
https://doi.org/10.1007/978-3-031-11898-2_217

surrounding swelling rock, the time are determined according to the results of monitoring measurement [1]. However, how to analyze the monitoring results and determine the timing of secondary lining has not been given.

Li and Ma [2] researched the tunnel structure of secondary lining by elasto-viscoplastic finite element, which is constructed when the deformation rate reached 2 mm/d. Liu et al. [3] obtained that it is reasonable and feasible to apply the secondary lining timely and in advance for the tunnel with large deformation of soft rock, through the analysis of secondary lining construction time about Wushaoling Tunnel. Wang et al. [4] recommended that the secondary lining be applied when the external loads of the segments are up to 83%–100% of the design load. Guo et al. [5] proposed the construction time of the secondary lining in the soft rock tunnel with different large deformation grades and different section.

Cheng et al. [6] conducted the numerical seismic analysis and shaking table test of tunnel in a loess soil considering rainfall and traffic load. Cheng et al. [7] compared the seismic response of the loess tunnel structure and the damping effect of the rubber damping layer under different seismic fields and rainfall conditions. Li et al. [8] analyse the dynamic response characteristics and failure process of a loess tunnel portal section under earthquake action and the transmission law of the seismic wave with consideration of tunneling elevation on the slope.

The previous studies about the construction time of secondary lining almost focus on the statics. This paper conducted the relative comparison analysis of statics and dynamics about the Qamdo tunnel of Sichuan-Tibet Railway with different construction times of secondary lining, which locals in strong earthquake region.

2 Introduction of the Project

2.1 The Sichuan-Tibet Railway

The Sichuan-Tibet Railway is the second "sky-road" into Tibet after the Qinghai-Tibet Railway (see Fig. 1). The whole line is divided into three sections: Chengdu to Ya'an section, Lhasa to Nyingchi section, and Ya'an to Nyingchi section. The total length of railway is about 1 700 km, more than 80% will be built in the way of tunnel and bridge, design speed of 200 km/h (speed limit of 160 km/h for some sections).

Fig. 1. The line of Sichuan-Tibet railway

The Ya'an to Nyingchi section is the just starting and most difficult project, which have the average elevation of more than 3 000 m, and have a accumulate climbing-up heighten of about 16 000 m. Optimal builders will have to face high seismic and geological faults, high geostress, high geotherm, high-density pebble bed, prone geological disasters and tunnel oxygen deficit, total six difficult problems. After completed, the travel time between Chengdu and Lhasa is expected to be shortened to 13 h, and the transportation conditions will be thoroughly changed in regions along railway.

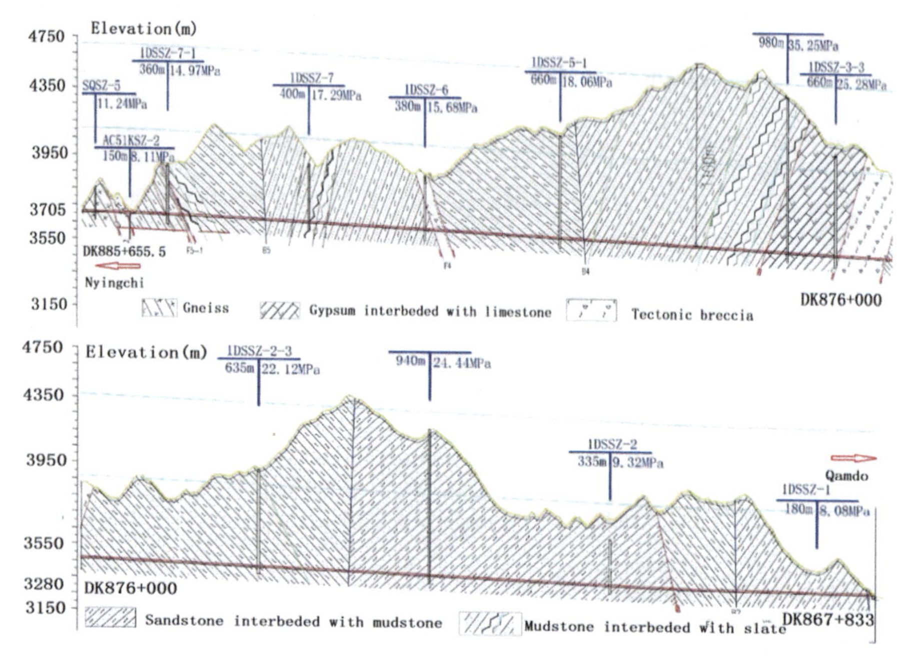

Fig. 2. The longitudinal profile of Qamdo Tunnel

2.2 The Qamdo Tunnel

The Qamdo Tunnel is just starting construction as a key controlling project, which have the length of 16 715 m and the largest burial depth of 1 150 m. The eastern access portion locals at the Jaka development zone of Qamdo city, the right bank of Lantsang river, and the exit portion locals at the Jitang town ChaYa county, which the line is showed in Fig. 3 [9]. The tunnel totally lies at the western of Hengduan Mountains on the Tibetan Plateau. According to the engineering geological survey report [10], the geology is shown as Fig. 2. The most portion of tunnel lies in the high geostress and extreme-high geostress regions, in which the ratio range of level stress to vertical stress is 1.64–1.98.

3 Analysis for Different Construction Times

3.1 Calculation Model

Select the DK870 + 920 section as the analyzing model, with the buried depth of 350 m and the sandstone interbeded with mudstone of grade V. It is assumed a plane strain problem, in which the surrounding rock is isotropic and elasto-viscoplastic material as the initial support, and the secondary lining is elastic material [2, 10]. Referring to the relevant engineering and the experiment results of similar rock, the material and support parameters are selected as in Tables 1 and 2 [2, 10].

The numerical analysis model is established as is showing in Fig. 4. The distances from the left/right edges to tunnel are 5 times of the tunnel span, and from top/bottom to tunnel is 4.5 times. The vertical geostress equaling the upper gravity is aligned through the pressure loading, in which the geostress ratio of horizontal to vertical is 1.64. The bottom boundary of model is fixed at the vertical direction, and the side boundaries are limited by 0 displacement at the level direction. The excavation method is the three benches seven steps method, through the element death and delay time to realize the excavation of tunnel face [11, 12].

Fig. 3. The line of Qamdo Tunnel

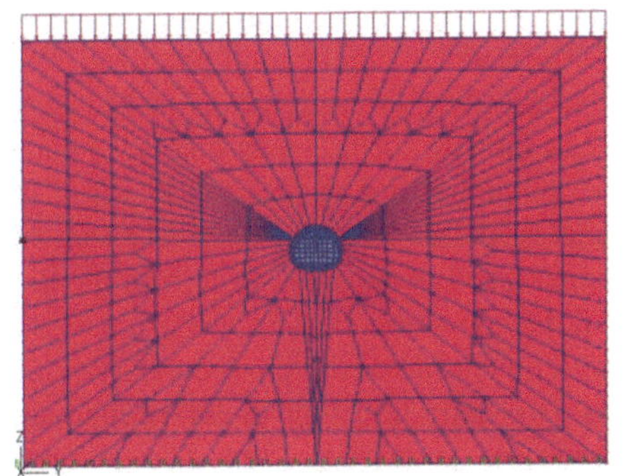

Fig. 4. The numerical analysis model

Table 1. The material parameters of numerical model

Material	Young's modulus (MPa)	Poisson's ratio	Density (kg/m^3)	Yield stress (MPa)	Fluidity parameter (d^{-1})
Surrounding rock	2 000	0.35	2 000	0.7	0.005 5
Initial support	26 000	0.28	2 400	20.0	0.001 2
Secondary lining	28 000	0.20	2 500	/	/

Table 2. The support parameters of tunnel

Material	Position	Specifications
C25 spray concrete	Full-circular	Thickness 35 cm
I25a section steel	Full-circular	Spacing 0.6 m
C35 second lining	Full-circular	Inver arch thickness 0.7 m, others 0.6 m
Steel bar	Full-circular	Φ22@20cm

3.2 Construction Time of Secondary Lining

For determining the suitable construction time, it is necessary to obtain the ultimate vault displacement. So the tunnel face is excavated with no secondary lining, and then the fitting curve of vault displacement (see Fig. 5) is got. Equation (1) is the fitting equation of vault displacement with no secondary lining.

$$z = -0.2451 \times \left(1 - e^{-0.1377t}\right), \quad (1)$$

where z is the vault vertical displacement (m); t is the time (day); the relation coefficient is $r^2 = 0.9864$; and the total predicted deformation is −0.2451 m.

According the suggestion of Liu et al. [3], the secondary lining are constructed when the ratio of previous displacement to ultimate displacement is to 65–80%. The deformation ratios of 65%, 80%, 90%, 95% and 100% are chose as the construction time, so the times are respectively 14d, 18d, 24d, 30d and 53d after the first bench of tunnel face is excavated. Referring the deformation rate $\leq$2 mm/d [2] and $\leq$0.2 mm/d [1], the construction times are 21d and 38d versus 84% and 98%. Finally the different work conditions about the secondary lining construction times are 14d, 18d, 21d, 24d, 30d, 38d and 53d behind the first bench of tunnel face excavated. The static analyses are conducted on secondary lining structure with the different work conditions.

3.3 Dynamic Response Analysis

According the basic seismic precautionary intensity of VIII (peak acceleration is 0.2 g) for Qamdo Tunnel, considering the influence of near-field pulse seismic, the Wolong wave recorded from the 5.12 Wenchuan Earthquake is selected, the acceleration time history curve is shown in Fig. 6. The Wolong wave is modulated that the peak acceleration is equal to 0.2 g, so as the seismic wave for dynamic response analysis, it is suitable for the seismic precautionary intensity of the tunnel.

Through the two times of integration for modulated wave, it is obtained that the seismic displacement curve as in Fig. 7. The seismic response is loaded by applying the horizontal displacement at the model boundary utilizing the seismic displacement curve. The seismic response analysis is conducted based on static analysis results under the different work conditions.

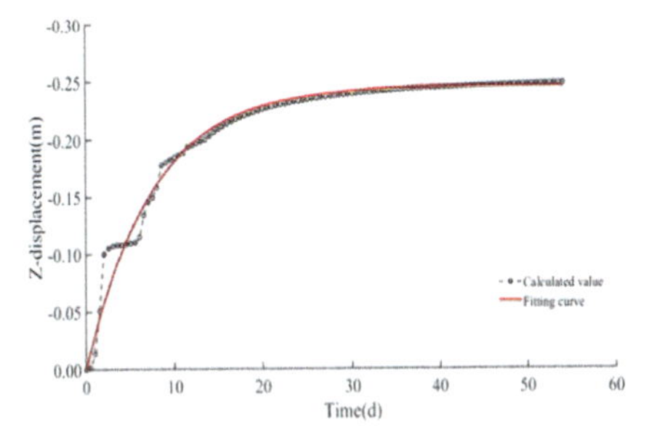

Fig. 5. The curves of vault displacement

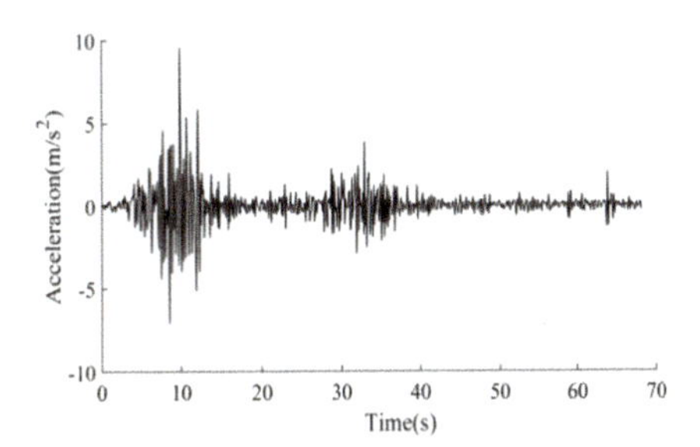

Fig. 6. The time history of ground acceleration recorded at wolong station

4 Result Analysis

4.1 Static Stress Analysis of Secondary Lining

According the static analysis result conducted under the different work conditions, the key positions' stresses of secondary lining concrete structure are listed out in Tables 3 and 4. The first principal stress (σ-p1) is listed in Table 3, and the third principal stresses (σ-p3) are listed in Table 4. The largest tensile stress and compressive stress are always at the joint of side-wall and invert-arch with different work conditions. As the time delay of installation time for secondary lining, they all have a clear decreasing tendency.

Table 3. The first principal stress of secondary lining (MPa)

Time Position	14	18	21	24	30	38	53
Vault center	−0.426	−0.255	−0.017	0.010	0.002	0.005	0.017
Arch-waist	0.340	0.344	0.033	0.280	0.191	0.116	0.013
Largest span	0.255	0.417	0.044	0.380	−0.129	−0.004	0.009
Side-wall foot	4.181	4.353	3.533	2.405	1.739	1.121	0.032
Invert-arch center	−0.183	−0.300	−0.313	−0.257	−0.200	−0.144	−0.006

Table 4. The third principal stress of secondary lining (MPa)

Time Position	14	18	21	24	30	38	53
Vault center	−3.776	−2.685	−2.125	−1.445	−0.670	−0.356	−0.139
Arch-waist	−0.816	−0.432	−0.262	−0.123	−0.095	−0.083	−0.098
Largest span	−5.497	−4.607	−3.785	−2.683	−1.335	−1.308	−0.360
Side-wall foot	−22.143	−14.917	−13.543	−8.997	−6.189	−3.807	−1.247
Invert-arch center	−9.256	−7.607	−5.778	−3.802	−2.727	−1.877	−0.849

4.2 Seismic Stress Analysis of Secondary Lining Structure

The seismic response analyses are conducted based on static analysis results under the different work conditions. The key positions' dynamic stresses of secondary lining are listed out in Tables 5 and 6. The dynamic first principal stresses (σd-p1) are listed in Table 5 and the dynamic third principal stresses (σd-p3) are listed in Table 6. The max dynamic tensile stress and max compressive stress are all at the joint of side-wall and invert-arch. They also have a decreasing tendency as the construction time of secondary lining.

The max stress comparison of different position under different work conditions are showed in Fig. 8. For σ-p3 and σd-p3 only the numerical value is showed in the figure. The obvious decreasing tendency as the installation time for secondary lining for σ_d-p1 and σ_d-p3 can be observed. For σ_d-p3 the tendency have a clear reduction when the installation time is to 30d, the 30d is the time of the 95% ratio of previous displacement to ultimate displacement. And for σ_d-p1 the clear tendency decreasing time is 38d, the time is the 98% displacement ratio, which is also the stabilized time of the code for design of railway tunnel (deformation rate $\leq$0.2 mm/d). In other words, the tunnel structure will have a low sensitivity to dynamic excitation when the displacement ratio is up to 98% to construct secondary lining [13].

Table 5. The first principal dynamic stress of secondary lining (MPa)

Time Position	14	18	21	24	30	38	53
Vault center	21.802	31.353	7.302	34.463	28.556	5.326	3.982
Arch-waist	33.196	35.924	16.845	18.284	4.913	2.759	4.705
Largest span	58.292	53.869	12.046	20.565	8.372	4.044	1.411
Side-wall foot	66.119	54.985	34.75	25.739	−0.918	0.109	1.092
Invert-arch center	45.382	40.426	44.111	30.952	−4.304	0.797	0.150

Table 6. The third principal dynamic stress of secondary lining (MPa)

Time Position	14	18	21	24	30	38	53
Vault center	−2.032	−4.482	−9.482	−3.842	−4.093	−5.655	−1.767
Arch-waist	−23.518	−49.476	−27.147	−26.037	−6.423	−14.359	−11.750
Largest span	−4.157	−9.335	−7.941	−8.947	−9.073	−2.568	−1.475
Side-wall foot	−60.032	−49.181	−47.745	−26.312	−2.279	−4.782	−10.050
Invert-arch center	−7.292	−5.778	−5.443	−16.043	−9.447	−6.811	−7.205

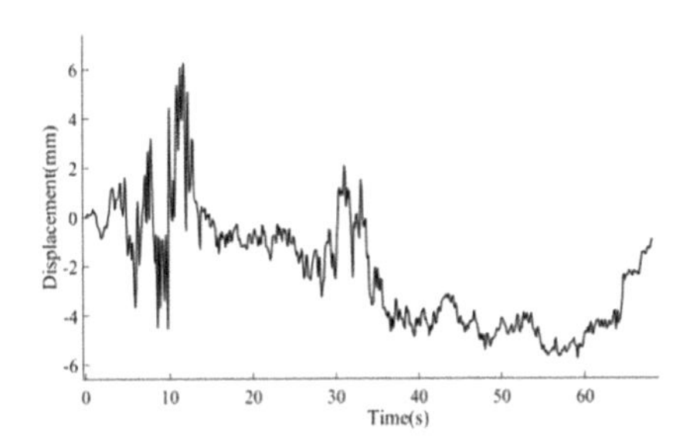

Fig. 7. The seismic displacement curve

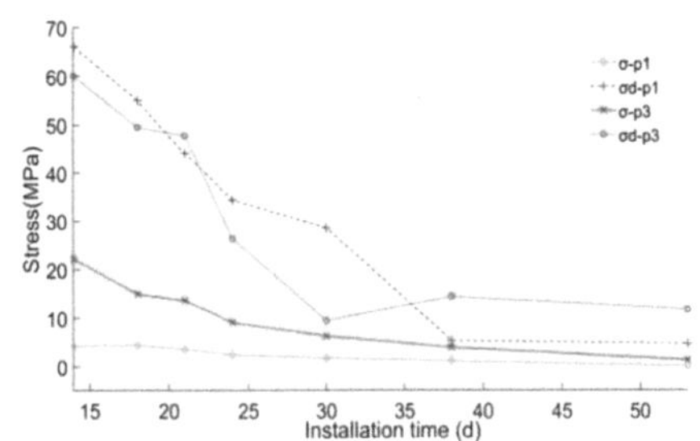

Fig. 8. Max stress comparison of secondary lining with different work conditions

5 Conclusion

According the above comparative analysis at the statics and dynamics through numerical calculation, the following viewpoint is obtained:

1. For large deformation of soft rock tunnel, it is very necessary to choose a appropriate construction time for secondary lining, no matter in terms of static and dynamic performance. The seismic response analyses show that the national code constructs the secondary lining when the ratio of previous displacement to ultimate displacement is up to 98%, that is more appropriate to structure safety.
2. The method of letting the secondary lining to bear more partial load of initial support is beneficial to the tunnel stability of large deformation, but the construction time must be comparatively calculated and determined according to the onsite monitoring measurement and the statics and dynamics analyses.
3. The strength growth as time for the concrete of secondary lining isn't considered in this paper. Considering the time effect of concrete strength, the schedule for constructing the secondary lining on site should be ahead of 1–2 days appropriately.
4. If it is must to bring forward the construction time of secondary lining to control the large deformation of surrounding rock and prevent collapse accident, the elastic modulus should be appropriately decreased, and the strength of concrete should be appropriately increased, especially the tensile strength, which will be beneficial for

structure safety neither at statics or dynamics. Because there are obvious dynamic-stress concentration phenomena at the joint point of side-wall and invert-arch, and at the large span, more attention should be given.

References

1. State Railway Administration: TB10003-2016. The Code for Design of Railway Tunnel. China Railway Publishing House, Beijing (2016)
2. Li, D.W., Ma, W.G.: Elasto-viscoplastic finite element analysis of the installation time of a secondary lining. Mod. Tunnel. Technol. **49**(4), 6–9 (2012)
3. Liu, Z.C., Li, W.J., Zhu, Y.Q., et al.: Research on construction time of secondary lining in soft rock of large-deformation tunnel. Chin. J. Rock Mech. Eng. **27**(3), 580–588 (2008)
4. Wang, S.M., Jian, Y.Q., Lu, X.X., et al.: Study on load distribution characteristics of secondary lining of shield under different construction time. Tunnel. Undergr. Space Technol. **89**, 25–37 (2019)
5. Guo, X.L., Tan, Z.S., Li, L., et al.: Study on the construction time of secondary lining in phyllite tunnel under high geostress. China J. Highway Transp. **33**(12), 249–261 (2020)
6. Cheng, X.S., Zhou, X.H., Liu, H.B., et al.: Numerical analysis and shaking table test of seismic response of tunnel in a loess soil considering rainfall and traffic load. Rock Mech. Rock Eng. **54**, 1005–1025 (2021)
7. Cheng, X.S., Zhou, X.H., Wang, P., et al.: Shaking table model test of loess tunnel. China J. Highway Transp. **34**(6), 136–146 (2021)
8. Li, S.J., Liang, Q.G., Fang, J., et al.: Seismic dynamic response characteristics of loess tunnel portal section with consideration of tunneling elevation on the slope. China Earthq. Eng. J. **42**(4), 996–1006 (2020)
9. Wu, S.S.: Study on Large Deformation Classification of Changdu Tunnel of Sichuan-Tibet Railway. Southwest Jiaotong University, Chengdu (2020)
10. China Railway First Survey and Design Institute Group Co. Ltd. Geological survey report of Qamdo Tunnel engineering. China Railway First Survey and Design Institute Group Co., Ltd., Xi'an (2019)
11. Zhao, Y., He, H.W., Li, P.F.: Key techniques for the construction of high-speed railway large-section loess tunnels. Engineering **4**(2), 254–259 (2018)
12. Liu, J.B., Gu, Y., Du, Y.X.: 3D consistent viscous-spring artificial boundaries and viscous-spring boundary elements. Chin. J. Geotech. Eng. **28**(9), 1070–1075 (2006)
13. Ma, W.G., He, W.B.: The influence evaluation of the waste slag above tunnel on the safety of the structure of the high-speed railway tunnel. In: The 7th Symposium on Innovation & Sustainability of Modern Railway, Nanchang, pp. 31–38 (2020)

Coupled Seismic Performance of Underground and On-Ground Structures

Juan Manuel Mayoral(✉), Daniel De La Rosa, Mauricio Alcaraz, and Enrique Barragan

Institute of Engineering at National Autonomous University of Mexico, Mexico, Mexico
JMayoralV@iingen.unam.mx

Abstract. Seismic performance evaluation of on-ground structures can be significantly affected by their interaction with underground structures. This effect becomes more relevant in densely populated areas where transportation systems such tunnel metro lines are located nearby urban overpasses, and buildings. This paper presents the findings of a numerical study carried out to conduct the seismic evaluation of a typical bridge-building-tunnel system, located in the high plasticity Mexico City clay. Through a series of three-dimensional finite difference numerical models developed with the software FLAC^{3D}, both detrimental and beneficial interactions effects among these structures are analyzed, establishing zones around each structure in which this interaction leads to ground motion variability, as well as the impact on both the building and bridge seismic performance. The parametric study was carried out considering both normal and subduction events for a return period of 250 years. Distances among each structure was varied to cover a wide range of scenarios. From the results gathered in here, the effect of this interaction on the modification of surface accelerations, seismic demand in on-ground structures, as well as deformations and structural stresses in the tunnel lining was established.

Keywords: Seismic · Soil-structure interaction · Tunnel · Bridge · Building

1 Introduction

Major earthquakes preparedness on densely populated urban areas located in seismically active regions, such as Mexico City, Los Angeles, and San Francisco, requires a proper assessment of the seismic demand expected to occur in buildings, bridges and tunnels. Thus, proper vulnerability assessment of the seismic interaction is a mandatory step to study strategies to increase strategic infrastructure resilience in cities during extreme events. Increasing awareness of vulnerability has led to a growing interest in the research community towards the quantification of risk and resilience. To date, however, there is a lack of knowledge of the seismic interaction among underground and on-ground structures. Thus, it is critical to study the complex interaction between the incident, reflected, and diffracted seismic waves in underground structures, and those generated by the surface structures moving back and forth (Fig. 1) [1]. Although various research

© The Author(s), under exclusive license to Springer Nature Switzerland AG 2022
L. Wang et al. (Eds.): PBD-IV 2022, GGEE 52, pp. 2331–2338, 2022.
https://doi.org/10.1007/978-3-031-11898-2_218

groups have studied the interaction between surface structures [e.g., 2, 3], they have not considered the interaction with underground structures. This leads to potentially expensive, and in some cases unsafe structure designs. These consequences can lead to negative effects on a structure such as reaching or exceeding a particular damage level or potential economic losses associated with these damages (e.g. costs due to required repair actions or due to loss of functionality).

Recently, Mayoral and Mosqueda, [4] conducted a numerical study to investigate the ground motion variability associated with the interaction of tunnels and buildings on soft clay deposits, such as those found in Mexico City, considering a normal and a subduction events. In this research, the detrimental effects of the interaction between both structures, as well as with surrounding structures, were established. In subsequent research [5], they analyze the effect of the variability of the ground motion, including a set of records with a broad range of duration, amplitude, and frequency content. This paper presents the comparison of the tunnel-soil-building and tunnel-soil-bridge interaction systems, including the frequency content, duration, and intensity effects in the seismic interaction. The results presented are part of an ongoing investigation.

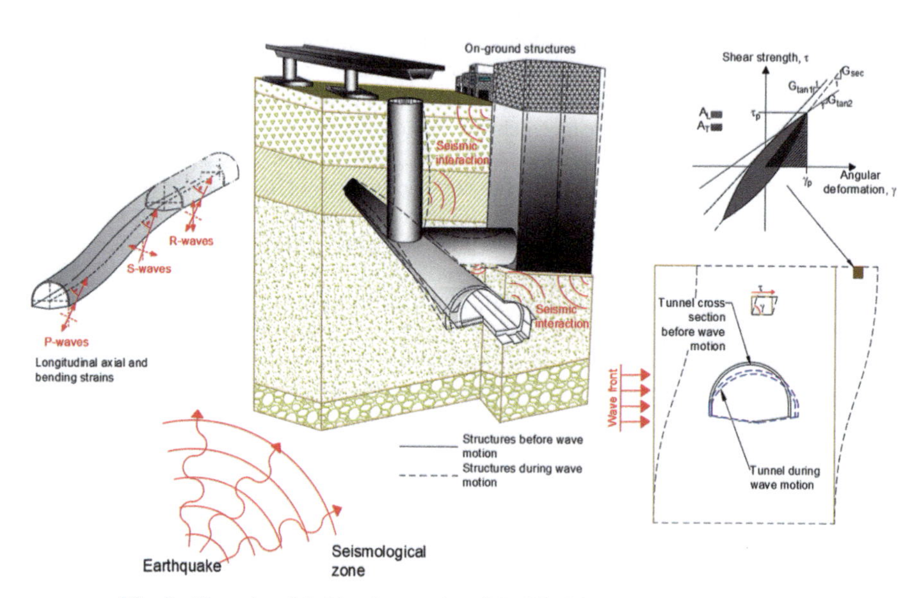

Fig. 1. Tunnel-soil-bridge interaction (Modified from Mayoral et al. [1]).

2 Problem Idealization

Tunnel-soil-building and tunnel-soil-bridge interaction in soft clays was studied considering the typology depicted schematically in Figs. 2a and 2b, using tridimensional finite difference models developed with the program FLAC^{3D} [6]. These correspond to common tunnels, buildings, and bridges typologies found in Mexico City. In the parametric studies, the tunnel width D, bridge height H_o, and building height H_b, were assumed to be 11, 12 and 21 m, respectively, and the tunnel cover remained constant and equal to 2D. The distance of interaction tunnel-building and tunnel-bridge L_i varied from 0 to 1D. Thus, including a reference case corresponding to the absence of the tunnel, three cases were analyzed for each scenario. The site is in the so-called Lake Zone of Mexico City, composed of highly compressible clay deposits separated by layers of compact to very compact silty or clayey sand. The shear wave velocity distribution was obtained by Seed et al. [7] using down-hole, and P-S suspension logging technique, and is also included in Fig. 2c. Gonzalez and Romo's model [8] was used to estimate the normalized modulus degradation and damping curves for clays. For sands, the upper and lower bounds proposed by Seed and Idriss [9] for normalized modulus degradation and damping curves, respectively, were deemed appropriated.

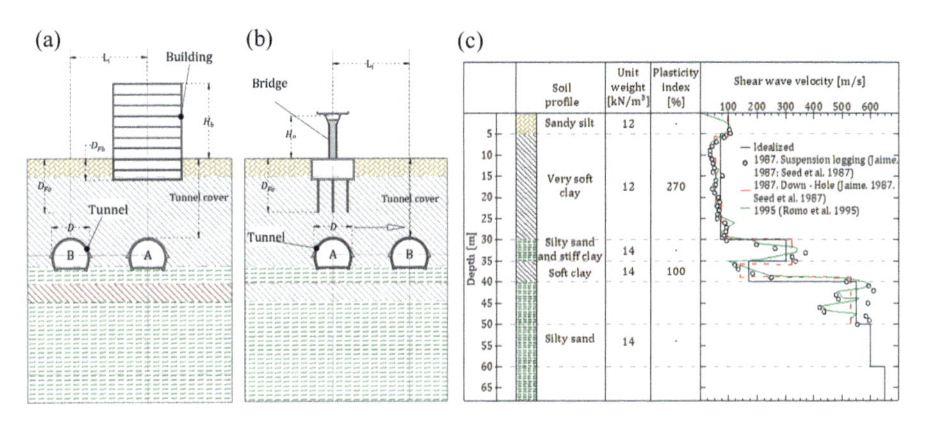

Fig. 2. (a) Tunnel-soil-building, (b) tunnel-soil-bridge interaction and (c) soil profile.

3 Numerical Model

Seismic tunnel-soil-building and tunnel-soil-bridge interaction analyses were carried out by a series of three-dimensional finite-difference models developed by the program FLAC^{3D} [6], assuming different positions of the tunnel with respect to the building and bridge (Fig. 3). A seven-story 20 by 20 m^2 square footprint building, with a compensated box-like foundation 6 m deep, was considered in the parametric study. The building was simplified as a shear beam comprised by solid elements, with equivalent stiffness, ki, and mass, mi, for each story i. The dimensions of the equivalent shear beam are the same as those of the building considered. The mass is evenly distributed on each floor, as well as the shear modulus, G. The details of this assumption can be found in Mayoral and Mosqueda, [4]. On the other hand, a common type of bridge found in Mexico City was considered in the numerical study. These bridges are precast pre-stressed concrete structures. The 9 m wide deck is composed of a hollow central girder and tablets and is structurally attached to the 12 m height column. The 15 m depth reinforced concrete foundation consists of a 6 m depth box with a 12.5 by 12.5 m^2 square footprint and 9 piles with a 50 cm side square section and a 9 m depth. The superstructure of the bridge was idealized by beam elements with an equivalent density in the upper deck to simulate the rocking effect of the real superstructure. The box foundation was modeled as a shear beam of solid elements, and the piles were modeled with pile elements, with spring parameters selected based on the surrounding soil. The tunnel has an 8 m height and 11 m width horseshoe section. The 20 cm thick primary lining is and made of shotcrete with steel fibers and the 40 cm thick secondary lining is made of reinforced concrete. The 60 cm thick lining was modeled with shell elements.

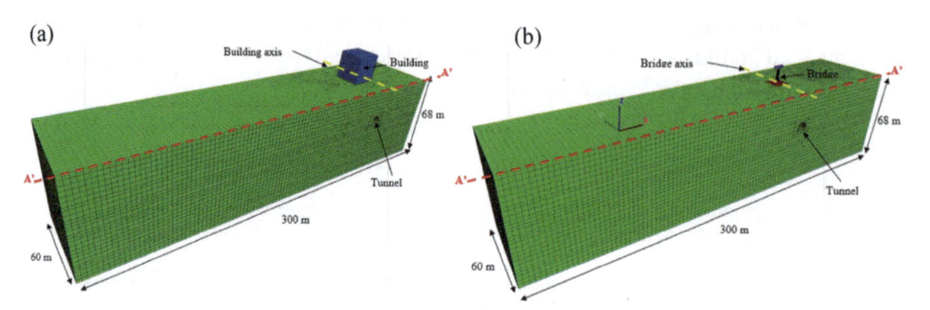

Fig. 3. Numerical models for (a) tunnel-soil-building, and (b) tunnel-soil-bridge interaction for case A (tunnel below the on-ground structure).

4 Seismic Environment

Mexico City's seismic hazard is controlled mainly by interplate earthquakes along the Mexican subduction zone, such as the September 19, 1985, earthquake, and intraplate earthquakes in the subducted Cocos plate (e.g., the event of September 19, 2017). Considering this, the Mexico City building code establish the seismic environment using a uniform hazard spectrum (UHS) with a return period of 250 years [10]. To include the effects of duration, amplitude, and frequency content, the seismic tunnel-soil-building and tunnel-soil-bridge interaction was analyzed using the set of six ground motions compiled in Table 1, with varying duration and frequency characteristics. These strong ground motions were recorded in firm soil or rock. According to the ASCE/SEI Standard 7–10, when the required number of recorded ground motion is not available, appropriate simulated ground motion is permitted to be used to make up the total number required [11]. To develop an acceleration time history which response spectrum reasonably matches the design response spectrum for the return period of analysis (i.e., T = 250 years), the selected time history, usually called seed ground motion, were modified using the method proposed by Lilhanand and Tseng [12] as modified by Abrahamson [13]. Table 1 summarizes the earthquakes considered in the analyses, where Mw, is the moment magnitude; PGA, is the peak ground acceleration, T_D, is the significant duration of ground motion defined as the difference of T-95 and T-5 which are respectively the times where 95% and 5% of Arias intensity is reached, Freq, is the dominant frequency, and AI, is the Arias intensity.

Table 1. Earthquakes considered in the analyses.

Seismogenic zone	Earthquake name	Year	Mw	PGA (g)	Freq (Hz)	T_D (s)	AI (cm/s)
Normal	Montenegro (former Yugoslavia)	1979	6.9	0.251	1.35	12.1	46.8
	Umbria (Gubbio-Piana, Italy)	1998	4.8	0.223	4.44	2.3	18.2
	CU17 (Puebla-Morelos, Mexico)	2017	7.1	0.059	0.56	29.6	12.7
Subduction	CU85 (Michoacan, Mexico)	1985	8.1	0.033	0.50	49.7	15.5
	Chile (Maule, Chile)	2010	8.8	0.638	1.25	71.2	1418.5
	Japan (Honshu, Japan)	2011	9.0	0.939	1.73	68.3	136.4

5 Seismic Tunnel-Soil-Building and Tunnel-Soil-Bridge Interaction

For the dynamic analyses, the synthetic ground motions were applied in two orthogonal directions x (i.e., transversal) and y (i.e., longitudinal), considering 100% of the ground motion in each component. To study the effect of the tunnel on the seismic tunnel-soil-building and tunnel-soil-bridge interaction, the results were presented in terms of the tunnel factor for the building, T_{rf} (Fig. 4a), and for the bridge, T_{df} (Fig. 4b), which are defined as the transfer function computed between the building and its foundation without the tunnel, and the upper bridge deck and its foundation without the tunnel respectively, divided by the transfer function between the roof and its foundation and the upper deck its foundation considering the presence of the tunnel. These results are presented for all the scenarios considered in Figs. 4a and b, for the transversal (X-direction) and longitudinal (Y-direction) components. T_{rf} and T_{df} allows to clearly show the effect of the amplification potential and energy distribution within the frequency content of interest. Values of T_{rf} and T_{df} above one corresponds to beneficial interaction (i.e., the building and bridge amplification potential decreases due to the tunnel presence), whereas values below one means detrimental interaction (i.e., the building and bridge amplification potential increases due to the tunnel presence). There is a clear amplification potential (i.e., detrimental interaction) on the building due to the tunnel for the frequency range of interest for the problem at hand (i.e., from 0.7 Hz to 1.2 Hz), which include the fundamental period of the structure, $Tf = 1.01$ s (0.99 Hz), and the characteristic elastic site period of the soil, $Tpe = 0.85$ s (1.17 Hz). This effect is evident in both normal and subduction events and appears to be more important for case A. The amplification potential (i.e., detrimental interaction) also is presented for the bridge, due to the tunnel for the frequency range from 1.3 to 2.6 Hz, which include the fundamental period of the structure, Tb = 0.4 and 0.6 s (2.5 and 1.6 Hz) for the transversal and longitudinal directions respectively. It's clearly noticed that these effects decrease when the tunnel gets away from the building and the bridge (Case B) for the frequency range of interest (i.e., 0.7 to 1.2 Hz and 1.3 to 2.6 Hz for tunnel-building and tunnel-bridge interaction respectively).

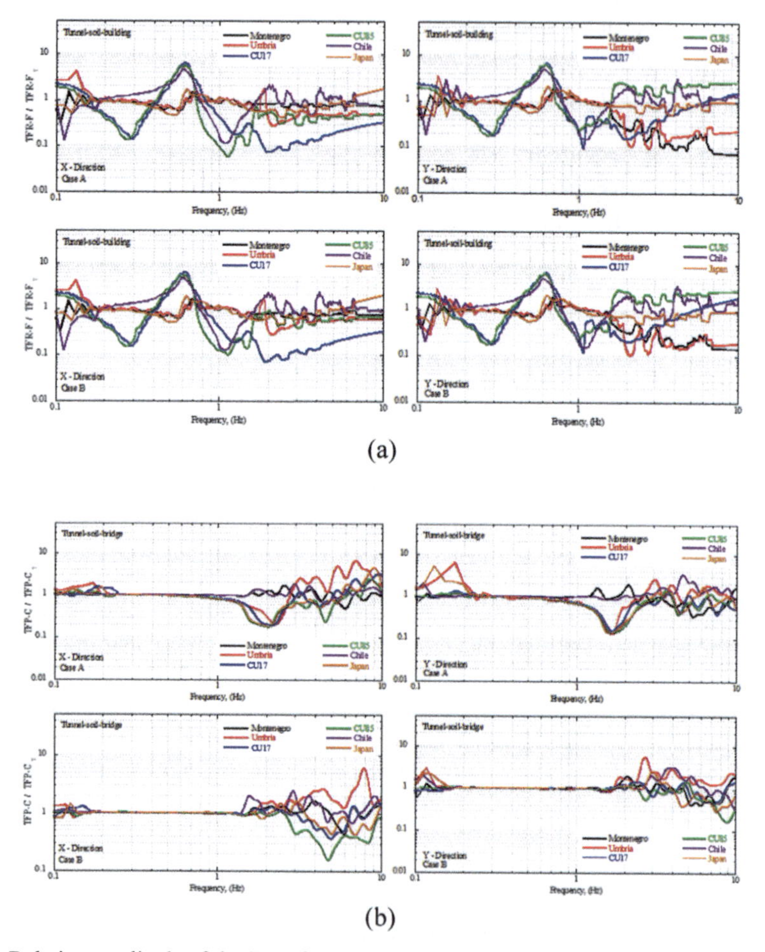

Fig. 4. Relative amplitude of the Transfer Functions between the foundation and the Roof (Case tunnel-building) and the Upper deck (Case tunnel-bridge) of the bridge for both transversal and longitudinal directions.

Overall, this increment in the amplification potential of the building and bridge associated to the tunnel, leads to an increment in the computed maximum acceleration at the building roof and at the upper deck, on the order of 10 and 8.5% respectively, as depicted in Fig. 5, for case A. As can be seen, the tunnel effect is stronger in the transversal direction, where the effect of the reduction in the mass and soil stiffness due to the excavation is higher than in the longitudinal direction. This amplification is also in agreement with the Housner spectral intensity, SI(ξ), computed for each scenario, as depicted in Fig. 6, for both transversal (x) and longitudinal (y) directions. The Housner spectral intensity was computed with and without tunnel, as a measured of the energy of the system. As can be seen, the effect is more important in the building-tunnel interaction in the transversal direction. For the bridge-tunnel interaction, the effect of the frequency content is stronger than in the building-tunnel interaction, where a lightly amplification is presented for the longitudinal direction for subduction events.

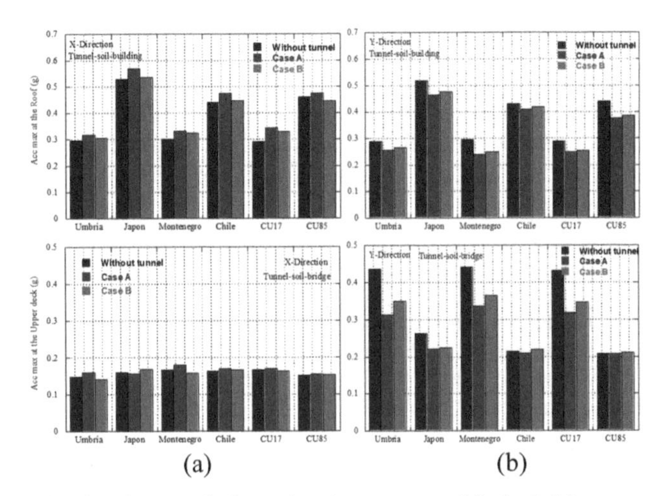

Fig. 5. Peak roof and peak upper deck accelerations computed for both (a) transversal, X, and (b) longitudinal, Y, directions

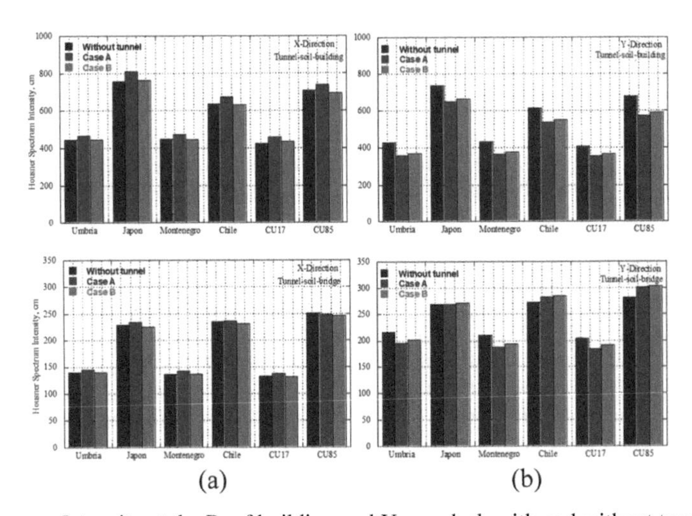

Fig. 6. Housner Intensity at the Roof building and Upper deck with and without tunnel for both (a) transversal, X, and (b) longitudinal, Y, directions

6 Conclusions

For the specific cases analyzed, there is an amplification (i.e., detrimental interaction) in the building due to the presence of the tunnel, increasing the maximum accelerations calculated on the roof of the building and the upper deck of the bridge. This effect is evident for both subduction and normal failure events and is more important for case A (i.e. tunnel under the building and bridge), and gradually reduces its impact as the tunnel moves away from the on-ground structure. The tunnel effect is stronger in the transversal direction, where the effect of the reduction in the mass and soil stiffness due to the excavation is higher than in the longitudinal direction. However, for the bridge-tunnel interaction, the effect of the frequency content is stronger than in the building-tunnel interaction, this is related to the period of the structure, and the stiffness of the bridge in that component. This fact must be considered to adequately estimate the seismic demand in the structures surrounding underground structures. The results presented are part of an ongoing investigation. The interaction of the three systems tunnel-soil-building-bridge will be published in the first part of 2022.

References

1. Mayoral, J.M., Mosqueda, G., De La Rosa, D., Alcaraz, M.: Tunnel performance during the Puebla-Mexico September 19, 2017. Earth Spectra **36**(S2), 288–313 (2020)
2. Isbiliroglu, Y., Taborda, R., Bielak, J.: Coupled soil-structure interaction effects of building clusters during earthquakes. Earthq. Spectra **31**(1), 463–500 (2015). https://doi.org/10.1193/102412EQS315M
3. Sahar, D., Narayan, J.P., Kumar, N.: Study of role of basin shape in the site–city interaction effects on the ground motion characteristics. Nat. Hazards **75**(2), 1167–1186 (2014). https://doi.org/10.1007/s11069-014-1366-2
4. Mayoral, J.M., Mosqueda, G.: Seismic interaction of tunnel-building systems on soft clay. Soil Dyn. Earthq. Eng. **139**, 106419 (2020)
5. Mayoral, J.M., Mosqueda, G.: Foundation enhancement for reducing tunnel-building seismic interaction on soft clay. Tunnel. Undergr. Space Technol. Incorp. Trench. Technol. Res. **115**, 104016 (2021)
6. Itasca Consulting Group. FLAC3D, Fast Lagrangian Analysis of Continua in 3 Dimensions. User's Guide (2009)
7. Seed, H.B., Romo, M.P., Sun, J., Jaime, A., Lysmer, J.: Relationships Between Soil Conditions and Earthquake Ground Motions in Mexico City in the Earthquake of September 19, 1985, College of Engineering, University of California, Berkeley, California (1988)
8. Gonzalez, C.Y., Romo, M.P.: Estimation of clay dynamic properties. Rev. Ing. Sísmica **2011**(84), 1–23 (2011)
9. Seed, H.B., Idriss, I.M.: Soil moduli and damping factors for dynamic response analysis. (UCB/EERC-70/10) (1970)
10. Ordaz, M.: Standards of earthquake design in Mexico City: some interesting news. Alternativas **17**(3), 106–115 (2016)
11. ASCE. Minimum Design Loads for Buildings and Other Structures. ASCE/SEI Standard 7–10. American Society of Civil Engineers (2010)
12. Lilhanand, K., Tseng, W.S.: Development and application of realistic earthquake time histories compatible with multiple damping response spectra. In: Proceedings of the 9th World Conference on Earthquake Engineering, Tokyo, Japan, vol. II, pp. 819–824 (1988)
13. Abrahamson, N.A.: State of the practice of seismic hazard evaluation. In: Proceedings of GeoEng 2000, Melbourne, 19–24 November 2000, pp. 659–685 (2000)

Simplified Numerical Simulation of Large Tunnel Systems Under Seismic Loading, CERN Infrastructures as a Case Study

A. Mubarak(✉) and J. A. Knappett

University of Dundee, Nethergate, Dundee DD1 4HN, UK
a.mubarak@dundee.ac.uk

Abstract. Seismic analysis of large tunnel systems using the continuum (Finite Element; FE) approach can be complex and computationally expensive. The inefficiency stems from the extended length of tunnels over very long distances, compared to the tunnel diameter, different terrain and lithological profiles, complex fixity conditions provided by the intermediate station boxes, and ground motion asynchronicity. This paper proposes an uncoupled numerical methodology to model and analyse the seismic response of large tunnel systems that is able to consider various tunnel alignments. The method is capable of simplifying the computationally intensive FE models into a lower-order, practically affordable numerical solution while still accounting for the aforementioned key features. This was achieved using a Beam-on-Non-linear Winkler Foundation (BNWF) model. The soil-structure interaction was considered using non-linear springs and frequency dependent dashpots. The springs were subjected to a free-field displacement time history obtained from 1-D wave propagation analysis. The proposed method is implemented for the case study of the circular Large Electron-Positron Collider (LEP) tunnel network at CERN in Geneva, Switzerland, the forerunner of the Large Hadron Collider (LHC). The tunnel system is 100m below the ground surface and completely embedded within a competent layered rock. The pre-LHC-upgraded tunnel complex contains four large underground cavern structures housing the particle detectors ('station-boxes') along its alignment. The study investigates forces developed along the circular tunnel alignment assuming a synchronous ground motion.

Keywords: Tunnel · Soil-structure interaction · Seismic analysis

1 Introduction

Owing to the rapidly increasing demand on underground space exploitation for various transportation and utility network systems, the assurance of structural integrity, safety and resilience due to natural hazards such as earthquakes is gaining an increased attention among researchers. Numerous analytical and numerical studies have been suggested over recent decades aiming to quantify the co-seismic behaviour of common tunnel sections under seismic excitations in different ground conditions (e.g., Wang, 1993; Hashash

© The Author(s), under exclusive license to Springer Nature Switzerland AG 2022
L. Wang et al. (Eds.): PBD-IV 2022, GGEE 52, pp. 2339–2347, 2022.
https://doi.org/10.1007/978-3-031-11898-2_219

et al. 2001; Yu et al., 2017; Zlatanović et al. 2017). Despite the importance of these studies, however, they either considered local plane-strain analyses with a site-specific ground profile or considered a limited tunnel length at a certain zone along the tunnel alignment. Such analyses cannot consider the global co-seismic response of large tunnel systems which usually extend over tens of kilometres. Such long tunnel alignments are likely to penetrate ground with different lithological constituents and properties, and the response may be affected by the lag-time (asynchronicity) of the propagated seismic waves along the alignment given its length. These key seismic features cannot be easily simulated using widely available and commonly used continuum-based computational methods (e.g., Finite-Elements 'FE') as the solution costs are very expensive for such large simulations.

This paper presents a methodology to model and analyse the co-seismic response of large tunnel alignments implemented on a real case study of the pre-LHC-upgrade tunnel network system (ring), shown in Fig. 1, of the Large Electron-Positron Collider (LEP) at CERN in Geneva, Switzerland. This analysis aims to identify the critical sections along the tunnel alignment that are most affected by the seismic shaking. The method provides a computationally effective tool to assess co-seismic performance that can be considered in planning and designing of large tunnel systems in earthquake-prone areas.

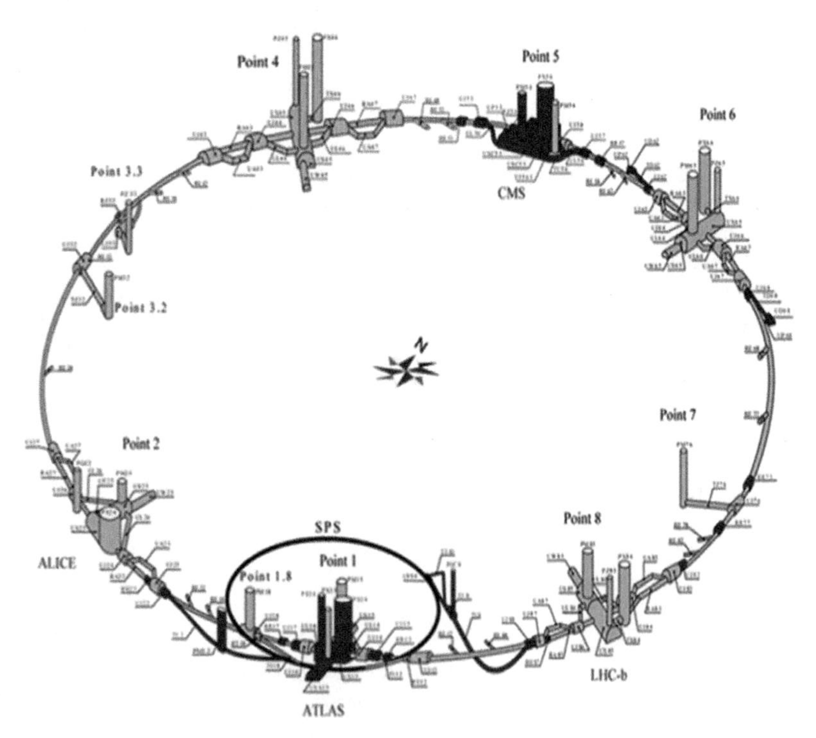

Fig. 1. Overview of CERN underground laboratory infrastructures including the existing LEP tunnel and caverns (at points 2, 4, 6, & 8) and the additionally constructed large underground caverns (CMS and ATLAS) as part of the upgrading project to the LHC accelerator. © CERN.

2 Methodology

An uncoupled numerical method is proposed to model and analyse the seismic response of large tunnel systems that is able to consider various tunnel alignments. This was achieved using a Beam-on-Nonlinear-Winkler Foundation (BNWF) model. The soil-structure interaction was represented by closely spaced discrete non-linear (*p-y*) springs derived from 2D plane-strain FE analyses and frequency-dependent dashpots. The tunnel axial resistance was modelled via frictional (*t-z*) springs. Figure 2 illustrates a hypothetical tunnel line modelled using a beam element attached with orthogonal springs and dashpots.

In this paper the boundary condition at each spring end was given an identical free-field displacement time history obtained from 1-D wave propagation analysis (i.e. synchronous ground motion). These displacements can be varied between springs to account for 2D and 3D motion components, synchronous and asynchronous ground effects, and varying ground conditions.

The methodology was implemented using two finite-element (FE) codes: the commercial FE geotechnical analysis software PLAXIS 2D 2019 was used to conduct plane-strain analyses to determine site-specific Winkler spring properties and the general FE analysis software ABAQUS/standard (2020) was used as a convenient platform to create and solve the BNWF interaction model.

2.1 FE Modelling of Underground Structures

The interaction springs in the BNWF model require two main inputs: (*i*) the non-linear stiffness response for a potentially non-symmetric and complex underground structure shape; and (*ii*) a boundary displacement time history amplitude vector for each earthquake component. A pseudo-static FE analysis was conducted for the former and

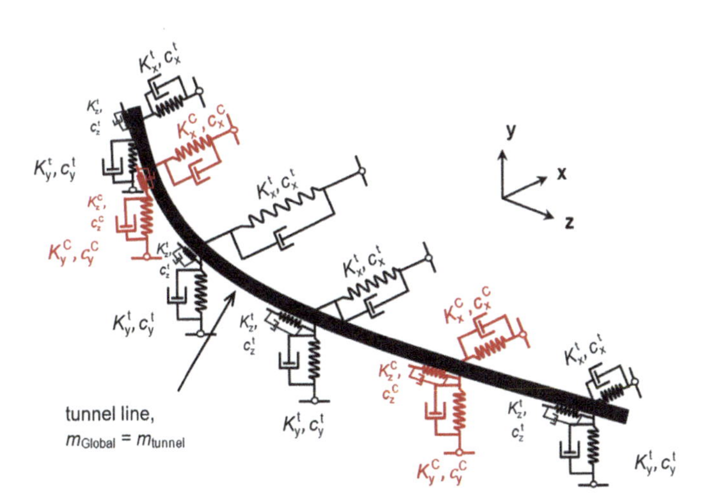

Fig. 2. Dynamic BNWF model to simulate the soil-structure interaction of large tunnel alignment, where K^{t}, c^{t} and K^{C}, c^{C} refers to the stiffness and dashpot constant for the tunnel and cavern respectively.

a non-linear dynamic analysis for the latter, respectively, using the same plane-strain model.

Analytical solutions were employed to determine appropriate axial non-linear response.

FE Model Description

Figure 3 shows the 2D plane-strain FE model for the CMS cavern that was used as a representative cavern at all stations around the LEP. The soil profile has a depth $y = 200$ m–$7H_{cavern}$ with soil and rock cover of approximately 70 m above the cavern crown, where the upper 45 m is moraine underlain by 25 m of layered rock. The width of the model $x = 600$ m–$12W_{cavern}$ is optimised such that the stresses and deformations at the boundaries of the model are equivalent to free-field conditions.

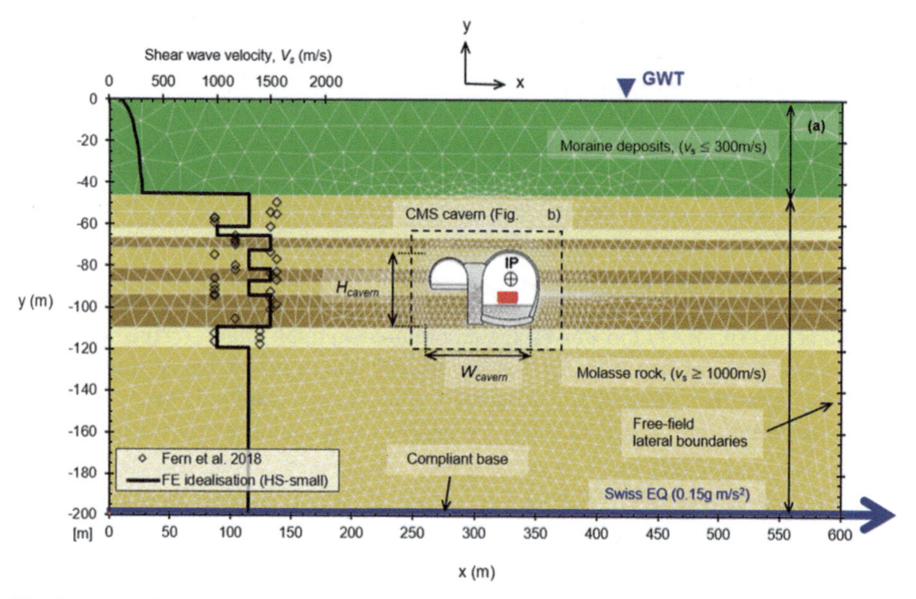

Fig. 3. FE model for the CMS cavern structure used for the pseudo-static and full dynamic analyses

The constitutive behaviour of the various materials in the ground profile was modelled using the non-linear elasto-plastic isotropic hardening hysteretic model developed by Benz et al. (2009), named the Hardening soil model with small strain stiffness (HS-small). Table 1 summarises the model parameters for the different geotechnical formations, where γ_{sat} is the saturated unit weight, E_{50}^{ref}, E_{oed}^{ref}, and E_{ur}^{ref} are the reference secant oedometer and unloading-reloading stiffnesses respectively at mean effective confining pressure of 100 kPa; G_0^{ref} is the reference shear modulus (also at 100 kPa), $\gamma_{0.7}$ is the shear strain at $G = 0.7G_0$, c'_{ref} is apparent cohesion, φ' is friction angle, ψ is the dilatancy angle, v is Poisson's ratio and m is the exponent for stress-level dependency of stiffness. For dynamic analysis, the model was subjected to an input horizontal earthquake motion

Table 1. HS-small model parameters for all geotechnical formations in the ground profile

Parameter	Unit	Moraine	Molasse rock		
			Weak	Med.-strong	Strong
γ_{sat}	kN/m^3	23	24	24	24
E_{50}^{ref}	kN/m^2	30×10^3	340×10^3	1.2×10^6	3.42×10^6
E_{oed}^{ref}	kN/m^2	24×10^3	272×10^3	960×10^3	2.73×10^6
E_{ur}^{ref}	kN/m^2	90×10^3	980×10^3	3.4×10^6	10.2×10^6
G_0^{ref}	kN/m^2	92×10^3	2.24×10^6	2.72×10^6	4.1×10^6
$\gamma_{0.7}$	[-]	0.09×10^{-3}	1.2×10^{-3}	1.2×10^{-3}	1.2×10^{-3}
c'_{ref}	kN/m^2	0	850	2000	3000
φ'	[°]	35	22	41	48
ψ	[°]	0	0	0	0
ν	[–]	0.3	0.25	0.2	0.2
m	[–]	0.5	0.7	0.7	0.7

at the base of the model with PGA = 0.15g m/s^2 according to the local seismicity of the Geneva area.

Non-linear Spring-Dashpot Properties

Due to the non-symmetric geometry of the CMS cavern structure and heterogeneity of the surrounding ground, the 2D FE model was employed to determine the soil-structure non-linear spring behaviour using a pseudo-static approach. The frictional resistance at the tunnel interface was estimated using an analytical solution for the load transfer mechanism by Bohn et al. 2017. Non-linear stiffness *p-y* and *t-z* curves are shown in Fig. 4a. Frequency-dependent damping coefficients (c) were determined for the cavern and tunnel assuming an equivalent circular section oscillating in an elastic space as suggested by Makris and Gazetas (1993) as shown in Fig. 4b.

2.2 BNWF Model for the LEP Tunnel Ring Alignment

The tunnel is 27 km in circumference, has a circular section with external radius of 2.6 m, thickness of 0.35 m, and runs at 100 m below the ground surface. The LEP tunnel alignment contains four large intermediate caverns (denoted as stations A to D), located symmetrically across the ring alignment. Figure 5a shows a simplified plan view for the LEP tunnel complex adopted for the BNWF model.

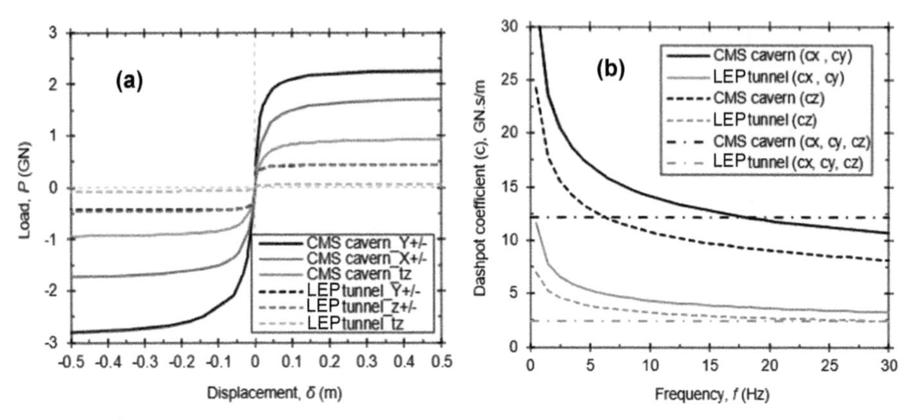

Fig. 4. (a) Soil impedance *p-y*, *t-z* curves and (b) frequency-dependent curves for the CMS Cavern and LEP tunnel structures in different directions,

This paper considers the co-seismic forces developed through the circular tunnel alignment assuming 2D synchronous motion. Due to the symmetry of the tunnel alignment around section A-A in Fig. 5a, a semi-circular BNWF model was built considering the lower half alignment with stations A and B respectively, as shown schematically in Fig. 5b. Appropriate boundary restraints were assigned with roller support at both ends (E_1 and E_2) to account for the circular tunnel continuity and free horizontal displacement. The tunnel line was modelled using 2-noded linear elastic shear flexible beam elements (type-B31 in ABAQUS) and has a section with an equivalent flexural stiffness of $EI = 432$ GN.m^2 to match the reinforced-concrete tunnel section.

The springs were assigned nonlinear stiffness and frequency-dependent dashpot properties obtained from previous sections (Fig. 4a&b). The free ends of the springs in the lateral, vertical and tangential directions were assigned displacement time-histories obtained from a point in the free-field of the 2D dynamic plane-strain analysis model at the running depth of the tunnel.

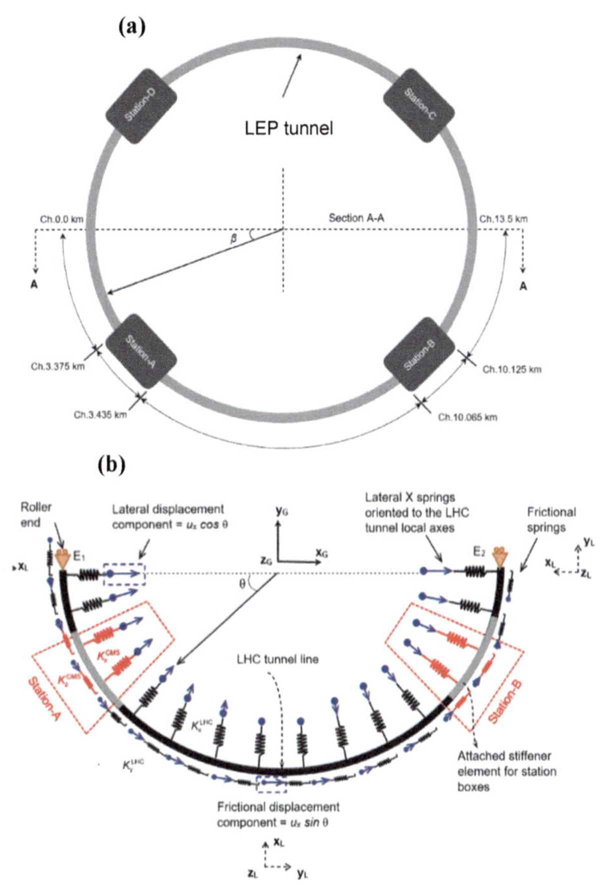

Fig. 5. Plane view for (a) the simplified model of the circular underground LEP tunnel complex adopted for the BNWF model.(b) the BNWF model of the semi-circular LHC tunnel subjected to 2D synchronous horizontal shaking. The subscripts G and L indicate the Global and Local axes respectively

3 Results

Figure 6 shows the computed peak seismic axial, moment and shear forces along the tunnel line around the semi-circular alignment (*a*, *c* & *e*) and a zoomed-in view at the tunnel-cavern-A connection zone (*b*, *d*, & *f*). These actions are based on the assumption of synchronous earthquake motion in the direction of x_G in Fig. 5b.

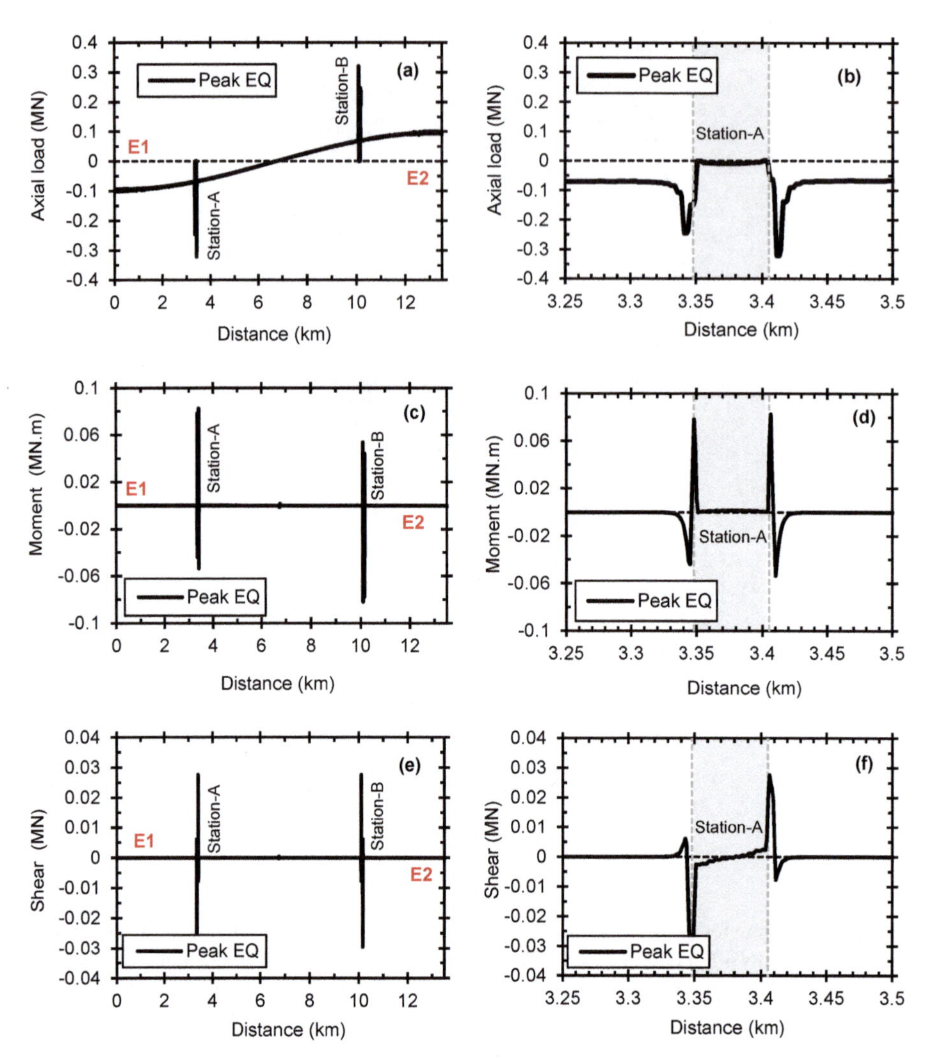

Fig. 6. Peak seismic axial, moment, and shear forces, where (*a*), (*b*) & (*c*) are forces distribution along the semi-circular tunnel alignment, and (*b*), (*d*) & (*f*) are a zoom in of forces at the tunnel-station -A connection points

From Fig. 6, it can be inferred that considering the simplest form of synchronous earthquake loading, the cavern (station-box) structures act as a restraint to the oscillation of the tunnel line at the zone near the tunnel-station connection points due to their much larger lateral ground-structure interaction stiffness, which creates additional boundary forces on the tunnel lining in these zones that should be considered when detailing the connection zones in the performance based seismic design of tunnels. Although not presented here, the model can be used to simulate asynchronicity in the ground motion through applying a time-lag to the ground displacement-time history inputs at each spring-dashpot pair and consider variation in ground conditions through varying the spring properties.

4 Conclusion

A simple uncoupled computationally efficient numerical methodology is proposed in this paper to simulate the co-seismic response of large tunnel systems under earthquake loading. The method can be applied to any tunnel alignment, ground conditions and considering synchronous and asynchronous effects. Despite the simplicity of the dynamic BNWF modelling approach, it captures the most significant features of the tunnel co-seismic response at the global scale through appropriate selection of spring and dashpot properties. The case study considered (the LEP ring at CERN, which is circular in layout) has indicated that the connection zones between parts of the system of strongly contrasting stiffness (tunnel-cavern here) have comparatively higher loading than along the rest of the tunnel, which has implications for seismic detailing.

References

Benz, T., Vermeer, P.A., Schwab, R.: A small-strain overlay model. Int. J. Numer. Anal. Meth. Geomech. **33**(1), 25–44 (2009)

Bohn, C., Lopes dos Santos, A., Frank, R.: Development of axial pile load transfer curves based on instrumented load tests. J. Geotech. Geoenviron. Eng. **143**(1), 04016081 (2017)

Hashash, Y.M., Hook, J.J., Schmidt, B., John, I., Yao, C.: Seismic design and analysis of underground structures. Tunn. Undergr. Space Technol. **16**(4), 247–293 (2001)

Makris, N., Gazetas, G.: Displacement phase differences in a harmonically oscillating pile. Geotechnique **43**(1), 135–150 (1993)

Yu, H., Yuan, Y., Bobet, A.: Seismic analysis of long tunnels: A review of simplified and unified methods. Underground Space **2**(2), 73–87 (2017)

Wang, J.N.: Seismic design of tunnels: A state-of-the-art approach. Monograph 7. New York: Parsons, Brinckerhoff, Quade and Douglas, Inc. (1993)

Zlatanović, E., Šešov, V., Lukić, D.Č., Prokić, A., Trajković-Milenković, M.: Tunnel-Ground Interaction Analysis: Discrete Beam-Spring VS. Continuous FE Model. Tehnicki vjesnik/Technical Gazette, 24 (2017)

Shaking Table Test of the Seismic Performance of Prefabricated Subway Station Structure

Lianjin Tao[1,4], Cheng Shi[1,4(✉)], Peng Ding[2,3], Linkun Huang[1,4], and Qiankun Cao[1,4]

[1] Key Laboratory of Urban Security and Disaster Engineering of the Ministry of Education, Beijing University of Technology, Beijing 100124, China
shi_cheng@emails.bjut.edu.cn
[2] Department of Hydraulic Engineering, Tsinghua University, Beijing 100084, China
[3] China Construction Science and Technology Group Co., LTD., Beijing 100195, China
[4] Key Laboratory of Earthquake Engineering and Structural Retrofit of Beijing, Beijing University of Technology, Beijing 100124, China

Abstract. In order to study the seismic performance of prefabricated subway station structure (PSSS), the shaking table test was carried out to analyze the seismic response of PSSS. According to the shaking table test results, a three-dimensional finite element model considering the interaction of PSSS, enclosure structure and soil was established. Numerical simulations of the seismic response of PSSS under different test conditions were implemented. Through the comparative analysis of numerical results and shaking table test results, the seismic response characteristics of PSSS were revealed, and the seismic damage mechanism of PSSS was summarized. The results showed that the numerical results and shaking table test results reflected similar regularities, indicating that the established finite element model and analysis method were reliable and effective. PSSS had good performance in earthquake resistance, prefabricated joints had outstanding performance in deformation resistance, which enabled the prefabricated components to work together. Under extreme earthquakes, the vault, the upper and lower ends of the side walls, the internal non-load-bearing structure, and the envelope structure of PSSS were the most severely damaged areas. It was reasonably predicted that the seismic damage mechanism of PSSS is divided into three stages: firstly, the enclosure structure suffered seismic damage; secondly, the non-load-bearing structure inside the structure loses stability; finally, the top arch structure degenerates into a three-hinged arch structure.

Keywords: Prefabricated subway station · Shaking table test · Numerical simulation · Seismic responses · Damage mechanism

1 Introduction

At present, the traditional cast-in-place technology for subway stations was relatively mature, but there were still unavoidable problems such as environmental pollution, resource consumption, long construction period, and poor project quality. In the construction of subway station projects, the application of prefabricated construction technology

© The Author(s), under exclusive license to Springer Nature Switzerland AG 2022
L. Wang et al. (Eds.): PBD-IV 2022, GGEE 52, pp. 2348–2361, 2022.
https://doi.org/10.1007/978-3-031-11898-2_220

could effectively eliminate many disadvantages of traditional cast-in-place technology, and it has become an important trend in the development of subway station construction technology in the future [1–3]. With the Kobe earthquake in 1995 [4], the Chichi earthquake in 1999 [5], the Wenchuan earthquake in 2008 [6], and the Great East Japan Earthquake in 2011 [7], the urban rail transit facilities were severely damaged. The seismic performance and safety of underground structures have become a hot spot for many scholars. The main methods for studying the seismic performance of underground structures were numerical simulation and model tests. Through research on the seismic performance of underground structures such as underground pipe corridors [8–10], long tunnels [11–14] and subway stations [15–20], the seismic response characteristics of underground structures have been qualitatively and semi-quantitatively summarized, and the damage mechanism of underground structures has been revealed.

In summary, the laws of seismic response of underground structures have been extensively studied by academic circles, which has facilitated the development and progress of seismic research on underground structures. However, the research on the seismic response of new prefabricated subway station structure (PSSS) was still in its infancy, and only a few scholars have made preliminary explorations. Therefore, it was urgent to carry out in-depth research to better promote the development of new prefabricated construction technology for subway stations. In this study, through shaking table tests and numerical simulations, the seismic response characteristics of PSSS, such as acceleration, strain, and joint deformation were analyzed, and the seismic damage mechanism of PSSS was revealed. The research work could provide reliable reference and guidance for the seismic design and seismic safety evaluation of PSSS.

2 Shaking Table Test Design

Based on the Buckingham π theorem, the shaking table test with the geometric ratio of 1:30 was carried out. The similar relationship of the physical quantities is shown in Table 1. The model structure production process is shown in Fig. 1, mainly including: firstly, the steel molds of the 7 prefabricated components were manufactured and then prefabricated components were poured. Secondly, the prestressed steel wire and solidified slurry were used for the assembly of standard rings and the assembly between rings. Finally, the non-load-bearing structure (middle plate, beam, column, bottom beam) and the enclosure structure on both sides of the prefabricated structure were post-poured to obtain a complete six-ring assembly structure. The prefabricated components were made of micro-concrete materials and galvanized steel wires, and the enclosure structure was made of gypsum materials. The joint solidified slurry was simulated by epoxy resin adhesive. The material parameters of the model structure are shown in Table 2. According to relevant research experience [21], the ultimate tensile strength of micro-concrete materials was taken as 1/10 of the ultimate compressive strength. The model foundation soil was simulated by representative fine sand in Beijing. The model soil was prepared by layering method. The mechanical parameters of the model soil are shown in Table 3. A representative Taft seismic wave was selected as the input ground motions for the shaking table test. The acceleration time history curves of the Taft seismic wave and its corresponding Fourier spectra are shown in Fig. 2. The input direction of the ground motion of the shaking table was horizontal. The seismic loading conditions of the shaking table test are shown in Table 4. The test was carried out on the shaking table of the Key Laboratory of Urban Security and Disaster Engineering of the Ministry of Education, Beijing University of Technology. The layered shear model box [17] independently developed by the research team was used as the container for the foundation soil, as shown in Fig. 3.The layout of monitoring equipment and sensors applied in the shaking table test are shown in Fig. 4. The shaking table test plan is detailed in the literature of the research group [15, 16].

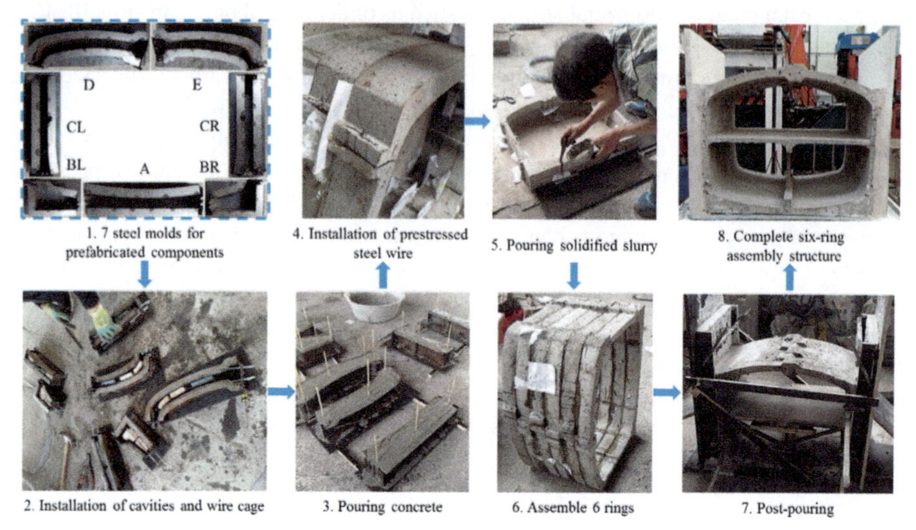

Fig. 1. Construction process of model structure.

Table 1. Similarity ratio of the model structure and soil.

Types	Physical quantity	Similarity	Similarity ratio	
			Model structure	Model soil
Geometric properties	Length l	S_l	1/30	1/4
	Linear displacement r	$S_r = S_l$	1/30	1/4
Material properties	Equivalent density ρ	$S_\rho = S_E/(S_l S_a)$	15/2	1

(*continued*)

Table 1. (*continued*)

Types	Physical quantity	Similarity	Similarity ratio	
			Model structure	Model soil
	Elastic modulus E	S_E	1/4	1/4
	Soil shear wave velocity V_s	S_V	–	1/2
Dynamic properties	Frequency ω	$S_\omega = 1/S_t$	5.4794	2
	Acceleration a	S_a	1	1
	Duration t	$S_t = \sqrt{S_l/S_a}$	0.1825	1/2
	Dynamic stress σ	$S_\sigma = S_l S_a S_\rho$	1/4	1/4
	Dynamic strain ε	$S_\varepsilon = S_l S_a S_\rho / S_E$	1	1

Table 2. Material properties of the model structure.

Material	Compressive yield strength (MPa)	Tensile yield strength (MPa)	Shear yield strength (MPa)	Elastic modulus (MPa)	Poisson ratio
Micro-concrete	7.5	0.75 (Estimated value)	–	7.9×10^3	0.17
Gypsum	1.31	–	–	7.6×10^2	0.36
Galvanized steel wire	–	1190	–	2×10^5	0.30
Epoxy resin adhesive	15	8	6	1×10^3	0.38

Table 3. Material properties of the fine sand.

Material	Density (kg/m^3)	Cohesion (kPa)	Friction angle (°)	Shear wave velocity (m/s)	Elastic modulus (MPa)	Poisson ratio
Fine sand	1750	0	25	230	276	0.36

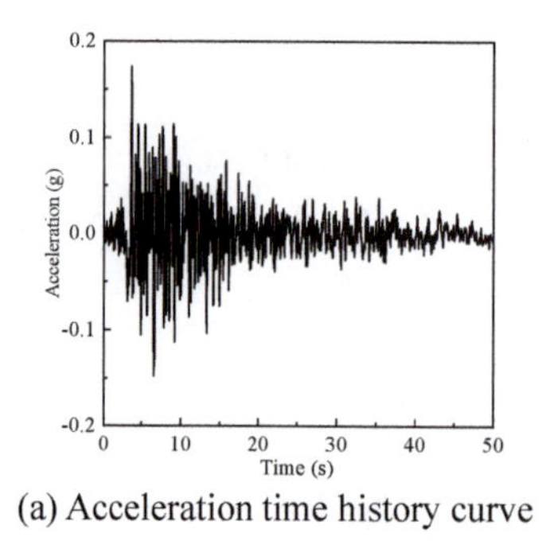

(a) Acceleration time history curve

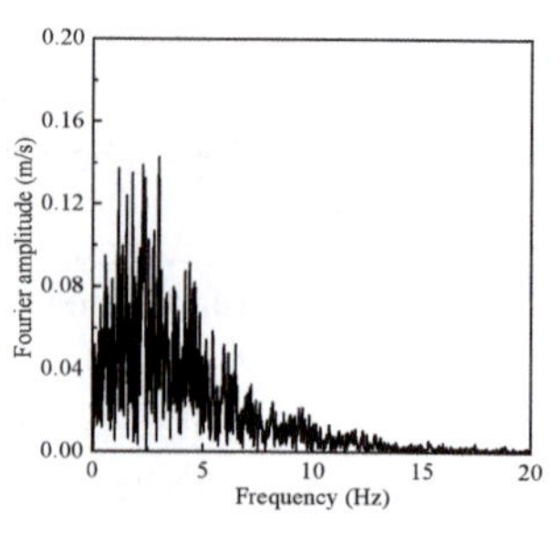

(b) Fourier spectra

Fig. 2. Taft seismic wave (TA).

Fig. 3. Shaking table system.

Fig. 4. Schematic diagrams of sensors layout.

Table 4. Seismic loading conditions for the shaking table tests.

Condition serial number	Input ground motion	Condition code	PGA (g)	Duration (s)
1	White noise	B-1	0.1	100
2	Taft wave (TA)	TA-1	0.1	27
3	White noise	B-2	0.1	100
4	Taft wave (TA)	TA-2	0.2	27
5	White noise	B-3	0.1	100
6	Taft wave (TA)	TA-3	0.4	27
7	White noise	B-4	0.1	100
8	Taft wave (TA)	TA-4	0.8	27
9	White noise	B-5	0.1	100

3 Analysis of the Test Results

3.1 Model Structural Seismic Damage

The seismic damage of model structure is shown in Fig. 5. The prefabricated structure and the enclosure structure work together to bear the seismic load. Since the strength of the enclosure structure is weaker than that of the main prefabricated structure, an obvious longitudinal crack occurs, appearing concrete spalling and exposed steel bars (Fig. 5(a)). In addition, through cracks appear on the middle beam-column joint (Fig. 5(b)) and the middle plate-bracket joint (Fig. 5(c)). In short, the model structure has been damaged to different degrees under the earthquake, and has entered the elastic-plastic working state.

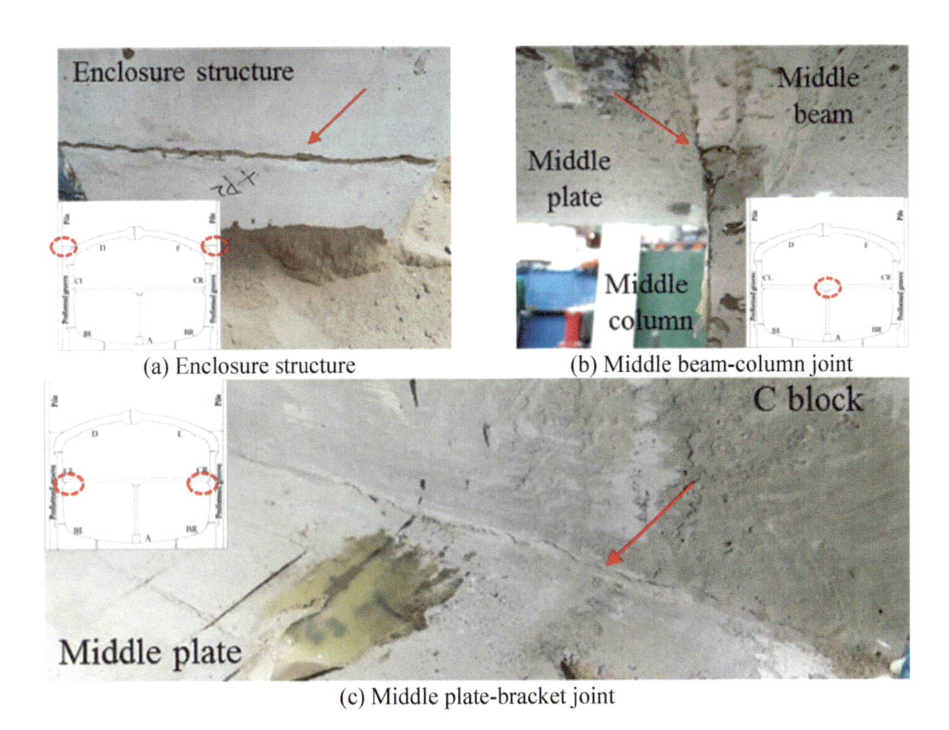

(a) Enclosure structure

(b) Middle beam-column joint

(c) Middle plate-bracket joint

Fig. 5. Seismic damage of model structure.

3.2 Model Structure Acceleration

The peak acceleration of the structural symmetry axis and side wall is shown in Fig. 6. The amplification factor is defined as the ratio of the measured peak acceleration amplitude of the soil to the input seismic peak acceleration Under a low-intensity earthquake, the maximum peak acceleration amplification factor of the symmetry axis and the side wall of the model structure gradually increases from bottom to top along the height direction, and the structure is in an elastic state. With the increase of the earthquake intensity, the maximum peak acceleration amplification factor of the symmetry axis of the model structure and the side wall showed a trend of first decreasing and then increasing. From the seismic damage in Sect. 3.1, the seismic damage occurs at the middle beam-column joint and middle plate-bracket joint. The acceleration reduction position of the symmetry axis and the side wall is consistent with the seismic damage position. In addition, as the input PGA increases, the deformation of the soil increases and the damping ratio increases. An excessively large damping ratio effectively weakens the amplification effect of the site. As a result, the acceleration amplification factor of the structure tends to weaken.

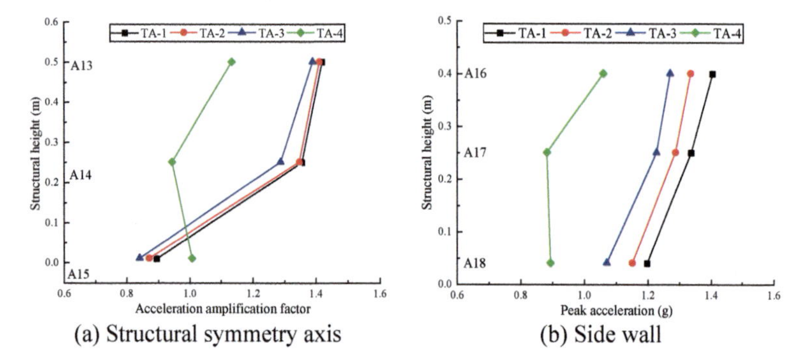

(a) Structural symmetry axis (b) Side wall

Fig. 6. The peak acceleration of the structural symmetry axis and side wall.

3.3 Model Structure Strain

The schematic diagram of the maximum tensile strain distribution of model structure is shown in Fig. 7, where the maximum tensile strain exceeding 95 $\mu\varepsilon$ is marked in red. The tensile strain response of the model structure could be described as follows:

a) When PGA = 0.1 g and 0.2 g, except that the maximum tensile strain at the middle column-bottom beam joint (S1-14) slightly exceeds the ultimate tensile strain of 95 $\mu\varepsilon$, the maximum tensile strain at the rest of the model structure is less than the ultimate tensile strain. Therefore, it can be considered that the entire PSSS is basically in an elastic working state under this earthquake intensity. The prefabricated subway station structure has good integrity and safety under earthquake action.
b) When PGA = 0.4 g, the maximum tensile strain at the variable section inside and outside of the vault (S1-1 and 16), the inside of the arch foot (S1-4), the B-C joint

(S1-9) and the A-B joint (S1-10) and the middle plate-beam-column joints (S1-12 and 13) exceeds the ultimate tensile strain. This shows that the PSSS is damaged and gradually enters the elastic-plastic working state. And with the increase of earthquake intensity, the area of structural damage has further expanded.

c) When PGA = 0.8 g, most areas of the model structures are damaged, which are mainly divided into six areas: 1) the variable section of the vault (circle 1), 2) the inside of the hance and arch foot (circle 2), 3) middle plate-bracket joint (circle 3), 4) the inner arc of B block (circle 4), 5) the middle plate-beam-column-bottom beam joints (circle 5 and circle 6). Among them, obvious cracks could be directly observed at the positions of circle 3 and circle 5 in Fig. 5. Although other areas exceed the extreme concrete tensile strain, the extreme strain might be relatively small, and no obvious cracks are observed. It is worth mentioning that the structural seismic damage area is mainly the variable section of the structure, the joint or the connection of the components, which is the weak part of the prefabricated structure for earthquake resistance.

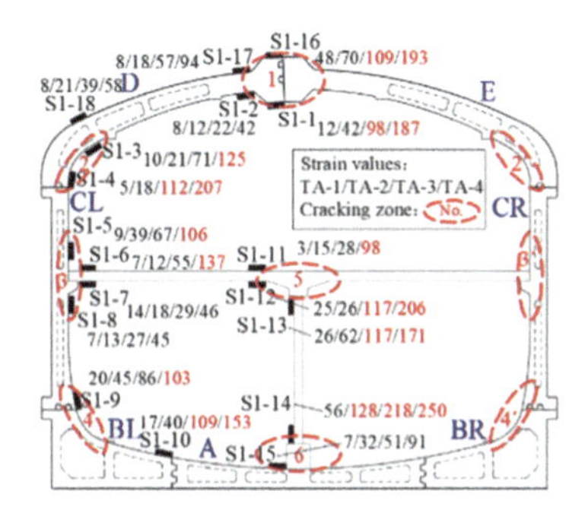

Fig. 7. Maximum tensile strains of model structure (unit: $\mu\varepsilon$).

3.4 Model Structural Joint Deformation

The schematic diagram of the tensile and compressive strain responses of model structure joints is shown in Fig. 8. As the input PGA increases, the tensile and compressive strains of each joint gradually increase. Compared with the extreme tensile strain of the structure in Sect. 3.3, the model structure joints deformation is smaller, which is in a safe and controllable range. This phenomenon is consistent with the experimental phenomenon of joint splitting failure (that is, the connection strength of the joint interface is greater than the strength of the surrounding concrete) in the joint static test [22–24]. It is verified that the integrity and synergy of prefabricated joints are favorable, which plays an important role in ensuring the overall stability of the structure and the continuous force transmission between components. The joint performance meets the seismic design goals of strong connections and weak components.

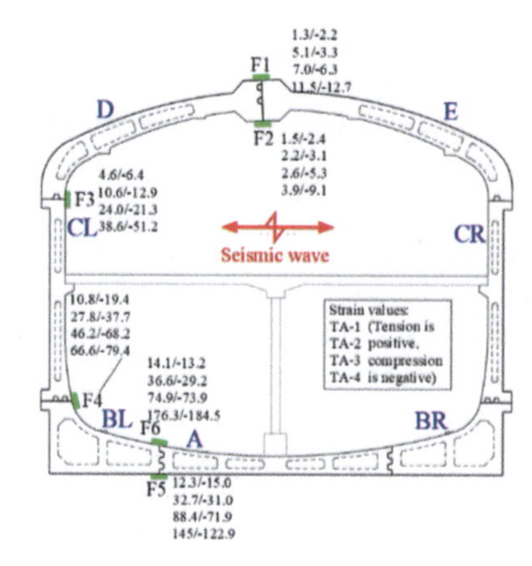

Fig. 8. The tensile and compressive strain of the model structure joints (unit: $\mu\varepsilon$).

4 Numerical Simulation Analysis of Model Structure

4.1 Numerical Model

The finite element model of the prefabricated subway station model structure is shown in Fig. 9. The material parameters of the model structure are shown in Table 2, and the material parameters of the model soil are shown in Table 3. The interaction between prefabricated joints, between prefabricated structure and enclosure structures, and between the soil and the station model were set as contact attributes. The contact normal behavior was simulated by the hard contact algorithm. The contact tangential behavior was simulated by Coulomb's law of friction. The friction coefficients of the three contact surfaces were respectively taken as 0.6, 0.6, and 0.4 [25]. The concrete structure and the soil were discretized by an eight-node linear reduced integration brick elements (C3D8R). The steel bars were discretized by three-dimensional two-node truss elements (T2D3) and embedded in the concrete structure without consideration for the slip property between the concrete and steel bar. The process of numerical simulation analysis was as follows: firstly, the model soil-PSSS system reached the initial in-situ stress equilibrium state under its own weight. Then, the seismic acceleration time history was input at the bottom boundary of the model for subsequent dynamic analysis. The normal direction of the bottom boundary of the model was fixed to simulate the supporting effect of the container on the soil. Meanwhile, the left and right boundaries were constrained to make the nodes of the same height move together without relative displacement, simulating the layered shear model box in a simplified manner [19].

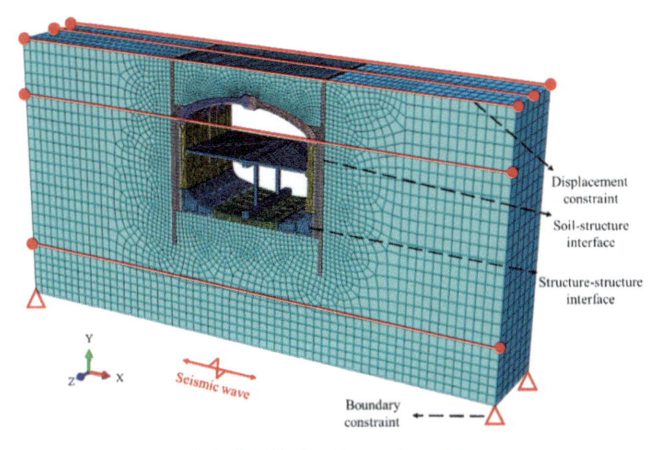

Fig. 9. Finite element model.

4.2 Comparative Analysis of Acceleration Response

The shaking table test results and numerical simulation (NS) results of the acceleration response of the monitoring points of the model structure bottom plate are shown in Fig. 10. The NS values of the monitoring points of the model structure bottom plate are in good agreement with the test values, and the red test values could almost completely cover the black simulation values. The slight difference between the NS values and the test values is mainly due to the simplified processing of boundary conditions, material constitutive models and dynamic parameters in the numerical analysis. In addition, the shaking table test is subject to inevitable manual measurement errors in the performance of test equipment and monitoring components. In general, the results of the two are mutually verified. It could be considered that the NS results are reasonable, and the test results are also reliable.

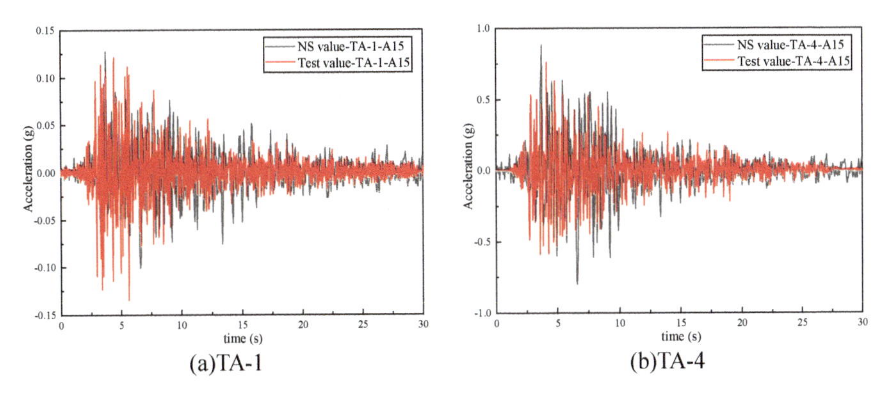

Fig. 10. Comparison of test and NS of the acceleration response of the bottom plate (A15)

4.3 Comparative Analysis of Structural Strain and Seismic Damage

Figure 11 shows the test and numerical simulation (NS) values of the structural strain amplitude under TA-1 and TA-4 conditions. As the earthquake intensity increases, the strain of the model structure shows the same increasing trend, and the tensile strain distribution of the numerical simulation values and the test values also have a higher degree of agreement. The seismic damage contour of the model structure under the TA-4 is shown in Fig. 12. It is determined again that the damage areas of the model structure are mainly: a) the vault, b) the upper and lower ends of the side wall, c) the internal non-load-bearing structure, d) the enclosure structure. The above four areas are weak areas of the PSSS for earthquake resistance.

By comprehensively analyzing the results of shaking table test and numerical simulation, the model structure seismic damage is reasonably predicted, as shown in Fig. 13. Under the design earthquake, the model structure is in an elastic state without seismic damage. Under the rare earthquake, the seismic damage occurs on the enclosure structure, the strengthening ability of the enclosure structure to PPSS gradually weakens. As a result, the non-load-bearing structure begins to lose stability after damage at connections such as the middle plate-bracket joint and the middle plate-beam-column-bottom beam joint. Afterwards, the variable section of the vault, the hence and the arch foot also suffered from seismic damage. Under the extreme earthquake, the top arch structure finally degenerates into a plastic three-hinge arch mechanism, and the degraded structure can still maintain a relatively stable self-equilibrium state without serious damage similar to the overall collapse. Therefore, under the action of non-extreme earthquakes, the PSSS has high safety and reliability.

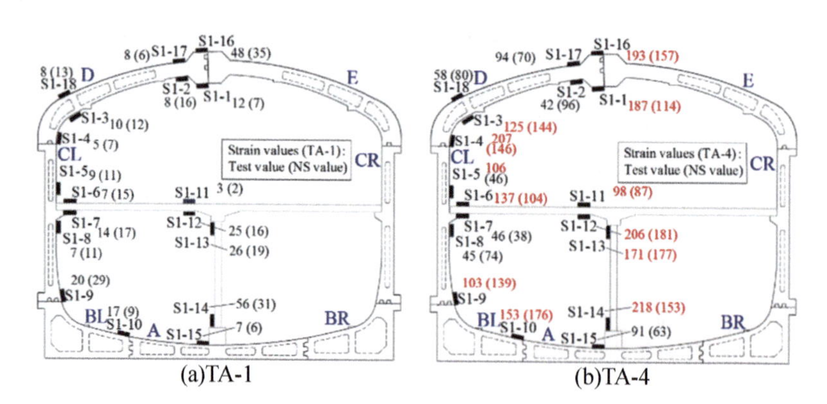

Fig. 11. Test and NS values of structural strain amplitude.

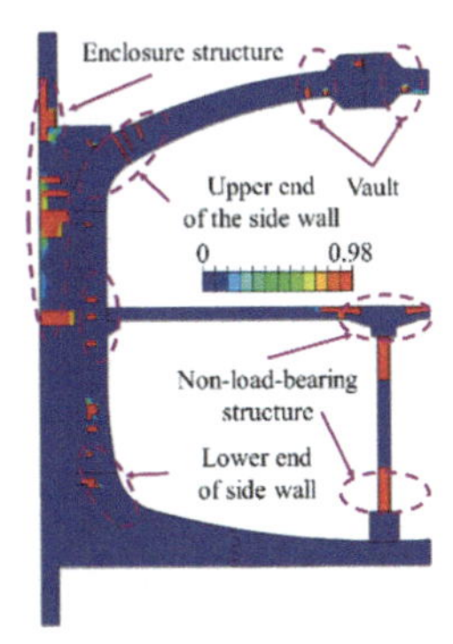

Fig. 12. Seismic damage under TA-4.

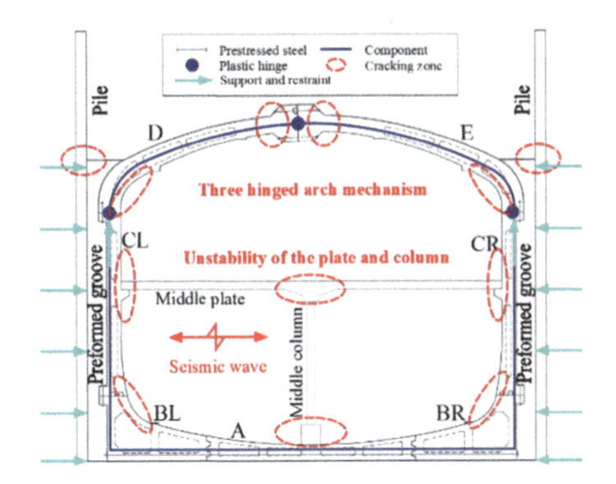

Fig. 13. Seismic damage mechanism

5 Conclusions

The seismic performance of PSSS under earthquake action is studied through shaking table tests and numerical simulations. The following conclusions could be obtained:

a) The numerical simulation results of the three-dimensional finite element model of the shaking table test are basically consistent with those obtained from the test, which verifies the reliability of the conclusions obtained from the test.
b) PSSS has good performance in anti-seismic and anti-deformation. The integrity of prefabricated joints is favorable, which meets the requirements of seismic design and enables the prefabricated components to work together.
c) The vault, the upper and lower ends of the side wall, the internal non-load-bearing structure and the enclosure structure are the weak parts of PSSS for earthquake resistance.
d) The seismic damage mechanism of PSSS is divided into the following stages: firstly, the enclosure structure suffered seismic damage; secondly, the non-load-bearing structure inside the structure loses stability; finally, the top arch structure degenerates into a three-hinged arch structure.

References

1. Steinhardt, D.A., Manley, K.: Adoption of prefabricated housing-the role of country context. Sustain. Cities Soc. **22**, 126–135 (2016). https://doi.org/10.1016/j.scs.2016.02.008
2. Kasperzyk, C., Kim, M.K., Brilakis, I.: Automated re-prefabrication system for buildings using robotics. Autom. Constr. **83**, 184–195 (2017). https://doi.org/10.1016/j.autcon.2017.08.002
3. Ferdous, W., Bai, Y., Ngo, T.D., Manalo, A., Mendis, P.: New advancements, challenges and opportunities of multi-storey modular buildings – a state-of-the-art review. Eng. Struct. **183**, 883–893 (2019). https://doi.org/10.1016/j.engstruct.2019.01.061
4. Iida, H., Hiroto, T., Yoshida, N., Iwafuji, M.: Damage to Daikai subway station. Soils Found. **36**, 283–300 (1996). https://doi.org/10.3208/sandf.36.special_283

5. Wang, T.T., Kwok, O.L.A., Jeng, F.S.: Seismic response of tunnels revealed in two decades following the 1999 Chi-Chi earthquake (Mw 7.6) in Taiwan: a review. Eng. Geol. **287**, 106090 (2021)
6. Yu, H., Chen, J., Bobet, A., Yuan, Y.: Damage observation and assessment of the Longxi tunnel during the Wenchuan earthquake. Tunn. Undergr. Space Technol. **54**, 102–116 (2016). https://doi.org/10.1016/j.tust.2016.02.008
7. Aydan, Ö.: Crustal stress changes and characteristics of damage to geo-engineering structures induced by the Great East Japan Earthquake of 2011. Bull. Eng. Geol. Env. **74**(3), 1057–1070 (2014). https://doi.org/10.1007/s10064-014-0668-7
8. Tao, L., et al.: Three-dimensional seismic performance analysis of large and complex underground pipe trench structure. Soil Dyn. Earthquake Eng. **150**, 106904 (2021). https://doi.org/10.1016/j.soildyn.2021.106904
9. Ding, X., Feng, L., Wang, C., Chen, Z., Han, L.: Shaking table tests of the seismic response of a utility tunnel with a joint connection. Soil Dyn. Earthq. Eng. **133**, 106133 (2020). https://doi.org/10.1016/j.soildyn.2020.106133
10. Chen, J., Jiang, L., Li, J., Shi, X.: Numerical simulation of shaking table test on utility tunnel under non-uniform earthquake excitation. Tunn. Undergr. Space Technol. **30**, 205–216 (2012). https://doi.org/10.1016/j.tust.2012.02.023
11. Yu, H., Yan, X., Bobet, A., Yuan, Y., Xu, G., Su, Q.: Multi-point shaking table test of a long tunnel subjected to non-uniform seismic loadings. Bull. Earthq. Eng. **16**(2), 1041–1059 (2017). https://doi.org/10.1007/s10518-017-0223-6
12. Yuan, Y., Yu, H., Li, C., Yan, X., Yuan, J.: Multi-point shaking table test for long tunnels subjected to non-uniform seismic loadings – Part I: theory and validation. Soil Dyn. Earthq. Eng. **108**, 177–186 (2018). https://doi.org/10.1016/j.soildyn.2016.08.017
13. Yu, H., Yuan, Y., Xu, G., Su, Q., Yan, X., Li, C.: Multi-point shaking table test for long tunnels subjected to non-uniform seismic loadings - Part II: application to the HZM immersed tunnel. Soil Dyn. Earthq. Eng. **108**, 187–195 (2018). https://doi.org/10.1016/j.soildyn.2016.08.018
14. Chen, J., Yu, H., Bobet, A., Yuan, Y.: Shaking table tests of transition tunnel connecting TBM and drill-and-blast tunnels. Tunn. Undergr. Space Technol. **96**, 103197 (2020). https://doi.org/10.1016/j.tust.2019.103197
15. Tao, L., et al.: Comparative study of the seismic performance of prefabricated and cast-in-place subway station structures by shaking table test. Tunn. Undergr. Space Technol. **105**, 103583 (2020). https://doi.org/10.1016/j.tust.2020.103583
16. Tao, L., Ding, P., Shi, C., Wu, X., Wu, S., Li, S.: Shaking table test on seismic response characteristics of prefabricated subway station structure. Tunn. Undergr. Space Technol. **91**, 102994 (2019). https://doi.org/10.1016/j.tust.2019.102994
17. An, J., Tao, L.: Shaking table tests for design and performance of a new type laminar shear model box. J. Vibr. Shock **39**, 201–207 (2020). https://doi.org/10.13465/j.cnki.jvs.2020.05.028
18. Iwatate, T., Kobayashi, Y., Kusu, H., Rin, K.: Investigation and shaking table tests of subway structures of the Hyogoken-Nanbu earthquake. In: Proceedings 12th World Conference on Earthquake Engineering, pp. 1–8 (2000)
19. Xu, C., Zhang, Z., Li, Y., Du, X.: Validation of a numerical model based on dynamic centrifuge tests and studies on the earthquake damage mechanism of underground frame structures. Tunn. Undergr. Space Technol. **104**, 103538 (2020). https://doi.org/10.1016/j.tust.2020.103538
20. Zhuang, H., Yang, J., Chen, S., Dong, Z., Chen, G.: Statistical numerical method for determining seismic performance and fragility of shallow-buried underground structure. Tunn. Undergr. Space Technol. **116**, 104090 (2021). https://doi.org/10.1016/j.tust.2021.104090
21. Chen, G., Chen, S., Qi, C., Du, X., Wang, Z., Chen, W.: Shaking table tests on a three-arch type subway station structure in a liquefiable soil. Bull. Earthq. Eng. **13**(6), 1675–1701 (2014). https://doi.org/10.1007/s10518-014-9675-0

22. Li, Z., Li, S., Su, H.: Study on the bending stiffness for double Tenon-Groove joints of metro station constructed by using prefabricated structure. China Civil Eng. J. **50**, 14–18 (2017). https://doi.org/10.15951/j.tmgcxb.2017.s1.003
23. Li, Z., Li, K., Lu, S., Su, H., Wang, C.: Experimental study on stress evolution rule of double Tenon-Groove joints for prefabricated metro station structure. China Railway Sci. **39**, 15–21 (2018). https://doi.org/10.3969/j.issn.1001-4632.2018.05.03
24. Li, Z., Su, H., Lu, S., Wang, C., Xu, X.: Experimental study on flexural mechanical properties of the double Tenon Groove joints of prefabricated subway station. China Civil Eng. J. **50**, 28–32 (2017). https://doi.org/10.15951/j.tmgcxb.2017.s2.005
25. Ding, P., Tao, L., Yang, X., Zhao, J., Shi, C.: Three-dimensional dynamic response analysis of a single-ring structure in a prefabricated subway station. Sustain. Cities Soc. **45**, 271–286 (2019). https://doi.org/10.1016/j.scs.2018.11.010

Numerical Study on Seismic Behavior of Shield Tunnel Crossing Saturated Sand Strata with Different Densities

Hong-Wu[1], Zhi-Ye[1], Hua-Bei Liu[1](✉), and Yu-Ting Zhang[2]

[1] Huazhong University of Science and Technology, Wuhan 430074, China
hbliu@hust.edu.cn

[2] Tianjin Research Institute for Water Transport Engineering, Tianjin 300456, China

Abstract. This study investigated the seismic response of a shield tunnel crossing two saturated sand stratums with different relative densities using a finite difference method. A bounding surface plasticity model was employed to model the liquefaction of saturated sand. A deformable force-displacement link model was proposed to simulate the interaction between circumferential joints, in which the face contact between adjacent lining rings was discretized into point contact and characterized by a series of multi-degree-of-freedom springs together with the bolted connections. A three-dimensional numerical model was then established to investigate the acceleration, pore pressure, and deformation of the tunnel–soil systems. The numerical results indicate that the seismic responses of tunnel structures are more significant near the soil interface. The liquefaction-induced horizontal and vertical displacements of the tunnel linings are coupled with each other, which results in the most unfavorable position of the tunnel section. In addition, different degrees of liquefaction between the two sand stratums lead to a large dislocation of circumferential joints and then contribute to a stress concentration on the segment joints. Therefore, the circumferential joints near the soil interface should be paid more attention to in the seismic design of tunnel structures.

Keywords: Tunnel · Joint pattern · Earthquake · Liquefaction · Soil interface

1 Introduction

Liquefaction is one of the major threats to underground structures buried in liquefiable sand strata. Moreover, the investigation and analysis of damage after earthquakes addressed that the seismic response of tunnels buried in adjacent different soil stratums are quite different, and the tunnel segments near the soil interface are more prone to severe damages [1–3]. Several studies [4–6] have demonstrated that the tunnel structures near the soil interface would suffer more serious seismic damage. In addition, a series of liquefaction constitutive models were developed to capture the dynamic response of sand stratums considering fluid-solid interaction. The Finn model [7] and the CycLiqCPSP model [8] were applied respectively to study the underground structure partly passing through liquefiable sand stratums. Their results indicated that stress/strain distributions

© The Author(s), under exclusive license to Springer Nature Switzerland AG 2022
L. Wang et al. (Eds.): PBD-IV 2022, GGEE 52, pp. 2362–2370, 2022.
https://doi.org/10.1007/978-3-031-11898-2_221

and the most unfavorable positions of the structure had changed, comparing with that buried in one soil layer.

Another critical consideration is how to identify the mechanical behavior of segmental joints and their contribution to the soil-tunnel system. The sophisticated shield tunnel model [9] is inapplicable in large-scale seismic analysis due to its high computational cost. Hence, many simplified soil-tunnel models were established. Chen et al. [10] focused on the bending deformations of segmental joints. Wu et al. [11] pointed out that shearing dislocation between lining rings was a significant aspect of longitudinal deformation of tunnels as well. A reasonable approach to represent the tunnel segmental joint pattern was applied in FLAC^{3D}, in which joints and their connections of three degree of freedom were considered [12].

In this paper, the bounding surface plasticity model [13] was adopted to simulate the soil liquefaction. Based on the previous works [12], a revised link model was proposed to simulate the mechanical behavior of circumferential joints. Then the three-dimensional fully coupled fluid-mechanical analysis model was built up in geotechnical code FLAC^{3D}, aimed to study the seismic behavior of long shield tunnels across liquefied and non-liquefied sand stratums.

2 Description of Numerical Method

A fully coupled fluid-mechanical analysis model was built up using FLAC^{3D} program, which is an explicit finite difference numerical modeling software commonly used for geotechnical analysis purposes [14]. The dynamic response of circumferential joints between adjacent segmental rings was carefully considered, and the dynamic constitutive model of the surrounding soil was investigated as well. More details of the proposed numerical methods are as follows.

A practice-oriented two-surface plasticity sand (P2PSand) model was adopted so that the dynamic deformation and liquefaction behavior of saturated sand under cyclic loadings can be captured. The P2PSand model is a modified extension of the fabric-dilatancy-related sand plasticity DM04 model [15] by adding a few parameters to improve the cyclic mobility of sand in order to make the simulation results match the laboratory tests or field observations better [13]. There are four different surfaces as defined: (1) yield surface; (2) dilatancy surface; (3) critical surface; and (4) bounding surface in P2Psand model. A unique set of parameters can be used for a special kind of sand with a wide range of initial stress and void ratio, videlicet, the P2PSand model is a unified elasto-plastic constitutive model.

In this study, the tunnel segments were modeled by shell-type elements and the joint bolts were considered by a series of springs by referring to Do et al. [12]. The mechanical behavior of circumferential joints was fully considered by the proposed force-displacement link model. For simplification, the longitudinal joints were not considered, and the tunnel-soil interface was bonded together. As shown in Fig. 1, the linking springs in the locations of 16 bolts are presented by bolt-type link (B-link), and the others are presented by concrete-type link (C-link).

The normal, shearing, and rotational deformation of ring joints were taken into account. The stress-strain relationships of these links are based on physical experimental tests, which are determined by the material characteristics, such as compression/tension rigidity (EA), shear rigidity (GA), and bending rigidity (EI).

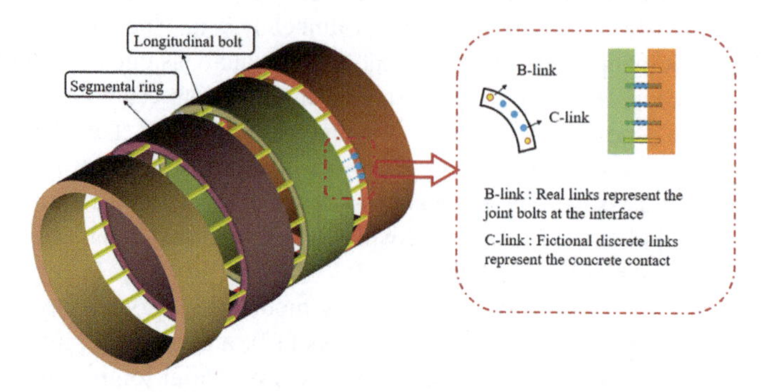

Fig. 1. The segment rings and segmental joints

3 Seismic Analysis of a Shield Tunnel crossing Saturated Sand Strata with Different Densities

3.1 Development of the Numerical Model

The numerical model introduced above was then implemented to investigate the seismic response of a shield tunnel crossing saturated sand strata with different densities. The basic parameters for the P2PSand model are listed in Table 1, which is calibrated for Toyoura sand similar to [13]. The relative densities (Dr) for the loose and dense sand layers are set as 30% and 80%, respectively.

Table 1. Main parameters of P2PSand model for Touyoura sand.

Properties of P2PSand model		Touyoura sand
Elasticity parameters	(g_0, C_{dr})	(171.45, 1.99)
Friction critical angle	ϕ_{cs}	31.5
Initial void ratio	(e_{max}, e_{min})	(0.977, 0.597)
Parameters of critical state	$(D_{rc0}, \lambda_c, \xi)$	(0.115, 0.05,0.7)
Factor of cycling	K_c	$0.63–0.25D_{r0}$
Factor of elasticity degradation	K_d	$0.66–0.58D_{r0}$

Based on the sub-sea shield-bored tunnel of Xiamen Rail Transit Line 2 [16], the parameters for tunnel segments and joint bolts are listed in Table 2. The circular segmental ring was divided into 64 shell elements by 64 links at the ring intersegment, 16 of which were bolt-type links. The link properties were calculated according to Sect. 2 except for

Table 2. Parameters of link model including B-type and C-type links

Degree of freedom	Link properties	B-link	C-link
Normal direction	Compression stiffness (MN/m)	2580	2580
	Initial tension stiffness (MN/m)	364	/
	Post-yield tension stiffness (MN/m)	3.64	/
	Tensile yield force (MN)	0.452	/
	Tensile ultimate force (MN)	0.566	10^{-6}
	Yield deformation (mm)	1.24	/
	Ultimate deformation (mm)	3.224	/
Shear direction	Radial shear stiffness (MN/m)	56	10.4
	Tangential shear stiffness (MN/m)	56	10.4
Rotation direction	Rotational stiffness (kN.m/rad)	8.19	/

the shear stiffness of B-links, which was derived from the previous related literatures [16].

Figure 2 shows that the numerical model has a dimension of 270 m (length) × 130 m (Width) × 25 m (height). The origin of coordinates is located in the bottom center of the numerical model. The shield tunnel consists of 114 segmental rings. The tunnel segment and soil medium are modeled by 16640 shell-type structural elements and 226200 hexahedral zone elements, respectively.

The lateral boundary in the vibration direction was considered in dynamic analysis. A 2-m wide relatively low elastic material, Duxseal, was laid at the left and right sides of the model respectively, and then the horizontal displacements of the Duxseal materials were tied together [12]. The calibrated Kobe motion scaled to various peak accelerations was applied on the base of the model. Additional 0.2% Rayleigh damping at the center frequency of 20 Hz was used to minimize the low-level noise at high frequencies.

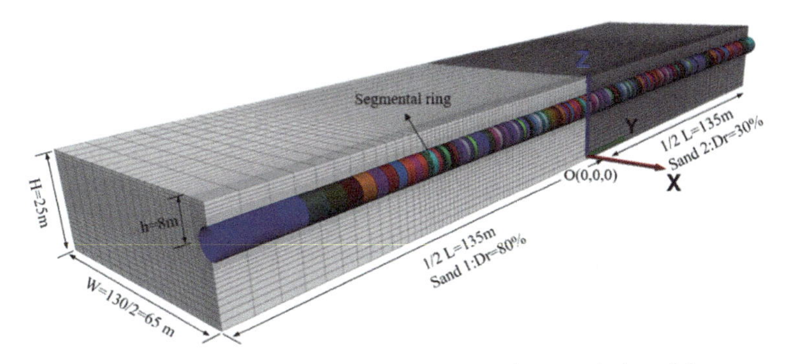

Fig. 2. Layout of tunnel and sand stratums in numerical model

3.2 Results and Discussions

Figure 3 depicts the typical residual deformation of tunnel and sand strata. There is a clear trend of increasing uplift of tunnel in loose sand (Dr = 30%). The cumulative soil deformation and displacement of the tunnel was consistent with previous studies. A more striking result from Fig. 3 is the torsion of the tunnel and dislocation of joints due to the coupling effect of horizontal seismic inertial force and the vertical floating force acting on the tunnel, which will be discussed later.

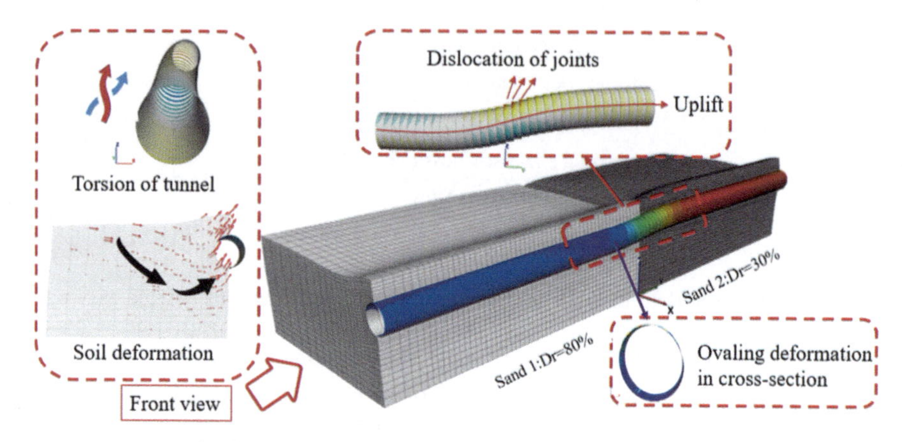

Fig. 3. Residual deformation modes of tunnel-soil system

Acceleration and Excess Pore Pressure. Figure 4 presents the acceleration amplification coefficient (β) on various locations at 6 cross-sections in the Y direction along the tunnel. It can be seen that there is more obvious earthquake-induced liquefaction ($\beta < 1.0$) observed with the increase of the input peak acceleration (IPA). However, some large values of β were got near the tunnel structure in loose sand. A reasonable explanation for this might be that there were larger shear strains in these positions due to the tunnel's uplift, leading to the sudden decrease of excess pore pressures and increase of effective pressure. Eventually, the sharp downward spikes called "dilation spikes" occurred which are similar to those derived from centrifuge experiments in the LEAP project [17]. For example, as shown in Fig. 5, the histories of acceleration and excess pore pressure ratio (EPPR) measured in point A(4,20,17) and point B(4,0,17) represent a typical "dilation spike" at about the 9th second.

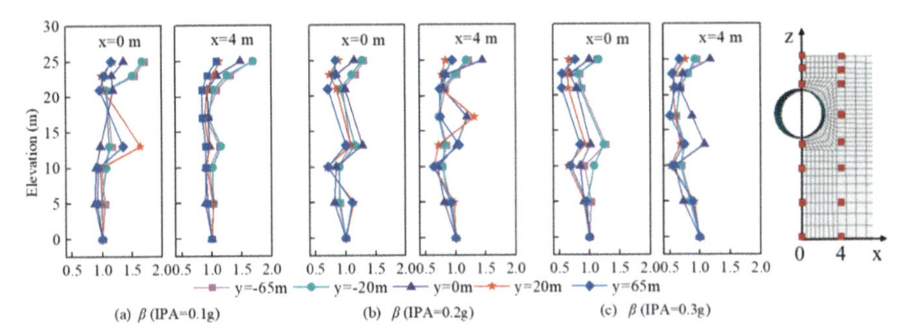

Fig. 4. The acceleration amplification coefficient β in various elevations and cross-sections when the input peak acceleration is (a) 0.1 g, (b) 0.2 g and (c) 0.3 g respectively

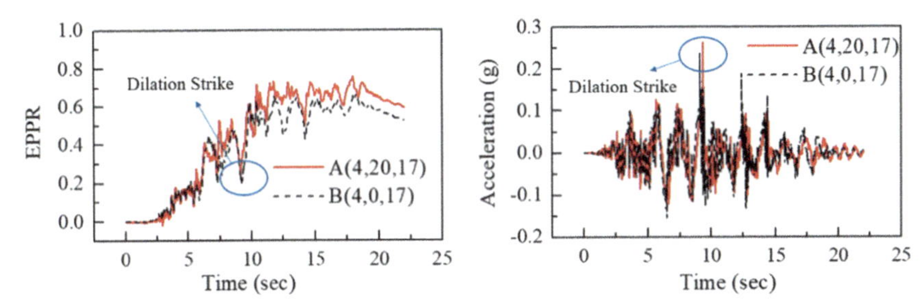

Fig. 5. The time history of acceleration and excess pore pressure ratio is measured point A(4,20,17) and point B(4, 0,17) when IPA = 0.2g

The Shear Displacements of Joints. Figure 6 shows the radial displacement (RD)and tangential displacement (SD) of segmental joints range from dense sand strata to liquefiable loose sand strata (−80 m to 80 m) in the Y direction, and the sequence 'd_0, d_1 ..., d_6, d_7' means various locations at each circular tunnel cross-section. Evidently, the RDs and SDs increase rapidly to the maximum values near the soil inter-face and gradually reduce to lower values away from the interface. Interestingly, there is one oblivious plateau of shear displacements in loose sand close to the soil interface except for RDs in d_2 and d_4, SDs in d_0 and d_6. A possible explanation for this might be that there are asymmetrical growth rates of uplift of the tunnel on each side of the soil interface, most of which were taken place in loose sand closed to the soil interface, posing the greater values of shear displacements in circumferential joints.

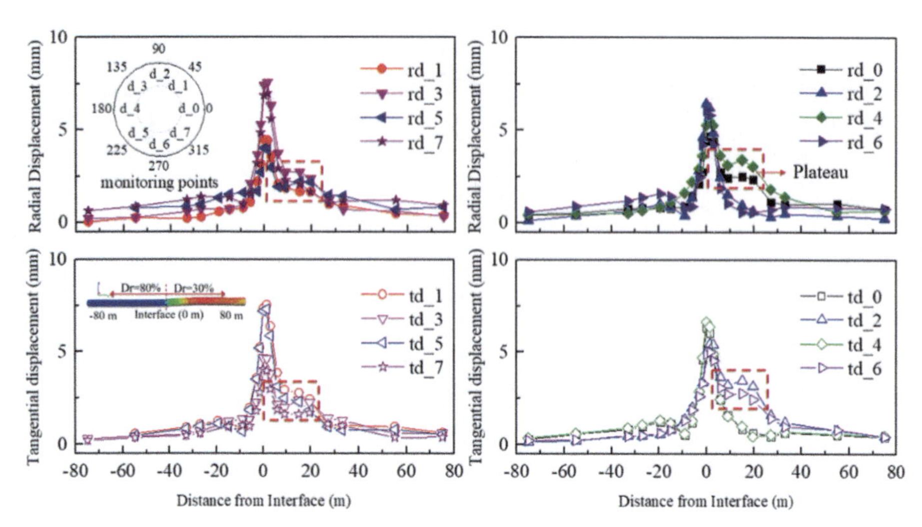

Fig. 6. The tangential and radial displacements of segmental joints along the shield tunnel when IPA = 0.2 g

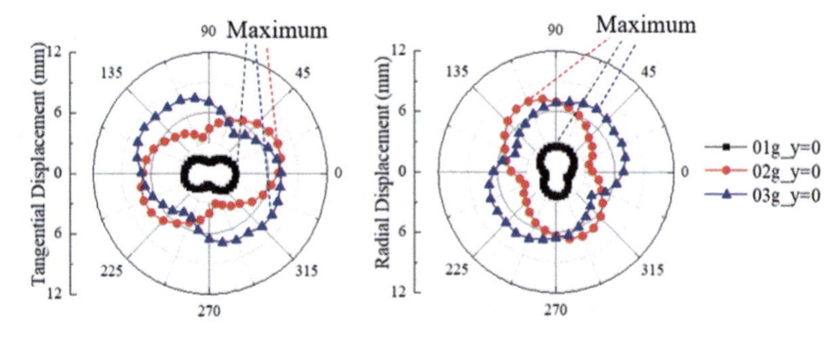

Fig. 7. The shear displacements of segmental joints at the cross-section in soil interface(y = 0)

In addition, the shear displacements in the same location vary with the magnitude of earthquake waves as can be derived from Fig. 7. The shear displacements are expected to increase with the rising of the magnitude of input earthquake waves. These findings suggest that the most unfavorable position of the tunnel section has been deflected due to the coupling relationship between the horizontal seismic inertial force and the vertical floating force applied on tunnel structure.

4 Conclusions

This study investigated the seismic response of a shield tunnel crossing two saturated sand stratums with different relative densities. Based on the numerical results, some main conclusions are drawn as follows:

(1) There are significant differences of seismic behavior of shied tunnel between saturated sand strata with different relative densities, and the most striking response of tunnel structure is near the soil interface.
(2) The shear displacements of tunnel segmental joints are large in the soil interface as well as in liquefiable sand strata closed to the interface. In other words, the boundary effect due to the soil interface in liquefiable loose sand strata is larger than that in dense sand strata.
(3) The most unfavorable position of the tunnel section is deflected due to the coupling relationship between the horizontal seismic inertial force and the vertical floating force acting on the tunnel. This study indicates that the deflection in the most unfavorable position should be taken into account in the seismic design of shield tunnels. However, the quantitative coupling relationship should be further investigated by considering more factors, such as the angle of soil interface, the relative density, and the depth of the tunnel.

References

1. Shrestha, R., Li, X., Yi, L., Mandal, A.K.: seismic damage and possible influencing factors of the damages in the Melamchi Tunnel in Nepal due to Gorkha earthquake 2015. Geotech. Geol. Eng. **38**(5), 5295–5308 (2020). https://doi.org/10.1007/s10706-020-01364-9
2. Wang, B., Li, T., He, C., et al.: Characteristics, causes and control measures of disasters for the soft-rock tunnels in the Wenchuan seismic regions. J. Geophys. Eng. **13**(4), 470–480 (2016)
3. Wang, T., Kwok, O.A., Jeng, F.: Seismic response of tunnels revealed in two decades following the 1999 Chi-Chi earthquake (Mw 7.6) in Taiwan: a review. Eng. Geol. **287**, 106090 (2021)
4. Yu, H., Zhang, Z., Chen, J., et al.: Analytical solution for longitudinal seismic response of tunnel liners with sharp stiffness transition. Tunn. Undergr. Space Technol. **77**, 103–114 (2018)
5. Liang, J., Xu, A., Ba, Z., et al.: Shaking table test and numerical simulation on ultra-large diameter shield tunnel passing through soft-hard stratum. Soil Dyn. Earthq. Eng. **147**, 106790 (2021)
6. Zou, Y., Jing, L., Li, Y.: Study of shaking table model test of tunnel through soil interface (in Chinese). Chinese J. Rock Mech. Eng. (s1), 3340–3348 (2014)
7. Zhang, Y., Wu, W., Li, G., et al.: Seismic response analysis of running tunnels in liquefiable site (in Chinese). China Earthq. Eng. J. **43**(02), 412–420 (2021)
8. Chen. R.: Underground structures in liquafiable ground: seismic response and numerical analysis method (in Chinese). Tsinghua University, Beijing (2018)
9. Ye, Z., Liu, H.: Mechanism and countermeasure of segmental lining damage induced by large water inflow from excavation face in shield tunneling. Int. J. Geomech. **18**(12), 4018161–4018163 (2018)
10. Chen, G., Ruan, B., Zhao, K., et al.: Nonlinear response characteristics of undersea shield tunnel subjected to strong earthquake motions. J. Earthq. Eng. **24**(3), 351–380 (2020)
11. Wu, H., Shen, S., Liao, S., et al.: Longitudinal structural modelling of shield tunnels considering shearing dislocation between segmental rings. Tunn. Undergr. Space Technol. **50**, 317–323 (2015)
12. Do, N., Dias, D., Oreste, P.: 3D numerical investigation of mechanized twin tunnels in soft ground – influence of lagging distance between two tunnel faces. Eng. Struct. **109**, 117–125 (2016)

13. Cheng, Z., Detournay, C.: Formulation, validation and application of a practice-oriented two-surface plasticity sand model. Comput. Geotech. **132**, 103984 (2021)
14. Itasca Consulting Group.: Fast Lagrangian Analysis of Continua in Three Dimensions. Minnesota (2019)
15. Dafalias, Y.F., Manzari, M.T.: Simple plasticity sand model accounting for fabric change effects. J. Eng. Mech. **130**(6), 622–634 (2004)
16. Sang, Y., Liu, X., Zhang, Q.: Stiffness analysis and application of segment annular joint based on bolt-concave and convex tenon connection in metro shield tunnel (in Chinese). Tunnel Constr. **40**(01), 19–27 (2020)
17. Kokkali, P., Abdoun, T., Zeghal, M.: Physical modeling of soil liquefaction: overview of LEAP production test 1 at Rensselaer Polytechnic Institute. Soil Dyn. Earthq. Eng. **113**, 629–649 (2018)

Seismic Response of a Tunnel-Embedded Saturated Sand Ground Subject to Stepwise Increasing PGA

Mingze Xu[1,2], Zixin Zhang[1,2], and Xin Huang[1,2](✉)

[1] Department of Geotechnical Engineering, Tongji University, Shanghai 200092, China
xhuang@tongji.edu.cn

[2] State Key Laboratory of Geotechnical and Underground Engineering, Ministry of Education, Shanghai 200092, China

Abstract. In this study, the interaction between a model tunnel and the saturated sand ground was investigated based on a series of shaking table tests. The model tunnel was made of stainless-steel tube and the model ground was composed of China ISO standard sand through water sedimentation technique. In order to reduce the boundary effect, polystyrene foam boards were used as a buffer layer to absorb vibration energy due to the reflection from the boundaries. A record from the 1995 Kobe earthquake in Japan was used as the input motion, of which the PGA (peak ground acceleration) was scaled from 0.1g to 0.6g. The input motion was applied in the direction perpendicular to the tunnel axis. The earth pressure, the pore water pressure, the accelerations of both the ground and tunnel, and the ground settlement were recorded during the tests. The model was settled between different testing groups until the recorded pore water pressure was stabilized. A new definition of pore pressure ratio was proposed which can be used to determine the degree of liquefaction at different PGAs. The liquefaction phenomenon was observed during the tests when the PGA reached 0.6g. It was also showed that the ground near the tunnel is more prone to liquefaction than in other regions.

Keywords: Saturated sand ground · Shaking table test · Liquefaction

1 Introduction

The seismic performance of underground structure has drawn wide attention since Kobe earthquake, in which both the super-ground structures and the underground structures, including subway stations and tunnels, were severely damaged. Since then, numerous studies have been conducted on the seismic performance of the underground structure. Shaking table test, as an important experimental technique, has been employed widely in these researches [1–5]. Previous shaking table tests can be divided into two categories: 1g shaking table tests and centrifuge tests. Although centrifuge tests can simulate real geo-stress environment, it is limited by size which makes 1g shaking table test a more popular choice.

© The Author(s), under exclusive license to Springer Nature Switzerland AG 2022
L. Wang et al. (Eds.): PBD-IV 2022, GGEE 52, pp. 2371–2380, 2022.
https://doi.org/10.1007/978-3-031-11898-2_222

A number of 1g shaking table tests have been conducted in the last two decades. Chen et al. [6] designed and fabricated a double-axis laminar shear box which was used to conduct a series of shaking table tests on scaled utility tunnels in the unsaturated clay soil. They studied the amplification factor of the soil with different PGAs and found that when PGA ≤ 0.2g, the amplification factor increased from the bottom to the top of the soil but it decreased bottom up while PGA > 0.2g. Yan et al. [7] conducted a series of shaking table tests using four shaking tables to simulate the immersed tunnel of Hong Kong-Zhuhai-Macau Bridge in the dry synthetic soil. Non-uniform excitations including different wave forms, peak accelerations, vibration directions and wave propagation directions were used in the tests. They also compared the test results with the simulation results to validate the reliability of the shaking table system [8]. Wang et al. [9] studied the interaction between the tunnel and the surface structure through a series of shaking table tests in the dry synthetic model soil and found that the existence of the tunnel reduced the rigidity of the model ground, which caused the amplification of seismic response of surrounding soil. Their experimental results were used to validate the numerical model by Liu et al. [10]. Haeri et al. [11] conducted a series of shaking table tests to study the response of the piles in the saturated sand soil when liquefaction happened. However, they only used the pore water pressure record to determine whether the ground liquefied or not, which seemed to be not so convincing.

Although numerous tests have been conducted, most of them used dry soil in their tests. Even though saturated sand was used in some researches, the method used to determine the onset of liquefaction was not convincing and the interaction between the tunnel and ground when liquefaction happened was missed. In this research, saturated sand has been used and a rational liquefaction determination method is proposed. Moreover, the tunnel influence on the ground liquefaction is studied.

2 Experimental Setup

2.1 Shaking Table

The shaking table tests were conducted in the State Key Laboratory for Disaster Prevention in Civil Engineering, Tongji University. As can be seen in Fig. 1(a), the shaking table is 4 m × 4 m in size with a payload capacity of 25 t in maximum. It can be input with a three-dimensional motion with six degrees of freedom in forms of simple harmonic vibration, shock or earthquake. More details of the shaking table can be found in Table 1.

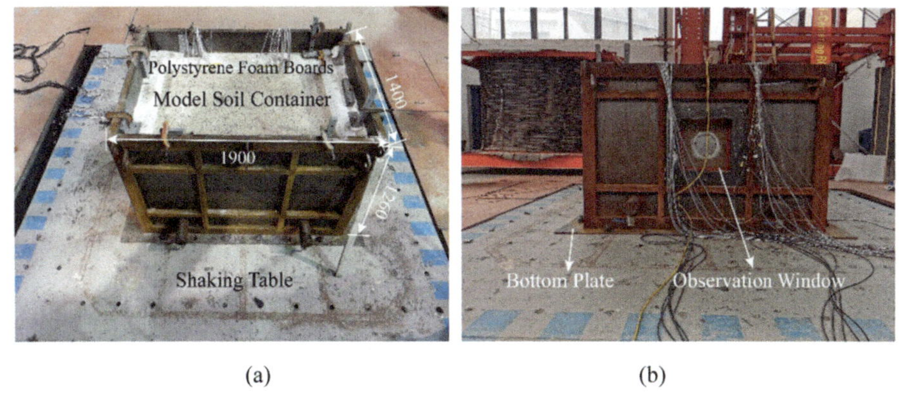

(a) (b)

Fig. 1. Shaking table and model soil container (unit: mm)

Table 1. Shaking table parameters

Characteristics	Index
Maximum bearing capacity	25 t
Size	4 m × 4 m
Excitation direction	X, Y, Z
Degree of freedom	6
Excitation form	Simple harmonic vibration, shock, earthquake
Maximum displacement	X: ±100 mm, Y: ±75 mm, Z: ±50 mm
Maximum velocity	±1000 mm/s, ±1000 mm/s, ±600 mm/s
Maximum acceleration	±4.0g
Working frequency	0.1–100 Hz
Test and data acquisition system	MTS469D, STEXPro, 256 channels

2.2 Model Soil Container

The model soil container was designed to meet the following requirement. Firstly, it should be strong enough to avoid any leakage or damage during the test. Secondly, it should meet the bearing capacity and size requirement of the shaking table. Rigid box was adopted instead of laminar box for its convenience in manufacture and quality control. The rigid container was 1.9 m in length (parallel to the excitation direction and perpendicular to the tunnel axis), 1.4 m in width (perpendicular to the excitation direction and parallel to the tunnel axis) and 1.26 m in height, as shown in Fig. 1(a). The main frame of the rigid container was welded by square steel pipe which was 50 mm × 100 mm in size and was enclosed by 10 mm thick steel plate. In order to improve the stiffness of the container, several 50 mm × 50 mm square steel pipes were welded around the container and also on the bottom of the container as cross braces. The 2.3 m × 1.6 m bottom plate was welded to the rigid container. Bolt holes with a diameter of

40 mm were set every 300 mm so that the container can be bolted to the shaking table (Fig. 1(b)).

Polystyrene foam boards, which were 140 mm in thickness were bonded to the side-walls to minimize the boundary effect (Fig. 1(a)). An observation window was set on the front of the rigid container so that the displacement of the tunnel can be recorded. The observation window was bolted to the rigid container as can be seen in Fig. 1(b).

2.3 Model Tunnel

Since the internal change of the model tunnel was not the focus of this research, the model tunnel was simplified to a cylindrical straight pipe with closed ends (Fig. 2). Stainless steel was selected to make the model tunnel considering its stability and ease of manufacture. The tunnel was 1m in length and has the width of the soil container with bonded polystyrene foam boards. Previous research shows that the tunnel has little influence on the dynamic response of surrounding soil if its radius is small [12]. However, the radius cannot be too large considering the boundary effect. Upon comprehensive consideration, the outer diameter of the model tunnel was set to be 200 mm. The wall thickness of the model tunnel was 4 mm so that the model tunnel may uplift when liquefaction happens.

Fig. 2. Model tunnel

2.4 Instrumentation

The sensors used in the shaking table tests included earth pressure cells, pore-water pressure gauges, accelerometers, strain gauges, a line laser displacement sensor and cameras. The abbreviations used to describe the sensors were as follows: Ac for accelerometers, W for pore-water pressure gauges, E for earth pressure cells, S for strain gauges and LVDT for line laser displacement sensors. Figure 3 presents the location of sensors. The locations of the two cameras are shown in Fig. 4: one sitting in front of the model soil container to record the tunnel displacement and the other one sitting obliquely above the soil container to monitor the earth surface change.

Since all the sensors were placed in the saturated sand, waterproof treatment was needed. In order to avoid the interference between the sensors, the accelerometers were placed along the central axis of the soil container with earth pressure cells and pore-water pressure gauges placed along the two sides (Fig. 3(b)). It is assumed that the pressures are the same over the same altitude as the excitation direction is perpendicular to the tunnel axis. Therefore, although for the earth pressure cell and pore-water pressure gauge with the same sensor number, for example W1 and E1, which were placed on the opposite two sides, their results can be compared to determine whether this location is liquefied.

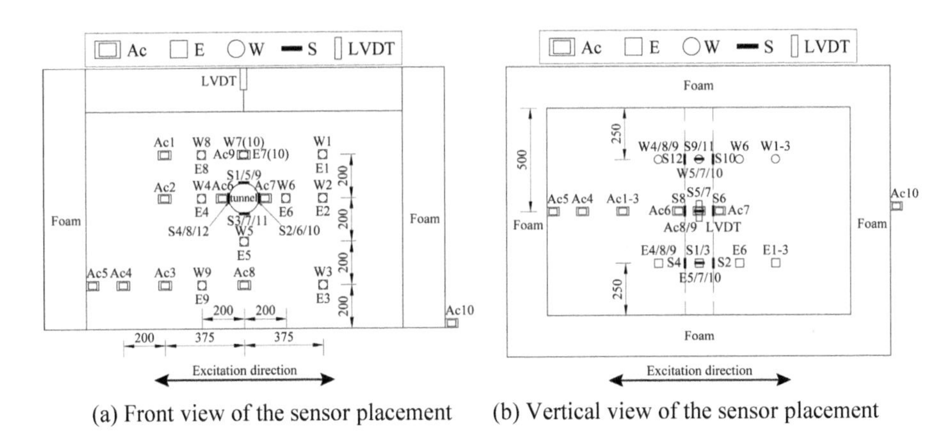

(a) Front view of the sensor placement (b) Vertical view of the sensor placement

Fig. 3. Schematic of sensor arrangement (unit: mm)

2.5 Ground Preparation

Fujian sand (China Standard Sand) was chosen to prepare the model ground using layered water sedimentation method. During the ground preparation, the water level was always kept higher than the ground level to guarantee all the sand was saturated. When the height of the ground reached 80 cm, sand was added in the soil container with no water injected until reaching the target height. The sand was completely saturated below 80 cm in height with no water on the ground surface after preparation work by using this method so that the loading level can be easily determined by monitoring the surface change. After settled for one night, the height of the ground is 88.68 cm with avoid ratio of 0.587 and a density of 1.7 g/cm^3.

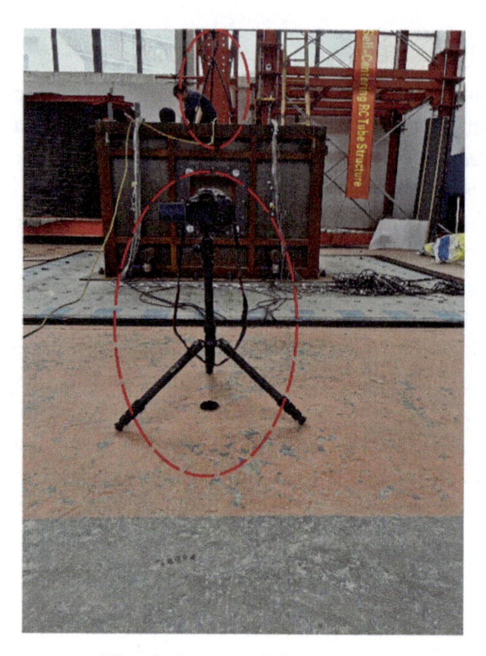

Fig. 4. Layout of the cameras

2.6 Test Case

In this test, as shown in Fig. 5, one Kobe seismic wave record [13, 14] was selected as the input wave. The input Kobe wave was scaled by its peak acceleration and the test cases are summarized in Table 2. The peak acceleration of the Kobe wave increased gradually to see under which case can the model ground liquefy. After finding the liquefaction case, the model would be reloaded using the same Kobe wave or the Kobe wave whose peak acceleration smaller than before to see the change of the liquefaction resistance of the model ground post liquefaction. Between different groups of Kobe wave cases, white noise was used to check the fundamental frequency of the model. After finishing one

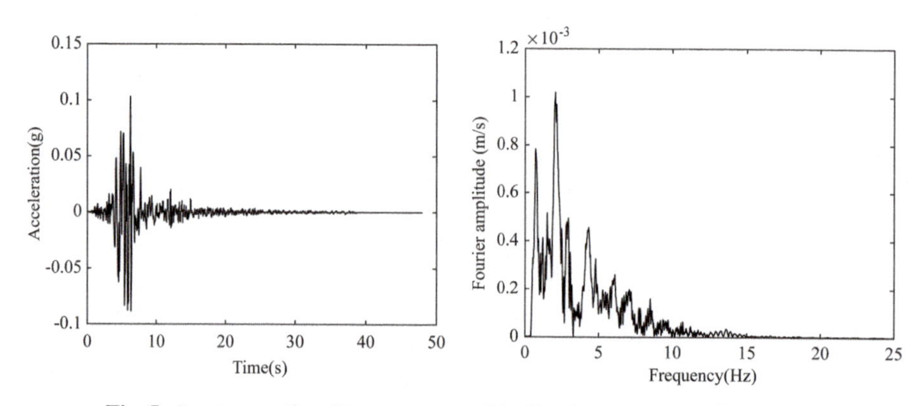

Fig. 5. Input wave time history curve and its Fourier spectrum (PGA = 0.1g)

test case, the model was settled for several minutes until the excess pore water pressure fully dissipated.

Table 2. Test cases

Test ID	Input wave form	Acceleration amplitude/g
WN1	White noise	0.07
Kobe0.1	Kobe	0.1
WN2	White noise	0.07
Kobe0.2	Kobe	0.2
WN3	White noise	0.07
Kobe0.3	Kobe	0.3
WN4	White noise	0.07
Kobe0.4-1	Kobe	0.4
WN5	White noise	0.07
Kobe0.5	Kobe	0.5
WN6	White noise	0.07
Kobe0.6-1	Kobe	0.6
Kobe0.6-2	Kobe	
Kobe0.6-3	Kobe	
Kobe0.6-4	Kobe	
WN7	White noise	0.07
Kobe0.4-2	Kobe	0.4

3 Test Results

Previous studies [2, 4, 15] defined the pore pressure ratio as:

$$r_u = \frac{\Delta u}{\sigma'} \tag{1}$$

in which, r_u was the pore pressure ratio, Δu was the excess pore water pressure and σ' was the initial vertical effective stress. Ground was considered to be liquefied when r_u reached 0.8. However, this critical value of r_u is subjective. Besides, in the real tests, the initial value of vertical effective stress cannot be determined precisely since the records of pore-water pressure gauges and earth pressure cells usually oscillate. An alternative definition of the pore pressure ratio is defined herein as:

$$f_u = \frac{u}{\sigma} \tag{2}$$

in which, f_u is the pore pressure ratio, u is the pore water pressure which is the record of pore-water pressure gauges, and σ is the vertical stress which is the record of earth pressure cells. The soil liquefies when f_u reaches 1, which means the vertical stress is fully contributed by the pore water pressure and the effective stress is 0. This definition makes the critical value of the pore pressure ratio has a solid theoretical basis. Besides, no initial vertical effective stress needs to be determined.

Figure 6 presents the evolution of the pore pressure ratio at monitoring point 4 (close to the left waist of tunnel) during seismic loading under different testing cases. The pore pressure ratio increased with the increase of PGA and when the PGA reached 0.6, the soil liquefied after loading for 6.74 s. Generally, the pore pressure ratio increases with increasing PGA and the largest pore pressure ratio occurs in accordance with the PGA. The excess pore pressure decreases sharply as the acceleration decreases.

The vertical distribution of the pore pressure ratio near tunnel can be seen in Fig. 7. Similar to Fig. 6, pore pressure ratios increased with the increase of PGA in every monitoring point. The pore pressure ratio is larger at the bottom than at the surface, while the pore pressure ratio of monitoring point 4 which is close to the tunnel is always the largest, indicating that the existence of a tunnel decreases the liquefaction resistance of the soil nearby. A possible explanation to this special phenomenon may be that the local vibration of tunnel will exert additional inertia force to the surrounding ground, which may lead to additional excess pore pressure.

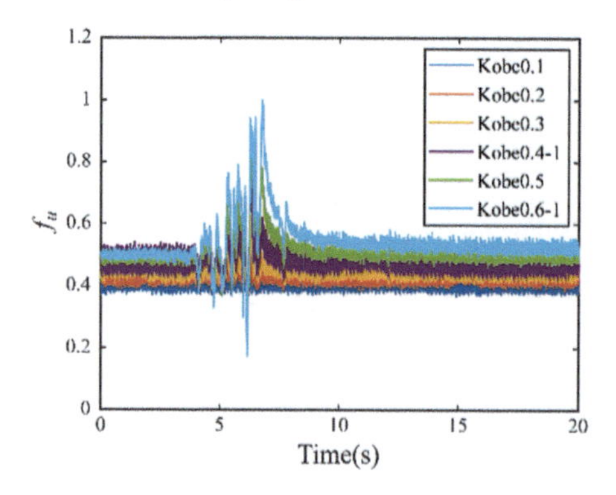

Fig. 6. Pore pressure ratio of monitoring point 4 under different PGAs

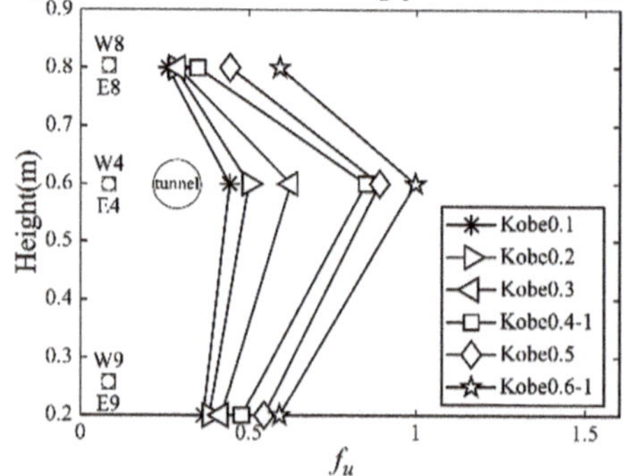

Fig. 7. Vertical distribution of pore pressure ratio near tunnel under different PGAs

The max pore pressure ratios around the tunnel can be seen in Fig. 8. Similar conclusions can be made that the max pore pressure ratios increase with the increase of PGA. Besides, the pore pressure ratio distributed nonuniformly around the tunnel. Pore pressure ratio of point 4 is the largest while point 10 is the smallest.

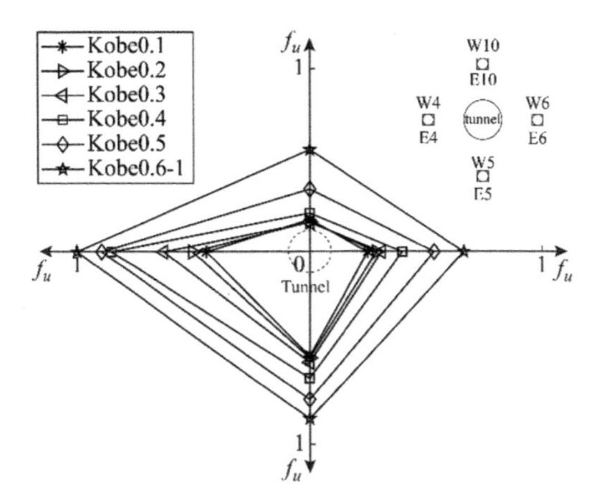

Fig. 8. Distribution of pore pressure ratio around tunnel under different PGAs

4 Conclusions

In this study, a series of shaking table tests considering tunnel-saturated sand ground interaction were conducted. The design and fabrication of the model soil container and model tunnel were given in detail together with the layout of the monitoring program. In particular, saturated sand was used in this test and liquefaction phenomenon was observed.

A new definition of pore pressure ratio with a solid theoretical basis is proposed to judge whether liquefaction occurs or not. Test results showed that the pore pressure ratio increased with the increase of PGA. When the PGA reached 0.6, the model ground liquefied. Besides, the existence of tunnel made the ground close by more prone to liquefaction, which may possibly be associated with additional inertia force exerted by the tunnel. Due to the limit of space, more details about the testing results will be reported elsewhere.

References

1. Liang, J., Xu, A., Ba, Z., Chen, R., Zhang, W., Liu, M.: Shaking table test and numerical simulation on ultra-large diameter shield tunnel passing through soft-hard stratum. Soil Dyn. Earthq. Eng. **147**, 106790 (2021). https://doi.org/10.1016/j.soildyn.2021.106790
2. Wu, Q., Ding, X.M., Chen, Z.X., Chen, Y.M., Peng, Y.: Seismic response of pile-soil-structure in coral sand under different earthquake intensities. Yantu Lixue/Rock Soil Mech. **41**, 571–580 (2020). https://doi.org/10.16285/j.rsm.2019.0122

3. Tsinidis, G., Rovithis, E., Pitilakis, K., Chazelas, J.L.: Seismic response of box-type tunnels in soft soil: experimental and numerical investigation. Tunn. Undergr. Sp. Technol. **59**, 199–214 (2016). https://doi.org/10.1016/j.tust.2016.07.008
4. Xu, C.S., Dou, P.F., Du, X.L., Chen, S., Han, J.Y.: Large-scale shaking table model test of liquefiable free field. Yantu Lixue/Rock Soil Mech. **40**, 3767–3777 (2019). https://doi.org/10.16285/j.rsm.2018.1339
5. Bao, Z., Yuan, Y., Yu, H.: Multi-scale physical model of shield tunnels applied in shaking table test. Soil Dyn. Earthq. Eng. **100**, 465–479 (2017). https://doi.org/10.1016/j.soildyn.2017.06.021
6. Chen, J., Shi, X., Li, J.: Shaking table test of utility tunnel under non-uniform earthquake wave excitation. Soil Dyn. Earthq. Eng. **30**, 1400–1416 (2010). https://doi.org/10.1016/j.soildyn.2010.06.014
7. Yan, X., Yu, H., Yuan, Y., Yuan, J.: Multi-point shaking table test of the free field under non-uniform earthquake excitation. Soils Found. **55**, 985–1000 (2015). https://doi.org/10.1016/j.sandf.2015.09.031
8. Yan, X., Yuan, J., Yu, H., Bobet, A., Yuan, Y.: Multi-point shaking table test design for long tunnels under non-uniform seismic loading. Tunn. Undergr. Sp. Technol. **59**, 114–126 (2016). https://doi.org/10.1016/j.tust.2016.07.002
9. Wang, G., Yuan, M., Miao, Y., Wu, J., Wang, Y.: Experimental study on seismic response of underground tunnel-soil-surface structure interaction system. Tunn. Undergr. Sp. Technol. **76**, 145–159 (2018). https://doi.org/10.1016/j.tust.2018.03.015
10. Liu, H., Liu, H., Zhang, Y., Zou, Y., Yu, X.: Coupling effects of surface building and earthquake loading on in-service shield tunnels. Transp. Geotech. **26**, 100453 (2021). https://doi.org/10.1016/j.trgeo.2020.100453
11. Haeri, S.M., Kavand, A., Rahmani, I., Torabi, H.: Response of a group of piles to liquefaction-induced lateral spreading by large scale shake table testing. Soil Dyn. Earthq. Eng. **38**, 25–45 (2012). https://doi.org/10.1016/j.soildyn.2012.02.002
12. Wang, G., Yuan, M., Ma, X., Wu, J.: Numerical study on the seismic response of the underground subway station- surrounding soil mass-ground adjacent building system. Front. Struct. Civ. Eng. **11**(4), 424–435 (2016). https://doi.org/10.1007/s11709-016-0381-7
13. Giardini, D., Woessner, J., Danciu, L., et al.: Seismic Hazard Harmonization in Europe (SHARE): online data resource. European Facilities Earthquake Hazard Risk (2013). https://doi.org/10.12686/SED-00000001-SHARE. Accessed 12 Jun 2021
14. Ancheta, T.D., Darragh, R.B., Stewart, J.P., et al.: PEER report on "PEER NGA-West2 Database". Pacific Earthquake Engineering Research Center, University of California, Berkeley (2013). https://apps.peer.berkeley.edu/publications/peer_reports/reports_2013/webPEER-2013-03-Ancheta.pdf. Accessed 12 Jun 2021
15. Xu, C.S., Dou, P.F., Du, X.L., Chen, S., Han, J.Y.: Study on solid-liquid phase transition characteristics of saturated sand based on large shaking table test on free field. Yantu Lixue/Rock Soil Mech. **41**, 2189–2198 (2020). https://doi.org/10.16285/j.rsm.2019.1716

Groundwater Response to Pumping Considering Barrier Effect of Existing Underground Structure

Xiu-Li Xue, Long Zhu, Shuo Wang, Hong-Bo Chen, and Chao-Feng Zeng(✉)

Hunan Provincial Key Laboratory of Geotechnical Engineering for Stability Control and Health Monitoring, School of Civil Engineering, Hunan University of Science and Technology, Xiangtan 411201, Hunan, China
cfzeng@hnust.edu.cn

Abstract. Existing underground structure may block the water flow or extend the seepage path, which should affect the groundwater response induced by pumping. In this study, finite element models are established by ABAQUS, based on a practical pre-excavation pumping test in Tianjin, to investigate the influence of existing underground structure on seepage flow under pumping. It is found that the existing underground structure has barrier effect in pumping. In addition, the effect of the spacing between the existing structure and the excavation on the groundwater response was revealed. This study can provide guidance for the design of groundwater recharge.

Keywords: Existing underground structure · Pumping · Barrier effect · Groundwater drawdown · Numerical simulation

1 Introduction

In deep foundation pit with a high groundwater level, there is often hydraulic connection between the inside and outside of a foundation pit. For example, the confined aquifer is not cut off by the waterproof curtain and the enclosure wall is defective [1, 2]. In this case, dewatering will lead to the drawdown of groundwater level outside the pit, which in turn will have an adverse impact on the environment [3–5].

In the existing studies, most scholars have focused on the impact of foundation pit dewatering on the environment. For example, Zeng [6] and Li [7] proposed that the deep soil settlement caused by pumping is greater than the ground surface settlement; Shen [8] conducted numerical simulations to reveal the characteristics of drawdown curve around an excavation where partial cut-off walls were utilized. However, the environment outside the foundation pit will also have some influence on the change of groundwater level during dewatering. For example, existing underground structure can block the groundwater seepage or prolong the seepage path, thus the groundwater response should be different in the process of foundation pit dewatering with/without subsurface structures. As shown in Fig. 1, groundwater seepage from outside of the pit to inside of the pit is caused by pumping; the underground structure has a barrier effect

© The Author(s), under exclusive license to Springer Nature Switzerland AG 2022
L. Wang et al. (Eds.): PBD-IV 2022, GGEE 52, pp. 2381–2390, 2022.
https://doi.org/10.1007/978-3-031-11898-2_223

on the water seepage and the water seepage path is extended, which makes the water recharge from the rear of the underground structure to the front (near the pit) slower compared to the condition without the underground structure. And the groundwater decline in this area is relatively larger. Therefore, when there is an underground structure, groundwater drawdown of the rear of the underground structure is relatively small and the water surface is relatively high. We define the above-mentioned barrier effect of the underground structure on groundwater as "water blocking effect".

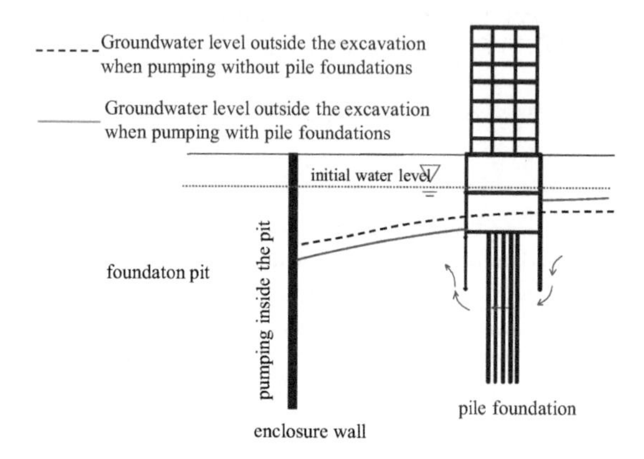

Fig. 1. Barrier effect of adjacent underground structure on groundwater seepage

As a result, groundwater response to pumping with/without underground structure outside the pit would be different,impacting on the number of pumping wells and flow rate in the pit. In this paper, we established a finite element model for foundation pit pumping by using ABAQUS in the context of actual foundation pit pumping test, and explored the groundwater response to pumping considering the barrier effect of existing underground structure under different spacing between existing pile foundation and foundation pit.

2 Project Background

2.1 Project Overview and Hydrogeological Condition

Figure 2 shows the foundation pit of a subway station in Tianjin, whose length, width and depth were 155 m, 40 m and 16.9 m respectively. The diaphragm wall was adopted as retaining structure that was 32.5 m deep and 0.8 m thick. The main soil parameters of each layer obtained from site investigation were exhibited in Table 1. The initial groundwater level of Aq0–AqIV was 2.0 m, 2.7 m, 3.0 m, 3.2 m and 3.7 m below the ground surface, respectively. Foundation pit dewatering was necessary to prevent potential safety of foundation pit. In addition, since the AqII depth ranges from 22 m to 35.5 m, and the AqII was not totally cut off by the diaphragm wall. In order to evaluate the dewatering effect on the environment and check the quality of dewatering well, the dewatering test was carried out before soil excavation.

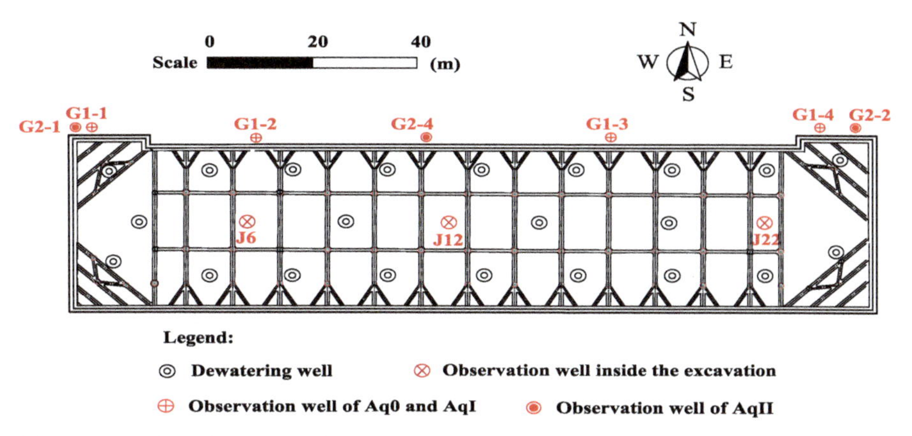

Fig. 2. Plan view of an excavation and instrumentation (adapted from Zeng et al. (2021)) [9]

Table 1. Strata distribution and main soil mechanical parameters

Hydrogeology	Soil classification	H/m	γ/ (kN/m^3)	K_0	ω /%	e	E_s/Mpa	V_s/(m/s)
Aq0	Silty clay	10	19.1	0.577	30.4	0.85	7.4	152
AdI	Silty clay	15	19.3	0.61	28.7	0.81	6.8	172
AqI	Silt	19	20.2	0.44	21.7	0.62	17.4	266
AdII	Silty clay	22	19.9	0.56	25.1	0.71	8	246
AqII	Silt	24.5	20.4	0.44	22.3	0.55	14.6	278
	Silt	29.5	20.6	0.41	20.9	0.58	15.9	278
	Silty clay	32.5	20.3	0.56	23.6	0.66	8.6	253
	Silty sands	35.5	20.6	0.398	16.3	0.521	18.9	300
AdIII	Silty clay	37	20.5	0.56	20.7	0.6	9.5	274.5
AqIII	Silt	41	20.7	0.44	18.2	0.54	19.9	328
AdIV	Silty clay	47	20.3	0.546	22.1	0.64	10.3	315
AqIV	Silty sands	50	20.6	0.384	17.5	0.53	25.3	360

Note: H = buried depth of soil bottom; γ = unit weight; K_0 = the coefficient of soil pressure at-rest; ω = moisture content; e = initial void ratio; E_s = compression modulus; V_s = shear wave velocity.

2.2 Dewatering Test

The first level strut was constructed before the dewatering test. In addition, Fig. 2 shows that 25 dewatering wells were installed inside the excavation and 7 observation wells were mounted outside the pit on the north side. There were 22 dewatering wells simultaneously operating to withdraw groundwater and others (i.e., J6, J12 and J22) were adopted as observation wells to monitor the water level change inside the excavation.

After the dewatering, the water level inside the pit dropped by about 15 m, causing groundwater drawdown of about 7 m in second confined aquifer and about 3 m in first confined aquifer outside excavation. The specific data of the experiment were detailedly discussed in the literature. Due to the limited data of actual project, it was impossible to discuss the barrier effect of existing underground structure on the groundwater response to foundation pit dewatering, and thus, ABAQUS was used to study the above content.

3 Numerical Simulation

3.1 Model Building

(1) Model Size and Element Type. Figure 3 shows the finite element model. Because the above foundation pit is very long, in our model only the central section of the foundation pit was modeled to facilitate modeling and reduce calculation. Considering the existence of hydraulic connection between inside and outside of the foundation pit, outer boundary was set at 800 m behind the foundation pit to eliminate the boundary effects on the calculation results, and this distance has already been over the influence radius of dewatering. In the model, the soil was simulated by C3D8P solid element, the retaining structure was simulated by C3D8I solid element, and the pumping well was simulated by S4 shell element. In the depth direction, the mesh size was between 0.4–2.5 m, and the mesh of the aquitard was relatively dense and that of the aquifers was coarse.

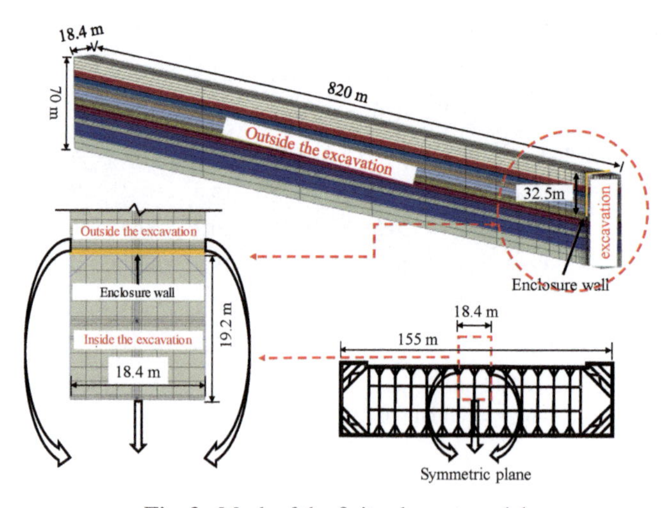

Fig. 3. Mesh of the finite element model

(2) Material Constitutive and Parameter Selection. Based on the calculation experience by former researcher, the Mohr-Coulomb constitutive model was used to simulate the deformation behavior of soil during pumping. The soil layers distribution and the physical parameters are shown in Table 2. Permeability coefficients (K_H, K_V) and elastic

modulus (E) were obtained by fitting the measured data with computed data [10]. The elastic model was adopted to simulate the retaining wall and dewatering well, and their elastic modulus was set as 30 GPa and 210 GPa, respectively. In addition, considering the friction between soil and structure, the Mohr-Coulomb's law was employed to control the friction behavior. The friction coefficient between soil and structure was set as 0.3.

Table 2. Soil distribution and parameters used in the model

Hydrogeology	Soil classification	c'/k Pa	φ'/°	μ	K_H/(m/d)	K_V/(m/d)	E/M pa
Aq0	Silty clay	17	25	0.33	0.03	0.003	43.5
AdI	Silty clay	18	23	0.33	0.025	0.001	56.3
AqI	Silt	10	34	0.3	0.2	0.1	137.6
AdII	Silty clay	19	26	0.33	0.006	0.001	118.6
AqII	Silt	8	34	0.3	2.5	0.5	151.8
	Silt	8	36	0.3	1	0.2	153.3
	Silty clay	17	26	0.33	1	0.16	128
	Silty sands	7	37	0.3	3	0.6	178.5
AdIII	Silty clay	19	26	0.33	0.002	0.0004	152.1
AqIII	Silt	10	34	0.3	3	0.9	214.5
AdIV	Silty clay	18	27	0.33	0.0005	0.0001	198.4
AqIV	Silty sands	7	38	0.3	3.5	1.5	257

Note: c' = effective cohesion; φ' = effective friction angle; μ = Poisson's ratio; K_H = horizontal hydraulic conductivity; K_V = vertical hydraulic conductivity; E = elastic modulus;

(3) Boundary Conditions and Pumping Simulation. On the three symmetrical planes, only deformation perpendicular to the direction of the plane was limited and these planes were set as impermeable boundaries. The deformation at the asymmetrical plane was limited in two horizontal directions. A constant head boundary was set on the asymmetric plane with the fixing head at the initial water head, which was assumed at the model surface Both the vertical and horizontal deformations were restricted in the bottom of the model, and meanwhile the model bottom was set as an impermeable boundary.

This section simulates the actual pumping test in Sect. 2, by setting a fixed water head boundary on the soil surface in contact with the well to simulate the pumping. Because the water level dropped by about 15 m in the actual pumping test, as exhibited in Fig. 4, zero pore water pressure was set in the depth range of 0–15 m(simulate the water level drop of 15 m). Because the dewatering well was 22 m deep, a fixed head boundary was set in the range of 15–22 m to form a hydrostatic pressure distribution in this range.

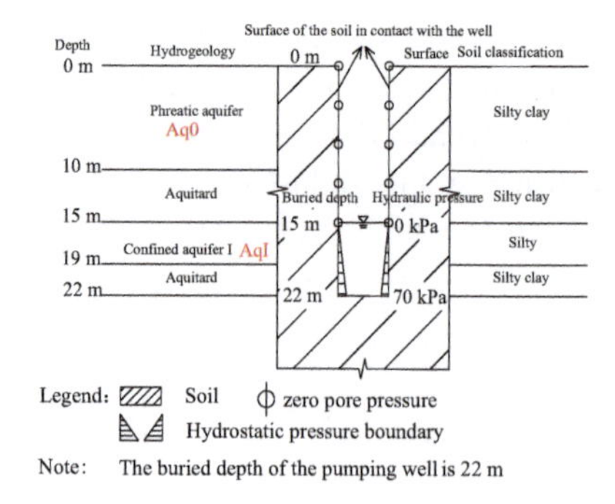

Fig. 4. Schematic diagram showing the simulation method of pumping from phreatic aquifer

3.2 Model Verification

The variation of pore water pressure in the model was derived and the water level drop at the corresponding position was calculated. Figure 5 compares the measured and calculated steady water level drop in pumping process. It can be seen that both measured and calculated water level declines increase with the pumping time, and the measured water level decline can be basically fitted by the computed ones. The use of the model to investigate the barrier effect of existing underground structure on the Groundwater response is justified.

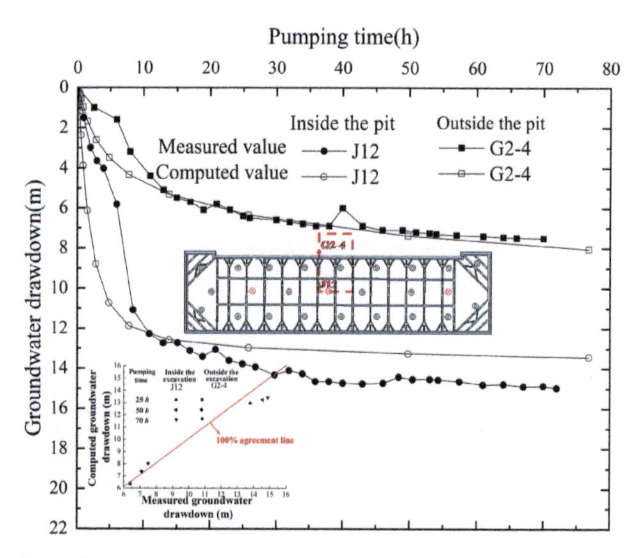

Fig. 5. Comparison of computed and measured groundwater drawdown (adapted from Zeng et al. (2021)) [9]

4 Barrier Effect of Existing Underground Structure

In order to investigate the influence of pile foundation on groundwater response to pumping, a building pile foundation and the related diaphragm wall were set outside the model pit. As exhibited in Fig. 6, the depth of the pile foundations and underground diaphragm wall were 37 m and 29.5 m, respectively. To simulate the pumping in the second confined aquifer, the pumping well was set 35.5 m; an impermeable boundary was set on the soil surface in the range of 0–22 m and a fixed head was set the range of 22–35.5 m, i.e., the pore-water pressure was set 120 kPa and 255 kPa at depth of 22 m and 35.5 m respectively, assuming a water level drop of 10 m(H_d)in the second confined aquifer, the pore-water pressure distribution within the range of 22-35.5 obeys a hydrostatic pattern (Fig. 7).

The pumping time was set to 200 days to stabilize the water level and soil deformation. Figure 8 shows the groundwater drawdown of the second confined aquifer during

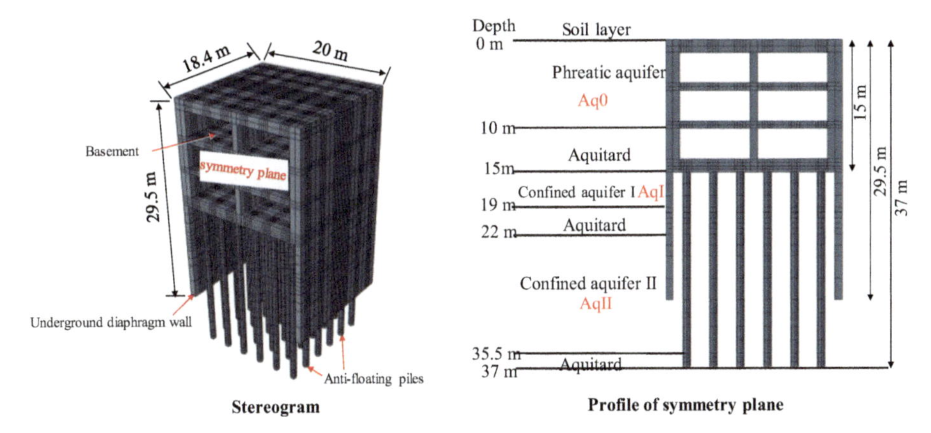

Fig. 6. Pile foundation in numerical model

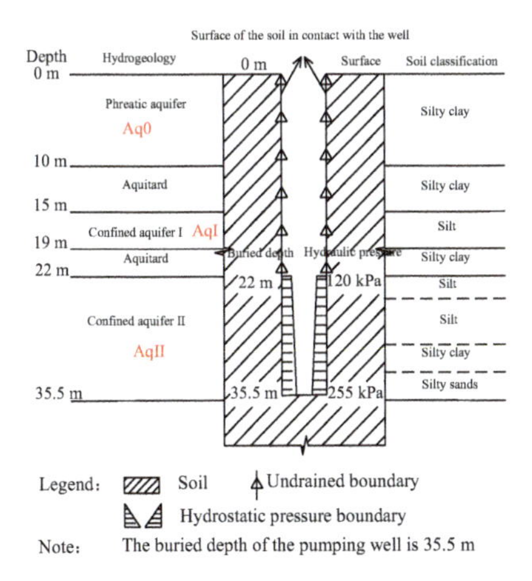

Fig. 7. Schematic diagram showing the simulation of pumping from confined aquifer

pumping under the condition of different spacing (D) between the existing pile foundation and foundation pit. It was observed that when there is no pile outside the foundation pit, the water level drop distribution was a continuous curve; when there is pile foundation outside the foundation pit, the water level drop distribution shows a stepped pattern. The abrupt change of the curve was at the position of the pile foundation; the water level drop in front of pile foundation (between the existing pile foundation and foundation pit) was larger than that without pile foundation, but the drop of water level behind pile foundation was smaller than that without pile foundation. This indicates that the existing piles produce a barrier effect on seepage of water outside the pit. With the increase of D, the maximum water level drop under the condition with pile foundation decreases continuously and approaches that without pile foundation, that is, the barrier effect of existing pile foundation decreases continuously with the increase of D.

Additionally, by comparing Fig. 8(a) and Fig. 8(b) can find that, in these two cases, the maximum difference between the water level drop with and without pile foundation was only 0.88 m, indicating that the barrier effect is not very obvious when the existing pile foundation does not continuously cut off the aquifer (the retaining wall of the building

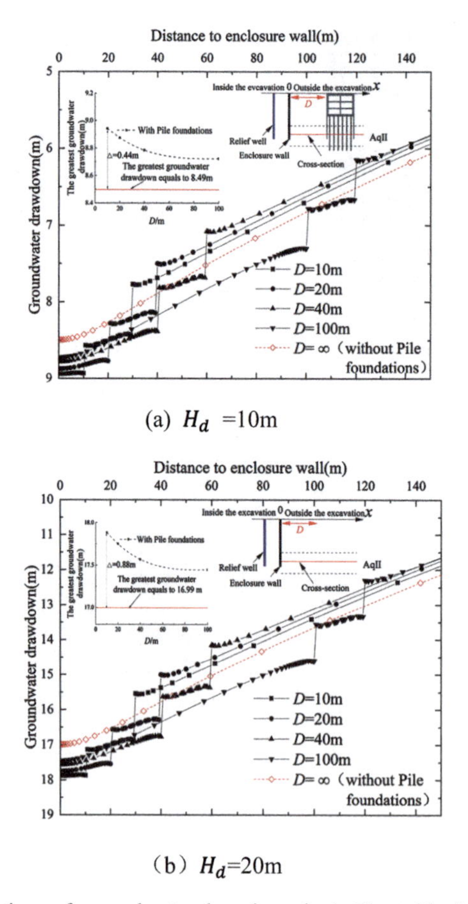

(a) H_d =10m

(b) H_d=20m

Fig. 8. Distributions of groundwater drawdown in AqII outside the foundation pit

basement does not completely cut off the aquifer). It can be inferred that if the existing pile foundation totally cut off the aquifer the barrier effect will be obvious.

5 Conclusions and Shortcomings

In this study, a finite element model was established by using ABAQUS with reference to the practical pumping tests in Tianjin, and pile foundation by which the aquifer was not continuously cut off are added to this model. Numerical simulation was used to study groundwater response to pumping considering barrier effect of existing underground structure. Based on this study, the conclusions and shortcomings are as follows:

(1) The existing pile foundation has a barrier effect on the groundwater response during foundation pit, which makes the groundwater drawdown between the existing pile foundation and the foundation pit greater compared to the condition without pile foundation.
(2) The dewatering barrier effect of the pile foundation decreases with the increase of the distance (D) between the existing pile foundation and the foundation pit. When there are buildings outside the pit (especially the existing buildings are close to the pit), the number of pumping wells inside the pit or the pumping volume of a single well can be appropriately reduced.
(3) This study only discusses the case that the aquifer was not totally cut off by pile foundation, i.e., the aquifer is not completely cut off by diaphragm wall, and the dewatering barrier is not very obvious at this time. However, when the aquifer was completely cut off by pile foundation, the dewatering barrier will be relatively more obvious.

References

1. Zhang, Y.-Q., Wang, J.-H., Chen, J.-J., Li, M.-G.: Numerical study on the responses of groundwater and strata to pumping and recharge in a deep confined aquifer. J. Hydrol. **548**, 342–352 (2017)
2. Pujades, E., Carrera, J., Vazquez-Sune, E., Jurado, A., Vilarrasa, V., Mascunano-Salvador, E.: Hydraulic characterization of diaphragm walls for cut and cover tunnelling. Eng. Geol. **125**, 1–10 (2012)
3. Roy, D., Robinson, K.E.: Surface settlements at a soft soil site due to bedrock dewatering. Eng. Geol. **107**(3/4), 109–117 (2009)
4. Zeng, C.F., Zheng, G., Xue, X.L., Mei, G.X.: Combined recharge: a method to prevent ground settlement induced by redevelopment of recharge wells. J. Hydrol. **568**, 1–11 (2019)
5. Zeng, C.-F., Song, W.-W., Xue, X.-L., Li, M.-K., Bai, N., Mei, G.-X.: Construction dewatering in a metro station incorporating buttress retaining wall to limit ground settlement: Insights from experimental modelling. Tunn. Undergr. Space Technol, **116**, 104124 (2021)
6. Zeng, C.-F., Zheng, G., Xue, X.-L.: Responses of deep soil layers to combined recharge in a leaky aquifer. Eng. Geol. **260**, 105263 (2019)
7. Li, M.-G., Chen, J.-J., Xia, X.-H., Zhang, Y.-Q., Wang, D.-F.: Statistical and hydro-mechanical coupling analyses on groundwater drawdown and soil deformation caused by dewatering in a multi-aquifer-aquitard system. J. Hydrol. **589**, 125365 (2020)

8. Wu, Y.-X., Shen, S.-L., Yuan, D.-J.: Characteristics of dewatering induced drawdown curve under blocking effect of retaining wall in aquifer. J. Hydrol. **539**, 554–566 (2016)
9. Zeng, C.-F., Wang, S., Xue, X.-L., Zheng, G., Mei, G.-X.: Evolution of deep ground settlement subject to groundwater drawdown during dewatering in a multi-layered aquifer-aquitard system: insights from numerical modelling. J. Hydrol. **603**, 127078 (2021)
10. Wang, X.-W., Yang, T.-L., Xu, Y.-S., Shen, S.-L.: Evaluation of optimized depth of waterproof curtain to mitigate negative impacts during dewatering. J. Hydrol. **577**, 123969 (2019). https://doi.org/10.1016/j.jhydrol.2019.123969

Ranking Method for Strong Ground Motion Based on Dynamic Response of Underground Structures

Wei Yu and Zhiyi Chen(✉)

Department of Geotechnical Engineering, Tongji University, Shanghai 200092, China
zhiyichen@tongji.edu.cn

Abstract. It is well known that different input ground motions result to significantly different underground structural response. Hence, rational selection of input Intensity Measures (IMs) is the basis for accurate seismic design of underground structures. In order to find out the most destructive ground motion for seismic design of underground structures, a ranking method for strong ground motions was presented to get their failure potential sequence which may result to maximum structural response of underground structures against ground motion excitations. In this paper, dynamic time-history analyses under a ground motion cluster were carried out on a typical subway station structure in Class IV site. In the analysis, the general finite element code ABAQUS was used to set up a refined numerical modeling, 64 real ground motion records were selected as seismic wave input, and 12 IMs of 3 groups were selected as original IMs. The moment, shear, axial force of central columns and relative displacement of central column of a subway station structure was used as the Structural Response Parameters (SRPs). Based on the Partial Least Squares (PLS) regression method, a series of composite IMs were developed and investigated. The results show that the composite IMs constructed by PLS regression method have better efficiency. This method can be used to rank the design input ground motion according to the degree of SRPs, which represents the failure potential of structure, and the results can provide a reference for the selection of input ground motions for seismic design of underground structures.

Keywords: Underground structures · Composite intensity measures · Dynamic time-history analysis · Partial least squares regression · Ranking method

1 Introduction

Earthquakes can cause great damage to building structures. Several major earthquakes near metropolis over the past few decades, such as the Kobe earthquake in Japan and the Chichi earthquake in Taiwan, China, had caused huge casualties and economic losses to human society. Among these, there are a large proportion were caused by the destruction of underground structures. In this environment, the seismic design of underground structures appears particularly necessary. When it comes to seismic design of underground structure, ground motion input is usually an important process, since different IMs have

© The Author(s), under exclusive license to Springer Nature Switzerland AG 2022
L. Wang et al. (Eds.): PBD-IV 2022, GGEE 52, pp. 2391–2398, 2022.
https://doi.org/10.1007/978-3-031-11898-2_224

great differences in evaluating the seismic response of the same structure at the same site. Therefore, selecting the input seismic records by IMs correctly and reasonably for structural seismic dynamic response analysis is a key step of safety seismic design of engineering structures. At present, El Centro or Taft ground motion records in 1940 are regarded as the preferred records in both design and scientific research in China. However, literature [1] suggests that it is lack of adequate reason to select El Centro or Taft ground motion records as design record. In recent years, many scholars have carried out various relevant studies. For example, Hariri et al. [2] studied the optimal IMs of concrete dams and found that the IMs based on spectral characteristics have higher efficiency and proficiency. Qian et al. [3] carried out a nonlinear time history analysis of expressway bridge, and evaluated the relationship between IMs and bridge SRP from various aspects comprehensively, and found that the average spectral acceleration has a higher overall acceptability. Zhong et al. [4] studied the correlation between 22 ground motion IMs and some SRPs based on nonlinear dynamic time-history analysis of Dakai station, and found that peak acceleration PGA, and acceleration spectrum intensity ASI are more suitable for predicting the dynamic response of single-story double-span subway station structure under earthquake excitation.

IMs contain a lot of information related to ground motion characteristics. Although the correlation analysis between IMs and SRPs above have get some achievements. However, only for a single ground motion IM will lead to the loss of a large number of ground motion information. In this paper, the composite IMs has been proposed based on the PLS regression method of multiple linear regression analysis, which considered multiple IMs synthetically. Taking a typical subway station at class IV site prototype to set up a numerical model as the research object, and 64 natural strong ground motion records were selected as seismic input, and 12 ground motion IMs and 4 SRPs are selected as the research parameters. The dynamic time-history analysis of subway station under the selected ground motion cluster were carried out. In the end, the logarithmic linear regression analysis between the composite ground motion IMs and the selected SRPs of underground structure were conducted. By comparing the results of these regression analysis above, the effectiveness of the composite ground motion IMs is verified.

2 Numerical Analysis of Subway Station

2.1 Numerical Model and Soil Parameters

The selected two-story and three-span subways structure prototype of our numerical model is located in Shanghai, China. The site ground elevation is 3.70–4.70 m, and the geomorphic type is coastal plain type with quaternary sedimentation stratum, and the site groundwater level is 0.5–2.6 m below the ground surface. The standard cross section size of subway station structure is given (see Fig. 1).

At the center of the standard section of the subway station, the excavation depth is 15.27 m, and the width is 19.9 m, and the main structure height is 12.37 m. The spacing between central columns along the longitudinal direction of the station structure is 8 m. while the spacing between central columns along the cross-sectional direction of the station structure is 5.3 m. The site where the subway station located is mainly composed of quaternary sedimentary soil. And the specific properties of the soil layer can be seen in literature [5].

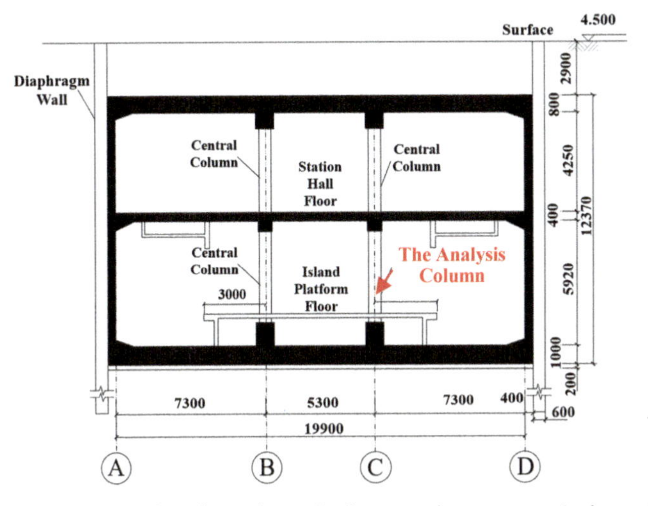

Fig. 1. Cross section dimensions of subway station structure (unit: mm)

A soil-structural dynamic interaction analysis model was established in the finite element code ABAQUS. The finite element structural model of soil-structure system is shown in Fig. 2. According to the Code for Seismic Design of Buildings (GB50011, 2010) [6], the width of each side of soil around the structure should be at least three times larger than the width of the structure. In the model, the width of each side of soil was 250 m (see Fig. 2).

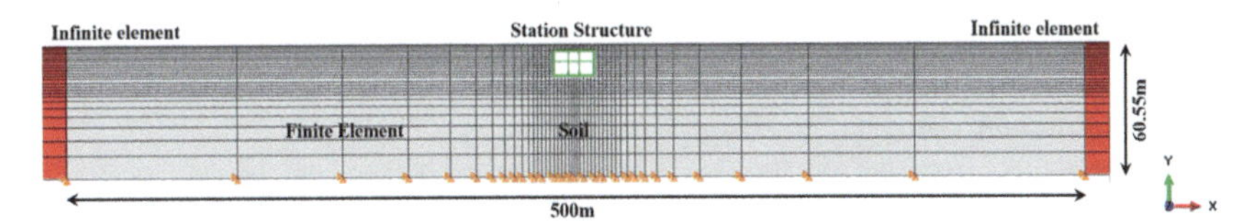

Fig. 2. Finite element analysis model

An infinite element boundary was used for the soil lateral boundary. As a results of the lack of experimental data concerning deep drilling soil samples, the soil was modeled down to a 60.55 m depth from ground surface. The depth of soil from the bottom plate was 45.28 m, which was more than three times the height of the station.

In order to ensure the composite ground motion IMs have ability to better characterize the structural failure potential, the selected input ground motions should lead to a wide value range of dynamic response of the structure as far as possible. In this paper, 64 ground motions were selected from NGA-west2 database released by Pacific Earthquake Engineering Research Center (PEER) through the screening condition that the peak acceleration is greater than 0.05 g.

2.2 The Selection of IMs and Dynamic SRPs

According to the common ground motion parameters in the seismic design of general underground structures, this paper also tries to avoid excessive information overlap in the parameters in the analysis process, and 12 IMs were selected. These selections were divided into three categories: (1) Peak IMs, including peak ground acceleration A_{pg}, peak ground velocity V_{pg}, peak ground displacement D_{pg}, effective peak acceleration A_{ep} and cumulative absolute velocity V_{ca}; (2) Spectral IMs, including spectral acceleration S_a, spectral velocity S_v and spectral displacement S_d; (3) Energy IMs, spectral intensity of acceleration A_{si}, spectral intensity of velocity V_{si} and spectral intensity of displacement D_{si} and Arias intensity I_a. The definitions of IMs were given in Table1.

Table 1. IMs used in analysis

Definition		
$A_{pg} = \max\|a(t)\|$	$A_{ep} = \frac{SA_m(\xi=0.05)}{2.5}$	$S_a = (T, \xi = 0.05)$
$V_{pg} = \max\|v(t)\|$	$A_{si} = \int_{0.1}^{0.5} S_a(T, 5\%)dt$	$S_v = (T, \xi = 0.05)$
$D_{pg} = \max\|d(t)\|$	$D_{si} = \int_{2.0}^{5.0} S_d(T, 5\%)dt$	$S_d = (T, \xi = 0.05)$
$V_{ca} = \int_0^{t_e} \|a(t)\|dt$	$V_{si} = \int_{0.1}^{2.5} S_v(T, 5\%)dt$	$I_a = \frac{\pi}{2g} \int_0^{T_d} a^2(t)dt$

where, $a(t)$, $v(t)$ and $d(t)$ is the acceleration, velocity and displacement, respectively. ξ is the damping ratio. Constant 2.5 is an empirical coefficient. T is the period. Where, T_d is the terminal time of ground motion; g is the acceleration of gravity.

In order to understand the failure strength of structure, it is necessary to know the failure mode and the key index of structure failure. Literature [7] suggests that the central column of subway station is the seismic weak segment. In this paper, the lower floor central columns of the structure were selected as the analysis members, and the SRPs, including moment M, shear force S, axial force F and relative displacement D of the selected column were selected (see Fig. 1).

2.3 Correlation Analysis Between IMs and SRPs

In order to study the relationship between the above 12 IMs and 4 SRPs, Pearson correlation coefficient was introduced. The correlation coefficients of the two random variables are calculated by the following formula:

$$\rho_{XY} = \frac{Cov(X, Y)}{\sqrt{D(X)}\sqrt{D(Y)}} \tag{1}$$

where, $Cov(X, Y)$ is the covariance of random variables X and Y; $D(X)$, $D(Y)$ is the variance of random variables X and Y, respectively. The calculation results of Pearson correlation coefficient are shown in Table 2..

As can be seen, the IMs with the highest correlation with SRPs M_{max}, S_{max}, F_{max} and D_{max} correspond to V_{si}, S_a, A_{si} and V_{si}, respectively. They are 0.83, 0.77, 0.89 and 0.93 respectively.

Table 2. The correlation coefficient of Pearson between IMs and SRPs

	M	S	F	D		M	S	F	D
A_{pg}	0.77	0.70	0.78	0.92	S_v	0.80	0.75	0.78	0.88
V_{pg}	0.79	0.75	0.78	0.81	S_d	0.59	0.56	0.61	0.69
D_{pg}	0.56	0.54	0.68	0.46	A_{si}	0.79	0.74	**0.89**	0.76
I_a	0.78	0.71	0.83	0.90	V_{si}	**0.83**	0.73	0.82	**0.93**
V_{ca}	0.66	0.64	0.71	0.45	D_{si}	0.62	0.59	0.62	0.73
S_a	0.82	**0.77**	0.80	0.83	A_{ep}	0.79	0.74	0.88	0.76

3 Structural Composite IMs

The purpose of constructing composite IMs is to linearly combine multiple IMs into a new IM. Generally, multiple linear regression method can be considered. However, in multiple linear regression analysis, if there exist a high correlation between variables and the least square method is still used to calculate the regression model, it will seriously affect the coefficient estimation in the regression model and increase the error of the model. In order to ensure the accuracy and reliability of the model, this paper adopts PLS regression method for statistical analysis.

3.1 The Results of PLS Regression

Cornell et al. [8] had pointed out that the relationship between SRPs and IMs meets the exponential relationship, approximately, as shown in Eq. (2). This relation has been widely accepted by researchers at home and abroad. Therefore, this study uses this relation to study the reasonable ground motion intensity parameters.

$$\ln SRP = a + b \ln IM \tag{2}$$

So, in this paper, the independent and dependent variables can be expressed as:

$$\ln IM = \ln\left[A_{pg},\, V_{pg},\, D_{pg},\, I_a,\, V_{ca},\, S_a,\, S_v,\, S_d,\, A_{si},\, V_{si},\, D_{si},\, A_{ep}\right] \tag{3}$$

$$\ln SRP = \ln\left[M_{\max},\, S_{\max},\, F_{\max},\, D_{\max}\right] \tag{4}$$

The 4 composite IMs were marked as $\ln I_{PLS1}$, $\ln I_{PLS2}$, $\ln I_{PLS3}$ and $\ln I_{PLS4}$, respectively. The regression results were given as follow:

$$\begin{aligned} \ln I_{PLS1} = {} & 0.047 \ln A_{pg} + 0.084 \ln V_{pg} - 0.033 \ln D_{pg} + 0.052 \ln I_a - 0.019 \ln V_{ca} \\ & + 0.025 \ln S_a + 0.116 \ln S_v - 0.005 \ln S_d + 0.008 \ln A_{si} + 0.154 \ln V_{si} \\ & - 0.022 \ln D_{si} + 0.009 \ln A_{ep} + 2.44 \end{aligned} \tag{5}$$

$$\ln I_{PLS2} = 0.040 \ln A_{pg} + 0.076 \ln V_{pg} - 0.031 \ln D_{pg} + 0.046 \ln I_a - 0.018 \ln V_{ca} + 0.020 \ln S_a + 0.106 \ln S_v - 0.004 \ln S_d + 0.005 \ln A_{si} + 0.141 \ln V_{si} - 0.020 \ln D_{si} + 0.005 \ln A_{ep} + 1.78 \quad (6)$$

$$\ln I_{PLS3} = 0.007 \ln A_{pg} + 0.004 \ln V_{pg} + 0.004 \ln I_a + 0.006 \ln S_a - 0.005 \ln S_v - 0.006 \ln A_{si} + 0.005 \ln V_{si} + 0.006 \ln A_{ep} + 4.046 \quad (7)$$

$$\ln I_{PLS4} = 0.123 \ln A_{pg} + 0.170 \ln V_{pg} - 0.0991 \ln D_{pg} + 0.106 \ln I_a - 0.064 \ln V_{ca} + 0.069 \ln S_a + 0.240 \ln S_v - 0.043 \ln S_d + 0.044 \ln A_{si} + 0.322 \ln V_{si} - 0.080 \ln D_{si} + 0.045 \ln A_{ep} + 0.018 \quad (8)$$

3.2 The Evaluation of Composite IMs

To compare the effectiveness of composite IMs compared with original IMs, the principle of least square method was used to carry out regression analysis on the obtained discrete points (SRP_i, IM_i), and then the logarithmic regression lines of variables $\ln(SRP_i)$ and $\ln(IM_i)$ were obtained as well as the conditional logarithmic standard deviation δ_ε,

$$\delta_\varepsilon = \sqrt{\frac{\sum\{\ln(SRP) - [a + b\ln(IM)]\}^2}{n-2}} \quad (9)$$

where, δ_ε is the efficiency of IMs, which is to ensure the reliability of the results in the process of nonlinear analysis, and can reduce the discreteness of structural response caused by the randomness of ground motion.

As for a single IM, the smaller the value of δ_ε, the smaller the discreteness of SRPs, the higher the effectiveness of the selected IM.

Since the SRPs of V_{si}, S_a, A_{si} and V_{si} with the highest Pearson correlation coefficient with the IMs of $M_{\max}$, $S_{\max}$, $F_{\max}$ and $D_{\max}$ respectively. Four control groups were set to verify the effectiveness of composite IMs by comparing, which come to the regression analysis of $\ln I_{PLS1}$ and $\ln V_{si}$, $\ln I_{PLS2}$ and $\ln S_a$, $\ln I_{PLS3}$ and $\ln A_{si}$, and $\ln V_{si}$. The regression results are given (see Fig. 3).

It can be summarized from Fig. 5 that the conditional logarithmic standard deviation δ_ε of composite IMs and SRPs are significantly lower than the correspond original IMs respectively. That represent the composite IMs can more efficiently in evaluating the seismic dynamic SRPs than the original single IM.

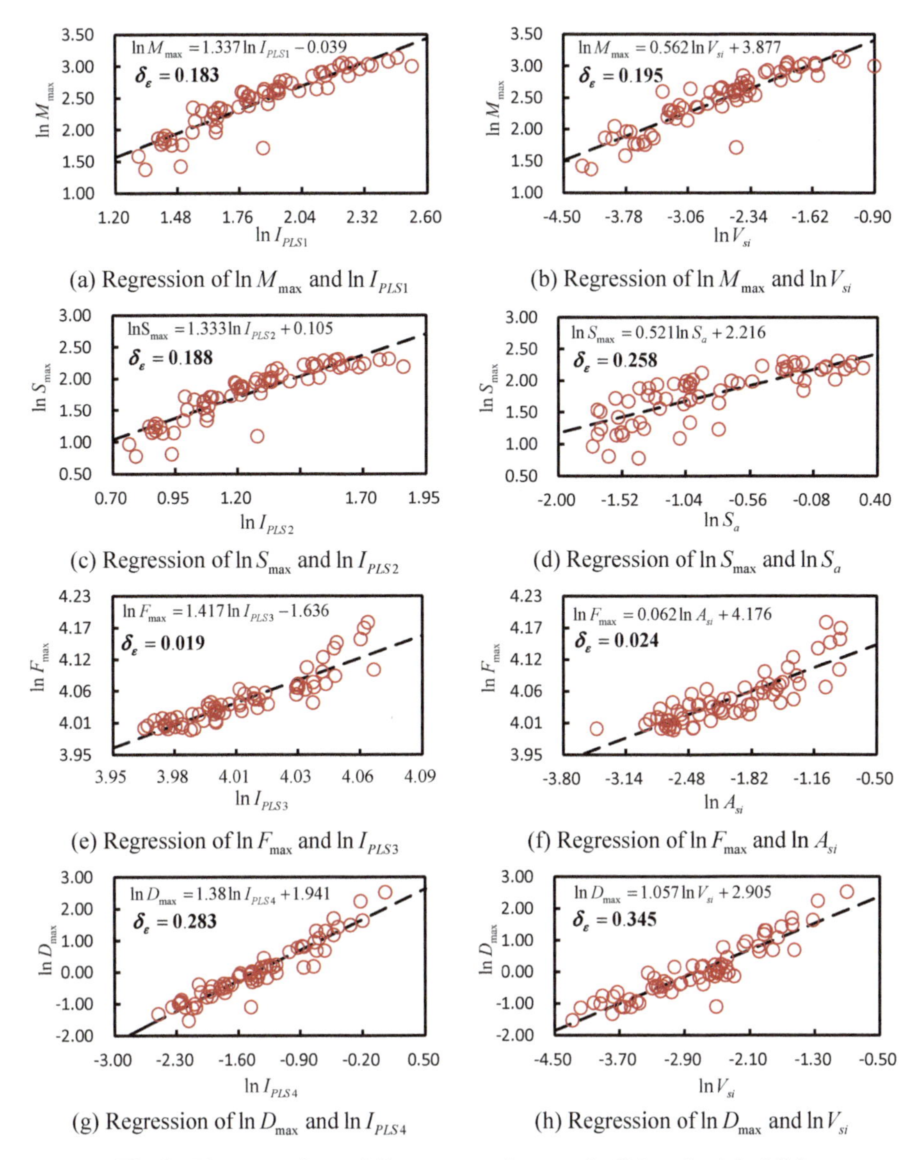

(a) Regression of $\ln M_{max}$ and $\ln I_{PLS1}$ (b) Regression of $\ln M_{max}$ and $\ln V_{si}$

(c) Regression of $\ln S_{max}$ and $\ln I_{PLS2}$ (d) Regression of $\ln S_{max}$ and $\ln S_a$

(e) Regression of $\ln F_{max}$ and $\ln I_{PLS3}$ (f) Regression of $\ln F_{max}$ and $\ln A_{si}$

(g) Regression of $\ln D_{max}$ and $\ln I_{PLS4}$ (h) Regression of $\ln D_{max}$ and $\ln V_{si}$

Fig. 3. The comparison of discreteness of composite IMs and original IMs

4 Conclusions

This paper has studied the IMs suitable for characterizing the dynamic structural responses of subway station structure. And the mainly conclusions of this paper were listed as follow:

(1) According to the Pearson correlation coefficient, the IMs of V_{si}, A_{si}, S_a are more relevant to the dynamic SRPs. And they are more suitable for predicting the dynamic

response of two-story and three span subway station structure in class IV soft soil site under earthquake.

(2) The composite IMs constructed by PLS regression method have better efficiency in evaluating the seismic dynamic structural responses than the original single IM. Thus, this method can be used to rank the design input ground motion according to the degree of structural responses, and the results can provide reference for the selection of input ground motion of relevant subway station structures.

In the analysis of this paper, only a single subway station was considered. As the underground structure of the subway becomes more and more complex, the relevant laws between the seismic response and ground motion parameters of subway stations with different structural forms and underground structures under different site conditions need to be studied in the follow up research to get a universal conclusion.

Acknowledgments. This research was supported by the National Natural Science Foundation of China (Grant No. 51778464), State Key Laboratory of Disaster Reduction in Civil Engineering (SLDRCE19-B-38) and the Fundamental Research Funds for the Central Universities (22120210145). All supports are gratefully acknowledged.

References

1. Xie, L.L., Zhai, C.H.: Study on the severest real ground motion for seismic design and analysis. Acta Seismol. Sin. **16**(3), 260–271 (2003) (in Chinese)
2. Hariri-Ardebili, M., Saouma, V.: Probabilistic seismic demand model and optimal intensity measure for concrete dams. Struct. Saf. **59**, 67–85 (2016)
3. Qian, J., Dong, Y.: Multi-criteria decision making for seismic intensity measure selection considering uncertainty. Earthquake Eng. Struct. Dynam. **49**(11), 1095–1114 (2020)
4. Zhong, Z.L., Shen, Y., Zhen, L.: Ground motion intensity measures and dynamic response indexes of subway station structures. Chin. J. Geotech. Eng. **42**(3), 486–494 (2020). (in Chinese)
5. Chen, Z.Y., Liu, Z.Q.: Effects of central column aspect ratio on seismic performances of subway station structures. Adv. Struct. Eng. **21**(1), 14–29 (2018)
6. GB50011 (2010) Code for Seismic Design of Buildings. Beijing, China: China Architecture & Building Press. (in Chinese)
7. Chen, Z.Y., Liu, W.B., Chen, W.: Performance experiment of a multi-story subway station. J. Tongji Univ. (Nat. Sci.) **48**(06), 811–820 (2020) (in Chinese)
8. Luco, N., Cornell, C.: Structure-specific scalar intensity measures for near-source and ordinary earthquake ground motions. Earthq. Spectra **23**(2), 357–392 (2007)

Author Index

© The Editor(s) (if applicable) and The Author(s), under exclusive license
to Springer Nature Switzerland AG 2022
L. Wang et al. (Eds.): PBD-IV 2022, GGEE 52, pp. 2399–2405, 2022.
https://doi.org/10.1007/978-3-031-11898-2

Printed by Printforce, the Netherlands